T0213341

AESS Interdisciplinary Environmental Studies and Sciences Series

Series editor

Wil Burns
Forum for Climate Engineering Assessment
School of International Service
American University
Washington, DC, USA

Environmental professionals and scholars need resources that can help them to resolve interdisciplinary issues intrinsic to environmental management, governance, and research. The AESS branded book series draws upon a range of disciplinary fields pertinent to addressing environmental issues, including the physical and biological sciences, social sciences, engineering, economics, sustainability planning, and public policy. The rising importance of the interdisciplinary approach is evident in the growth of interdisciplinary academic environmental programs, such Environmental Studies and Sciences (ES&S), and related 'sustainability studies.'

The growth of interdisciplinary environmental education and professions, however, has yet to be accompanied by the complementary development of a vigorous and relevant interdisciplinary environmental literature. This series addresses this by publishing books and monographs grounded in interdisciplinary approaches to issues. It supports teaching and experiential learning in ES&S and sustainability studies programs, as well as those engaged in professional environmental occupations in both public and private sectors.

The series is designed to foster development of publications with clear and creative integration of the physical and biological sciences with other disciplines in the quest to address serious environmental problems. We will seek to subject submitted manuscripts to rigorous peer review by academics and professionals who share our interdisciplinary perspectives. The series will also be managed by an Editorial board of national and internationally recognized environmental academics and practitioners from a broad array of environmentally relevant disciplines who also embrace an interdisciplinary orientation.

More information about this series at http://www.springer.com/series/13637

Peter Saundry • Benjamin L. Ruddell
Editors

The Food-Energy-Water Nexus

Editors
Peter Saundry
Energy Policy and Climate
Advanced Academic Programs, Krieger School of Arts and Sciences
Johns Hopkins University
Washington, DC, USA

Benjamin L. Ruddell
Northern Arizona University
Flagstaff, AZ, USA

ISSN 2509-9787 ISSN 2509-9795 (electronic)
AESS Interdisciplinary Environmental Studies and Sciences Series
ISBN 978-3-030-29916-3 ISBN 978-3-030-29914-9 (eBook)
https://doi.org/10.1007/978-3-030-29914-9

This Springer imprint is published by the registered company Springer Nature Switzerland AG
The registered company address is: Gewerbestrasse 11, 6330 Cham, Switzerland

Preface

Human societies require a constant supply of The Food-Energy-Water Nexus (FEW). These are critical consumable commodities for human life. Each has been the subject of extensive research and action aimed at providing secure access for the 7.7 billion people on the Earth at this time. Yet, billions of people today do not experience food, energy, and water security. As human population and demand for food, energy, and water have grown, the challenge of meeting the demand has become ever harder and resulted in ever more consequences on the natural world. The food-energy-water nexus has become the embodiment of many of the most important practical challenges of sustainability. Food, energy, and water systems are complex coupled natural-human systems with many dependencies and interactions. It is, therefore, unsurprising that efforts to look at the food, energy, and water as a "system of systems" have advanced over the past decade, including via the energy-water and FEW-land-environment nexus conversations.

On January 19–21, 2016, the National Council for Science and the Environment (NCSE) focused its 16th National Conference and Global Forum on Science, Policy, and the Environment on the opportunities and challenges of advancing science on the food-energy-water nexus. At the time, Peter Saundry was the Executive Director of the NCSE and edited a special issue of the *Journal of Environmental Studies and Sciences* (JESS) which included 22 papers on the subject.[1] In summer 2016, many of the authors of the JESS papers agreed to create this book.

While a number of books and compilations of papers on the FEW nexus have appeared over the past decade, we believe that this is the first attempt to publish a textbook specifically for the FEW nexus. The scope of this book is broad and introductory, and it is intended to be accessible to advanced undergraduate students, graduate students, practitioners, and also those researchers and scholars new to the field and who seek a scholarly introduction to the nexus issues, tools and applications.

---

[1] Saundry. P. (2016) The food-energy-water nexus. *Journal of Environmental Studies and Sciences*, Volume 6, Issue 1, March 2016, Springer. Online at: https://link.springer.com/journal/13412/6/1/page/1.

This book reflects the diverse interdisciplinary scholarship that contributes to the nexus, including biophysical sciences, economics, agriculture, social sciences, business, political science, engineering, modeling, computing, and data science, and their applications to real-world problem-solving. We assume that some readers may have substantial disciplinary expertise in one or more of these areas. However, few, if any, readers will come to this subject with a background in all aspects of the nexus. Therefore, we have included introductory material in all chapters. The reader may choose to skip introductory material in areas where they have expertise.

Each chapter's end matter includes a summary of key points, a limited set of discussion points and exercises to facilitate learning, references, and suggested further readings. Appendices A–D contain a list of nexus institutions, a list of treaties, declarations and laws referred to in the book, data sources that are utilized in exercises, and a rich set of supplemental FEW educational resources that can be utilized by teachers and students. Finally, a glossary is provided. Glossary terms are in bold when first used in the text.

While this book is the result of the work of many authors who were responsible for different chapters, considerable effort has been made to integrate the text into a seamless whole. The authors have reviewed each other's work and revised their chapters to integrate with the rest of the book. Most chapters include contributions of text and insight from the authors of other chapters. Chapters 12, 17, and 21 are team-writing efforts. All chapters were subject to careful review and editing. We thank those who contributed to the integration and editing.

Because the food-energy-water nexus is a subject where scholarship is relatively new and changing rapidly, we do not claim that this is the "definitive" text on the subject. Rather, this is the first step in an exciting and profoundly important area of education, science, and practice.

Germantown, MD, USA
Flagstaff, AZ, USA

Peter Saundry
Benjamin L. Ruddell

Acknowledgments

The editors and authors of this book would like to acknowledge the National Council for Science and the Environment, whose 16th National Conference and Global Forum on January 19–21, 2016 addressed the Food-Energy-Water Nexus and catalyzed the collaboration that led to this book. We also thank the Association for Environmental Studies and Sciences and Springer for providing support for bringing this book to print. The Nexus work of many of us has been supported by grants from the National Science Foundation, the U.S. Department of Agriculture, and other sources. The opinions expressed in this book are, of course, those of the authors and not necessarily the funding agencies.

The editors and authors have benefited from scholarly support and encouragement from a host of institutions and individuals to whom we wish to express our gratitude to those who have helped us get to where we are.

On a personal note, Peter Saundry wishes to express his gratitude and appreciation to his wife Claire, who has both supported and endured the considerably greater-than-expected effort required to bring this project to fruition. There would be no book without you.

Contents

Part II Tools

Part III Applications

About the Authors

Antti Belinskij, LL.D. is a Professor of Environmental Law at the University of Eastern Finland and a Research Professor at the Finnish Environment Institute SYKE. He is specialized in water, natural resources, and nature conservation law. His contribution to this book was supported by the Strategic Research Council of Finland under project Winland (Chap. 6).

Andrew Reid Bell is Assistant Professor of Environmental Studies at New York University. His research uses surveys, behavioral experiments, and agent-based models to understand decision-making in the areas of agricultural development, water management, migration, and livelihoods (Chap. 9).

Robert T. Caccese is an attorney focused on administrative law, water policy, and fisheries issues. He serves as Assistant Counsel for the Pennsylvania Fish and Boat Commission and received his B.S. in Wildlife and Fisheries Science and J.D. from Penn State University. He also serves as an adjunct instructor for Penn State Law, teaching environmental law and policy (Chap. 20).

Michael Carbajales-Dale heads the Energy-Economy-Environment (E3) Systems Analysis research group at Clemson University. He joined Clemson in August 2014 as an Assistant Professor in the Environmental Engineering and Earth Sciences Department. Before joining Clemson, Mik was an Energy Systems Analyst with Stanford's *Environmental Assessment and Optimization Lab* and with the *Global Climate and Energy Project* (GCEP). Twitter:@EconomyE3 (Lead Author for Chaps. 12, 13 and 17)

Mary Doidge is an Assistant Professor of Agricultural Economics at McGill University. Her contribution to this book occurred while a postdoctoral researcher in the Ohio State University School of Environment and Natural Resources. Dr. Doidge has expertise in behavioral economics, land use change, and crop insurance (Chap. 4).

Emre Eftelioglu is a Data Scientist at Cargill Inc. where he focuses on analytics on spatial, spatiotemporal, and sensor datasets to improve food production efficiency in a sustainable way. His research is focused on identifying and analyzing patterns from spatial datasets to understand the effect of location on phenomena. He received his Ph.D. in Computer Science from the University of Minnesota (Chaps. 12, 16, and 17).

Lara B. Fowler, J.D. is a Senior Lecturer at Penn State Law and Assistant Director at the Penn State Institutes of Energy and the Environment. With a background as a practicing mediator and attorney, she focuses on interdisciplinary research and managing conflicts that arise in water, energy, and agriculture (Chap. 20).

Hongkai Gao is a Postdoctoral Research Fellow at Arizona State University working to develop a Green Infrastructure Support Tool (GIST) for decision-making on water resources management (Chap. 10).

Craig Harris is an Associate Professor in the Department of Sociology of Michigan State University. He focuses on the behavior of tightly and moderately coupled social-natural systems, including agriculture, fisheries, and food safety (Chap. 18).

Kaisa Huhta is a senior lecturer in European Union law at the University of Eastern Finland Law School. Her research focuses on EU energy law; in addition to which she actively carries out consultancy work in the field of EU and energy law (Chap. 6).

Elena Irwin is an environmental and urban economist specializing in land use, coupled with human-natural systems, and sustainability. She is a Professor in the Department of Agricultural, Environment, and Development Economics and the Faculty Director of the Sustainability Institute at Ohio State University (Chap. 4).

Carey W. King is a Research Scientist and Assistant Director of the Energy Institute at the University of Texas at Austin. Dr. King performs interdisciplinary research related to energy systems integration as well as the macro-scale interaction of energy within the economy and environment. Twitter: @CareyWKing (Chaps. 12, 13, and 17)

Marko Keskinen is a Senior University Lecturer at Aalto University's Water and Environmental Engineering Research Group in Finland. His research focuses on water resources management, water-energy-food security nexus as well as linkages between sustainability, security, and resilience. His contribution to this book was supported by the Strategic Research Council of Finland under project Winland (Chap. 6).

Beth Kinne, J.D., LL.M. is an Associate Professor of Environmental Studies at Hobart and William Smith Colleges and the Environmental Studies Chair of the

Finger Lakes Institute in Geneva, NY. She formerly practiced water, land use, and municipal law. Her current research focuses on watershed management collaborations among state and local governments and third sector organizations (Chaps. 8 and 21).

Natalie Lyon is an urban planner and trainee in the NSF Transformative Research in Urban Sustainability Program at Wayne State University. She conducts interdisciplinary research on urban sustainability issues including green stormwater infrastructure, the interface of natural and urban systems, and invasive species management in cities. This material is based upon work supported by the National Science Foundation under Grant No. 1735038 (Chap. 18).

Darrin Magee is an Associate Professor of Environmental Studies at Hobart and William Smith Colleges in Geneva, New York. A specialist in environmental issues in China, he holds degrees in geography, international studies, mathematics, and foreign languages. His research and teaching focus primarily on water, energy, and waste issues (Chaps. 8 and 21).

Nathanial Matthews is an international development practitioner with experience spanning 35 countries. He also holds positions as a Senior Visiting Fellow at King's College London, a Visiting Fellow at the University of East Anglia Water Security Centre, and a Fellow of the Royal Geographical Society. Twitter: @Nate_Matthews_ (Chap. 9)

Bruce A. McCarl is University Distinguished Professor and Regents Professor of Agricultural Economics at Texas A&M University. His recent research efforts have largely involved policy analysis as well as the proper application of quantitative methods to such analyses. He was part of the 2007 Nobel Peace Prize winning Intergovernmental Panel on Climate Change. This material is based upon work partially supported by the National Science Foundation under Grant 1739977, Addressing Decision Support for Water Stressed FEW Nexus Decisions, and Grant 1639327, INFEWS/T1: A Modeling Framework to Understand the coupling of Food, Energy, and Water in the Teleconnected Corn and Cotton Belts (Chap. 5).

Carol Miller is a Professor of Civil and Environmental Engineering at Wayne State University in Detroit. She supervises research and curriculum in hydraulics, urban water resources, and the energy/water interface. She is Director of the Healthy Urban Waters Initiative and the U.S. Co-Chair of the Great Lakes Science Advisory Board, International Joint Commission (Chap. 18).

Fernando R. Miralles-Wilhelm is a hydrologist with research interests in modeling of surface and groundwater systems, climate-hydrology-vegetation interactions, water quality, and modeling of the water-energy-food nexus. He is Chair and Professor of Atmospheric and Oceanic Science at the University of Maryland as well as Lead Scientist for Freshwater at The Nature Conservancy (Chaps. 11, 15, and 17).

Rabi Mohtar is Professor and Dean, Faculty of Agricultural and Food Sciences, American University of Beirut, and TEES Research Professor, Texas A&M University. He founded A&M's Water-Energy-Food (WEF) Resource Nexus Initiative and coordinates AUB's Water-Energy-Food-Health Nexus Renewable Resources Initiative: both work toward WEF securities by implementing integrated solutions through holistic, systems-level thinking (Chaps. 17 and 21).

Okan Pala is an expert in disaster management for critical infrastructures performing research at the Center for Geospatial Analytics and teaching for the Computer Science Department at NC State University. His research applies geospatial analytics along with cutting-edge computing techniques to solve grand challenges in energy, water, and transportation systems. Dr. Pala received his Ph.D. from the College of Computing at UNC Charlotte (Chap. 10).

Kami Pothukuchi is an expert in community food systems planning, with a specific interest in planning for social justice and sustainability. She is Distinguished Service Professor in the Department of Urban Studies and Planning at Wayne State University and the founding director of SEED Wayne, a campus-community food system collaborative (Chap. 18).

Outi Ratamäki, D.Soc.Sci. works as a senior lecturer at the Law School University of Eastern Finland. She is specialized in analyzing natural resources governance and conflicts at the interfaces between law and policy (Chap. 6).

Robert B. Richardson is an ecological economist and Professor in the Department of Community Sustainability at Michigan State University. His research interests are related to the contribution of ecosystem services to socioeconomic well-being. He uses methods from the behavioral and social sciences to study decision-making and the values of ecosystem services (Chaps. 19 and 21).

Benjamin L. Ruddell is Professor and Director of Northern Arizona University's (NAU) School of Informatics, Computing, and Cyber Systems (SICCS). He is the lead PI on the FEWSION project, an NSF-funded effort to produce meso-scale datasets describing national food-energy-water systems (NSF/USDA ACI-1639529). He served for 8 years on the faculty of the Fulton Schools of Engineering at Arizona State University before serving at NAU. He holds a Ph.D. in Civil and Environmental Engineering from the University of Illinois at Urbana-Champaign (Editor; Chaps. 1, 2, 3, 7, 10, 12, 14, 17, and 21).

Richard Rushforth is an Assistant Research Professor in the School of Informatics, Computing, and Cyber Systems at Northern Arizona University. His research focuses on big data modeling of food, energy, and water systems to further the understanding of complex, coupled natural-human systems. He also works on the National Water-Economy Project and the NSF-funded FEWSION project (Chap. 10).

John Sabo is an ecologist who studies the importance of water in determining the viability and resilience of animal and plant populations in river and riparian ecosystems. He also has projects that examine the effects of dams on energy flow through aquatic food webs. He is a Professor and Senior Sustainability Scientist at Arizona State University's Wrigley Global Institute of Sustainability (Chap. 10).

Peter Saundry is an Adjunct Professor of Energy at the Johns Hopkins University and a Senior Fellow at the National Council for Science and the Environment. His scholarship focuses on how to effectively integrate energy and climate science into decision-making and public policy. He has a Ph.D. in physics from the University of Southern California. Twitter: @psaundry (Editor; Chaps. 1, 2, 3, 6, 7, 11, 12, 17, and 21)

Shashi Shekhar a McKnight Distinguished University Professor at the University of Minnesota, is a leading scholar of spatial computing and Geographic Information Systems (GIS). His contributions include algorithms for evacuation route planning and mining spatial patterns (e.g., colocation, linear hotspots), an *Encyclopedia of GIS*, and a *Spatial Databases* textbook (Chaps. 12, 16, and 17).

Nicole Sintov is an environmental psychologist. She currently serves as Assistant Professor of Behavior, Decision-Making, and Sustainability in OSU's School of Environment and Natural Resources. Her work aims to advance psychological theory while producing insights that can be applied to benefit the environment and society (Chap. 4).

Lara Treemore-Spears is Program Coordinator for Healthy Urban Waters Initiative at Wayne State University where she manages interdisciplinary research projects that combine physical science and engineering with human behavior and health relating to the natural and built environment, water resources and infrastructure, and the transport and fate of pollutants (Chaps. 9, 18, and 21).

Robyn S. Wilson is a Professor in the School of Environment and Natural Resources at the Ohio State University. She is a behavioral decision scientist, focused on decisions under risk. She is also interested in the development of messaging and decision support tools that support more informed choices (Chap. 4).

Yingqian Yang is conducting research on food-water-energy Nexus. She has a Ph.D. in Agricultural Economics from Texas A&M University. This material is based upon work partially supported by the National Science Foundation under Grant Addressing Decision Support for Water Stressed FEW Nexus Decisions Numbered 1739977 (Chap. 5).

Wei Zhang is a senior research fellow at the International Food Policy Research Institute in the Environment and Production Technology Division. She leads the research program on ecosystem services under the Natural Resource Management theme. She has a Ph.D. in Agricultural, Food, and Resource Economics from Michigan State University (Chap. 9).

Yongli Zhang is a faculty member in the Department of Civil and Environmental Engineering at Wayne State University. Currently, she serves as the co-Director of Graduate Certificate Program in Sustainable Engineering. Her research focuses on the water-environment-energy nexus through system modeling/optimization and technology development (Chap. 18).

Part I
Framing the Nexus

Chapter 1
Introduction

Peter Saundry and Benjamin L. Ruddell

The "Blue Marble" image of Earth taken from Apollo 17, December 7, 1972 (Source: NASA; Apollo 17 astronauts Eugene Cernan, Ronald Evans and Harrison Schmitt)

P. Saundry (✉)
Energy Policy and Climate, Advanced Academic Programs, Krieger School of Arts and Sciences, Johns Hopkins University, Washington, DC, USA
e-mail: psaundr1@jhu.edu

B. L. Ruddell
School of Informatics, Computing, and Cyber Systems, Northern Arizona University, Flagstaff, AZ, USA
e-mail: Benjamin.Ruddell@nau.edu

P. Saundry, B. L. Ruddell (eds.), *The Food-Energy-Water Nexus*, AESS Interdisciplinary Environmental Studies and Sciences Series,
https://doi.org/10.1007/978-3-030-29914-9_1

1.1 The Structure of This Book

Given the many critical challenges facing society, why select the particular challenges of **food**, **energy**, and **water** for integration? Why not air, land, climate, population, or economics? Why now at the beginning of the twenty-first century? This chapter sets up the rest of the book by answering this fundamental question.

The reader is asked not to assign any importance to the order of the words "food," "energy," and "water." This order is used primarily because the resulting acronym "FEW" is easy to remember. Others prefer the equally valid "WEF." Any order is equally valid because the nexus of FEW **systems** is, by definition, where all three components and their interactions are all essential to understanding a challenge and developing solutions.

The integration of FEW systems is necessary because human beings and their societies require all three all the time, and because each of the three depends on the others. Food requires water and energy; energy requires water, and (in the case of **biofuels**) food; water requires energy (which may require food). Practical challenges such as how human beings adapt to limitations in their access to food, energy, and water; how they develop policies and laws to ensure and/or restrict access; how the regulations in each system affect adaptation and access in the other two systems; how they resolve disagreements; and whether and how they will do so in a sustainable manner in the future represent some of the underlying motivations for nexus studies. In Chaps. 3–11, we will explore the human demands on FEW systems and how the critical roles played by ecosystems, **infrastructure**, **climate change**, law and **policy**, economics, and culture shape the resulting systems and their interactions.

In the second part of this book, Chaps. 12–17, we will explore tools for measuring and modeling integrated food, energy, and water systems individually. We will address how to ask questions, with spatial and temporal **boundaries**, that can lead to results useful to decision-makers in the real world as they respond to the challenges of meeting the needs for food, energy, and water simultaneously, challenges that are both scientific and human.

While the nexus of food, energy, and water systems is still an emerging field of study, it is not new. Nearly every chapter in this book includes case studies related to the focus of that chapter. Chapters 18 and 19 will explore two areas where the importance of integrating FEW systems has already been recognized by significant research carried out over several decades—cities and watersheds. Both of these areas lend themselves to certain spatial boundaries which helps define questions at the nexus. Work in these contexts has been varied and is responsible for many of the tools presented in this book. These tools and others have been applied to manage various types of human **conflict** at the nexus (Chap. 20).

We conclude this book (Chap. 21) with a brief, but speculative, look into the future, exploring how advances in the study of the integration of FEW systems have the potential to foster the development of many solutions to specific challenges.

1.2 Fundamental Challenges

The study of integrated food, energy, and water systems requires the application of physical, biological, and social sciences; engineering; scholarship in policy, law, and public health; and awareness of relevant aspects of arts and letters. It is not the intention of the authors to prioritize the relative importance of different areas of scholarship. All contribute significant value, and the absence of any one area can result in unsuccessful outcomes in any particular situation. Just as the FEW nexus results from the interaction of many biological, physical, and social systems, effective FEW scholarship results from the integration of the many relevant areas of scholarly work.

Readers are urged to keep in mind throughout this book the many challenges to scholarly work in this area. Food, energy, and water systems are challenging topics in their own right. The integration of all three into a system of systems is still an emerging field because it is especially scientifically challenging, requiring careful consideration of the following:

1. The fundamental questions which drive all later scientific considerations and the utility of outcomes.
2. The appropriate spatial and temporal **scales** to balance the requirements of science, geography, **data**, and the needs of decision-makers and other end-users of results.
3. The choices and consequences of selecting metrics that have utility to end-users of results and are capable of being addressed with available data, **models**, and computational **resources**.
4. How to select and integrate multiple data sets with diverse attributes.
5. How to choose the right type of model and develop the specific model, based on all relevant laws of nature, that is capable of integrating all aspects of an integrated system, recognizes *all significant* internal and external interactions, and can produce results useful to end-users with available computational resources.
6. The human dimensions of FEW systems which are affected by perceptions, cultural and economic motivations, laws and policies operating at different scales, and the conflicts that arise between various parties with different values, power, and aspirations.

Communities of **Practice** that engage scientific and non-scientific **stakeholders** are key to the successful application of nexus science to practical problems (Chap. 17). This requires the engagement of a wide range of stakeholders (broadly defined as anyone who is affected by a particular topic).

When all these factors are recognized, scholars can be tempted to throw their hands up in despair. However, scientists working with stakeholders can help frame research questions can help lead to more integrated work that leads to actual implementation. Further, the development of better data sets, models, and computational resources combined with more sophisticated ways of connecting science to non-scientific stakeholders will likely transform Nexus studies over the coming decades. Indeed, the authors of this book and many others devote their time to this area in large part because they are optimistic that their work will lead to positive outcomes and innovative solutions. We hope that the readers of this book will reach a similar conclusion, and, perhaps, will be at least tempted to join in the ongoing process.

This book attempts to provide the reader with a **survey** of issues, methods, and tools that reflects best practices in nexus work at the time of writing and looks ahead to what the authors believe will be important in the future.

Note that issues of scale will arise throughout this book and be explored in detail in Chap. 15. Most generally we will use (with many important caveats) the following terms to describe phenomena at different scales:

- "Micro" for small, fine, or local scales;
- "Meso" for intermediate or regional scales;
- "Macro" for large, national, international, or **global scales**.

1.3 Why Food, Energy, and Water?

1.3.1 Criteria

Nexus studies recognize the importance of integrating FEW systems based upon approximately five criteria that are interwoven with each other. Food, energy, and water are critical consumable **commodities**, require massive infrastructure, are currently footprint-heavy, must be extremely accessible and affordable, and are the focus of high-level decision-making and policy. We study and manage FEW systems, rather than other parts of the **FEW-everything** (FEWe) system, because FEW systems usually, and uniquely, meet most or all of these five criteria.

1.3.1.1 Criterion 1: Critical Consumable Commodities for Human Life

Human beings require a continuous supply of food and water to consume to live and energy to support most aspects of comfortable living. Food, energy, and water are essential **commodities** for human life level and require constant replenishment. These commodities are essential in both sufficient quantity and quality (i.e., food nutrition, energy in useful forms, and clean water).

The phrase "society is only three meals away from anarchy" (sometimes "nine meals" or "from revolution") captures the critical need for food. The human body can survive without water for 2–3 days, without food for perhaps 30–40 days. In large cities, the human economy and society are less resilient. Energy sustains water and food supplies and multiple additional life-dependent aspects of human societies. As is all too frequently demonstrated during electricity blackouts, communities can quickly become dysfunctional, even disorderly, when there are disruptions in modern energy supplies. Disruptions of water supplies have proven critical in war-torn areas of Syria and Iraq (see Sect. 20.3.5), and disruptions in energy supplies are disrupting Venezuela in 2018–2019.

The criticality of these "three great consumables" is reflected in the commonly used terms "**food security**," "**energy security**" and "**water security**" referring to individuals and communities having affordable access to sufficient food, energy, and water to meet their basic needs for healthy and productive lives (see Sect. 3.2). While these terms may seem abstract to some in affluent societies like the USA and other economically prosperous communities, these terms are very immediate and personal for billions of people who do not have reliable access to an adequate supply of nutritious food, clean water, or modern energy sources. For these people, gaining access to these three great consumables occupies a central part of their daily lives and labor.

Even within affluent societies like the USA, food, energy, and water are recognized by emergency planners as "community lifelines" which provide indispensable services that enable "the continuous operation of critical business and government functions, and is critical to human health and safety, or economic security" (FEMA 2008).

FEW commodities are bought and sold in large quantities in markets at many economic levels. Many FEW products are also produced for direct consumption and are never traded.

Food

In the global North, the majority of households get most of their food from retail stores and commercial food service establishments and **institutions** (e.g., hospitals, dormitories, prisons). These facilities are situated at the endpoints of long complex **supply chains**. As noted in Chap. 18, approximately 5% of the **urban** population of the USA live in a "**food desert**" where they do not have access to a full-line grocery store; the same is true in **rural** areas of the USA. Both in urban and rural areas of the USA, according to the National Gardening Association (NGA) 36% of households grow food either at home or in a community garden.

The major cities in the global South also have retail stores and commercial food service establishments and institutions supplied by long complex supply chains, although the performance may be somewhat less complete and less timely than in the global North. In addition, urban food production is much more prevalent in neighborhoods and in **peri-urban** areas, small-scale farmers' markets are more prevalent in local neighborhoods, and many households receive a significant amount of food from relatives in rural areas.

In the rural areas of the global South, the vast majority of households produce most of their own food. As with water and energy, much of the burden of food production is borne by women and children, especially girls. These rural farmers may also produce some crops for local sale and some commodities for sale to regional or national or global markets, but the majority of their production provisions their own households, sometimes including one or more draught animals. While the harvesting of seafood is more likely done by men than women, it also goes both for household provisioning and for local and regional markets (often processed and sold by female members of the household). The water and energy collected by women and children are very important for household food preparation.

Energy

In the global North, most residences and institutional and commercial facilities are connected to the electrical grid; this can provide the energy needed for lighting, cooking, electronics, and heating. In the USA, approximately 5% of households heat with coal or oil, a little more than half heat with natural gas, and roughly 5% of US households heat with propane. Two percent of US households use wood as the primary heating fuel.

Although electricity is generally available in the major cities in the global South, the service is not highly dependable, and the current is highly variable. Outside the major cities, much of the energy for heating is wood or peat or dung, and much of the energy for cooking is charcoal or dung or wood. The use of these fuels causes significant indoor air pollution and high levels of respiratory disease. While charcoal is produced locally and distributed by small-scale vendors, wood and dung are (like water) collected by women and children often on a daily basis. Both in the global North and in the global South, residential solar facilities are increasingly used for cooking, electricity, and heating.

Water

In the global North, most residences have running water service for personal and household uses and sanitation. But even in the USA, roughly 15% of households rely on private wells for drinking water and septic tanks for wastewater treatment; approximately 1% of the population lacks basic plumbing facilities. In Mexico **City**, 70% of the city has fewer than 12 h of running water per day, and 18% of the population receives water for only a couple hours every several days. The lack of dependability and affordability and safety of the municipal water supply leads roughly half of the households to purchase additional water privately.

Although water and sanitation service in the major cities in the global South resembles that in Mexico City, outside the cities, the services are much scarcer. In 2015, according to the World Health Organization (WHO), approximately 30% of the global population did not have a safely managed drinking water service (on-premises, as needed, and contamination-free). Members of these households (usually women and children) have to transport water every day. One-third of these households rely on untreated **surface water**; both these surface waters and shallow wells provide two-thirds of these households with water that is contaminated with feces. This water is important for the FEW nexus because many of the meals in these households are stewed meats or seafood, mashed vegetables or fruits, or porridges. So water is essential for cooking, as is the fuel that is locally obtained. The cooking techniques also provide the benefit of treating some of the contaminants in the water. Also, some of the water that is collected is used to irrigate crops in home gardens.

As noted, outside the major cities in the global South, sanitation services are much less developed. According to the WHO, 2.3 billion people (30% of the world's population) still do not have basic sanitation facilities such as toilets or latrines. Of these, 892 million still defecate in the open, for example, in street gutters, behind bushes or into open bodies of water. This leads to high levels of mortality for children under 5 years old, and for women and children, it leads to a high **risk** of sexual assault.

Nations attempt to provide secure access to these resources for their citizens and achieve some degree of FEW self-sufficiency, even when it comes at an economic or environmental cost. This includes a wide range of financial incentives for domestic production of FEW commodities or tariffs on imports. Environmental laws are also commonly structured to avoid impacting the production of FEW commodities.

1.3.1.2 Criterion 2: Heavy on Physical Infrastructure

Maintaining a reliable and consistent supply of food, energy, and water has resulted in the building of extensive physical **infrastructure** to store, transport, distribute, and deliver them to all economically advanced human communities remote from their sources. While most communities have access to local water supplies, and many have access to some locally produced foodstuffs and energy sources, all cities currently rely on extensive physical infrastructure to sustain the regular importation of huge masses, **watts**, and volumes of food, energy, and clean water. Human-built physical infrastructures supplement, and in some cases, replace or dominate, natural physical infrastructures that provide FEW commodities in some degree of quality and quantity. Significant physical infrastructure also exists for the disposal of waste materials and water.

In addition to physical infrastructure, humanity has communication infrastructures, service infrastructures, and social infrastructures that guide the functioning of physical infrastructure and convey values, culture, and money.

Food, energy, and water systems each include infrastructure that extends from production to processing to distribution to consumption.

For Food

- Food production, for much of the world's population, comes from a global network of farms and fisheries which receive inputs of human labor, seeds, animal feeds, fertilizers, herbicides, fuels, pesticides, and irrigation water; may produce waste byproducts such as manure and contaminated water that require processing and management; and utilize a wide range of farming and fishing equipment. In many countries, agriculture is one of the largest users of water. In the USA and the nations of the European Union, agricultural production is guided through governmental policy infrastructure that influences what food, fiber, and fuel will be produced in what ways. These influences include when crops are planted, how much water is used, how much fuel is produced, how global food markets **function**, and more.

- Food processing transforms raw products into edible products using technology that ranges from handheld tools to large factories. Which edible products are produced is influenced by the social infrastructure of government dietary guidelines and the communication infrastructure of commercials and advertisements?
- Food distribution ranges from local transportation to international **trade routes** that utilize stationary and mobile **storage** equipment as well as roads, railways, ships, and aircraft. A myriad of distribution nodes connects farms and processing facilities to restaurants, markets, and shops that make the food available to the plates of end consumers. Thus, food distribution relies on social infrastructure ranging from commodity exchanges to marketing contracts to food assistance programs.
- Electrically-intensive refrigeration in residences, food services, retail, and wholesale, is one of the largest users of anthropogenic energy in the food system.

For Energy

- Energy production comes from a global network of oil and gas wells, coal mines, hydropower dams, nuclear power plants, wind turbines, solar cells, and fields of sugar and corn and soybeans. For many individuals and communities, energy comes from the gathering of wood, peat, and animal waste. Each energy source requires particular equipment. Even locations with onsite energy sources generally need regular supplies of fuel from offsite or to be connected to an electrical grid to balance supply and demand.
- Energy processing includes refineries and electric power plants of many kinds, as well as charcoal kilns and inverters on rooftop solar panels. Refineries and power plants rely on marketing infrastructure to arrive at the necessary contractual arrangements and on government policy infrastructure to regulate air and water and **soil** pollution.
- Energy distribution includes oil and gas pipelines, barges and oil tankers, road and rail, terminals, and the most complicated machine in the world, the modern electric grid. These pipelines and tankers and grids are guided and regulated by governmental and other social infrastructure to ensure their **security** and equity.

For Water

- Water production comes from lakes, rivers, **groundwater**, and increasingly, rain gathering, direct and indirect potable (drinkable) reuse of wastewater, or desalination of seawater. These processes use both "grey" built infrastructure and engaging natural or "**green**" **infrastructure**.
- The twentieth century's largest producers of electrical power, thermoelectric and hydroelectric power plants, require copious **withdrawals** of water, particularly for cooling.
- In economically developed areas, massive and energy-intensive physical and social infrastructure process and treat and distribute water with the expressed aim of protecting consumers from water-borne pathogens, toxic chemicals, and other contaminants.
- Water distribution requires aqueducts and pipe infrastructure that are second only to the electrical grid in their extent and pervasiveness. Storing water often requires dams, tanks, wells, and other human infrastructure. Distributing water through large physical infrastructure requires economic arrangements for the maintenance of the physical system ("user fees"), and governmental policies determining which users have a right to water.

The sheer complexity of FEW infrastructures is possibly the most significant "wonder" of the modern world, a wonder often unnoticed while in plain sight. One challenge is to make this "invisible" set of systems visible.

Such systems are much more visible in societies where such infrastructure is missing or dysfunctional, and the impact shapes the daily lives of people who toil each day to gather sufficient food, energy, and water to sustain themselves and their families. In most such societies, a disproportionate share of this burden falls to women, who must pay much higher real costs for these resources in terms of time consumed, education forgone, and health impacts.

The "hard" metal and concrete infrastructures that provide the physical functions of modern societies are quite visible. However, a **soft infrastructure** of human institutions and activities enable the building, maintenance, and functioning of **hard infrastructure**. These and other infrastructure issues are explored in Chap. 10.

1.3.1.3 Criterion 3: Footprint-Heavy

The direct and indirect **biophysical** impact of consumption of **natural resources** on Earth is commonly referred to collectively as the "**footprint**" of consumption. The footprint concept is particularly useful when applied to natural resources that have a "planetary boundary" on their availability.

A **planetary boundary** is a human-determined threshold beyond which, with a margin of safety, there is a significant risk that an important biophysical process undergoes an irreversible decline or collapse. Nine planetary boundaries that have received particular attention (Rockström et al. 2009) are climate change, ocean acidification, stratosphere ozone depletion, global phosphorus and nitrogen cycles, rate of **biodiversity** loss, global freshwater use, land system change, aerosol loading, and chemical pollution. Food, energy, and water are related to nearly all of these issues. Note that biophysical boundaries exist at all geographical scales ranging from local to planetary, and local and regional boundaries often have a global impact. Understanding biophysical, or **biogeophysical**, boundaries is a challenging scientific subject that includes many of the issues related to the production and movement of food, energy, and water; their many interrelationships to each other and to **ecosystem services** (Chap. 9). Establishing a margin of safety for a biophysical process is a normative exercise based on decisions about risk and uncertainty.

We noted above that FEW commodities are essential in both sufficient quantity and quality. A significant aspect of footprint related to FEW commodities is their form and qualities. For example, the footprints associated with meat production are significantly different from those associated with grains. Similarly, the footprints associated with coal-based energy are very different from that associated with natural gas-based or wind-based energy. And water obtained from a nearby well has a very different footprint than water delivered in plastic bottles far from its source. Throughout this book, we will focus on the energy and water footprints of food, the land and water footprints of energy, and the land and energy footprints of water.

Examination of footprints also brings into focus inefficiency and waste in the production, use, and discarding of FEW commodities. Later in the book, we will explore the concept of **Life Cycle Assessment** as a methodological framework for assessing the environmental impacts associated with a FEW systems (see Sect. 13.2.1).

A concept closely related to that of planetary boundaries is that of **carrying capacity**. Carrying capacity is the estimated maximum population of a species that an environment can sustain indefinitely. The commonly asked question, "How many people can the earth support?" is thus based on understanding planetary boundaries. A book titled with this question by Joel Cohen (1996) demonstrates the difficulties of providing a clear answer.

Food, energy, and water are responsible for the majority, and arguably, the vast majority of humanity's footprint on the earth. twenty-first-century humanity is exceeding the Earth's local, regional, and even global carrying capacities as measured by terrestrial photosynthetic productivity and usable land, replacement rates of ocean fish, renewable fresh surface water availability, or the ability of the **biosphere** and oceans to absorb **greenhouse gases** like **carbon dioxide**.

The fundamental long-term **sustainability** of these consumption rates is in doubt. Solving the sustainability problem will require some combination of increased efficiencies of production, the redistribution of consumption from more affluent consumers to less affluent consumers, and reduced per capita consumption by larger consumers. Affluent city dwellers tend to outsource their FEW supplies and wastes, and the associated **externalities** and footprints, to their rural neighbors.

1.3.1.4 Criterion 4: Extreme Affordability

By definition, to support life, all people need reliable supplies of food, energy, and water that must be of at least adequate quantity and quality. Because a significant fraction of humanity has modest, or even minimal, financial resources, quality FEW products are required at near-zero marginal price, even during **droughts** or floods, harsh winters and hot summers, and economic downturns—**extreme affordability**.

Extreme affordability may necessitate massive **economies of scale**, household-level subsistence economies, government subsidies to production and storage, policies providing FEW accessibility and human rights, and government regulation of minimum quality standards. The affordability, subsidy, and regulation for FEW commodities can often be a barrier to **private sector** innovation and solutions because maintaining low-cost access results in small (or nonexistent) profit margins to producers and disincentives to investment. Thus, one of humanity's most potent problem-solving tools, private for-profit innovation, is often not fully engaged in solving FEW problems.

1.3.1.5 Criterion 5: Governance-Heavy

The word **governance** refers to the processes by which groups of people make and implement decisions, policies, and rules. As such, governance refers to the policies, laws, institutions, and actions made by formal governments at all levels of society; between and among governments (e.g., **international treaties**); and with entities outside of formal government (e.g., private corporations and **civil society** organizations). Throughout this book, it is important to not confuse the words "governance" and "government." It has been said that governance is "governing with or without a government, policy-making with or without politics" (Colombi and von der Pfordten 2011). Governance will be explored more deeply in later chapters, especially Chaps. 6 (International Governance) and 8 (United States Governance).

Given that FEW commodities are critical for all human societies, it not surprising that they are subject to significant governance. Nations establish domestic policies on production and consumption, support public and private investment in infrastructure, engage in international diplomacy, and have strategies to address disruptions in supplies. Policies include:

- Recognition of many FEW commodities as a human right (see Sect. 6.2.4). For example, through the United Nations, countries have asserted the right to an adequate standard of living in the International Covenant on the Economic, Social and Cultural Rights (ICESCR), which also refers to the right to food as a vital element of an adequate standard of living (Article 11).
- Investment and incentives for the extraction/production and storage of food, energy, and water.
- International arrangements to ensure external supplies from foreign nations through bilateral and multilateral trade agreements (Chaps. 6 and 7). Natural

flows of water across international boundaries are a particular source of potential conflict and motivation for cooperation. Domestic policies that minimize dependence on foreign supplies are also standard; take, for instance, the USA's emphasis on energy self-sufficiency following the oil shocks of the 1970s, or the emphasis of Israel, Iran, or, until recently, China on FEW self-sufficiency despite strained local natural resources.

- Distribution of food, energy, and water via hard (physical) and soft (organizational) infrastructure (see Chap. 10) and in a manner that balances the essential need of consumers with the needs and rights of producers and intermediaries.
- Quality control for healthy consumption leads to policies regarding contaminants, nutrition, and health impacts related to food production and preparation, energy use, and water supply systems, as well as the disposal of residues and waste.
- **Externalities** must be regulated to avoid adverse impacts on human health and natural ecosystems. While FEW systems have historically been considered topics of local or regional policy, they are now a source of tension between local, regional, national, and international policies. Chapters 5 (Economics), 6, 8 (International and US Governance), 9 (Ecosystems), 11 (Climate Change), and 20 (Human Conflicts) will all explore this tension in greater depth.
- Failure to provide access to minimum supplies of food, energy, and water can lead to the breakdown of civil order. Therefore, public policies exist everywhere to ensure sufficient volumes of supply, and address issues of cost and services to poor and vulnerable communities are critical; issues related to food, energy, and water security are developed in Sect. 3.2.
- Programs supporting research, development, education, and outreach aim to empower participants in different parts of FEW systems to achieve societal goals ranging from production to healthy consumption.

Separate or conflicting policies for food, energy, and water frequently lead to conflicts between communities with a primary interest in one of the food, energy, or water components. For example, in the USA currently, there are conflicts between farmers whose fields generate contaminated runoff to streams versus downstream communities who bear the negative impacts of the contaminated water. While international policies usually involve diplomacy and trade arrangements, nations can and have resorted to a military conflict to address food, energy, and water crises. Shortages in food, energy, or water can also lead to innovative ways of managing these shortages (see Chap. 20).

Each of these five criteria motivates the integrated study of food, energy, and water systems. Each criterion must be addressed by societies for each critical consumable in a manner that does have significant adverse effects on the other two. Thus, the many conflicts and trade-offs between food, energy, and water systems that meet the demands of people motivate the integration of study, governance, and actions.

While the five criteria define the motivations for integrated FEW studies, these systems operate in the context of important processes that are fundamental to their functioning regarding environmental and human concerns. These processes are essential components of any consideration at the nexus of FEW systems and receive our attention now.

1.3.2 Core Processes Influencing the Nexus

Several processes that influence and interact with the nexus are commonly discussed, each with the potential for a variety of nexus impacts depending on their characteristics, functions, and interactions with other processes. Each has varying degrees of responsiveness and importance to the five criteria introduced above. The importance of these processes has led some to frame the nexus as including them as components equivalent to food, energy, and water; an alternative to the FEW framings described above. Here, based upon the criteria that we have just discussed, they are viewed as critical processes shaping the FEW nexus. A fundamental understanding of these processes and their underlying components will be helpful to a scholar of nexus studies, who may find in the Further Reading and Educational Resources throughout the book additional background for FEW nexus topics outside their primary discipline.

1.3.2.1 Population Growth and Societal Development

Recognizing food, energy, and water as "three critical consumables" anchors the nexus in considerations of people, their demographics and locations, the conditions in which they live, and their aspirations for nutrition, energy services, and water—and a certain "quality of life."

The total human population on earth, 7.7 billion in 2019, while profoundly important, is a superficial **indicator**. The consequences of how human societies function concerning food, energy, and water are complicated and related to far more than the numbers of people within them. Factors such as geography, local resources, and ecosystems affect how people live. The age structure, wealth, education, social inequality, fertility, political and economic structures, and many other factors are every bit as necessary as the absolute number of people for an understanding of the structure and functioning of the FEW nexus.

In Chap. 3, we will explore Demographics and Sustainable Development. We will do so with an eye to the billions of people on this planet who aspire to live like their wealthier neighbors. We will focus on the concepts of resource "security," "development," and initiatives like the Millennium Development Goals (2000–2015) and the Sustainable Development Goals (2015–2030) which seek to bring about greater access to food, energy, and water, as well as other aspects of human development.

While the focus of development is usually on the poorest countries on the planet where simple access to adequate supplies of food, energy, and water are critical and ethical objectives, it is important to remember that development never stops, even in the wealthiest countries. Therefore, how development occurs in the more developed nations is also essential. Development in those countries includes increasing sustainability, and increasing quality of life while decreasing the ecological footprint. Thus, in this book, we present case studies from countries across the full range of human development. We will explore this issue as it applies to cities in Chap. 18, which includes case studies on Portland and Detroit (USA), Curitiba (Brazil), and Tianjin (China).

1.3.2.2 Air Pollution

Human beings require a continuous supply of clean air to live, and so it is a critical consumable. Significantly, energy production and energy consumption to process and maintain food and water supplies, among other societal demands, may threaten air quality. While specific infrastructure is required to limit air pollutants entering the atmosphere, physical infrastructure that is protective of the environment and public health may interact substantially with FEW systems. For example, field crop production may release both dust and pesticides, and animal production may release noxious chemicals and pathogens into the air. Water and wastewater treatment can also emit noxious chemicals.

Rightly or wrongly, societal actions to ensure clean air have been secondary to activities on food, energy, water, and other forms of economic development. Most nations have achieved, or are reaching, significant levels of development while enduring very unhealthy air quality, as discussed further in Chap. 18 on Cities at the Nexus. While western countries began to address air issues in the 1950s, 1960s, and 1970s, China has only recently mobilized politically to start tackling air pollution, and India continues to suffer from severe air quality degradation. Thus, air has been a significantly lower policy priority compared to food, energy, and water.

1.3.2.3 Ecosystem Services

While humans need land (and importantly its soil and hydrological qualities, in the context of its climate and biogeography), productive waters (oceans, and especially estuaries because of their necessity for fisheries production), freshwater **aquifers** and surface waters, and broader ecosystem services for the production of FEW resources, ecosystem services are not consumable commodities in the same sense as land and water.

They are, however, degradable, which can significantly impair their ability to contribute to the production of FEW commodities. Recognizing degradation can tempt one to consider land and ecosystem services as consumable. By definition, ecosystem services can be restored, and land can be made available for productive use, but perhaps not on the scale or timeframe or having the quality or quantity needed to address societal demands and preserve socio-political-economic stability. A core motivation for integrating the study of FEW systems is to ensure that all three consumables are provided simultaneously and sustainably—in a manner that requires management and maintenance of land, soils, waters, air, and ecosystems.

Hard (physical) infrastructure has a nuanced relationship to ecosystem services. Superficially, one can say that hard infrastructure for food and water delivers the products of ecosystem services. However, while energy from hydroelectricity and other renewables draw on ecosystem services, energy from fossil fuels is the result of geological actions on biomass over millions of years rather than recent **ecosystem functions**.

Nexus studies often treat ecosystem services in a similar way to infrastructure, or as "natural extensions" of hard infrastructure. However, in practice, ecosystem services are different from hard infrastructure because of their broader values and services to society.

Ecosystems can mediate changes within individual food, energy, and water systems, and between systems. For example, hydroelectricity production impacts sediment flow in rivers, which can affect downstream agriculture and water use. Several case studies in this book explore land and ecosystems as mediums for nexus interactions.

1.3.2.4 Climate Change

Climate averages and cycles and the weather patterns that they give rise to are central to where human communities exist and how they operate. It is primarily the impact of climate change on where and how human communities live, which makes it a significant concern.

In some ways, the earth's climate operates similarly to land, soils, waters, air, and ecosystems, providing environmental services rather than being a consumable commodity. Like land, soils, waters, air, and ecosystems, the climate has proven to be subject to profound impact by human activity, which can alter climate averages, increase and decrease rainfall, increase the frequency of extreme events, and bring about sea-level rise.

Climate also resembles air in that it arrives naturally without the aid of human-built physical infrastructure. Rather, like air, human-built physical infrastructure related to climate is primarily to protect humans from negative impacts related to weather.

Climate change as a high-level policy issue arrived in the 1980s, several decades later than air regulations in economically advanced nations. However, climate change has risen to the top of the global policy agenda in a manner in which local and regional air pollution has not because climate is a global, rather than local or regional, problem.

Human impact on the earth's climate change is primarily the result of the dominance of carbon fuels in nearly all energy systems, and, to a lesser degree, to land-use changes related to food production. Both result in the intensification of the earth's natural greenhouse effect, warming the planet with many consequences. Therefore, efforts to mitigate climate change require profound shifts in energy and food production, hopefully without reducing energy services or nutrition.

Changes in climate patterns significantly impact the supplies of food, energy, and water on which societies depend, and, to a lesser degree, the demand for these consumables. Climate shapes what types of crops can be grown and where, and what kinds of seafood can be harvested and where. Changing growing seasons and **regions** suitable for specific food crops is a significant concern. Climate defines how much water is available and where. Climate impacts many kinds of energy production, such as hydropower, wind, and solar. Climate also influences the demand for water and energy, as when more water is needed to grow food in a desert, like California, or less energy is needed for shelter in a mild climate like San Diego.

The impacts of climate change are seen as a **threat multiplier**: floods, droughts, extreme heat, and other climate-related issues can multiply underlying threats and increase tensions. For example, the Syrian Civil War is traced in part back to a 7-year drought that caused massive crop failure and exacerbated civil unrest from Syrians lacking sufficient food, energy, and water. However, as explored in Chap. 20, the sheer necessity of a critical resource like water can bring people together to address these tensions.

Because fossil fuels and agricultural changes in land use are drivers of climate change, it is appropriate to think of climate as mediating modifications between food, energy, and water systems. Climate as a medium for nexus interactions is explored in Chap. 11 and case studies throughout this book.

1.3.2.5 Sociopolitical Economics

Sociopolitical economics is one of the fields that study how societies choose to allocate scarce resources to satisfy their unlimited wants. Food, energy, and water are, for practical purposes, physically limited resources for which human societies have unlimited wants (especially for energy). Sociopolitical economics at the nexus is not just about quantifying the value of the three consumable commodities and in their many forms along with **inputs and outputs**, but, critically, about quantifying the value of impacts related to FEW systems and their interactions. However, sociopolitical economics struggles in practice with commodities and markets that are heavily regulated and subsidized, subject to intangible valuation, or laden with externalities. The integration of FEW systems is impossible to do without the application of economics and political science and sociology and anthropology. Thus, in Chap. 5, we will explore many aspects of sociopolitical economics at the nexus and apply social and political and cultural and economic considerations throughout our study of the nexus. Chapter 18 discusses the **dynamics** of sociopolitical economics relating to cities.

Now that it is clear why food, energy, and water are the three prioritized components of nexus studies, it should also be clear that population and development; air, land, soils, water, and ecosystem services; climate and climate change; and economics are core contextual components of nexus studies. These additional topics do not bound the scope of essential elements of nexus studies. Other critical issues like biodiversity are also crucial in specific nexus contexts.

1.3.3 Grand Challenges

The relevance of the five criteria described in Sect. 1.3.1 is reflected in several initiatives to articulate the "grand challenges" facing humanity in different areas.

In Chap. 3, we will explore the challenges of people everywhere to having affordable access to their basic food, energy, and water needs to live healthy and productive lives. This issue of food, energy, and water security is a significant part of sustainable development. Food and water were explicit parts of the United Nations (UN—see Sect. 6.2.2) Millennium Development Goals adopted in 2000 (see Sect. 3.5). The 2012 UN Sustainable Energy for All program brought energy security to the fore (see Sect. 3.6). The 15-year UN Sustainable Development Goals adopted in 2015 includes specific goals on food, energy, and water. Such high-level needs have also been compiled by others.

In 2018, the U.S. National Academy of Engineering identified 14 grand challenges for engineering in the twenty-first-century, emphasizing technical solutions needed to address the national economy and general welfare. Several of these grand challenges directly implicate FEW:

- Make solar energy economical
- Restore and improve urban infrastructure
- Provide access to clean water
- Provide energy from fusion
- Manage the nitrogen cycle
- Develop carbon sequestration methods

The U.S. National Science Foundation's Mathematical and Physical Sciences Advisory Committee published a 2014 report on global opportunities for research in FEW, emphasizing basic science and applied science opportunities. Six foundational knowledge areas were identified as opportunities, with an emphasis on the need for transformative and non-incremental solutions:

- Ensuring a sustainable water supply for agriculture
- Closing the loop for nutrient life cycles
- Crop protection
- Innovations to prevent waste of food and energy
- Sensors for food security and safety
- Maximizing biomass conversion to fuels, chemicals, food, and materials

Many other high-level planning efforts are currently underway. Additionally, several prior efforts have addressed related topics, most notably the "energy–water nexus" reports from the first decade of the millennium. The U.S. Department of Energy released in 2014 a water-energy nexus report identifying the following issues and priorities in the USA:

- Interdependency of energy and water at regional scales
- Use of energy for water production and vice versa
- Impact of drought on **thermoelectric power plant** cooling

- **Cascading failures** where a power system failure causes a water system failure
- Changing climate drives changing water demand and availability
- Massive population growth in water-stress regions like the Southwestern USA
- Interactions of water and energy efficiency via life cycles and footprints
- Water rights and water requirements for energy production including hydraulic fracturing
- Tribal water rights and energy production
- Water demands by irrigated agriculture

These examples illustrate that food, energy, and water challenges have been recognized from many perspectives and further demonstrate the importance of integrating them.

1.4 FEW System Framings in the Literature

1.4.1 The Macroscope

In studying the nexus of FEW systems, we are integrating three systems into one more extensive system. While traditional studies of FEW systems have similarities, they also have some significant differences. These differences must be carefully understood when attempting to treat the three components in a consistent and balanced manner to undertake an integrated study. Therefore, before we take the step of integrating the three systems, we will consider some aspects of individual food, energy, and water systems.

Complex systems are named this way by scientists because they are too complex for a human being to perceive, understand, or reduce in their entirety. As a result, we need to use “lenses,” “slices,” or other thinking tools to get inside these systems’ **structure** and function and to grab hold of some key aspect of the system—only one part of the system, but a useful and accurate perception of that part. Joel de Rosnay in 1979 described the strategy that we use to perceive and sense complex systems as “The Macroscope.” A **macroscope** is a tool that lets people see complexity in perspective, and is an analogy to the much older concepts of microscopes and telescopes. The FEW nexus concept generally, and its many specific conceptualizations and implementations, is viewed through the lens of a macroscope (Fig. 1.1).

The literature since 2014 has exploded with ideas and concepts for the framing of FEW systems. A handful of these systems concepts are summarized below. These experts have attempted to build macroscopes for the complex FEW system. There are counterarguments that this effort to build a FEW macroscope has not been successful to date. However, given the criticality of the FEW problem space, this is a good reason to think deeply and continue searching for frameworks that will allow us to accurately sense the context of this complexity and act.

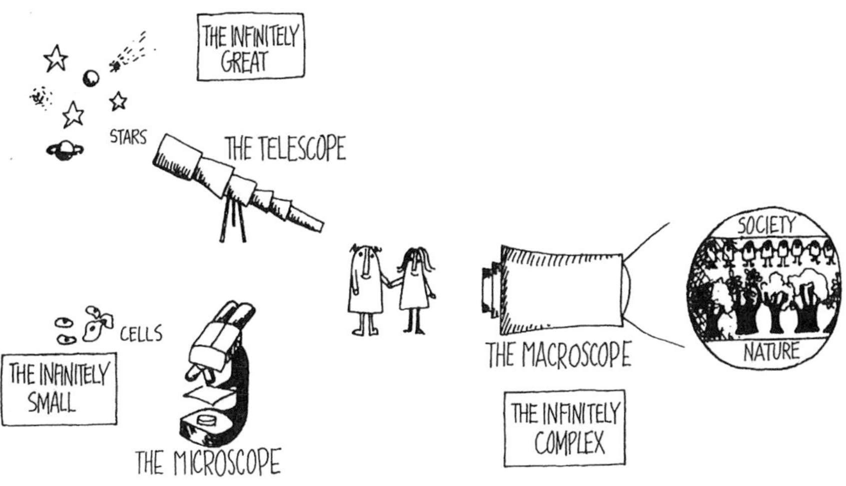

Fig. 1.1 The macroscope. (Source: de Rosnay 2014)

1.4.2 Lant Framing

Lant et al. (2018) identified the following characteristics of the FEW system:

1. It has a clear "core," but fuzzy system boundaries because of the many connections that exist between natural systems and human societies—the **FEW-everything system**.
2. There are four FEW system facts should inform our work:
   (a) It dominates the human footprint on earth's carrying capacities.
   (b) It operates primarily at the **meso-scale** of cities, counties, and local watersheds (see Sect. 12.3 for a discussion of meso- and other scales).
   (c) Cities are its hubs of processing, transit, and consumption (but not production).
   (d) It forms a network and can be understood best with network concepts (Fig. 1.2).

1.4.3 D'Odorico Framing

D'Odorico et al. (2018) emphasized that the FEW system implicates the economic factors of land, labor, and capital. The system can be understood by focusing on the transitions that need to be undertaken in its components (e.g., renewable energy transitions), and by focusing on the interdisciplinary problems and solutions that touch on all three subsystems (e.g., climate change) (Fig. 1.3).

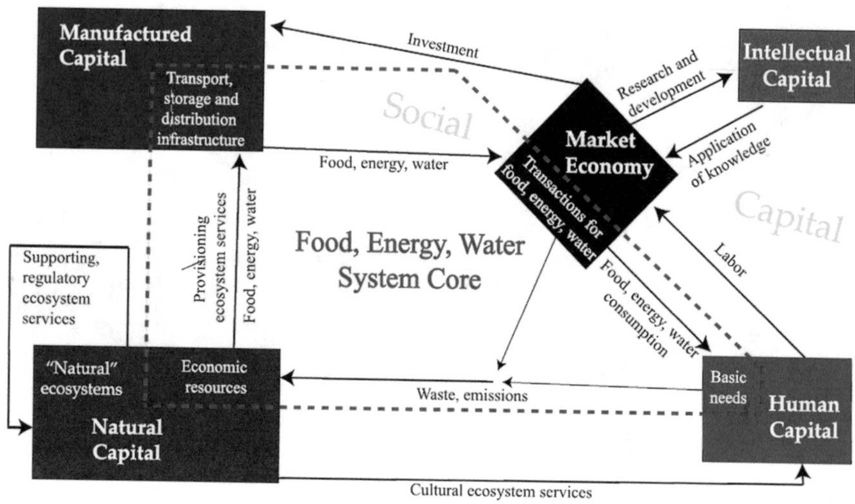

Fig. 1.2 Diagram of how food–energy–water (FEW) systems are embedded within the broader ecological economy. Note that FEW systems are important for ecosystem services. (Source: Lant et al. 2018)

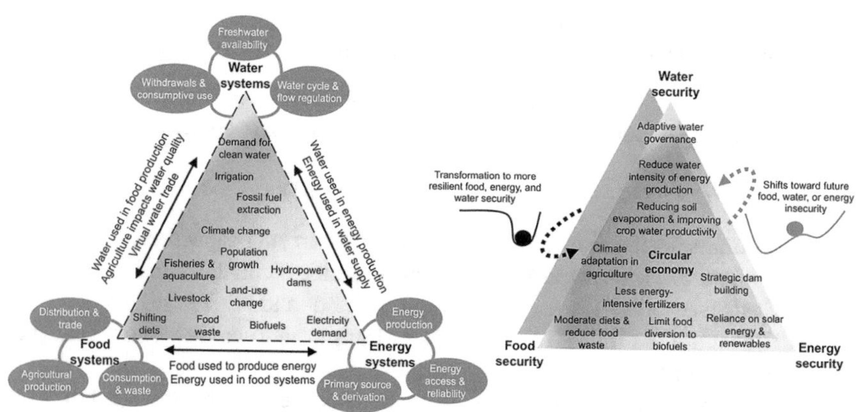

Fig. 1.3 The inherent linkages between individual food, energy, and water systems, including the competition in demand for water between food and energy production (adapted from UN Water, 2013). The right panel shows a conceptual depiction of resilience in the food–energy–water nexus. (Source: D'Odorico et al. 2018)

1.4.4 Scanlon Framing

Scanlon and colleagues (2018) join D'Odorico and colleagues in emphasizing the interdisciplinary nature of the nexus and the importance of focusing on problems and solutions at the nexus, as a strategy for making sense of the complex system.

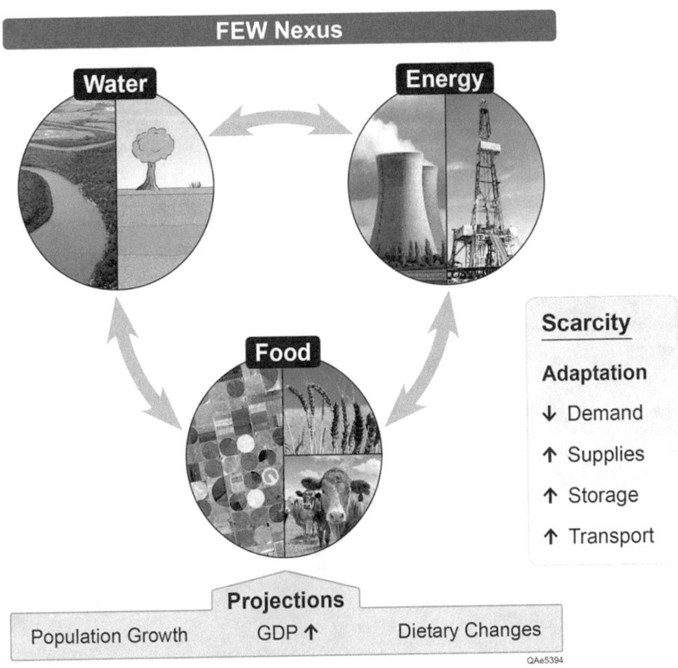

Fig. 1.4 The food–energy–water nexus showing the interconnection among the components and adaptation strategies for scarcity. (Source: Scanlon et al. 2018)

In general, the goals are straightforward. We need to decrease FEW demands, increase supplies, increase storage buffers, increase transportation and trade capacity and connectivity, and do so in the presence of climate change, population growth, growing wealth and consumption, and dietary changes. An emphasis is placed on the need for improved monitoring and information on the system, especially at the micro- (or fine-) to meso-scales where most of the system's impacts and decisions happen; national-level data is insufficiently actionable to provide solutions (Fig. 1.4).

1.4.5 Bazilian Framing

Bazilian et al. (2011) conceptualized the FEW system as a modeling problem that should be addressed in a reductionist manner using an integrated assessment framework. This paper emphasizes the importance of measuring and modeling the interrelated security of FEW systems and accurately resolving the causes and effects of changes in FEW securities. This focus makes sense because key attributes of FEW systems include their criticality and their inherent insecurity (Fig. 1.5).

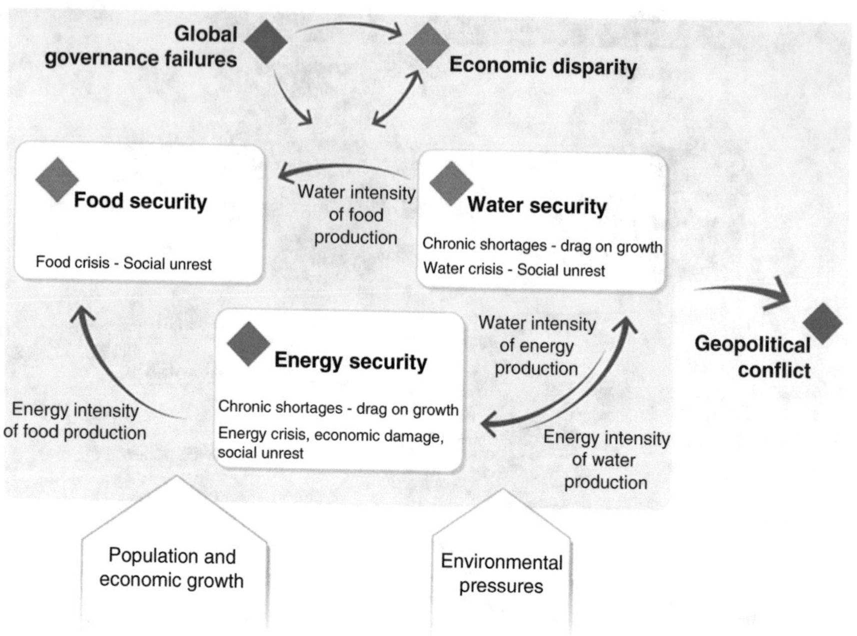

Fig. 1.5 System diagram for risks associated with the macroeconomic imbalances nexus. (Source: World Economic Forum 2011b used by Bazilian et al. 2011)

1.4.6 Ringler Framing

Ringler et al. (2013) include land as a fourth component of the FEW nexus. This study emphasizes the importance of focusing on aspects of FEW systems that are directly helpful to addressing the U.N. Sustainable Development Goals (see Sect. 3.5), whose goals can be used as indices for the success of our efforts to transform the FEW system. Additionally, this study emphasizes the synergistic efficiencies that can be gained when increases in efficiency or reductions in consumption of one of the three commodities (e.g., food) results in a reduction in consumption of the other two (water, energy) (Fig. 1.6).

1.4.7 California Framing

The US State of California is one of the world's largest and most advanced economies and is also a microcosm for many of the world's FEW problems, given its status as a major food-producing region located in a desert. The California Department of Water Resources is responsible for managing water deliveries to ecosystems, cities, farms, and hydropower facilities and has attempted, therefore,

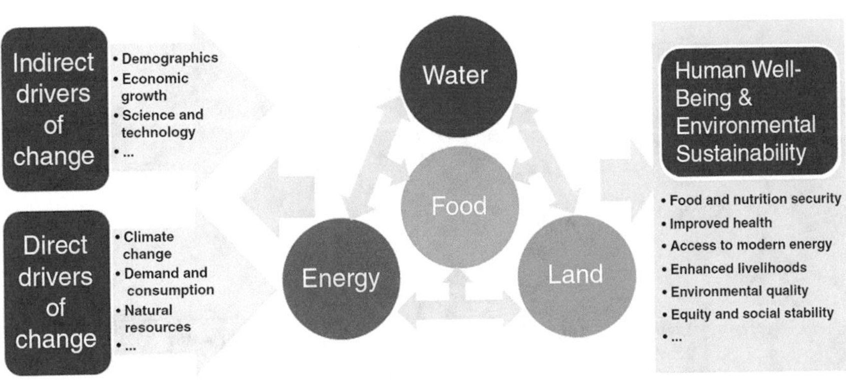

Fig. 1.6 Extended water, land, energy, and food nexus framework. (Source: Ringler et al. 2013)

to develop an understanding of the FEW nexus in this state. One of the three legs of the FEW system—in this case, water—can provide an appropriate lens through which to see the system. Climate change, drought, and its forcings on water supply and demand figure prominently in the California Department of Water Resources (CADWR) conceptualization of the FEW system, as do ecosystem water requirements and life cycle costs and efficiencies of FEW consumption (Fig. 1.7).

1.5 Solving Problems at the Nexus

1.5.1 Objectives of Studies of FEW systems

The criteria that lead us to integrate FEW systems are human-centric. Thus, it is not surprising that the ultimate purpose of studying the nexus of FEW systems is to guide human decision-making so that obtaining food, energy, and water for human use can be achieved in a manner that is both sustainable and consistent with other objectives related to environment (e.g., ecosystems, biodiversity, climate, etc.) and human development (e.g., equity, self-actualization).

Decision-making on food, energy, and water occurs at every level of social organization, both public (i.e., village, town, city, county, state/province, region, nation, international, and global) and private (i.e., household, small businesses like farmers, large companies, nonprofit and nongovernmental organizations, and many more).

FEW systems nearly always include a large number of independent actors of different kinds (individual, public, and private) of different sizes and power with different objectives and different values and cultures. Problems between different actors may be said to be "tractable" when their goals overlap, and compromise is desired or pushed. Where there is no overlap in the objectives of the different actors,

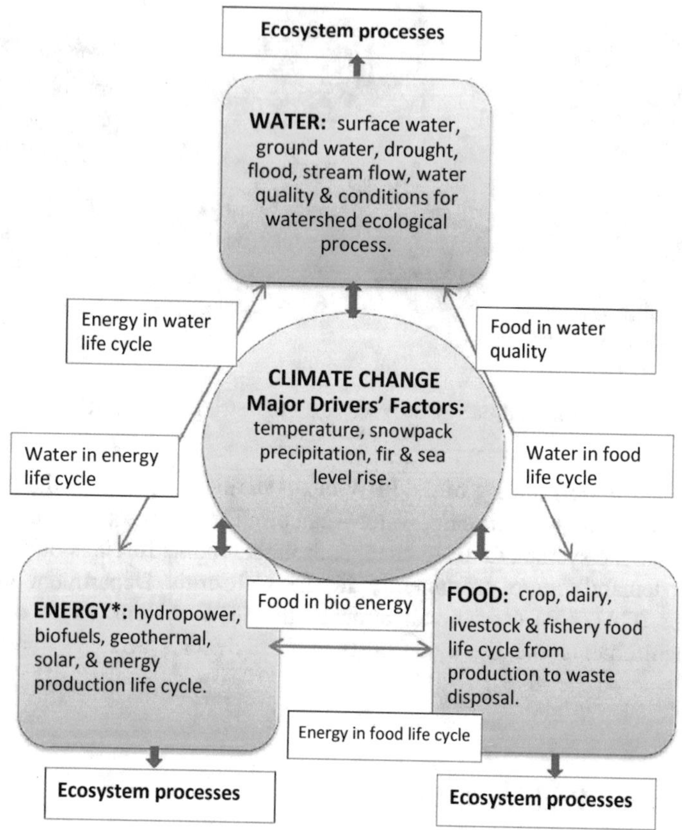

Fig. 1.7 Conceptual framework to connect climate change, water, energy, and food to ecosystems. (Source: California Department of Water Resources (CADWR) 2017)

no willingness to compromise, or big differences in framing the challenges, problems may be viewed as intractable or "wicked."

Power imbalances, where one actor, or set of actors, has significantly greater power than others can often lead to outcomes that prioritize their objectives over those of others. Some actors have a preference for culture-based or values-based interpretation of facts and science. Societies have different types of formal or informal conflict resolution, including political action and judicial decision-making; a range of conflict resolution methods are explored in Chap. 20.

Risk is a function of multiple factors and has been discussed as a function of (variously) total exposure to events, event severity, armoring or **vulnerability** to the event, probability of the event, and the number of potential events that may occur. Specifically, **risk** is an estimate of likely impact or damage (sometimes measured in dollars ($)), and is a function of roughly three factors: (1) the likelihood (or frequency) of an event, (2) the vulnerability of a system to the event (or, how protected a system is from the event), and (3) potential impact or severity of an event if it occurs and overwhelms the system's protection; for events with high likelihood, high vulnerability, and high severity, risk is high for a system.

Against this backdrop, studies at the nexus of food, energy, and water systems seek to understand the integrated system well enough to assess important attributes such as the following:

- The ability to provide essential food, energy, and water resources over select scales of time (**reliability**);
- The capacity to recover from disruptions (**resilience**);
- The vulnerability of communities to rapid, abrupt, nonlinear, or cascading changes which "tip" part of the natural world into a new mode of behavior;
- Estimate systemic risk;
- The evolution of stressors on the system that reduces reliability and resilience over given scales of space and time and function.

Further, studies of FEW systems seek to do the following:

- Identify and develop options for meeting essential needs for food, energy, and water simultaneously with achieving other decision-maker objectives within a given context.
- Provide methods, decision-making tools, and technical guidance to guide the management resolution of conflicts and develop effective policies.

1.5.2 Decision-Making Context

It is critical to understand FEW system solutions are shaped by decision-making processes that are contextual in two crucial ways—the biophysical environment, and the socio-economic-political system. There is tremendous variation in geographic factors such as environmental conditions, and the natural resources present within a system. In addition, time considerations, especially political and economic timescales, are usually a significant factor influencing different decision-makers. Further, decision-making occurs within a context of the culture, the social and political and economic structures, the supporting technology, and the motivations and characteristics of the decision-makers. For example, a farmer has to make annual decisions about planting crops, a politician may be under a longer re-election time scale for making decisions, and the ecosystem itself may respond over decades to changes on the land through a longer "lag" time. Context means that studies of FEW systems often lead to very different conclusions about how to apply the results in different places.

Of course, some areas of decision-making provide greater impetus and opportunity for integrated FEW science and governance. For example, when multiple FEW commodities are acutely scarce, major systemic actions are often contemplated. For example, water scarcity might lead to careful consideration of the trade-offs in water use by people directly against that used by agriculture and for cooling power plants. Another example is where the actions of some people cause significant externalities that impact the welfare of others via FEW system connections. For example, the largest emissions of greenhouse gases come from affluent, high energy-use societies, while the adverse **externality** (climate change effects) fall disproportionately on less wealthy, low-energy-use societies.

A more positive example of where there are greater impetus and opportunity for integrated FEW science and governance is where potential benefits to many communities exist as a result of coordinated actions. For example, the international trade of FEW commodities creates systemic benefits when certain products can be produced and transported with a smaller footprint and at a lower cost in one area compared to another. In such cases, both parties to the exchange benefit.

We will return to these examples of greater impetus and opportunity for integrated FEW science and governance in the final chapter.

1.5.3 Projections, Predictions, Assumptions, and "Well-Known" Solutions

The behavior of complex systems is, by their nature, difficult to understand and model accurately. When systems are studied in the present to help understand the future, models of complex systems make **projections**, not **predictions**. Projections are estimates of future outcomes, based on specified **assumptions**.

A set of assumptions is a **scenario**; that is, the assumptions constitute a stated version of what the key inputs, conditions, and functioning will be for the system under study. For example, assumptions about resource availability, economics, technologies, policies, as well as about the relationships between those and other factors and outcomes like consumption, **technology adoption**, and changes in behavior, collectively constitute a scenario for the future of a system.

While projections are statements about what "would" happen under certain assumptions, predictions are **forecasts** of what "will" happen. Human actors in systems introduce uncertainty and make predictions are, at their core, informed guesses. Good forecasts include clearly stated margins of uncertainty.

Models, as explored in this book, are based on assumptions and therefore produce projections best summarized as: "If such and such happen, this will be the outcome. However, if such and such does not happen, the outcome will be different."

The simplest and most common assumption, but rarely the most accurate, is a "Business as Usual" (BAU) assumption where trends in resource availability, economic change, and technological advancement occur as close to the same rate and direction as they have in the recent past while policies remain unchanged. Given that policies change, sometimes significantly, in accordance with political actions that are very hard, if not impossible to know in advance, policy assumptions are, by their very nature, highly uncertain. The future availability of natural resources is often based on expectations about the depletion of known reserves, assuming present rates of consumption unchanged by discoveries of additional resources or changes in technology. The future of infrastructure is often based upon assumptions about the operating lifetime of the infrastructure (e.g., a power plant) in the absence of changes in economics, technologies, and policies that might shorten or lengthen the operating lifetime.

Assumptions about how technologies and policies will evolve are intrinsically uncertain. Like sporting events, while everyone knows that there will be a particular score, no one knows for certain what it will be. The careful selection and communication of assumptions are critical to successful studies of the FEW nexus. Assumptions and models are often tested through monitoring what is actually happening on the ground through networks of gages and sensors (**sensor systems**), market analysis, and so on.

The public and decision-makers commonly believe that assumptions in scientific models are declarations of what scholars "believe will happen," and model projections are quantified predictions of what will happen. This is incorrect and is frequently the source of frustrations on both sides.

A second common belief is that there are clear and straightforward answers to food, energy, and water challenges. As the journalist, satirist, and cultural critic Harold Louis Mencken wrote,

> Explanations exist; they have existed for all time; there is always a well-known solution to every human problem—neat, plausible, and wrong. The ancients, in the case at bar, laid the blame upon the gods: sometimes they were remote and surly, and sometimes they were kind. In the Middle Ages lesser powers took a hand in the matter, and so one reads of works of art inspired by Our Lady, by the Blessed Saints, by the souls of the departed, and even by the devil. [from H. L. Mencken (1920), Prejudices: second series, Alfred A. Knopf]

Throughout the remainder of this book, the reader should always remember that the systems being addressed are complex and that projections about how FEW systems will evolve will usually contain significant uncertainties. Uncertainties require decision-makers to acknowledge risk and adopt strategies that can be adapted as understanding develops. Different decision-makers and different stakeholders will have different perceptions and tolerance of risk.

Key Points

- The integration of food, energy, and water systems is essential because of five criteria:
  - FEW systems deliver critical consumable commodities;
  - They are heavy on physical infrastructure;
  - They are footprint-heavy;
  - They must be extremely affordable; and
  - They are policy-heavy at all levels of government.
- Several critical processes influence the FEW nexus, including population growth and societal development; air pollution; ecosystem services; climate change; and socio-political-economic processes.

- The academic literature provides several alternative "macroscopes" or frameworks. Applied resilience, sustainability, development, and security problems are the usual motivation for different FEW nexus framings.

Discussion Points and Exercises

1. Discuss the strengths and weakness of the five criteria for selecting food, energy, and water systems for analysis.
2. Describe a situation where an alternative framing of a nexus using one or more of the core processes influencing the nexus described in Sect. 1.3.2. What are the strengths and weaknesses of such an approach?
3. Discuss why the FEW Nexus has become an important topic for science and policy in the early twenty-first century—and why it did not emerge before.
4. Develop a list of five "grand challenges" for the Twenty-first Century with implications for food, energy, and water for one of the following countries: (a) Chile, (b) Fiji, (c) Japan, (d) Nigeria, (e) Poland, and (f) Saudi Arabia.
5. Discuss the strength and weakness of the FEW nexus framing proposed by Lant and colleagues.
6. Describe an example where the Lant framing is very effective and an example where it is not effective.
7. Discuss the strength and weakness of the FEW nexus framing proposed by D'Odirico and colleagues.
8. Describe an example where the D'Odirico framing is very effective and an example where it is not effective.
9. Discuss the strength and weakness of the FEW nexus framing proposed by Scalon and colleagues.
10. Describe an example where the Scalon framing is very effective and an example where it is not effective.
11. Discuss the strength and weakness of the FEW nexus framing proposed by Bazilian and colleagues.
12. Describe an example where the Bazilian framing is very effective and an example where it is not effective.
13. Discuss the strength and weakness of the FEW nexus framing proposed by Ringler and colleagues.
14. Describe an example where the Ringler framing is very effective and an example where it is not effective.
15. Discuss the strength and weakness of the FEW nexus framing proposed by the California Department of Water Resources.
16. Describe an example where the California DWR framing is very effective and an example where it is not effective.
17. Compare how factors important to FEW nexus considerations differ between a highly urbanized region and a rural region in the same country.
18. Compare how factors important to FEW nexus considerations differ between a highly industrialized country and one where large parts of the population have limited access to basic levels of food, energy, or water.
19. Discuss how changes in the FEW nexus might serve as a threat multiplier for underlying societal dynamics. Can you identify a country where such a situation exists?

References

Colombi, C. & von der Pfordten, D. (2011). *Governance meets law: Exploring the relationship between law and governance: A proposal*. Netherlands Institute for Law and Governance. Retrieved from http://nilg.nl/wp-content/uploads/2011/06/ProjectomschrijvingGovernanceMeetsLaw.pdf.

FEMA. (2008). *Community lifelines. National Response Framework update* (4th ed.). Washington, DC: Author. Retrieved from https://www.fema.gov/media-library-data/1544471807449-68999c40805e98f058822c55b1fef233/Revised_Community_Lifelines_Information_Sheet.pdf.

Mencken, H. L. (1920). *Prejudices: Second series*. Alfred A. Knopf ASIN: B00088H4QM

U.S. Department of Energy. (2014). *The water-energy nexus: Challenges and opportunities*. Washington, DC: DOE. Retrieved September 7, 2018, from https://www.energy.gov/sites/prod/files/2014/07/f17/Water%20Energy%20Nexus%20Full%20Report%20July%202014.pdf.

World Economic Forum. (2011b). *Global risks 2011*. Geneva: Author.

Further Reading

Aldaya, M. M., et al. (2012). *The water footprint assessment manual: Setting the global standard*. London: Routledge.

Bakke, G. (2016). *The grid: The fraying wires between Americans and our energy future*. New York, NY: Bloomsbury.

Bazilian, M., et al. (2011). Considering the energy, water and food nexus: Towards an integrated modelling approach. *Energy Policy, 39*(12), 7896–7906.

California Department of Water Resources (CADWR). (2017). *Connecting the dots between water, energy, food, and ecosystems issues for integrated water management in a changing climate*. Sacramento, CA: Author Retrieved September 7, 2018, from https://water.ca.gov/-/media/DWR-Website/Web-Pages/Programs/All-Programs/Climate-Change-Program/Climate-Program-Activities/Files/Reports/Connecting-the-Dots-between-Water-Energy-Food-and-Ecosystems-Issues-for-Integrated-Water.pdf.

Cohen, J. E. (1996). *How many people can the earth support?* New York, NY: W. W. Norton & Company ISBN: 0393314952.

D'Odorico, P., et al. (2018). The global food-energy-water nexus. *Reviews of Geophysics, 56*, 456.

De Rosnay, J. (2014). *Le macroscope. Vers une vision globale*. Paris: Le seuil Translated. Retrieved September 7, 2018, from http://pespmc1.vub.ac.be/macroscope/default.html.

Dodds, F., & Bartram, J. (Eds.). (2016). *The water, food, energy and climate nexus: Challenges and an agenda for action. Earthscan studies in natural resource management*. London: Routledge.

Evenson, R. E., & Gollin, D. (2003). Assessing the impact of the green revolution, 1960 to 2000. *Science, 300*(5620), 758–762.

Flammini, A., et al. (2014). *Walking the nexus talk: Assessing the water-energy-food nexus in the context of the sustainable energy for all initiative*. Rome: Food and Agriculture Organization of the United Nations.

Gleick, P. H. (2014). *The world's water: The Biennial report on freshwater resources* (Vol. 8). Washington, DC: Island Press.

Lant, C., et al. (2018). The U.S. food–energy–water system: A blueprint to fill the meso-scale gap for science and decision-making. *Ambio, 2018*, 1–13. https://doi.org/10.1007/s13280-018-1077-0.

McKinsey Global Institute. (2011). *Resource revolution: Meeting the world's energy, materials, food, and water needs*. New York, NY: McKinsey & Company.

Meadows, D., & Randers, J. (2012). *The limits to growth: The 30-year update*. London: Routledge.

NAE. (2018). *Grand challenges for engineering*. Washington, DC: U.S. National Academy of Engineering. Retrieved September 7, 2018, from http://www.engineeringchallenges.org/challenges.aspx.

NSF. (2014). *Food, energy, and water transformative research opportunities in the mathematical and physical sciences*. Alexandria, VA: National Science Foundation, Mathematical and Physical Sciences Advisory Committee. Retrieved September 7, 2018, from https://nsf.gov/mps/advisory/mpsac_other_reports/subcommittee_report_food_water_energy_nexus_final.pdf.

O'Riordan, J., & Sanford, R. W. (2015). *The climate nexus: Water, food, energy and biodiversity in a changing world*. Calgary, AB: Rocky Mountain Books.

Postel, S., & Richter, B. (2012). *Rivers for life: Managing water for people and nature*. Washington, DC: Island Press.

Ringler, C., Bhaduri, A., & Lawford, R. (2013). The nexus across water, energy, land and food (WELF): Potential for improved resource use efficiency? *Current Opinion in Environmental Sustainability, 5*(6), 617–624.

Rockström, J., et al. (2009). Planetary boundaries: Exploring the safe operating space for humanity. *Ecology and Society, 14*(2), 32 Retrieved March 11, 2019, from http://www.ecologyandsociety.org/vol14/iss2/art32/b.

Ruddell, B. L., & Kumar, P. (2009). Ecohydrologic process networks: 1. Identification. *Water Resources Research, 45*(3), W03419.

Sarni, W. (2015). *Beyond the energy-water-food nexus: New strategies for 21st-century growth*. London: Routledge.

Scanlon, B., et al. (2018). The food-energy-water nexus: Transforming science for society. *Water Resources Research, 53*, 3550–3556. https://doi.org/10.1002/2017WR020889.

Skaggs, R., et al. (2012). *Climate and energy-water-land system interactions* (Technical Report to the U.S. Department of Energy in Support of the National Climate Assessment, PNNL-21185). Richland, WA: Pacific Northwest National Laboratory.

U.S. Environmental Protection Agency. *Learn about sustainability*. Washington, DC: Author. Retrieved April 21, 2018, from https://www.epa.gov/sustainability/learn-about-sustainability.

Vörösmarty, C. J., et al. (2013). Global water, the Anthropocene and the transformation of a science. *Current Opinion in Environmental Sustainability, 5*(6), 539–550.

Webber, M. E. (2016). *Thirst for power: Energy, water, and human survival*. New Haven, CT: Yale University Press.

World Commission on Environment and Development. (1987). *Our common future*. Oxford: Oxford University Press.

World Economic Forum. (2011a). *Water security: The water-food-energy-climate nexus*. Washington, DC: Island Press.

Chapter 2
Systems Science

Peter Saundry and Benjamin L. Ruddell

2.1 Introduction to Systems Science

A system is a set of things connected in a way that creates some unified whole. The nature of a system is, to some degree, simple, complicated, or complex. This distinction between "complicated" and "complex" is important and subtle and is addressed below (Sect. 1.4.2).

Simple systems typically have a small number of parts with usually linear "cause–effect" interactions between the parts. For example, imagine a system of pulleys connected by a rope where a force pulling on the rope turns one pulley and then additional pulleys in a linear succession through the simple application of force imparted through friction between the rope and the pulleys. As an output, one of the pulleys is attached to a weight and lifts that weight. If the pulleys are suitably arranged, an applied input force moving the rope a considerable distance can result in a much larger force moving the weight a much shorter distance. This simple system operates under the application of physics to a small number of parts. There is an independent variable X (or a small number of independent variables), a dependent variable Y, and changes in X explain changes in Y through some function $Y = f(X)$. Simple systems have clear boundaries and are fundamentally predictable without much effort (if you know the calculus).

P. Saundry (✉)
Energy Policy and Climate, Advanced Academic Programs, Krieger School of Arts and Sciences, Johns Hopkins University, Washington, DC, USA
e-mail: psaundr1@jhu.edu

B. L. Ruddell
School of Informatics, Computing, and Cyber Systems, Northern Arizona University, Flagstaff, AZ, USA
e-mail: Benjamin.Ruddell@nau.edu

P. Saundry, B. L. Ruddell (eds.), *The Food-Energy-Water Nexus*, AESS Interdisciplinary Environmental Studies and Sciences Series,
https://doi.org/10.1007/978-3-030-29914-9_2

A food, energy, or water system might be defined by the set of sources, movements, uses, and sinks that constitute a way of understanding the unified whole in the context of a particular place and time. There are many parts, and the interactions between them are by no means simple; the relationships may be

For example:

- Nonlinear (e.g., water withdrawals may have thresholds beyond which significant changes in ecosystem function occur or where certain uses are prohibited).
- Multivariate (e.g., changes in energy demand depend on weather conditions, the rate economic growth, building size and location, demographic changes, and other factors).
- Multiscalar (e.g., food production occurs at the local level; domestic markets at a regional level; and trade at an international level; with each affected by factors at that level).

Further, a system's boundaries may be multifaceted (e.g., a food–energy–water system may have boundaries associated with an agricultural region, a regional electric grid; a watershed; and several political jurisdictions).

Studying such systems requires the careful use of science and much effort, but complicated systems are still fundamentally predictable in principle.

As the number of simply interacting subsystems within a system increases, the number of interactions increases geometrically, and systems become complicated very quickly. For example, a water system in isolation might have N interactions, an energy–water system might have $2N + 2$ interactions (energy and water separately, plus each of their effects on the other), and a food–energy–water system might have $3N + 6$ interactions following the same pattern. Water systems are determined by processes of supply and demand for water, water balances, and water quality, but when energy is included, every change to the water system cascades to affect the demand for energy to produce water and the demand for water to produce energy.

Complicated systems are predictable in practice if you can afford the workforce, data collection, and computing power necessary. Engineers are specialists in designing and managing complicated systems—like the space shuttle, the power grid, a fuel refinery, or a computer.

Complex systems are different because although they may have many parts or only two, they are fundamentally unpredictable to some degree, and chaotic because feedback renders the traditional idea of cause and effect meaningless. "Interaction" is a general term for all kinds of connections, correlations, feedbacks, and cause-effect relationships—both biophysical and human. Forcings or controls are interactions by which one subsystem causes effects in another subsystem. Feedback involves loops of causes and effects. For example, when increased demand for water increases demand for energy to produce water which reduces the supply of water and increases the cost of both the energy and the water. More broadly, weather conditions, policy decisions, ecological impacts, and economic activities are all difficult to predict and have two-way dependencies and impacts on the demand and production of FEW commodities over different timeframes.

The principles or processes in a complicated system might include the application of physics, chemistry, botany, hydrology, engineering, and many other physical and life sciences, as well as the application of social sciences applied to human beings, social organizations, and societies which operate under various economic, political, and sociocultural rules.

For FEW systems, these principles and processes of "system science" are applied to the sources, movements, transformations, uses, and sinks of food, energy, and water—a combination of the functioning of the biophysical world and the demands and impacts of the social world.

Engineering, in particular, is the discipline that focuses on the quantitative analysis, optimization, and control of real-world systems, including the infrastructure underlying FEW systems. Historically, engineering has focused on complicated systems, not complex systems, but this is changing presently.

Applied areas within the social sciences (psychology, economics, political science, sociology, anthropology) focus on the analysis, optimization, and management of real-world human-based systems.

System science is the scientific study of a unified whole composed of many parts:

1. It is defined by some unifying identity or macroscopic framework (e.g., food, energy, or water).
2. It exists within certain **boundaries** of space, time, or institution.
3. It relates to **external or exogenous factors** or "forcings" (e.g., sources and sinks of matter or energy and drivers) that may be parts of other systems (e.g., the climate system interacting with a water system).
4. It has **structural relationships** or "networks of relationships" among its parts (e.g., the relationship of water flows between rainfall, reservoirs, aqueducts, and consumers) and between its parts and external systems. **Structure** establishes the potential for function and the pathways of functional interaction.
5. It has internal or endogenous **functional relationships** between the parts which are governed by natural and anthropogenic principles or processes (laws of thermodynamics, economics, engineering of infrastructure, public policy, etc.). Function is distinct from, and constrained by, structure. Function is what matters, but structure enables function. Infrastructure is structure while commodity flow is function.
6. It often involves **agents** that are not entirely rational or predictable.
7. It changes **dynamically** in response to external and **internal interactions**.
8. It may be described over space and time by mathematical **models** which attempt to recognize and incorporate all crucial factors.

The objective of system science is to understand the entire system holistically, and with as much precision as needed for purposes of analysis and decision-making (i.e., excess detail can be ignored). At the very least, system science is needed to establish the nature of the system (simple, complicated, complex) so that the limits of its predictability can be clearly understood.

How a system is defined and studied is usually shaped by balancing important human dimensions, e.g., the scale and boundary of the decision-making process or institution, such as a farm, or a political jurisdiction, such as a city (Chap. 18) with consideration of physical scales and boundaries, e.g., an environmental boundary like a watershed (Chap. 19), or the boundaries of important material or energy inputs and outputs. These boundaries are not just defined by space but also time (e.g., growing seasons or political cycles). This is a **Coupled Natural-Human System** (CNH).

The application of system science is to predict system behavior in order to (1) design systems and (2) guide decision-making to maximize benefits and minimize adverse impacts.

Models of systems (Chap. 15) are often considered in two ways: "bottom-up" and "top-down."

Bottom-up models start by experimentally isolating and understanding the individual components of a system and then adding them together (or linking them) to construct the system. For FEW systems, bottom-up models tend to emphasize the environmental and technological aspects of a system. The challenge with bottom-up models is that the whole is greater than the sum of the parts; that is, the isolated parts do not add up to explain the whole, due to the complex interactions between the parts. For example, efforts to model how water moves through the system may capture environmental factors like precipitation (snow, rain), movement of water through the hydrologic cycle, and even the built environment like dams and networks of water distribution, but miss the legal and policy structures that also control water flows.

The main problem with bottom-up models is, therefore, that they are never complete or detailed enough to understand the system's behavior as a whole—although they may be very accurate for one subsystem or component.

A secondary problem with bottom-up models is that their representation of the whole system's behavior may be poor despite a good representation of the behavior of the subsystems. For example, a weather model of a hurricane could get the energy of the ocean surface precisely correct, and also its rainfall totals, but still fail to accurately predict the trajectory of the hurricane as a whole.

Top-down models "deconstruct" a whole system into a few essential components, and then proceed to disaggregate each of the components into a hierarchy of finer subsystems. For FEW systems, top-down models tend to emphasize economic and policy aspects of a system and global or national processes. The problem with top-down models is their limited predictability because of complicated and complex systems where the large-scale pattern emerges from the interactions of many atomic (small) parts; this yields surprises. For example, a top-down model of regional water stress might be based on demographics and prosperity, which motivate financing and policy, which leads to infrastructure, and withdrawals. This approach might accurately project long-term water shortages and economic problems of a water-scarce arid region by evaluating aggregated supply and demand for water. However, this model could not tell you much about whether any individual city or family is going to run out of water. One city might be in serious trouble, and another immune to the water stress, based on details that are only available at a finer level of

disaggregation. Such an approach also has limited ability to identify and quantify ecological impacts and environmental trade-offs.

The pros and cons of different approaches to modeling systems are addressed later in the book (especially Chap. 15).

2.2 Complex Systems

Complex systems have attributes that distinguish them from simpler systems; including:

- **Heterogeneity**: The many parts of the system are diverse (heterogeneous) in their characteristics and modes of operation. In the systems that we are considering, there can be both many distinct biophysical and human elements operating in diverse ways. Subsystems are of many types; operate at many scales; can process mass, energy, or information; and can be quantified with many different units. For example, the biophysical aspects of food production are quite different to the financial and policy aspects. A second example is that electrical power production typically occurs at a few large generation facilities of a few types, but food production is widely distributed across the landscape and takes on near-infinite forms.
- **Interconnections**: Components (subsystems) of the system are interdependent. That is, the behavior of subsystems is dependent on the behavior of other subsystems. Components can act on each other directly and indirectly through other parts of the system. There can be interactions operating under the laws of nature and interactions operating under the influence of cultural norms, governmental laws, human motivations, and economic principles as applied by independent decision-makers. Physical and human elements are interrelated because of the way they impact each other and depend on each other (recognizing that natural ecosystems could function without human intervention while human activities shape how many ecosystems function). For example, policies and laws governing natural water bodies are connected to food consumption through a series of interconnections; water law > water body > water use by farmers > food availability and price > food retail > food consumer.

  Often interconnections can be described by a set of mathematical expressions. This can allow systems to be described by a computational model where many mathematically described interactions between subsystems are calculated simultaneously and influence the next set of projected interactions between subsystems. These types of models are referred to as **Process Networks**.

  Process Networks are typically represented by graphs of nodes (representing subsystems) connected by "edges" (lines representing interactions) and studied within a field of mathematics called "Network Theory." Social networks, communication systems, and FEW infrastructures are subtypes of process networks.

  An electric grid is an excellent example of a process network because generation and demand must be kept in balance at all times to the grid to function. As a result, an extensive network of technology measure conditions at a large number of nodes on the grid and provide feedback to electricity sources (e.g., power plants or energy storage devices) to increase or decrease generation to match demand. The application of electric sensors and internet communication to the grid constitutes the so-called "smart grid." As the diversity of energy sources increases along with more distributed variable generation sources, more energy storage devices, and demand management tools, the need for ever more sophisticated tools to ensure a reliable and resilient electric grid.

The critical need for balancing supply and demand for all FEW commodities and the existence of nodes where commodities flow in and out on a continuous basis mean that process networks are a very useful tool for FEW systems.

Network theory provides many tools for the analysis of complex systems; most network theory applies to simple networks like internet-based social networks, but more sophisticated network methods are being developed to address the more complicated types of real-world Process Networks found in FEW systems. Scientists and engineers have done the most science on communication and computer networks, so the fields of **Information Theory** and computer science are particularly valuable sources of methods for Process Network study.

- **Distributed Natural and Distributed Human Controls**: The combination of the complex interactions between different parts of a system, and changes to individual elements, causes changes to ripple through the entire system. A complex system is not controlled by one force or by one component but by multiple forces and components that are distributed throughout the system. Ecosystems and the laws of nature provide a number of controls on how systems operate. Distributed natural controls for FEW systems include such factors as soil conditions, annual climate cycles, seasonal precipitation, and wind speed.

  Similarly, human systems usually have many independent decision-makers and actors (**agents**) who have different priorities and objectives and frequently work toward different (and often conflicting) outcomes. For example, the individual choices of billions of people determine the demand for food products, which in turn drives production patterns and natural resource consumption. However, control is not equitable. Hierarchies and hubs for control exist; there is a net flow of information from "controlling" to "controlled" parts of the system, even when both parts are exerting some control.

  Distributed human controls for FEW systems include such factors as fertilizer application by farmers and government agricultural policies for food; drilling of new oil wells and wholesale electricity markets for energy; groundwater pumping rates and water pricing for water; and for all aspects of FEW systems, consumers, investors, distributors, and regulators in diverse locations.
- **Hierarchy**: Complex systems still have hierarchies of scale, importance, and control. Despite their heterogeneity, interdependency, and emergent properties, some parts of the system exert more control than others, and some scales are more important than others. Complex systems have distributed control, but there are centers and hubs of control. At the same time, a more complete understanding of a complex system includes recognition of the free parameters of

smaller-scale elements. For example, farming systems operate with regional agricultural ecosystems that define what types of farming are practical, and agricultural ecosystems exist within larger economic systems, which operate within sociopolitical systems that shape investment and policy drivers. A second example is the U.S. Power Grid, which has three physical "interconnections" (Eastern, Western, and Texas), but within each interconnection, there are various "balancing regions" that govern power quality, and within those multiple power utilities produce and distribute the power.

- **Emergence**: The characteristics of the whole system cannot be adequately understood from the separate study of individual parts and the bottom-up **aggregation** of the properties of disconnected components. Instead, characteristics of the entire system "emerge" from the interconnections between the fine-scale parts of the system. For example, epidemics that destroy food crops or livestock emerge from a combination of bad luck and bad management practices at individual farms and processing facilities and then spread widely only if enough facilities follow bad management practices.

  A simple way to visualize emergent properties is to consider how the features of a building are distinct from the separate properties of the various elements of construction, such as the joists, bricks, windows, doors, wiring, and paint. One might say that the properties of the building, its rooms, its controlled environment, and so forth emerge from how the building elements interact with each other.

  Food, energy, and water markets, where they exist, are emergent properties that result from the interaction between consumers, policymakers, energy producers, and the technologies and infrastructure required to produce, move, and utilize the various forms of food, energy, and water.
- **Feedback (Coevolution, Synchronization)**: As one part of the system changes or "evolves" over time, other parts of the system will change or evolve as a result. That change will influence the change of the first part, a phenomenon known as "feedback." As a result, parts of the system "coevolve" based on their interactions with each other, yielding synchronized or partially synchronized subsystem states. In the presence of feedback, "cause" and "effect" lose their classical or original simplistic meaning. Complex systems may exhibit forms of relative stability or equilibrium even as they include dynamic processes. However, slow or small changes may lead to rapid or abrupt changes, which can sometimes occur at "tipping points" where nonlinear change can "cascade" through a system.

  For example, food production coevolves with energy and water systems because of the importance of water for irrigation and energy for fertilizers and machinery. In a second example, decreased electrical power demand decreases demand for water to generate power and then decreases the demand for power to pump the water.
- **Self-organized criticality**: The dynamics of complex systems often grow toward one or more critical limits where "catastrophe" (rapid and large-scale change) is just a small step away, and where a small disturbance to push them over that edge into a new system state. Forest fires, earthquakes, and avalanches are examples

in natural systems, but in human systems, we see "trigger events" that mobilize action and fundamentally shift the landscape. Trigger events can include, for instance, major disasters, "viral" cultural moments, key elections, successful terrorist attacks, the opening of a new communication or transportation route, or the introduction of disruptive technology. For example, a 5-year drought in Syria is thought to be a trigger event that led, in part, to the Syrian Civil War (see Chap. 20). The dramatic shifts in the energy systems of many countries as a result of the energy crisis of 1973 is another example.

- **Sensitivity to Initial Conditions**: How a system evolves is dependent on the starting conditions of a system. In many cases, the evolution of a system is dramatically different based on small changes in the initial state. The widely known "butterfly effect" illustrates this point, but is commonly misunderstood. The classic "butterfly effect" is the best example of the importance of initial conditions—and of chaos in systems; when a butterfly flaps its wings in China, the tiny alteration in the system's condition can produce dramatically different weather in the USA—through a series of amplifying feedback loops and processes in the atmosphere.

  In practice, most chaotic systems tend to fall into one or more relatively stable states regardless of their initial conditions. The precise state of these systems cannot be predicted or controlled, but the general "ballpark" state of the system (the attractor) can be predicted and controlled. The important difference between projections and predictions was noted above (Sect. 1.5.3). Estimates of future outcomes (projections or forecasts) are based on specified **assumptions** relevant to a question and decision option.
- **Complex Adaptive Systems**: Inherent in human decision-making, natural evolutionary processes, especially ecosystems, physics, and recently AI-based **machine learning** is the ability of complex systems to learn from experience, experimentation, observation, and study, and to adjust system structure and control to achieve a more preferred or optimal outcome. Often these changes are thought to occur in pursuit of some optimality principle, such as maximization of information. Thus, complex systems that connect human and natural biophysical elements, including many that will be examined in this book, have the additional attribute of being adaptive. Adaptive systems sense, anticipate, learn, and act; complex adaptive systems must do these things rapidly and skillfully because they change so frequently and unpredictably.

  The deployment of new technology in all sectors results in a period of learning and adaptation. Changes in how consumers understand FEW commodities also results in adaptation. For example, a better understanding of the nutritional value of food products, or household energy consumption, or the environmental consequences of certain products or practices frequently results in changes in consumer behavior and adaptations to FEW systems.
- **Sentience**: Human beings and social organizations are particularly adept at the invention of new ideas, memes, and values. Sentient behavior can include the pursuit of idea-based and value-based objectives that appear to be maladaptive but which nevertheless have a rationale. With sentient agents controlling a system,

its behavior is not necessarily predictable, because the principles and values guiding the system's function can change rapidly, unlike, for instance, the Law of Gravity which is stable over time. For example, while people may be concerned about the impacts of climate change may understand the contribution of driving large fuel-inefficient vehicles or having a diet heavy in meat consumption, only some will change their driving or eating behavior and then only modestly. However, when such concern becomes widely shared in a society, cultural shifts can occur, leading to larger changes in perception and behavior.

2.3 Food Systems

In the narrow sense used in this book, food systems bring together the components of the "food chain" or food supply chain path from production to processing, distribution, and consumption of nutritional substances that humans and their household animal pets eat and drink. Generally, this is understood to include feed for agricultural animals. With the recent expansion of biofuels, and because our subject is the FEW nexus, we include in food systems the production of plant ethanol and biodiesel. Because natural fibers are produced by the same plants and animals that produce foodstuffs, as part of the same farming operations, food systems are considered to produce fiber also. In a broader sense, food systems integrate all of the inputs, processes, conversions, infrastructure, outputs, uses, wastes, allocations, and *impacts* of food, feed, fuel, and fiber. For example:

1. Food **production** affects what foods are produced, how they are produced, and where. Food production (including seafood harvesting and aquaculture) integrates soils; land-use; ecosystem functioning; water movements and use; seeds and animal **stocks**; fertilizers, herbicides, and pesticides; climate; nutrient cycles; energy use; agricultural practices and economics; labor relations; agricultural and food processing machinery; farm management and operations; production wastes and pollutants; manufacturing and processing corporations; public agricultural policies and food policies; farmer and consumer organizations; processing systems; and environmental policies and consequences.
2. Food **distribution** affects how certain foods are moved to where, who gains access to them, and who benefits from the distribution. Food distribution includes the economics and social organization and cultures of actors from farmers to retailers and institutional food services; roads, rail; ports, refrigerated rail cars and trucks, and other forms of transportation infrastructure; food marketing and food service corporations and institutional food services; trade policies; and other issues which shape how certain foods are moved and where to, who gains access, who benefits, and how much.
3. Food **consumption** affects who gains access to what foods and the associated nutrition. Food consumption includes issues of types of nutritional needs based on **demography** and public health; food quality and preservation; affordability

and issues impacting access to food; culture and social equity; food preservation, preparation and homemaking roles; food policies; food waste; and recycling.

With the exception of seafood harvesting and some aquaculture, and emergent vertical farming (including hydroponics and aquaponics), most contemporary food production is land-based. Because of the foundation of most contemporary food production in land-use, including soils and ecosystem functions and decisions related to them, some scholars prefer land-energy–water integration to food–energy–water integration. While such an approach does have some benefits from a food perspective, it can obscure, or de-emphasize non-food aspects of land use and ecosystems, and all of the non-land aspects of food, as well as aquatic food systems such as fisheries.

Food production, distribution, and consumption each raise important issues about where to define the **boundaries** of food systems and how to address **external factors**. A consideration of the global food system requires explicit recognition of the geochemical cycles of water and nutrients like nitrogen and phosphorus and the operations of transnational food corporations. At smaller scales, considerations lead toward local flows of water and nutrients and pesticides and labor and farm operator decisions as inputs and outputs, sometimes resulting in imbalanced conditions. For example, importation of feed from the Midwestern region of the USA results in a nutrient imbalance in the Chesapeake Bay.

Food systems studied at various scales are explored in different parts of this book. Here it suffices to note that boundaries for studies of food systems can include subparts of larger systems such as the following:

- Individual production facilities (e.g., gardens, greenhouses fields, farms, hydroponic and aquaponic systems);
- Facilities that process or convert raw agricultural products into food products;
- Food processing and manufacturing corporations;
- Storage and stockpiling systems for food including refrigeration;
- Landscape systems encompassing many farms or agricultural communities;
- Crop systems which look a particular crop or set of crops across multiple regions or nations;
- "Foodsheds" that serve a particular population;
- Both raw agricultural products and food products (because the differences can be subtle);
- Market and nonmarket economics of trade and exchange of food;
- Restaurants, grocery stores, and food markets;
- Wholesale, warehouse, and retail supply chains for a particular food product or location;
- National and international government agencies for food policy and regulation;
- National and international producers and consumers civil society organizations, and
- Local, state/provincial, national, and international food policy systems.

Food systems include some significant complications as compared with water and energy systems. For example, there is an immense variety of food product types, brands, and qualities, each with its nutritional attributes, some of which are commoditized and some not, while water and energy feature a smallish and relatively well-defined set of types and properties and (especially in the global north) are heavily commoditized.

There are multiple conversion processes from agricultural commodities to food commodities that require careful attention. These can be simple like the conversion of wheat into flour, or the use of one product like corn as a feedstock for a secondary product like beef, or the combination of products into a process to produce an output like a frozen dinner.

These complexities lead to a diversity of **structural relationships** between different parts of food systems, with some structural arrangements more dominated by anthropogenic factors like agricultural practices, economics, industrial labor relations, and diet choices than others, which orient more toward biological, chemical, and environmental factors. These different structural relations lead to various **internal interactions** between the parts of the food system and a wide range of approaches to **modeling**. Figure 2.1 shows an illustrative example of an integrated model of a farm system with select flows and interactions.

Modern approaches to food systems are typically oriented toward efficiency, standardization, and quantity of production and/or delivery, such as how to maximize outputs (tonnage, **calories**, nutritional value, delivered products, etc.) while minimizing inputs (seed, land, fertilizers, herbicides, pesticides, irrigation, processing, preservation requirements, etc.) and achieving a consistent and regulation-compliant quality (if not high quality) product, to maximize profit and marketability of the food.

A common metric for modern food systems is the price per unit of foodstuffs paid to the farmer by the food processor; or the price paid by the retailer or the consumer, all of which are frequently impacted by subsidies and other forms of public policy. This is a "value chain" economic model for food that adds value and price at

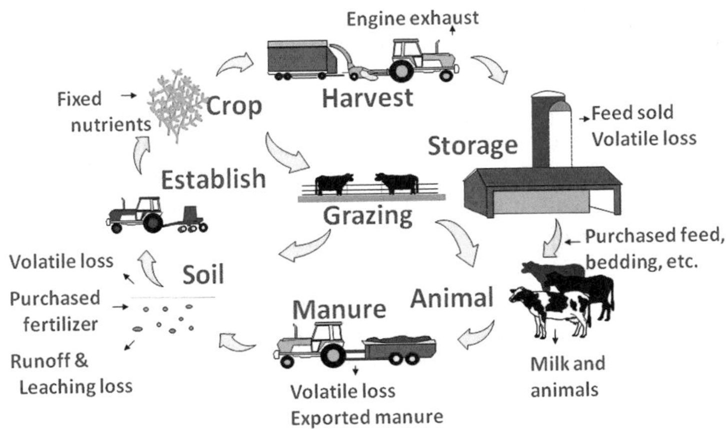

Fig. 2.1 Integrated farm system model. (Source: USA Department of Agriculture Integrated Farm System Model website)

each step, much like the factory model for industrial goods. To some degree, price entrains attributes of cost, perceived value, quality, and regulatory compliance, along with the timing of delivery. Thus, many food system models, especially those in the global north, address the system from an economic perspective using **econometric** models that use statistics and economic data to understand human system behavior.

However, alternative metrics for food systems emphasize non-economic factors such as environmental and social concerns or food quality. Examples include: "local food" systems which seek to minimize transportation and develop food community; organic food systems which seek to minimize synthetic chemical inputs; approaches which emphasize the nutritional qualities of the food; "fair trade" systems which seek to maximize compensation to farmers in poor or disadvantaged regions of the world; "slavery-free" systems which seek to ensure humane and equitable relations of production; farm animal welfare systems that seek to ensure the humane treatment of farm animals; and "footprint" or "life cycle" metrics that measure environmental impacts.

One notable challenge to the study of food systems is that they are often profoundly integrated into both particular aspects of local and regional and national geography, such as land availability and climate, as well as culture which strongly influence what foodstuffs can be produced and in what ways, what foodstuffs can be marketed profitably, and what types of foods are desired locally and regionally and globally (e.g., vegetables versus meats).

The emergence of affordable long-distance transportation and storage of foodstuffs aided by refrigeration has reduced the "distance from production" issue to some degree, allowing ever more urbanized consumers to become ever more separated from food production which can occur in different parts of the world, even as they enjoy the increased convenience and diversity and year-round availability of food options in their stores and restaurants and food service institutions. The same trend raises issues of equity as people local to production can lose (1) control over, and access to, traditional food supplies, and (2) market share as they increasingly have to compete with producers in different societies around the world. In some cities, urban agriculture and innovative programs integrating FEW elements at the city scale have been used as strategies to address the disconnect between low-income communities and local food access (see Sect. 18.5 for illustrative case studies in cities). Further, as food production is concentrated in fewer areas, there is increased vulnerability to problems impacting those areas such as climate change.

As societies develop, cultural attitudes toward food also change. Perhaps the most notable change is the growing desire for protein, especially meat protein, in societies emerging from relative poverty and transitioning to relative prosperity. Because meat protein is produced one step higher on the value chain and food web than vegetable protein, its costs, and environmental impacts tend to be an order of magnitude higher.

The effect of development on food systems is often treated in a straightforward three-phase evolution from traditional to intermediate to modern.

- Traditional food systems describe the approach of indigenous people to produce foods locally or gather them based on local environmental conditions, including locally available animal power, surface water or rainwater, and natural fertilizer inputs, and to consume them in accordance with local cultural customs and local seasons. Subsistence and hunter-gatherer models are forms of traditional food systems, alongside traditional "city-hinterland" agrarian models. Traditional systems do not require much capital intensive or specialized machinery, chemicals, GMO (genetically modified organism) seeds, imported technology, or non-animal energy inputs. Traditional food systems are highly local, yielding extremely diverse system types. While traditional food systems are commonly viewed as "sustainable" due to their modest ecological impacts, scaling up production to feed large urban populations can be challenging, and localized droughts and disasters easily propagate to cause local famine due to a lack of access to food from other regions. Note, however, that some ancient societies used surprisingly modern food systems, with the massive irrigation projects of ancient Egypt, Mesopotamia, or China as examples.
- Modern food systems are a complex network of industrial-scale food production occurring in diverse environments with significant chemical and engineering inputs, processed in a variety of ways and transported over vast distances to consumers. Modern food systems are marked by "industrial" characteristics of high levels of inputs, economies of scale, specialization of producers, branding of products, both "just in time" production and large-scale storage, corporate ownership and management, separation of (mostly rural) producers from (mostly urban) consumers, separation of the local growing season from the timing of consumption, and increasing global homogeneity of crops, agricultural practices, policies, and diets emphasizing the most commercially successful, profitable, and efficient types. Efficiency is typically defined in terms of cost, volume, or mass and (usually) not in terms of nutritional values and environmental costs.
- Intermediate food systems combine local production with a connection to larger systems.

The simplistic application of these categories lends itself to ideological, rather than practical, thinking. In the real world, food systems tend to fall into a grey area blending these categories in ways that reflect subtle contextual trade-offs and constraints. Much of this book is oriented toward recognizing and engaging with the complications of systems in a manner that promotes nuanced decision-making about trade-offs and integrates food, energy, and water aspects in a balanced way without idealizing one component or model over others.

Food systems change dynamically as a result of varying soil conditions, environments, climate and weather, crop decisions, agricultural practices and innovations, availability of inputs, population, changes in diet and culture, political and economic conditions, the market power of food corporations, and numerous other factors. **Climate variability**, seasonality, and disturbances from extreme events are natural sources of dynamics, but technological and policy change, market changes, consumption habits, and conflicts also drive dynamics.

Water and energy are often considered as inputs to food system models reflecting the demands that food production makes on water and energy and subject to possible supply constraints. Examples of how food systems place demands on water have already been given. Examples of how food systems place demands on energy include:

- Energy embedded "virtually" in the life cycle of agricultural inputs such as fertilizers, pesticides, and irrigation water.
- Energy demand (fuels) for operating agriculture equipment, transportation, and distribution, and (electricity and natural gas) for the processing and preservation of food products.
- Energy demand for agricultural labor.

Food systems are extremely nutrient-intensive because of the need for Nitrogen (N) and Phosphorous (P) fertilizers on crops and because of the transportation of nutrients and carbon embodied within food products. N and P limitations may become critical for some food systems in the twenty-first century. N and P helped create the "Green Revolution" in food production, but they are not unlimited resources (especially P), and they contribute dramatically to freshwater pollution via **"nonpoint source" pollution** of waterways and "dead zones" where oxygen has been depleted from waters by oxygen-eating microorganisms feeding on N and P, so fish cannot live.

In studies of food systems, the use of food crops as feedstocks for biofuels brings competing demands for water with food production for human consumption. In the USA, a large fraction of corn is used for ethanol production; Brazil is also a leader in biofuel production from corn and sugar cane. Biofuel production is controversial because "first generation" biofuels like corn ethanol compete with human food for land and water resources. This is in contrast to "second generation" advanced biofuels-based crops like algae, willow, switchgrass, and other woody products. See Sect. 8.2.2 for more on biofuels.

Changes in land use and ecosystem functioning because of water- and energy-related uses of land (e.g., mining, reservoirs, wind farms, pipelines, solar farms) may or may not raise food production issues. The addition of solar and wind production has allowed landowners to "produce" renewable energy as a crop and maintain their other crop productions as well. At the same time, dams on rivers provide both hydroelectric generation and supplies of water for agriculture.

Major considerations for modern food systems include the following:

- Concerns over genetic modification, pesticides and herbicides, and industrial-scale food systems.
- Local food movements and farmer-to-consumer linkages.
- Organic food movements.
- Food cultures that emphasize authenticity or specific diets.
- Food self-sufficiency as national policy, with its consequences for water demand in dry regions.
- The right to food as a local and national policy.

- **"Virtual" water**-embedded crops and products that are traded.
- Drought and famine in low development status countries and subsistence farming economies.
- Access to out-of-season food via trade and long-distance transportation.
- Food waste as a major inefficiency (over 30% is wasted).
- Government policies of overproduction and subsidy.
- Transitions (and declines) of farm communities.
- Smart agriculture technology.
- Bioengineering for higher yields (the green revolution).
- Nutrient and energy input management, including extra use of fertilizer as "**insurance**."
- Nonpoint source farm pollution, oxygen depletion (hypoxia), and aquatic ecosystems.
- Industrial-scale food supply chains and food safety.
- Changing the nutritional content of foods as a result of breeding and/or heavy processing.
- Changing diet and its health implications in high development status countries.
- Growing meat consumption and its footprint implications.
- Refrigeration and its fragility and electrical demands.
- Humans as the largest users of the terrestrial land surface.
- Volatility in water supplies from both drought and flood (sometimes in proximity).
- Volatility of food prices in low development status countries.
- Land competition between crops for first-generation biofuels and other agricultural products.
- Land use for crops that are exported from less developed countries to more developed countries.
- Impact of global markets and trade.

2.4 Energy Systems

Energy systems, at the largest scale, integrate into a whole the various components of energy resources, including their form (solid, liquid, gas, etc.), production, conversion processes (and efficiencies), long-distance transmission, short-distance distribution, end-use, and wastes. Conversion processes include such actions as the refining of gasoline and the generation of electricity and the conversion of biomass into biofuels.

Decision-making defined by political borders, energy processes, or end-user communities typically defines the boundaries of energy system studies and leads to a demarcation between external factors and internal interactions, and to the identification of the structural or functional relationships between different parts of the system being considered. For example, electrical power grids have no internal physical boundaries, but the wires cross many different corporate, State,

National, and regulatory jurisdiction boundaries. This is especially significant because some aspects of energy systems are highly regulated (e.g., electric power) and/or dominated by large public and private corporations (e.g., integrated oil and gas companies).

In the USA, generation, local distribution, and retail sales of electricity have historically been regulated at a state level. Thus, until the 1990s, vertically integrated electric utilities generated, transmitted, distributed, and sold electricity to retail customers within states (see Sect. 10.3.5) at rates set by each state. Each state would decide what power plants would be constructed by which utility, with (nuclear power plants excepted) modest oversight from the federal government. However, the US power system has become more distributed, utilizing power generated at a distance, especially wind and solar farms built in locations best suited to them, with more wholesale power crossing state lines regulated by the federal government. The outcome has been a restructuring of electricity markets in most locations. In 2018, two-thirds of Americans received their electricity via competitive, usually multi-state, wholesale markets. The restructuring of US electricity markets is an ongoing process.

The decision-making perspective also leads to a choice about taking a bottom-up (i.e., starting with individual components) or a top-down (i.e., beginning with a whole system) approach to viewing energy systems. Bottom-up approaches to energy start with specific technologies of energy production, conversion, or use with internal interactions dominated by physical, environmental, engineering, subcultural, and microeconomic factors. Top-down approaches are dominated by macroeconomic and policy and cultural considerations. They may also be dominated by decisions made in **arbitration** unseen to most of society. For example, after the Fukushima earthquake and tsunami that impacted one of Japan's major nuclear facilities, Germany decided to phase out nuclear power, thus ending contracts early and subjecting it to major arbitration cases. In the example of US electricity markets described above, historically, states have top-down projected demand and approved new power plant construction and what type of plants are constructed, with costs inserted into electricity rates charged to customers. Restructured electricity markets are far more open to bottom-up power plant decisions by independent actors. Chapter 20 includes many examples of decision-making processes and tools to address conflicts that arise.

Energy systems can be studied at the scale of the following:

- Individual devices and machines.
- Buildings.
- Facilities ranging from a power plant to an industrial facility like a refinery or a factory.
- Human communities such as cities or metropolitan regions (see Sect. 18.2).
- Regional transmission systems that connect multiple states or regions together.
- Particular energy resources, fuels, and energy products (e.g., electricity systems).
- National and international multi- and total-energy systems.

Notable examples of tools to explore large-scale energy systems that will be addressed in Chap. 15 (Modeling) include the following:

- MARKAL (derived from "Market Allocation") developed by the International Energy Agency is a model widely applied at many scales to project the evolution of energy systems over 40–50 years under certain assumptions. This is a bottom-up model that allows the assessment of different techno-economic assumptions about the future.
- TIMES (The Integrated MARKAL-EFOM System) is a successor to MARKEL (Fig. 2.2).
- National Energy Modeling System (NEMS) developed by the U.S. Energy Information Administration to project the future of the U.S. energy system and support an "Annual Energy Outlook."
- Model for Energy Supply Strategy Alternatives and their General Environmental Impact (**MESSAGE**) developed by the International Institute for Applied Systems Analysis (IIASA) used to explore energy scenarios related to several large-scale analyses including the Intergovernmental Panel on Climate Change (IPCC).

An interesting tool worth mentioning in the same context is the Long-Range Energy Alternatives Planning System (**LEAP**) developed by the Stockholm Environment Institute to explore scenarios of energy use (in all sectors) with emissions of greenhouse gases.

The ability of many primary energy sources to be converted to electricity and the significance of electricity as an end-use energy source (nearly 40% of global

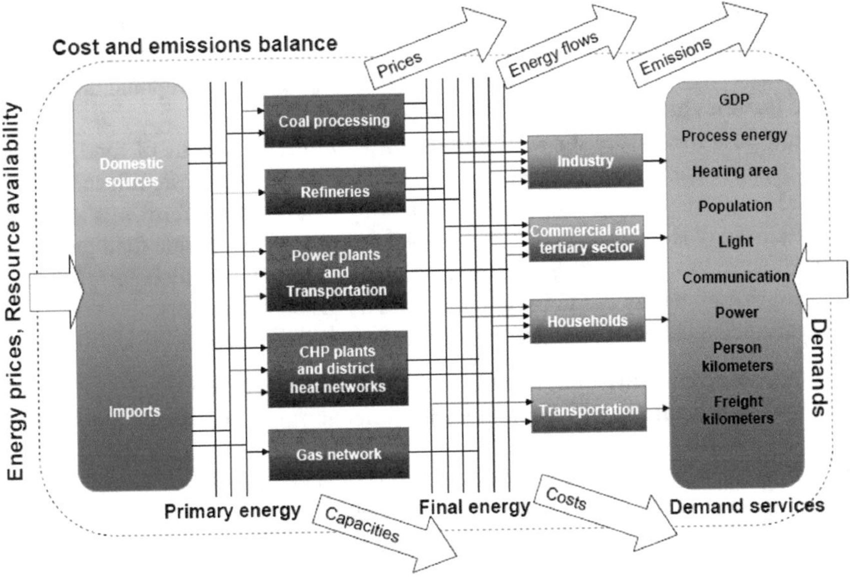

Fig. 2.2 Schematic of TIMES energy system model inputs and outputs. (Source: Remme et al. 2001)

Total Primary Energy consumption) has resulted in many studies of energy systems looking just at the electricity sector, so the knock-on impacts of electricity on water use, on other energy sources, and on the environment were often ignored. Similarly, the dominance of **petroleum** as the primary input for transportation leads to designs and models of transportation that ignored the availability of electricity, which is the transportation fuel of the future. CO_2 and greenhouse gas emissions were not considered during the design of the modern energy system, and this has turned out to be a serious problem.

As should be expected, water and food should be considered when developing models of energy systems, and vice versa.

The water demands of energy are significant. Some uses, such a hydroelectric generation, are largely **non-consumptive**, in that the water is returned after use (except for the increased evaporation from the impoundment). Other uses, such as thermo electric power plants (41% of all the water withdrawal in the USA in 2015), irrigation (38%), and public water supplies (12%) are **consumptive** users of water (USGS 2018). Although much of the water withdrawn for thermo electric power is returned to the water body, significant evaporation occurs, and the returned water is a higher temperature than the withdrawal. After 2007, a major shift in US electricity generation away from coal to natural gas and renewables dramatically reduced power plant water consumption. This has been offset somewhat by the increased use of water in hydraulic fracturing for unconventional oil and gas production.

Conversely, the energy demands of water are also significant. Power is used to pump water, move it, clean it for use, heat and cool it, and, to clean wastewater before return to the environment. In a move toward recognizing the value of integrating water and energy systems, the considerable energy use to heat water has led the State of California to prioritize the efficient use of heated water during a recent drought because of its added benefits for energy efficiency. Some cities have created hot water districts around thermo-electric plants, and some hydroponic and aquaponic facilities are co-locating with thermo-electric plants.

Similarly, energy use for food production is primarily a focus of food system studies and is merely one more demand for energy use. However, significant energy is used in agricultural operations and for creating inputs such as fertilizers and pesticides. Energy is required to preserve, move, process, and distribute food products, as well as address waste by-products. Refrigeration to preserve foodstuffs is a significant part of energy use in the commercial and residential sectors.

Some key considerations for modern energy systems include:

- greenhouse gas emissions, treaties, and climate change (Chap. 11);
- rapid advancement in renewable energy technology and economics (solar, wind, etc.);
- access to electricity and other modern forms of energy;
- aspirational development of renewable biofuels;
- the transition to electrical power from other energy sources;
- increasing use of geothermal energy for residential heating;
- rapid evolution and reliability problems of the massive and complex electrical power grid (Chap. 10);

- managing power grid peak demands, including with battery technologies;
- power grid peak demands driven by urban air conditioning;
- energy independence, dependence, and geopolitics, especially in oil and natural gas, but increasingly with wind and solar siting;
- vehicle fuel efficiency increases and vehicle electrification;
- energy for mass transportation;
- air pollution and health from fossil fuel burning, including power plants, and vehicles;
- air pollution and health from biomass and charcoal burning for cooking fires and residential heating;
- the problems with establishing safe and agreeable disposal of wastes from nuclear energy;
- the Faustian bargain of "normal accidents" with nuclear energy;
- aging energy infrastructure;
- falling energy prices and disincentives for developing new cleaner technologies; and
- economic and technological approaches for increasing energy efficiency.

2.5 Water Systems

The water system most familiar to many readers is the hydrologic/water cycle of the earth, defined with planetary boundaries between the upper atmosphere and subsurface water tables. The water cycle of the planet is also one of many geochemical cycles that are studied at a planetary level, including oxygen, carbon, phosphorus, and nitrogen (Fig. 2.3).

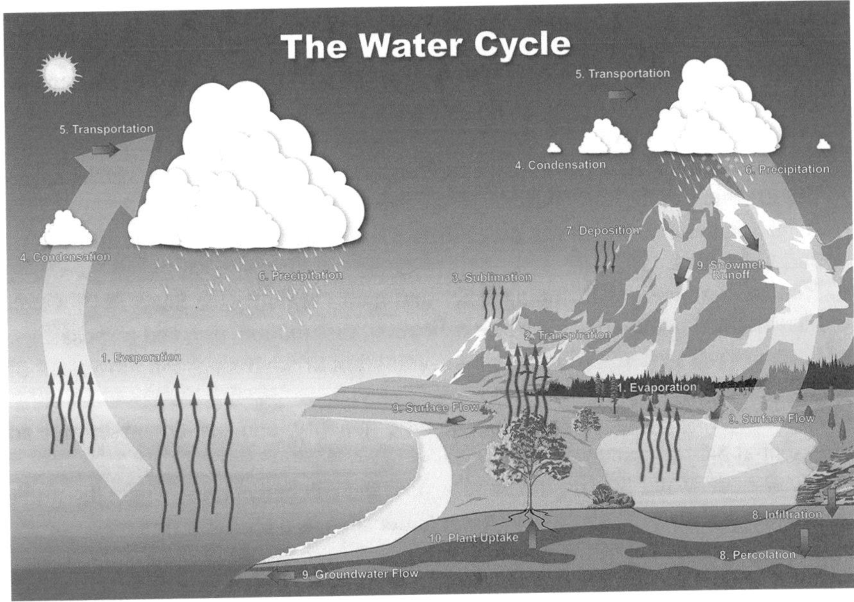

Fig. 2.3 Schematic of the water cycle. (Source: NOAA)

The energy provided by the sun is the most significant external factor impacting the system, although the earth's gravitation force also operates as an external force. There are structural relationships between the water contained in oceans, ice, groundwater, soils, lakes, atmosphere, swamps, rivers, and biology.

Internal interactions between these parts of the system are mediated by the laws of nature governing the hydrosphere and further mediated through the climate system, oceans, the biosphere (the living parts of the earth systems), and the **cryosphere** (the frozen water part of the Earth system) and manifest them as more straightforward processes such as evaporation, condensation, precipitation, transpiration, sublimation, surface and subsurface flows, percolation, and plant uptake.

Human actions include the creation of reservoirs, irrigation, and multiple types of consumption that have more or less impact on the larger natural process in different locations. The water cycle changes dynamically on many time scales from hours (storm intensity) to years (seasonal changes) to hundreds of thousands of years (ice age cycles).

Modeling the earth's water cycle with Global Hydrological Models (GHMs) is a major scientific discipline in its own right and is a significant part of efforts to understand the climate system through Global Circulation Models (GCMs) or the earth system as a whole through "Earth System Models" (ESMs). In this book, we will be looking at FEW systems integrations primarily at smaller scales.

The famous image of the entire earth taken from the Apollo 17 spacecraft in 1972 (page 1) has given rise to the view of the earth as the "water planet," the "blue planet," and the "blue marble." However, from a human perspective, it is 1% of the planet's water that is fresh and accessible, which is most important. Thus, water systems at sub-planetary scales tend to focus on the freshwater systems that can meet human needs, their capture/extraction, distribution, pretreatment, use or consumption, post-use treatment, and disposal. Examples of traditionally studied water systems at various scales include the following:

- Hydroponic systems where water acts as a medium for transporting nutrients to plants.
- Water facilities such as pre- and post-use treatment facilities, hydroelectric power plants, thermoelectric power plant cooling, and a wide variety of industrial facilities in which water flows have a critical function.
- Irrigation systems which can range from a single field to a farm to an entire agricultural region.
- Human communities including cities and metro regions (see Sect. 18.2) which have defined boundaries for water collection, distribution, use, and disposal.
- Groundwater systems which drive the evolution of aquifers or the movement of pollutants.
- Aquatic ecosystems where water quality, quantity, and movement impact an essential natural resource.
- Watersheds, water **basins**, drainage basins, and catchment areas where the water flows in a given area go to a common outlet such as a reservoir or a bay.

The spatial and temporal scale of these types of water systems is primarily driven by human decision-making processes and conflicting demands on limited supplies of available water. These conflicts are often rooted in, quantified by, and sometimes resolved by infrastructure and economic considerations that are incorporated into models (see Chap. 15). Such systems often depend on assumptions of what the hydrologic cycle has been versus what it is projected to be.

The decision-making process will often suggest an important metric that may be simple such as the volume of production or consumption, or a more complex metric such as the efficiency of use, or intensity of water demand, or a metric of a side effect such as the greenhouse gas emissions related to the water flows. The decision-making process will usually suggest a timescale for the parameter. Metrics will be explored in depth in Chap. 13.

Food and energy are often considered when developing models of water systems. Traditionally, the consideration is where food and energy are two types of demands on water.

Food demand on water includes the following:

- Agricultural production uses of water such as crop irrigation, water for animals, and water for farming practices.
- Impacts of runoff from agricultural fields and production.
- Production of agricultural inputs such as fertilizers.
- Food processing water use.
- Changes in land use and ecosystem functioning from agricultural practices impacting water flows.

Energy demand on water includes the following:

- Production processes such as hydroelectric power, oil and gas drilling (especially hydraulic fracturing), and irrigation of biomass crops.
- Energy transformation processes such as thermoelectric power plant cooling systems and biofuels production.
- Changes in land use and ecosystem functioning from energy production such as the creation of reservoirs, adding heat to rivers with water used for cooling.
- Impacts on the water cycle, such as through the release of greenhouse gases.

Other demands on water are also considered, such as demographics and human consumption, ecosystems, industrial uses, and non-consumption uses for human recreation. There is also a movement to ensure that rivers have a right to water as well.

Thoughtful approaches to water systems also recognize the land and energy needs of water, including the following:

- Land required for water capture and storage diverted from agricultural use.
- Energy use to pumping groundwater, transport and distribute water.
- Energy use to treat water before and after use and for the heating and cooling of water.

However, disciplinary hydrology and water resource (HWR) studies of water systems typically focus on the impacts on the water system rather than the effects of water's land and energy use. When your tool is a water balance equation, your analysis tends to ignore factors that do not appear directly in that equation. The commitment to core methods, concepts, and theories is both the greatest strength and greatest weakness of the traditional disciplinary approach; it is a weakness for systems work.

Some key considerations of the modern water system include the following:

- Massively centralized infrastructure dependency (Chap. 10);
- Growing global demand for water, especially for irrigated agriculture;
- Growing importance of managing life cycle water use and water footprints;
- The growth of human populations and economies in desert regions;
- The transition from a water-abundant world to a water-scarce world;
- Regional and planetary boundaries and carrying capacities for water;
- The conflict between environmental flow requirements and human demands;
- Humans as a major, or dominant, part of the water cycle;
- Groundwater mining and depletion;
- Outsourcing of water-intensive food production via virtual water (see Sect. 7.5);
- Informal water systems and water quality problems in low development status countries;
- Water pollution and water quality;
- The impact of both floods and droughts, often in proximity;
- The impact of existing and new hydropower development.

A study by the U.S. Department of Energy (DOE 2014) represented the estimated US energy and water flows in 2011 (Fig. 2.4). This study illustrates the connections and trade-offs that come with the interactions between food, energy, and water systems, using a Sankey flow diagram. The energy flows into the transportation, industrial, residential, and commercial sectors include energy for water and food systems. The water flows into thermoelectric cooling, and agriculture are very large.

It is important to note the difference between **water consumption** and **water withdrawal**. Consumption is different between water withdrawn from the immediate aquatic environment as compared with the quantity of water that is returned (discharged) to the same immediate environment at a similar time, place, and quality. Generally, water consumption is due to evaporation and evapotranspiration or its embodiment in some products (e.g., food). However, the water returned to a watershed may be altered by its use. For example, water use for cooling in a thermoelectric power plant typically raised the temperature of the water. In another example, water use in agriculture may result in the addition of nutrients. Both of these examples can result in significant ecological impact when non-consumed water is returned to a watershed, via thermal or chemical pollution.

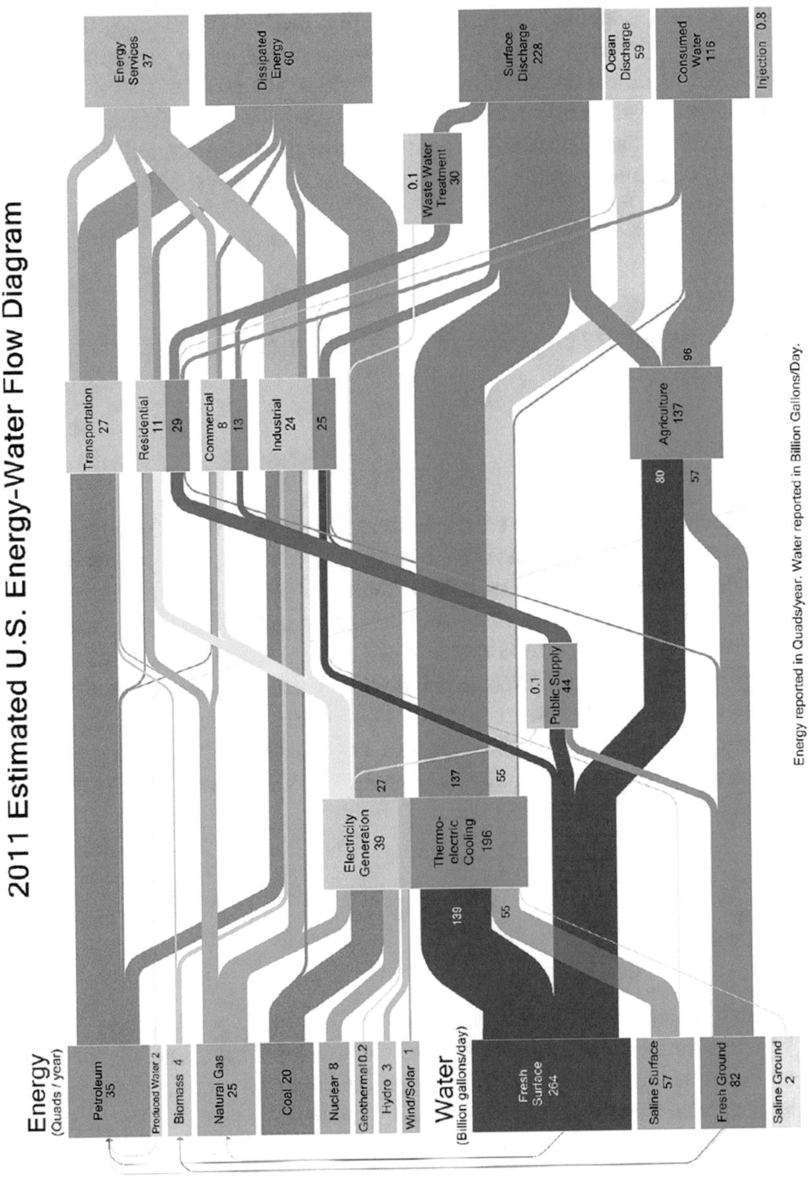
2011 Estimated U.S. Energy-Water Flow Diagram
Energy (Quads / year)
Petroleum 35
Produced Water 2
Biomass 4
Natural Gas 25
Coal 20
Nuclear 8
Geothermal 0.2
Hydro 3
Wind/Solar 1
Water (Billion gallons/day)
Fresh Surface 264
Saline Surface 57
Fresh Ground 82
Saline Ground 2
Electricity Generation 39
Thermo-electric Cooling 196
139
55
27
137
55
0.1 Public Supply 44
Transportation 27
Residential 11
29
Commercial 8
13
Industrial 24
25
Agriculture 137
80
57
96
0.1 Waste Water Treatment 30
Energy Services 37
Dissipated Energy 60
Surface Discharge 228
Ocean Discharge 59
Consumed Water 116
Injection 0.8
Energy reported in Quads/year. Water reported in Billion Gallons/Day.

Fig. 2.4 2011 estimated U.S. energy–water flow diagram. (Source: DOE 2017)

It is also important to note that much (most?) energy does not go into providing energy services but is dissipated as heat before its use. This is especially true in thermal power plants using turbines and internal combustion engines.

Also note that the flow diagram in Fig. 2.4 is not a model, but rather data that is visualized using a specific type of visualization method (the Sankey diagram). While Fig. 2.4 represents how energy and water flows go to different sectors and how much is consumed and discarded, it does not explain the intention of the flows, embody scientific and engineering concepts, show interactions and causation, or allow the user to experiment with changes to inputs and interactions to see what outcomes result. This diagram is descriptive and is based on empirical data.

2.6 From Separate Systems to an Integrated System of Systems

2.6.1 Science

In the previous sections, we have seen how food, energy, and water each play a part in careful studies of systems of the other two components. As such systems become ever more comprehensive, the treatment of the other elements become ever more detailed, until those systems become subsystems embedded in a more extensive system. Advanced studies of food systems, for example, will include many water and energy interactions, both direct and indirect. Where sets of those interactions are connected, they begin to be recognized as components of the food system with internal attributes of a system—mini-systems. Thus, advances in the study of food, energy, and water systems separately (**separated FEW systems**) lead to considerations of an integrated "system of systems."

2.6.2 Sustainability

There is also a sustainability path to the same outcome. Societies have thought about how to sustainably provide themselves with essential food, energy, and water services for millennia. The conflicts between efforts to provide the three critical consumables have also been long recognized. The environmental consequences of meeting the rapidly expanding demand for each of the three came to the fore in the 1960s and received serious attention in 1972, at the United Nations Conference on the Human Environment in Stockholm, Sweden, now referred to as the Stockholm Conference (The UN is explored in see Sect. 6.2.2).

Later, the United Nations World Commission on Environment and Development, chaired by former Norwegian Prime Minister Gro Harlem Brundtland, famously define sustainable development in its 1987 report, Our Common Future, as follows:

> Sustainable development is development that meets the needs of the present without compromising the ability of future generations to meet their own needs.

While the Brundtland Report, as Our Common Future is now commonly referred to, contains the most widely known definition of sustainable development, there are many others.

The U.S. Environmental Protection Agency states that

> Sustainability is based on a simple principle: Everything that we need for our survival and well-being depends, either directly or indirectly, on our natural environment. To pursue sustainability is to create and maintain the conditions under which humans and nature can exist in productive harmony to support present and future generations.

Such general definitions of sustainability are, unsurprisingly, subject to debate, and challenge. Regardless of the definition of sustainability, the core question of sustainability is how to provide essential services to human societies without causing long-term (decades or centuries) degradation to natural ecosystems and the services that they provide to human communities.

As a scholarly field, sustainability science has come to broadly encompass the study of interactions between the natural environment and human societies, classified as "human-environment systems" or "**social-ecological systems**," and recognized as "coupled systems."

One core scientific challenge is how to measure sustainability. This vital issue, in the context of FEW systems, will be addressed in detail in Chap. 13 (Metrics).

One core practical challenge is how to address sustainability when human societies vary dramatically in their resources, population growth, social and economic development, and values.

Studies of food, energy, and water systems have helped bring into focus the scientific and practical challenges of sustainability.

With growing demands for food, energy, and water-related to both population growth and economic prosperity, political and policy conflicts between different claims and demands on food, energy, and water resources have become ever more frequent. As a result, for some policy-makers, the question of how to provide constituents with food, energy, and water in a manner that can be sustained for decades to come is the practical definition of sustainability. While environmental conditions are not explicit in such a question, large-scale ecological degradation makes an answer to such a question impossible and places environmental considerations at the core. Such questions can also be critical in bringing disparate parties together to find solutions (see Chap. 20).

While sustainability is much broader than sustainable food, energy, and water systems, the necessity of simultaneously providing all three critical consumables has provided a human- and ecosystem-focused impulse for integrating FEW systems as an essential practical application of sustainability.

2.6.3 Principles of a System of Systems

Studying an integrated "system of systems" utilizes the similar principles of system science described above in Sect. 1.4 and of the emerging field of sustainability science. The attributes of a system of systems include:

2.6.3.1 A Question (or Problem) as a Macroscope Which Defines Boundaries and Scales

If we are integrating three components, what is the unifying identity (or macroscope) by which we perceive the system of systems? In practical situations, the unifying identity is the question (or problem) that is being addressed: "how do we manage a system in a certain context in which food, energy, and water are critical components to achieve desirable outcomes?"

The keyword and phrase in such a question are "context" and "desirable outcomes." The question will also define issues of spatial and temporal scale (defining its boundaries and external factors) metrics, data, modeling, and computing.

The issue of defining or framing the question will be addressed in Chap. 12; its relationship to metrics, data, modeling, and computing will be explored in Chaps. 13–16; and the application of science to practical questions addressed in Chap. 17.

2.6.3.2 Heterogeneous Parts Which Have Mutual Relationships

While in FEW systems, the parts include food, energy, and water elements; there are usually other parts like population, economics, infrastructure, ecosystem services, and biodiversity to include.

Interactions are both direct and indirect and usually operating in both directions, so that we can think of elements of an integrated system having complex interactions embodied in mutual relationships. Further, when one element changes, the interactions with other parts of the system result in additional interactions on the first element, that is, reciprocal relationships usually include **feedback interactions**.

Direct, or first-order, interactions are the influence on a system by another system; for example, the demands on water by the energy system and on energy by the water system. Many examples of this type of interaction were given above in our consideration of water, food and energy systems separately.

Indirect interactions include the impacts of one element on another to which it is not in direct contact. Rather the impact is mediated through other intermediate parts of the system or factors external to the system. Here are three examples where there is one step mediating the indirect interaction:

- Energy use of crops for biofuels makes demands on water because of the irrigation needs of those crops.

- Particular crops can alter ecosystem functioning and the change the percentage of rainfall that accumulates in groundwater or flows in streams and leaves a certain area.
- Energy used to heat water can lead to emissions of greenhouse gases that alter the climate, which in turn impacts the growth of food crops in a variety of ways.

Where there is one step to mediating the interaction, we can call this a "second order" interaction. If two steps are mediating an indirect interaction, we can have "third order" interaction. If more steps are mediating the interaction, we have "higher order" interactions. Each additional indirect step in an interaction makes a system more complex to study.

Indirect interactions are frequently folded into direct bilateral interactions or a two-way mutual relationship. For our three examples, this might be done in the following manner:

- The water demands of crops for biofuels are considered within the bilateral or mutual relationship of agricultural water use.
- The impacts of particular crops on groundwater storage are regarded within the bilateral relationship of land use/cover and water.
- The effects of water energy use on crops are considered within the reciprocal relationship between climate and crops.

2.6.3.3 Structural Arrangements

The parts of an integrated system are usually grouped into a system of dynamic "components" or "modules" based on strong interactions rooted in biophysical and societal relationships (e.g., resource flows, a shared decision-making process, or a robust economic relationship).

Parts usually reflect distributed control of a system, indicating opportunities or change their operation and management. Parts can sometimes also be seen as the units of coevolution.

In the most straightforward approach to this, we might think of food, energy, and water components managed to achieve the desired outcome. However, in practice, integrated systems are far more complicated.

Models describing structural arrangements must address the systems simultaneously and their mutual relationships consistently (both biophysical and societal.) An integrated energy–water system model must recognize the direct interactions of the demands on water by the energy system and on energy by the water system. It must also acknowledge the indirect interactions mediated through the food system, land use, climate, ecosystems, economics, and social changes.

2.6.3.4 Emergence

The purpose of studying an integrated system of systems is to identify patterns and find solutions to severe problems that "emerge" from understanding the complex interactions between the parts of the subsystems.

2.6.4 System of Systems and Models

Given this evolution of individual food, energy, and water systems towards an integrated analysis via a system-of-systems approach, one emerging area of challenge is modeling tools to go along with this approach. Modeling of food, energy, and water systems is the subject of ample literature, including textbooks. However, when modeling these systems individually, typically the other two are assumed to be unchanged or unimpacted. This simplifies the analysis and enables solutions to problems in each of these systems to be found, and to some extent, these solutions may work, at least under some constrained conditions that satisfy this assumption. But in a more general sense, and particularly when these systems are intensely stressed (e.g., by human activity such as urbanization, and expansion of services in food, energy, and water sectors), it is intuitively easy to understand that these systems will interact and be dependent upon each other (nexus). This interacting **coupling** is illustrated in Fig. 2.5. We explore this system of systems approach in more detail from a modeling perspective in Chap. 15.

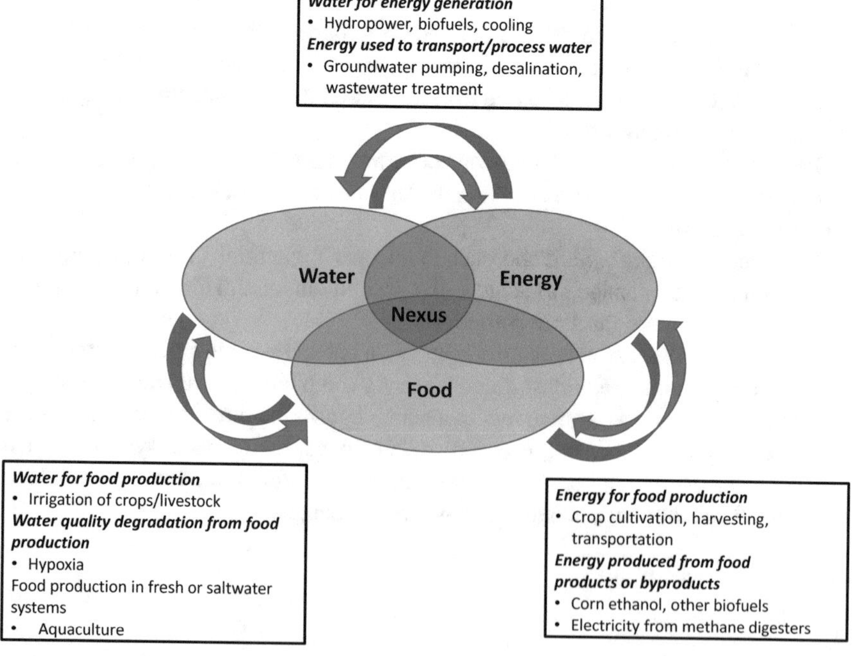

Fig. 2.5 Examples of modeling interactions and feedbacks among the FEWS nexus. (Source: Fernando R. Miralles-Wilhelm)

Key Points

- Systems describe a whole composed of many interacting parts with a unifying framework; boundaries in space and time; external forcing factors; structural arrangements; functions; change and variability; and humans as both forcings and participants in the system.
- Complex systems have heterogeneous parts with complex interactions that make them interdependent, coevolving, and subject to distributed control. The characteristics of the whole system emerge from the interactions between the components of the system giving rise to stability and changes that are often hard to predict and sensitive to initial conditions.
- Because complex systems are influenced by external factors such as human actions, models of complex systems make projections based upon certain assumptions rather than predictions about actual future outcomes.

Discussion Points and Exercises

1. The FEW system is an excellent example of all four system-of-systems attributes. Now that you have read this chapter, describe how this is so.
2. Discuss what makes a system complex, and why the FEW system IS or IS NOT complex.
3. Describe a food system as a process network.
4. Describe heterogeneity in a FEW system at the local, regional, and national levels.
5. Describe hierarchy in an energy system.
6. Describe three emergent properties in a: (a) food system; (b) energy system; (c) water system; (d) FEW system
7. Develop a Process Network of a (a) food system; (b) energy in transportation system; (c) water system; (d) farm system including FEW components; (e) household system including FEW components.
8. Describe a Complex Adaptive System in: (a) the agricultural sector; (b) the electricity sector; (c) a urban FEWs context; (d) a rural FEW context.

   For Exercises 9–17, consider modeling a system. Describe the following:

   (a) The boundaries of the system;
   (b) The main components;
   (c) The structural arrangement of the components;
   (d) The most significant functional interactions between components;
   (e) External factors and their interactions with the system;
   (f) The most significant (distributed) controls on the system including human actions;
   (g) Issues that might alter the stability of the system; and
   (h) An emergent property of the system.

9. The water system for a city. Specify the location of the city and think about geographic variation.
10. The water system for a farm growing an irrigated crop.
11. The food system of a landscape encompassing many farms or agricultural communities.

12. The food system of a facility that processes or converts food products.
13. The energy system of a building.
14. The energy system of a coal-fired power plant.
15. The integrated food, energy, and water systems of a house.
16. The integrated food, energy, and water systems of a college campus.
17. The potential differences between a food, energy, and water system in the USA, India, Ghana, and an island country like Trinidad and Tobago.

References

Agricultural Research Service. (2017). *Integrated farm system model*. Washington, DC: U.S. Department of Agriculture. Retrieved from https://www.ars.usda.gov/northeast-area/up-pa/pswmru/docs/integrated-farm-system-model/.

Buchanan, M. (2003). *Nexus: Small worlds and the groundbreaking theory of networks*. New York, NY: WW Norton & Company.

Capra, F. (2002). *The hidden connections: Integrating the biological, cognitive, and social dimensions of life into a science of sustainability*. New York, NY: Doubleday.

DOE. (2017). Environment Baseline Vol. 4: Energy-Water Nexus prepared by the Office of Energy Policy and Systems Analysis, U.S. Department of Energy https://www.energy.gov/sites/prod/files/2017/01/f34/Environment%20Baseline%20Vol.%204--Energy-Water%20Nexus.pdf

Midgley, Gerald (2016) *Four domains of complexity*. Emergence: Complexity and organization. Retrieved September 26, 2018, from https://journal.emergentpublications.com/article/four-domains-of-complexity-2/.

Pikovsky, A., et al. (2003). *Synchronization: A universal concept in nonlinear sciences* (Cambridge nonlinear science series). Cambridge: Cambridge University Press.

Remme, U., et al. (2001). *MESAP / TIMES — Advanced Decision Support for Energy and Environmental Planning in Operations Research Proceedings* (pp. 59–66). Springer

U.S. Department of Energy. (2014). *The water-energy nexus: Challenges and opportunities*. Washington, DC: Author. Retrieved September 7, 2018, from https://www.energy.gov/sites/prod/files/2014/07/f17/Water%20Energy%20Nexus%20Full%20Report%20July%202014.pdf.

USGS. (2018). Estimated Water Use in the United States in 2015, Circular 1441, U.S. Geological Survey https://pubs.usgs.gov/circ/1441/circ1441.pdf.

von Bertalanffy, L. (2015). *General system theory: Foundations, development, applications*. New York, NY: George Braziller Inc.

Westhoek, H., et al. (2016). *Food systems and natural resources* (A report of the Working Group on Food Systems of the International Resource Panel). Nairobi: United Nations Environment Programme.

Wolfram, S. (2002). *A new kind of science*. Champaign, IL: Wolfram Media.

Further Reading

Bale, C. S. E., Varga, L., & Foxon, T. J. (2015). Energy and complexity: New ways forward. *Applied Energy, 138*, 150–159.

Capra, F. (2004). *The hidden connections: A science for sustainable living*. New York, NY: Anchor.

Chase, L., & Grubinger, V. (2014). *Food, farms and community: Exploring food systems*. Lebanon: University Press of New England.

Gunderson, L. H. (2001). *Panarchy: Understanding transformations in human and natural systems*. Washington, DC: Island Press.
Jørgensen, S. E., et al. (2011). *A new ecology: Systems perspective*. Amsterdam: Elsevier.
Liu, J., et al. (2007). Complexity of coupled human and natural systems. *Science, 317*(5844), 1513–1516.
Meadows, D. H. (2008). *Thinking in systems: A primer*. White River Junction, VT: Chelsea Green Publishing.
Menzel, P., & D'Aluisio, F. (2005). *Hungry planet: What the world eats*. Berkeley, CA: Material World Books/Ten Speed Press.
Mobus, G. E., & Kalton, M. C. (2015). *Principles of systems science* (No. 5) (Vol. 7). New York, NY: Springer.
Weinberg, G. M. (1975). *An introduction to general systems thinking* (Vol. 304). New York, NY: Wiley.
Zeeman, E. C. (1979). *Catastrophe theory. Structural stability in physics* (pp. 12–22). Berlin: Springer.

Chapter 3
Development

Peter Saundry and Benjamin L. Ruddell

3.1 Introduction

In Chap. 1, we examined the reasons for focusing on food, energy, and water, three critical consumables requiring extensive infrastructure, and a high-level policy. In Chap. 2, we explored what this means on a global scale, especially for the billions of people for whom access to these resources is severely limited and needs improvement.

In this chapter, after analyzing the idea of food, energy, and water security, we will look through the lens of demographics to see the complexities of engaging the world's population in our study of the nexus. Next, we will examine the concept of development in countries and the human development of people within a country. Based on these elements, we will explore efforts to guide policy and action on a global level to meet the basic need for food, energy, and water through the Millennium Development Goals (2000–2015), the Sustainable Energy for All initiative (2012–2030), and the Sustainable Development Goals (2016–2030).

Aspects of this chapter foreshadow later chapters, such as international governance (Chap. 6) and the challenges of metrics, data, and modeling (Chaps. 13–16).

P. Saundry (✉)
Energy Policy and Climate, Advanced Academic Programs, Krieger School of Arts and Sciences, Johns Hopkins University, Washington, DC, USA
e-mail: psaundr1@jhu.edu

B. L. Ruddell
School of Informatics, Computing, and Cyber Systems, Northern Arizona University, Flagstaff, AZ, USA
e-mail: Benjamin.Ruddell@nau.edu

P. Saundry, B. L. Ruddell (eds.), *The Food-Energy-Water Nexus*, AESS Interdisciplinary Environmental Studies and Sciences Series,
https://doi.org/10.1007/978-3-030-29914-9_3

3.2 Food, Energy, and Water Security

Security, including "food security," "energy security" and "water security" refers to the ability of people to have affordable, reliable, and high-quality access to their basic FEW needs so they can live healthy and productive lives unconstrained by existential resource limitations. Security implicates attributes of availability, access, **utilization**, stability, sustainability, reliability, and resilience. Separate definitions of these terms are as follows:

Food Security exists when "all people, at all times, have physical and economic access to sufficient safe and nutritious food that meets their dietary needs and food preferences for an active and healthy life" (FAO 1996).

Energy Security requires "the uninterrupted availability of energy sources at an affordable price" (International Energy Agency).

Water Security is "the capacity of a population to safeguard sustainable access to adequate quantities of acceptable quality water for sustaining livelihoods, human well-being, and socio-economic development, for ensuring protection against water-borne pollution and water-related disasters, and for preserving ecosystems in a climate of peace and political stability" (UN Water).

What constitutes an individual's "basic needs" for each consumable is related to factors such as age, physical characteristics, level of activity, and their geographic and community context. However, four common attributes can be identified.

1. Physical **availability** of the resource brings a focus on the following issues:

    (a) Production of raw commodities (e.g., agricultural products, energy sources, and untreated water) and processed products (e.g., food products, electricity and fuels, and treated water).
    (b) Stocks of commodities and products in locations controlled by the country and available for distribution.
    (c) Trade balance (imports–exports) of commodities and products.
    (d) The infrastructure required to make products available (e.g., refrigeration, electric grid, storage, pipes).
    (e) Capacity of systems and infrastructures that produce and deliver the resource.

2. Physical, legal, and economic **access** to the resource which determines the ability of individuals to obtain the resource when it is available. Access brings a focus on the following issues:

    (a) Affordability (e.g., cost of products compared to incomes)
    (b) Markets (e.g., socioeconomic arrangements which can limit access even if the person has the money for the resource)
    (c) Support programs that provide access to vulnerable populations and at times of crisis (e.g., during an emergency, conflict, drought, flood, famine, or severe heat/cold)

3. **Utilization** referring to the ability of a person to make use of the resource productively. For example:

(a) Utilizing food requires the ability of a person to be able to hygienically store, preserve, refrigerate and prepare a variety of foods and consume them in a manner that provides them with both sufficient energy for their age, gender, weight and activity level, and to receive sufficient nutrition to develop healthily and/or maintain appropriate body weight.
(b) Utilizing energy requires the ability to obtain energy services through the use of machines, vehicles, devices such as clean cooking equipment.
(c) Utilizing water requires the capability to use it in a manner that prevents the spread of waterborne diseases and to produce food, energy, and other goods and services with the water. This includes such factors as having piping to bring water to the user or the ability to keep the water clean, soap for hygiene, and toilets, or other systems to remove waste.

It is important to recognize that utilization is not just about having the necessary equipment but also about having the necessary knowledge and/or culture to utilize a resource healthily and efficiently that maximizes its benefits.

4. **Stability and Reliability** refer to the short-term preservation of availability under shocks and stresses. These bring a focus on the following issues:

(a) Vulnerability to disaster events such as droughts, storms, earthquakes, and flooding.
(b) Political or military conflict.
(c) Economic instability such as loss of income and volatility of prices.
(d) Reliability of systems that produce and deliver the resource.

Sustainability refers to the long-term preservation of security as time passes. Sustainability brings a focus on the following issues:

(a) Infrastructure deterioration.
(b) Climate change.
(c) Social capital and well-being.
(d) Efficiency of consumption and impact.
(e) Resource stock depletion or resource scarcity.
(f) Degrading ecosystem services.

5. Resilience thinking is beginning to replace and supersede sustainability thinking because **resilience** emphasizes our ability to prevent, recover from, adapt to, and thrive under both short term and long-term shocks and stresses. There are at least four schools of thought about resilience in a FEW systems context:

(a) Emergency Management, emphasizing quick recovery of function after disruptions.
(b) Engineering, emphasizing prevention of disruptions through **redundancy**, **buffering**, armoring, and control.
(c) Ecological, emphasizing adaptation of structure and connections (especially green infrastructure) to preserve function.

(d) Reversibility, emphasizing the avoidance of overcommitment of variable levels of available water to preserve adaptability and avoid stranding capital in light of an uncertain future.

When one or more of these attributes are lacking for FEW systems, it is appropriate to speak of food, energy, and water insecurity. There are, however, gradations of insecurity. Many experience chronic insecurity of one or more of the three commodities, even as much of humanity is relatively secure in one, two, or all three commodities at the moment. All of humanity is threatened with FEW insecurities on long timescales due to sustainability problems. The differences and distinctions are consequential.

1. Chronic insecurity exists when one or more of the attributes are persistently lacking. Examples include the following:

    (a) Natural resources or ecosystem services for food, energy, and water are insufficient.
    (b) Absent or degraded infrastructure prevents sufficient resources from reaching a community.
    (c) Systemic, long-term poverty, and other forms of social **exclusion** severely limit access.
    (d) Political and governance systems are structured in a manner that prevents addressing solutions.
    (e) People lack the equipment or knowledge to utilize a resource effectively.

2. Periodic/Seasonal insecurity exists when a repeating cycle in conditions exists, which causes insecurity. This type of insecurity is, to some degree, predictable. Examples include the following:

    (a) Agricultural practices or crops result in insufficient food at certain times of the year.
    (b) Seasonal weather patterns like monsoons disrupt access to clean water.
    (c) Seasonal employment leads to regular periods of economic hardship.

3. Temporary/Emergency insecurity exists when a triggering event causes short-term and temporary insecurity. Examples include the following:

    (a) Loss of electricity for much of the population of Puerto Rico following Hurricane Maria in September 2017.
    (b) Disruptions in food imports to Yemen due to civil conflict that began in 2015.
    (c) Loss of drinking water supplies in the city of Flint, Michigan (USA) because of lead contamination (2015) (see Sect. 18.4).

Appropriate solutions to insecurity depend on the context. Emergency aid can provide temporary assistance to large populations in a crisis. However, many populations experience chronic insecurity. This can be rooted in poverty, disability, discrimination, a number of other issues. “Safety net” programs often exist to mitigate chronic challenges. In countries where large portions of the population experience chronic insecurity, the concept of country-wide "development" is critical (see Sect. 3.4 below).

All kinds of security problems can be helped using a resilience analysis framework where we choose to proactively anticipate, sense, adapt, and respond to stresses and shocks to get ahead of the problem. However, this can be challenging when what could be seen as a short-term shock turns into longer chronic insecurity. For example, the California drought that began in 2012 stretched into 2017. Tree rings indicate that this region is subject to multi-decadal droughts. However, water storage systems are built to withstand 2–3 year droughts, not multi-decadal droughts.

Even currently-secure communities are at risk of becoming insecure. Thus, much effort on food, energy, and water security is risk-based—that is, devoted to identifying and managing the most significant risks. In Chaps. 9–11 (Ecosystems, Infrastructure, and Climate), we will explore issues of risk and vulnerability at different scales of time and space.

There are three primary **responses** to vulnerability:

1. **Mitigation** to reduce the drivers underlying the risk.
2. **Adaptation** to reduce the impact of a hazardous event or outcome.
3. **Crises Response** that helps populations endure an event or outcome without long-term negative consequences.

The principles of system science (Chap. 2) can be applied to resource security, and the attributes of boundaries of space and time, external factors, structural arrangements, internal interactions, human actions, dynamic changes, and models all take on particular applications to populations and their sources of food, energy, and water.

In Chap. 4, we will explore how individuals, communities, and nations respond and adapt to food, energy, and water insecurity. In Chap. 20, we will examine how food, energy, and water insecurity can create conflict and tools for reducing or managing these conflicts. In Sects. 3.4–3.7, and Chaps. 6 and 7, we will examine how the international community attempts to address the issues of insecurity.

3.3 Population

3.3.1 Demography

We have already noted that an individual's "basic needs" for each consumable is related to many factors that are unique to that person. It is rarely possible to survey the conditions and needs of every individual when determining the needs and prevalence of insecurity in a community or a country. Instead, data is gathered on a sample population. That data is then combined with a statistical profile of the total population in a model to produce a quantitative indicator or metric of the condition being explored. We will look at various development metrics below, and explore metrics generally in Chap. 13.

A statistical profile of a population includes all relevant characteristics of that population and any vulnerable subpopulations like infants, both in the present and in the future. This is the field of **demography**. The following list illustrates, in a non-comprehensive way, some of the demographic factors that usually require consideration:

1. Absolute population numbers: how many people there are.
2. The age distribution of the population. An individuals' demand for critical consumables changes over a human lifespan. It is, therefore, important to know the population in different age groups and how they will likely evolve.
3. Fertility, reproductive health, public health, and mortality, which affect future absolute population and age profile.
4. The wealth of a population which shapes both the ability to gain access to consumables and the ability to invest in producing them for a population as a whole.
5. Distribution of wealth which shapes which parts of a population consume and how much and how different parts of the population engage in food, energy, and water systems. It also reflects differentials in the capacity to participate in or influence decisions related to the structure and governance of systems.
6. Education, skill development, gender and social inequality, cultural and social influences on consumption choices and the efficiency of use of the consumables, and related factors that influence decision-making.
7. Location of populations concerning the availability of resources, ecosystem functions, infrastructure requirements; vulnerability to disruptions in supplies of commodities; and rural-urban context.
8. Migration of populations into and out of areas.
9. Political and institutional capacity which mediate the ability of a population to produce or import, distribute, gain access to, and utilize food, energy, and water resources stably.
10. Considerations of political- and income-inequality, culture, priorities, habits, and values among the population.

As this list illustrates, demographic issues provide a statistical context for people within societies while also adding a layer of diversity and complexity, particularly if including them in modeling efforts.

The diversity of human societies is significant at all scales. While much of this chapter will look at large aggregations of people within countries, it is important to remember that significant diversity exists at the regional and local levels of countries and between different socioeconomic groups within any given community. The most basic diversity criteria, gender, and age apply everywhere there are people.

3.3.2 Population and the Legacy of Malthus

Modern awareness of food security rose to prominence with the 1798 book "An Essay on the Principle of Population" by Thomas Malthus (1766–1834). Malthus asserted that when resources such as food were available, the human population grew geometrically; that is, the population grew at some fixed rate, resulting in an ever-larger absolute increase in population number each year. However, Malthus also asserted that food production would only increase arithmetically, that is by some constant amount each year.

Malthus concluded that "[t]he power of population is indefinitely greater than the power in the earth to produce subsistence for man" and that "positive checks" to population growth such as famine and starvation would operate as an inevitable check on human population growth unless "negative checks" to population growth such as practices to limit childbirth occurred.

While Malthus' conclusions have been influential for over two centuries, they were immediately controversial and contested, and have remained so. It is now clear that population need not grow geometrically, or even at all, as demonstrated by such countries as Italy, Japan, and Russia, where populations are declining. Of course, some of the reasons for low birth rates, such as modern contraception, were unknown to Malthus.

Further, food production has, during many periods of time, increased at rates far more rapid than the linear arithmetic growth postulated by Malthus. In particular, the widespread use of new crop varieties, irrigation, fertilizers, and other practices in the second half of the twentieth century, known as the Green Revolution, has resulted in waves of dramatic increases in food production.

While many of the assumptions upon which Malthus based his reasoning have proven to be inaccurate, the core challenge of providing sufficient food and other resources to a growing human population remains. Those who emphasize efforts to limit population growth are grouped under the label "**neo-Malthusian**" while those who emphasize the ability of humanity to find innovative solutions, especially market and technological solutions, are grouped under the label "**cornucopian**." Human development requires, among other things, that we avoid a "Malthusian trap" that is, a situation where food, energy or water cannot be provided in sufficient quantities to feed a large human population made possible by prior success in expanding their production, whether that be 10, 15, 20, or more billion people.

In 1972, the Club of Rome published a landmark model of the earth's resource consumption dynamics in its report, "The Limits to Growth." This model describes the Earth as an ecosystem with fixed sustainable resource carrying capacities, geometrically growing populations, and resource demands, and resource "stockpiles," or buffers, that can be consumed when resource demands exceed carrying capacities. This model predicts overshoot, which is where large consumable buffers enable the growth of population and consumption far past the sustainable carrying capacity, followed by a collapse of population and consumption when the resource buffers are degraded. The Earth's human population and is ecological resources roughly fit this model. In 2004, an update to the "The Limits to Growth," concluded that the 1972 model's predictions remained sound.

In his 2005 book, "Collapse: How Societies Choose to Fail or Succeed," Jared Diamond provided a collection of historical examples of how many past human societies had faced of resource constraints with catastrophic outcomes. These were isolated and smaller regional societies prior to the creation of the globally connected societies of the current time.

Ehrlich (Ehrlich 1968; Ehrlich and Ehrlich 2009) provides the most notable modern interpretation of Malthus. Ehrlich (1968) famously predicted widespread famine and violence due to the "Population Bomb" of the 1970s. A revisitation of the "Population Bomb" in 2009 concludes that the predicted famines and resource wars did occur, albeit at a lower intensity than expected due primarily to the Green Revolution, to declining birthrates in the rich world, and to massive and unsustainable groundwater mining to support expanded irrigated agriculture in arid regions. The same revision concludes that the worst Malthusian impacts have merely been postponed, not avoided, because the global population will soon triple from the 1968 baseline, demand for land-intensive meat continues to rise, climate change will erode yield and productivity gains, and groundwater will run out.

The laws of economics suggest that resource scarcity and rising real prices for FEW commodities (along with other commodities) are harbingers of a Malthusian crisis and an ecological and economic collapse. By contrast, successful development policies and sustainable growth should yield steadily declining real prices for commodities, as an accumulated wealth of infrastructure, policy, efficiencies, supply chains, and knowledge steadily reduces FEW scarcities and frees up those resources for higher pursuits. The Simon-Ehrlich Wager in 1980 was a bet between business professor Julian Simon and Paul Ehrlich, chancing whether the real price of copper, chromium, nickel, tin, and tungsten (five nonrenewable mined metals, not FEW resources) would rise or fall. Prices fell on all five commodities over the next decade. More significantly, the market prices of food and energy have fallen in the USA since 1980, both in real terms and as a share of the average household's living expenses. The price of water has remained relatively constant. However, water is not a market good in most cases. Further, water infrastructure is aging and requires significant investment in many locations. Is sustainable FEW development occurring in the USA? Will it in the future?

Human conflict is thought to be a symptom of the Malthusian Trap, both because it leads to resource scarcity relative to the human population and because growing human populations are known to be politically and socially volatile. The median age of a society's population is an extremely effective predictor of civil war and revolutionary political activity. A "youth bulge" in the demographics yields a high likelihood of armed conflict, genocide, starvation, and collapse of democratic institutions, whereas demographically balanced populations (balanced age distribution and stable population) are likely to experience peace, economic growth, and transitions to stronger democratic institutions. Wealthier societies tend to transition toward lower birthrates and balanced age demographics. Will large, volatile, young, rapidly growing, low-development nations and regions increasingly threaten the security and prosperity of their stable and wealthy neighbors—and themselves—in the Twenty-first century? The present immigration crisis in Europe and the USA suggests that this trend is already underway.

Which predictions turn out to be true, time will tell, but in light of the catastrophic consequences of famine and resource wars, we had best focus serious attention on this development challenge before it is too late. We have already spent 50 years of our lead time since the warning issued by The Population Bomb… is it ticking? Development in the Twenty-first century takes on a new urgency because we may be in a race to avoid the Malthusian Trap.

With a balanced view of the legacy of Malthus in mind, we can consider the growth of the human population from 2.5 billion in 1950 to 7.4 billion in 2015, and its projected growth to 9.8 billion in 2050 and 11.4 billion in 2100. Figure 3.1 shows population aggregated in three categories:

- **More developed countries** include the nations of Europe, Northern America, Australia, New Zealand, and Japan.
- **Less developed countries** are those in all regions of Africa, Asia (except Japan), Latin America and the Caribbean plus Melanesia, Micronesia and Polynesia.
- **Least developed countries (LDCs)** defined as "low-income countries confronting severe structural impediments to sustainable development. They are highly vulnerable to economic and environmental shocks and have low levels of human assets." In 2018, 47 nations were listed as LDCs: 33 in Africa, 9 in Asia, 4 in Oceania, and one in Latin America and the Caribbean. Nations are evaluated every 3 years and may be moved from the LDC list.

It is no less important for being well-known that the nations with the least capacity to meet the food, energy, and water needs of their citizens also possess the highest

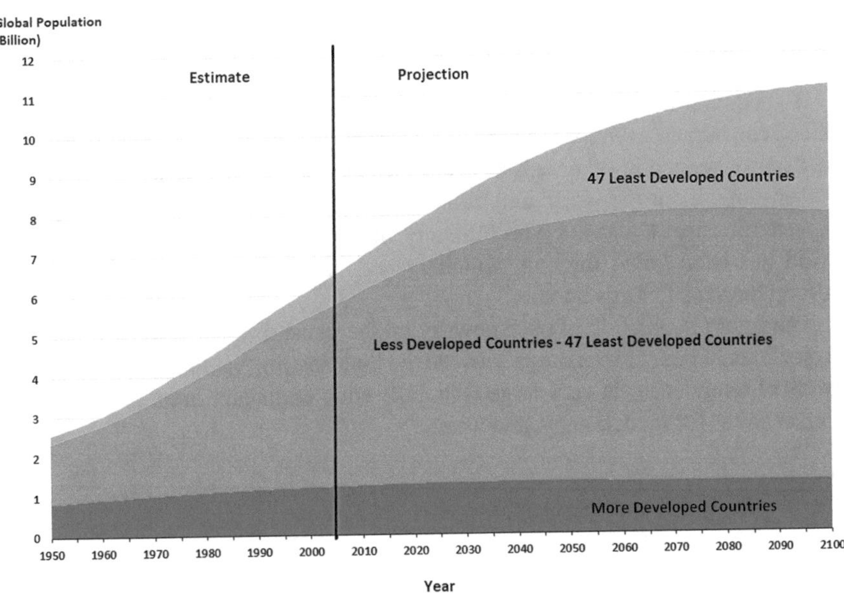

Fig. 3.1 World population estimates (1950–2015) and medium-variant projection (2015–2080). (Source: United Nations, Department of Economic and Social Affairs, Population Division, 2017. World Population Prospects: The 2017 Revision. New York: United Nations)

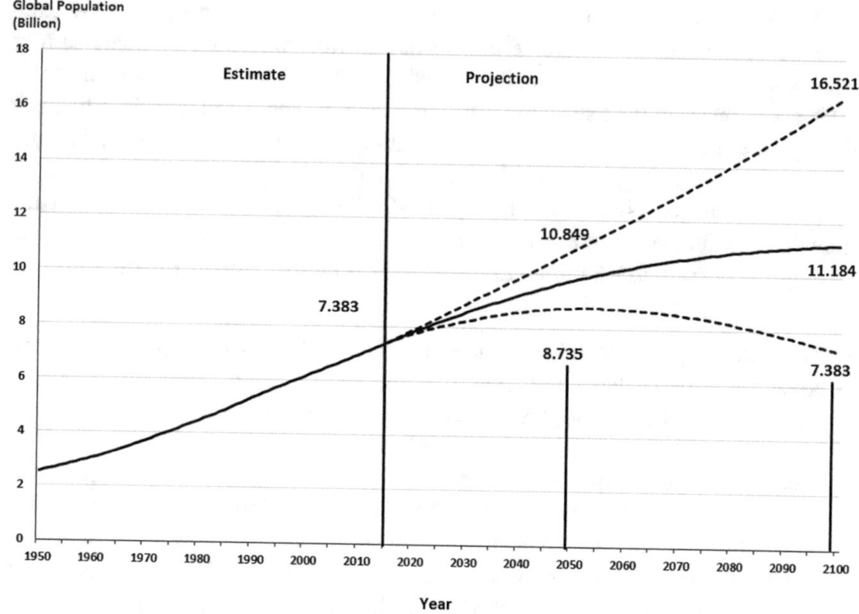

Fig. 3.2 The low, medium, and high variants (95% prediction intervals) for global population projections. (Source: United Nations, Department of Economic and Social Affairs, Population Division 2017. World Population Prospects: The 2017 Revision. New York: United Nations)

rates of population growth. In 1950, the more developed nations possessed 32% of the global population; in 2015, 17% and are projected to have 12% of the global population in 2080.

It is equally important to recognize that during the twentieth century, humans gained considerable ability to control their fertility through the use of contraception, later marriage, and in other ways. As a result, the rate of growth in population has slowed throughout the world and has slowed more rapidly and dramatically than expected in most countries. The full story of the demographic transition that the world has been going through, voluntary and involuntary in some instances, is beyond the scope of this book.

The future population of each country on the planet is subject to many factors. Projections of populations range between the high and low variants (defined as 95% levels of confidence) is very large (Fig. 3.2) with significant implications for the human needs for food, energy, and water.

3.4 Development

Issues of food, energy, and water security, along with land distribution and economic, and housing security, are embedded in the concept of "development" with its meanings of maturing, growth, and progress. What constitutes development is subjective. Further, the deeper one explores the idea, the more difficult a simple definition becomes.

The founding Charter of the United Nations (UN—see Sect. 6.2.2) committed the organization to promote "higher standards of living, full employment, and conditions of economic and social progress and development" (Article 55). The Charter did not define "social progress and development."

Development is viewed as encompassing widely accepted ideas of what constitutes progress in human societies at a particular time. The meaning of development has evolved and continues to do so.

Early measures of "international development" focused on national averages related to economics, such as the following:

1. **Gross Domestic Product** (GDP) and GDP per capita, based on the market value of goods and services produced *within* a country's geographic borders over a specified period (typically 1 year).
2. **Gross National Product** (GNP) and GNP per capita, based on the market value of goods and services produced (i.e., the "output") by the citizens and corporations of a country regardless of where that activity takes place. For some countries, significant economic activity takes place within its borders by foreign citizens and corporations. The value of that economic activity is counted toward the host country under GDP and toward the home country of the citizen/corporation under GNP.
3. **Gross National Income** (GNI) and GNI per capita, is based on the sum of a nation's GDP (inside its geographic borders) and the net income it receives from outside its geographic borders. GNI differs from GNP in that it focuses on income rather than the economic outputs of production. Some organizations like the World Bank prefer GNI to GNP because it better captures the financial wealth within a country beacuse much of the economic value of production can move to third countries regardless of the home country of the corporation that owns that production.

Such measures, although useful, fail to capture many of the attributes that relate to the condition and well-being of individuals within societies. From the 1960s onwards, many explored and proposed alternatives to financial measures of development. The Indian economist Amartya Sen (1933–) wrote many influential papers on the concept of "Human Capabilities," which promoted the ideas of development being measured in terms of the conditions of human lives.

In 1990, the United Nations Development Programme (UNDP) began to produce Human Development Reports, including measures of the average human conditions within a society, such as:

- life expectancy at birth (years);
- expected years of schooling (years);
- mean years of schooling (years); and
- Gross National Income (GNI) per capita (measured in dollars adjusted for "purchasing power parity").

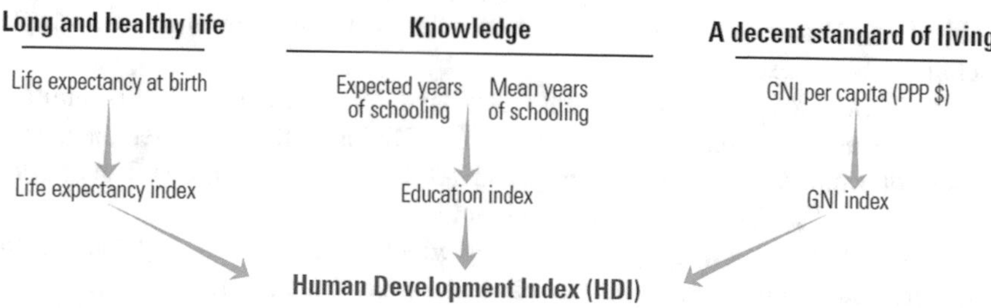

Fig. 3.3 The human development index. (Source: UNDP, http://hdr.undp.org/sites/default/files/hdr2016_technical_notes.pdf)

Table 3.1 Human development indices for 2017

Human development index range	Category	Number of countries with this category
>0.800	Very high human development	59
0.700–0.799	High human development	53
0.550–0.699	Medium human development	39
<0.549	Low human development	38

Source: UNDP (2018) Human Development Report 2018 Human Development for Everyone, United Nations Development Programme

These indices are aggregated into a **Human Development Index (HDI)** (Fig. 3.3) on a scale of 0–1 and applied to countries as a measure of their level of development.

The 2017 HDIs for 189 countries (contained in the Human Development Report 2018) are summarized in Table 3.1. The three countries with the highest HDI were Norway (0.953), Switzerland (0.944), and Australia (0.939). The three countries with the lowest HDI were South Sudan (0.388), Central African Republic (0.367), and Niger (0.354).

Both financial metrics or the broader Human Development Index are problematic measures of development because they only address specific attributes of the conditions of people and omit many other factors that most would consider important to the human condition or the world in which they live, including the concept of sustainability or sustainable development.

The UNDP acknowledges that the HDI "simplifies and captures only part of what human development entails. It does not reflect on inequalities, poverty, human security, empowerment, etc." Therefore, the UNDP has developed other "composite indices," including:

- **Inequality-adjusted Human Development Index** (IHDI), which modifies the three main metrics within the HDI in accordance with the degree of inequity in that index.

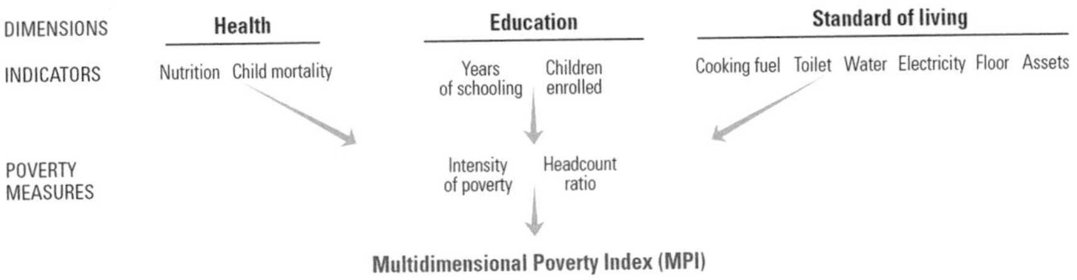

Fig. 3.4 The multidimensional poverty index. (Source: UNDP, http://hdr.undp.org/sites/default/files/hdr2016_technical_notes.pdf)

- **Gender Development Index** (GDI), which calculates separate HDIs for males and females.
- **Gender Inequality Index** (GII), which utilizes metrics differentiated by gender on health, empowerment, and labor.
- **Multidimensional Poverty Index** (MPI), which measures "overlapping deprivations suffered by individuals at the same time" in terms of health (nutrition and child mortality), education (years of schooling and children enrolled) and standards of living (cooking fuel, sanitation, drinking water, electricity, housing, and assets) (Fig. 3.4).

While the Human Development Report presents data on a national level, such country-level aggregation obscures essential differences within societies that the geographic and socioeconomic variations that occur within countries, which the composite indices only begin to illuminate.

There is a substantial body of literature that indicates that women's literacy is an especially important indicator of overall human development.

It should also be recognized that "development" is not a concept reserved for less-developed countries. All nations are developing all the time. The relative importance of food, energy, and water in each country compared to other aspects of development is different.

All countries are continually changing their demands for food, energy, and water, and changing how they make them available, how their citizens obtain access to them, how they are utilized, and how they address issues related to stability and sustainability. Moreover, the balance struck in each country between its concerns about food, energy, and water security and concerns about the consequences of that security changes with time.

Broadly speaking, the least developed countries are concerned about achieving basic levels of food, energy, and water security for their citizens, emphasizing accessibility and reliability over other dimensions of security. Wealthier nations put a greater emphasis on the quality, sustainability, and impact criteria. However, situations like the exposure of high levels of lead in water pipes in Flint, Michigan, and other US cities show the impact of failure to continue investing in infrastructure in developed countries.

3.5 Millennium Development Goals

3.5.1 Creation of the Millennium Development Goals (2000–2015)

Over many decades the United Nations and its constituent organizations (see Sect. 6.2.2) have explored many aspects of development and environmental deterioration. For many decades, such events produced speeches and grand goals, but limited follow through and impact. In the 1990s, more effective work before and after UN meetings, and the growing power of nongovernmental organizations (NGOs) resulted in greater follow through.

In 1996, the Organization for Economic Cooperation and Development (OECD), which brings wealthier, more developed countries together (then 24, now 35), released a list of International Development Goals (IDGs) derived mainly from various UN summits. While many OECD member countries did little to follow up on the IDGs, they were advanced as the basis of a significant development agenda to be adopted at the 2000 Millennium Assembly of the United Nations along with targets and indicators (metrics) that would quantify progress.

In September 2000, a Millennium Declaration was adopted by the 189 members of the UN General Assembly. The Declaration included language on development goals somewhat different from the OECD's IDGs.

In 2001, the two sets of goals were merged by a working group of individuals from the UN, OECD, World Bank, and International Monetary Fund into a list of eight formal Millennium Development Goals (MDGs) with targets and metrics (called indicators) that had primarily been developed for the IDGs. The eight MDGs, approved at a 2002 Finance for Development Conference with a target date of 2015, were as follows:

1. To eradicate extreme poverty and hunger
2. To achieve universal primary education
3. To promote gender equality and empower women
4. To reduce child mortality
5. To improve maternal health
6. To combat HIV/AIDS, malaria, and other diseases
7. To ensure environmental sustainability
8. To develop a global partnership for development

Food, an explicit component of MDG 1, was given Target 1.B, to "halve, between 1990 and 2015, the proportion of people who suffer from hunger." Two metrics were ultimately adopted to measure progress toward this target:

1.8 Prevalence of **underweight** children under five years of age
1.9 Proportion of population below minimum level of dietary energy consumption

The two metrics for food address aspects of utilization: one broad (dietary energy consumption) and one specific to a vulnerable population (underweight children under 5 years of age).

Water, included under MDG 7, was given Target 7.C, to "halve, by 2015, the proportion of people without sustainable access to safe drinking water and basic sanitation." Two metrics were ultimately adopted to measure progress toward this target:

7.8 Proportion of population using an improved drinking water source
7.9 Proportion of population using an improved sanitation facility

The two metrics for water are also utilization metrics (using an improved drinking water source and using an improved sanitation facility).

3.5.2 *Outcomes for Food and Water 2000–2015*

Before assessing the impact of the Millennium Development Goals, we will look at the four metrics over the period of 2000–2015.

3.5.2.1 Prevalence of Underweight Children Under 5 Years of Age

Figure 3.5 shows the prevalence of underweight children under 5 years of age, as estimated by the World Health Organization (WHO) and the United Nations Children's Fund (UNICEF) based on data from national surveys.

Underweight is defined by the WHO as "less than two standard deviations below the median weight for age groups in the international reference population." WHO/UNICEF estimates that the percentage of underweight children dropped from 20.9% (1.28 billion of 6.13 billion) to 14.4% (1.06 billion of 7.35 billion) between 2000 and 2015, a reduction of 31%.

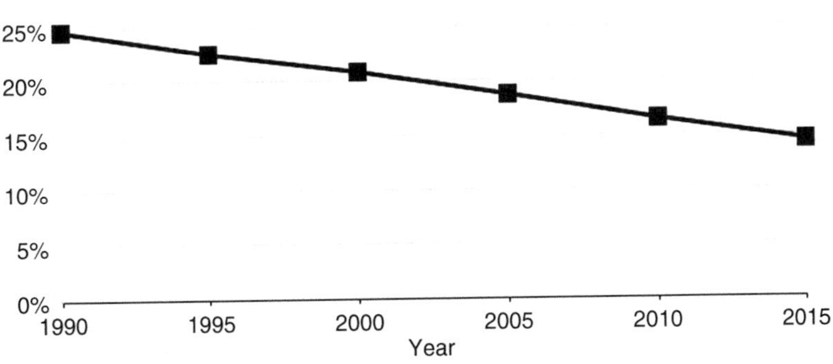

Fig. 3.5 Prevalence of underweight children under 5 years of age. (Source: WHO/UNICEF. World Bank Indicator Code: SH.STA.MALN.ZS)

3.5.2.2 The Proportion of Population Below Minimum Level of Dietary Energy Consumption

While energy intake affects the ability of a person to maintain appropriate body weight and their ability to perform work, it does not measure nutritional intake, which has a significant impact on health. Children who do not receive healthy nutritional diets have impaired growth and physical development, a phenomenon known as "stunting."

Figure 3.6 shows the "prevalence of **undernourishment**" as a percentage of the population. Undernourishment is defined as people receiving dietary energy from their usual food consumption that is below "minimum energy requirement norms" for their age, gender, body weight, and level of activity. Such people are termed "underfed."

For the world as a whole, the proportion of undernourished people dropped from 14.8% (907 million) to 10.7% (786 million) between 2000 and 2015, a 28% drop. LDCs saw a higher (32%) drop. Because of their substantial populations, China (40% decrease) and India (a 16% increase) had a significant impact on global averages. It is also worth noting the persistence of about 2.7% prevalence of undernourishment in high-income nations that was mostly unchanged during the period of the Millennium Development Goals.

Although the trend had been good, the number of undernourished people increased by at least 30 million in 2016. The FAO reports that "After a prolonged decline, this recent increase could signal a reversal of trends. The food security situation has worsened in particular in parts of sub-Saharan Africa, South-Eastern Asia, and Western Asia, and deteriorations have been observed most notably in situations of conflict and conflict combined with droughts or floods" (FAO 2017).

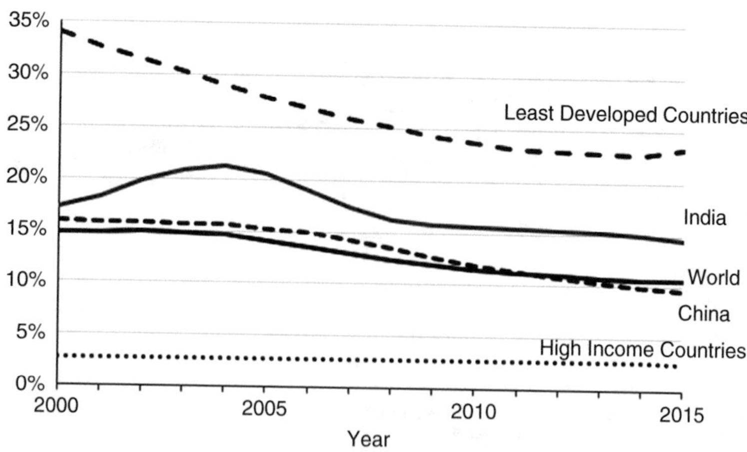

Fig. 3.6 Prevalence of undernourishment (% of the population). (Source: UN Food and Agriculture Organization. World Bank Indicator Code: SN.ITK.DEFC.ZS)

3.5.2.3 The Proportion of the Population Using an Improved Drinking Water Source

Improved drinking water sources are defined by the World Health Organization/ UNICEF Joint Monitoring Programme (JMP) to be "those which, by nature of their design and construction, have the potential to deliver safe water." However, the JMP provides a more sophisticated metric in the form of a "drinking water ladder," which allows deeper insight:

- Surface Water/No Service: Drinking water directly from a river, dam, lake, pond, stream, canal, or irrigation canal. This has a high risk of contamination.
- Unimproved: Drinking water from an unprotected dug well or unprotected spring. This risks contamination.
- Limited: Drinking water from an improved source for which collection time exceeds 30 min for a round trip, including queuing. This is safe but laborious and interferes with other living activities and productivity.
- Basic: Drinking water from an improved source, provided collection time is not more than 30 min for a round trip, including queuing. This satisfies the basic human need.
- Safely Managed: Drinking water from an improved water source, which is located on-premises, available when needed, and free from fecal and priority chemical contamination. This maximizes a person's chances of being healthy and productive and is the typical practice in the wealthiest countries. Considering the costs of a person being sick and spending their valuable time gathering water, and the economic value of productive activities requiring access to water, this is usually the most affordable type of water source—but it requires financial capital, professional expertise, and political stability to provide this type of water access.

Additional JPM metrics for improved water sources include the following:

- Accessible on-premises
- Available when needed
- Free from contamination
- Piped
- Non-piped

Figure 3.7 shows the percentage of the world's population without access to at least basic improved sources of drinking water. The percentage of the world's population without access to at least basic improved sources of drinking water dropped from 20.0% (1224 million) in 2000 to 11.5% (848 million) in 2015, a 42% improvement. Note the significant difference between rural and urban populations.

3.5.2.4 The Proportion of the Population Using an Improved Sanitation Facility

The Joint Monitoring Programme defines sanitation services as "the management of excreta from the facilities used by individuals, through emptying and transport of excreta for treatment and eventual discharge or reuse." Improved sanitation facilities

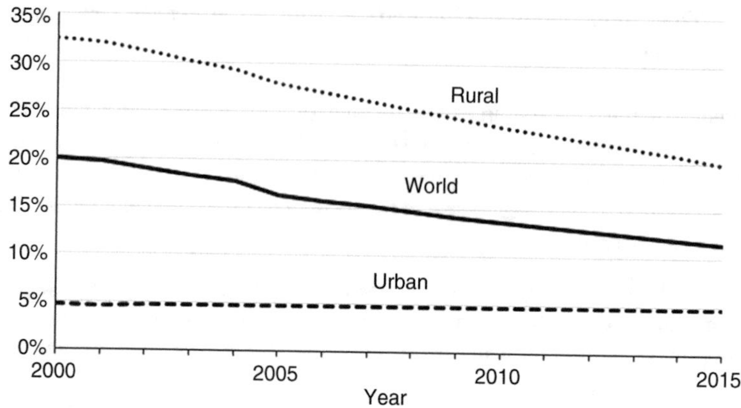

Fig. 3.7 Percentage of the world population without access to basic improved sources of drinking water. (Source: World Health Organization/UNICEF Joint Monitoring Programme (JMP) for Water Supply, Sanitation, and Hygiene. World Bank Indicator Codes: SH.H2O.BASW.ZS, SH.H2O.BASW.RU.ZS [rural], and SH.H2O.BASW.UR.ZS [urban])

are "those designed to hygienically separate excreta from human contact." The JMP also has a "sanitation ladder" metrics:

- Open defecation: Disposal of human feces in fields, forests, bushes, open bodies of water, beaches and other open spaces or with solid waste
- Unimproved: Use of pit latrines without a slab or platform, hanging latrines, or bucket latrines
- Limited: Use of improved facilities shared between two or more households
- Basic: Use of improved facilities which are not shared with other households
- Safely Managed: Use of improved facilities which are not shared with other households and where excreta are safely disposed in situ or transported and treated off-site

Figure 3.8 shows that the percentage of the world's population without access to basic sanitation services dropped from 41.5% (2.54 billion) in 2000 to 31.9% (2.35 billion) in 2015, a 23% improvement. Again, note the significant difference between rural and urban populations.

3.5.2.5 Metrics, Data, and Models

The metrics used in the Millennium Development Goals did not address many aspects of food, energy, and water security because of both the political desire to have a small number of metrics and for the practical reason that the data underlying many desirable metrics is limited on a global scale or inconsistently gathered in different countries. The complexities associated with metrics, data, and modeling and computation illustrated here are explored in later Chapters.

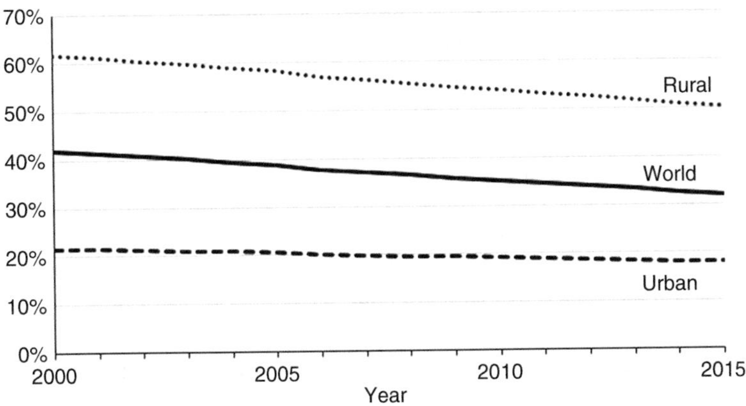

Fig. 3.8 Percentage of the world population without access to basic sanitation services. (Source: World Health Organization/UNICEF Joint Monitoring Programme (JMP) for Water Supply, Sanitation, and Hygiene. World Bank Indicator Codes: SH.STA.BASS.ZS, SH.STA.BASS.RU.ZS [rural] and SH.STA.BASS.UR.ZS [urban])

3.5.3 *Impact of the Millennium Development Goals*

While the halving or 50% reductions of the MDGs were not achieved for the four metrics, significant progress was made between 2000 and 2015:

- 31% reduction in the prevalence of underweight children under 5 years of age
- 28% reduction in the proportion of population below minimum level of dietary energy consumption
- 42% reduction in the proportion of the population using an improved drinking water source
- 23% reduction in the proportion of the population using an improved sanitation facility

How much did the process of setting the Millennium Development Goals and follow up that happened afterward contribute to these achievements?

While it is tempting to view the MDGs as a form of coordinated global public policy facilitated by the United Nations system (see Sect. 6.2.2), and implemented in a systematic way by national governments, aid agencies, NGOs and the UN system itself, the reality is far more complicated.

The process of global policymaking has been likened to an ancient Greek marketplace (an “agora”) where individuals and organizations mix their politics, economics, and culture in “a domain of relative disorder and uncertainty where institutions are underdeveloped and political authority [is] unclear and dispersed through multiple institutions and networks” (Stone 2008).

Significant efforts were made to confront all of the issues addressed by the MDGs before 2000. How did the MDGs change those efforts? While many programs continued as before, many of the significant actors in development publicly committed to the MDGs as a shared vision and framework for their efforts. Funds were raised using the MDG frameworks, and actions under the framework became targets for efforts.

Some viewed the MDGs as leading the way in lifting billions out of poverty. Jeffrey Sachs of Columbia University declared that the "Millennium Development Goals (MDGs) mark a historical and effective method of global mobilization to achieve a set of important social priorities worldwide."

However, others were far more critical. William Easterly of New York University noted that "[m]easuring social and economic progress is not at all as straightforward as the discussion of the MDGs makes it seem. Setting targets in a particular way will make some regions look better, and others look worse depending on a number of choices that any target-setting exercise must make."

It is challenging to provide a robust quantitative correlation between the MDGs and the improvements in the four metrics that took place. One approach by John McArthur and Krista Rasmussen of the Brookings Institution measured the rate of improvement of MDG metrics before 2000 to the rate of improvement in the period 2000–2015. They found that most rates of progress accelerated after the adoption of the MDGs, especially in low-income countries and sub-Saharan African countries, while "middle-income countries typically registered larger cumulative gains but less acceleration over the period."

Further, "the greatest advances were in matters of life and death. At least 20.9 million and as many as 30.3 million additional lives were saved due to accelerated rates of progress, with sub-Saharan Africa accounting for approximately two-thirds of the total." Note that this analysis was applied to all the MDGs, not just the food and water metrics, and so other factors, like improvements in health services, also played an essential role in declines in mortality rates.

However, showing a correlation between the MDGs and increases in rates of improvements in different aspects of development is not the same as proving direct causation. It is impossible to know how institutions and organizations would have behaved had there been no MDGs.

From a public policy perspective, the Millennium Development Goals were perceived as successfully mobilizing action on development issues, even they did not specifically address energy. This naturally led to consideration of what would succeed them following 2015. Addressing energy and creating a more ambitious set of "Sustainable Development Goals" became the focus of the third United Nations Conference on Environment and Development held in 2012.

Significantly, the Millennium Development Goals did not address energy despite the connection between energy and human well-being. In 2005, an unsuccessful attempt was made to add energy targets to the MDGs. However, in 2010, the UN General Assembly declared that 2012 would be the International Year of Sustainable Energy for All.

3.6 Sustainable Energy for All

The relationship between energy use and human well-being is well known and takes a logarithmic form (see Fig. 3.9). Energy is essential for the provision of most modern human services provided at scale, including mobility; lighting; food production and storage; water extraction, purification and delivery; sanitation; education; health care; construction; and, of course, most forms of economic activity.

While energy was mostly an implicit, rather than explicit, part of the development agendas of the international aid community, it has always been a very high priority for national governments and international investment.

Following the first international energy crisis of late 1973, the OECD established the International Energy Agency (IEA) to provide, among other things, comprehensive statistics and analysis on a wide range of energy issues, including energy poverty and development.

Energy poverty includes many issues in addition to the low levels of energy consumption. In poor communities, the primary sources of energy are traditional forms of biomass—wood, charcoal, leaves, agricultural residue, animal/human waste, and urban waste. These forms of energy, while "renewable" have many problematic issues, including:

- They often entail a significant burden on people (particularly women and children) to collect and use, reducing the time available for other productive purposes and education.
- **Traditional biomass** fuels (wood, charcoal, leaves, agricultural residue, animal/ human waste, and urban waste) are very inefficient ways of obtaining energy,

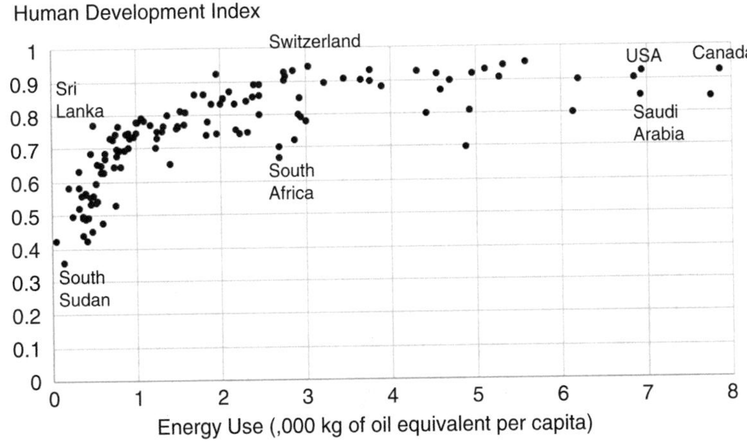

Fig. 3.9 Human development index (HDI) as a function of energy use per capita of 125 countries in 2015. (Source: UNDP 2016 for HDI data and World Bank (2018) for energy data, Indicator Code: EG.USE.PCAP.KG.OE)

typically only 10% of the potential energy is utilized for an energy service such as cooking. Traditional biomass is considered renewable.
- They result in harmful levels of air pollution in homes. The WHO estimates that 3.8 million deaths occur each year as a result of exposure to dirty cookstoves and the indoor use of unclean fuels.
- Gathering traditional biomass fuels on a large scale can have adverse environmental consequences such as increasing deforestation, soil loss, and degrading of ecosystem services.

Low-income countries typically obtain a large part of their energy from **traditional biomass** sources. Mid-income countries typically obtain most of their energy from fossil fuels. In contrast, many high-income countries obtain a growing share of their energy from **modern renewable energy** such as hydropower, wind, solar, and modern biomass (Fig. 3.10)—along with (arguably) third-generation and newer nuclear sources. For example, in 2017, the USA obtained 11% of its total primary energy from renewable sources (EIA 2018).

In the 1980s, at the same time as the term "sustainable development" was becoming popular, rising concern about climate change brought a new perspective and tension to the challenge of providing energy security. The new question became how to provide the energy services required to advance human development without dramatically increasing emissions of greenhouse gases?

While initiatives to address energy security increased during the 1990s, the issue finally came to the fore in 2010, when the UN General Assembly declared that 2012 would be the International Year of Sustainable Energy for All.

In 2012, the Sustainable Energy for All (SEforAll) initiative was launched with a declaration that "Energy is the golden thread that connects economic growth,

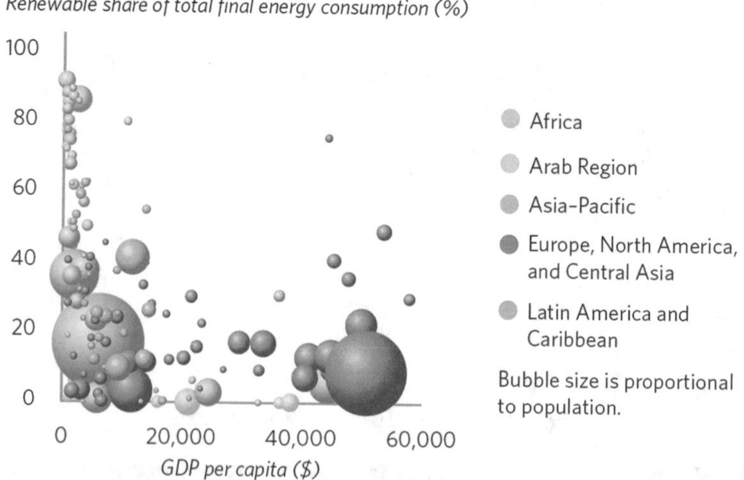

Fig. 3.10 Renewable energy (traditional and modern) as a percentage of total final energy consumption. (Source: Global Tracking Framework, Progress *toward Sustainable Energy* 2017)

increased social equity, and an environment that allows the world to thrive. Development is not possible without energy, and sustainable development is not possible without sustainable energy." SEforAll had three objectives to be achieved by 2030:

- Ensuring universal access to modern forms of energy like electricity
- Doubling the share of renewable energy
- Doubling the rate of improvement in energy efficiency

These objectives were quickly subsumed within the 2015 Sustainable Development Goals.

3.7 Sustainable Development Goals

The 2012 United Nations Conference on Environment and Development focused on establishing a new round of development goals. A 3-year "Post-2015 Development Agenda" process culminated in the adoption by the UN General Assembly in 2015 of a document "Transforming Our World: the 2030 Agenda for Sustainable Development."

Much bigger and bolder than the Millennium Development Goals, the 17 Sustainable Development Goals (SDGs) addressed: (1) poverty; (2) hunger; (3) health and well-being; (4) education; (5) gender equality; (6) water and sanitation; (7) energy; (8) work and economic growth; (9) industry, innovation and infrastructure; (10) reduced inequalities; (11) cities and communities; (12) consumption and production; (13) climate change; (14) oceans, seas and marine life; (15) terrestrial ecosystems, forests, desertification, land degradation and biodiversity; (16) peace, justice and social institutions; and (17) cooperation, finance, data and monitoring and other activities to achieve the goals.

As of summer 2018, the 17 SDG includes 169 targets and 243 indicators/metrics.

Here food, energy, and water are explicit in SDGs 2, 6, and 7 and arguably implicit in many of the other goals. Within these three SDGs are specific objectives with some additional strategies. Each includes a diverse set of objectives and indicators.

3.7.1 SDG 2 Food

Sustainable Development Goal 2 aims to "end hunger, achieve food security and improved nutrition and promote sustainable agriculture." To achieve this goal, SDG 2 has the following targets and indicators/metrics:

Target 2.1. By 2030, end hunger and ensure access by all people, in particular, the poor and people in vulnerable situations, including infants, to safe, nutritious and sufficient food all year round.

Metric 2.1.1: Prevalence of undernourishment. This is the same metric as used in the MDGs and is shown in Fig. 3.6 above.

Metric 2.1.2: Prevalence of moderate or severe food insecurity in the population, based on the **Food Insecurity Experience Scale** (FIES).

Target 2.2. By 2030, end all forms of malnutrition, including achieving, by 2025, the internationally agreed targets on stunting and wasting in children under 5 years of age, and address the nutritional needs of adolescent girls, pregnant and lactating women and older persons.

Metric 2.2.1: Prevalence of stunting (height for age <−2 standard deviation from the median of the WHO Child Growth Standards) among children under 5 years of age.

Metric 2.2.2: Prevalence of malnutrition (weight for height >+2 or <−2 standard deviation from the median of the WHO Child Growth Standards) among children under 5 years of age, by type (wasting and overweight). This is a variation on the metric used in the MDG and is shown in Fig. 3.5 above with the addition of overweight.

Target 2.3. By 2030, double the agricultural productivity and incomes of small-scale food producers, in particular women, indigenous peoples, family farmers, pastoralists, and fishers, including through secure and equal access to land, other productive resources and inputs, knowledge, financial services, markets and opportunities for value addition and non-farm employment.

Metric 2.3.1: Volume of production per labor unit by classes of farming/pastoral/forestry enterprise size.

Metric 2.3.2: Average income of small-scale food producers, by sex and indigenous status.

Target 2.4. By 2030, ensure sustainable food production systems and implement resilient agricultural practices that increase productivity and production, that help maintain ecosystems, that strengthen capacity for adaptation to climate change, extreme weather, drought, flooding, and other disasters and that progressively improve land and soil quality.

Metric 2.4.1: Proportion of agricultural area underproductive and sustainable agriculture.

Target 2.5. By 2020, maintain the genetic diversity of seeds, cultivated plants and farmed and domesticated animals and their related wild species, including through soundly managed and diversified seed and plant banks at the national, regional, and international levels, and promote access to and fair and equitable sharing of benefits arising from the utilization of genetic resources and associated traditional knowledge, as internationally agreed.

Metric 2.5.1: Number of plant and animal genetic resources for food and agriculture secured in either medium or long-term conservation facilities.

Metric 2.5.2: Proportion of local breeds classified as being at risk, not-at-risk, or at an unknown level of risk of extinction.

Three additional targets address implementation issues: finance and trade and markets:

Target 2.A Increase investment, including through enhanced international cooperation, in rural infrastructure, agricultural research and extension services, technology development, and plant and livestock gene banks in order to enhance agricultural productive capacity in developing countries, in particular, least developed countries.
Metric 2.A.1: The agriculture orientation index for government expenditures.
Metric 2.A.2: Total official flows (official development assistance plus other official flows) to the agriculture sector.

Target 2.B Target 2.B. Correct and prevent trade restrictions and distortions in world agricultural markets, including through the parallel elimination of all forms of agricultural export subsidies and all export measures with equivalent effect, in accordance with the mandate of the Doha Development Round of the World Trade Organization (see Sect. 6.2.3).
Metric 2.B.1: Producer Support Estimate.
Metric 2.B.2: Agricultural export subsidies.

Target 2.C Target 2.C. Adopt measures to ensure the proper functioning of **Food Commodity (agricultural)** markets and their derivatives and facilitate timely access to market information, including on food reserves, in order to help limit extreme food price volatility.
Metric 2.C.1: Indicator of food price anomalies.

SDG 2 is a far cry from the hunger component of MDG 1. The two MGD food metrics are replaced by 14, which move toward a food system perspective. Each of the four attributes of food security (availability, access, utilization and stability, and sustainability) have some metrics. Each year, the Food and Agricultural Organization of the United Nations (FAO) leads an annual global review of the state of food security, which provides some tracking of SDG 2.

3.7.2 SDG 6 Water

Sustainable Development Goal 6 aims to "ensure availability and sustainable management of water and sanitation for all." To achieve this goal, SDG has the following targets and indicators/metrics:

Target 6.1. By 2030, achieve universal and equitable access to safe and affordable drinking water for all.
Metric 6.1.1: Proportion of population using safely managed drinking water services. Note that this metric is the highest level of the JMP "drinking water ladder" described above (Sect. 3.5.2). Figure 3.7 shows those without access to basic improved sources of drinking water. This

was 11.5% of the world in 2015. 28.2% of the world did not have access to safely managed drinking water in 2015.

Target 6.2. By 2030, achieve access to adequate and equitable sanitation and hygiene for all and end open defecation, paying special attention to the needs of women and girls and those in vulnerable situations.
Metric 6.2.1: Proportion of population using safely managed sanitation services, including a hand-washing facility with soap and water. Note that this metric is the highest level of the JMP "sanitation ladder" described above (Sect. 3.5.2). Figure 3.8 shows the percentage of the world population without access to basic sanitation services. This was 32.0% of the world in 2015. 60.7% of the world did not have access to safely managed sanitation services in 2015, and 12.1% of the world still relied on open defecation.

Target 6.3. By 2030, improve water quality by reducing pollution, eliminating dumping and minimizing the release of hazardous chemicals and materials, halving the proportion of untreated wastewater and substantially increasing recycling and safe reuse globally.
Metric 6.3.1: Proportion of wastewater safely treated.
Metric 6.3.2: Proportion of bodies of water with good ambient water quality.

Target 6.4. By 2030, substantially increase water-use efficiency across all sectors and ensure sustainable withdrawals and supply of freshwater to address water scarcity and substantially reduce the number of people suffering from water scarcity.
Metric 6.4.1: Change in water-use efficiency over time.
Metric 6.4.2: Level of water stress: freshwater withdrawal as a proportion of available freshwater resources.

Target 6.5. By 2030, implement integrated water resources management at all levels, including through transboundary cooperation as appropriate.
Metric 6.5.1: Degree of integrated water resources management implementation (0–100).
Metric 6.5.2: Proportion of transboundary basin area with an operational arrangement for water cooperation.

Target 6.6. By 2020, protect and restore water-related ecosystems, including mountains, forests, wetlands, rivers, aquifers, and lakes.
Metric 6.6.1: Change in the extent of water-related ecosystems over time.

Two additional targets address implementation issues; one addressing development assistance and one addressing community engagement:

Target 6.A By 2030, expand international cooperation and capacity-building support to developing countries in water- and sanitation-related activities and programs, including water harvesting, desalination, water efficiency, wastewater treatment, recycling, and reuse technologies.
Metric 6.A.1: Amount of water- and sanitation-related official development assistance that is part of a government-coordinated spending plan.

Target 6.B Support and strengthen the participation of local communities in improving water and sanitation management.
Metric 6.B.1: Proportion of local administrative units with established and operational policies and procedures for participation of local communities in water and sanitation management.

As with food, SDG 6 represents a dramatic expansion in targets and metrics. The two MGD water and sanitation metrics are replaced by 11, which again move toward a water system perspective and apply to the four attributes of water security (availability, access, utilization and stability, and sustainability). Target 6.6 directly addresses ecosystems and ecosystem services.

The WHO/UNICEF Joint Monitoring Programme (JMP) provides annual reporting on progress on SDG 6.

3.7.3 SDG 7 Energy

Sustainable Development Goal 7 draws directly on the Sustainable Energy for All initiative with a goal to "ensure access to affordable, reliable, sustainable and modern energy for all." Note the use of the term "modern energy," which directly aims at a transition away from tradition biomass fuels and the problems associated with them. To achieve this goal, SDG adopts the same three targets as SEforAll with the following indicators/metrics:

Target 7.1. By 2030, ensure universal access to affordable, reliable and modern energy services.
Metric 7.1.1: Proportion of population with access to electricity (see Fig. 3.11).

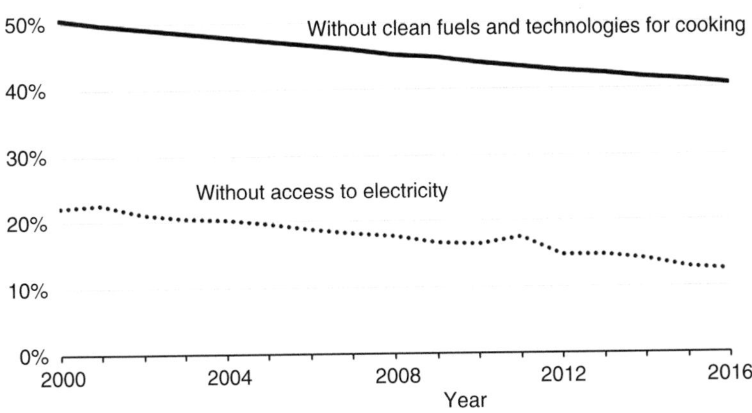

Fig. 3.11 Percentage of the world population without access to electricity (Metric 7.1.1) and without access to clean fuels and technologies for cooking (Metric 7.1.2). (Source: World Bank, Indicator codes EG.ELC.ACCS.ZS [electricity] and EG.CFT.ACCS.ZS [cooking])

Metric 7.1.2: Proportion of population with primary reliance on clean fuels and technology (see Fig. 3.11).

Target 7.2. By 2030, increase substantially the share of renewable energy in the global energy mix.

Metric 7.2.1: Renewable energy share in the total final energy consumption (see Fig. 3.12).

Target 7.3. By 2030, double the global rate of improvement in energy efficiency.

Metric 7.3.1: Energy intensity measured in terms of primary energy and GDP (see Fig. 3.13).

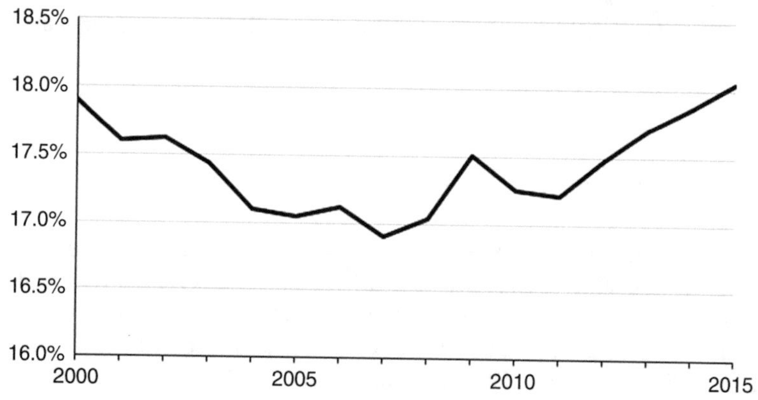

Fig. 3.12 Renewable energy share in the total final world energy consumption. (Source: World Bank (Indicator code EG.FEC.RNEW.ZS). Note that the declining trend before 2007 represents a reduced share of energy from traditional biomass and greater use of fossil fuels. The increasing trend after 2007 represents the acceleration in the deployment of modern renewables, mainly wind. It is important to remember the SDG 7 calls for a reduction in the use of traditional biomass, and therefore modern renewables must both replace traditional biomass and displace fossil fuels and nuclear)

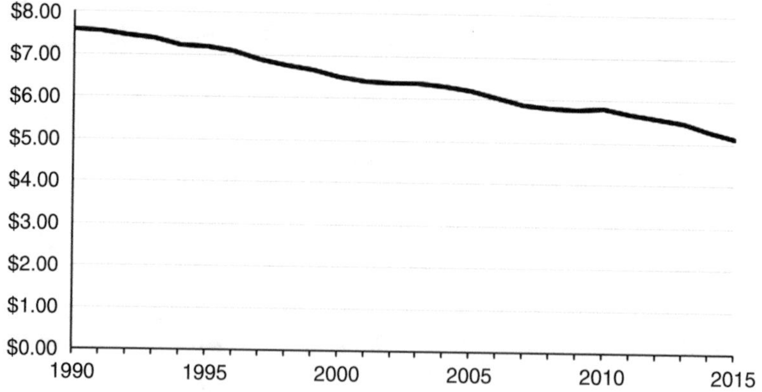

Fig. 3.13 Energy intensity level of primary energy (MJ/$2011 PPP GDP). (Source: World Bank, Indicator code EG.EGY.PRIM.PP.KD)

Two additional targets address implementation issues; one addressing finance and one addressing investment in infrastructure and technology:

Target 7.A By 2030, enhance international cooperation to facilitate access to clean energy research and technology, including renewable energy, energy efficiency and advanced and cleaner fossil-fuel technology, and promote investment in energy infrastructure and clean energy technology.
Metric 7.A.1 Mobilized amount of US dollars per year starting in 2020 accountable towards the $100 billion commitment.

Target 7.B By 2030, expand infrastructure and upgrade technology for supplying modern and sustainable energy services for all in developing countries, in particular, least developed countries, small island developing States, and land-locked developing countries, in accordance with their respective programs of support.
Metric 7.B.1 Investments in energy efficiency as a percentage of GDP and the amount of foreign direct investment in financial transfer for infrastructure and technology to sustainable development services.

The World Bank leads a lead an annual review related to the SEforAll initiative (see World Bank, 2017 Global Tracking Framework in Further Reading). The IEA, IRENA, UN Statistics Division, World Bank, and WHO are the custodian agencies for SDG 7 and put out a tracking report each year (see Further Reading).

3.7.4 Other SDGs

All SDG goals have some relevance to food, energy, and water.

- Achieving SDG 1 (poverty) and SDG 8 (work) will enable the poor of the world to be able to afford more resources and address the access component of security.
- Achieving SDG 3 (health and well-being) and SDG 11 (cities and settlements) requires addressing hunger and nutrition, water and sanitation, and energy for such services as refrigerating medicines. See Chap. 18 for FEW issues at the city-scale relevant to SDG 11.
- Achieving SDG 4 (education) will result in less poverty and better utilization of resources and services.
- Gender quality (SDG 5) and reduced inequality (SDG 10) will lower fertility rates and allow women and others greater access to resources which can be utilized more efficiently (resources for basic need are used more efficiently than resources for needs beyond basic)
- SGD 12 (consumption and production) includes a strong focus on the efficient use of resources (including waste reduction, reuse, and recycling), which will lower the demand for energy and water.

Specific to food, SDG 12 includes: Target 12.3 By 2030, halve per capita global food waste at the retail and consumer levels and reduce food losses along production and supply chains, including post-harvest losses. This is measured by the **Global Food Loss Index** (Metric 12.3.1).

Specific to energy, SDG 12 includes Target 12.C. Rationalize inefficient fossil-fuel subsidies that encourage wasteful consumption by removing market distortions, in accordance with national circumstances, including by restructuring taxation and phasing out those harmful subsidies, where they exist, to reflect their environmental impacts, taking fully into account the specific needs and conditions of developing countries and minimizing the possible adverse impacts on their development in a manner that protects the poor and the affected communities. Metric 12.C.1 measures the level of fossil-fuel subsidies per unit of GDP (production and consumption) and as a proportion of total national expenditure on fossil fuel.

Specific to water, SDG 12 includes Target 12.4, which aims to reduce the release of chemicals and wastes into air, water, and soil.

- SDG 13 (climate) addresses the topic with a focus on supporting the ability of poor and vulnerable countries to adapt, adopt effective policies, support education, finance clean energy, and take other steps consistent with the United Nations Framework Convention on Climate Change without encroaching on its primary authority on climate change (Chap. 11).
- SDG 14 (oceans, seas, and marine life) and SDG 15 (terrestrial ecosystems, forests, desertification, land degradation, and biodiversity) address a variety of traditional environmental issues, many of which impact food, energy, and water. For example, reducing pollution and the degradation of ecosystem functioning has significant implications for food production on land and in fisheries.

  Specific to food are the following:

  (a) Target 14.2 addresses destructive and overfishing practices and calls for science-based management plans to restore fish stocks to levels that can be sustainably harvested. This is measured by the proportion of fish stocks within biologically sustainable levels (Metric 14.4.1).
  (b) Target 14.6 address subsidies which contribute to destructive overfishing practices. This is measured by progress by countries in the degree of implementation of international instruments aiming to combat illegal, unreported and unregulated fishing (Metrics 14.6.1).
  (c) Target 14.7 addresses fisheries and Small Island Developing States, while Target 14.B addresses access for small-scale artisanal fishers to marine resources and markets.
  (d) Targets 15.1–15.5 address freshwater ecosystems, forests, desertification, mountain ecosystems, and biodiversity loss respectively. These targets and related issues will be addressed in Chap. 9 (Ecosystems).

- SDG 16 (peace, justice, and social institutions) and SDG 17 (cooperation, finance, data and monitoring, and other activities to achieve the goals) address the framework of societies and their interactions that will enable the SDG to be successful.

Key Points

- Food, energy, and water security implicate four key attributes, including availability; access; utilization; and, stability and reliability.
- A wide array of quantitative and qualitative factors are required to understand food, energy, and water security and (more generally) human development.
- Development is a general and subjective concept to describe the evolution of aggregate human well-being.
- Economic metrics such as GDP, GDP and GNI, and the Human Development Index are useful but limited measures of development.
- The Millennium Development Goals (2000–2015) provided a framework for mobilizing development efforts in a small number of areas.
- The Sustainable Development Goals (SDGs) (2016–2030) are a much larger enterprise, including 17 goals, 169 targets, and 243 indicators/metrics. SDG 2 addresses food; SDG 6 addresses water and sanitation; and, SDG 7 addresses energy. Many other SDGs have essential interlinkages to food, energy, and water.
- SDGs and other goals have implications for FEW Nexus, but exactly in what manner or magnitude are hard to outline in the current context. Much work still is needed to identify Nexus connections to goals, develop Nexus-related goals and benchmarks, and devise strategies and policies.

Discussion Points and Exercises

1. Discuss how the attributes of complex systems such as heterogeneity, interconnections, distributed natural and distributed human controls, hierarchy, emergence, feedback, self-organized criticality, sensitivity to initial conditions, adaption, and sentience map on to attributes of resource security.
2. Discuss the role of demographics in three countries—one LDC country, one less developed non-LDC, and one high developed country.
3. Discuss the strengths and weaknesses of the Human Development Index and suggest an index of your own.
4. Discuss relationships between targets and metrics in SDG 2 (hunger) and SDG 6 (water and sanitation).
5. Discuss relationships between targets and metrics in SDG 2 (hunger) and SDG 7 (energy).
6. Discuss relationships between targets and metrics in SDG 6 (water and sanitation) and SDG 7 (energy).
7. Map the metrics of SDG 2 on to the four attributes of food security. What additional metrics might provide a fuller understanding of food security?
8. Map the metrics of SDG 4 onto the four attributes of water security. What additional metrics might provide a fuller understanding of water security?
9. Map the metrics of SDG 7 onto the four attributes of energy security. What additional metrics might provide a fuller understanding of energy security?
10. Map the metrics of SDG 6 on to the attributes of a water system. What parts of a water system are not represented in the indicators?
11. Using the World Bank's World Development Indicators, graph prevalence of undernourishment against GDP for countries where data is available (see

Appendix D). Identify countries that have relatively high or low prevalence of undernourishment compared to their relative GDP (i.e., relatively poor/wealthy countries that have relatively low/high undernourishment.)

12. Using the World Bank's World Development Indicators (see Appendix D), graph percentage of population without access to basic improved sources of drinking water against GDP for countries where data is available. Identify countries that have relatively high or low use of improved sources of drinking water compared to their relative GDP (i.e., relatively poor/wealthy countries that have relatively high/low use of improved sources of drinking water.)
13. Using the World Bank's World Development Indicators (see Appendix D), graph percentage of population without access to electricity against GDP for countries where data is available. Identify countries that have relatively high or low access to electricity their relative GDP (i.e., relatively poor/wealthy countries that have relatively high/low access to electricity.)
14. Write and discuss an opinion on whether, and in what specific sense, your home community, home country, or the world face a Malthusian trap in the twenty-first century.

References

Allan, J. A. (1998). Virtual water: A strategic resource global solutions to regional deficits. *Groundwater, 36*(4), 545–546.

Bakker, K. (2012). Water security: Research challenges and opportunities. *Science, 337*(6097), 914–915.

Bakker, K., & Morinville, C. (2013). The governance dimensions of water security: A review. *Philosophical Transactions of the Royal Society A, 371*(2002), 20130116.

Beddington, J. R., et al. (2012). *Achieving food security in the face of climate change: Final report from the Commission on Sustainable Agriculture and Climate Change*. Copenhagen: CGIAR Research Program on Climate Change, Agriculture and Food Security (CCAFS).

Bhaduri, A., et al. (2015). Sustainability in the water-energy-food nexus. *Water International, 40*, 723–732.

Bruckner, M., et al. (2012). Materials embodied in international trade–Global material extraction and consumption between 1995 and 2005. *Global Environmental Change, 22*(3), 568–576.

D'Odorico, P., Laio, F., & Ridolfi, L. (2010). Does globalization of water reduce societal resilience to drought? *Geophysical Research Letters, 37*(13), L13403.

D'Odorico, P., et al. (2014). Feeding humanity through global food trade. *Earth's Future, 2*(9), 458–469.

Daily, G., et al. (1998). Food production, population growth, and the environment. *Science, 281*, 1291–1292.

Diamond, J. (2005). *Collapse: How societies choose to fail or succeed*. Camberwell, VIC: Penguin.

Easterly, W. (2009). How the millennium development goals are unfair to Africa. *World Development, 37*(1), 26–35. Retrieved from https://papers.ssrn.com/sol3/papers.cfm?abstract_id=2066805.

Ehrlich, P. R. (1968). *The population bomb*. New York, NY: Sierra Club, Ballantine.

Ehrlich, P. R., & Ehrlich, A. H. (2009). The population bomb revisited. *The Electronic Journal of Sustainable Development, 1*(3), 63–71.

Famiglietti, J. S. (2014). The global groundwater crisis. *Nature Climate Change, 4*(11), 945.

FAO (1996) Report of the World Food Summit 13-17 November 1996, WFS 96/REP, Food and Agriculture Organization of the United Nations, Rome, 1996 http://www.fao.org/3/w3548e/w3548e00.htm

FAO, IFAD, UNICEF, WFP, WHO. (2017). *The state of food security and nutrition in the world 2017. Building resilience for peace and food security*. Rome: FAO. Retrieved from http://www.fao.org/3/a-i7695e.pdf.

Folke, C., et al. (2002). Resilience and sustainable development: Building adaptive capacity in a world of transformations. *Ambio: A Journal of the Human Environment, 31*(5), 437–440.

Folke, C., et al. (2010). Resilience thinking: Integrating resilience, adaptability, and transformability. *Ecology and Society, 15*(4), 20.

Garnett, T. (2014). Three perspectives on sustainable food security: Efficiency, demand restraint, food system transformation. What role for life cycle assessment? *Journal of Cleaner Production, 73*, 10–18.

Garrick, D., & Hall, J. W. (2014). Water security and society: Risks, metrics, and pathways. *Annual Review of Environment and Resources, 39*, 611–639.

Godfray, H. C. J., et al. (2010). Food security: The challenge of feeding 9 billion people. *Science, 327*, 812.

Gustafson, D., et al. (2016). Seven food system metrics of sustainable nutrition security. *Sustainability, 8*(3), 196.

Haddow, G., Bullock, J., & Coppola, D. P. (2017). *Introduction to emergency management*. Oxford: Butterworth-Heinemann.

Hanjra, M. A., & Qureshi, M. E. (2010). Global water crisis and future food security in an era of climate change. *Food Policy, 35*(5), 365–377.

Hashimoto, T., Stedinger, J. R., & Loucks, D. P. (1982). Reliability, resiliency, and vulnerability criteria for water resource system performance evaluation. *Water Resources Research, 18*(1), 14–20.

Hirsch, R. L., Bezdek, R., & Wendling, R. (2005). *Peaking of world oil production: Impacts, mitigation, & risk management* (No. DOE/NETL-IR-2005-093; NETL-TPR-2319). Pittsburgh, PA; Morgantown, WV; Albany, OR: National Energy Technology Laboratory (NETL).

Hoekstra, A. Y., et al. (2012). Global monthly water scarcity: Blue water footprints versus blue water availability. *PLoS One, 7*(2), e32688.

Holling, C. S. (1996). Engineering resilience versus ecological resilience. *Engineering Within Ecological Constraints, 31*(1996), 32.

Hollnagel, E., Woods, D. D., & Leveson, N. (2007). *Resilience engineering: Concepts and precepts*. Aldershot: Ashgate Publishing, Ltd.

IEA, IRENA, UN Statistics Division, World Bank Group, WHO. (2018). *Tracking SDG7: The energy progress report 2018: A joint report of the Custodian Agencies*. Washington, DC: Bank. Retrieved from https://irena.org/publications/2018/May/Tracking-SDG7-The-Energy-Progress-Report.

Jones, A. D., et al. (2013). What are we assessing when we measure food security? A compendium and review of current metrics. *Advances in Nutrition, 4*(5), 481–505.

Kasprzyk, J. R., et al. (2013). Many objective robust decision making for complex environmental systems undergoing change. *Environmental Modelling & Software, 42*, 55–71.

Kennedy, C., Cuddihy, J., & Engel-Yan, J. (2007). The changing metabolism of cities. *Journal of Industrial Ecology, 11*(2), 43–59.

Kijne, J. W., Barker, R., & Molden, D. J. (Eds.). (2003). *Water productivity in agriculture: Limits and opportunities for improvement* (Vol. 1). Wallingford: CABI.

Lambin, E. F., & Meyfroidt, P. (2011). Global land use change, economic globalization, and the looming land scarcity. *Proceedings of the National Academy of Sciences, 108*(9), 3465–3472.

Ljunberg, D., & Lundh, V. (2013). *Resilience engineering within ATM - Development, adaption, and application of the Resilience Analysis Grid (RAG)* (LiU-ITN-TEK-G--013/080--SE). Linköping: University of Linköping.

Mayer, A., Mubako, S., & Ruddell, B. L. (2016). Developing the greatest Blue Economy: Water productivity, fresh water depletion, and virtual water trade in the Great Lakes basin. *Earth's Future, 4*(6), 282–297.

McArthur, J., & Rasmussen, K. (2018). Change of pace: Accelerations and advances during the Millennium Development Goal era. *World Development, 105*, 132–143. https://doi.org/10.1016/j.worlddev.2017.12.030.

Meadows, D., & Randers, J. (2012). *The limits to growth: The 30-year update*. London: Routledge.

Meadows, D. H., et al. (1972). *The limits to growth: A report to the Club of Rome*. New York, NY: Universe Books.

Mekonnen, M. M., & Hoekstra, A. Y. (2016). Four billion people facing severe water scarcity. *Science Advances, 2*(2), e1500323.

Miller, F., et al. (2010). Resilience and vulnerability: Complementary or conflicting concepts? *Ecology and Society, 15*(3), 1.

Molden, D., & Sakthivadivel, R. (1999). Water accounting to assess use and productivity of water. *International Journal of Water Resources Development, 15*(1-2), 55–71.

Rosegrant, M. W., & Cline, S. A. (2003). Global food security: Challenges and policies. *Science, 302*(5652), 1917–1919.

Rushforth, R. R., Messerschmidt, M., & Ruddell, B. L. (in review). *The water SRVC model for metropolitan water portfolio water security*.

Rushforth, R. R., & Ruddell, B. L. (2015). The hydro-economic interdependency of cities: Virtual water connections of the Phoenix, Arizona Metropolitan Area. *Sustainability, 7*(7), 8522–8547.

Rushforth, R. R., & Ruddell, B. L. (2016). The vulnerability and resilience of a city's water footprint: The case of Flagstaff, Arizona, USA. *Water Resources Research, 52*(4), 2698–2714.

Sovacool, B. K., & Mukherjee, I. (2011). Conceptualizing and measuring energy security: A synthesized approach. *Energy, 36*(8), 5343–5355.

Sovacool, B. K., et al. (2011). Evaluating energy security performance from 1990 to 2010 for eighteen countries. *Energy, 36*(10), 5846–5853.

Stone, D. (2008). Global public policy, Transnational policy communities, and their networks. *Policy Studies Journal, 36*(1), 19–38.

The Secretary-General's High-Level Group on Sustainable Energy for All. (2012). *Sustainable Energy for All: A Global Action Agenda*. Vienna: Sustainable Energy for All. Retrieved from https://www.seforall.org/sites/default/files/l/2014/01/SEFA-Action-Agenda-Final.pdf.

U.S. Geological Survey. (2018). *Summary of estimated water use in the United States in 2015*. Reston, VA: Author. Retrieved from https://pubs.usgs.gov/fs/2018/3035/fs20183035.pdf.

UNDP. (2016). *Human development report 2016. Human development for everyone*. New York, NY: Author. Retrieved from http://hdr.undp.org/en/content/human-development-report-2016-human-development-everyone.

United Nations. (1992). *United Nations Conference on Environment & Development* (Agenda 21). New York, NY: Author. Retrieved from https://sustainabledevelopment.un.org/content/documents/Agenda21.pdf.

United Nations. (2010). *Resolution adopted by the General Assembly on 20 December 2010, 65/151. International year of Sustainable Energy for All*. New York, NY: Author. Retrieved from https://www.seforall.org/sites/default/files/l/2013/10/a.res_.65.151-int-year-of-sefa.pdf.

United Nations. (2015). *Transforming our world: The 2030 agenda for sustainable development* (A/RES/70/1). New York, NY: Author. Retrieved from https://sustainabledevelopment.un.org/.

UN-Water. (2013). *Water security & the global water agenda. United Nations University, 2013*. Geneva: Author. Retrieved from http://www.unwater.org/publications/water-security-global-water-agenda/.

UNDP. (2018). *Human development indices and indicators: 2018 Statistical update*, United Nations Development Program. Retrieved from http://hdr.undp.org/en/2018-update

Urdal, H. (2004). *The devil in the demographics: The effect of youth bulges on domestic armed conflict, 1950-2000* (Social development papers 14). Washington, DC: World Bank.

Vogt, K., et al. (2012). *Sustainability unpacked: Food, energy and water for resilient environments and societies*. London: Routledge.

Walker, B., & Salt, D. (2012a). *Resilience practice: Building capacity to absorb disturbance and maintain function*. Washington, DC: Island Press.

Walker, B., & Salt, D. (2012b). *Resilience thinking: Sustaining ecosystems and people in a changing world*. Washington, DC: Island Press.
Walker, R. V., et al. (2014). The energy-water-food nexus: Strategic analysis of technologies for transforming the urban metabolism. *Journal of Environmental Management, 141*, 104–115.
Wheeler, T., & Von Braun, J. (2013). Climate change impacts on global food security. *Science, 341*(6145), 508–513.
Wiedmann, T. O., et al. (2016). The material footprint of nations. *Proceedings of the National Academy of Sciences, 112*(20), 6271–6276.
Yergin, D. (2006). Ensuring energy security. *Foreign Affairs, 85*, 69–82.
Zimmerman, R., Zhu, Q., & Dimitri, C. (2016). Promoting resilience for food, energy, and water interdependencies. *Journal of Environmental Studies and Sciences, 6*(1), 50–61.

Further Reading

Easterly, W. (2009). How the millennium development goals are unfair to Africa. *World Development, 37*(1), 26–35 Retrieved from https://papers.ssrn.com/sol3/papers.cfm?abstract_id=2066805.
FAO, IFAD, UNICEF, WFP, WHO. (2017). *The state of food security and nutrition in the world 2017. Building resilience for peace and food security*. Rome: FAO Retrieved from http://www.fao.org/3/a-i7695e.pdf.
HDRO Outreach. (2015). *What is human development?* New York, NY: United Nations Development Programs Retrieved from http://hdr.undp.org/en/content/what-human-development.
Hulme, D. (2009). *The millennium development goals (MDGs): A short history of the world's biggest promise* (BWPI working paper 100). Manchester: University of Manchester Brooks World Poverty Institute Retrieved from https://www.unidev.info/Portals/0/pdf/bwpi-wp-10009.pdf.
IEA, IRENA, UN Statistics Division, World Bank Group, WHO. (2018). *Tracking SDG7: The energy progress report 2018: A joint report of the custodian agencies*. Washington, DC: World Bank Retrieved from https://irena.org/publications/2018/May/Tracking-SDG7-The-Energy-Progress-Report.
Malthus, T. R. (1798). *An essay on the principle of population*. London: J. Johnson Reprinted by Dover Publications in 2007. Retrieved from http://www.esp.org/books/malthus/population/malthus.pdf.
McArthur, J., & Rasmussen, K. (2018). Change of pace: Accelerations and advances during the Millennium Development Goal era. *World Development, 105*, 132–143. https://doi.org/10.1016/j.worlddev.2017.12.030.
Meadows, D., & Randers, J. (2012). *The limits to growth: The 30-year update*. London: Routledge.
Reisner, M. (1993). *Cadillac desert: The American West and its disappearing water*. New York, NY: Penguin.
Sachs, J. (2012). From millennium development goals to sustainable development goals. *Lancet, 379*, 2206–2211 Retrieved from http://jeffsachs.org/wp-content/uploads/2012/06/From-MDGs-to-SDGs-Lancet-June-2012.pdf.
Stone, D. (2008). Global public policy, Transnational policy communities, and their networks. *Policy Studies Journal, 36*(1), 19–38.
The Secretary-General's High-Level Group on Sustainable Energy for All. (2012). *Sustainable energy for all: A global action agenda*. Vienna: Sustainable Energy for All Retrieved from https://www.seforall.org/sites/default/files/SEFA-Action-Agenda-Final.pdf.
UN. (2010). *Resolution adopted by the General Assembly on 20 December 2010, 65/151. International year of Sustainable Energy for All*. Washington, DC: United Nations Foundation Retrieved from https://seforall.org/sites/default/files/l/2013/10/GA-resolution-A-67-215-SE4ALL-DECADE.pdf.

UNDP. (2016). *Human development report 2016: Human development for everyone*. New York, NY: Author Retrieved from http://hdr.undp.org/en/content/human-development-report-2016-human-development-everyone.
United Nations. (1992). *United Nations Conference on Environment & Development* (Agenda 21). New York, NY: Author Retrieved from https://sustainabledevelopment.un.org/content/documents/Agenda21.pdf.
United Nations. (2015a). *Transforming our world: The 2030 agenda for sustainable development* (A/RES/70/1). New York, NY: Author Retrieved from https://sustainabledevelopment.un.org/.
United Nations. (2015b). *The millennium development goals report 2015*. New York, NY: Author Retrieved from http://www.un.org/millenniumgoals/2015_MDG_Report/pdf/MDG%202015%20Summary%20web_english.pdf.
UN-Water. (2013). *Water security & the global water agenda*. Hamilton, ON: United Nations University Retrieved from http://www.unwater.org/publications/water-security-global-water-agenda/.

Chapter 4
Human Behavior and Adaptation

Mary Doidge, Elena Irwin, Nicole Sintov, and Robyn S. Wilson

4.1 Introduction

Human behavior plays an important role in food, energy, and water systems. The complexity of these systems (see Chap. 2) means that studying components in isolation will not allow for a complete understanding of system dynamics, and will potentially lead to solutions that omit important elements. The coupled nature of FEW systems means the human and natural systems are linked such that human behavior impacts natural processes, and the outcomes of these natural processes influence human behavior. Effectively managing FEW systems requires understanding how humans behave and interact with their natural and social environments, as well as understanding the environmental impacts of social changes (e.g., population growth and urbanization).

M. Doidge (✉)
Department of Natural Resource Sciences, McGill University, Ste-Anne-de-Bellevue, QC, Canada
e-mail: mary.doidge@mcgill.ca

E. Irwin
Department of Agricultural, Environmental, and Development Economics, The Ohio State University, Columbus, OH, USA
e-mail: irwin.78@osu.edu

N. Sintov · R. S. Wilson
School of Environment and Natural Resources, The Ohio State University, Columbus, OH, USA
e-mail: sintov.2@osu.edu; Wilson.1376@osu.edu

P. Saundry, B. L. Ruddell (eds.), *The Food-Energy-Water Nexus*, AESS Interdisciplinary Environmental Studies and Sciences Series,
https://doi.org/10.1007/978-3-030-29914-9_4

Studying either the natural or human system in isolation omits important drivers of change in FEW systems and how people respond and adapt to these changes. This can potentially lead to an incomplete understanding of these integrated systems.

Researchers may approach the role of human behavior and adaptation within FEW systems using different frameworks or methods, drawing from and incorporating many social science disciplines. Despite the variety of approaches, researchers in this context seek to address how human behavior impacts the natural environment, and how individual, community, and political decisions impact these systems.

In this chapter, we consider the role of human behavior and adaptation in FEW systems. We discuss the importance of including more sophisticated models of human behavior in the study of FEW systems and provide examples of how past research has incorporated complexity in human behavior into models of these systems. We do so from the perspectives of psychology, economics, and **decision science**—all social sciences with well-developed theories and models of human behavior that are useful in informing models and policies. We present two case studies as examples of how research can explicitly account for human behavior in these systems. Finally, we discuss challenges in incorporating human behavior and adaptation into models of these systems and identify future directions for work in this field.

4.2 Consequences of Overlooking Complexities in Human Behavior

The complexities of FEW systems are apparent throughout the chapters of this book. Section 1.4 discusses the macroscope of the FEW nexus concepts and outlines several framings of this nexus. The complexity of the system presents many challenges to studying it as a whole, often leading researchers to focus on a single aspect of the system and omit other drivers of change within these systems. Many models of FEW systems may neglect or simplify human behavior to the point that it does not accurately represent how people behave in certain situations. Not accounting for behavioral responses has led to mismanagement of resources ranging from fisheries to forests to endangered species.

As an illustration of the complex feedbacks between human behavior and FEW systems, the American **Renewable Fuel Standard** (RFS—see Sects. 9.2.1 and 9.3.1) demonstrates how a policy can cause people to alter their behavior, potentially leading to changes in FEW systems beyond the stated goals of the policy.

The RFS requires the incorporation of renewable fuels, such as corn ethanol, in commercial gasoline in an effort to reduce greenhouse gas emissions. One potential outcome of this requirement is increased demand for domestically produced corn to meet the demand for ethanol. An increase in demand for corn will likely impact commodity markets nationally, potentially leading to increased food prices (see Sect. 5.2.1). An increase in corn prices may provide an incentive for farmers to change their crop production and land allocation on their farms. These individual decisions by farmers may impact water demand due to changing crop production patterns and have downstream effects on water quality from farmers' input use. Farmers' production decisions may also affect international commodity markets through price responses to changes in production levels.

Thus, a decision at the national level may affect individual decisions, with consequences for environmental quality and resource availability at multiple scales. Neglecting any of these behavioral responses will lead to an incomplete understanding of the potential effects of a policy such as the RFS on local, national, and international systems.

As the above example illustrates, ignoring human behavior can lead to inaccurate models of FEW systems. Simplifying human behavior can lead to similar results. Much of the work that incorporates human behavior and decision-making includes assumptions about how people behave that may not fully reflect reality. Such assumptions may facilitate modeling these systems and are often necessary to make models tractable to understand how individual decision-making influences system-level change and vice versa. However, simplifying how people respond to changes in their social and natural environments can also lead to inaccurate conclusions about how such changes can impact human and natural systems.

Economic models of individual decision-making are a common approach to representing human behavior in integrated human-natural models of FEW systems. Standard economic models assume that people are rational, using optimization rules such as profit or **utility maximization**[1] to guide their behavior. Using simplified models such as these in FEW system research has advantages for researchers: under these frameworks, decision-making processes are fairly straightforward to model with clean decision rules (i.e., make the decision that will maximize the individual's profit or utility). Models that can be easily quantified are more easily integrated into models of the natural systems.

These models can incorporate some amount of individual or landscape heterogeneity—for example, differences in income can lead to differences in the demands for food, energy or other goods and services embedded in FEW systems. Economic models can indirectly account for interactions among individuals through changes in prices that result from changes in demand and can account for non-market interactions as well, e.g., pollution caused by individual activities that in turn induces individual adaptations to the pollution. However, because the models rely on quantifying equilibrium conditions to characterize such system feedbacks, they are

[1] Utility maximization is a theoretical framework often used to model human behavior in economics. Utility maximization assumes people have a well-defined function that determines their utility (called a utility function). In economics, this function has certain properties that make it easy to model. For example, utility functions are assumed to be decreasing in the price of an object, so that an individual experiences less utility if the price of that object increases. These functions and their defined properties make them easy to incorporate into economic models, although some of the assumptions they make may not be accurate in describing someone's well-being, and the behavioral predictions made by using such models have often found to be lacking in their ability to model people's actual behavior.

limited in the variety and complexity of individual-level heterogeneity and interactions that can be considered.[2]

In addition to being constrained in terms of the amount of heterogeneity that can be considered, there are other potential limitations of simplifying human decision-making. Optimization assumes that agents are maximizing or minimizing a known objective, which may not reflect the underlying behavior. Assumptions of optimizing behavior can be made more realistic by imposing constraints, e.g., by accounting for uncertainty and then assuming a process by which individuals form expectations over future unknown conditions or states of the world. However, this expectation formation process is often assumed to be highly simplified and may not account for key processes, such as learning over time or through interactions with others.

Furthermore, while profit and other economic factors often play an important role in individuals' decision-making processes, people consider other factors in their decision-making process as well. For example, farmers may use their land in such a way that takes the environmental consequences of their decisions into account, even if their land use strategy is not the one that maximizes profit. Similarly, people may choose gasoline-inefficient vehicles over smaller, more efficient vehicles to signal status or other aspects of their identities, even when inefficient vehicles are economically irrational choices.

As discussed in Sect. 1.3.1, the FEW system domain is subject to policy intervention at all levels of government. Formal models of these coupled human and natural systems often inform policy recommendations for FEW system management. As a result, relying on inaccurate representations of human behavior may lead to policies that fail to achieve their stated aims or policies that are not implemented at all. For example, economic models of policies intended to mitigate the effects of climate change often present the economic costs of strategies to address climate change without accounting for the potential benefits of new technologies, such as economic growth or increases in employment. Additionally, models designed to inform global climate change policies have traditionally assumed a market **discount rate** (see Sect. 5.3.3), the rate at which the market discounts future economic returns (typically the prevailing interest rate), rather than a social discount rate, which would reflect individuals' sense of moral obligation and concern for future generations. The use of a market discount rate underweights future environmental costs and can have substantially different implications for policy.

Furthermore, studies that predict policy impacts often do not consider **distributional effects**, such as whether and how the policy change may be felt by high-versus low-income populations, or how people in urban centers and rural areas may be differentially impacted (See Sects. 5.3.2 and 18.4 for case studies that illustrate distributive effects in FEW policy at the city-scale).

Policies that have different distributional impacts may result in conflict as groups compete for increasingly scarce resources such as water or land, triggering human

[2] See Irwin and Wrenn (2014) for a discussion of equilibrium-based and other modeling approaches, including agent-based models, in the context of land use decision-making and land change systems.

migration among certain groups if resources are no longer available. For example, migration among residents of rural Mexico to the USA has been shown to be affected by the incidence of significant drought events. These drought events do not have the same effect on the more affluent urban Mexican population (Hunter et al. 2013).

4.3 Towards More Realistic Models of Human Behavior in the Study of FEW Systems

Decision science and its related disciplines have grown substantially since the early 1990s. This growing body of work has generated a better understanding of the impacts of environmental changes on human behavior and adaptations and vice versa. While many of these insights are generic across many different types of systems, some are particularly germane to human adaptations within FEW systems. Here we highlight and discuss some of the key insights.

4.3.1 FEW System Dynamics

Studying human behavior and adaptation in FEW systems requires understanding the changes in these systems to which people react. The nature of these changes, including the time horizon over which these changes occur, will impact how people respond. Changes in FEW systems can occur gradually, resulting in so-called **"press" events**, which alter the system incrementally, or as sudden shocks or **"pulse" events**. Gradual rising temperatures and sea levels, or gradual population decline, are examples of press events. Pulse events, such as sudden weather events like hurricanes or floods, or economic shocks like a sudden increase in food prices, are likely to affect people and communities over a short period of time and are more difficult to predict.

The ways in which people respond to gradual or sudden changes in their social and natural environments are likely to differ. Press events are more gradual and long-term in nature and are therefore likely to require adaptation strategies that are incremental. Farmers may respond to gradually increasing temperatures by adjusting their crop mix on a year-by-year basis, or altering when they plant and harvest crops. Families may move inland in response to increased frequency and severity of coastal flooding. In contrast, pulse events require people to respond more quickly, such as evacuating an area due to an imminent hurricane threat.

Studying responses to press events may be challenging because of the gradual nature of the changes. Pulse events provide their own challenges, as these events are less frequent and may not provide researchers with enough data to draw conclusions about how people adapt to rapid changes in their environments. It is vital to understand how these changes will differentially impact affected populations, and how people and communities are likely to adapt to their changing environment.

4.3.2 Behavioral Heterogeneity

Recognizing that environmental and social changes may cause people to respond in different ways is important in accurately accounting for human behavior and adaptation in FEW systems. Allowing for **behavioral heterogeneity** recognizes that individuals may have different motivations for their actions and that how they react to changes in their social or natural environments (e.g., due to policy changes, climate change, etc.) may also be different.

Models that incorporate behavioral heterogeneity can be complex, and may therefore be more computationally difficult than those that model the behavior of a single representative agent. However, methods have been developed that allow for this heterogeneity in agents' decision-making strategies, and that can incorporate interactions between different agent types.

Models that incorporate types of individuals with different decision rules (beyond those based on profit maximization), and interactions between individuals, allow for behavioral heterogeneity within a population. These models demonstrate that allowing for different decision rules and direct interaction between individuals has implications for environmental quality, such as resource depletion and pollution levels. (See Jager et al. 2000 for a more detailed discussion of incorporating heterogeneity in models of human behavior.)

van Duinen et al. (2015) provide an example of the potential implications of introducing behavioral heterogeneity in FEW systems research. They use an **agent-based model (ABM)** to study farmers' adoption of irrigation systems in response to increased drought risk. Behavioral heterogeneity and social interaction are introduced into the model by including types of agents with different decision-making strategies. Agents differ in what they consider when deciding whether to adopt irrigation technology and the degree to which they consider others' adoption decisions. The model simulates the effects of drought from climate change on regional agricultural income, adaptation rate, water demand, and behavioral strategies. The results show that allowing for decision-making heterogeneity and interactions between farmers results in slower adoption than when farmers' decision-making is based purely on expected profit maximization. See Sect. 4.4.3 for more discussion of the use of ABMs in FEW systems research.

4.3.3 Technology Adoption

Given the essential role that new technologies play in improving resource efficiencies and reducing environmental impacts of human consumption, technology adoption can be a key component of human adaptation in FEW systems. However, technology will be ineffective in improving FEW system sustainability without widespread adoption. It is therefore critical to consider drivers of technology adoption, such as consumer decision-making and the social impacts of technology adoption, to reach a more complete understanding of the role that technology can

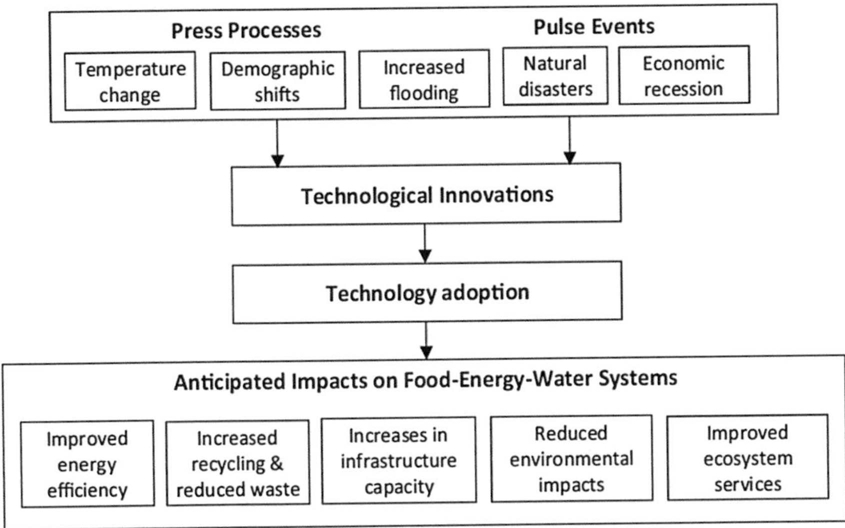

Fig. 4.1 A simplified coupled human-natural systems illustrating how failing to consider human responses and feedbacks in FEW systems predict the impact of technological innovations in response to press and pulse events. (Source: Irwin et al. 2016a)

play in FEW systems. For example, adoption of solar panels by homeowners has been shown to be partially driven by adoption by peers, where one additional solar panel unit in a zip code increases the probability of another unit being adopted by almost 1% (Bollinger and Gillingham 2012).

Figures 4.1 and 4.2 illustrate how explicitly accounting for human behavior and adaptation in FEW systems can more accurately describe the role of technology in addressing challenges in these coupled human and natural systems.

Fig. 4.1 shows how technological innovation may be modeled assuming that technological innovation will directly lead to technology adoption. However, this assumption may neglect important behavioral feedbacks within the system. In contrast, Fig. 4.2 shows the important feedback mechanisms between technological innovation, technology adoption, and human adaptation. The figures also illustrate how the pulse and press events may directly affect human adaptation, as well as the indirect impacts of these events through technological innovations. When these important feedback mechanisms are ignored or omitted, an incomplete understanding of the system can emerge. This may lead to inaccurate predictions of how changes within the system will impact other elements of the system.

The integration of natural subsystems (e.g., land, primary energy, and water resources) and human subsystems (e.g., food production, power generation, and water supply) in FEW systems offers opportunities to explore issues of human behavior and adaptation in a predictive manner, using tools of data analytics and modeling. These integrative tools that enable to explore patterns and forecasting in how FEW systems interact and coevolve are explored in Chaps. 14 (Data) and 15 (Modeling).

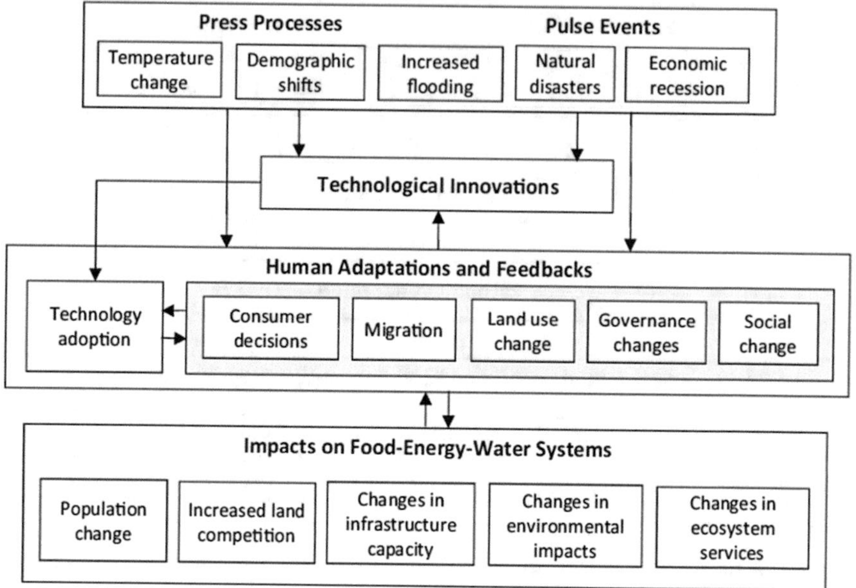

Fig. 4.2 A more holistic coupled human-natural systems framework, illustrating how the consideration of human responses and adaptation to press-pulse events can mediate the relationship between technological innovation and changes in FEW systems. (Source: Irwin et al. 2016a)

4.3.4 Behavioral Responses and Feedbacks

People can respond to changes in FEW systems or to new FEW policies with desirable changes in their behavior, such as reducing water consumption in response to a water conservation policy. However, people may also alter their behavior in undesirable ways. Specifically, rebound and **boomerang effects** can occur (See Sect. 5.2.2), stemming from market signals or from socio-psychological processes.

Rebound effects are unintended responses to changes in the social system, such as the introduction of new technology or policy, which results in a reduction or reversal of the intended impact. The introduction of a new, more energy-efficient technology may result in a decrease in household energy use and subsequent cost. In response to these energy savings, the household may respond by using more energy so that the energy reduction impact of the new technology is not as great as initially projected. Similarly, the introduction of more efficient irrigation systems has been found to result in an *increase* in water use (Pfeiffer & Lin 2014). Not accounting for potential rebound effects may cause researchers to over-estimate the impact of certain technological advances and assume environmental benefits that may not materialize.

Boomerang effects describe the unexpected ways in which people may respond to efforts intended to promote a particular behavioral change. Boomerang effects are observed when the promotion of a particular behavioral or attitudinal change has an effect opposite of what was intended. For example, messaging aimed at increasing people's awareness about the potential impacts of climate change has been found to *decrease* support for policies aimed at addressing climate change among certain populations (Hart & Nisbet 2012). There are many potential mechanisms to explain this effect, but the idea of psychological reactance is one such mechanism. Reactance occurs when someone feels threatened by new information or the actions of others, and so they choose to act out in ways to restore their sense of freedom or autonomy (e.g., acting out of spite).

4.3.5 Behavioral Spillover

In addition to accounting for potential unanticipated behavioral responses to policy, it is important to acknowledge the potential connections among FEW behaviors themselves. **Behavioral spillover** occurs when engaging in a target behavior is linked to the performance of another, seemingly unrelated behavior.

Spillover may be positive when an increase in one sustainable FEW behavior is associated with an increase in another. For instance, restrictions on outdoor irrigation may lead to a family reducing water used for irrigation as well as conserving water in other household activities, like reducing shower times.

Spillover may also be negative, such as when a homeowner increases air conditioning use after adopting water conservation measures, often as a result of moral licensing or the idea that she has already done her part. Spillover can occur within one given FEW domain, such as when engaging in one water conservation behavior influences another water conservation behavior, or across FEW domains, such as water conservation impacting energy conservation behavior.

Because spillover is defined as a causal process whereby the performance of one behavior causes a subsequent behavior, spillover studies must establish temporal order of behaviors to make a case for causality. The link between the target and spillover behaviors is mediated by the spillover pathway—the psychosocial mechanisms, such as moral licensing, that account for the relationship between seemingly independent behaviors. The second case study below provides an example of behavioral spillover.

4.3.6 Individual and Collective Decision-Making in FEW Systems

Decisions made collectively, such as at the community level or through policies instituted by various levels of government, have the potential for significant environmental impact and for addressing the many resource problems within FEW systems. These decisions are likely to be different from those made by individuals, and thus the effects of these decisions on the broader food, energy, and water systems will also differ. Both collective and individual decisions have implications for FEW systems. As such, it is important to consider the different scales at which decision-making takes place within these systems.

Communities are complex social systems, made up of individuals who may have multiple motivations for their actions. Communities may also have common interests that differ from those of the individuals that make up the community. Systems of governance often constrain the behavior of individuals within these groups and also determine the actions that the community itself can take (e.g., the laws or policies that a government can enact).

Theories of community decision-making can be generally categorized into those that assume that decisions are made pluralistically, with power and influence diffuse across the community without a dominant actor, and those that assume that decisions within communities are made by an elite group of individuals. Thus, characteristics of individuals that determine their power, influence, and role in collective decision-making, and the structures of the networks in which individuals are embedded, are key attributes for understanding organizational dynamics and decision-making.

The impact a community has on its environment and the decisions it makes about FEW system management are affected by several characteristics, including the community's population and its economic resources. Urban centers may have access to more resources and therefore may be better able to protect themselves from climate hazards (Chap. 11), but high population densities may also make adaptation more complex (Chap. 18). Rural areas may be more vulnerable to changes in FEW systems due to their closer links to industries that rely on the natural resources (i.e., agriculture, timber, fishing, etc.), and they may be less able to invest in infrastructure and development.

Participation and involvement of the individuals within the community can also impact its ability to respond to FEW system challenges. For example, communities with higher levels of community participation were better able to put policies in place to limit tree mortality (Flint & Haynes 2006). In contrast, absentee land ownership has been shown to impact the ability of local governments to respond to threats to forestland (Bailey & Majumdar 2014).

4.3.7 Temporal Scales

There is often a **temporal mismatch** between ecological and social responses to changes in FEW systems. (See Sect. 12.3.1 for a more comprehensive discussion of temporal scales of FEW systems and Case Study 1 in this chapter for a practical application of differing temporal scales in FEW systems.) Information and resources available to make decisions over the long term are limited, and there are often trade-offs between short-term costs and long-term gains. Interventions by individuals and communities to address potential threats like climate change or water quality may have significant costs in the short term, but the benefits are not likely to occur until well into the future. Research has shown that people are often less willing to incur immediate costs for benefits that will occur in the medium or long term, especially when there is some uncertainty about the benefits (Frederick et al. 2002). This mismatch in temporal scales can often lead to negative impacts in FEW systems and policies that fail to address them.

Wilson et al. (2016) present a decision framework for conservation and environmental management for decisions made by individuals within an institution, such as a governmental agency. This framework includes five elements: objectives of stakeholders and the decision-maker; actions of the decision-maker (e.g., policy); responses of the social and ecological systems; and learning by the decision-maker (see Fig. 4.3). The elements of the framework have their own timescales, which may differ from one element to another.

Mismatches in these time scales can occur both within and between individual decision-makers within an organization, actors in the social (human) system, or the ecological (natural) system. Not accounting for these temporal mismatches may lead to incorrect predictions of potential interventions aimed at conservation and management. These mismatches may require behavioral interventions to adjust the timescales of human systems. Given it is not often possible to adjust the timescales of the natural system, this may be the only clear way to obtain the desired outcomes. For example, placing a tax on gasoline will increase its short-term cost, potentially causing people to drive less often and decreasing carbon emissions from personal transportation. Alternatively, desirable behaviors like the use of solar energy can be subsidized to encourage uptake when market forces are ineffective in the immediate term. The benefits of these behaviors (driving less and using solar energy) are likely to occur in the long term, and interventions can be used to impose short-term costs and benefits to more closely align with the short-sighted nature of human decision making.

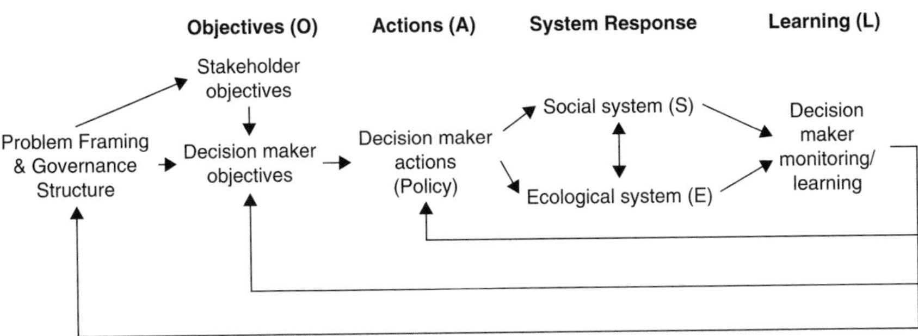

Fig. 4.3 A decision framework for conservation and environmental management for decisions made by individuals within an institution. (Source: Wilson et al. 2016)

4.4 Data and Methods for Studying Human Decision-Making and Behavior

Including models of human behavior and adaptation in the study of FEW systems necessitates the collection of data beyond those required for studying the natural systems in isolation. Data from the natural sciences, including data on climate, soils, hydrology, and land cover, are needed to specify physical models of environmental systems (i.e., the agricultural, energy, and hydrological systems). Likewise, data on individuals and their decision-making processes are necessary to inform models of human behavior and adaptation in FEW systems. The population of interest, the context in which human behavior is being studied, and the scale at which decisions are made, among others, are all important considerations in determining the appropriate data collection and research approach. (See Chap. 14 for a more comprehensive discussion on the types of data used in various FEW system research and the integration of data on the physical and human systems.)

4.4.1 Primary Data Collection

4.4.1.1 Surveys

Data collected directly from people about their behavior and decision-making processes can be informative about how they are likely to behave in particular settings and scenarios. A common method of behavioral and decision-making data collection involves surveying individuals from the population of interest. Asking people about their past behavior, or how they would react and what decisions they would make in certain hypothetical situations, is one method of collecting these data to inform models of human decision-making. Surveys can be tailored to the context of interest, asking a sample of people from a population of interest how they make decisions in different scenarios.

For example, to learn about how farmers are likely to respond to changing climate conditions or if new policies were put in place, researchers may conduct surveys that ask what crops they might plant, or how they might use their land if certain changes were to occur. Researchers wanting to learn about likely technology adoption may survey consumers about how much they would pay for new energy-efficient appliances, or under what conditions they would use new technologies.

Although surveys can be a valuable source of data and provide more detailed information than secondary sources (see below), they are not without limitations. For instance, when people are asked how they would behave in a particular scenario, their response is hypothetical. They may report that they would take a particular action, or pay a certain amount for new technology, but their behavior may differ when actually faced with that decision, resulting in a **hypothetical bias**. Similarly, asking people about their past behavior may introduce memory recall issues, as people are unlikely to remember their past decisions with perfect accuracy. Surveys must be carefully designed to ensure that the information collected will adequately address questions that researchers are attempting to answer. (See Krosnick 1999 for a more in-depth discussion of survey design for the study of human behavior and Rust and Golombok (2014) for an in-depth discussion of **measurement** or psychometrics.)

4.4.1.2 Experiments

Conducting experiments is another way to address these potential response biases and inform models of human behavior and decision-making. Experiments allow researchers to reduce variation in external factors to isolate the effect of a particular variable of interest, such as prices, information provided, or another experimental treatment. Importantly, experimental designs allow researchers to draw causal conclusions. They often allow researchers to observe participants' actual choices and behavior, rather than relying on hypothetical choices or reporting past behaviors. Experimental methods are varied and include experiments conducted in a controlled laboratory setting as well as in a more natural setting (e.g., field experiments).

Laboratory experiments involve randomly assigning some participants to one type of treatment (e.g., having participants set goals, or exposing them to a persuasive message), and others to a control group, in order to enable comparisons across these groups. Often, participants are then asked to make choices, such as which product they would purchase or whether or how they would dispose of waste. These procedures are conducted in a controlled setting such as a computer lab, reducing external variables that may provide alternate explanations for results, and allow researchers to directly observe participants' behavior rather than relying on participants to report on past or hypothetical behavior.

Similar to lab experiments, field experiments allow researchers to control variation to observe how participants make decisions. However, field experiments are conducted in an environment that more closely resembles the decision-maker's natural choice setting. For example, researchers may alter the information presented to supermarket customers about the conditions in which animals were raised and observe whether they choose to purchase conventionally or organically grown products.

Despite the advantages of conducting experiments, they also have some limitations. As with surveys, it may be difficult to recreate the decision context of interest exactly, and some hypothetical bias may still exist. Participants may be aware of the fact that their behavior is being observed, and they may alter it to align with the study objectives more closely. Additionally, sample sizes for experiments are often significantly smaller than those for surveys, potentially limiting the scope of conclusions that can be made from experimental evidence.

4.4.2 Secondary Data

Data from secondary sources can be valuable in developing models of human behavior and adaptation. Economic and demographic data collected from secondary sources, such as data collected as part of a national census or from government agencies like the USA Department of Agriculture (USDA), can be used to inform and test models of human behavior and capture trends in how people respond to changes in their social and natural environments. Often, **secondary data** can be used to complement **primary data** collection such as surveys and experiments discussed above.

Using data from secondary sources has several advantages. These data are often readily available at no or little cost to researchers. Data from these sources are often collected from a population over a long time period, enabling researchers to analyze trends in human behavior and adaptation. For example, census data can be used to study population movements in response to increased flood risk in coastal areas over time. The USDA collects yearly data on the crops produced in the USA, allowing researchers to study how farmers' crop mix and land use has changed over time.

Secondary data often has some advantages over primary data in that it is much more readily available and can often be collected for an entire population, e.g., all households living within a region. As we discuss below, larger numbers of observations enable statistical models with more power and a greater opportunity for identifying a significant relationship that may be a key parameter governing FEW system dynamics.

While data from secondary sources is a valuable resource in studying how people behave and adapt to changing environmental or economic conditions, users of these data are limited by the data collection strategies of these agencies or organizations. If answers to particular questions are needed, researchers may have to adapt their research questions and approach to fit the data available to them or collect data from the particular populations of interest themselves. Secondary data sources that contain measures of psychological constructs, such as values, attitudes, and beliefs, are also much more difficult to come by (as opposed to sociodemographic or economic variables; see The World Values Survey (Inglehart et al. 2014), for an example).

4.4.3 Using Data in Decision-Making Models for FEW Systems

Data on human behavior and adaptation are necessary to empirically specify the underlying decision-making process or rules that are hypothesized to determine behavioral outcomes. Once specified, these decision-making rules can be incorporated into FEW systems models to understand the implications of these decisions for FEW systems and management of these systems. There are several ways in which data may be used to specify these decision-making rules and the different ways in which these rules may be incorporated into FEW systems models. Here we briefly summarize a handful of approaches for developing such models. This is not intended as an in-depth or exhaustive review. For further details see, for example, Irwin and Wrenn (2014) and Robinson et al. (2007).

Primary data can be used to estimate the key parameters of individual decisions. Zhang et al. (2016) use survey data and discrete choice modeling to estimate the parameters of farmers' decisions to adopt a specific best management practice. Alternatively, secondary data may be used to estimate a key decision-making parameter. For example, data on vehicle purchases and registration can be used to learn about how changes in gas prices affect consumer demand for fuel-efficient vehicles.

Because of data limitations, it is often not possible to fully estimate the relevant set of parameters for the population or context of interest. In these cases, descriptive data analysis may be used to identify patterns among one group that can be applied to different segments of a population. Alternatively, parameters that have already been estimated in the literature may be used. For example, economic simulation models commonly specify plausible demand and supply elasticity parameters based on the estimates that have been reported in the literature.

Once the decision-making rule has been specified, simulation models can be used to incorporate these decisions into **FEW system models**. Different types of simulation models are possible, but two of the most common approaches are structural economic models and agent-based models. As Irwin and Wrenn (2014) discuss in more detail, structural economic models are primarily focused on modeling consumption and production decisions and price feedbacks for one or more key sectors of the economy. While these models focus on economic outcomes, they may also account for key environmental externalities, such as carbon emissions or natural land degradation.

These models assume some kind of optimizing behavior (e.g., profit or utility maximization or cost minimization) subject to resource and other constraints. They have become increasingly sophisticated by incorporating multiple types of uncertainty and environmental tipping points that can lead to large discontinuous changes in the coupled human-natural system.[3] However, these models are computationally intensive, making it difficult to incorporate many sources of behavioral heterogeneity.

Agent-based models offer an alternative that, because they do not rely solely on optimization, are more flexible and do not require the same approach to "solving" the model. Instead of assuming market equilibrium in which prices instantaneously adjust to the cumulative actions of individuals, these models are ad hoc in their treatment of the price mechanism, requiring them to specify alternative assumptions about how buyers and sellers interact and how prices emerge from these interactions. While this is often viewed as a limitation of these models, their advantage is

[3] See Irwin et al. (2016b) for more discussion of these and other dynamic coupled models of human-natural systems.

that they can incorporate much more heterogeneity in the model, including differences in individuals and their behaviors.

Researchers define decision rules for particular subsets of agents (individuals) and the proportion of agent types within a population. With environmental and social conditions specified, the model simulates how agents react to changes in their environments and the effects of their actions on the system. Behavioral heterogeneity can be incorporated by introducing multiple decision-making strategies, allowing for different responses by agents to changes in their environments. Interactions between agents can also be modeled, allowing for agents' behavior to be directly impacted by the behaviors of other agents.

While the use of agent-based models to study behavior and adaptation in FEW systems has allowed researchers to introduce different agent types with different decision strategies, these models present additional challenges to researchers. Because they are more flexible and allow for a greater variety of behaviors, the amount and type of data needed to calibrate and verify these models is substantial and may not always be available. In addition, the predicted outcomes of these models depend on the particular scenario, which may limit the generalizability of these models and their usefulness for policy.

4.5 Case Studies: Key Adaptations and Their Implications for FEW Systems

Two case studies are provided here as concrete examples of how accounting for human behavior and adaptation can inform the study of FEW systems, and how behavioral interventions can aid in FEW system management. The first case study uses the example of eutrophication in the Great Lakes to demonstrate how behavioral heterogeneity can influence the adoption of fertilizer management strategies. The second examines behavioral spillover, showing how the adoption of one FEW-related behavior can lead households to engage in other FEW behaviors.

4.5.1 Case Study 1: Incorporating Behavioral Heterogeneity into FEW System Models and Policies in the Lake Erie Watershed

Harmful algal blooms (HABs) in Lake Erie provide an important example of how misalignment of temporal and **spatial scales** may contribute to problems in FEW systems (See Sect. 8.2.3 for material on HABs, and Sect. 12.3 for a discussion on the spatial and temporal scales of FEW systems), and how they can be addressed by interventions aimed at altering human behavior within these systems. HABs in Lake Erie are a particularly negative example of eutrophication. They can be harmful to animal and human health, and toxins from the algae have entered into cities' water systems, making the water undrinkable. Severe blooms also limit recreation activities on the lake (see Chap. 19).

HABs in Lake Erie are due in large part to increased phosphorus entering the lake. These blooms have become more frequent and severe in recent years. This is due to several factors, including more frequent and intense rainfall events that release soil nutrients from surrounding agricultural land to streams and rivers that flow into Lake Erie (Bosch et al. 2014; Michalak et al. 2013). Farmers' application and management of fertilizers containing phosphorus, therefore, play a significant role in these HABs.

In many watersheds where downstream eutrophication is a problem, the focus of the coupled human-natural system is on agency decision-makers, farmers, and the interconnectedness between soil, nutrient management, water quality, and regional climate patterns (Michalak 2013; Wilson et al. 2014). Agency decision-makers grapple with the inherent conflict between annual economic goals (e.g., maximizing crop production within a growing season) and long-term protection for both the downstream ecological system and the regional economy. Individual farmers grapple with a similar conflict between their short-term goals of maximizing profits and personal and societal goals of protecting soil and water quality over larger time scales (i.e., decades or possibly generations).

Such temporal mismatches can delay action while actors within the human system struggle to deal with these difficult trade-offs. Even when individual and societal objectives are aligned across time, there might be delays between policy implementation and individual compliance, especially where new capital-intensive technologies or equipment are required, or new and unfamiliar nutrient management practices are recommended. Further, farmers' intentions to implement best management practices (BMPs) such as injecting fertilizer beneath the surface of the soil or using cover crops on their farms are not always translated into action, as demonstrated by Fig. 4.4. The figure shows that a majority of farmers intend to and use several recommended practices (soil testing and applying fertilizer at the right time). However, other practices are used at much lower rates despite good intentions to use the practice by a majority of motivated farmers (cover crops, subsurface placement). Several recent surveys of farmers in the western Lake Erie basin provide some insight into the heterogeneity that exists in farmer land management decisions, as well as to whether or not current policies are effectively leading to behavioral change.

Several regulations and programs have been implemented in an effort to reduce the loss of nutrients from agricultural land in the western Lake Erie basin. Farmers in Ohio must now become certified fertilizer applicators to ensure that they understand how best to minimize nutrient loss, and applying fertilizer on saturated or frozen ground is now prohibited (Ohio Senate Bill 150, 2014). Despite these regulatory changes, HABs in the lake continue to occur. BMPs have been promoted through both educational efforts and incentive-based programs. Despite the fact that these BMPs show some promise in decreasing the harmful algal blooms in Lake Erie (Scavia et al. 2017), uptake has been relatively low over time (Wilson et al. 2019).

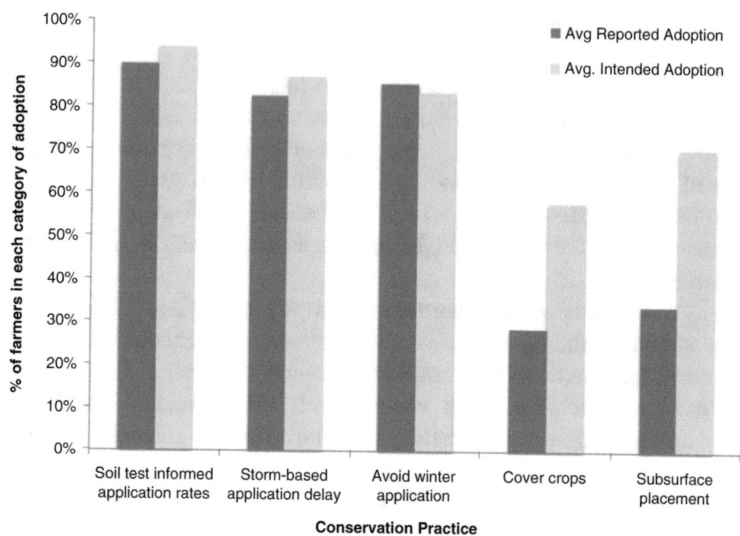

Fig. 4.4 Farmers' reported behavior on a representative field (average reported use from 2015 and 2017) and willingness to perform behaviors in the future (average intended future use from 2016 and 2018) for five practices aimed at reducing phosphorus loading into waterways from agricultural fields. (Source: Wilson et al. 2019)

Researchers investigating farmers' land management decisions have found that not all farmers have the same motivations for adopting BMPs and that policies that endeavor to target their behavior may not be taking this heterogeneity into account. Characteristics, such as how effective farmers believe the BMPs to be, have been found to impact their willingness to adopt certain recommended practices; however, these perceptions do not impact all farmers in the same way. Wilson et al. (2014) found that the perceived effectiveness of the practices is more likely to affect the decisions of older farmers more concerned with their farm's profitability. Similarly, work by Zhang et al. (2016) found that farmers can be classified into different groups based on their propensity to adopt certain BMPs and that these groups may have different motivations for their land management decisions. For example, farmers who self-identify as conservationists were more likely to adopt BMPs on their land than those who identify primarily as agricultural producers. Farmers may also place differential importance on the BMPs employed by other farmers in their area and therefore respond differently to social pressure to engage in particular practices.

4.5.1.1 Practical Implications

Management of this complex system, where the goal is to produce food in an economically viable manner while protecting critical ecosystem services, demonstrates that understanding human behavior is critical. The work being done in this area highlights the importance of accounting for behavioral heterogeneity in FEW management strategies. Different actors may have different motivations for the decisions they make; programs and policies may, therefore, have to be tailored to incorporate these different attitudes if they are to be successful in motivating behavioral change.

Despite the millions of dollars being spent in the Great Lakes through a combination of incentive-based programs to offset BMP implementation costs and extensive education and outreach,[4] adoption of conservation practices is largely static over time, and water quality is not improving. It is clear that intervening more strategically to encourage more thoughtful trade-offs across time will be key to addressing this challenge. For example, many incentive-based programs that offset the costs of cover crops do not last long enough for farmers to experience the ecological, let alone economic benefits of cover crops. This leads to just as many farmers dropping the practice as are adding the practice on an annual basis (Wilson et al. 2019). A more effective policy would be one that extends the payments for a minimum of 5 years to ensure that the on-farm benefits can be realized (e.g., decreased fertilizer inputs, improved yield), allowing farmers to continue the practice despite the added cost.

4.5.2 *Case Study 2: Behavioral Spillovers Among Household FEW Consumption Behaviors*

Households represent a fundamental nexus of FEW systems. More than 67% of food calories are consumed at home (Lin and Guthrie 2012), more than 75% of direct energy use is residential (Energy Information Administration 2018), and more than 50% of water withdrawals go directly to households or create food and thermoelectric power consumed by households (Dieter 2018).

As discussed in Sect. 4.3.2, there is the potential for spillover effects in FEW-related household behaviors. Several community field experiments provide examples of spillover effects. For example, following a Welsh policy charging a plastic shopping bag fee, Thomas et al. (2016) observed positive spillover from increased use of reusable bags to sustainable transportation choices and home water and energy conservation. Another study found that Australian residents reporting curtailed water use were more likely to subsequently install water efficient appliances (Lauren et al. 2016).

The current case study examines spillover among FEW behaviors as part of a larger intervention project that encouraged composting through a curbside collection program. (For more details on this case study, see Sintov et al. 2017.) The primary objective of this study was to test whether adopting a new FEW management

[4]Approximately $32 million per year is spent in the Lake Erie basin by the federal government on agricultural conservation.

behavior—composting—would increase the performance of other household FEW management behaviors. Participants were 284 adult residents of single-family homes in Costa Mesa, California who were not already composting at baseline.

In an effort to divert waste from landfills, the Costa Mesa Sanitary District provided homes in its territory with 64-**gallon** organics carts, weekly collection service, and informational flyers describing the types of food and yard waste to be placed in the organics cart (see Fig. 4.5). Residents were asked to keep food and yard waste intended for organics carts separate from landfill waste intended for trash carts. Organic waste was then to be placed in the organics cart, which was to be placed at the curb on collection day.

Study participants completed a baseline survey at recruitment and a follow-up survey 6 months later. Change in composting was assessed by comparing follow-up survey data to behavior at baseline. This provided for evaluation of the effect of performing a new FEW behavior on subsequent FEW spillover behaviors.

The follow-up survey also assessed other household FEW management (spillover) behaviors, including two water behaviors (taking shorter showers, turning water off while brushing teeth), five energy behaviors (unplugging devices not in use, refraining from using heater when cold, walking or biking instead of driving, encouraging others to save energy, turning lights off in unoccupied room), and three food behaviors (planning out meals, assessing and using food at home before shopping).

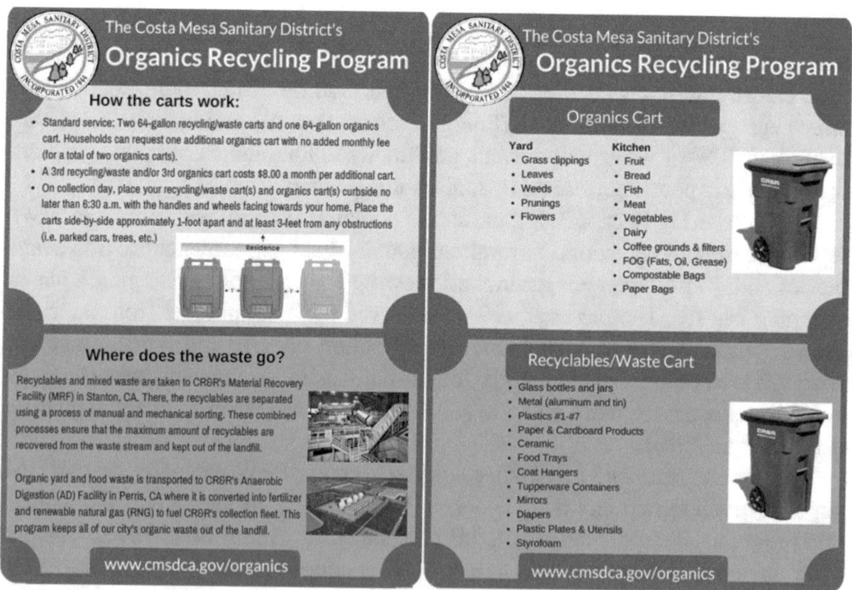

Fig. 4.5 Procedural information delivered as part of the intervention

4.5.2.1 Results

At baseline, none of the 284 respondents were composting. By follow-up, approximately 72% of respondents had begun composting.

Beginning to compost was associated with significantly more energy and water management spillover behaviors at follow-up. This significant positive spillover was observed for both water and all five energy management behaviors. Effect sizes were small, ranging from $r = 0.15$ for using the heater (less) to $r = 0.26$ for letting water run (less). None of the three food spillover behaviors differed significantly among those who began composting vs. those who did not.

4.5.2.2 Discussion

This study found that beginning to manage household FEW behaviors by composting can spill over to other FEW behaviors. Furthermore, across all water and energy behaviors, results showed positive, rather than negative, spillover.

Although effect sizes were small, the positive spillover effect was relatively robust across types of behaviors, ranging from taking shorter showers to walking/biking to conserve gasoline, to encouraging others to save energy. Importantly, these represent different quite behaviors that occur in unique contexts, suggesting spillover is not limited to a particular setting or FEW domain.

Surprisingly, no significant spillover was observed among any of the food waste management behaviors. Ceiling effects may have contributed to these null results, particularly given that these behaviors were quite prevalent among those who did not begin composting (see Fig. 4.6). Alternatively, preventing food waste often

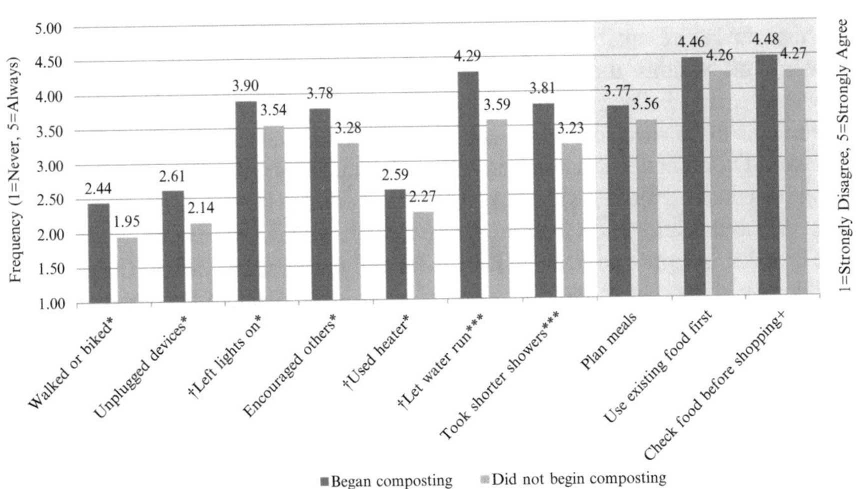

Fig. 4.6 Mean scores on spillover behaviors among participants who did vs. did not begin composting over the course of the study (N = 284). †Reverse scored item. Holm-Bonferroni-corrected p-values from Wilcoxon-Mann-Whitney tests: $^{+}p_{HB} < 0.10$, $*p_{HB} < 0.05$, $**p_{HB} < 0.01$, $***p_{HB} < 0.001$

occurs temporally and spatially distant from the point at which food becomes waste, whereas preventing energy and water waste occurs more proximally to the points at which waste is recognized (e.g., leaving water running) and waste management acts can occur.

4.5.2.3 Practical Implications

These findings suggest that local policies can have unintended yet desirable consequences. Results suggest that municipalities should take into account target *and* spillover behaviors when quantifying the impacts resulting from such programs. This information can be used to understand how the pursuit of one goal, such as landfill diversion, may support or impede the achievement of others, such as water conservation. A better understanding of spillover can aid **practitioners** in designing real-world interventions to improve FEW sustainability, illuminating opportunities to maximize program benefits (i.e., positive spillover) and avoid unintended consequences (i.e., negative spillover).

4.6 Conclusions

As discussed in this chapter, incorporating human behavior and adaptation in FEW system research is important for our understanding of these complex systems. Allowing for an accurate representation of human behavior requires drawing on social science disciplines and data that may not otherwise be employed in research of the natural systems alone. Researchers have made progress in incorporating more sophisticated representations of behavior and adaptations in FEW system models by drawing on various theoretical and empirical approaches of social sciences. However, many challenges remain.

First, as there are many social sciences, there are many ways to model human behavior. Different disciplines often approach human behavior and adaptation from different perspectives. The approaches of different disciplines often vary in their level of formalization, and therefore the ease with which they can be incorporated into broader, integrated models of human-natural systems also varies. These frameworks may also differ in their scope (e.g., narrow and detailed vs. broad and general), and whether they allow for feedback within the systems, such as social dynamics or learning from past experiences. The diversity of social sciences and their theoretical approaches mean that there are many candidates for modeling human behavior, and not always a clear-cut method for determining the most appropriate framework.

Second, incorporating behavioral heterogeneity presents both computational and conceptual challenges for modeling. Accounting for multiple dimensions of heterogeneity quickly leads to models that are not analytically tractable and pushes the limits of simulation-based models. In addition, models that seek to be more realistic can become overly detailed or specified, making it impossible to generalize and draw policy implications for a larger population, e.g., a community or region.

Third, accounting for how individual decisions relate to higher-level decisions made by communities or governments, and the proper scaling of models from an individual level to a regional or global scale, is a continuing challenge. As human agency varies across different scale levels, the ways that human agency and human decision-making are represented should similarly depend on the scale of analysis. In addition, behavioral interactions and feedbacks that occur across scales, such as many of the rebound and spillover effects discussed earlier, greatly increase the complexity of a model and thus may be difficult to incorporate.

By definition, models are simplifications of reality. Given the pace with which advances in behavioral research are occurring, the above and other challenges to incorporating behavioral realism into FEW system research are likely to be addressed. Paradoxically, as we discover more about the anomalies and differences of human behavior, determining the right amount of detail and which details to track becomes harder, not easier. What is needed is to scale up our current level of understanding, making the focus of research more about determining the sources of heterogeneity that matter at an aggregate or systems level, which is the scale that is relevant for policy. This is perhaps the greater challenge, as policymakers do not require perfect models to develop well-informed policies. Hence, for research to be policy relevant, researchers need to produce models that achieve these appropriate levels of detail and accuracy.

Key Points

- Studying either the natural or human system in isolation omits important drivers of change in FEW systems and how people respond and adapt to these changes. This can potentially lead to an incomplete understanding of these integrated systems.
- Relying on inaccurate representations of human behavior may lead to policies that fail to achieve their stated aims or policies that are not implemented at all.
- Allowing for an accurate representation of human behavior requires drawing on social science disciplines and data that may not otherwise be employed in research of the natural systems alone, and often requires drawing on various theoretical and empirical approaches of social sciences.
- Recognizing that environmental and social changes may cause people to respond in different ways is important in accurately accounting for human behavior and adaptation in FEW systems. Allowing for behavioral heterogeneity recognizes that individuals may have different motivations for their actions and that how they react to changes in their social or natural environments may also be different.
- People can respond to changes in FEW systems or to new FEW policies with desirable changes in their behavior, or with undesirable changes (e.g., rebound or boomerang effects). It is also important to consider potential connections among FEW behaviors themselves (e.g., behavioral spillover).
- Including human behavior in the study of FEW systems requires collecting of data on individuals and their decision-making processes to inform models of

human behavior and adaptation in FEW systems. The population of interest, the context in which human behavior is being studied, and the scale at which decisions are made, among others, are all important considerations in determining the appropriate data collection and research approach.
- Researchers have made progress in representing human behavior and adaptations in FEW system models, drawing on various theoretical and empirical approaches of social sciences. However, challenges remain, including determining the most appropriate social science framework, computational and conceptual challenges for modeling, and the proper scaling of models from an individual level to a regional or global scale, among others.

Discussion Points and Exercises

1. Why is it important to consider human behavior in the context of FEW systems? What are the potential consequences of not considering human behavior?
2. Discuss the challenges of incorporating human behavior in the study of FEW systems. How are researchers working to overcome these challenges?
3. Provide an example of a technology whose impact may be overstated by using the framework in Fig. 4.1 to model its adoption. How might the framework presented in Fig. 4.2 help researchers more accurately predict the impact of the technology you have chosen? What elements of the figures are important when considering the impacts of new technologies on FEW systems?
4. The EPA estimates that the average passenger vehicle emits 4.6 metric tons of carbon per year (https://www.epa.gov/greenvehicles/greenhouse-gas-emissions-typical-passenger-vehicle). Suppose that electric vehicle manufacturers want to estimate the potential impact of their cars on carbon emissions in the USA. They use data on electric and plug-in vehicles from the EPA's Alternative Fuels Data Center (https://afdc.energy.gov/data/categories/afvs-and-hevs) to predict electric vehicle sales 20 years from now, assuming similar sales growth into the future. Using their prediction for the number of cars sold and the amount of carbon used by the average vehicle, they report the estimated reduction in carbon from electric vehicles in 20 years. What factors might have been overlooked in this estimate? Do you think this estimate is understated or overstated? What factors would you consider to provide a more accurate estimate of the future potential impacts of electric vehicles on carbon emissions?
5. What is the **rationality assumption**? Why might it lead to inaccurate depictions of human behavior and adaptation in FEW systems?
6. Provide examples of press and pulse events in FEW systems. Why is considering human behavior and adaptation to these events important in studying their effects?
7. What are some challenges of studying human behavior and adaptation to press events? To pulse events? In your answer, please use specific examples of types of press and pulse events.
8. What is meant by the term "behavioral heterogeneity"? What challenges does behavioral heterogeneity pose to the study of human behavior and adaption in FEW systems? What are some ways that these challenges have been addressed?

9. How can rebound effects affect the impact of a new technology on FEW systems? Find examples from the literature.
10. What are spillover effects? Find examples in the literature of positive and negative behavioral spillover in FEW systems, and describe the impact of behavioral spillover on these systems.
11. Discuss how **timescale mismatches** could contribute to negative consequences within a FEW system, using examples not discussed in this chapter. What are some ways that these mismatches have been addressed?
12. Looking at Fig. 4.3, how might the timescales of decision-makers and the social and ecological systems differ? Provide an example of these different time scales using a specific FEW system. Propose a policy instrument that could address these timescale mismatches in your chosen FEW system.
13. Discuss the advantages and disadvantages of primary and secondary data to study human behavior in different FEW system domains. Provide examples of when primary data should be used, and others of when secondary data may be more appropriate.
14. Suppose you want to study the impact of a new technology that increases the efficiency of household water use. How might you design an experiment to test the impact of the technology? What would your target population be? How would you measure the impact of the new technology on water use? What challenges would you expect in running your experiment?
15. How could you use secondary data to study the adoption of electric vehicles? What existing data sources could you use? Be specific.

References

Bailey, C., & Majumdar, M. (2014). Absentee forest and farmland ownership in Alabama. In *Rural wealth creation* (Vol. 134). London: Routledge.

Bollinger, B., & Gillingham, K. (2012). Peer effects in the diffusion of solar photovoltaic panels. *Marketing Science, 31*(6), 900–912.

Bosch, N. S., Evans, M. A., Scavia, D., & Allan, J. D. (2014). Interacting effects of climate change and agricultural BMPs on nutrient runoff entering Lake Erie. *Journal of Great Lakes Research, 40*(3), 581–589.

Dieter, C. A. (2018). Water Availability and Use Science Program: Estimated Use of Water in the United States In 2015. Geological Survey.

EIA. (2018). 2015 Residential Energy Consumption Survey. https://www.eia.gov/consumption/residential/data/2015/index.php?view=consumption

Flint, C. G., & Haynes, R. (2006). Managing forest disturbances and community responses: Lessons from the Kenai Peninsula, Alaska. *Journal of Forestry, 104*(5), 269–275.

Frederick, S., Loewenstein, G., & O'Donoghue, T. (2002). Time discounting and time preference: A critical review. *Journal of Economic Literature, 40*(2), 351–401.

Hart, P. S., & Nisbet, E. C. (2012). Boomerang effects in science communication: How motivated reasoning and identity cues amplify opinion polarization about climate mitigation policies. *Communication Research, 39*(6), 701–723.

Hunter, L. M., Murray, S., & Riosmena, F. (2013). Rainfall patterns and U.S. migration from rural Mexico. *International Migration Review, 47*(4), 874–909.

Inglehart, R., Haerpfer, C., Moreno, A., Welzel, C., Kizilova, K., Diez-Medrano, J., Lagos, M., Norris, P., Ponarin, E., Puranen, B., et al. (2014). *World values survey: Round Six - Country-pooled datafile version*. Madrid: JD Systems Institute. Retrieved from http://www.worldvaluessurvey.org/WVSDocumentationWV6.jsp.

Irwin, E., Campbell, J., Wilson, R., Faggian, A., Moore, R., & Irwin, N. (2016a). Human adaptations in food, energy, and water systems. *Journal of Environmental Studies and Sciences, 6*, 127–139.

Irwin, E., Gopalakrishnan, S., & Randall, A. (2016b). Welfare, wealth and sustainability. *Annual Review of Resource Economics, 8*, 77. https://doi.org/10.1146/annurev-resource-100815-095351.

Irwin, E. G., & Wrenn, D. (2014). An assessment of empirical methods for modeling land use. In J. M. Duke & J. Wu (Eds.), *The Oxford handbook of land economics*. London: Oxford University Press.

Jager, W., Janssen, M. A., De Vries, H. J. M., De Greef, J., & Vlek, C. A. J. (2000). Behaviour in commons dilemmas: Homo economicus and Homo psychologicus in an ecological-economic model. *Ecological economics, 35*(3), 357–379.

Krosnick, J. A. (1999). Survey research. *Annual Review of Psychology, 50*(1), 537–567.

Lauren, N., Fielding, K. S., Smith, L., & Louis, W. R. (2016). You did, so you can and you will: Self-efficacy as a mediator of spillover from easy to more difficult pro-environmental behaviour. *Journal of Environmental Psychology, 48*, 191–199.

Lin, B.H., & Guthrie, J. (2012). Nutritional quality of food prepared at home and away from home, 1977-2008 Washington, DC US Department of Agriculture, Economic Research Service (Economic Information Bulletin Number 105.)

Michalak, A. M., Anderson, E. J., Beletsky, D., Boland, S., Bosch, N. S., Bridgeman, T. B., ... & DePinto, J. V. (2013). Record-setting algal bloom in Lake Erie caused by agricultural and meteorological trends consistent with expected future conditions. *Proceedings of the National Academy of Sciences, 110*(16), 6448–6452.

Pfeiffer, L., & Lin, C. Y. C. (2014). Does efficient irrigation technology lead to reduced groundwater extraction? Empirical evidence. *Journal of Environmental Economics and Management, 67*(2), 189–208.

Robinson, D. T., Brown, D. G., Parker, D. C., Schreinemachers, P., Janssen, M. A., Huigen, M., Wittmer, H., Gotts, N., Promburom, P., Irwin, E., Berger, T., Gatzweiler, F., & Barnaud, C. (2007). Comparison of empirical methods for building agent-based models in land use science. *Journal of Land Use Science, 2*(1), 31–55.

Rust, J., & Golombok, S. (2014). *Modern psychometrics: The science of psychological assessment*. London: Routledge.

Scavia, D., Kalcic, M., Muenich, R. L., Read, J., Aloysius, N., Bertani, I., Boles, C., Confesor, R., DePinto, J., Gildow, M., & Martin, J. (2017). Multiple models guide strategies for agricultural nutrient reductions. *Frontiers in Ecology and the Environment, 15*(3), 126–132.

Sintov, N., Geislar, S., & White, L. V. (2017). Cognitive accessibility as a new factor in proenvironmental spillover: Results from a field study of household food waste management. *Environment and Behavior, 51*, 50. https://doi.org/10.1177/0013916517735638.

The Ohio Legislature. (2014). *To revise the law governing the abatement of agricultural pollution, to require a person that applies fertilizer for the purposes of agricultural production to be certified to do so by the Director of Agriculture, to make other changes to the Agricultural Additives, Lime, and Fertilizer Law*. Ohio Senate Bill 150. 113th General Assembly.

Thomas, G. O., Poortinga, W., & Sautkina, E. (2016). The Welsh single-use carrier bag charge and behavioural spillover. *Journal of Environmental Psychology, 47*, 126–135.

van Duinen, R., Filatova, T., Jager, W., & van der Veen, A. (2015). Going beyond perfect rationality: Drought risk, economic choices and the influence of social networks. *Annals of Regional Science, 57*(2-3), 335–369.

Wilson, R. S., Beetstra, M., Reutter, J., Hesse, G., Fussell, K., Johnson, L., King, K., LaBarge, G., Martin, J., & Winslow, C. (2019). Commentary: Achieving phosphorus reduction targets for Lake Erie. *Journal of Great Lakes Research, 45*(1), 4–11.

Wilson, R. S., Hardisty, D. J., Epanchin-Niell, R. S., Runge, M. C., Cottingham, K. L., Urban, D. L., Maguire, L. A., Hastings, A., Mumby, P. J., & Peters, D. P. (2016). A typology of time-scale mismatches and behavioral interventions to diagnose and solve conservation problems. *Conservation Biology, 30*(1), 42–49.

Wilson, R. S., Howard, G., & Burnett, E. A. (2014). Improving nutrient management practices in agriculture: The role of risk-based beliefs in understanding farmers' attitudes toward taking additional action. *Water Resources Research, 50*(8), 6735–6746.

Zhang, W., Wilson, R. S., Burnett, E., Irwin, E. G., & Martin, J. F. (2016). What motivates farmers to apply phosphorus at the "right" time? Survey evidence from the Western Lake Erie Basin. *Journal of Great Lakes Research, 42*(6), 1343–1356.

Further Reading

Arneth, A., Brown, C., & Rounsevell, M. D. A. (2014). Global models of human decision-making for land-based mitigation and adaptation assessment. *Nature Climate Change, 4*(7), 550–557.

Collins, S. L., Carpenter, S. R., Swinton, S. M., Orenstein, D. E., Childers, D. L., Gragson, T. L., Grimm, N. B., Grove, J. M., Harlan, S. L., Kaye, J. P., & Knapp, A. K. (2011). An integrated conceptual framework for long-term social-ecological research. *Frontiers in Ecology and the Environment, 9*(6), 351–357.

Irwin, E., Campbell, J., Wilson, R., Faggian, A., Moore, R., & Irwin, N. (2016). Human adaptations in food, energy, and water systems. *Journal of Environmental Studies and Sciences, 6*, 127–139.

Irwin, E. G., & Wrenn, D. (2014). An assessment of empirical methods for modeling land use. In J. M. Duke & J. Wu (Eds.), *The Oxford handbook of land economics*. London: Oxford University Press.

Millner, A., & Ollivier, H. (2016). Beliefs, politics, and environmental policy. *Review of Environmental Economics and Policy, 10*(2), 226–244. https://doi.org/10.1093/reep/rew010.

Shogren, J. F., & Taylor, L. O. (2008). On behavioral-environmental economics. *Review of Environmental Economics and Policy, 2*(1), 26–44. https://doi.org/10.1093/reep/rem027.

Wilkinson, N., & Klaes, M. (2017). *An introduction to behavioral economics*. London: Macmillan International Higher Education.

Chapter 5
Economics

Bruce A. McCarl and Yingqian Yang

5.1 Introduction

As the previous chapters have illustrated, the FEW sectors are often strongly linked. Actions that seek to optimize outcomes for food, energy or water separately often lead to less than optimal outcomes for the other sectors. Bazilian et al. (2011) conclude that treating the three areas of the FEW Nexus holistically "would lead to a more optimal allocation of resources, improved economic efficiency, lower environmental impacts, and better economic development conditions, in short, overall, optimization of welfare." Indeed, much Nexus work arises from an underlying assumption that by better managing the resources overall societal benefits arising from those resources can be increased.

Economics is frequently defined as the study of how people allocate scarce resources when needs and wants of those resources are unlimited. Economics can provide metrics which can be utilized when evaluating the benefits and costs arising from potential actions.

Economics also permits one to look at not only total regional welfare but also at the welfare of participants in the Nexus identifying who gains and who loses under alternative scenarios. This provides a way to understand the incentives needed to attain full cooperation in strategy implementation.

B. A. McCarl (✉)
Department of Agricultural Economics, College of Agriculture and Life Sciences, Texas A&M University, College Station, TX, USA
e-mail: mccarl@tamu.edu

Y. Yang
Agricultural Economics, Texas A&M University, College Station, TX, USA
e-mail: yingqianyang@tamu.edu

P. Saundry, B. L. Ruddell (eds.), *The Food-Energy-Water Nexus*, AESS Interdisciplinary Environmental Studies and Sciences Series,
https://doi.org/10.1007/978-3-030-29914-9_5

In this chapter, we address economic aspects of the Nexus and broader issues regarding the analysis of the Nexus. We will cover the following:

1. Concerns about incorporating market reactions and prices.
2. Behavioral reactions of individuals given nexus actions.
3. Non-market valuation.
4. Welfare analysis.
5. The value of water in alternative uses.
6. Economic influences on observed nexus strategies.
7. The transfer of results between studies.
8. Induced innovation.
9. Adding consideration of limits.
10. Designing incentives.

In broader terms, we will cover externalities, income distribution and inequality effects, dynamics, uncertainties and risk aversion, public–private roles, and **cost–benefit analysis**. We will also use a case study as an example to illustrate economic considerations concerning FEW Nexus metrics, data, and modeling which will be explored in greater detail in Chaps. 13–16.

5.2 Economic Aspects of the Nexus

Here we discuss major economic issues when considering potential FEW Nexus actions, which we will refer to as "projects." Projects may refer to any type of activity designed and undertaken to achieve specific outcomes at the Nexus. In doing this, we will both reveal theoretical concepts and ground them in practical FEW Nexus domains to illustrate why consideration of these concepts is essential.

5.2.1 Incorporation of Demand and Supply Relations

A FEW project can both add extra supply to the market and alter input usage leading to market price changes. In turn, such price changes can alter the revenue and cost outcomes of the project. However, it is common for project evaluation to assume prices of outputs, by-products, and inputs remain unchanged. Thus, price reactions are essential considerations in Nexus project evaluation.

Bioenergy provides several illustrations of this effect. In 2005, the USA adopted a Renewable Fuel Standard or "RFS" which requires the blending of renewable biofuels into traditional hydrocarbon fuels. In practice, this primarily involves corn-based ethanol blended with gasoline and biodiesel blended with diesel. Over time the volumes required to be blended have increased, and both the price of the ethanol and the price of corn have increased exhibiting a response along the supply curve.

In fact, corn prices in 2011 were triple those in 2005 (note other forces contributed as discussed in Abbott et al. 2011). In 2017, corn prices were still more than

50% greater than 2005 levels. These significant price changes have a significant impact on the economics of corn-based ethanol production.

In the bioenergy world, by-products are often advanced as valuable items that can help support bioenergy production. However, one must be careful to consider the price changes brought about by increased supply.

Consider one such by-product: glycerol arises as a by-product when producing biodiesel. The expansion in biodiesel production resulted in additional glycerol production so large (2.8 million tons) that it exceeded the market volume before the expansion (2 million tons). As a result, the market saturated, and the price crashed. Moreover, the crude glycerol by-product from biodiesel production contained toxic elements and exhibited a substantial difference in color, decreasing its market value. Consequently, the glycerol by-product became worth less than the cost of selling it and now it is an item for disposal.

This example shows the need for considering by-product demand relations when evaluating Nexus projects. Moreover, this is not always done, for example, Wooley (1999) identified glycerol as a by-product contributing to profitability while Ciriminna et al. (2014) identified glycerol as a disposal issue that costs money to dispose of.

The anticipation of price changes requires broad industry level consideration. For any single **firm**, the input use or by-product quantity is typically not large enough to stimulate price changes. However, when many firms pursue the same actions, then the quantities are large enough, and price changes occur. Such a process is called the **fallacy of composition** by economists. The lesson here is that industry trends also need to be considered when evaluating Nexus projects.

Demand relationships are also key when Nexus actions cause product prices to increase. For example, in an energy-only context, in the early 1970s, the state of Washington Public Power Supply System (WPPSS) responded to a 7% annual growth rate in electricity demand by initiating construction on five nuclear-generating facilities. The cost of constructing nuclear power plants was deemed too high to be covered by funds just raised by issuing bonds and WPPSS began to raise electricity rates paid by customers. This led to a demand response in the form of lowered electricity consumption. The needed revenue for the financing did not materialize and contributed to the abandonment of four of the new plants and the largest bond default in US history.

Collectively, one must incorporate product demand and input supply relations in Nexus project evaluations. If this is not done, there is a high likelihood of a biased evaluation and an unanticipated result.

5.2.2 *The Rebound Effect*

We noted in Sect. 4.3.4 that rebound effects are unintended responses to changes in the social system, such as the introduction of new technology or policy, which results in a reduction or reversal of the intended impact. For example, economists have noted in many cases that subsidizing conservation, like water conserving technology, can stimulate decision-makers to increase usage—the "Rebound Effect." However, Nexus projects are often analyzed under a strong assumption that the current economic and technical characteristics will be unchanged. However, this may not hold as increasing refrigerator efficiency and thus lowering the cost of say refrigerating food may cause people to buy more refrigerators resulting in less to no savings in energy.

As a specific example, several western US states subsidized water conserving irrigation technologies, and assumed that only the equipment would change. However, Pfeiffer and Lin (2014) analyzed such a case in Kansas and found this lowered water costs to farmers and stimulated a higher production of irrigated crops, including expansions onto previously unirrigated lands. The end result was increased overall water use.

Thus, one needs to consider possible rebound effects when Nexus projects are implemented to lower water or energy usage and costs.

5.2.3 Non-market Valuation

Nexus projects often alter abundance or characteristics of items that do not trade in marketplaces such as ecosystem functions, air and water quality, recreational access, climate, and other phenomena. Economic appraisals of the effects on these items involve what economists called nonmarket valuation.

Valuing non-market items involves determining how much society would need to be paid to live with the adverse effect (diminished air or water quality, lost recreation, or climate change), or conversely, how much they would be willing to pay for the more desirable outcome.

Placing a value on such items may make a project more or less desirable. For example, replacing a coal-fired generating plant with a solar farm may not be cost-efficient in some locations. However, the impact of the solar farm on reducing air pollution and emissions of greenhouse gases may make the project more desirable.

Many techniques of non-market have been introduced. Techniques are usually divided into two approaches: stated and revealed preference methods. **Stated preference** method measures individuals' value for environment quality directly, by asking them to state their preference for the environment. **Revealed preference** method seeks to recover estimates of individuals' will to pay for environment quality by observing their behavior in related markets (see the detailed discussion in the estimated "cost of carbon" or, various revealed or, stated preference approaches).

As reviewed in Chap. 9 (Ecosystem), economic valuation reveals the trade-offs within the Nexus and the impacts of various agricultural systems on the ecosystem services. Estimating the valuation of an ecosystem involves estimating the way that the ecosystem is involved with increasing production of market goods then estimating the marginal effect of a change in ecosystem services (i.e., altered water quantity or quality as discussed in Hanley et al. 2002). These ecosystem services can involve many diverse items such as changes in spawning **habitat** for commercial fish species or water yields or clean water. Other ecosystem services which could be valued similarly include nutrient cycling and the provision of genetic resources. The value of the ecosystem can then be indirectly inferred from the changes in the value of goods produced due to changes in the supply of these services because of, say, a loss in the area of the ecosystem.

5.2.4 Welfare

Economic welfare is a monetary measure of the gains (or losses) achieved by consumers from having cheaper, more abundant (more expensive, scarcer) Nexus products and services. It also measures the gains to producers in the form of profits from having more Nexus resources available or losses under the converse. Welfare estimates the willingness to pay to avoid some negative force like pollution, or the willingness to accept compensation in the face of a Nexus management practice being adopted that worsens their well-being. The producer component is called **producers' surplus**. The consumer component is called **consumers' surplus**.

Thus, the welfare effects of adopting a Nexus practice involve both producers and consumers welfare. This is an important distinction because many Nexus type studies only estimate the effect on producers without considering any consumer effects. The consumer benefits arise from lower product prices or greater product availability at a given price. For example, when considering climate change effects on crop production one should not only consider changes in commodity prices and producer revenue but also in the cost of consumer food purchases (see Adams et al. 1990 for an example).

In general, treatment of consumer effects means incorporating demand curves and assumptions other than fixed prices for commodities. In particular, as more is produced then, assuming that the market share is significant, this will cause prices to go down giving consumers more for their money or the converse occurs with prices going up.

Overall, it is useful to do a welfare analysis in conjunction with the evaluation of a Nexus project on recognized groups of producers and consumers (i.e., farmers, electricity producers, low-income consumers, urban dwellers, rural parties, overseas parties, etc.) as opposed to aggregate analysis. Such a welfare analysis is commonly called a **benefit–cost analysis**.

Benefit–Cost Analysis

Benefit–cost analysis is founded on a branch of economics known as **welfare economics**. That is, what are the benefits and costs arising when implementing an action when the action affects welfare across elements of the economy? Who benefits and who bears costs when a project is built and, considering those who benefit and those who bear costs, are the benefits larger than the costs are the questions where benefit–cost analysis can be applied. The formula below simply speaks of benefits and costs, that is,

$$\sum_{i,t} B_{i,t} \cdot (1+s)^{-t} - \sum_{i,t} C_{i,t} \cdot (1+s)^{-t},$$

where t is time, i is the ith individual, $B_{i,t}$ are benefits to the ith individual in period t, and $C_{i,t}$ are corresponding costs while s is the discount rate. Thus, the first term is summation of benefits to individuals and the second term is the summation of costs to individuals. The basic decision rule for judging whether a project may be desirable occurs is when the results from the above formula are greater than zero. The discount rate refers to the rate at which the value of something in the future is reduced to obtain a value in the present. For example, a discount rate of 10% means that something with a value of 100 a year from now is valued at 90 today. Thus, future costs and benefits are discounted to some reduced value today.

There is also one result that is often confusing for some that merits explanation. Often actions that increase supply decrease producer welfare but benefit consumers, while actions that reduce supply are beneficial to producers but not consumers. This occurs since agriculture and energy both typically face an **inelastic demand curve**. In economics, inelastic demand curve means when the percentage change in price exceeds the change in quantity demanded, thus, for example, a 10% quantity increase could cause commodity price to fall by say 20%. The more inelastic the demand, the steeper the curve and the more price will react to a quantity change. More supply typically lowers prices substantially reducing producers' net incomes but causing consumers to increase the amount of goods they buy, and causing the consumer dollar to go farther.

5.2.5 *Value of Water in Alternative Uses*

Water has differential value in alternative uses, such as irrigation, cooling in power plants, ecological support, human consumption, and direct use, pollution dilution, hydroelectric power generation, and unconventional oil and gas production (fracking). These differential values exist because of the high costs of moving water and historical water allocation procedures (like prior appropriation which give certain users water rights regardless of use value).

Consumers also derive value from water and actions that lower its costs as this allows them to consume more water or divert money to buy other goods again causing the dollar to go farther.

Consistent methods of estimating and comparing water values are required when examining the implications of Nexus projects. Approaches to value water are discussed in Young and Loomis (2014).

Market-based methods (Colby 1989) include:

1. A **comparable sales approach** which involves a comparison of specific water, one is trying to value with the prices and characteristics of similar water that has been recently sold or leased;
2. A **capitalization approach** which involves taking the net present value of the stream of income arising from the water;
3. A **land value differentials approach** which assumes the value of water is capitalized into land values and involves a comparison of the values of agricultural land with and without water access;
4. A **replacement approach** which involves the estimation of the **replacement cost** when replacing the water with the lowest cost alternative water supply source; and
5. An **econometric estimate** of water demand can be formed in some situations, where trading data can be attained along with sufficient information on other characteristics of the trade (i.e., is the water conveyed from a senior or junior right? (senior or junior water right here means, for example, priority in water use when scarcity arise with senior water rights having first priority.) Or, is the transfer permanent or a lease? And what are the lease terms?).

5.2.6 Economic Influences on Observed Nexus Strategies

One way to identify possible Nexus strategies that might be undertaken in the future in target region is to observe the impact of prior actions that addressed the Nexus either in the target region or similar regions. In such a case, there is an inherent bias in what can be observed that arises due to economic prices. In particular, the range of prices that have been observed for both output and inputs restrict what Nexus opportunities may have been chosen and thus can be observed. Let us look at theoretical and practical examples of this.

In setting up this example, we use the classical production possibilities curve as in Fig. 5.1. The production possibilities curve is a graphical representation of the

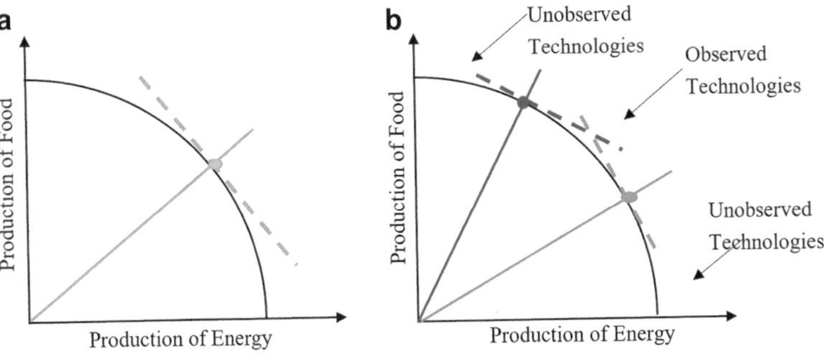

Fig. 5.1 (**a**) Optimal choice given a price. (**b**) Domain of strategies given a price range

alternative combinations of goods an economy can produce showing how an expansion of production of one good may cause contraction of production of another. The bold black line in panels A and B gives a continuous set of energy-food production possibilities using different technologies. Each point on the curve represents a choice of a technology or a particular resource allocation which results in a certain level of food production (y-axis) and the corresponding level of energy production (x-axis). Thus, lower energy production correlates to higher food production or vice-versa.

In panel A, the solid green line gives the ratio of the food price to the energy price at a point in time. According to economic theory, the production technology chosen will be the one where there is a tangency between the line giving the price ratio and the production possibilities curve. This means production in our case will occur where the green dashed line is tangent to the bold black line—at the green dot.

Now given this basic setup, consider panel B where we have a solid red line representing the highest observed ratio of food price to energy price in recent history, and the blue solid line representing the lowest ratio again in recent times. Then, in this case, the only Nexus technologies we have observed fall between the red and blue dotted points and those outside will not have been seen. This means for example if the ratio of say ethanol to gasoline prices has been in one interval that we would not have seen strategies appearing that allowed use of much more ethanol in cars which would only happen if much lower relative prices appeared.

A second example can be cast in terms of inputs using the classical **isoquant** that explains the relative use of two inputs given their prices as in Fig. 5.2. Therein, assume the bold line in panel A gives a continuous set of possible quantities of energy and water used across the set of possible technologies. Also, assume the solid line gives the ratio of the energy price to the water price.

According to economic theory, given the input price, the production technology chosen is the one at the point where there is a tangency between the line giving relative prices of energy and water and the isoquant, as occurs at the green point in panel A. Now given this basic setup, consider panel B where we have a solid red line representing the highest ratio of energy price to water price we have ever seen, and a solid blue dotted line representing the lowest ratio. Then, in this case, the only Nexus possibilities we have observed fall between the two-colored dots, and again

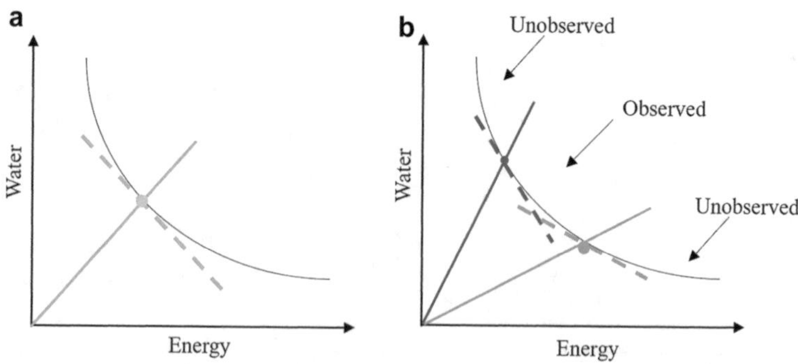

Fig. 5.2 Production Isoquant and strategies chosen. Panel (**a**): Optimal choice given a price. Panel (**b**): Domain of strategies given a price range

there are a lot of unobserved possibilities that never got chosen because the prices did not favor them. This means, for example, if the ratio of say natural gas generated electricity to coal prices has been in one interval that we would not have seen strategies appearing that used much more natural gas in electricity generation which would only happen if much lower relative prices appeared.

However, Nexus actions or external forces can alter production possibilities, isoquants, and relative prices. Under such shifts, previously unattractive Nexus related production or resource usage strategies can become desirable. Thus, not all possible strategies will be observed and thus new never before seen strategies may arise.

As a consequence, identifying Nexus related strategies through surveys, interviews or other means will not generally describe the full set of possible strategies that may arise in the future.

5.2.7 *Can I Transfer Results from Other Assessments into This One?*

As discussed in Sect. 5.2.6 (Economic influences on observed nexus strategies), there exists prior actions that address the Nexus either in the target region or similar region. Frequently, results from other studies are used in a Nexus evaluation rather that developing primary estimates. A big issue in such a setting raises what economists called "**benefits transfer**" which refers to the transfer of benefit estimates from some other location into the differing project location. For example, Hansen and Ribaudo (2008) provide dollar-per-**ton** estimates for 14 categories of soil conservation projects while Young and Loomis (2014) contain an estimate of the value of water from various regions. However, using such numbers needs to be done with caution. Brouwer (2000) argues that most transfers appear to result in substantial transfer errors.

In covering benefits transfer, the Ecosystem Valuation website developed by King and Mazzotta (2000) states "The more similar the sites and the recreational experiences, the fewer biases will result" and then presents a discussion and cites the benefits of such an action as:

1. Reductions in the cost of carrying out an appraisal;
2. Speed of attaining the information;
3. Ability to use the transferred estimate in constructing a rough estimate of the value of a project to see if more effort on it is justified; and
4. Ability to use in making a gross estimate of the total item value (i.e., cost of water or reduced **erosion**).

It also cites the limitations as:

1. The transfer may not be accurate, except for gross estimates unless the sites share all characteristics;
2. Good estimates for the item at hand may not be available;

3. Appropriate studies may not be published and are hard to access;
4. Reporting in the studies found may not give one enough information to allow transferring the information with appropriate adjustments;
5. Quality of the other studies may be difficult to assess;
6. Extrapolation beyond what is covered in the initial study is questionable;
7. The accuracy of the transferred item is limited by that of the item itself; and
8. Estimates may be out of date.

5.2.8 *Induced Innovation*

New technologies are likely to evolve as input or product prices change. This involves **induced innovation**. The theory indicates that when the price of a particular input used in production increases or falls significantly relative to the price of other inputs, society will innovate by developing technologies that reduce or increases usage of that particular factor.

In a Nexus setting, an example is that when labor prices dramatically increase due to scarcity, society will invent ways of substituting other factors for labor, like going to the more capital-intensive harvesting practices. Similarly, if a fee is charged for GHG emissions, this will induce industry and others to develop strategies that produce goods with less emissions.

Induced innovation has also been observed in corn to ethanol conversions. In particular, when processing corn into ethanol, a by-product called distillers dry grains (DDGS) is produced. The rise in biofuel production resulted in a large increase in DDGS production and a fall in price. DDGS were initially only used in wet form (up to 70% moisture) and, due to transport costs, usage was limited to cattle feeding near the refinery. Eventually, innovation was induced, which transformed the wet DDGS into more valuable forms. Today high valued oil is extracted from DDGS, and the remainder is mixed with low-value corn stalks for animal feed. This product can now be processed into dry pellets, allowing long-distance shipping.

5.2.9 *Adding Consideration of Limits*

All strategies have factors which can limit their adoption. Such limits involve the following:

1. Financial capital availability, such as capital constraints and lending practices.
2. Human education and abilities, including sheer labor availability, labor skill, leadership capabilities, and educational attainment.
3. Resources available, including regional land, water, equipment, and infrastructure resources.

4. Consistency with cultural practices, such as societal values, worldviews, cultural norms and behaviors, perceptions of needs for action, and compatibility of strategies with lifestyles.
5. Availability of technology.
6. Knowledge of new practices, such as knowledge and awareness of water-conserving irrigation strategies as elaborated on in Chambwera et al. (2014).

Limits may be alleviated through educational programs, extension programs, loan programs, grants, and other actions. They may also render some strategies useless.

5.2.10 The Role of Incentives

It is rarely the case that a FEW Nexus action will make everyone better off (what economists commonly call an action that is **Pareto optimal**, and others call a win-win situation). Generally, at least one participant in Nexus or a group thereof will be made worse off by the given action.

In judging action desirability, economists generally utilize the **compensation principle** finding that the action is desirable if those gaining from the implementation have gains large enough to compensate those who lose (for discussion see Just et al. 2008).

When an action is implemented, there is no guarantee that the compensation will actually occur, that is, if consumers benefit from a production increase while producers lose consumers generally do not compensate producers. However, if the individuals that would need to implement the Nexus action can choose whether or not to implement, then some form of direct compensation would be needed to get the Nexus action implemented. For example, suppose an action involves the construction of a reservoir in a valley containing a small number of farms, and that the purpose of the reservoir is supplying water to a nearby town, but it takes water away from the farmers. If the people who benefit from the reservoir, that is, the urban population could in principle fully compensate the losers, that is, the farmers, and farmers feel that they will be worse off as a result of the reservoir, then the reservoir may not be built. However, if compensation is actually paid, then the farmers may cooperate. Such a situation is common with water trades from agriculture to urban interests where water markets are a way that farmers can be compensated.

Compensation in the form of incentives needs to stimulate target entities to adopt costly practices which generally do not yield benefits to themselves even though others receive gains. Steps can also be taken to make current practices undesirable steering decision-makers to shift toward the Nexus action. Many different forms of incentives or steering disincentives are possible, including:

1. The introduction of markets for Nexus items, like a water market, where cities can buy the water from farmers who would lose if they lost access to the water, but the price for the water could be high enough to compensate the farmers for any losses.

2. The introduction of subsidies for equipment that a group would need to use to achieve the desired result, reducing the costs of the equipment directly or reducing the cost of money borrowed to buy the equipment. This might involve cities subsidizing the cost of more water efficient cooling equipment for use in electricity generation.
3. The introduction of taxes on equipment crucial to the continuation of current undesirable practices, such as one could tax conventional tillage equipment when one is interested in higher water use efficiency, and lower greenhouse gas emissions.
4. The use of technology standards which mandate an upgrade in technology to the desirable actions such as the automobile café standards on vehicle miles per gallon.
5. The imposition of some form of regulations such as banning appliances that do not meet certain water or energy efficiency characteristics such as high water shower heads.
6. The development of differentiated markets favoring products from Nexus implementing parties, for example, opportunities to purchase electricity only from renewable wind sources.

In implementing such incentives, one naturally needs to be careful about inducing such things as the rebound effect as discussed above and also will need to be flexible potentially increasing prices and decreasing prices to get the amount of resource transferred that is desirable.

5.3 Broader Items

Some other economic concepts merit mention including externalities, income distribution, dynamic concerns, uncertainty and risk aversion, public and private roles, and cost–benefit analysis.

5.3.1 Externalities

Frequently, activities have positive or negative impacts that damage others for which they are not held liable. Such a situation is called an **externality** by economists where production or consumption of one of the Nexus items imposes negative impacts on other parties.

Here we will generally deal with negative externalities where the impacts on the other parties are adverse. Examples of such externalities commonly involve polluting emissions, as follows:

1. Where applications of nitrogen fertilizers on food crop impact local rivers and aquifers but is not reflected in the costs of the fertilizer appliers or their resultant crop product price.

2. GHG emissions from fossil fuel-based power generation.
3. Pollutants entering aquifers associated with infiltration of produced fracking water.

Non-market valuation is often applied to value externality effects as discussed (see Chaps. 4 and 5).

A related concept is **maladaptation** where Nexus actions that meet the needs of one sector increase the vulnerability of sectors elsewhere or in the future. This is an important issue recognized in climate change adaptation literature (see Barnett and O'Neill 2010). Examples of maladaptation include: diverting floodwaters away from a city may result in flooding of other citizens along the path of diversion canal; and extensive use of groundwater today depletes the resource so that it cannot be used in the future.

In dealing with negative externalities economists often state that the externality should be internalized. There are many ways of doing this including some of the incentives below:

1. Assignment of property rights, such as allocating grazing use permits on federal lands providing an incentive for those using land for grazing to avoid overgrazing.
2. Instituting markets for rights to pollute (commonly called cap and trade. For example, one can reduce GHG emissions below current levels by allocating rights to emit but than allowing the purchase and sale of those rights).
3. Imposing performance standards like the use of denitrification inhibitors or the institution of a Renewable Fuel Standard which requires certain volumes of renewable fuel.
4. Providing subsidies for equipment that reduces the negative impact, for example, lowering the cost of precision agriculture equipment to reduce nitrogen runoff, or lowering existing subsidies to fossil fuel production to increase price and lower consumption reducing air pollution and greenhouse gas emission externalities.

In doing this, one must be careful about adopting strategies with minimum implementation—**transaction costs**, as the magnitude of such costs, have caused some schemes to fail (Stavins 1995; Tietenberg 2003).

Externalities are critical concepts in the Nexus arena. Hoff (2011) argues Nexus thinking is concerned with addressing externalities across multiple sectors, decreasing adverse effects of some sectoral actions on others with a focus on system efficiency rather than on the productivity of individual sectors.

5.3.2 *Income Distribution and Inequality Effects*

The Nexus also needs to be contextualized within the debate around social justice. Food/water poverty indicate the presence of a strong relationship between levels of poverty and levels of resource consumption—with resource consumption and resource availability decreasing as poverty indices increase.

As discussed in Chap. 3, water security involves access to, water for human and ecosystem uses; energy security involves access to clean, reliable and affordable energy services for daily uses; food security involves physical and economic access to sufficient, safe and nutritious food to meet dietary needs and food preferences for an active and healthy life. The Bonn2011 conference specifically aimed to explore "how a nexus approach can enhance water, energy, food security by increasing efficiency, reducing trade-offs, building synergies and improving governance across sectors" (Hoff 2011). Nexus decisions can alter food/water/energy availability and prices, in turn, altering the welfare of groups outside the FEW production region or domain under study including disadvantaged ones. Such potential effects increase the importance of making informed and balanced choices not only on our natural resources but also in working toward the achievement of Sustainable Development Goals as discussed in Chap. 3.

Section 4.2 noted the importance of considering distributional effects, such as whether and how the policy change may be felt by high- versus low-income populations, or how people in urban centers and rural areas may be differentially impacted. Section 18.4 includes four case studies that illustrate **distributional (distributive) effects** in FEW policies at the city-scale.

5.3.3 Incorporating Dynamic Concerns

Nexus interrelationships and demands for Nexus commodities are changing over time. Therefore, decision-making processes need to be proactive and consider the dynamic evolution of the FEW arenas. (See Sect. 4.3.1 for a discussion of FEW system dynamics and human behavior.) For example:

1. Growing populations alter FEW demands.
2. Climate change alters water supplies plus regional FEW demands.
3. Aquifer and fossil fuel reservoir depletion is ongoing.
4. Evolving technology influences FEW supplies and demands.
5. Regional FEW infrastructures and resource stocks/availabilities may be depreciating or being depleted.

The incidence of growing populations and climate change forces economists to consider how to balance current versus future resource allocation properly so as to maximize society's welfare over a long time horizon.

Economic effects at different points in time are not usually valued the same. The promise of payment of a dollar today is more valuable than payment 10 years from now since one could buy an interest-bearing bond returning say a dollar plus 25 cents interest by 10 years from now. Economists use the concept of a discount **rate** to place economic effects across different times on an equal footing. This reduces the future value by an accounting of the compound interest one would achieve by that period from investing the same amount of money currently.

Choice of the discount rate is crucial when making comparisons between decisions or impacts at different times. The use of too high a discount rate will result in too little value placed on avoiding damaging future events and too little investment in technologies that enhance sustainability. Conversely, applying too low a discount rate will result in too much investment in items that benefit the future at the expense of the current.

Investment in climate change mitigation is a clear instance where the discount rate causes different conclusions to be reached. Debate in the literature on discounting has often focused on how to select the correct discount rate (Stern 2007; W. D. Nordhaus 2007; Weitzman 2007; Zhuang et al. 2007). Regardless of the rate chosen, it is important to remember that the discount rate is a critical determinant in the outcome of an analysis, and for each project, a single rate must be applied to all future benefits and costs. For example, Stern (2007) advocates more ambitious greenhouse gas mitigation than Nordhaus does, and this is in part influenced by Stern's use of a much lower discount rate.

5.3.4 Uncertainty and Risk Aversion

Uncertainty adds complexity to Nexus systems. Uncertainty may be represented by year-to-year variations in water supplies and commodity prices caused by drought or floods plus an uncertain future for the rate of population growth, climate change incidence, technological progress or aquifer/fossil fuel reservoir depletion. Collectively, such uncertainties raise needs for stochastic modeling and scenario analysis. Stochastic modeling involves considering multiple possible say water availability situation and their probabilities. For example, in the South-Central Texas EDSIMR model (see Sect. 5.4), shorter run uncertainty was addressed by having nine levels water availability and their historical probabilities. For long run uncertainly, the model was run under alternative scenarios involving population growth and future climate change.

Broadly following Moschini and Hennessy (2001), the main sources of uncertainty in the Nexus system as being from:

1. Production uncertainty which refers to the variation in levels of production like crop yields where the amount and quality of output that will result from a given bundle of inputs are typically not known with certainty.
2. Price uncertainty, where production decisions are made in advance of the time when the final product becomes available, so that market price for the output is typically unknown when these decisions have to be made.
3. Technology improvement uncertainty which acknowledges that increases in production output and input usage efficiency are uncertain across all sectors.
4. Policy uncertainty where one is unsure of the persistence and enactment of economic policies that significantly impact sectors like renewable energy subsidies or requirements.

Risk preferences and management have been widely addressed both analytically and numerically in the economic literature. Many stakeholders such as farmers or firms are typically risk-averse. The more significant the variability they experience in their profits or service supply, the more willing they will be to adopt measures that reduce risks.

In the face of climate variability, vulnerable farmers employ ex-ante (forecast-based) strategies, to protect against the possibility of catastrophic loss in the event of a climatic shock. Farmers' precautionary strategies include the following:

1. Selection of a portfolio including less risky but less profitable crops.
2. Overuse of fertilizers.
3. Diversifying income sources.
4. Avoiding investment in production assets and technology.

Arguments have been made that if farmers can trade away part of the risks on their farm at an acceptable cost, the expected utility of the farmer will increase and this provides another incentive direction—development of risk sharing mechanisms like insurance or crop share arrangements when Nexus practice adoption increases risk exposure.

Although sharing risks can increase utility, individuals are not likely to share all risks. Factors that may influence this decision include the following:

1. An individual's degree of risk aversion.
2. The costs involved in risk sharing.
3. The relative size of a risk.
4. The correlation of the risk with other risks.
5. Other sources of indemnity.
6. An individual's perception of the nature of risk.
7. An individual's income and wealth.

5.3.5 *Public–Private Goods, Incentives, and Roles*

Some strategic responses to Nexus issues involve adaptation strategies that occur autonomously by private individuals and through planned public actions.

McCarl (2015) presents a list of possible adaptation categories in a climate change adaptation context with an indication of whether the actions will be public or private. Individuals serving their personal interests take private actions. For example, altering crop, livestock mix, or modifying irrigation practices are primarily exercised by private individuals who manage the land. However, other strategies are not feasible or desirable for implementation by individuals (called **public goods**) which, in turn, bring in a public role. Public entities may alter policy, provide incentives, provide information, develop certain classes of technology, or build infrastructure. Examples of such public actions include:

1. technology development on FEW components through research, for example, developing ways to generate energy with less water, grow food with less irrigation and energy and utilize less energy and water for municipal use;
2. building a more efficient public utility generation and distribution system or reducing leaks in water distribution;
3. providing extension information on FEW situations, improving practices like water conserving agricultural practices;
4. financing insurance coverage for cases where for example water is limited and operations are curtailed; and
5. creating public goods that benefit many without excluding those who have not financed implementation, for example, capital investment in reservoirs or canals.

In public cases, an individual does not capture all of the benefits and is likely to choose not to pay for the investment. Thus, in cases where the Nexus action creates a public good then broader involvement is needed to achieve implementation as private actions will underinvest in such actions.

5.3.6 Cost–Benefit: Not Just Economics

Finally, let us deal with the broad issue of benefit–cost analysis inclusiveness. One quite frequently hears people say that for a project to be justified it must have a ratio greater than one of the benefits divided by costs. In fact, it is often a requirement that a benefit–cost ratio is constructed for almost any considered project.

However, it is also important to note that the benefit–cost ratio will generally not be all-inclusive relative to the items considered in making a decision. In particular, there may well be many nonmarket impacts which cannot be quantified in dollars and cents. Nexus project induced items such as reductions in the amount of erosion getting into the water, or a reduced amount of air pollution emissions, or an alteration in biodiversity in the region are difficult to represent in general and certainly in monetary terms. In such cases, one may do both a benefit–cost and a simultaneous environmental analysis and emphasize that both should be considered, not just the cost–benefit ratio as could be implemented within the systems approach discussed in Chap. 2.

5.4 FEW Nexus Metrics, Data, and Modeling

In this section, we use an ongoing Texas FEW Nexus case study to illustrate the complexity and challenges regarding the economic considerations related to FEX Nexus metrics, data, and modeling but also see the detailed discussion in Chaps. 13–16.

The Nexus model is named the Edwards Aquifer and River system Simulation Model (EDSIMR). The model depicts regional dryland and irrigated farming, water diversion/pumping, river flows, environmental indicators, aquifer elevation status, thermal energy cooling, hydropower, and hydraulic fracturing. The model when solved generates output on water prices, water use, and allocation, farming crop mix, agricultural production, irrigation strategy, aquifer levels, spring flow discharge into rivers, farm incomes, municipal and agricultural pumping, pumping lifts, energy generation, and energy use, among other items.

A FEW Nexus model needs to be based on high-quality data. For the EDSIMR analysis we needed to integrate data from:

1. Regional aquifer simulations that employed the groundwater model (GAM) to simulate aquifer level, pump lift and spring flow discharge given alternative amounts of pumping in the region;
2. Crop growth simulations using EPIC under different irrigation strategies and climate conditions to develop estimates of dryland and irrigated crop yields along with water use plus erosion and nutrient flows;
3. River flow and groundwater infiltration where we used **SWAT** to simulate levels of aquifer recharge, net inflows at river locations, evaporation, reservoir operations, and water quality characteristics, given changes in agricultural production, climate and the typical regional distribution of rainfall;
4. Econometric based urban water demand equations that show water demand as a function of water price and climate conditions;
5. Engineering models of electrical power generation cooling with which we estimated the alternative cooling methods and their implications for water use, cost and power generation; and
6. Calculations of energy use and water loss associated with many water development alternatives (reservoir construction, pipelines from distant locations, desalination, aquifer storage, and recovery and conservation incentives among others).

EDSIMR is formulated as a unifying component that includes modeling of rivers, aquifers, agricultural water use, irrigated and dryland cropping, water project development, energy generation, cooling water use, cooling water retrofits and non-agricultural water use among other things. That model is used to look at a regional welfare-maximizing allocation of water across urban, industrial, electrical generating and agricultural users coupled with an optimal choice among the water development and power cooling alternatives. As shown in Fig. 5.3, the total project encompasses data collection, model development, and feedback from stakeholders.

Key Points

- Nexus analyses can be biased if one neglects product demand and input supply price-quantity relationships. For example, using US corn for biofuel was a significant force behind corn price increases while the supply of the by-product glycerol from biodiesel refining reduced glycerol prices. Also, demand quantity projections may fall if prices are increased as expensively discovered by WPPSS power suppliers.

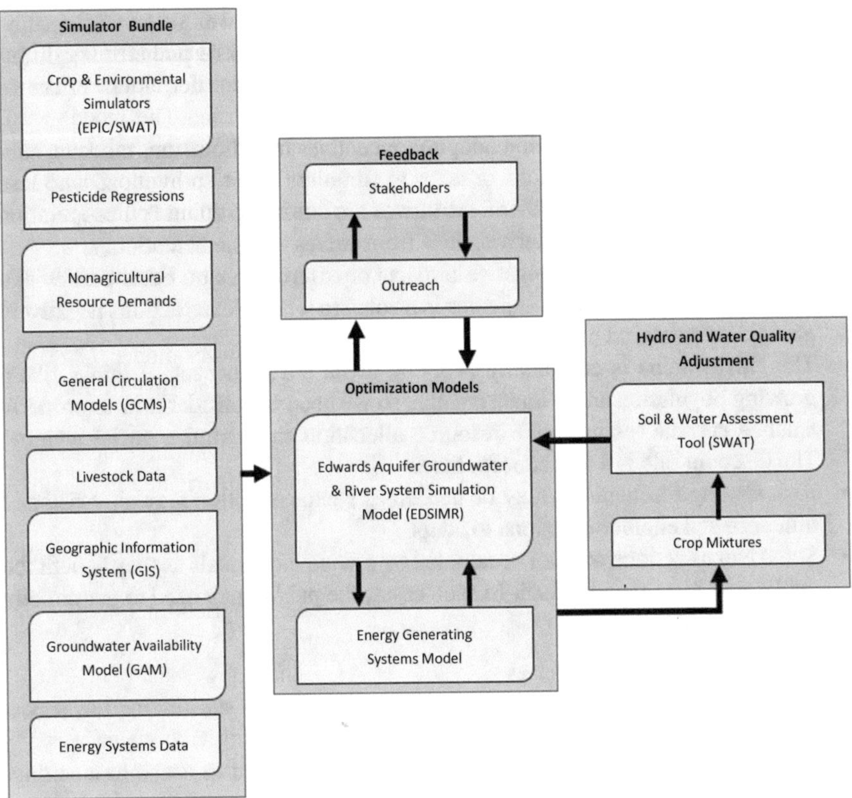

Fig. 5.3 Edwards aquifer and river system simulation model (EDSIMR) framework

- Incentives to conserve can have perversely increased usage as demonstrated by the rebound effect.
- Projects may affect non-market items like water quality which may, if valued, influence an appraisal.
- Transferring evaluation estimates from elsewhere have both advantages and drawbacks.
- Water has a different value in different uses. This requires both quantification and consideration when making decisions. The consideration of water movement costs and water allocation patterns between users is a crucial consideration. Techniques exist for estimating the value of water in alternative uses so one can look at the implications of Nexus–based reallocations or new water development activities.
- When identifying strategies based on those currently employed, one must realize this is biased by historical prices and that other strategies could be used if prices change.
- Economics provides welfare metrics that can be utilized when examining making decisions about Nexus possibilities in a benefit–cost analysis. The welfare

metrics can describe not only total welfare but also the welfare of participating groups identifying who gains and who loses. One should be prepared for different effects across producers and consumers when a practice alters prices of commodities.

- One can provide Nexus action adoption incentives by introducing markets, subsidies, performance standards, or taxes to stimulate adoption by those who lose in the interest of gains to others. Incentives are needed to attain full cooperation especially to those who otherwise lose from strategy implementation.
- Nexus actions can cause negative impacts or externalities on other parties that need to be considered. Social justice is a concern when Nexus actions negatively affect disadvantaged populations.
- The Nexus arena is continually evolving given the influences of things like a growing population and climate change, so we need to consider how to properly balance current versus future resource allocation to maximize social welfare. The discount rate is a key concept here.
- Risk sharing mechanisms may be needed as Nexus adoption may alter risk and influence stakeholder decisions to adopt.
- Some Nexus actions will not be adopted by private individuals as they benefit the public, not just the individual. In such cases, the public may need to get directly involved in adoption.

Discussion Points and Exercises

1. Discuss why the rebound effect might occur if you are subsidizing lower cost lawn irrigation practices in a growing municipality.
2. Discuss why not all nexus alternatives can be observed in usage in a setting. Discuss the role of historical commodity and water prices along with induced innovation.
3. Discuss why incentives may be needed to implement FEW Nexus actions by listing two to three examples.
4. Discuss the consequences of changing to more water efficient cooling practices for power generators in a scarce water setting where the change makes more water available for municipal use and reduces water purchases from agriculture. Could power generators lose? What might happen to agricultural producers? What would be the consequence of having this lower priced water for municipalities? Would some form of compensation possibly be needed and if so how could it be implemented as a way of resolving negative effects?
5. Discuss how reducing discount rates will affect the desirability of investments with significant upfront costs and benefits occurring later.
6. Suppose you are doing a study of changing cooling water alternatives and the one being considered eliminates the need for a cooling pond that provides spawning grounds for an endangered fish species and also reduces the discharge of pollutants into a nearby river. Would there be any need for non-market valuation in such a setting? Could benefits transfer be used and if so what qualifications might you state on the resultant estimates?

7. In the San Antonio region, two 140+ mile long pipelines are being considered to deliver water from one place to another. If you were appraising their FEW consequences what types of effects would you consider?
8. Suppose you are evaluating a US case where to generate more energy: 5% of US cropland and 5% of current grasslands will be diverted to producing energy crops which use less fertilizer than existing crops but much more than pasture. Conceptually, could this influence US and global commodity markets? What about water use and water quality?
9. Suppose we have several Nexus water-related alternatives: build a 150-mile pipeline, have agriculture shift to more water conserving crops, increase home appliance water efficiency and change fracking techniques to use less water. Who makes the decision to adopt the practice in each case and is a public role needed to provide funds or disseminate information?
10. When you are examining issues regarding water what should you use, mean water availability, drought availability, flood availability or the distribution of water availability?
11. Suppose a pipeline is to be built that crosses private lands and some government lands. Could any externalities be involved?
12. Suppose a consulting firm needs to appraise whether water could be transferred from agriculture to cities. How would you estimate the value of water to both parties?
13. Energy prices rose substantially between 2000 and 2011. During that period, we saw smaller cars and more miles per gallon from new models. Prices have now gone back down some, and larger cars are again selling. Which two economic concepts above are relevant to these observations?
14. Why should one not only look at the cheapest cost water when making an urban decision on water supply sources?
15. Why do we worry about policies that increase food prices? Who is affected by food prices? Also, should we only worry about the average consumer?
16. The formula for the elasticity of demand is $e = \frac{\Delta q}{\Delta p} \cdot \frac{p}{q}$, where p is commodityprice and q is quantity consumed, and Δq and Δp are changes in quantity and price due to firm actions respectively. This formula can be manipulated to express the change in product prices or the change in quantity sold as follows:

$$\Delta q = \Delta p / p * q * e$$
$$\Delta p = \Delta q / q * p / e$$

Assume $p = q = 1$ and answer the questions in the following two paragraphs:

How much would the sale price at which more production could be sold change if they produced 10% more under an elasticity of −0.1 versus one of −10? Does this mean FEW actions that expand production can do this more safely if they face elastic demand ($e < -1$ (the −10 above)) or inelastic demand ($e > -1$ (the −0.1 above))? How does this relate to the Washington Public Power case described above?

How much would the quantity they could sell change if their cost of production increased by 10% under an elasticity of −0.1 versus one of −10? Does this mean FEW actions that expand the price they need to charge to cover the cost of production can be implemented more safely if they face elastic demand ($e < -1$ (the −10 above)) or inelastic demand ($e > -1$ (the −0.1 above))? How does this relate to the glycerol case described above?

References

Abbott, P., Hurt, C., & Tyner, W. E. (2011). *What's driving food prices in 2011?* (Issue report). Oak Brook, IL: Farm Foundation.

Adams, R. M., Rosenzweig, C., Peart, R. M., Ritchie, J. T., McCarl, B. A., Glyer, J. D., Curry, R. B., Jones, J. W., Boote, K. J., & Allen, L. H., Jr. (1990). Global climate change and U.S. agriculture. *Nature, 345*(6272), 219–224.

Barnett, J., & O'Neill, S. (2010). Maladaptation. *Global Environmental Change, 20*(2), 211–213.

Bazilian, M., Rogner, H., Howells, M., Hermann, S., Arent, D., Gielen, D., Steduto, P., Mueller, A., Komor, P., & Tol, R. S. J. (2011). Considering the energy, water and food nexus: Towards an integrated modelling approach. *Energy Policy, 39*, 7896–7906.

Brouwer, R. (2000). Environmental value transfer: State of the art and future prospects. *Ecological Economics, 32*, 137–152.

Chambwera, M., Heal, G., Dubeux, C., Hallegatte, S., Leclerc, L., Markandya, A., McCarl, B. A., Mechler, R., & Neumann, J. E. (2014). Economics of adaptation. In *Climate change 2014: Impacts, adaptation, and vulnerability. Part A: Global and sectoral aspects. Contribution of Working Group II to the Fifth Assessment Report of the Intergovernmental Panel on Climate Change*. Cambridge: Cambridge University Press. Chapter 17.

Ciriminna, R., Pina, C. D., Rossi, M., & Pagliaro, M. (2014). Understanding the glycerol market. *European Journal of Lipid Science and Technology, 116*, 1432–1439.

Colby, B. G. (1989). Estimating the value of water in alternative uses. *Natural Resources Journal, 29*, 511.

Hanley, N., Shogren, J. F., & White, B. (2002). *Environmental economics in theory and practice*. New York, NY: Palgrave Macmillan.

Hansen, L., & Ribaudo, M. (2008). *Economic measures of soil conservation benefits* (Technical bulletin number 1922). Washington, DC: USDA Economic Research Service.

Hoff, H. (2011). *Understanding the nexus: Background paper for the Bonn2011 Nexus Conference*. Stockholm: SEI.

Just, R. E., Hueth, D. L., & Schmitz, A. (2008). *Applied welfare economics*. Cheltenham: Edward Elgar Publishing.

King, D. M., & Mazzotta M. (2000). *Benefit transfer method*. Ecosystem valuation. Retrieved from http://www.ecosystemvaluation.org/benefit_transfer.htm#advant.

McCarl, B. A. (2015). Elaborations on climate adaptation in U.S. agriculture. *Choices, 30*, 1–5.

Moschini, G., & Hennessy, D. A. (2001). Uncertainty, risk aversion, and risk management for agricultural producers. *Handbook of Agricultural Economics, 1*, 87–153.

Nordhaus, W. D. (2007). A review of the stern review on the economics of climate change. *Journal of Economic Literature, 45*, 686–702.

Pfeiffer, L., & Lin, C. C. (2014). Does efficient irrigation technology lead to reduced groundwater extraction? Empirical evidence. *Journal of Environmental Economics and Management, 67*, 189–208.

Stavins, R. N. (1995). Transaction costs and tradeable permits. *Journal of Environmental Economics and Management, 29*, 133–148.

Stern, N. H. (2007). *The economics of climate change: The Stern review*. Cambridge: Cambridge University Press.

Tietenberg, T. (2003). The tradable-permits approach to protecting the commons: Lessons for climate change. *Oxford Review of Economic Policy, 19*(3), 400–419.

Weitzman, M. L. (2007). A review of the stern review on the economics of climate change. *Journal of Economic Literature, 45*, 703–724.

Wooley, R. (1999). Process design and costing of bioethanol technology: A tool for determining the status and direction of research and development. *Biotechnology Progress, 15*(5), 794–803.

Young, R. A., & Loomis, J. B. (2014). *Determining the economic value of water: Concepts and methods*. London: Routledge.

Zhuang, J., Liang, Z., Lin, T., & De Guzman, F. (2007). *Theory and practice in the choice of social discount rate for cost-benefit analysis: A survey*. Metro Manila: Asian Development Bank.

Further Reading

Abbott, P., Hurt, C., & Tyner, W. E. (2011). *What's driving food prices in 2011?* (Issue report). Oak Brook, IL: Farm Foundation Retrieved from https://www.farmfoundation.org/news/articlefiles/1742-FoodPrices_web.pdf.

Chambwera, M., Heal, G., Dubeux, C., Hallegatte, S., Leclerc, L., Markandya, A., McCarl, B. A., Mechler, R., & Neumann, J. E. (2014). Economics of adaptation. In *Climate change 2014: Impacts, adaptation, and vulnerability. Part A: Global and sectoral aspects. Contribution of Working Group II to the Fifth Assessment Report of the Intergovernmental Panel on Climate Change*. Cambridge: Cambridge University Press Chapter 17. Retrieved from http://pure.iiasa.ac.at/11126.

Colby, B. G. (1989). Estimating the value of water in alternative uses. *Natural Resources Journal, 29*, 511 Retrieved from http://www.jstor.org/stable/24883538.

Greening, L. A., Greene, D. L., & Difiglio, C. (2000). Energy efficiency and consumption -- The rebound effect -- A survey. *Energy Policy, 28*, 389–401. https://doi.org/10.1016/S0301-4215(00)00021-5.

Heal, G., Barbier, E. B., Boyle, K. J., Covich, A. P., Gloss, S. P., Hershner, C. H., Hoehn, J. P., Pringle, C. M., Polasky, S., Segerson, K., & Shrader-Frechette, K. (2005). *Valuing ecosystem services: Toward better environmental decision-making*. New York, NY: National Academies Press Chapter 4. Retrieved from https://www.nap.edu/read/11139/chapter/6.

McCarl, B. A. (2015). Elaborations on climate adaptation in U.S. Agriculture. *Choices, 30*, 1–5 Retrieved from http://www.choicesmagazine.org/UserFiles/file/cmsarticle_432.pdf.

Moschini, G., & Hennessy, D. A. (2001). Uncertainty, risk aversion, and risk management for agricultural producers. *Handbook of Agricultural Economics, 1*, 87–153. https://doi.org/10.1016/S1574-0072(01)10005-8.

Myers, E., & Myers, D. L. (2011). *Lessons from WPPSS*. Langley, WA: Context Institute Retrieved from https://www.context.org/iclib/ic07/myers/.

Nordhaus, W. D. (2007). A review of the stern review on the economics of climate change. *Journal of Economic Literature, 45*, 686–702 Retrieved from http://www.jstor.org/stable/27646843.

Pfeiffer, L., & Lin, C.-Y. C. (2014). Does efficient irrigation technology lead to reduced groundwater extraction? Empirical evidence. *Journal of Environmental Economics and Management, 67*(2), 189–208. https://doi.org/10.1016/j.jeem.2013.12.002.

Rasul, G., & Sharma, B. (2016). The nexus approach to water–energy–food security: An option for adaptation to climate change. *Climate Policy, 16*, 682–702. https://doi.org/10.1080/14693062.2015.1029865.

Ruttan, V. W., & Hayami, Y. (1984). Toward a theory of induced institutional innovation. *The Journal of Development Studies, 20*, 203–223. https://doi.org/10.1080/00220388408421914.

Chapter 6
International Governance

Antti Belinskij, Kaisa Huhta, Marko Keskinen, Outi Ratamäki, and Peter Saundry

6.1 Introduction

One of the criteria for considering the nexus of food, energy, and water systems are that they are governance-heavy, as noted in Chap. 1. The word *governance* refers here generally to the processes by which organizations and other groups of people make and implement decisions, policies, and rules.

While governance has many definitions, it can generally be seen to consist of two key elements, that is, actors and institutions. Actors include variety of organizations, groups, and individuals that participate and/or have an interest in the governance processes at different scales. They can thus include formal government-related actors such as ministries, international actors such as the United Nations organizations as well as other actors such as private corporations, academia, and civil society organizations. **Institutions**, on the other hand, can be generally defined as persistent social arrangements that shape and regulate actors' behavior and actions. Institutions include both formal institutions (such as policies and laws) and informal institutions (such as norms and traditions).

A. Belinskij (✉) · K. Huhta · O. Ratamäki
Social Sciences and Business Studies, Law School, University of Eastern Finland, Joensuu, Finland
e-mail: antti.belinskij@uef.fi; kaisa.huhta@uef.fi; outi.ratamaki@uef.fi

M. Keskinen
Water & Development Research Group, Aalto University, Espoo, Finland
e-mail: marko.keskinen@aalto.fi

P. Saundry
Energy Policy and Climate, Advanced Academic Programs, Krieger School of Arts and Sciences, Johns Hopkins University, Washington, DC, USA
e-mail: psaundr1@jhu.edu

P. Saundry, B. L. Ruddell (eds.), *The Food-Energy-Water Nexus*, AESS Interdisciplinary Environmental Studies and Sciences Series,
https://doi.org/10.1007/978-3-030-29914-9_6

This chapter focuses on a key formal institution, that is, international law, and also discusses briefly some key actors related to international governance related to food, energy, and water security. While formal laws created and enforced by formal governmental actors (**hard law**) are an important governance tool, it is important to note that many other governance processes and tools exist inside and outside of these more formal government arrangements. Given the context of this book, this chapter looks particularly at the interlinkages between food, energy, and water security in international law. It examines how international treaties and **customary international law** in respect of each of these sectors take into account the other two sectors.

Much of international governance related to food, energy, and water is motivated by the interlinked concepts of food, energy, and water security (see Sect. 3.2). A nexus approach to food, energy, and water security aims to integrate the many different aspects of management and governance of these resources across sectors and scales (Hoff 2011). Such scales range from local to regional and even to a global scale, and an increasing amount of literature looks at the nexus from perspectives of local sociopolitical structures like cities (Chap. 18) and natural structures like watersheds (Chap. 19) which can encompass all or parts of several countries.

While policy and decision-making in relation to each sector should therefore always consider the effects on the other two, international FEW governance has traditionally taken place in separate food, energy, and water "silos." Binding legal instruments and concrete obligations (hard laws) directed at all three sectors are scarce, typically provisions in one sector having only implicit connections to the other two sectors. The importance of the nexus approach and cross-sectoral linkages are more explicitly expressed in non-legally binding (**soft law**) instruments and policy papers.

We argue that, while explicit interlinkages between food, energy, and water are largely missing from international law, legal instruments leave much room for interpretation to enhance the nexus approach. A nexus approach allows for the possibility of achieving synergies and enhancing trade-offs between the water, energy, and food sectors. This is by no means easy and often leads to conflicting views on the most sustainable and equitable uses of our limited natural resources. Integrated FEW governance is thus also inherently political.

The structure of the chapter is as follows. Section 6.2 describes some critical elements of contemporary international FEW governance. Sections 6.3, 6.4, and 6.5 look at these elements from the perspective of each sector along with interlinkages. Section 6.6 presents the conclusions, placing the analysis in a broader context. In addition to the recommendations for further reading and the chapter references, readers should explore treaties and declarations (Appendix C) and relevant case law (included within a section at the end of the chapter).

6.2 Key Elements of International Governance

Governance of our societies has its roots in small, localized groups of people who had their own rules, customs and norms. Over time, larger governance structures at higher scales emerged, creating more stable and formalized social arrangements within areas where they were in effect: this ultimately also lead to the establishment of different nations and their governance. While such larger social arrangements helped to reduce conflicts between localized groups of people, they could remain biased toward some groups and individuals. Further, such nation-wide governance structures could not prevent conflicts between nations, with the two World Wars of the first half of the twentieth century representing the most destructive examples.

As a result, considerable efforts have been made globally to establish international institutions (namely, international norms of behavior, codified into rules and laws) and related international organizations that seek to minimize the main sources of conflict while maximizing the benefits of peaceful cooperation and exchange of goods and services. The League of Nations (1920–1946) can be seen as the first major effort for such a global governance initiative. Yet, its failure to achieve wide acceptance, meaningful influence and, ultimately, to prevent World War II led to a renewed effort to create both organizations and institutions that would be more successful. Since the World War II, an extensive set of international arrangements have reduced (though far from eliminated) conflict and facilitated rapid increase in production of FEW commodities and their trade between nations (Chap. 7). While the United Nations system created in 1945 is the most visible representation of international governance, it is only a part of myriad treaties, institutions, and other arrangements that seek to establish a generally accepted approach for international governance.

6.2.1 *Concepts of International Law*

Governance encompasses more than formal laws created and enforced by formal governments (**hard law**). This is also reflected to the core concepts of international law. International law refers to the "rules and principles governing the relations and dealings of nations with each other, as well as the relations between states and individuals, and relations between international organizations" (Legal Information Institute). As such, international law recognizes:

1. International treaties, conventions and agreements which have legal binding obligations which are considered "**hard law**" because there are specific or "hard" consequences if breached.
2. Agreements without legally binding components to them which are considered "**soft law**." One example of soft law is the use of voluntary commitments called Nationally Determined Contributions under the 2015 Paris Agreement on Climate Change (see Sect. 11.3). In addition to non-legally binding treaties and

conventions, there are many other examples of international soft law such as the resolutions and declarations frequently adopted by the United Nations General Assembly, and most "action plans," "statements of intent," "codes of conduct," and "guidelines."

While soft law obligations are frequently criticized for have "no teeth," they have the advantages of:

(a) providing States greater flexibility in fulfilling commitments;
(b) being more readily agreed to by States; and
(c) creating goals to which governments may be held accountable to by domestic political actors.

For example, it is unlikely that the USA would have accepted hard law requirements under the Paris Agreement because of opposition within the legislative branch of its national government. However, the US commitments had, and have, clear political consequences.

Many international obligations are rooted in established or "customary" practices, referred to as **customary international law**. The International Law Commission states that customary international law is "unwritten law deriving from practice accepted as law. It remains an important source of public international law" (ILC 2018). How customary law is understood and accepted differs between States. Customary law is frequently referred to by courts, jurists, legal scholars, diplomats, and governments as a source of international law. International treaties often seek to codify customary law to bring about formal, clear, and broad agreement about such law.

Both hard and soft international law are frequently (but not always) expressed in international treaties. The 1980 Vienna Convention on the Law of Treaties defines a treaty as "an international agreement concluded between States in written form and governed by international law."

Treaties that can be considered hard law typically include provisions on how a State consents to be bound by the agreement; territorial scope of and reservations to the agreement; how it enters into force; monitoring compliance and consequences of non-compliance; allowing other States to join the agreement; and revoking or modifying (amending) the agreement.

Many international agreements and declarations include non-binding (i.e., soft law) normative statements and goals (that is desirable behaviors, practices, outcomes and objectives) which, over time, can become customary law, and, ultimately, hard law either domestically or internationally.

Note that it is customary to refer to independent countries as "**Nation States**" or simply "States." The use of the word "State" in this Chapter should not be confused with sub-national regions in certain countries like Australia, Brazil, Mexico, and the USA. States are considered "sovereign" in that their governing body of the State has supreme, or ultimate, authority for political decision-making within the territory of the State. There exists a diverse array of relationships between the governing bodies of States and the people who are citizens of the State, some democratic in nature, some not.

The principle of sovereign equality asserts that every sovereign State has the same legal rights as every other sovereign State. According to this principle, a State has a sovereign right to use its territory but, at the same time, it must respect the territorial integrity of other States and their "**correlative rights**" to resources shared with other States, like transboundary groundwater resources.

International law is sometimes divided into public and private components, with **public international law** referring to laws governing the relationship between States, and **private international law** referring to laws governing individuals and organizations (like corporations) when they move across borders and operate in countries other than their home. Because many international treaties have a significant impact on private international law and domestic law (laws internal to countries), it is important not to consider different areas of law to be entirely separate and distinct.

6.2.2 The United Nations System

Following the creation of the United Nations (UN) in 1945, at the end of the second World War, the organization quickly became the most significant body for facilitating and overseeing international cooperation and governance. However, it is essential to remember that the UN:

1. Is just one source of international governance.
2. Has little direct authority of member states; rather, it is best viewed as a "club" where members discuss issues and establish arrangements to address those issues; and whose coercive power only exists when members voluntary enforce agreements individually, or cooperatively on others.

As of 2019, the UN recognizes 193 sovereign States as members and which are represented in the United Nations General Assembly. Two States have observer status, the Holy See and Palestine.

The General Assembly is one of six "principal organs" of the United Nations. The others are the Economic and Social Council, the International Court of Justice, the Security Council, the Trusteeship Council, and the UN Secretariat. There are also a large number of agencies and organizations established by and reporting to different parts of the UN such as the World Bank Group, International Monetary Fund (see Sect. 6.2.3), the World Food Programme, the United Nations Framework Convention on Climate Change, the International Law Commission, and many more. All of these entities collectively are referred to as the United Nations System.

In the United Nations system, the General Assembly adopts multilateral treaties, the International Law Commission promotes the development of international law, and the International Court of Justice, based in city of The Hague, Netherlands, settles disputes between States.

Article 38 of the governing Statute of the International Court of Justice brings together many of the core concepts of international law by what types of laws it applies:

1. The Court, whose function is to decide in accordance with international law such disputes as are submitted to it, shall apply:
    (a) international conventions, whether general or particular, establishing rules expressly recognized by the contesting states;
    (b) international custom, as evidence of a general practice accepted as law;
    (c) the general principles of law recognized by civilized nations;
    (d) subject to the provisions of Article 59, judicial decisions and the teachings of the most highly qualified publicists of the various nations, as subsidiary means for the determination of rules of law.

The International Court of Justice has occasionally been asked to address questions related to the food, energy, or water system (see, e.g., Sect. 6.3.1 below); however, addressing the FEW nexus itself is something most courts have not yet fully grappled with. More discussion of food, energy, or water has often occurred in different pieces of the UN system (Fig. 6.1).

The UN system includes a number of sub-organizations, including, but not limited to:

- Food and Agriculture Organization of the United Nations (created in 1960) "works in partnership with governments and other development actors at global, regional and national levels to develop supportive policy and institutional environments" to lead "international efforts to defeat hunger." (FAO) The FAO includes work on fisheries and agriculture.
- World Food Programme (1961) delivers food assistance in emergencies and works with communities to improve nutrition and build resilience.
- International Atomic Energy Agency was created to support the peaceful use of atomic energy, including nuclear power.
- United Nations Development Programme (1965) "works to eradicate poverty and reduce inequalities through the sustainable development of nations" (UNDP).
- United National Environment Program (1972) seeks to set "the global environmental agenda, promotes the coherent implementation of the environmental dimension of sustainable development within the United Nations system, and serves as an authoritative advocate for the global environment." (UNEP).
- United National Framework Convention on Climate Change (1992) facilitates international negotiations on climate change (see Chap. 11).
- UN-Energy coordinates the activities of many UN agencies and offices on energy.
- UN-Water coordinates the activities of many UN agencies and offices on water.
- UN Human Rights Council promotes and protects human rights around the world.

In addition, the World Bank Group and International Monetary Fund (see Sect. 6.2.3) have significant impacts on the FEW nexus.

6.2.3 World Trade Organization and GATT

In Chap. 7, we explore the important role that trade plays in supporting food, energy, and water security. Trade governance is therefore an important part of international governance related to FEW systems.

The modern approach to regulating international trade was profound shaped by the same process that led to the creation of the United Nations system following World War II which included a belief that the way in which had nations managed their trade and currencies during the interwar years were a significant factor in making the Great Depression as harmful and long-lasting as it was, and created conditions that help lead to the second global war within a quarter century of the first. In particular, trade barriers such as tariffs, import quotas, and subsidies were viewed as exacerbating economic and social instability.

At the famous 1944 Bretton Woods conference, proposals were endorsed to create three international organizations to facilitate economic growth:

1. The *International Monetary Fund* (IMF) to promote financial stability by having countries cooperate in how they managed their currency exchange rates and financial flows; and to support nations which face problems in their payments of international debts, thereby avoiding broader international financial crises.
2. The International Bank for Reconstruction and Development, to provide debt-financing and aid reconstruction following the war. This Bank subsequently became one of five institutions within the *World Bank Group* (WBG). The World Bank expansion included additional components primarily aimed at supporting development in low-income countries, including the:

   (a) International Finance Corporation (1956) to work with the private sector;
   (b) International Development Association (1960) to provide interest-free loans or grants;
   (c) International Centre for Settlement of Investment Disputes (1965); and
   (d) Multilateral Investment Guarantee Agency (1988) to provide insurance for investments.

   Data aggregated and disseminated by the World Bank has been used throughout this book.
3. The International Trade Organization (ITO) to establish rules for international trade that would reduce barrier to international trade. Because the ITO was viewed as a threat to domestic policy-making in the USA, the treaty to establish it failed to be ratified in that country. However, a weaker **General Agreement on Tariffs and Trade** or "GATT" was endorsed and has provided the most significant framework for regulating international trade.

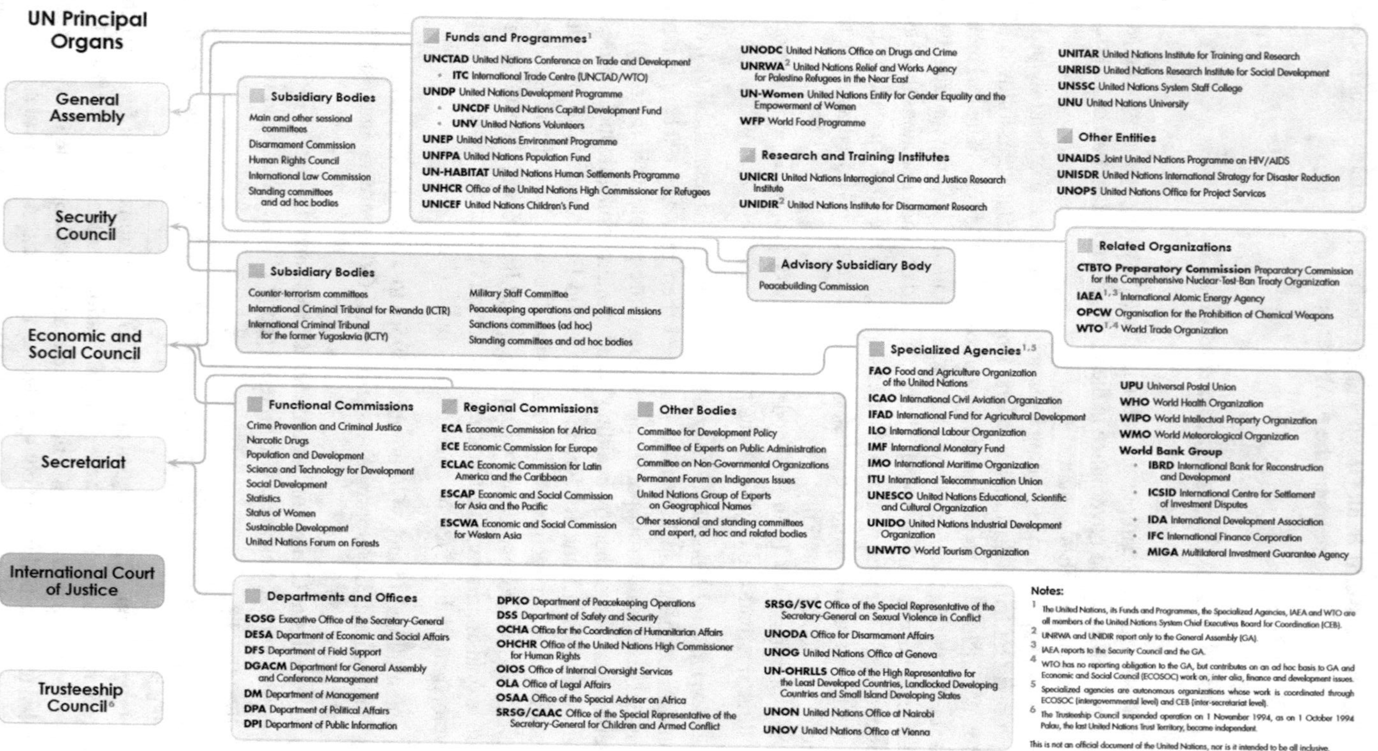

Fig. 6.1 The United Nations system. (Source: United Nations)

Following the original 1947 legal agreement between 23 countries, GATT was revised seven times in "rounds" of international negotiations in 1949, 1951, 1955–1956, 1960–1962, 1962–1967, 1973–1979, and 1986–1994 (called the "Uruguay Round"). Each round of GATT negotiations addressed a wide array of issues, further reduced tariffs, and included more countries.

In 1995, GATT was subsumed within the **World Trade Organization** which became, in practice, a club for the GATT signatories. As of 2019, the 164 member governments of the WTO represent 98% of world trade. The WTO facilitates member fulfillment of pledges and conflict resolution, as well as ongoing negotiations (the "Doha Round" began in 2001 and is ongoing).

When the United Nations was established in 1945, the IMF and World Bank became agencies of the UN. The World Trade Organization is not an agency of the UN.

At the core of the GATT/WTO arrangement is that all members commit to treating imports from all other members equally. That is, every member must subject every other member to the same trade conditions. This is referred to as "most favored nation" treatment because every GATT/WTO member should be allowed to trade under the same conditions as the most favorable conditions of any member. Further, members should have access to other markets on a reciprocal basis. Finally, members are required to be transparent about their trade restrictions.

Membership in the WTO does not preclude countries being involved in other trade agreements such as the European Union and the **North American Free Trade Agreement** (NAFTA) provided that they do not conflict with their GATT/WTO commitments.

GATT and WTO have significant impacts on food, and more recently energy. Each of these is addressed below.

6.2.3.1 Food

From the outset, agriculture has been treated differently to other products under the GATT/WTO. Because food insecurity (see Sect. 3.2) has often been associated with social instability, nations have long treated agriculture differently to other areas of their economy. Nations have used tariffs, import quotas, and subsidies to achieve stable food prices and to mitigate large swings in crop prices harmful to domestic farmers. In addition, many countries have used the same tools to support rural economies impacted the movement of capital and people to industrialized urban centers—a social policy goal.

The importance of food and agricultural policies, especially in the USA, resulted in major exceptions to the application of GATT to agricultural and fishery products. In particular, the use of subsidies and import restrictions were allowed under GATT in most circumstances. The effect of such exemption led to subsidized agriculture in wealthier countries and frequently to large surpluses of certain crops. These surpluses were then exported with the help of export subsidies simultaneously providing cheap food to developing nations and economically damaging the agriculture sector by lowering the cost of domestically produced products. The USA and the European Union (because of its Common Agricultural Policy "CAP") were the chief users of such policies and GATT exemptions and were the most common litigants in GATT disputes.

While GATT negotiations attempted to address the exceptions for agriculture throughout its first forty years, it was not until the Uruguay Round (1986–1994) that serious changes were advanced by a group of agricultural exporting nations known as the **Cairns Group**. This effort, ultimately led to the adoption of the Agreement on Agriculture (1995). We will return to the Agreement on Agriculture in Sect. 6.5.

Since 2001, the WTO has been facilitating the Doha Development Round of trade negotiations with a focus on the needs of developing nations. In 2008, negotiations stalled over disagreement between developed nations and large developing nations about agricultural subsidies.

While there may be sufficient food for the global population, there remains tremendous inequity where some populations struggle with obesity while others starve. Thus, world trade in food supplies has a long way to go to facilitate food security for all.

6.2.3.2 Energy

Energy was not a priority in early GATT negotiations. This was likely connected to the fact that energy trading was dominated by a combination of a small number of large multinational corporations and state-owned corporations.

Following the first oil crisis of 1973, the member nations of the Organisation for Economic Co-operation and Development (OECD) established the International Energy Agency (IEA) to provide data and analysis to help them coordinate a collective response to such crisis. Today the IEA "examines the full spectrum of energy issues including oil, gas and coal supply and demand, renewable energy technologies, electricity markets, energy efficiency, access to energy, demand side management and much more. Through its work, the IEA advocates policies that will enhance the reliability, affordability and sustainability of energy in its 30 member countries and beyond." (IEA).

Following the political transitions in Eastern Europe (1989–1991) that marked the end of the Cold War, the significance of the energy resources of those countries to those in Western Europe resulted in the adoption of the European Energy Charter (1991) which provided a political framework for developing a treaty. An Energy Charter Treaty (ECT) was adopted in 1994 and entered into force in 1998. A subsequent International Energy Charter was adopted in 2015 by 64 states to provide the framework process to develop a larger treaty. The Treaty addresses:

- the protection of foreign investments;
- non-discrimination in trading conditions (based on WTO rules);
- ensuring stable movements of energy across international borders;

- dispute resolution;
- advancing energy efficiency; and
- minimizing negative environmental impacts.

We will return to the Energy Charter in Sect. 6.4 and recognize its importance to energy trading in Sect. 7.4, and its role in Energy arbitrations in Chap. 20.

6.2.4 *Human Rights*

Before addressing international law with respect to food, energy, and water, we will explore a foundational issue for many laws, that of human rights.

Over the past several centuries, the concept of "rights" has developed both as a human construct and as a social technology. The difference is manifested in the distinction between the U.S. Declaration of Independence ("all men are endowed by their creator with certain unalienable rights") versus the Bill of Rights in the US Constitution that enumerates the rights which citizens enjoy under the federal government.

The difference continues to be a source of tension as activist advocate a "right to food" or a "right to water" or a "right to energy," and politicians debate how to embody such a right in laws or public programs.

6.2.4.1 Food

Food has been explicitly recognized as a human right in various ways (Kent 2005; Mechlem 2004). Article 25 of the Universal Declaration of Human Rights adopted by the United Nations General Assembly in 1948 includes the following language:

> Everyone has the right to a standard of living adequate for the health and well-being of himself and of his family, including food, clothing, housing and medical care and necessary social services, and the right to security in the event of unemployment, sickness, disability, widowhood, old age or other lack of livelihood in circumstances beyond his control.

These rights were expanded in Article 11 of 1966 International Covenant on the Economic, Social and Cultural Rights (ICESCR) which entered into force in 1976:

1. The States Parties to the present Covenant recognize the right of everyone to an adequate standard of living for himself and his family, including adequate food, clothing and housing, and to the continuous improvement of living conditions. The States Parties will take appropriate steps to ensure the realization of this right, recognizing to this effect the essential importance of international cooperation based on free consent.
2. The States Parties to the present Covenant, recognizing the fundamental right of everyone to be free from hunger, shall take, individually and through international cooperation, the measures, including specific programs, which are needed:

(a) To improve methods of production, conservation, and distribution of food by making full use of technical and scientific knowledge, by disseminating knowledge of the principles of nutrition and by developing or reforming agrarian systems in such a way as to achieve the most efficient development and utilization of natural resources;
(b) Taking into account the problems of both food-importing and food-exporting countries, to ensure an equitable distribution of world food supplies in relation to need.

Further, the World Conference on *Human Rights* declared that "*human rights* derive from the *dignity* and worth *inherent* in the *human person*."

While these statements provide an important foundation for international FEW governance, it is important to note that energy and water were not explicitly recognized, and that no enforceable obligations are included (i.e., they constitute soft law).

In the 1990s, it was observed that the increase in global food production per person made possible by the Green Revolution had not reduced the total number of people without adequate food supplies. Amartya Sen (see Sect. 3.4) and others argued that the problem was a result of poor governance associated with social injustice, and failures in governments and other important institutions. The 1996 World Summit of Food Security prosed that tools be created to allow people to force their governments to take concrete steps to enforce the right to food.

In 1999, the UN Committee on Economic, Social and Cultural Rights (CESCR), a group of experts that monitored implementation of the Covenant by countries that were parties to the Covenant issued statement on what the right entailed. They stated,

> States have a core obligation to take the necessary action to mitigate and alleviate hunger as provided for in paragraph 2 of article 11, even in times of natural or other disasters…
>
> Every State is obliged to ensure for everyone under its jurisdiction access to the minimum essential food which is sufficient, nutritionally adequate and safe, to ensure their freedom from hunger…

> The right to adequate food, like any other human right, imposes three types or levels of obligations on States parties: the obligations to respect, to protect and to fulfil. In turn, the obligation to fulfil incorporates both an obligation to facilitate and an obligation to provide. The obligation to respect existing access to adequate food requires States parties not to take any measures that result in preventing such access. The obligation to protect requires measures by the State to ensure that enterprises or individuals do not deprive individuals of their access to adequate food. The obligation to fulfil (facilitate) means the State must proactively engage in activities intended to strengthen people's access to and utilization of resources and means to ensure their livelihood, including food security. Finally, whenever an individual or group is unable, for reasons beyond their control, to enjoy the right to adequate food by the means at their disposal, States have the obligation to fulfil (provide) that right directly. This obligation also applies for persons who are victims of natural or other disasters.

The obligations to respect, protect, and fulfill form the core obligations of governments. These obligations were elaborated upon in a 2004 FAO document titled Voluntary Guidelines to support the Progressive Realization of the Right to Adequate Food in the Context of National Food Security (also called "Right to Food Guidelines", Food and Agricultural Organization 2005).

Civil society organizations have used these statements of government obligations to press for actions in courts. In one notable case, in 2001, responding concerns that the state government of Rajasthan in a northern Indian was not using stockpiles of food reserves to address shortages and high prices, several civil society organizations petitioned the Supreme Court of India to issue directives to the national and state governments of how they should address problems of malnutrition. The Supreme Court of India issued several ordinances in response to the case.

6.2.4.2 Water

In 2002, The UN Committee on Economic, Social and Cultural Rights (CESCR) issued a statement on the right to water, based upon Article 11 and 12 of the International Covenant on the Economic, Social and Cultural Rights (ICESCR), as well as other treaties. Article 12 of ICESCR addresses "the right of everyone to the enjoyment of the highest attainable standard of physical and mental health." CESCR stated,

> Water is a limited natural resource and a public good fundamental for life and health. The human right to water is indispensable for leading a life in human dignity. It is a prerequisite for the realization of other human rights. The Committee has been confronted continually with the widespread denial of the right to water in developing as well as developed countries…
>
> The human right to water entitles everyone to sufficient, safe, acceptable, physically accessible and affordable water for personal and domestic uses. An adequate amount of safe water is necessary to prevent death from dehydration, to reduce the risk of water-related disease and to provide for consumption, cooking, personal and domestic hygienic requirements…
>
> The right to water contains both freedoms and entitlements. The freedoms include the right to maintain access to existing water supplies necessary for the right to water, and the right to be free from interference, such as the right to be free from arbitrary disconnections or contamination of water supplies. By contrast, the entitlements include the right to a system of water supply and management that provides equality of opportunity for people to enjoy the right to water.

CESCR also stated that "[w]hile the adequacy of water required for the right to water may vary according to different conditions, the following factors apply in all circumstances:" availability, quality, and accessibility. Accessibility includes physical accessibility, economic accessibility, non-discrimination, and information accessibility. Here we see many of the elements of water security discussed in Sect. 3.4.

As with the right to food, the obligations to respect, protect, and fulfill form the core obligations of governments. CESCR goes further is asserting an international obligation:

> To comply with their international obligations in relation to the right to water, States parties have to respect the enjoyment of the right in other countries. International cooperation requires States parties to refrain from actions that interfere, directly or indirectly, with the enjoyment of the right to water in other countries. Any activities undertaken within the State

> party's jurisdiction should not deprive another country of the ability to realize the right to water for persons in its jurisdiction

This includes remaining from using water "as an instrument of political and economic pressure" such as preventing the supply of water reaching another country; or, allowing communities of citizens from violating the water rights of their counterparts in other countries.

Finally, CESCR also connected the right to water to the right to food:

> The Committee notes the importance of ensuring sustainable access to water resources for agriculture to realize the right to adequate food. Attention should be given to ensuring that disadvantaged and marginalized farmers, including women farmers, have equitable access to water and water management systems, including sustainable rain harvesting and irrigation technology.

In addition, food and water are very closely linked with each other in human rights law. For example, the 1979 Convention on the Elimination of all Forms of Discrimination against Women (Article 14) calls on parties to ensure that women have equal access to "adequate and healthy living conditions" where adequate living conditions also include electricity.

Finally, there is a growing movement under human rights law to give rivers themselves the status of "legal persons." For example, New Zealand's parliament passed an act giving the Whanganui River and ecosystem legal standing, guaranteeing its "health and well-being." Also building of human rights law, Ecuador enshrined in its constitution the rights of nature. A court in India ruled that the Ganges and Yamuna rivers have the "status of a legal person, with all the corresponding rights, duties and liabilities… in order to preserve and conserve them."

6.2.4.3 Energy

While discussion of the right to energy is in its infancy compared to that of the right to food or the right to water, a growing body of literature addresses the concepts of energy poverty, energy justice, and access to energy as a precondition for ensuring socio-economic human rights.

The connection between access to energy and human rights was first acknowledged in the World Commission on Environment and Development report in 1987 (Brundtland report), which stated that energy services are a crucial input to the primary development challenge of providing, for example, adequate food and water. Among other things, energy enables cooking, piped water and sewerage facilities, and fuels agriculture.

Although there is no explicit mention of energy in the global human rights treaties, it has been strongly argued that access to energy is a fundamental precondition for the fulfillment of human rights obligations (Bradbrook et al. 2008).

Regulating energy is a matter of reconciling mutually conflicting interests of security, affordability, and sustainability. The food–energy–water nexus in international energy law can be seen as an aspect of global sustainability efforts. However, pursuing sustainability as an objective becomes challenging when it conflicts with the objectives of energy security or the affordability of energy. This is acknowledged in the ECT, for example, which states that contracting parties should strive to minimize environmentally harmful impacts but should do so economically efficiently and take safety into account (Article 19(1) of the ECT). Interruptions in energy supply are considered politically unacceptable and, as a result, governments are unlikely to allow interruptions in the interest of pursuing sustainability objectives.

6.3 International Water Law

International water law applies to the uses and protection of international freshwater resources.

Approximately half of the earth's surface, containing about 40 percent of the world's population (3.1 Billion), is contained within 263 transboundary lake and river basins. There are hundreds of transboundary groundwater aquifers. International waters constitute an estimated 60% of global freshwater flow and have a great impact on economic development, poverty reduction, and the attainment of the Sustainable Development Goals (see Sect. 3.7).

Early international water agreements concentrated primarily on the regulation of navigation and fishing. However, today, international water law mainly relates to water uses such as hydropower production and irrigation.

6.3.1 Two Conventions

International water law was codified by the 1966 International Law Association (ILA) non-binding **Helsinki Rules on the Uses of the Waters of International Rivers** (ILA 1967).

Currently, there are two global water conventions in force:

1. The 1992 Convention on the Protection and Use of Transboundary Watercourses and International Lakes (**ECE Water Convention** or ECEWC). This Convention was established as a regional convention under the auspices of the United Nations Economic Commission for Europe (UNECE) and went into effect in 1996. It was amended from a regional convention to a global convention in 2013.
2. The 1997 Convention on the Law of the Non-Navigational Uses of International Watercourses (**United Nations Watercourses Convention** or "UNWC") which entered into force in 2014.

In addition, States have concluded hundreds of bilateral and multilateral water agreements dating back centuries, although many transboundary water bodies remain outside the scope of these agreements. The provisions of the two global water conventions are compatible and mostly complementary (McCaffrey 2014).

Despite the global scope of their application, the Water Convention and the Watercourses Convention have major gaps in their geographical coverage. The former has only 43 and the latter 36 parties as of mid-2018. However, it is often considered that the conventions provide authoritative terms of reference for customary international water law (McCaffrey 2001).

6.3.2 Principles of International Water Law

Three principles, largely based on the principle of limited territorial **sovereignty**, provide the substantive and procedural basis of international water law:

6.3.2.1 The Principle of Equitable and Reasonable Utilization

The principle of equitable and reasonable utilization requires States to utilize and develop an international watercourse in an equitable and reasonable manner to attain optimal and sustainable utilization thereof and benefits therefrom. For that purpose, States have to take into account all relevant factors and circumstances such as factors of a natural character, the social and economic needs of the States concerned, and the effects of its water use on other States and reach a conclusion on the basis of the whole.

The principle of equitable and reasonable utilization is embodied in Article 2(2)(c) of the ECE Water Convention and Article 5 of the UN Watercourses Convention. Article 6 of the Watercourses Convention recognizes this principle and

> "requires taking into account all relevant factors and circumstances, including:
>
> (a) Geographic, hydrographic, hydrological, climatic, ecological and other factors of a natural character;
> (b) The social and economic needs of the watercourse states concerned;
> (c) The population dependent on the watercourse in each watercourse state;
> (d) The effects of the use or uses of the watercourses in one watercourse State on other watercourse states;
> (e) Existing and potential uses of the watercourse;
> (f) Conservation, protection, development and economy of use of the water resources of the watercourse and the costs of measures taken to that effect;
> (g) The availability of alternatives, of comparable value, to a particular planned or existing use."

In 1997, the International Court of Justice (ICJ) referred to the principle of equitable and reasonable utilization in its resolution of a dispute concerning a stretch of the Danube River which defines the border between Hungary and Slovakia. Under the 1977 Budapest Treaty, Hungary and Czechoslovakia agreed to jointly build two dams on the river, near Gabčikovo, Slovakia and Nagymaros, Hungary, to provide flood control and hydroelectric power. The Treaty was inherited by of the Slovak Republic following the 1993 dissolution of Czechoslovakia. When Hungary withdrew from the agreement because of environmental concerns, the Slovak Republic decided to proceed independently with a modified project. The IJC ruled that:

> "The Court considers that Czechoslovakia, by unilaterally assuming control of a shared resource, and thereby depriving Hungary of its right to an equitable and reasonable share of the natural resources of the Danube—with the continuing effects of the diversion of these waters on the ecology of the riparian area of the Szigetkoz—failed to respect the proportionality which is required by international law." (ICJ 1997).

6.3.2.2 The No-Harm Rule

According to the no-harm rule, States must take all appropriate measures to prevent the causing of significant harm to other watercourse States (Article 2 of the ECE Water Convention and Article 5 of the UN Watercourses Convention). The no-harm rule provides an obligation of conduct (take all appropriate measures), not an obligation to reach a fixed result. For example, a State may need to take legislative measures or prevent illegal activities in its territory to fulfill its obligation. The no-harm rule requires a level of care expected from a reasonable government and must be balanced with the degree of risk of transboundary harm (UNECE 2013).

6.3.2.3 Principle of Cooperation

Last but definitely not least, the **principle of cooperation** aims to enhance cooperation between watercourse States to attain the substantive objectives of the principle of equitable and reasonable utilization and the no-harm rule. According to the water conventions, States sharing international waters have to cooperate on the basis of sovereign equality. In order to cooperate, States need bilateral and multilateral agreements and joint bodies or other arrangements. In concrete, cooperation may include joint monitoring and action programs, alarm procedures and the exchange of information on existing and planned uses.

Both the ECE Water Convention and the UN Watercourses Convention require cooperation between States that share international water resources and provide guidance toward for that cooperation (Articles 2 and 9 of the ECE Water Convention and Articles 7 and 8 of the UN Watercourses Convention). While the UN Watercourses Convention provides a general obligation to cooperate, the Water Convention is more detailed and demanding in this respect.

In 2006, Argentina instituted proceeding before the International Court of Justice against Uruguay for violating a 1975 treaty between the countries concerning the Uruguay River which constitutes the boundary between the two countries. Under the Statute of the River Uruguay, the two nations agreed to coordinate with each other on activities impacting the river. Article 60 of the Statute stated that any disagreement which could not be resolved through direct negotiations could be referred to the ICJ. Argentina asserted that Uruguay had unilaterally approved the building of two pulp mills on the river in 2003 and 2005 without appropriate notification and consultation. The dispute between to two countries escalated with public protests by Argentinian citizens blockading a bridge across the river connecting the two countries. One of the two pulp mills was canceled. In its 2010 ruling in the case of Pulp Mills on the River Uruguay (Argentina v. Uruguay), the ICJ declared that Uruguay has not notified the Commission established under the Statute to monitor the river. In addition, the Court ruled that general international law requires an environmental impact assessment when there is a risk that an industrial project may cause a significant adverse impact in a transboundary context (McIntyre 2011). However, the Court concluded that Uruguay had appropriately studied pollution issues, that they were not sufficient to violate the Statute between the two countries, and the second pulp mill could continue to operate.

6.3.3 Interlinkages to Energy and Food

Neither the ECE Water Convention nor the UN Watercourses Convention explicitly mentions energy or food.

However, international water law and various bilateral and multilateral water agreements aim to balance the uses of international water resources between different States and different interests such as hydropower production and irrigation for agriculture. As a result, the FEW-ecosystem nexus has been one of the main areas of focus within the ECE Water Convention regime for many years.

The UN Watercourses Convention provides a non-exhaustive list of factors relevant to the application of the principle of equitable and reasonable utilization. These factors encapsulate many interests related to the nexus approach, such as the social and economic needs of watercourse States, the population dependent on the watercourse, the existing and potential uses of the water, as well as the availability of alternatives to a planned or existing use.

In determining what constitutes a reasonable and equitable use between States sharing an international watercourse, all relevant factors must be considered. For example, the production of hydropower might endanger agricultural uses relevant to the social and economic needs of other watercourse States, and alternative energy sources may be viable.

The UN Watercourses Convention does not provide for any specific order of priority between different uses of international waters. However, conflict between different water uses must be resolved with particular regard being given to the requirements of vital human needs (Article 10). Vital human needs include drinking water and water for food production (ILC 1994).

The no-harm rule requires that States take all appropriate measures to prevent causing significant harm such as pollution or thermal consequences from energy production to other watercourse States (ILC 1994). This rule must be applied in tandem with the principle of equitable and reasonable utilization and requires diligence on the part of a State to conduct itself in a way that is generally considered appropriate and proportional, taking into account the case-specific risks of transboundary impacts. Although the no-harm rule involves flexible and broadly defined phrases such as "significant harm" and "all appropriate measures" (UNECE 2013), it directs decision-makers to consider how water uses affect food energy and water security in other watercourse States.

All in all, the substantive basis of international water law—comprising the principle of equitable and reasonable utilization and the no-harm rule—leaves wide discretion to watercourse States to decide on the reconciliation of water uses. From the perspective of the nexus approach, this is both an advantage and a disadvantage. On the one hand, a wide discretion offers the possibility to integrate the energy and food sectors in the decision-making processes of international water law. The UN Watercourses Convention encourages this integration by listing various factors relevant to equitable and reasonable utilization. On the other hand, wide discretion and flexible concepts also make it possible to concentrate only on the water sector despite its obvious interlinkages with energy and food.

Climate variability and climate change (Chap. 11) further underline the need for the nexus approach in the management of international watercourses (UNECE 2015, p. 2). (See also Chap. 20 on managing conflict). The Paris Agreement—which enhances the implementation of the UN Framework Convention on Climate Change—aims, for example, to increase the ability to adapt to climate change, in which water management plays a pivotal role (Article 2 of the Paris Agreement). In this regard, watercourse States need to find common ground to reconcile their different sectoral and development priorities.

Joint river basin commissions and other joint bodies for cooperation in the management of international waters may well provide a platform for the implementation of the nexus approach between neighboring countries (see Sect. 8.1.1, and the case studies of Chap. 19).

In order to support the implementation of the nexus approach in international water cooperation, the UNECE has developed procedural footsteps for the nexus assessment in transboundary basins. They consist of:

1. the identification of basin conditions and its socioeconomic context as well as the key sectors and stakeholders;
2. an analysis of the key sectors;
3. the identification of intersectoral issues;
4. nexus dialogue; and
5. the identification of synergies across the sectors and countries.

The nexus assessment is well in line with the principle of cooperation under international water law, which obligates States to cooperate on the basis of sovereign equality, territorial integrity, mutual benefit, and good faith and with the help of joint bodies or other joint mechanisms. The two global water conventions also include specific provisions on knowledge mobilization between watercourse States, such as the requirements on the exchange of data on the condition of an international watercourse and information on issues covered by international water law.

All in all, States should consider the intersectoral implications of water uses in transboundary water cooperation in order to implement the nexus approach. The challenge is, however, how to reconcile water uses between the food, energy, and water sectors and in line with the principle of equitable and reasonable utilization to achieve benefits, capitalize synergies and address trade-offs. This requires close collaboration between watercourse States as well as the identification of community interests and the mutual benefits of cooperation. Watercourse States should be able to negotiate and enter into mutual gain agreements in order to develop joint opportunities in the management of international watercourses (Grzybowski et al. 2010). The nexus assessment methodology contained in the UNECE presents a promising path to support States in achieving the objectives of the nexus approach within the context of international water law.

6.4 International Energy Law

6.4.1 Main Characteristics

Energy law address a variety of issues, including resources. Regulating markets and energy efficiency as well as on ensuring security of supply, i.e., the uninterrupted availability of affordable energy. The importance of resources is captured in one definition of energy law as the 'allocation of rights and duties concerning the exploitation of all energy resources between individuals, between individuals and the government, between governments and between States' (Bradbrook 1996, p. 194).

The sovereignty over and the exploitation of energy resources has traditionally been a national matter and subject to strong national protection. The permanent sovereignty over natural resources was established by United Nations General Assembly Resolution 1803 (XVII) in 1962 (Declaration on Permanent Sovereignty over Natural Resources). Furthermore, issues such as energy security and energy efficiency have commonly been regulated nationally and subjected to strong national control. Therefore, the traditional legal approach to energy has also been national or subnational, not international.

In parallel with the national development of energy law, the growing internationalization of trade in energy (see Sect. 7.4) has resulted in the internationalization of energy law and growing international energy governance. This body of legal norms is not, however, a clearly distinguishable area of law but a rather fragmented array of customary, regional, national and international rules that borrow well-established concepts from international investment law, international trade law, and international environmental law and contract law, just to name a few sources (Talus 2014).

A limited number of sector-specific international legal instruments directly address energy, of which the Energy Charter Treaty (ECT) is globally the most well-known. As noted above (see Sect. 6.2.3), the ECT is a multilateral trade treaty, the purpose of which is to promote long-term cooperation in the energy sector (Article 2 of ECT). In common with the vast majority of contemporary energy trade, it relies on open and competitive markets to ensure the objectives of energy security, sustainable development, and sovereignty over energy resources (Article 3 and 6). The ECT not only introduces well-established rules of international investment law such as non-discrimination and the protection of foreign investment to the energy sector but also provides for a system of dispute resolution. The increasing body of case law generated through this dispute resolution system is a significant source of international energy law along with the ECT itself.

In addition to the ECT, there are international treaties that address specific issues within the energy sector. Safety and liability in the nuclear energy sector have been addressed through numerous international legal instruments and institutions like the International Atomic Energy Agency. The International Convention for the Prevention of Pollution from Ships (MARPOL) includes significant provisions on the design and operation of oil tankers. Furthermore, the international climate regime directly addresses energy efficiency, the promotion of renewable energy sources and the progressive reduction of public subsidies to greenhouse-gas-emitting sectors, for example.

Alongside these sector-specific legal instruments, there is a body of international law, referred to as *lex petrolea* which can be understood as both the application of international law to the petroleum sector and as sector-specific international rules adapted to the specificities of the petroleum industry (Martin 2012).

In addition to the international law that specifically addresses the energy sector, international investment law and trade law have also proved to have global significance in the energy sector. The application of WTO law to the energy sector, in particular, has attracted a great deal of academic attention. Overall, the share of energy-related disputes in international arbitration is significant, which indicates the economic importance of trade and investment in the energy sector (see Chap. 20 for more on arbitration).

In addition to the multilateral investment treaties, there is an abundance of bilateral investment treaties that apply to energy.

As demonstrated in the discussion above, it is clear that there is no single international legislative instrument or a clear body of international law that could be categorized as constituting international energy law. However, there is firstly a fragmented body of international law that addresses specific issues in the energy sector, and secondly a large body of private and public international law that is applicable to and has particular relevance in the energy sector, albeit it does not necessarily directly address energy issues.

Many of the energy-specific legal instruments currently in force aim to balance the oft-conflicting interests of securing the availability of energy on affordable and equitable terms without compromising sustainability (Wyman 2015). However, the number of provisions that address issues of sustainability within these legal instruments is not abundant, irrespective of whether the instruments in question relate to investment, exploration, production, trade or the transmission of energy. Against this background, it is perhaps not surprising that the existing texts of international energy law contain almost no explicit legal references to the food–energy–water nexus. Nevertheless, it is possible to identify legal linkages with water and food in international energy law, as discussed below.

6.4.2 Interlinkages to Water and Food

The prima facie absence of references to the water–energy–food nexus in international energy law does not by any means indicate that its substantive interconnections with the water and food sectors are sparse. Extracting, refining, and processing fossil fuels require large amounts of water, and as governments look further afield to find new reserves of oil and gas, questions arise on such issues such as how to project the Arctic should energy resources be developed there (see Sect. 6.4.2.2).

Global trends in the energy markets give rise to new connections between energy activities and the water and food sectors. For example, the potential impacts of hydraulic fracturing on groundwater have been the subject of heated global debate.

In Sect. 6.2.4, we noted the role of human rights with respect to energy. Two additional areas of international law further illustrate how legal obligations can be imposed on energy governance with impact on water and food.

6.4.2.1 Environmental Law

Environmental provisions included in international legal instruments concerning energy are the connecting legal element between energy, food, and water. For example, Article 19 of the ECT imposes an obligation on the parties to strive to minimize harmful environmental impacts, either within or outside of their territory, that are caused by operations within their energy cycle. ECT refers to the energy cycle as "the entire energy chain, including activities related to prospecting for, exploration, production, conversion, storage, transport, distribution and consumption of the various forms of energy, and the treatment and disposal of wastes, as well as the decommissioning, cessation or closure of these activities" (Article 19(3)(a)).

Environmental impacts are defined as any effects on the environment caused by a given activity, including human health and safety, soil and water as well as interactions among different environmental factors (Article 19(3)(a)).

Furthermore, Article 5 of the Energy Efficiency Protocol to the ECT specifies that the parties are under an obligation to formulate strategies and policy aims for improving energy efficiency and thereby reducing environmental impacts of the energy cycle. Although the formulation 'strive to minimize harmful environmental impacts' imposes no concrete obligation to include considerations in respect of water or agriculture, Article 19, nevertheless, enables contracting parties to take into account the potential adverse effects of energy cycles on the water and food sectors.

Similar provisions can also be found in general trade agreements. The GATT/WTO allows exemptions from the fundamental rules of the agreement in the interests of human, animal or plant protection. These exemptions can be invoked under certain terms to conserve exhaustible natural resources or to prevent critical shortages in foodstuffs (Articles XI and XX of the GATT).

6.4.2.2 International Climate Regime

The international climate regime addresses climate change and its impacts and, therefore, is the overarching legal discipline that connects the effects of increasing global energy consumption—and associated greenhouse gas emissions—to the availability of and access to food and water. The United Nations Framework Convention on Climate Change (UNFCCC) entered into force on 21 March 1994 (see Chap. 11).

Global warming is leading to loss of Arctic sea ice. As a result, new shipping channels are opening up, resulting in more opportunities for fishing and navigation. In addition, opportunities are opening up for oil and gas drilling. However, given the challenging conditions, concerns are mounting about how to address hazards of more people, more ships, and more industrial activity in the region. Recent rescue operations of grounded ships off both the coasts of Alaska and Norway highlighted the challenging of even rescuing people, let alone cleaning up an oil spill. The risk of an oil spill is to local areas but also to global fishing stocks.

The global move towards the decarbonization of the energy sector raises new issues in connection with the nexus, such as ensuring a balance between biofuels, food production and water use or between sufficient hydropower production and the protection of surrounding ecosystems. For example, following the approval of two small hydropower plants on the Schwartz Sulm River in Austria, the European Commission filed a case with the European Court of Justice. The Commission argued that the projects violated Austria's obligations under the EU Water Framework Directive to "prevent deterioration of the status of all bodies of surface water." In 2016, the Court ruled that sufficient environmental study had taken place to fulfill the obligations of the Directive (see C-346/14 European Commission v Republic of Austria 2016).

This connection between climate, energy and food was most recently addressed in the Paris Agreement, which entered into force in late 2016, but the matter had already been acknowledged in the 1994 United Nations Framework Convention on Climate Change. Article 2 of the Paris Agreement, for example, aims to strengthen the global response to climate change by, among other things, increasing the ability to adapt to climate change, foster climate resilience and low greenhouse gas emissions development in a manner that *does not threaten food production* (Article 4 of the UNFCCC).

6.5 International Food Law

6.5.1 Main Characteristics

Food is one of the most heavily regulated social and economic sectors. Much of this regulation falls under the areas of international trade law and human rights law (see Sect. 6.2.4 above). Food production, especially livestock farming, is a very intensive business from the point of view of land-use, water, and energy. For example, direct and indirect emissions from livestock sector globally are responsible for approximately 14.5% of greenhouse gas emissions measured in CO2 equivalent (Gerber et al. 2013). Agriculture, including irrigation, livestock watering and cleaning, and aquaculture, accounts for 69 percent of the world's water withdrawal, and is the largest user of water (Food and Agricultural Organization 2016a, b), and accounts for approximately 70% of total water withdrawal and about 90% of virtual water (the water that was consumed in the production of a good or service—see Sect. 7.5) flows globally relate to trade in agricultural products (Chico et al. 2014). Yet, at the same time, the UN has identified a dietary change towards meat-based diets as a global trend with diffuse economic and political implications. These issues make food law a highly topical area of international law.

Most legal structures related to food on an international level are the creations of the United Nations (see Sect. 6.2.2) and the WTO (see Sect. 6.2.3). Food falls under the mandate of several UN organizations such as: the World Food Programme (WFP); the UN Food and Agricultural Organization (FAO); and the World Health Organization (WHO).

Legal tools created by these organizations include the 2001 FAO International Treaty on Plant Genetic Resources for Food and Agriculture (International Seed Treaty). The Treaty is approved under the provisions of Article XIV of the FAO Constitution. Its objectives are the conservation and **sustainable use** of all plant genetic resources for food and agriculture and the fair and equitable sharing of the benefits arising out of their use, in harmony with the Convention on Biological Diversity, for sustainable agriculture and food security.

In addition, the 2005 WHO International Health Regulations aim to prevent the international spread of diseases, including foodborne diseases.

WTO law has three major areas: trade in goods, trade in services, and intellectual property rights. From the perspective of food law, the agreements on trade in goods are most significant. The following four WTO agreements are especially relevant to food law:

(a) The General Agreement on Tariffs and Trade (GATT);
(b) The Agreement on the Application of Sanitary and Phytosanitary Measures (SPS);
(c) The Technical Barriers to Trade Agreement (TBT); and
(d) The Agreement on Agriculture (AoA).

In addition, the Agreement on Trade-Related Aspects of Intellectual Property Rights (TRIPs) governs food trade by regulating geographical indicators, e.g., to prevent misleading consumers about the origin of goods.

While the GATT/WTO aims to liberalize international trade in goods by setting equal treatment of all trading partners as the main rule, it also recognizes exceptions (Articles XX and XXI). On several occasions, countries have blocked trade in food, citing human health concerns. Concerns have been raised that national authorities are too eager to invoke such exceptions with negative impact on international trade.

The WTO's SPS and TBT agreements are important tools with which to harmonize these oft-conflicting interests.

The SPS agreement specifies the exceptions listed in the GATT. It enables WTO members to adopt measures to protect the health of humans, animals, and plants. However, these measures have to be based on a scientific risk assessment. The SPS agreement refers to three international standard-setting organizations (the "Three sisters") for food, plant protection, and animal health. The Codex Alimentarius (Food Code) is the most important international point of reference for food safety standards. The Codex Alimentarius was established by the FAO and the WHO in 1963, and it currently consists of standards, guidelines, codes of practice and advisory texts, which are non-binding in nature.

The TBT agreement complements the SPS. It aims to ensure that technical regulations, standards, and conformity assessment procedures are non-discriminatory in nature and do not create unnecessary obstacles to trade. It further specifies the WTO members' right to take measures to protect consumers and other public interests but also requires that these protective measures are transparent and non-discriminatory. With regard to food, the requirements related to packaging and labeling are the most relevant in the TBT agreement.

As noted above, agriculture received significant exemptions under GATT due to its connections with national food security. The Agreement on Agriculture (AoA) was established in 1995 to boost fair competition and to reduce distortion in agricultural trade. The agreement contains provisions for the granting of preferential treatment to producers in developing countries to promote their access to international markets.

All in all, even though food is regulated through various agreements and two quite separate legal domains in trade law and human rights law, it is commonly observed that there is no incompatibility between the different provisions of international food law. On the contrary, international trade law includes various provisions aimed at ensuring human health that also enhance the right to food and food security. However, there are various policy-driven problems in the food sector in relation to unjust trade practices that may endanger food security (van `der Meulen 2010).

6.5.2 Interlinkages to Water and Energy

As is the case for international water and energy law, the food–energy–water nexus is not explicitly addressed in international food law. However, all the treaties in the area recognize the need to protect natural resources or the environment, which allows for the consideration of energy and water issues in relation to food.

As noted above, the GATT provides possibilities for WTO members to be exempted from its rules (Article XX). These exemptions include measures necessary to protect human, animal or plant life or health (paragraph b) or to conserve exhaustible natural resources (paragraph g). However, the exemptions cannot create unjustifiable discrimination between countries or disguised restrictions on international trade. The protection of human, plant, and animal health or the environment is also acknowledged in the TBT and SPS Agreements as well as in the AoA. The WTO Committee on Trade and Environment (CTE) has a mandate to contribute to identifying and understanding the relationship between trade and the environment in order to promote sustainable development. However, the CTE has been widely criticized in terms of the level of success it has achieved.

When observed from the perspective of food law, much of the food–energy–water nexus is about the connection between trade and the environment, which is a highly contested topic. It also goes beyond the food–energy–water nexus per se, since the international food trade has also been linked with severe threats to biodiversity, habitats, and species (see Chap. 9).

The fact that WTO law is not designed to address issues relating to the production of products due to the principle of mutual recognition is a major source of dispute in WTO law. In fact, so-called PPM (processing and production measures) rules prohibit member States from restricting trade based on the ways in which goods have been produced (Falkner and Jaspers 2012). Yet, in many cases, the environmental concerns associated with certain products arise from the methods used in their processing and production, for instance through the intensive use of water and energy (Esty 2001). The 2003 EC–Biotech case, in which the USA, Canada, and Argentina brought WTO proceedings against the EU's restrictions on the marketing of genetically modified organisms offers a good example of such concerns (though not specifically focused water or energy law).

Even though there are shortcomings in how water and energy issues are integrated into international trade law, it has been shown that WTO jurisdiction is gradually moving towards a more generous interpretation of environmental exemptions (Falkner and Jaspers 2012).

In this context, one of the most frequently cited cases is the 1998 Shrimp-Turtle case in which the WTO Appellate Body discussed the meaning of conservation of exhaustible natural resources. The Appellate Body recognized that textually, Article XX(g) of the GATT is not limited to the conservation of "mineral" or "non-living" natural resources but also extends to "living" natural resources and thus applies also to "renewable" natural resources, such as animals (including fish). The Appellate Body also referred in its ruling to the preamble of the WTO Marrakesh Agreement, which explicitly acknowledges the objective of sustainable development.

In addition to the trade-environment nexus discussed above, the question of whether water is food is another interesting issue associated with the water-food nexus. Water is included in the definition of food in the Codex Alimentarius, for example. Accordingly, food means any substance, whether processed, semi-processed or raw, which is intended for human consumption, and includes, for example, drink.

Food and water are closely related in the Convention on the Rights of the Child, under which the Parties must combat disease and malnutrition, through, inter alia, the provision of adequate nutritious foods and clean drinking-water, taking into consideration the dangers and risks of environmental pollution (Article 24).

Overall, the above-mentioned linkages between food law and the clauses on human health and environmental protection in the trade agreements enable interpretations that support the food–energy–water nexus. Furthermore, references to sustainable development in the preambles of these treaties allow their interpretations to evolve as required by growing societal needs, offering further opportunity to take the nexus approach into account.

6.6 Integrating Food, Energy, and Water Law

When comparing international food, energy, and water law, water law provides the most practical point of entry for the nexus approach. There are two reasons for this.

First, it is in the essence of international water law to allocate international water resources between different interests such as energy production, agriculture, and vital human needs, whereas international energy and food law are much more fragmented areas of law in which balancing international trade against national protectionism plays a key role.

Second, international water law requires transboundary cooperation between States sharing international water resources, while international energy and food law do not have a similar emphasis on transboundary cooperation from the point of view of the sustainable management of resources.

In terms of the nexus, one interesting way to find common legal ground would be to look at nexus integration in a clearly defined geographical area such as a region including countries sharing the same transboundary river basin. While the discussion about transboundary cooperation in such contexts often focuses on water flows, the countries involved are also very commonly linked by transboundary flows of energy and food (e.g., Keskinen et al. 2016), and related international and regional laws and regulations. This also means that while international water cooperation may provide a starting-point for the food–energy–water nexus, watercourse States should understand the community of their interests within and beyond international water law. This would provide an opportunity for cross-sectoral bilateral and multilateral agreements to enhance regional water, energy, and food security and sustainable development in the long term (see Sadoff and Grey 2002; Grzybowski et al. 2010; Belinskij 2015).

Finally, human rights law seems to provide a point of connection between all three sectors. While this chapter does not analyze the human rights system in detail, it can be concluded that human rights law supports the conclusion that water, energy, and food should be first allocated to meet vital human needs. Both the nexus approach and human rights law can, therefore, be seen to share common ground: both aim to address fundamental aspects of human life on this planet: sustainability and equity. It can even be argued that one cannot exist without the other: human rights law helps to address the politics inherent to management of the nexus, while the nexus brings important aspects of sustainability and resource scarcity to the discussion of human rights.

At the same time, it is important to remember that the nexus still lacks the internationally recognized status that is enjoyed, for instance, by Integrated Water Resources Management. In order to change this, the nexus should be more clearly implemented in international treaty regimes, agreements, and processes. While the ECE Water Convention regime has led the way in this respect and transboundary water agreements can provide such a context at a regional level, it appears that on a global scale the Sustainable Development Goals (see Sect. 3.7) provide a common platform for advancing both the cross-sectoral role of law and the objectives of the nexus. The 2030 Agenda for Sustainable Development emphasizes the importance of international law and regional cooperation for sustainable development and—despite its sectoral structure—includes several interlinked goals relating to water, energy, and food (UN 2017).

Key Points

- Governance can be understood as a process by which organizations and other groups of people make and implement decisions, policies, and rules. International governance for FEW consists of key actors (such as the United Nations organizations and WTO) and key institutions (such as international laws and treaties) that together form the general framework for FEW governance.
- International law does not include many explicit interlinkages between the water, energy, and food sectors. Instead, the food–energy–water nexus consists of a fragmented body of provisions in different sectors, in different areas of law and with different legal functions.
- While different legal provisions in international water, energy, and food law often have only implicit connections to other two sectors, the importance of the nexus approach and cross-sectoral linkages are more explicitly expressed in soft law instruments and policy papers. In this regard, perhaps the best example is the United Nations Economic Commission for Europe guidance on the nexus assessment in transboundary basins.
- International law on water differs in scope and scale from that on energy and food. International water cooperation may provide a starting-point for the food–energy–water nexus but watercourse Stated should understand the community of their interests within and beyond international water law.
- The human rights regime can be seen as a common element connecting all three themes and providing a general frame in a similar manner to that of the nexus.

Discussion Points and Exercises

1. Which of the following are hard international laws and which soft international laws?

   (a) Agreement on Agriculture.
   (b) Convention on the Law of the Non-Navigational Uses of International Watercourses.
   (c) Convention on the Protection and Use of Transboundary Watercourses and International Lakes.
   (d) Energy Charter Treaty.

(e) General Assembly resolution on 28 July 2010: The Human Right to Water and Sanitation, GA/10967.
(f) FAO International Treaty on Plant Genetic Resources for Food and Agriculture.
(g) United Nations Framework Convention on Climate Change.
(h) Describe examples of customary international law as applied to (a) food; (b) energy; (c) water; and (d) two or more of food, energy, and water.

2. Discuss the strengths and weaknesses of the UN System to integrating food, energy, and water concerns.
3. Describe how the human rights obligations of respect, protect, and fulfill discussed in the context of food and water might be applied to energy.
4. Describe the trade-offs and challenges of applying the human rights obligations of respect, protect, and fulfill to food, energy, and water simultaneously.
5. Describe an example of international law related to shared resources in the area of (a) food; (b) energy; (c) water; and (d) two or more of food, energy, and water.
6. Describe one or more examples of the principle of equitable and reasonable utilization in the area of food, energy, and water.
7. Describe one or more examples of the no-harm rule in the area of food, energy, and water.
8. Describe one or more examples of the principle of cooperation in the area of food, energy, and water.
9. Discuss the competing principles and trade-offs of the ICJ decision in the Gabčikovo-Nagymaros case.
10. Discuss the competing principles and trade-offs of the ICJ decision in the Pulp Mills case.
11. Discuss the competing principles and trade-offs of the EU Court of Justice decision in the Schwartz Sulm River case.
12. Discuss the challenges of international food governance.
13. Discuss the challenges of international energy governance.
14. Discuss the challenges of international water governance.
15. Discuss the challenges of integrating international FEW governance.

References

Belinskij, A. (2015). Water–energy–food Nexus within the framework of international water law. *Water, 7*, 5396–5415.

Bradbrook, A. J. (1996). Energy law as an academic discipline. *Journal of Energy and Natural Resources Law, 194*, 194.

Bradbrook, A. J., Gardam, J. G., & Cormier, M. (2008). A human dimension to the energy debate: Access to modern energy services. *Journal of Energy & Natural Resources Law, 26*(4), 526–552.

Chico, D., Aldaya, M. M., Flachsbarth, I., & Garrido, A. (2014). Virtual water trade, food security and sustainability: Lessons from Latin America and Spain. In P. Martinez-Santos, M. M.

Aldaya, & M. R. Llamas (Eds.), *Integrated water resources management in the 21st century: Revisiting the paradigm* (pp. 79–98). Boca Raton: CRC Press.

Committee on Economic, Social and Cultural Rights. (2002). *Substantive issues arising in the implementation of the international covenant on economic, social and cultural rights. General comment no. 15: The right to water (Arts. 11 and 12 of the covenant)*. Retrieved January 30, 2017, from http://tbinternet.ohchr.org/_layouts/treatybodyexternal/Download.aspx?symbolno=E%2fC.12%2f2002%2f11&Lang=en

Esty, D. C. (2001). Bridging the trade-environment divide. *Journal of Economic Perspectives, 15*(3), 113–130.

Falkner, R., & Jaspers, N. (2012). Environmental protection, international trade and the WTO. In K. Heydon & S. Woolcock (Eds.), *The Ashgate research companion to international trade policy* (pp. 245–260). Farnham: Ashgate.

Food and Agricultural Organization. (2005). *Voluntary guidelines to support the progressive realization of the right to adequate food in the context of national food security*. Adopted by the 127th Session of the FAO Council November 2004, http://www.fao.org/3/a-y7937e.pdf

Food and Agricultural Organization. (2016a). *Codex Alimentarius. Understanding Codex*. Retrieved February 6, 2017, from http://www.fao.org/3/a-i5667e.pdf

Food and Agricultural Organization. (2016b). *Water withdrawal by sector, around 2010, AQUASTAT*. http://www.fao.org/nr/aquastat. Update: November 2016.

Gerber, P. J., et al. (2013). *Tackling climate change through livestock—A global assessment of emissions and mitigation opportunities. Food and Agriculture Organization of the United Nations (FAO), Rome*. Retrieved April 19, 2019, from http://www.fao.org/3/i3437e/i3437e.pdf

Grzybowski, A., McCaffrey, S. C., & Paisley, R. K. (2010). Beyond international water law: Successfully negotiating mutual gains agreements for international watercourses. *Pacific McGeorge Global Business & Development Law Journal, 22*, 139–154.

Hoff, H. (2011). *Understanding the Nexus, background paper for the Bonn 2011 conference: The water, energy and food security nexus*. Stockholm: Stockholm Environment Institute Accessed October 3, 2018.

Hunter, T., & Paterson, J. (2014). Prevention of and response to shale gas well incidents: An assessment of the current U.K. regulatory framework for shale gas activities. *Oil, Gas & Energy Law Intelligence, 12*(3).

IEA, Our Mission. (n.d.). (website). International Energy Agency. Retrieved January 29, 2019, from https://www.iea.org/about/ourmission/

International Law Association (ILA). (1967). *The Helsinki rules on the uses of the waters of international rivers, adopted by the international law association* (pp. 7–55). London: International Law Association.

International Law Commission (ILC). (1994). Draft articles on the law of the non-navigational uses of international watercourses and commentaries thereto and resolution on transboundary confined groundwater. In *Yearbook of the International Law Commission, 1994* (Vol. II, Part Two, pp. 89–135).

International Law Commission (ILC). (2018). Draft conclusions on identification of customary international law, with commentaries. In *Yearbook of the International Law Commission, 2008* (Vol. II, Part 2). United Nations.

Kent, G. (2005). *Freedom from want. The human right to adequate food*. Washington, DC: Georgetown University Press.

Keskinen, M., Guillaume, J. H. A., Kattelus, M., Porkka, M., Räsänen, T. A., & Varis, O. (2016). The water–energy–food nexus and the transboundary context: Insights from large Asian rivers. *Water, 8*(5), 193.

LII, International Law. (n.d.). (website). *Legal information institute, Cornel University*. Retrieved January 29, 2019, from https://www.law.cornell.edu/wex/international_law

Martin, T. (2012). Lex petrolea in the international oil and gas industry. In R. King (Ed.), *Dispute resolution in the energy sector: A practitioner's handbook*. London: Globe Law and Business.

McCaffrey, S. C. (2001). The contribution of the UN convention on the law of the non-navigational uses of international watercourses. *International Journal of Global Environmental Issues, 1*, 250–263.
McCaffrey, S. C. (2014). International water cooperation in the 21st century: Recent developments in the law of international watercourses. *Review of European Community and International Environmental Law, 23*, 4–14.
McIntyre, O. (2011). The World Court's ongoing contribution to International Water Law: The pulp mills case between Argentina and Uruguay. *Water Alternatives, 4*, 124–144.
Mechlem, K. (2004). Food security and the right to food in the discourse of the United Nations. *European Law Journal, 10*(5), 631–648.
Sadoff, C., & Grey, D. (2002). Beyond the river: The benefits of cooperation on international rivers. *Water Policy, 4*, 389–403.
Talus, K. (2014). Internationalization of energy law. In K. Talus (Ed.), *Research handbook on international energy law* (p. 4). Camberley: Edward Elgar.
United Nations (UN). (2017). *Transforming our world: The 2030 agenda for sustainable development*. United Nations A/RES/70/1.
United Nations Economic Commission for Europe (UNECE). (2013). *Guide to implementing the water convention*. ECE/MP.WAT/39, United Nations.
United Nations Economic Commission for Europe (UNECE). (2015). *Reconciling resource uses in transboundary basins: Assessment of the water-food-energy-ecosystems nexus*. Geneva: United Nations.
United Nations Economic Commission for Europe (UNECE). (2003).
van der Meulen, B. (2010). The global arena of food law: Emerging contours of a meta-framework. *Erasmus Law Review, 3*(4), 217–240.
Wyman, O. (2015). *World energy trilemma: Priority actions on climate change and how to balance the trilemma*. London: World Energy Council.

Further Reading

Ajanovic, A. (2011). Biofuels versus food production: Does biofuels production increase food prices? *Energy, 36*(4), 2070–2076.
Allouche, J., Middleton, C., & Gyawali, D. (2015). Technical veil, hidden politics: Interrogating the power linkages behind the nexus. *Water Alternatives, 8*, 610–626.
Barral, V. (2016). National sovereignty over natural resources: Environmental challenges and sustainable development. In E. Morgera & K. Kulovesi (Eds.), *Research handbook on international law and natural resources* (pp. 3–25). Camberley: Edward Elgar.
Boas, I., Biermann, F., & Kanie, N. (2016). Cross-sectoral strategies in global sustainability governance: Towards a Nexus approach. *International Environmental Agreements, 16*, 449–464.
Bodansky, D., Brunnée, J., & Rajamani, L. (2017). *International climate change law*. New York: Oxford University Press.
Bradbrook, A. J., & Gardam, J. G. (2006). Placing access to energy services within a human rights framework. *Human Rights Quarterly, 389*(28), 389–390.
Brownlie, I. (1990). *Principles of public international law* (4th ed.). Oxford: Clarendon Press.
Carlarne, C., Gray, K., & Tarasofsky, R. (Eds.). (2016). *The Oxford handbook of international climate change law*. New York: Oxford University Press.
Charnovitz, S. (2007). The WTO's environmental progress. *Journal of International Economic Law, 10*(3), 685–706.
Childs, T. C. (2011). Update on Lex Petrolea: The continuing development of customary law relating to international oil and gas exploration and production. *Journal of World Energy Law & Business, 4*(3), 214–259.

Eckersley, R. (2004). The big chill: The WTO and multilateral environmental agreements. *Global Environmental Politics, 4*(2), 24–50.

Endo, A., et al. (2015). Methods of the water-energy-food nexus. *Water, 7*, 5806–5830.

Flammini, A., Puri, M., Pluschke, L., & Dubois, L. (2014). *Walking the Nexus talk: Assessing the water-energy-food nexus in the context of the sustainable energy for all initiative*. Environment and natural resources working paper no. 58, FAO 2014.

Food and Agricultural Organization. (2006). *Livestock's long shadow. Environmental issues and options. Rome, Italy*. Retrieved January 30, 2017, from ftp://ftp.fao.org/docrep/fao/010/a0701e/a0701e.pdf

Food and Agricultural Organization. (2013). *Reviewed strategic framework (FAO C 2013/7)*. Retrieved February 7, 2017, from http://www.fao.org/docrep/meeting/027/mg015e.pdf

Hawkes, C., Delia, G., & Thow, A. M. (2015). Trade liberalization, food, nutrition and health. In R. Smith, C. Blouin, Z. Mirza, P. Beyer, & N. Drager (Eds.), *Trade and health: Towards building a National Strategy*. Geneva: World Health Organization.

Healy, S., Pearce, R., & Stockbridge, M. (1998). *The implications of the Uruguay round agreement on agriculture for developing countries—A training manual (Training materials for agricultural planning—41)*. Rome: FAO. Retrieved February 8, 2017, from http://www.fao.org/docrep/w7814e/w7814e00.htm#Contents.

Heffron, R. J., & Talus, K. (2016). The development of energy law in the 21st century: A paradigm shift? *Journal of World Energy Law & Business, 9*(3), 189–202. Oxford Academic.

Horn, H., & Mavroidis, P. C. (2014). Multilateral environmental agreements in the WTO: Silence speaks volumes. *International Journal of Economic Theory, 10*, 147–166. https://doi.org/10.1111/ijet.12033.

Hunter, T. (2014). Regulating the environmental impact of tight and shale gas tight gas projects in Western Australia: An assessment of the existing regulatory framework. *Oil, Gas & Energy Law Intelligence, 3*, 1–17.

International Energy Agency. (2017). *Energy security*. Retrieved February 1, 2017, from https://www.iea.org/topics/energysecurity/

International Law Association (ILA). (2004). *Berlin conference, water resources law*. Retrieved February 22, 2017, from http://internationalwaterlaw.org/documents/intldocs/ILA_Berlin_Rules-2004.pdf

Jackson, L. A., Pene, C., Martinez-Hommel, M.-B., Hofmann, C., & Tamiotti, L. (2014). Water policy, agricultural trade and WTO rules. In P. Martinez-Santos, M. M. Aldaya, & M. R. Llamas (Eds.), *Integrated Water Resources Management in the 21st Century: Revisiting the paradigm* (pp. 59–75). Boca Raton: CRC Press.

Lawford, R., Bogardi, J., Marx, S., Jain, S., Wostl, C. P., Knüppe, K., Ringler, C., Lansigan, F., & Meza, F. (2013). Basin perspectives on the water-energy-food security nexus. *Current Opinion in Environment Sustainability, 5*, 607–616.

Leal-Arcas, R., & Abu Gosh, E. S. (2014). Energy trade as a special sector in the WTO: Unique features, unprecedented challenges and unresolved issues. *Indian Journal of International Economic Law, 6*(1), 25–35.

Lenzen, M., Moran, D., Kanemoto, K., Foran, B., Lobefaro, L., & Geschke, A. (2012). International trade drives biodiversity threats in developing nations. *Letter. Nature, 486*, 109–112. https://doi.org/10.1038/nature11145.

McCaffrey, S. C. (1992). A human right to water: Domestic and international implications. *Georgetown International Environmental Law Review, 5*(1), 1–24.

McCaffrey, S. C. (2007). *The law of international watercourses* (2nd ed.). Oxford: Oxford International Law Library.

McIntyre, O. (2012). The human right to water as a creature of global administrative law. *Water International, 37*, 654–669.

Neumayer, E. (2001). *Greening trade and investment: Environmental protection without protectionism*. London: Earthscan.

Oregon State University, the College of Earth, Ocean, and Atmospheric Sciences. (2017). *International freshwater treaties database*. Retrieved February 6, 2017, from http://www.transboundarywaters.orst.edu/database/interfreshtreatdata.html

Rathmann, R., Szklo, A., & Schaeffer, R. (2010). Land use competition for production of food and liquid biofuels: An analysis of the arguments in the current debate. *Renewable Energy, 35*(1), 14–22.

Rieu-Clarke, A. (2015). Remarks of the drafting history of the convention. In A. Tanzi, O. McIntyre, A. Kolliopoulos, A. Rieu-Clarke, & R. Kinna (Eds.), *The UNECE convention on the protection and use of transboundary watercourses and international lakes: Its contribution to international water cooperation* (International Water Law Series) (Vol. 2015, pp. 1–14). Leiden: Brill Nijhoff.

Rieu-Clarke, A., & Kinna, R. (2014). Can two global UN water conventions effectively co-exist? Making the case for a 'Package Approach' to support institutional coordination. *Review of European, Comparative & International Environmental Law, 23*(1), 15–31.

Rieu-Clarke, A., Moynihan, R., & Magsig, B-O. (2012). UN watercourses convention: User's guide. University of Dundee: IHP-HELP Centre for Water Law, Policy and Science, Dundee.

Romppanen, S. (2015). *New governance in context: Evaluating the EU biofuels regime*. University of Eastern Finland Ph.D. thesis.

Sampson, G. P. (2005). *The WTO and sustainable development*. Tokyo: United Nations University Press.

Shelton, D. (2009). *Soft law in handbook of international law* (pp. 68–80). Oxford: Routledge Press ISBN-10:0415418763.

Sovacool, B. K., & Dworkin, M. H. (2014). *Global energy justice*. New York: Cambridge University Press.

Stokke, O. S. (2007). A legal regime for the Arctic?: Interplay with the law of the sea convention. *Marine Policy, 31*(4), 402–408.

Tanzi, A. (2010). Reducing the gap between international water law and human rights law: The UNECE protocol on water and health. *International Community Law Review, 12*, 267–285.

Tanzi, A. (2015). *The Economic Commission for Europe Water Convention and the United Nations Watercourses Convention: An analysis of their harmonized contribution to international water law*. ECE/MP.WAT/42, United Nations.

Tanzi, A., & Arcari, M. (2001). *The United Nations convention on the law of international watercourses: A framework for sharing* (International and National Water Law and Policy Series). London: Kluwer Law International.

Tscherning, R. P. (2014). Renewable energy for food and water security projects in dry-land countries: Towards a model legal framework for the Qatar national food security programme. In K. Talus (Ed.), *Research handbook on international energy law*. Camberley: Edward Elgar.

UNDESA. (2014). *From silos to integrated policy making*. HLPF issue briefs 5.

United Nations (UN). (2018a). *Uphold international law*. Retrieved September 3, 2019, from http://www.un.org/en/sections/what-we-do/uphold-international-law/

United Nations (UN). (2018b). *International law and justice*. Retrieved September 3, 2019, from http://www.un.org/en/sections/issues-depth/international-law-and-justice/index.html

United Nations Economic Commission for Europe (UNECE). (2009). *Capacity for water cooperation in Eastern Europe, Caucasus and Central Asia: River basin commissions and other institutions for transboundary water cooperation*. ECE/MP.WAT/32, United Nations.

United Nations Economic Commission for Europe (UNECE). (2016). *Report of the meeting of the parties on its seventh session* (ECE/MP.WAT/49).

United Nations Economic Commission for Europe (UNECE). (2017). *Implementation: Status of bilateral and multilateral agreements*. Retrieved February 6, 2017, from https://www.unece.org/env/water/partnership/part621.html

United Nations Treaty Collection (UNTC). (2018a). *Chapter XXVII, Environment, 5. Convention on the protection and use of transboundary watercourses and international lakes*. Retrieved September 12, 2018, from https://treaties.un.org/Pages/ViewDetails.aspx?src=TREATY&mtdsg_no=XXVII-5&chapter=27&clang=_en

United Nations Treaty Collection (UNTC). (2018b). *Chapter XXVII, Environment, 12. Convention on the law of the non-navigational uses of international watercourses*.

Retrieved September 12, 2018, from https://treaties.un.org/pages/ViewDetails.aspx?src=TREATY&mtdsg_no=XXVII-12&chapter=27&clang=_en

UN-Water. (2008). *Transboundary waters: Sharing benefits, sharing responsibilities*. UN-Water thematic paper.

UN-Water. (2013). *Water security & the global development agenda*. UN-Water analytical brief, United Nations University.

UN-Water. (2014). *Water and energy. The United Nations World Water development report 2014, Volume 1*, UNESCO.

Villarroel Walker, R., et al. (2014). The energy-water-food nexus: Strategic analysis of technologies for transforming the urban metabolism. *Journal of Environmental Management, 141*, 104–115.

World Commission on Environment and Development. (1987). *Our common future*. New York: Oxford University Press.

World Food Programme. (2017). *What is food security?* Retrieved February 1, 2017, from https://www.wfp.org/node/359289

World Trade Organization. (n.d.). *Agriculture: Fairer markets for farmers*. Retrieved October 3, 2017, from https://www.wto.org/english/thewto_e/whatis_e/tif_e/agrm3_e.htm

Human Rights Law Network. (n.d.). (website). *Right to food initiative*. https://hrln.org/initiatives/right-to-food/

Case Law

Arbitral Tribunal. (1957). *Lake Lanoux arbitration (France V. Spain). 12 R.I.A.A. 281; 24 I.L.R. 101.*

C-346/14 Commission v Austria. (2016). *Judgment of the Court of 4 May 2016.*

Government of the State of Kuwait v. American Independent Oil Co. (AMINOIL). (1982). *Award of 24 May 1982, 21 ILM 976.*

International Court of Justice (ICJ). (1998). *Gabčikovo-Nagymaros project (Hungary/Slovakia), Judgment.*

International Court of Justice (ICJ). (2010). *International court of justice. Pulp mills on the River Uruguay (Argentina v. Uruguay), Judgment.*

WTO, European Communities. (n.d.). *Measures affecting the approval and marketing of biotech products, DS291, DS292 and DS293, 29 September 2006.*

WTO, European Union and a Member State. (n.d.). *Certain measures concerning the importation of biodiesels, DS443, 17 August 2012.*

WTO, European Union and Certain Member States. (n.d.). *Certain measures affecting the renewable energy generation sector, DS452, 5 November 2012.*

WTO, European Union. (n.d.). *Certain measures on the importation and marketing of biodiesel and measures supporting the biodiesel industry, DS459, 15 May 2013.*

WTO, European Union. (n.d.). *Anti-dumping measures on biodiesel from Argentina, DS473, 19 December 2013.*

WTO, European Union. (n.d.). *Cost adjustment methodologies and certain anti-dumping measures on imports from Russia, DS474, 23 December 2013.*

WTO, European Union. (n.d.). *Certain measures relating to the energy sector, DS476, 30 April 2014.*

WTO, European Union. (n.d.). *Anti-dumping measures on biodiesel from Indonesia, DS480, 10 June 2014.*

WTO, USA. (n.d.). *Import prohibition of certain shrimp and shrimp products, WT/DS58/AB/R, 12 October 1998.*

Chapter 7
Trade

Peter Saundry and Benjamin L. Ruddell

7.1 Introduction

Trade is a key feature of the FEW Nexus. The D'Odorico and Scanlon framings described in Sect. 1.4 emphasize the importance of focusing on problems and solutions at the nexus, as a strategy for making sense of complex FEW systems. Typically, the practical goals are straightforward—decrease demands for FEW commodities, increase supplies, increase storage buffers, increase transportation and trade capacity and connectivity, and do so in the presence of climate change, population growth, growing wealth and consumption, and dietary changes. Trade is a powerful tool in achieving practical goals in FEW systems.

In Chap. 2, we noted the importance of inputs and outputs in system science. Trade is one of the major classes of system inputs and outputs. Imports and exports of commodities through trade (the buying or selling of goods and services) are frequently important inputs and outputs in FEW systems.

In Chap. 3, we noted that food, energy, and water security require four attributes: availability, access, utilization, and reliability (see Sect. 3.2). Trade can facilitate all four attributes.

0. Trade makes commodities physically **available** at locations where local supplies do not meet demand.

P. Saundry (✉)
Energy Policy and Climate, Advanced Academic Programs, Krieger School of Arts and Sciences, Johns Hopkins University, Washington, DC, USA
e-mail: psaundr1@jhu.edu

B. L. Ruddell
School of Informatics, Computing, and Cyber Systems, Northern Arizona University, Flagstaff, AZ, USA
e-mail: Benjamin.Ruddell@nau.edu

P. Saundry, B. L. Ruddell (eds.), *The Food-Energy-Water Nexus*, AESS Interdisciplinary Environmental Studies and Sciences Series,
https://doi.org/10.1007/978-3-030-29914-9_7

1. For many reasons, commodities can be produced at a lower price in locations far from demand centers, and can be produced or stored far from a local shock that has disrupted local supply, and thus trade can lower the cost of a product and increase its **accessibility**.
2. Trade can support **utilization** (the ability of a person to make use of the resource productively) through the import of equipment such as refrigeration, cookers, vehicles, water treatment, waste management, and expertise (human services), and by allowing producers in low-population and resource-rich areas to more fully utilize available land and energy resources to produce goods for export.
3. Trade allows for **the stable and reliable** provision of commodities that are subject to variability because of growing seasons, disruption because of natural disasters, and other factors.

Trade is a broad concept that generally refers to buying and selling of goods and services, but in this chapter, as is conventional in the academic economics literature, trade refers specifically to the exchange between nations, but more generally to trade between regions (Chap. 5).

Trade is one of the main adaptive behaviors by which humanity has historically responded to local resource scarcity, and to shocks and stresses. Trade has been a major tool used by humanity to raise its productivity and standard of living (Chap. 4). FEW trade policy tends to balance self-sufficiency objectives against the need to access less expensive commodities or access commodities during local shocks like drought or severe winter weather.

In general, trade increases resilience to disruptions in food, energy, and water supplies by increasing a region's ability to access diverse suppliers in times of need. It is unheard of in the modern global world for a region or nation to suffer severe food, energy, or water supply shocks if that nation is wealthy and well-connected to global suppliers via trade. However, it remains tragically routine for severe food energy and water supply shocks to impact traditional peoples and economies that lack access to the global trade network (e.g., famine in the Horn of Africa).

The net benefits of trade to FEW resilience are strongly positive, global producers of a commodity are generally diversified, mitigating major disruptions. However, trade shocks can be and have been, used as a geopolitical weapon when the international trade markets are overly dependent on a small number of exporters. This was powerfully illustrated by the 1973 disruptions in oil world petroleum supplies by certain members of the Organization of Petroleum Exporting Countries (OPEC, more below).

Trade is an ancient institution that has for millennia moved geographically rare goods like salt, obsidian, metal tools, spices, and gems thousands of miles. Every historical civilization engaged in trade across long distances. Trade was arguably the primary motivating force behind the exploration of the Earth. However, in the past 100 years, trade has accelerated exponentially in both absolute and relative terms. Five historical turning points that notably accelerated trade include:

1. The medieval Islamic Expansion and subsequent Crusades which connected the Old World from Europe to China (632–1291),

2. The Columbian Exchange between the Old World and New World in the century following the opening of the Americas by the Spanish in 1492 (Mann 2011),
3. The creation of the International Monetary Fund and the World Bank as part of the United Nations system (1945) to promote economic stability and development (see Sect. 6.2.3),
4. The launch of the internet by the USA (1983), and
5. The Pax Americana following the collapse of the USSR and the lifting of the Iron Curtain, including the founding of the World Trade Organization (1991 and 1995).

Globalization is the trend toward a dramatically increased exchange between cultures and nations. Today, except for a few exceptions like North Korea, every global culture and nation on Earth is strongly connected via trade and communication (Baldwin 2016).

Nations have widely differing trade policies that reflect a blend of openness versus protectionism (e.g., producer subsidies, price controls, quotas, and tariffs). **Neoliberal economics** argues that free and open trade enriches all parties and maximizes resilience (Wolf 2004). However, there are real-world consequences of globalization, including a loss of local social and political control and the "**race to the bottom**," which occurs when a nation with low social or environmental standards undercuts a more responsible and healthy nation on price (Sassen 1999).

Trade and globalization of trade have created stark winners and losers among nations and demographic segments, and have arguably benefitted high-skilled specialist labor and also those with mobile capital, that is, banks and the "rich," to the detriment of most "ordinary" laborers. Globalization has dramatically benefitted "developing" countries that have adopted an export-oriented manufacturing strategy, including notably the **petro-states** of the Middle East and the "Asian Tiger" economies like Japan, South Korea, and especially China.

Those involved in exporting industries expand their enterprise and increase their revenue. However, those involve in those same industries at the site of import, see the price for their products lowered by the presence of cheap imports. For example, exports of many grain crops by US farmers increases their income. However, any producer of that same crop in a country importing it from the USA sees its revenue decline. As a result, each nation, when setting up trade relations, balances the benefits of low prices to its citizens against any negative impacts on domestic industries. For example, countries balance the benefits of lower-cost food products for their citizens against negative economic impacts on their farmers. For better or worse, globalization has created a world where a region's economic growth—and economic collapse—benefit and harm everyone around the world.

In prior centuries, the colonial trade model of "**mercantilism**" was perfected by nations including the Netherlands and England for the purpose of extracting natural resources, monopolizing valuable trade routes, and enriching the home country through lending and value-added manufacturing monopolies. More recently, cartels like the Organization of Petroleum Exporting Countries (OPEC) have attempted to manipulate trade to their members' advantage by forming oligarchies and monopolies to control the supply of rare goods or services. One of the original and most important functions of the US government was to create a single-currency open trading block, and this strategy of open trading blocks has been replicated to some extent via the North American Free Trade Agreement (NAFTA) and the European Union (EU).

In the twenty-first century, the **World Trade Organization** (WTO) creates a framework for open trade between nations, and this framework has encouraged ever-increasing globalization and trade. The modern international trade system was engineered by Europe, and then especially the USA, following World War 1 and 2. The primary motivation for this liberalization of international trade was the promotion of international peace and security by engineering global prosperity and economic interdependency. You are less likely to go to war with a country when their economy is tightly integrated with your own and when there is a high level of communication, travel, and cultural exchange between the countries.

The US Dollar and the **U.S. Federal Reserve System** are central to the international trade regime because most international trade is priced in US Dollars, even trade between countries outside of North America and Europe. This U.S. central bank, therefore, remains the key financial institution for trade and for economic stability worldwide, because it sets the supply and price of the Dollar. The internet has become the key cultural and communication institution for trade and globalization. The internet was founded by and is partly controlled by the US government and by US companies, but the internet is not a strictly national institution.

We will explore international laws in greater detail in Chap. 6, including how they affect trade issues. In Chap. 8, we will examine laws and policies in the USA, and note how they affect domestic and international trade. Finally, it is important to restate that food, energy, and water systems impact many things that do not trade in marketplaces. These include ecosystem functions (see Chap. 9), air and water quality, recreational access, climate, and other phenomena. Nonmarket valuation was introduced in Sect. 5.2.3 and is revisited in Chap. 9.

In the following sections, we will briefly review some conceptual fundamentals of trade, then we will discuss the specifics of trade in food, energy, and water, along with "virtual" trade, trade regulations, and the role of trade in the FEW Nexus framing. In the coming sections of this chapter, we will explore how these factors and others that impact trade in food, energy, and water separately and together. We will introduce some international policies and organizations important to FEW trading in this chapter.

7.2 Rationale for Trade

There are many reasons why trade occurs between two locations. The most common reason is that one location has some kind of **comparative advantage** in the production of a product, which allows it to provide that product for sale/trade at a lower cost than another location. For trade to occur, the comparative advantage must be sufficient to overcome both transportation costs and any barriers to trade like a tariff.

There are many reasons for a comparative advantage to exist, including:

1. **Natural resources**. Many places do not have sufficient natural resources to provide food, energy, and water for their populations. Within countries, cities and large urban areas must bring FEW resources from their surrounding areas and other areas inside and outside their home country to make sufficient FEW commodities available to their resident populations. Rural areas are often able to produce more FEW resources than their people require. Chapter 18 explores some of the complex FEW relationships of cities. Countries may have an abundance of arable land or in one or more minerals, while other countries have limitations on one or more such resources.
2. **Product Preferences**. Even when locations can produce their own FEW resources, populations may have preferences different from those that they can provide for themselves. For example, an area may produce more than enough grains but demand more meat, while another location may produce wind power but demand oil. A difference in preferences may be rooted in cultural differences that shape dietary choices or in aspirations such as lower carbon emissions.
3. **Capital goods**. This term refers to the machinery, infrastructure, communications systems, and other physical assets of a location that facilitate its ability to produce products. Farm equipment, power plants and electric grid infrastructure, water treatment and waste management plants, and all physical systems that enable them to operate are examples of capital goods that affect a location's ability to produce food, energy, and water commodities. A location with limited capital good is limited in its ability to utilize natural resources.
4. **Human resources**, such as the skills and abilities of a workforce, has traditionally been a major factor in their ability to produce diverse products. These are often a reflection of investment in education, skill development, and dissemination of new ideas and skills. However, increasing education standards throughout the world, combined with the mobility of many skilled workers, has reduced the very major differences within countries and between countries that have existed historically.
5. **Technological capacity** combines capital goods with natural and human into the ability to produce specific outputs (goods and services) and is influenced by additional factors such as economics and government policies.
6. **Economies of Scale** result in a lower unit price for a commodity. This can occur in two quite different ways:

   First, within a single process, the output of many goods can be achieved without requiring an equivalent increase in inputs, because of increasing the efficiency of use of one or more input or by spreading fixed costs (e.g., the cost of a piece of equipment) over a larger amount of production. For example, historically, larger power plants have been able to produce electricity more efficiently than a smaller power plant. As a result, one large power plant could generate electricity at a lower cost than two smaller power plants of equivalent capacity.

   The second way that economies of scale can occur is through the amalgamation of industries that are supportive of each other. For example, a petrochemical industry can, in part, produce critical inputs to agriculture like fertilizers and insecticides. Thus, in a region with both a petrochemical industry and agriculture can have more efficient food production. Efficiencies through amalgamation mean that larger economies with industries that are supportive of each other and closely linked geographically or economically realize greater efficiencies.

Large economies with large domestic demand often experience both kinds of economies of scale. This is sometimes referred to as "large country advantage." However, smaller countries that closely connect their economies in free trade zones like the European Union can also experience such advantages. For economies of scale, production tends to become concentrated in that location at the expense of other locations.

7. **Government Policies** can dramatically influence the production and trade in many ways. As noted already, national governments often seek to ensure FEW security for its citizens. Domestically, policies can stimulate production to shape consumption. Internationally, policies usually involve diplomacy and trade arrangements, or sometimes the ownership of resources in other countries, to guarantee supplies. While nations can and have resorted to a military conflict to address food, energy, and water crises, trade is preferable.

While these seven factors are important within countries, there is a profound difference between domestic trade and international trade. Domestic trade is subject to largely uniform policies, although significant regional policy differences do sometimes occur. Further, the movement of people, natural resources, and capital within countries occur a lot more easily and rapidly, that across national borders.

While the past century has seen a significant convergence in the trade policies of different countries, facilitated by many multilateral treaties and organizations, profound differences domestic policies and other attributes of different sovereign countries, results in significant challenges to the free movement food, energy, and water cross national boundaries.

7.3 International Food Trade

The importance of food trade to food security is recognized in Sustainable Development Goal 2.B (see Sect. 3.7.1), which aims to "correct and prevent trade restrictions and distortions in world agricultural markets" through the World Trade Organization (WTO) (see Sect. 6.2.3). In addition, the Food and Agricultural Organization of the United Nations (FAO) supports developing nations in achieving trade agreements that improve their food security.

In the 50 years ending in 2017, the value of international trade in food in current dollars rose 40-fold (Fig. 7.1). This dramatic rise is the result of growth in agricultural production and the global population, as well as shifts in diets throughout the world away from locally grown food stuffs to diets that reflect the global diversity of food crops.

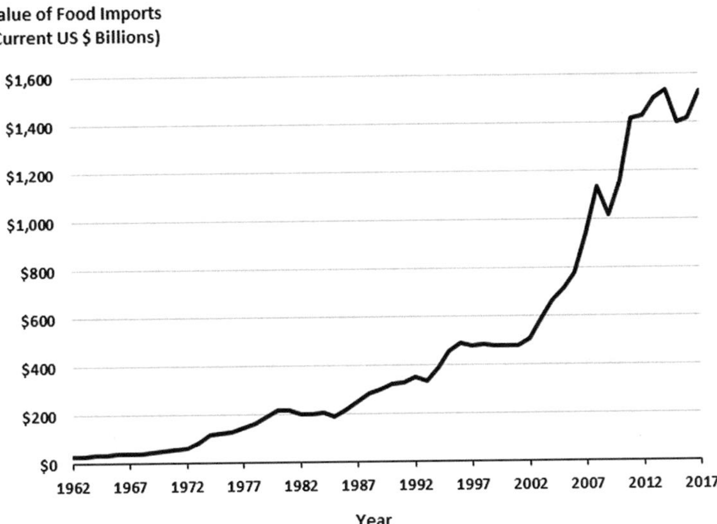

Fig. 7.1 Global food imports. (Source: World Bank Data Bank (Indicator codes TM.VAL.MRCH.WL.CD [Merchandise imports in current US$]))

The international food trade is essential for global food security in most nations and regions. "Just three crops—maize, wheat and rice—account for around 60% of global food energy intake. A fourth crop, soybean, is the world's largest source of animal protein feed, accounting for 65% of global protein feed supply. Each year, the world's transport system moves enough maize, wheat, rice, and soybean to feed approximately 2.8 billion people. Meanwhile, the 180 million **tonnes** of fertilizers applied to farmland annually play a vital role in helping us grow enough wheat, rice and maize to sustain our expanding populations." (Bailey and Wellesley 2017)

Nearly all countries are both importers and exporters of food (Table 7.1). However, the importance of food imports to a nation's food security is dependent on several factors. Wealthy countries have the financial ability to purchase foods as a matter of preference rather than need. This is reflected in the fact that the members of the European Union (collectively) and the USA are the top two importers and exporters of food by value and account for half of all such trade.

FEW systems that address countries and subregions must acknowledge flows of food into and out of the system boundaries—flows that are increasing with time and changing structurally in terms of food types and locations, as illustrated in Fig. 7.2. Note that North America and Europe are net exporters (exports-imports) of food.

Imports are also a reflection of population. The past two decades have seen a rise in both imports and exports from large emerging economies such as Brazil, China, India, and Indonesia. The growing wealth of China, in particular, saw its share of food imports rise from 2.3% of global imports in 2000 to 8.2% in 2016 whiles its share of food exports increased from 3.0% to 4.2%. These changes reflect part of the overall advances in human development achieved (see Sects. 3.4 and 3.5).

Table 7.1 Top importers and exporters of food by economic value 2016

	Top importers of food		Top exporters of food	
1	European Union	39.1%	European Union	41.1%
2	USA	10.1%	USA	11.0%
3	China	8.2%	Brazil	5.7%
4	Japan	4.2%	China	4.2%
5	Canada	2.7%	Canada	2.4%
6	Mexico	2.0%	Argentina	2.8%
7	China	1.9%	Australia	2.5%
8	India	1.9%	Indonesia	2.4%
9	Republic of Korea	1.9%	Mexico	2.3%
10	Russian Federation	1.9%	India	2.2%
	Other	26.1%	Other	23.4%

Source: Food and Agriculture Organization of the United Nations, The State of Agricultural Commodity Markets (2018)

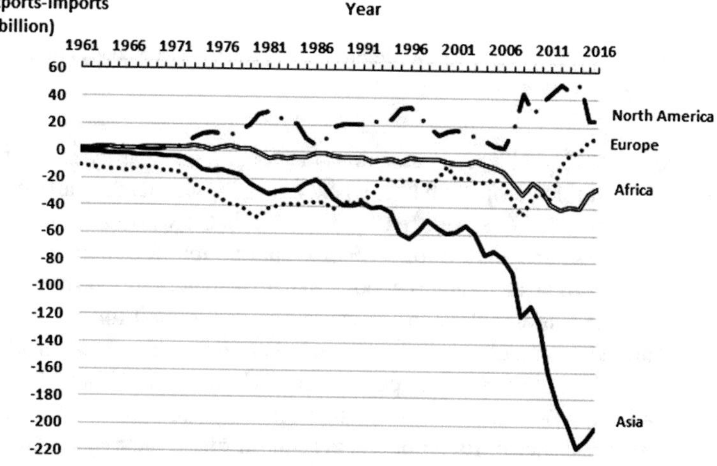

Fig. 7.2 Net exports (exports—imports) of aggregate agricultural (crops and livestock) products. (Source: Food and Agriculture Organization of the United Nations, FAOSTAT)

The connections between the most important factors related to trade in food are illustrated in a model developed by the International Food Policy Research Institute (IFPRI) known as the International Model for Policy Analysis of Agricultural commodities and Trade or "IMPACT." We look at IMPACT here to both gain insight into food trade and to introduce some basic elements of modeling, a topic that will be addressed in depth in Chap. 15.

IMPACT was developed in the 1990 "to address a lack of long-term vision and consensus among policymakers and researchers about the actions necessary to feed the world in the future, reduce poverty, and protect the natural resource base. Over time, this economic model has been expanded and improved, and IMPACT

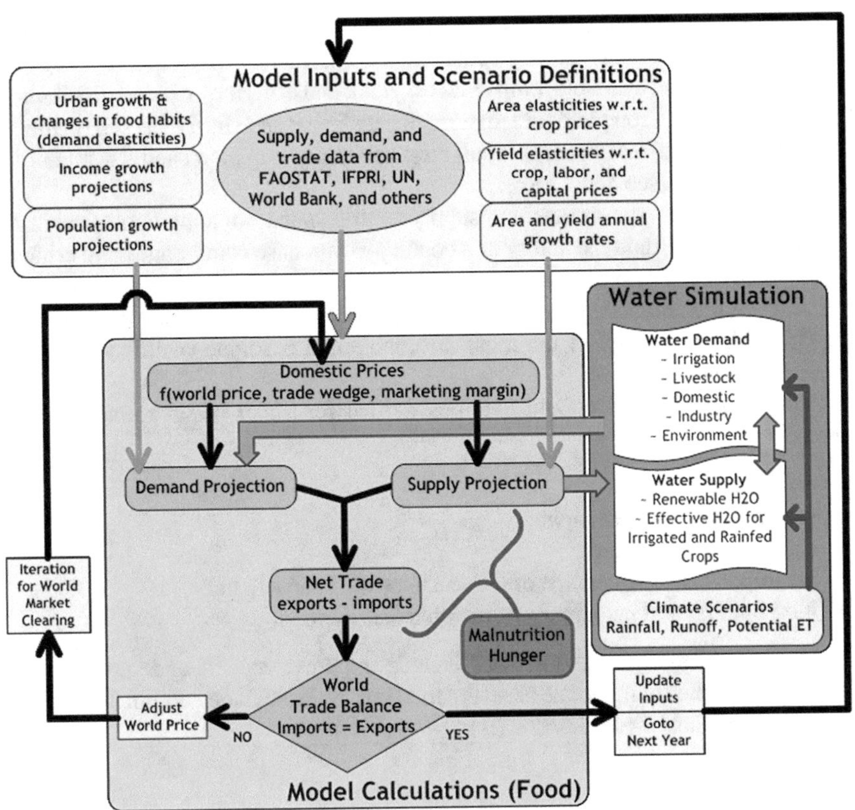

Fig. 7.3 The international model for policy analysis of agricultural commodities and trade (IMPACT). (Source: Rosegrant et al. 2012)

is now a system of linked models around a core multimarket economic model of global production, trade, demand, and prices for agricultural commodities." (Robinson et al. 2015)

IMPACT (Fig. 7.3) is a set of mathematical relations between different parts of a food system connecting inputs (e.g., population, diet, production, prices, etc.) to outputs (e.g., demand, exports, imports, etc.) (Robinson et al. 2015).

IMPACT is a **macro-scale** model (see Sect. 12.3) that covers 159 countries, 154 water basins, and 320 food production units based on 62 agricultural commodities. The agricultural commodities are staple crops (e.g., wheat, corn, rice, barley, maize, sugar, fruits, and vegetables) and animal products (e.g., fish, meat, and dairy products).

When considering IMPACT, it is important to remember the difference between projections and predictions (see Sect. 1.5.3). Models provide projections (or estimates) or what will happen under specified scenarios, defined by a set of assumptions about a system such as its key inputs, conditions, and functioning.

The dependency of a country on food trade is thus rooted in demand and supply factors. Demand factors include population, income, and food habits. Supply factors include arable land available, crop choices, yield, and efficiency of food distribution and preservation. Crop yields are, in turn, dependent on such factors as soil qualities, crop choices, labor and nutrient input, crop irrigation, irrigation capital availability, and other factors (see Sect. 2.3).

Trade depends on more than supply and demand in a particular country. Commodity prices and the ability of a country to purchase commodities are critical. High prices on world markets can result in food exports for cash, even in the presence of high domestic need. Conversely, low world prices without tariff barriers result in higher imports of the more affordable and desirable (within a domestic market) food stuffs.

Common metrics (see Data Sets below) that form part of such consideration include:

- arable land per capita;
- food productivity per capita;
- net imports per capita;
- net imports as a percentage of Gross Domestic Product; and
- food imports as a percentage of merchandise imports.

Metrics of impacts and outcomes include the prevalence of undernourishment and the prevalence of moderate or severe food insecurity, included within Sustainable Development Goal 2 (see Sect. 3.7.1).

7.4 International Energy Trade

The importance of energy security to nations is reflected in policies to support the production of energy sources that exist domestically and policies to minimize reliance on energy sources from other countries that might be disrupted. The 1973 oil crisis mentioned above was the first of a number of "oil shocks" in the 1970s and 1980s that had huge impacts on economies throughout the world (Huang et al. 1996; Lutz 2008). In response, many nations implemented significant policies to reduce their vulnerability to OPEC-led disruptions to oil flows.

According to the Energy Information Administration, the USA is 88% energy self-sufficient, compared with Japan at 8%, China at 80%, or Russia and Canada at 100% (IEA 2018, Fig. 7.4).

However, for purposes of trade, it is the differences between the production and consumption of specific energy sources that are critical. The USA, for example, has traditionally consumed more oil and gas than it produced, and was a major importer of both commodities. However, in recent years, the USA has become an exporter of natural gas and will, in the 2020s, be a net export of petroleum too.

If a nation can substitute one energy source for another, the implications of a shortage in the initial fuel are significantly mitigated.

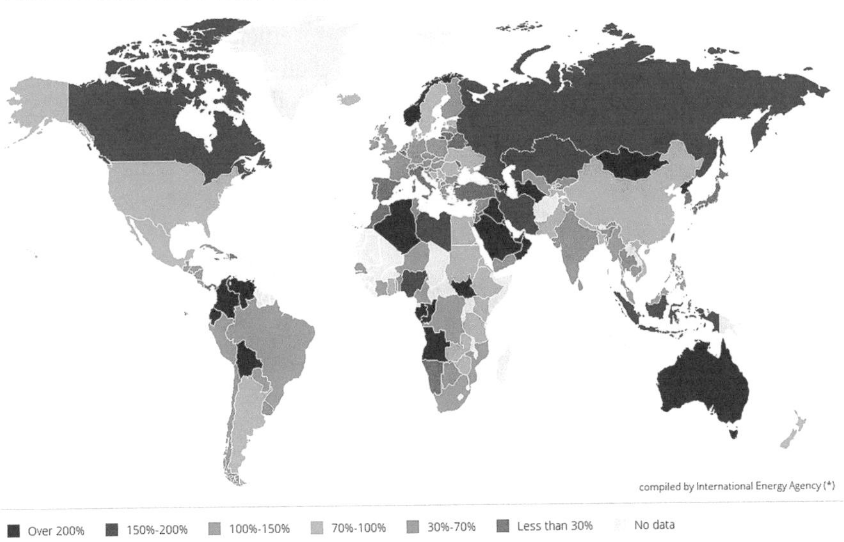

Fig. 7.4 Energy self-sufficiency in 2016 (IEA 2018)

The two largest uses of primary energy are as fuel for the generation of electricity and as fuel for transportation. While nearly all primary energy sources can be converted into electricity, transportation fuels are dominated (over 95%) by petroleum-based fuels. The near-monopoly of petroleum-based fuels for transportation means that it is impossible to mitigate significant shortages, except through reduced consumption. Thus, petroleum is considered a strategic commodity for nearly all societies, and its trade significant to energy security. However, there may be a significant transition to electrification of transportation during the coming decades. If that occurs, petroleum will lose much of its strategic importance.

Figure 7.5 shows primary energy consumption by modern fuel source. Consumption of traditional forms of biomass such as wood, charcoal, peat, and animal waste are not included. Prior to the industrial revolution, almost everyone on the planet received nearly all of their energy from traditional biomass. Today traditional biomass still accounts for an estimated 8% of primary energy consumption.

In 2017, fossil fuels accounted for 85% of primary energy consumption (oil 34%, coal 28%, and natural gas 23%), renewables accounted for 11% (hydropower 7%, others 4%), while nuclear power provided 4%. The climate change consequences of such dominance of fossil fuels are explored in Chap. 11.

The non-transportation energy use of countries is strongly influenced by the sources of energy that are domestically available. China, India, and Indonesia all produce and consume large amounts of coal. However, Indonesia also produces large amounts of oil and natural gas, which are reflected in high levels of use and

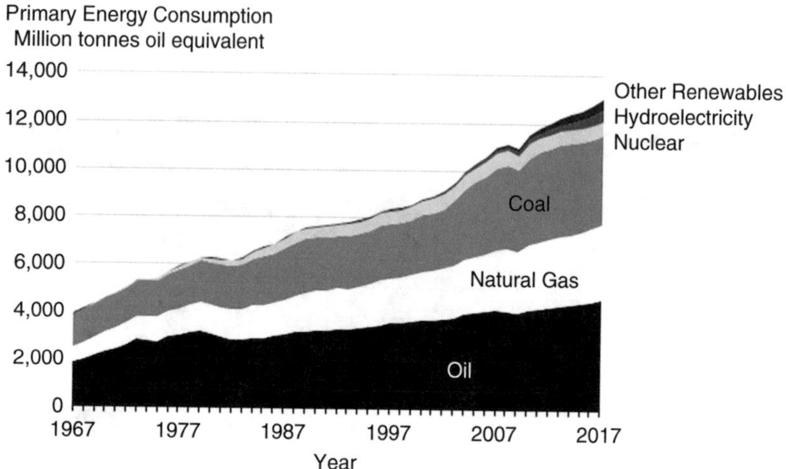

Fig. 7.5 Total primary energy consumption by modern fuel source, 1967–2017. (Source: BP Statistical Review of World Energy 2018)

limited use of other sources. In response to the oil crisis of 1973, nations such as Japan, France, Germany, and the USA all expanded their use of nuclear power to reduce their dependence on petroleum imports. Countries with large natural hydropower capacity, such as Brazil and Canada, have developed them, and they reflect a high share of their energy use.

Petroleum and natural gas are naturally created in the same manner are found in closely associated locations. Production of crude oil and natural gas usually includes some level of production of the other. In 2018, 28% of US petroleum field production was in the form of **Natural Gas Liquids**.

Production of petroleum and natural gas is dominated by a relatively small number of countries (Fig. 7.6). Three countries accounted for 39% of crude oil production in 2017 (USA 14%, Saudi Arabia 13%, and the Russian Federation 12%). Three-quarters of crude oil production occurred in just 13 countries, and 39 countries accounted for 98% of production. The USA (20%) and the Russian Federation (17%) account for 37% of global natural gas production. Three-quarters of natural gas production occurred in 13 countries, and 41 countries accounted for 98% of production.

Table 7.2 summarizes global trade in petroleum and natural gas in 2017. A critical point to understand about oil trading is that crude oil is a blend of **hydrocarbons** and, therefore, a range of products. Fuels for automobiles, large trucks, aircraft, and ships are all derived from different hydrocarbons contained in crude oil. Each source of crude oil has a unique blend of hydrocarbons. Therefore, even when a country produces as much crude oil as it consumes, it will be still be involved in trading—exporting the hydrocarbons that it has a surfeit of and importing those in which it has a deficit. Further, some trading is associated with refining capacity. For example, the USA has significant excess capacity to refine heavy

Fig. 7.6 Crude oil production (IEA 2018). Crude oil production is concentrated in Middle East, Russia, and North America, although many nations have substantial reserves

Table 7.2 Share of global trade in petroleum and natural gas for select countries and regions. (Source: BP Statistical Review of World Energy 2018)

	Petroleum (Crude + Products)		Natural gas (Pipeline + LNG)	
	Exports	Imports	Exports	Imports
US	8%	15%	7%	7%
Canada	6%	2%	7%	2%
Europe	5%	21%	18%	43%
Russian Federation	13%	0%	20%	2%
Middle East	36%	2%	14%	3%
China	2%	15%	0%	8%
India	2%	7%	0%	2%
Japan	0%	6%	10%	0%
Rest of World	32%	32%	33%	24%

crude oil (that is crude oil with a blend that includes a high proportion of long "heavy" hydrocarbons) and therefore imports some crude oil simply to refine and export finished petroleum products.

The term "**petrostate**" refers to countries like Saudi Arabia whose economies are dominated by oil exports and which are extremely vulnerable to changes in oil price and oil demand on the international market. The Organization of Petroleum Exporting Countries (OPEC) attempts to facilitate coordination of petroleum policies of 14 countries, many of whom are considered petrostates.

The USA has experienced a dramatic increase in production of both petroleum and natural gas since 2007 as a result of developments in hydraulic fracturing, which has allowed the production of "unconventional" resources that were previously inaccessible. These are commonly referred to as "tight oil" and "shale gas." The scale of this development has moved the US importer of oil (60% of consumption in 2006) and natural gas (16% of consumption in 2007) to become a net exporter of gas in 2017 and (soon to be) petroleum (expected in 2020–2022). Unconventional production of oil and gas is being deployed worldwide.

Coal deposits are more widely distributed in the world, leading producers of with some notable exceptions, different from leading producers of oil and gas. Further, because coal is primarily used for electricity generation, which can also be generated from other primary energy sources, it is not as critical to the energy security modern economies as petroleum.

Petroleum and natural gas can be moved efficiently over large land distances (and shallow seas) by pipeline. Coal is primarily moved overland by rail or river barge. Electricity can be moved efficiently over several hundred miles using about-ground, High Voltage Alternating Current (HVAC) transmission lines. Longer distances require more costly High Voltage Direct Current (HVDC) transmission lines.

Moving energy across seas and oceans typically requires ships. Because oil has a high energy density (by mass and volume) and requires no treatment before or after movement by ship, it is a relatively easy commodity to move. As a result, transportation costs are relatively low, and oil prices are very similar in all major markets around the world—moving up or down in unison.

In contrast, coal has a lower energy density (approximately 24 MJ/kg compared to approximately 45 MJ/kg for oil) and, therefore, higher transportation costs per unit of energy. Thus, islands that require energy to be shipped to them generally utilize petroleum to a large extent (e.g., for electricity generation and heating as well as transportation) than other locations.

Natural gas has a very low energy density and requires liquefaction before being shipped. Thus, transportation of liquefied natural gas (LNG) has a high transportation cost compared to oil and markets in different parts of the world typically have notable differences in prices.

The movement of all energy sources requires significant infrastructure.

Energy is also "embedded" in the trade of other products. That is, energy is used to produce products that are then traded to other countries. The location of the energy used to produce a product is separated from the location of the final use or consumption of the product. This increases energy use (and) in the country producing the product (known as "on-shoring") and decreases the energy use in that country receiving and using/consuming the product ("off-shoring"). Energy-intensive industries include bulk chemicals, refining, mining, agriculture, and iron, steel, and aluminum production. **Embedded energy** is analogous to virtual water trade, as discussed in the next section.

An important aspect of embedded energy is that any emissions of greenhouse gases or other pollutants associated with the traded products are also on-shored or off-shored.

The Energy Charter Treaty (see Sect. 6.2.3) began as a framework for Western Europe to respond to the changes in eastern Europe following the end of the Cold War, before being expanded. An Energy Charter Treaty was adopted in 1994 and went into force in 1998. A subsequent International Energy Charter was adopted in 2015 by 64 states to provide the framework process to develop a larger treaty. The ECT includes provisions on non-discrimination in trading conditions (based on WTO rules); ensures stable movements of energy across international borders; and plays a significant role in trade dispute resolution.

7.5 International Water Trade (and Virtual Water Trade)

Water is not formally traded in any significant volume between nations at the present time, with only limited exceptions, such as bottled water and beverages. This is because water is too heavy and is priced so low that water tankers and water pipelines are not yet economically feasible. The fact that water infrastructure tends to be both very expensive and also nationally financed likewise discourages the construction of the transboundary water infrastructures that would convey traded water.

On the other hand, **transboundary** rivers and other fresh waters that cross or straddle national boundaries are incredibly important for many nations' water supplies. Transboundary waters are shared, not traded, although an implicit water trade of a sort exists in the treaties and de-facto agreements between nations. Transboundary waters are common among the world's nations, owing to the tendency of rivers to disrespect political boundaries, with roughly 263 transboundary river basins and 300 transboundary aquifers in the world, involving nearly all of the world's nations (UNEP 2016). Few of these transboundary water flows have a cooperative management framework (see Sects. 6.3, 8.1.1, and the case studies of Chap. 19).

A notable cooperative management framework is the **Great Lakes Compact** (see Sects. 8.1.1 and 19.2.3) between the USA and Canada (and their States and Provinces) governing the Great Lakes, one of the world's great fresh waters. The GLC explicitly allows a preapproved list of water transfers, and also allows bottled water exports as long as small containers are used. Bottled water could be considered trade, although it is not a large trade in terms of volume, mass, or value. The GLC regulates any activities that create cumulative impacts on the Great Lakes water resources, and prohibits diversions and transfers that are not explicitly approved by the parties to the compact. The GLC is a very conservative cooperative management framework in the sense that it is legally binding and that it is comprehensive in the protections it guarantees.

Another notable cooperative management framework is the Colorado River Compact (CRC, U.S. Congress 1921) between the USA and Mexico, governing the sharing of the water supply on one of the world's most important arid region rivers. This is a complicated agreement that divides the waters of the Colorado River between the less fully developed and populous Upper Basin States (Colorado, Utah, Wyoming, New Mexico, and portions of Arizona), the more fully developed and populous Lower Basin States (California, most of Arizona, and Nevada), and Mexico. Nearly all of the Colorado River's water originates in the snowcapped peaks of Wyoming and Colorado, but far-downstream California and Arizona were the first to exploit its waters on a large scale. The CRC's rules ensure that water is shared equitably according to strict pre-negotiated rules, especially during times of drought. Pre-negotiated diversions are explicitly allowed, such as the large diversions of river water to Denver, Phoenix, Tucson, Las Vegas, and Los Angeles, and their surrounding agricultural operations. Water is not traded or sold in this compact, but the compact does ensure that downstream users far from the original source of the water—especially California, Arizona, and (internationally) Mexico—receive water.

Oil and grain are easy to move and trade and are valuable enough to make this trade profitable. On the other hand, water is relatively massive and voluminous, and its price is very low (usually zero), so water is not normally traded in quantity. Hard-to-trade Resources (like water) may be traded "virtually" by trading goods that require a lot of that resource to produce. This is analogous to the concept of embedded energy discussed above.

Virtual Water is the water that was consumed in the production of a good or service. For instance, a lot of virtual water is traded via grain exports, because grain is easy to produce in massive quantities in arid regions if you have irrigation water. Irrigated agriculture accounts for approximately 70% of total water withdrawal, and about 90% of virtual water flows globally relate to trade in agricultural products.

Nations with large populations and scarce water and farmland tend to be virtual water importers, and nations with small populations, lots of water, and abundant farmland tend to be virtual water exporters. Interestingly, the USA is among the largest net virtual water exporters due to its massive bulk grain exports (Fig. 7.7). This virtual water trade is so large and important that the study of this trade has

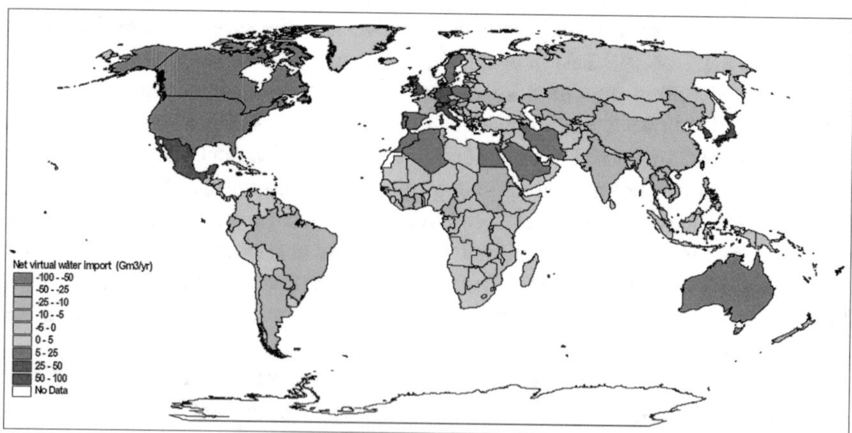

Fig. 7.7 Net virtual water imports in the period 1997–2001; green is a net exporter, and red is a net importer (Fig. 4.5 from Chapagain and Hoekstra 2004). Green nations tend to be rich in agricultural capacity; red nations tend to be populous, rich, and/or relatively poor in agricultural capacity

spawned a niche academic subdiscipline (Allan 2003; Chapagain and Hoekstra 2004; Konar et al. 2011).

Energy use, greenhouse gas emissions, minerals, land use, photosynthetic capacity, and other goods and materials are likewise "embedded" virtually in every international trade (Bruckner et al. 2012). The field of life cycle analysis generally, and "footprint" analysis specifically, is concerned with measuring these embedded virtual flows. Because a great deal of energy and water use is embedded in nearly every kind of good or service, it is generally true that nations with limited domestic energy or water resources rely heavily on trade in virtual energy or water resources as substitutes for scarce local resources. This virtual trade reflects the comparative advantages of energy-rich and water-rich nations, especially in the production of energy-or-water-intensive bulk grains and raw materials. The comparative advantage driving virtual trade is real, although it tends to be modest (Debaere 2014).

7.6 Modeling Trade in FEW Systems

In addition to IMPACT, which focuses on the trading of agricultural commodities, there are other modeling tools and approaches to analyze FEW systems, including trade at the local, regional, and global scales. In Chap. 15, we discuss modeling approaches that include the trade of food (directly and virtually through land and water) and energy at various scales. In one of these approaches, referred to as Integrated Assessment Modeling (see Sect. 15.2.4), trade of food and energy commodities is explicitly included in the calculations of supply and demand of land, extractive, and water resources that need to be considered in improved accounting efforts of FEW.

Another approach worth mentioning is Multi-Region Input/Output (MRIO). MRIO analysis is a widely used modeling approach, which enables analysts to explore the entire supply chain and the associated ("embodied") emissions or natural resource use. At its core, it is an accounting procedure relying on regional economic input–output (IO) tables and interregional trade matrices, depicting the flows of money to and from each sector within and between the interlinked economies and thus revealing each sector's entire supply chain. The MRIO modeling approach has been frequently used in water and land footprint and virtual land/water studies by utilizing the IO ability to quantify direct and indirect (upstream supply chain) land and water consumption for sectorial production at regional, national, or global scales.

The integration of trade modeling with modeling of FEW systems remains an important area of research and innovation. Given the intensity of the exchanges of FEW brought by trade, this is likely to remain an important area of activity in the years ahead.

Key Points

- Trade is an ancient institution, but globalization has dramatically accelerated trade.
- Nations trade to buy goods and services that are not available locally or that are available at lower prices from other nations.

- Nations trade to sell goods and services in greater quantity than could be sold locally.
- Differences in the prices of food and energy commodities reflect a combination of policy-driven distortions (like subsidies) and genuine comparative advantages.
- Because it is unlikely for human or natural shocks to simultaneously impact the entire group of global producers of a commodity, global trade markets increase FEW resilience.
- Grain and Petroleum products are among the most-traded commodities by volume.
- Trade policy tends to balance FEW self-sufficiency objectives against the need to access less expensive commodities or access commodities during local shocks like drought or severe winter weather.
- Hard-to-trade Resources (like water) may be traded "virtually" by trading goods that require a lot of that resource to produce; for instance, a lot of virtual water is traded via grain.
- FEW trade shocks have been used as a geopolitical weapon during the global era.
- Nations have widely differing trade policies that reflect a blend of openness versus protectionism (e.g., tariffs).
- Neoliberal economics argues that free and open trade enriches all parties and maximizes resilience, but there are real-world limitations to this philosophy.
- Cartels like the Oil and Petroleum Exporting Countries (OPEC) attempt to manipulate trade to their members' advantage.

Discussion Points and Exercises

1. Is trade in food and energy a good thing, or a bad thing? Why or why not, or in what cases?
2. List, in descending order, the top ten food and energy importing and exporting nations in the world, in a recent year of record (four lists, 40 total nations).
3. Is your nation is self-sufficient in terms of food, energy, and/or water? Is your nation a net importer, exporter, or neutral?
4. Has there ever been an event where your nation's food, energy, or water supply was unexpectedly interrupted? What happened; what was the impact? How did your leaders respond?
5. Consider the Oil Shocks of the 1970s and 1980s. What changes did your nation make to its FEW policies as a result of those shocks?
6. What characteristics, policies, and/or production advantages do the world's major FEW exporting nations share?
7. What characteristics, policies, and/or needs to the world's major FEW importing countries share?
8. Why do you think some countries like the USA pursue a policy of overproducing food and energy commodities, despite the costs of this policy?
9. Are there any countries that specialize specifically in food or energy production and export? What are the risks involved in being heavily concentrated in this type of export, or in depending heavily on that exporter?
10. Should trade in food and energy be totally free? Why or why not?
11. Can you think of an example where water is traded directly (not virtually)?

References

Allan, J. A. (2003). Virtual water-the water, food, and trade nexus. Useful concept or misleading metaphor? *Water International, 28*(1), 106–113.

Bailey, R., & Wellesley, L. (2017). *Chokepoints and vulnerabilities in global food trade. Chatham House*. Retrieved February 20, 2019, from https://reader.chathamhouse.org/chokepoints-vulnerabilities-global-food-trade#

Baldwin, R. (2016). *The great convergence*. Cambridge: Harvard University Press.

Benton, T. (n.d.). *Food security, trade and its impacts, ResourceTrade.Earth, Chatham House*. Retrieved February 20, 2019, from https://resourcetrade.earth/stories/food-security-trade-and-its-impacts#top

BP (2018) BP Statistical Review of World Energy, 67th Edition, British Petroleum Co, London. https://www.bp.com/content/dam/bp/business-sites/en/global/corporate/pdfs/energy-economics/statistical-review/bp-stats-review-2018-full-report.pdf

Bruckner, M., et al. (2012). Materials embodied in international trade–Global material extraction and consumption between 1995 and 2005. *Global Environmental Change, 22*(3), 568–576.

Cazcarro, I., Duarte, R., & Sanchez-Choliz, J. (2013). Water footprints for Spanish regions based on a multi-regional input-output (MRIO) model. In J. Murray & M. Lenzen (Eds.), *The sustainability practitioner's guide to multi-regional input-output analysis* (pp. 119–132). Champaign, IL: Common Ground.

Chapagain, A.K. and Hoekstra, A.Y. (2004) Water footprints of nations, Volume 1: Main Report, Value of Water Research Report Series No. 16 UNESCO-IHE Delft https://waterfootprint.org/media/downloads/Report16Vol1.pdf

D'Odorico, P., et al. (2014). Feeding humanity through global food trade. *Earth's Future, 2*(9), 458–469.

Debaere, P. (2014). The global economics of water: is water a source of comparative advantage? *American Economic Journal: Applied Economics, 6*(2), 32–48.

Dietzenbacher, E., & Velazquez, E. (2007). Analysing Andalusian virtual water trade in an input-output framework. *Regional Studies, 41*, 185–196.

EIA. (n.d.). *The changing U.S. energy trade balance is still dominated by crude oil imports*. Retrieved February 20, 2019, from https://www.eia.gov/todayinenergy/detail.php?id=37253

Feng, K., Chapagain, A., Suh, S., Pfister, S., & Hubacek, K. (2011). Comparison of bottom-up and top-down approaches to calculating the water footprints of nations. *Economic Systems Research, 23*, 371–385.

Feng, K., Hubacek, K., Pfister, S., Yu, Y., & Sun, L. (2014). Virtual scarce water in China. *Environmental Science & Technology, 48*, 7704–7713.

FAO (2018) *The State of Agricultural Commodity Markets 2018*. Agricultural trade, climate change and food security. Food and Agriculture Organization of the United Nations Rome. http://www.fao.org/3/I9542EN/i9542en.pdf

Hathaway, D. (1997). *Agriculture and the GATT: Rewriting the rules*. Washington, DC: Institute for International Economics, Policy Analysis in International Economic.

Huang, R. D., Masulis, R. W., & Stoll, H. R. (1996). Energy shocks and financial markets. *Journal of Futures Markets: Futures, Options, and Other Derivative Products, 16*(1), 1–27.

IEA. (2018). *World energy balances 2018*. International Energy Agency, 978-92-64-30155-9.

International Energy Charter. (n.d.). Retrieved February 20, 2019, from https://energycharter.org/process/overview/

Keynes, J. M. (2017). *The economic consequences of the peace*. New York: Routledge.

Kilian, L. (2008). The economic effects of energy price shocks. *Journal of Economic Literature, 46*(4), 871–909.

Konar, M., Dalin, C., Suweis, S., Hanasaki, N., Rinaldo, A., & Rodriguez-Iturbe, I. (2011). Water for food: The global virtual water trade network. *Water Resources Research, 47*(5).

Lenzen, M., Moran, D., Bhaduri, A., Kanemoto, K., Bekchanov, M., Geschke, A., & Foran, B. (2013). International trade of scarce water. *Ecological Economics, 94*, 78–85.

Mann, C. C. (2011). *1493: Uncovering the new world Columbus created*. New York: Vintage.
Muñoz-Castillo, R., Feng, K., Sun, L., Guilhoto, J., Pfister, S., Miralles-Wilhelm, F., & Hubacek, K. (2019). The land-water nexus of biofuel production in Brazil: Analysis of synergies and tradeoffs using a multiregional input-output model. *Journal of Cleaner Production, 214*, 52–61.
Robinson, S., et al. (2015). *The international model for policy analysis of agricultural commodities and trade (IMPACT): Model description for version 3, IFPRI discussion paper 01483, International Food Policy Research Institute*. Retrieved from http://www.ifpri.org/publication/international-model-policy-analysis-agricultural-commodities-and-trade-impact-model-0
Rosegrant, M. W., et al. (2012). *The international model for policy analysis of agricultural commodities and trade (IMPACT) model description, International Food Policy Research Institute (IFPRI)*. Retrieved from http://www.ifpri.org/publication/international-model-policy-analysis-agricultural-commodities-and-trade-impact
Sassen, S. (1999). Globalization and its discontents: Essays on the new mobility of people and money. The New Press. ISBN-10: 1565845188
Scanlon, B. R., Ruddell, B. L., Reed, P. M., Hook, R. I., Zheng, C., Tidwell, V. C., & Siebert, S. (2017). The food-energy-water nexus: Transforming science for society. *Water Resources Research, 53*(5), 3550–3556.
Serrano, A., Guan, D., Duarte, R., & Paavola, J. (2016). Virtual water flows in the EU27: A consumption-based approach. *Journal of Industrial Ecology, 20*, 547–558.
Torreggiani, S., Mangioni, G., Puma, M. J., & Fagilo, G. (2018). Identifying the community structure of the international food-trade multi network. *Environmental Research Letters, 13*(5), 054026. https://doi.org/10.1088/1748-9326/aabf23.
U.S. Congress. (2008). *Senate joint resolution 45, 110th Congress 2nd session*. Retrieved March 11, 2019, from https://dnr.wi.gov/topic/GreatLakes/documents/Congress_Compact_Consent.pdf
UNEP. (2016). *Transboundary waters systems—status and trends: Crosscutting analysis*. Nairobi: United Nations Environmental Programme (UNEP).
US Congress. (1921). *Colorado river compact, U.S. Congress, 42 statutes at large, p. 171*. Retrieved March 11, 2019, from https://www.usbr.gov/lc/region/g1000/pdfiles/crcompct.pdf
Wolf, M. (2004). *Why globalization works (No. 3)*. New Haven, CT: Yale University Press.
World Factbook. (2019). *Country comparison: Natural gas exports*. U.S. Central Intelligence Agency. Retrieved March 8, 2019, from https://www.cia.gov/library/publications/the-world-factbook/rankorder/2251rank.html

Further Reading

Allan, J. A. (2003). Virtual water-the water, food, and trade nexus. Useful concept or misleading metaphor? *Water International, 28*(1), 106–113.
Bailey, R., & Wellesley, L. (2017, June). *Chokepoints and vulnerabilities in global food trade, Chatham House*. Retrieved April 10, 2019, from https://www.chathamhouse.org/publication/chokepoints-vulnerabilities-global-food-trade
Bouet, A., & LaBorde, D. *Building food security through international trade agreements, IFPRI*. Retrieved February 20, 2019, from http://www.ifpri.org/blog/building-food-security-through-international-trade-agreements
D'Odorico, P., et al. (2014). Feeding humanity through global food trade. *Earth's Future, 2*(9), 458–469.
Distefano, T., Laio, F., Ridolfi, L., & Schiavo, S. (2017). *Shock transmission in the International Food Trade Network. A data-driven analysis*. SEEDS working papers 0617. SEEDS, Sustainability Environmental Economics and Dynamics Studies.
Friedman, M., & Friedman, R. (1997). The case for free trade. *Hoover Digest, 4*.
Hamilton, J. D. (2003). What is an oil shock? *Journal of Econometrics, 113*(2), 363–398.

Headey, D. (2011). Rethinking the global food crisis: The role of trade shocks. *Food Policy, 36*(2), 136–146.
Keynes, J. M. (2017). *The economic consequences of the peace*. New York: Routledge.
LEVIN Institute. (n.d.). Globalization 101: Trade and globalization, SUNY LEVIN Institute. Retrieved February 20, 2019, from http://www.globalization101.org/
Ortiz-Ospina, E., Beltekian, D., & Roser, M. Trade and globalization, our world in data. Retrieved February 20, 2019, from https://ourworldindata.org/trade-and-globalization
Sassen, S. (1999). Globalization and its discontents: Essays on the new mobility of people and money.
UN Water. (n.d.). Transboundary waters. Retrieved February 20, 2019, from http://www.unwater.org/water-facts/transboundary-waters/
Worstall, T. (2014). *The benefits of international trade in food; or the limits to localism, Forbes, August 14*. Retrieved April 12, 2019, from https://www.forbes.com/sites/timworstall/2014/08/14/the-benefits-of-international-trade-in-food-or-the-limits-to-localism/#6fd1fef31691
Yergin, D. (2006). Ensuring energy security. *Foreign Affairs, 85*, 69–82.

Chapter 8
US Governance

Beth Kinne and Darrin Magee

8.1 Introduction

In Chap. 1, we noted that high-level decision-making and policy in many countries tend to treat food, energy, and water (the "three great consumables") as strategically important *separately*, but not in an integrated fashion. Failure to address the critical interlinkages between food, energy, and water systems leads to conflicts among users and producers whose primary interests lie in one of the "Three Great Consumables."

Creating legal, social and economic structures to promote responsible stewardship of the Three Great Consumables at all geographic scales—from the individual and community to the nation and the entire globe—is challenging. Chapter 3 explored how international law grapples with nexus challenges. Policies at different levels of government need to address specific nexus problems and relationships that arise with particular geographic configurations. Chapter 12 provides definitions of three particular scales of interest to nexus analyses and policies: micro-, meso-, and macro-.

Chapter 1 frames the nexus issue and explicitly identifies policy as one of the critical criteria for improving the way we manage nexus resources. This chapter examines US law and policy at the FEW nexus. Most laws regulating nexus resources are primarily directed at protecting or promoting the development of one resource sector, including the related human health and safety impacts. Nevertheless,

B. Kinne
Environmental Studies, Hobart and William Smith Colleges, Geneva, NY, USA
e-mail: kinne@hws.edu

D. Magee (✉)
Environmental Studies, Hobart and William Smith Colleges, Geneva, NY, USA

Asian Studies, Hobart and William Smith Colleges, Geneva, NY, USA
e-mail: magee@hws.edu

P. Saundry, B. L. Ruddell (eds.), *The Food-Energy-Water Nexus*, AESS Interdisciplinary Environmental Studies and Sciences Series,
https://doi.org/10.1007/978-3-030-29914-9_8

laws primarily governing one or another nexus resource can require intentional consideration of *environmental* impacts, which often include impacts on nexus resources.

In this chapter, we provide an overview of US law vis-à-vis international law and in light of the US federalist system, which gives individual states greater leeway than is found in the sub-national jurisdictions of many other countries. We then examine key laws at the national, state, and local levels that address one or more FEW components. This chapter demonstrates that attention to FEW impacts is currently rather fragmented in the USA. In order to change this, state and federal regulatory agencies in the USA must develop procedures that require the integrated assessment of nexus impacts across all three FEW sectors for every action taken on one of those sectors.

There is a pressing need for information and quantitative tools to support integrated planning of FEW resource development and use, in an effort to avoid unwanted and unsustainable scenarios in coming years. Although the FEW nexus is fairly evident, these three sectors have historically been regulated and managed separately; and despite growing concern over these trends, decision makers often remain ill-informed about their drivers and ill-equipped to deal with possible outcomes. Such quantitative tools (data and models) that can support more effective governance of FEW systems are explored in further details in Chaps. 14 and 15, respectively. At the heart of this discussion lies the question of the role science plays in informing policy, and the responsibility policymakers have to consider scientific expertise when formulating laws and policies. We return to that question in the final chapter of this textbook.

8.1.1 Framing International and US Governance at the Nexus

The relationship between the US and international governance is a two-way, iterative one. In some cases, domestic law shapes international norms, while in other cases, the reverse is true. For example, the Oil Pollution Act of 1990 passed in the aftermath of the Exxon Valdez oil spill required double hulls for new vessels carrying oil in US waters. This motivated the adoption of an equivalent requirement internationally under the 1992 MARPOL Convention.

Once the US Senate approves international treaties and the President ratifies them, those treaties become domestic law.[1] However, from a practical perspective, implementation in the domestic sphere often requires the creation of new laws and regulations consistent with treaty obligations.

Examples of international agreements that impact the food–energy–water nexus in the USA include:

1. The Great Lakes-St. Lawrence River Basin Sustainable Water Resources Agreement between eight states and two Canadian provinces to regulate

[1] It is common to speak of Congress "ratifying" a treaty, but it is the president who ratifies the treaty only after two-thirds of the Senate has provided "advice and consent."

diversions from the Great Lakes basin. The agreement was approved by legislation in all states and a Joint Resolution of US Congress in 2008.
2. The Colorado River "Law of the River," a collection of over a dozen treaties, interstate compacts, federal acts of legislation, and court cases that govern rights to the water in the Colorado River Basin as it is shared among the seven basin states and Mexico.
3. Trade agreements such as the 1993 North American Free Trade Agreement (NAFTA) and the agreements underpinning the World Trade Organization (WTO), which are adopted by acts of legislation rather than ratified as international treaties.
4. The 1992 United Nations Framework Convention on Climate Change (UNFCCC), which was ratified after a two-thirds vote of the US Senate in 1992.

In many cases, international and domestic laws covering natural resources such as energy, water, and food share the same goals: conservation, long-term planning, **facilitation** of resource development and trade, or promotion of human health and safety. The concurrent need to protect interstate relationships, national sovereignty, and the priorities of key powerful industries, results in the use of different mechanisms to achieve similar goals (Hall 2007). Many of the principles applied in international disputes—the geographical locus of the action that caused the dispute or contractual agreements to submit to the jurisdiction of a particular sovereign, for example—are also relevant to adjudicating disputes among US states, which retain a degree of sovereignty vis-à-vis the federal government.

The USA's interests in international law at the food, energy, water nexus relate most directly to relationships between the USA, Canada, and Mexico, due to geographical connectivity, in addition to other key agricultural trading partners such as the European Union. Concerning water, the USA enjoys the privilege of being Mexico's upstream neighbor. Therefore, the Colorado River Compact failed to incorporate Mexico's interests for the first 23 years, treating this vital river as something to be shared only among seven US states. Only in 1944, long after the river's annual flow had been over-allocated, was Mexico recognized as a party to the Compact. In contrast, the USA is the downstream neighbor on about half the streams along the US-Canadian border, providing an incentive for the USA to promote equitable apportionments of water resources across this—and arguably other—international boundaries.

According to the U.S. Department of Agriculture, Canada, Mexico, and the European Union are the largest suppliers of foreign agricultural products to the USA. At the same time, Canada has consistently remained one of the top two importers of food from the USA. NAFTA frames the international relationships among the USA, Mexico, and Canada with respect to water, food, and energy to some degree. Reforms to NAFTA could enhance or inhibit (or prohibit) domestic policies promoting the development of low-carbon energy sources, economically sustainable food production, and environmental rules and regulations that protect water. The direction of such changes will depend on the degree to which decision-makers recognize the importance of FEW interconnectivity.

8.1.2 Federalism and the Food–Energy–Water Nexus

Within the USA, the federalist structure creates shared authority between the federal government and the states, resulting in a varied landscape of natural resource regulation. While federal environmental and other laws may preempt or constrain state and local regulations in some cases, but state and local (municipal) governments retain authority over many FEW resources through water resource allocation and land use decisions, resulting in a great variation of regulation of FEW resources.

At the federal level, the **National Environmental Policy Act (NEPA)** provides for consideration of significant environmental impacts of regulations and projects that are created, funded, or permitted by federal government agencies. State analogs to NEPA require a similar assessment of state-level laws and projects found likely to have significant environmental impacts. At the municipal level, various local land-use laws provide a mechanism for protecting water resources, promoting food production, and governing impacts of energy infrastructure and production. These laws often reflect local social and economic priorities as well as place-specific environmental sensitivities, but the resulting mosaic of regulations that are inconsistent from state to state and municipality to municipality creates some unwanted effects, such as hot spots for pollution or resource development.

The federal regulatory system rarely explicitly addresses the food–water–energy nexus, despite extensive research showing strong interconnection among the three sectors. Most decisions impacting water pollution, energy production, and agriculture are made by specialized administrative agencies such as the Environmental Protection Agency (EPA), the Federal Energy Regulatory Commission (FERC), and the U.S. Department of Agriculture (USDA), and in some cases the Food and Drug Administration (FDA). Agency actions are subject to judicial review under limited circumstances, and such legal challenges form a substantial body of **precedent**—that is, past court decisions that guide and constrain courts in future cases—in the field of administrative, environmental, and natural resource law.

Legal precedent grants federal agencies discretion to make their best expert judgment when they are acting within the scope of the authority granted by Congress. Where conflicts arise between laws at different scales, agency rulings sometimes seek to mitigate them. For example, after a 2004 case found that back-pumping water into the Florida Everglades required a National Pollutant Elimination Discharge System (NPDES) permit, the EPA issued a Water Transfer Rule in 2008. This rule explicitly stated that discharge permits would not be required for transfers of water from one navigable water body to another, and all navigable waters would be treated as one, the "unitary waters" theory. This rule prevents disruption of water law in western states, which manage water resources by extensive transfers of water from one watershed to another, avoiding conflict between the NPDES system and state water rights systems. However, it risks allowing lesser quality water to be transferred to bodies of higher quality water.

For practical reasons, many federal environmental statutes, such as the Clean Air Act (CAA) and the **Clean Water Act** (CWA), delegate implementation to the states. Upon EPA's approval of a state implementation plan, the state enjoys **primacy** with respect to the federal law and is in charge of implementing federal standards. Granting states primacy has two significant impacts: within the parameters of approved plans, states can devise funding and management strategies appropriate for their own natural resources; and the federal government is largely removed from making specific decisions and funding their implementation.

Under the doctrine of **preemption**, states may not regulate in ways that contradict federal law, though often they may pass regulations that are *stricter* than the floors set by federal laws. Similarly, when states choose to regulate in a certain arena, such as unconventional oil and gas development, local governments in that state may be prohibited from enacting local regulations that conflict with state law. A thorough treatment of the doctrine of preemption is beyond the scope of this chapter, but it plays a critical role in the interaction among federal, state, and local agencies on issues such as energy resource development, water resource protection, and regulation of production and transportation of food.

The federal government has not been completely silent on nexus issues. Recent energy developments, in particular, have catalyzed concern about impacts on water and food. During the Obama Administration (2009–2017), which coincided with the boom in unconventional shale gas development through hydraulic fracturing ("fracking"), the Government Accountability Office (GAO) published several reports aimed squarely at the energy–water nexus, motivated partly by the fracking revolution and its water impacts (See Further Reading). However, the majority of regulation of fracking remains within the purview of the States, particularly after industry first challenged federal rules governing fracking on federal and Indian lands finalized in 2015, and then saw those rules rescinded by the Trump Administration in 2017.

Existing environmental laws mitigate some of the hazards from the erosion, brine, radioactive cuttings, and fugitive emissions of greenhouse gases that result from unconventional oil and gas development but not all. Those laws often fail to address the various stages of resource development and production comprehensively. For instance, while the wastes associated with exploration, development or production of crude oil, natural gas, or geothermal energy are themselves exempt from regulation under the Resource Conservation and Recovery Act (RCRA), the landfills where those rock and dirt cuttings from drilling are routinely disposed of *are* subject to RCRA. Similarly, the leachate from landfills containing similar wastes is transferred to municipal wastewater treatment plants that discharge into rivers and lakes, even though direct discharges from unconventional oil and gas production facilities to municipal wastewater treatment facilities are prohibited. The effluent from wastewater treatment plants is, however, is subject to CWA discharge permits.

Produced water—water that comes back out of the wellbore after fracking—is often injected into abandoned oil and gas wells as a means of disposal. Deep well injection of wastes is typically regulated by the Safe Drinking Water Act's Underground Injection Control (UIC) regulations. However, under what is commonly referred to as the "Halliburton Loophole," encoded in the Energy Policy Act of 2005[2] wastes resulting from oil and gas exploration and production are exempt from UIC requirements "unless such requirements are essential to assuring that underground sources of drinking water will not be endangered by such injection."

Other key exceptions include exemptions from CWA stormwater permitting requirements for oil and gas exploration and production, processing and transmission facilities These exemptions from federal law, however, do not preclude state law from requiring stormwater permits for construction activities associated with oil and gas development. This is just one example of how the regulatory landscape for FEW resources is complex and multi-scalar.

8.1.3 *Private Property Rights and FEW Resource Regulation*

There is a tradition of strong private property rights in US jurisprudence. Water law, for example, has historically favored individuals who harness and divert water away from natural channels to private uses, even if doing so harms the ecological health of watersheds. Similarly, conventional oil and gas law is based on the rule of capture—the party who removes oil or gas from the ground gains a property right in that resource. With respect to food or any other vegetable matter, the farmer who owns the land the crop is grown on owns the crop unless a contract stipulates otherwise. However, regulations to protect the public interest routinely impinge on individual property rights and alter market forces. Governments can regulate the location and manner of energy exploration and production, and the placement and operation of a larger infrastructure for transmission of oil gas and electricity, for example.

Federal and state financial assistance and insurance programs help stabilize agricultural markets. But the production of food and energy is still largely under the control of individuals and corporations, and production rates respond to price signals in the market, which can result in industry trends that are not conducive to the long-term sustainability of FEW resources. According to the USDA, 90–99% of crops grown in southwestern US states (based on 2012 market value) depended on irrigation.[3] However, data from the U.S. Geologic Survey (USGS) shows that total

[2] Critics named the loophole after a Wellfield Services company that benefited from it. Richard (Dick) Cheney, US vice-president at the time of the 2005 Act, had previously served as CEO of Halliburton.

[3] See USDA, National Agricultural Statistics Service (NASS), 2012 United States Census of Agriculture, AC 12-A-51, Washington, DC, May 2014, http://www.agcensus.usda.gov/Publications/2012/Full_Report/Volume_1,_Chapter_1_US/usv1.pdf.otal.

US water use, and use in all sectors except agricultural, declined between 2005 and 2015, while agricultural use increased only slightly.[4] This increase is largely due to the increased reliance on irrigation in US states east of the Mississippi River.

8.2 The U.S. Regulatory Framework by FEW Sector

The following sections outline some important regulatory frameworks at the federal level for the water, energy, and food sectors. The list is by no means exhaustive and focuses on critical regulations designed primarily to regulate one sector but which turn out to have significant impacts on other sectors at the food–energy–water nexus. It may be helpful here to recall the discussion on the micro, meso, and macro scales of nexus effects and policies in Chap. 12.

8.2.1 Water

In Sect. 2.5, we introduced the concept of water systems at various scales. Similarly, some US laws and policies also are based on hydrologic units such as the **watershed** or **catchment basin**; some are based on political, jurisdictional boundaries such as states, counties, towns, and cities; and still, others are based on "hydro-economic units" such as irrigation districts. The term, hydro-economic unit is often used in modeling. In this text, it denotes a group of people who share, for a distinct economic purpose, water that is in one or more hydrological units.

Water governance in the USA is complicated by the division between water quantity laws and water quality laws and the historical lack of recognition of the intimate hydrological connection between surface water and groundwater flows. Water quantity law prioritizes obtaining maximum utility from water resources, promoting multiple users and beneficial uses of any given water body. Water quality law, meanwhile, focuses on preventing degradation and improving impaired waters, goals arguably more subjective and difficult to measure. The two regulatory systems are sometimes at odds with one another. For example, maximizing use of water quantity in a stream can degrade its quality.

Water quantity regulation is critical to food and energy production; conversely, production of food and energy substantially impacts water resources. Of all freshwater withdrawals in the USA in 2010, irrigation and thermoelectric power generation accounted for 36% and 40%, respectively, both of which are much larger than the 14% used for public water supplies (Maupin et al. 2014). This contrasts with the global average, where irrigation comprises about 70% of water usage (Black 2016). These decreases may be attributed to increased efficiency in irrigation

[4] See USDA, NASS, Quick Stats, http://quickstats.nass.usda.gov/.

technologies, improvements in power plant cooling technologies, and advances in industrial wastewater recycling. Also, it is important to note that some withdrawals are a poor proxy for water use; much of the water withdrawn for agricultural use is **consumptive**, whereas much of that withdrawn for power plant cooling is not.

Water quantity law is state law and varies considerably among states. Those with abundant fresh water tend to have less regimented systems governing water rights, while those with more limited supplies tend to have elaborate water rights regimes. Two predominant models have shaped state water quantity law: riparian rights and prior appropriation rights, although over time many states have adopted modifications and combinations of the two.

A pure riparian rights doctrine grants use rights in water to landowners whose property abuts a water body such as a river or a lake. In times of shortage, all users participate in reducing use. This doctrine originated in England and was transplanted to the USA where the "reasonable use" overlay, sometimes called the "American Rule" was established, restricting riparian users from making unreasonable use of the water.

"Reasonable use" is, of course, a slippery term, and the determination of reasonable use is multifaceted. It includes consideration of the suitability of the water body for a particular use, such as drinking water, irrigation, fish propagation, or navigation, along with the social and economic values of that use, and the extent to which that use causes harm to others.

The appropriative rights doctrine was introduced in the arider western states where cultivating crops required regular irrigation. One who diverts water from a stream and puts that water to *beneficial use* establishes a water use right. The right is usually limited in scope by purpose (e.g., irrigation of a precise number of **acres**), the total amount of water or flow rate, and sometimes by the time of year.

Priority in time trumps proximity in space; an established right is protected against any other user who might assert the use of the water at some later time. Water rights with earlier appropriation dates are more valuable than rights with later dates, because in times of shortage, later (more "junior") appropriations are curtailed first. Early ("senior") water rights, particularly non-consumptive water rights for hydroelectric plants, can have significant impacts on downstream users. For example, the 1902 water right for the relatively small (15MW) Shoshone power plant on the upper Colorado River creates substantial, reliable downstream flows, protecting rights of municipal and agricultural users lower on the river.

Watersheds frequently span two or more states, creating the need for interstate compacts or other instruments governing water sharing. These agreements usually focus on three areas: water apportionment, pollution control, and flood control (see Chap. 19).[5]

For example, the Mississippi River watershed, one of the largest in the world, drains about 40% of the land mass in the continental USA and includes 31 states and two Canadian provinces. Therefore, the watershed is impacted by the laws and regulations of all 33 state/provincial jurisdictions and two national governments. In addition, the Mississippi is subject to two basin-specific interstate compacts governing water pollution: The Louisiana-Mississippi Tangipahoa River Waterway Compact and the Mississippi River Interstate Pollution Phase-out Compact.

---

[5] An exception to the three focal areas noted above is the Atlantic Salmon Compact of 1983 (Public Law 98-138) which focuses on restoration of the Atlantic salmon fishery in the Connecticut River Basin.

Interstate organizations governing water attempt to bridge the gap between the hydrological system and the political systems that impact it. Those organizations can take several forms. The most formal is the commission, which is governed by a compact to which the federal government is a party. Examples include:

(a) The Delaware River Basin Commission (1961), which includes the states of Delaware, New Jersey, Pennsylvania, and New York as well as the Army Corps of Engineers as the federal representative;
(b) The Susquehanna River Basin Commission, (1970), which includes the states of Maryland, Pennsylvania and New York and a representative from the federal government;
(c) The Upper Colorado River Basin Commission (1948), which includes the states of Colorado, Wyoming, Utah, and New Mexico, and an appointee from the federal government.

One important interstate water organization not governed by a compact to which the federal government is a party is the Upper Mississippi Basin Association (UMBA). While the UMBA includes no formal representative from the federal government, representatives from the U.S. Army Corps of Engineers, the Departments of Agriculture, Transportation, and Interior, and the EPA all serve in advisory, non-voting capacities to inform the governance of the river basin. These institutions all have their own particular missions and priorities that may differ for the river.

Advancing scientific understanding of hydrological connections, therefore, creates challenges for water rights and water pollution law. The interpretation of the federal–state jurisdictional boundary is challenged, as seen in the heated negotiations over the legality and practical implications of the **Clean Water Rule of 2015**. The debate over the extent of federal jurisdiction over water resources maps the alternate interpretations by U.S. Supreme Court Justice Scalia and Justice Stevens in the 2006 *Rapanos* v. the *United States 547 U.S. 715* case. Justice Scalia found that federal jurisdiction applied to "relatively permanent, standing or flowing bodies of water," but not "occasional," "intermittent," or "ephemeral" flows, and that a "hydrological connection" between an intermittent body of water and a navigable body of water was insufficient to confer federal jurisdiction on the former, and a "continuous surface connection" with navigable waters was required to confer federal jurisdiction over any non-navigable body of water.

The Obama Administration promulgated the **Clean Water Rule of 2015**, sometimes referred to as the *WOTUS Rule* (Waters of the USA). The Clean Water Rule of 2015 followed Justice Kennedy's reasoning in *Rapanos*, in which Kennedy found that federal jurisdiction reached any water body that had a "significant nexus" to navigable waters or seas. The rule effectively expanded federal jurisdiction over water, particularly over wetlands, and spurred litigation around the country in attempts to define the boundary between state and federal control in this area. The Trump Administration stayed the application of the rule for 2 years, effectively reinstating the 1986 interpretation, which followed Scalia's opinion in *Rapanos*. That stay was vacated (canceled) by a federal judge in August 2018, impacting its interpretation in 26 states, but not in the 24 states where other litigation over the rule is, as of January 2019, still pending.

In the following paragraphs, we discuss the role of the Clean Water Act and the Safe Drinking Water Act in regulating surface water and groundwater resources.

8.2.1.1 The Clean Water Act

The Clean Water Act (CWA) is the primary water regulation mechanism that also affects the energy sector. The CWA governs **surface water** quality but is largely silent on groundwater. The CWA has been interpreted by courts to govern quantity indirectly, in that **minimum in-stream flows** have been found to be a valid water quality criterion. In-stream flow requirements impact permitting of removal of water from river systems for a variety of uses, from hydroelectric power plants to municipal and irrigation withdrawals.

At the federal level, the EPA is charged with enforcing the provisions of the CWA, which it does by issuing **pollutant discharge** permits for **point-source** polluters through the **National Pollution Discharge Elimination System** (NPDES). In keeping with the architecture of the federal system, the CWA also provides an avenue for states to assume **primacy** for drafting and enforcing regulations in compliance with the CWA.

Primacy is based on the notion that state regulators, rather than federal ones, are more familiar with local conditions and therefore better able to adapt federal laws to those conditions while implementing enforcement regimes. In states like California, where the linkages between water services and energy use are clear and significant, primacy has enabled state regulators to tailor programs serving local needs for water and energy conservation. However, shifting the enforcement authority to state regulators also shifts the costs of enforcement to state budgets. Asserting primacy may also expose state regulators to more direct pressure from users wishing to skirt or soften those regulations.

Most states have applied for and been delegated primacy with respect to the CWA, many in the years immediately following the Act's initial passage in 1972. As of January 2019, the only states *not* authorized to administer the program were Idaho, Massachusetts, New Hampshire, and New Mexico, although Idaho began transitioning to state administration of the system in 2018 and expects to take complete control by 2021. The EPA also remained the jurisdictional authority for the District of Columbia, Puerto Rico, and US territories in the Pacific.

The CWA encompasses both **pollutant discharges** and **thermal discharges**. The justification for linking thermal and pollutant discharges is straightforward: increases in water temperature can reduce the ability of a water body to hold dissolved oxygen, which in turn can affect the water body's suitability for aquatic life or its ability to promote bacterial decomposition of organic matter. According to the USGS, over 40% of all US water withdrawals in 2015 were used for thermoelectric power; most of those withdrawals were from surface sources.

Figure 8.1 shows the geographic breakdown of water withdrawals used for thermoelectric power. The CWA's recognition of the synergistic effects between thermal pollutants and other contaminants provides critical leverage for increasing water

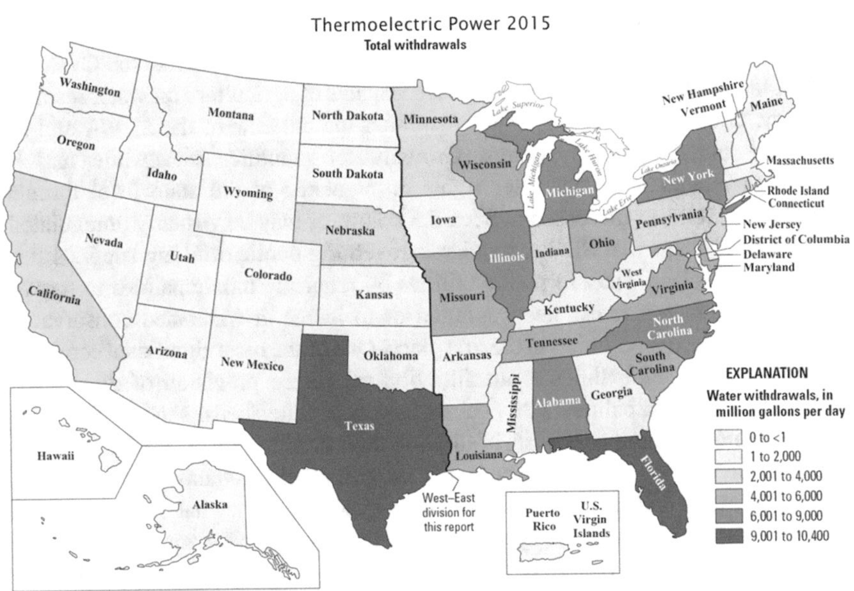

Fig. 8.1 Map of water withdrawals for thermoelectric power. (Source: U.S. Geological Survey, https://water.usgs.gov/watuse/wupt.html)

quality standards over time, but limitations in enforcement, along with the complexities in understanding how different pollutants and polluters interact, can impede such improvements.

One final point on the Clean Water Act bears mention: its emphasis on **point-source pollution** means that regulation for drinking water supplies in urban areas polluted by point sources subject to NPDES permits is far tighter than regulation for water users near farms, where **non-point** sources prevail. Regulation of non-point sources of pollution falls to the states, creating a "*de facto* fifty-state experiment in regulation—or rather, non-regulation –" of non-point source pollution (Craig and Roberts 2015, p. 2). Agricultural operations are a major contributor. Craig and Roberts note relatively few cases where states have found political support for regulating non-point source pollution: the fate of salmon in the Pacific Northwest, the "blue-baby syndrome" resulting from nitrates, and the degradation of the Chesapeake Bay, for example. More recently, health impacts of harmful algal (cyanobacteria) blooms have garnered political attention in watersheds throughout the country (see Sect. 5.2.3).

8.2.1.2 Safe Drinking Water Act

The Safe Drinking Water Act (SDWA), enacted in 1974 and amended in 1996, was created to protect both surface and subsurface drinking water sources. According to Pollans (2016), almost one-third of the pollutants regulated by the SDWA enter waterways through non-point source pollution. As with the CWA, the EPA shares enforcement responsibilities with state and local agencies. Also like the CWA, the SDWA largely fails to prevent the negative impacts of agriculture on water resource integrity. Groundwater quality thus falls under the purview of the SDWA and the act's sole source aquifer provisions. Groundwater quantities, meanwhile, may be subject to federal reserved water rights or regulated at the state level through adjudicated wells, as discussed in Sect. 5.3 below, or may be virtually unregulated.

Pollans explains that SDWA standards governing public drinking supplies shift the burden of clean water to public utilities by requiring municipalities to pay for cleaning water after it has been polluted or to invest in watershed conservation measures to prevent pollution in the first place. One of the most significant economic incentives for municipalities is the filtration avoidance provision of the SDWA, which allows municipalities with heavily protected watersheds, such as New York City, to avoid the significant costs of filtering their drinking water. Yet, the law does not mandate protection of watersheds, nor provide other mechanisms to attain it.

In contrast, SDWA groundwater protection plans are somewhat more structured. Section 1424(e) of the SDWA specifies protective measures for aquifers qualifying as sole source aquifers (SSA), or those which provide 50% or more of the drinking water in a service area and for which there are no alternatives. SSA designation prevents federal financial support for any project within the aquifer recharge zone that could contaminate the aquifer but does not prevent all contamination.

8.2.2 Energy

As described in Chap. 2, Energy and water are intimately connected. Their use converges at three critical junctures where efficiency policies can make significant sustainability gains. First, pumping and treating water for drinking, and removing contaminants from wastewater, require significant energy. The California Energy Commission (n.d.) estimates that 30% of non-power plant use of natural gas goes to meet water services directly. Similarly, one study of major municipalities in southern California tied roughly one-fifth of electricity use to "water-related services" (Stokes and Horvath 2009). Even in the simplest and most straightforward of cases, energy costs can account for 30–50% of a wastewater treatment plant's budget.

When technologies such as reverse osmosis, currently the dominant technology in desalination plants, are employed, energy (and financial) costs increase even further. The Yuma Desalting Plant on the lower Colorado River in Arizona, designed to reduce the concentrations of salts and farm chemicals in the river before it passes into Mexico, is a prime example; since its completion in 1992, it has operated only intermittently due to costs related to energy and brine (i.e., concentrated saltwater waste) disposal. While higher-than-average precipitation in the US Southwest during the first decade of the twenty-first century made it geopolitically and hydrologically feasible to idle the plant, such wet conditions were anomalous, and the costs of water treatment to meet treaty obligations will likely only rise in the coming years.

The second critical confluence of water and energy resource use occurs in the conversion of primary energy sources such as coal, natural gas, and uranium into electricity. Conventional thermal power plants burn fossil fuels to create high-temperature, high-pressure steam that then spins turbines, which in turn spin generators to produce electricity. Nuclear plants derive their heat source from the fissioning of uranium nuclei, but otherwise, their process is essentially the same as conventional thermal-electric plants. In both cases, water is used in two ways. First, some circulates in the steam loop, where it is alternately heated to steam and condensed back to water in a **non-consumptive** process; the water in the steam loop is almost entirely reused. Second, water is drawn into the plant on a continual basis for cooling, to provide the cold sink necessary for re-condensing the steam after it has flowed across the turbine-generator assembly and done its work. After absorbing the steam's excess heat, the cooling water—now significantly warmed—is discharged to the environment, usually into the same body of water from which it was first withdrawn. It is precisely this type of **thermal discharge** that falls under the purview of Section 316 of the CWA, as discussed above.

As Fig. 8.2 shows, water use for thermoelectric power generation from 2005 to 2015 declined. This is largely due to more efficient cooling mechanisms and the fact that gas-fired plants, which comprise the majority of new thermal power additions, require less cooling. However, it remains to be seen whether that trend will continue.

Not all scholars view the water-energy nexus as a useful concept for developing strategies that simultaneously address water and energy demand growth. Ackerman

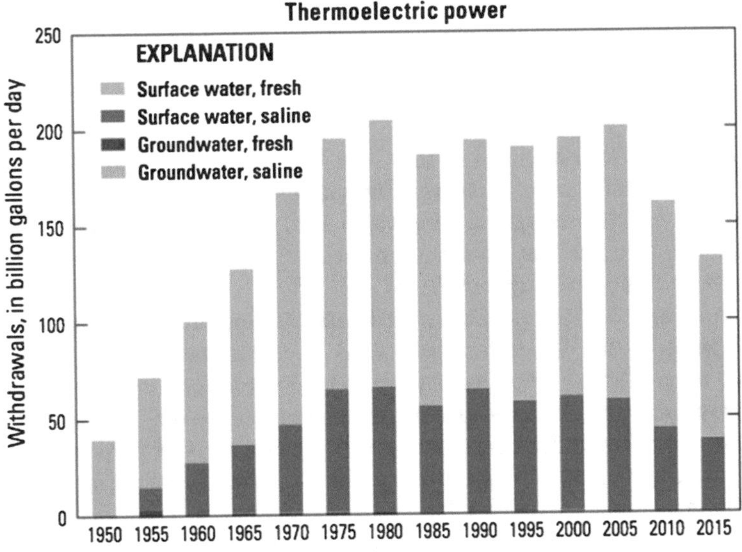

Fig. 8.2 US water withdrawals for thermoelectric power generation. (Source: U.S. Geological Survey, https://water.usgs.gov/watuse/wupt.html)

and Fisher (2013) argue that the economics of carbon reduction will likely be much more effective drivers of change, at carbon prices plausible by 2100, than will the economics of water savings. That is, policies that push the electricity sector to reduce carbon emissions by putting a price on carbon will become effective at carbon prices one can reasonably expect to see in the near future. Similar policies aimed at reducing water usage, they argue, would only become effective at water prices *far higher* than one could reasonably expect in the coming century, even in the desert southwest. For them, "'water-energy nexus'…might be better understood as two distinct problems that intersect, quite asymmetrically, with energy planning, and call for quite different responses." They conclude that "the [US west's] water crisis…does not originate in, and cannot be solved in, the electricity sector."

Berkman (2015) echoes this perspective, noting that "alarmist" or "crisis" studies (his terms) focusing solely on the electricity sector's share of water withdrawals fail to account for the fact that "as much as 96% of water withdrawn by power plants is returned to the water supply," or non-consumptive. In a review of the water planning documents of four states that have faced persistent water resource uncertainty, he finds that "electricity generation demand is not a primary source of concern, but that infrastructure (reservoirs and conveyance), conservation (especially agriculture) and regional cooperation are of primary concern."

Biofuels development is one area where consideration of the food–energy–water nexus is particularly relevant. Promoters of biofuels recognize that meeting growing global energy demands will require developing significant reserves of oil and gas, which, while available and increasingly technologically feasible, will exacerbate climate change. Increasing biofuels production, they argue, could mitigate climate damage while decreasing US dependence on oil-exporting countries, not all of which are supportive of US political goals. According to the Energy Information Administration, US net petroleum imports totaled 25% of total US consumption in 2016 and 20% in 2017. Just over a decade prior, at the time the USA adopted Renewable Fuel Standards (RFS), oil imports represented 60% of American's petroleum use.

Amendments to section 211 of the Clean Water Act introduced by the **Energy Policy Act of 2005** and the **Energy Independence and Security Act of 2007** require the EPA to set annual Renewable Fuel Percentage Standards with the purpose of "driving the market to overcome constraints in renewable fuel infrastructure" (Federal Register 2016).

Biofuels are classified as first-, second-, and third-generation, depending on the source material and the technologies used to create the fuel. First-generation biofuels are made from crops that otherwise could be used as human or animal feed, such as soy and corn, and oilseed crops. Second-generation biofuels are made from the waste vegetable matter from crops primarily grown for other purposes, non-food crops (switchgrass), or municipal waste or food processing waste. Third generation biofuels (such as algae-derived fuel) have the least impact on food production because they do not compete with food crops or occupy land used to grow food crops.

From a nexus perspective, the production of biofuels can occupy land and resources that might otherwise be used for food production, and conversion of land to agriculture typically increases water use and pollution. First-generation biofuels make up over 99% of the biofuels feedstock, and expansion of demand for them in the USA resulted in food price volatility in 2008 and 2010-2011 (Wise and Cole 2015). Corn prices reported by the USDA oscillated from a high of $7.60 per **bushel** in August 2012 to a low of $3.15 per bushel in November 2017.[6] In 2018, the USDA recorded that 38% of US corn production volume was used for ethanol production, up from 14% in 2007 when demand first began increasing to meet the RFS.[7] However, in 2018 the amount of corn dedicated to ethanol production declined for the first time since 2012. In 2015, the Department estimated that the RFS and similar targets in other countries with aggressive biofuel goals could result in some 32–42 million additional acres of land devoted to first-generation biofuel production worldwide. Not surprisingly, biofuels mandates can impact domestic production practices, as farmers respond to price signals in the commodities markets and plant more crops for use as biofuels feedstocks.

A second reason biofuels remain controversial is that their ability to reduce greenhouse gas emissions is questionable, especially with current technology. **Energy balance**, a common metric for measuring the energy benefits of biofuels, compares the total energy needed to produce one unit of fuel to the total energy output of that unit of fuel. As Saundry (2019) noted recently, the energy balance of corn ethanol, while somewhat better now than 20 years ago, is not impressive. The fossil fuels used to produce, harvest, and transport corn cancels out much of the energy savings resulting from the ethanol. Biofuels produced from waste materials such as sugar cane waste (bagasse), might improve the energy and economic balance, but true breakthroughs remain elusive.

One of the most obvious connections between water, food, and energy is dams. Dams are constructed for hydropower, irrigation, water supply, flood control, aquaculture, navigation, and recreation. Hydropower, regulated mainly by the Federal Energy Regulatory Commission, is technically a non-consumptive water use, but can nevertheless displace water resources in time and space, altering the natural flow regime of a river. Reservoirs created by dams often flood fertile valleys, removing those lands from cultivation. But the water held by them may be used to propagate fish or irrigate crops, thereby improving agricultural production in other ways.

Thermal power plants tell a different story. While the water intensity—how much water it takes to produce a kilowatt-hour of electricity—of US thermal power plants has declined in recent years, nearly one-third of recent thermal capacity additions (new power plants) requiring cooling has occurred in Texas and California, two dry western states. The decline in water intensity is motivated not so much by forward-looking nexus thinking as by economic and pollution control dynamics that make older, less efficient power plants less economically competitive. Thermal discharges from most of these plants, some 1100 in total, are covered by the EPA's Steam

[6] See USDA, Prices Received for Corn by Month at https://www.nass.usda.gov/Charts_and_Maps/graphics/data/pricecn.txt 11/29/2018.

[7] See USDA, World Agricultural Supply and Demand Estimates, WASDE 584 (2018) p 12, U.S. corn and grain supply and use, https://www.usda.gov/oce/commodity/wasde/latest.pdf.

Electric Power Generating Effluent Guidelines and Standards, enacted in 1974 and most recently revised in 2015.

Assessing the nexus impacts of new development in any FEW sector requires a sound understanding of the initial baseline conditions of all three sectors prior to development. Several GAO reports since 2010 have cited a simple lack of comprehensive data on the water impacts of energy development, making effective regulation all the more problematic. Others highlight the lack of coordinated policy-making among different agencies. The Energy Policy Act does not specifically refer to the nexus per se but does contain clear guidance that the DOE should research and develop linked water and energy resource needs. In addition, the Act directed the DOE to coordinate work with other relevant federal agencies including but not limited to the EPA, the Interior Department, the Army Corps of Engineers, the Commerce Department, and the Defense Department. The Act also includes provisions governing energy impacts on water, for example, targeting leaking tanks that risk contaminating groundwater. As the GAO reports show, much work remains to be done to meet the coordination goals in the Energy Policy Act.

8.2.3 Food

In Chap. 2, we noted that "food systems integrate all of the inputs, processes, conversions, infrastructure, outputs, uses, wastes, allocations, and impacts of food." The relationship between food systems and water is complex: water use for food production can be consumptive (water integrated into the product, for example) or non-consumptive (water used in washing, or return flows from irrigation). Saundry points out that our twenty-first-century food system is global in nature, which makes it difficult to draw boundaries around systems for analysis, let alone regulation. Not surprisingly, the global nature of the food system brings certain difficulties in terms of law and policy.

Food law and policy scholars note a need to "view food and agriculture as part of an integrated system," and to broaden food and agriculture law to account for other priorities such as public health, the environment, and economic development (Beyranevand and Leib 2017). Food regulation implicates the nexus in four key areas: food production, processing, transportation, and disposal. All these require energy and water inputs. Poor management of food production and waste can also pollute water resources through nutrient runoff. Therefore, increases in efficiency that reduces food use and waste *overall* can reduce negative impacts on energy and water resources. Food waste, such as manure, grain stalks, vegetable clippings, and even post-consumer waste, can also be a source of energy if the legal and physical infrastructure permits collection and anaerobic processing of the waste to produce methane.

According to the Food and Agricultural Law Program at the University of Arkansas, the USA wastes 96 billion **pounds** of food annually.[8] A significant

[8] See "Food Recovery: A Legal Guide" at http://law.uark.edu/documents/2013/06/Legal-Guide-To-Food-Recovery.pdf.

fraction of that—some 38 million tons in 2014, according to the EPA—gets landfilled. In landfills, that waste produces methane when it decomposes anaerobically. If not captured and burned as fuel, the methane is released into the atmosphere, where it has a **global warming potential** over a 100-year timescale more than 20 times as powerful as that of carbon dioxide. Composting food waste in high-oxygen conditions drastically reduces methane production. Processing it in specially designed anaerobic facilities called digesters, on the other hand, allows for the capture and potential beneficial reuse of that methane.

In addition to water and climate impacts, food waste has social justice impacts. In 2015, the U.S. EPA initiated a Food Recovery Challenge (FRC) as part of its Sustainable Materials Management Program. The FRC includes a "Food Loss and Waste Reduction Goal" that aims to cut food waste to half of 2010 levels by 2030. As part of the strategy, the EPA developed a Food Recovery Hierarchy, which prioritizes tactics for food waste reduction in the following order:

- source reduction (not producing surplus food);
- redirecting unused food to feed hungry people;
- diverting food not used by people to feed animals;
- industrial uses of food waste;
- composting; and
- (as a last resort) landfilling.

The EPA's food recovery hierarchy mirrors the prioritization seen in RCRA, which applies to both hazardous and household wastes. RCRA Subtitle C incentivizes producers of hazardous wastes to minimize waste creation by reducing or substituting raw materials or changing manufacturing processes (source reduction); followed by recycling or repurposing waste materials, followed by disposal after appropriate treatment. Under the Food Recovery Challenge (FRC) the concept of source reduction seen in RCRA becomes reducing over-consumption/over-production of food; repurposing becomes distribution to other users or animals and composting/digestion, and landfilling is a last resort.

Hamilton (2013) describes the current era in food policy beginning in the mid-1990s as the "post-industrial food democracy period," characterized by new, often smaller, agricultural businesses as well as animal welfare and environmental sustainability movements, and the use of ballot initiatives to achieve food-related goals at the state and local levels. But the lack of a national food strategy results in counter-productive initiatives and conflicting or disconnected laws and policies implemented by state and federal agencies. A 2016 GAO report to the U.S. Senate pointed out such inefficiencies, noting that in addition to federal agencies governing food, some 3000 non-federal agencies also do (Government Accountability Office 2016) .[9]

One such inconsistency arises between the federal Farm Bill, authorized by Congress every 5 years, and the U.S. Dietary Guidelines, published by the Department of Health and Human Services. The Farm Bill prioritizes funding of commodity crops such as corn, soy, and wheat and cotton over specialty crops

[9] See Government Accountability Office (2016)

(fruits and vegetables). That prioritization, however, led to a situation in 2005 where US farmers were, by a USDA estimate, some 15 million acres short of producing sufficient specialty crops to meet the recommended ratio of fruits and vegetables to grains and starches given in the dietary guidelines.[10]

In 2011, The Food Safety Modernization Act (FSMA) shifted the focus of regulatory efforts to the prevention of foodborne illness by addressing multiple steps in the production and supply chain. It directs the FDA to balance environmental concerns with food safety concerns by standards enacted by environmental and conservation agencies when making rules under the Act. However, as Pollans (2016) notes, this approach is ineffective at minimizing negative environmental impacts of farming practices, including those on water, because it "puts the onus of considering environmental effects on the regulatory agency (FDA) despite the fact that the structure of the regulatory scheme shifts substantial regulatory authority to private parties."

Some laws meant to promote and protect domestic agriculture can create adverse impacts on water resources. **Right-to-farm** laws, originally intended to protect farmers from nuisance suits—for smells or dust, for example—by neighbors, have sometimes been applied in manners that effectively given farmers, as Hamilton (2013) argues, a "right to commit nuisance." Hamilton also notes that many farming activities, such as the installation of drain tile in poorly drained fields, are virtually unregulated, even though they have significant impacts on local hydrology.

8.2.4 Food–Energy–Water Nexus Approaches

Food–energy–water Nexus approaches can be seen in a limited number of federal initiatives. One example is the U.S. EPA's NetZero program, designed to reduce the use of energy and water and eliminate solid waste production in military and non-military facilities with the goals of "protecting human health and the environment while generating societal and economic benefits." To promote a better understanding of the interconnections between FEW resources, a collaboration between the USDA's National Institute of Food and Agriculture and the National Science Foundation is funding new research into FEW nexus projects.

8.2.5 Microorganisms in the Spotlight at the FEW Nexus

One significant, visible, and dangerous linkage between food production, energy, and water quality is the phenomenon of **Harmful Algal Blooms (HABs)**, an unfortunate misnomer since the blooms are actually composed of cyanobacteria. A key

[10] See Buzby and Wells (2007)

catalyst in HABs events appears to be nutrient loading (e.g., nitrogen and phosphorus) of water bodies from over-application of nitrogenous fertilizer (mainly manure) to fields, which then washes into lakes and streams during precipitation events. Other factors include increasing temperatures and still water. In areas where climate change is resulting in warmer temperatures and intensification of weather events, conditions favoring cyanobacteria blooms and the cyanotoxins they produce will likely increase, resulting in the need to carefully assess how to reduce nutrient loading from non-point sources in the watershed (Paerl et al. 2016). Cyanotoxins produced by extreme HABs events can be difficult for drinking water treatment plants to remove, and exposure via skin contact, inhalation, or drinking can even result in renal failure and death, making them of significant political interest.

The Finger Lakes region of Western New York, home to roughly 21% of the state's agricultural land, is a prime example of this type of food–water nexus impact. There, farm runoff contributes to water quality degradation in the lakes, which, together with the region's pastoral scenery and vibrant wine and culinary scene, drive roughly $3 billion in tourism income annually. Figure 8.3 shows a drone photo of a HAB on a lake in upstate New York, taken in the summer of 2017. HABs are particularly troubling because it is difficult to know when the algae start to produce toxins or when the toxins have dissipated. HABs can threaten the health of boaters, swimmers, and municipal water customers, as well as the economies of communities near water bodies where they occur.

Fig. 8.3 Drone photo of a HAB occurrence in September 2017 on a lake in upstate New York. The intake for the municipal water system lies not far from the breakwater; the clear wake was caused by a boat carrying divers to inspect the intake pipe. (Photo credit: Tim Schneider, Cayuga County Water Inspector, reproduced here with permission)

Watershed associations address this issue by promoting integrated management across FEW sectors and municipal boundaries, through education, volunteer coordination, and data collection efforts that inform watershed decision-making. Voluntary initiatives can be particularly successful when linked to regulatory targets or mandates. In the case of impaired waters, Section 303 of the Clean Water Act requires states to establish a Total Maximum Daily Load (TMDL) for given pollutants such as nutrients, mercury, for waters that do not meet state regulatory standards. States attempt to bring the impaired water body back into compliance with CWA water quality standards through a two-prong approach: reducing loading by permitted point-source polluters by decreasing allowances in National Pollutant Discharge Elimination System (NPDES) permits (a stick approach), and implementing a variety of voluntary, often grant-funded, non-point source pollution mitigation strategies (a carrot approach).

Agricultural policies and incentives to improve **soil quality** also impact the nexus. Healthy soil sequesters carbon and holds water available for plants. One USDA study found that cover crops, which can reduce topsoil loss and improve the soil's water retention capacity, were used on only 1% of US cropland in 2013. The Natural Resources Defense Council (NRDC) estimates that increasing by 1% the organic matter on merely half the acreage used in the USA to grow soy and corn—the two largest crops grown in the USA by acreage—could result in the retention of approximately a trillion gallons of water. As a result, the NRDC recommends incentives such as discounts on crop insurance to promote cover crop use.

8.3 Nexus Regulations at the State and Local Levels

8.3.1 Water

In the absence of any applicable over-arching federal framework such as the Endangered Species Act or interstate compacts, the burden for ensuring in-stream flows remain within the specified acceptable minimum, and maximum levels fall mostly to individual states. Therefore, the degree to which state regulatory agencies value and legally recognize the maintenance of minimum in-stream **flows** or **environmental flows** as much as withdrawals or impoundments for economic purposes such as irrigation or hydropower matters greatly. Some western states such as Oregon, Washington, Texas, and Colorado have made significant strides in this arena, mainly through **water trusts** that allow in-stream water rights to be held by state entities, countering the norm of private use rights and prioritization of non-ecosystem uses.

As explained in Sect. 8.2.1, energy and agricultural production practices can pollute or deplete water resources. Innovative methods to prevent these negative impacts are often seen at the state and local level.

For example, in Pennsylvania, the boom in shale gas development between 2007 and 2015 resulted in the construction of thousands of natural gas drilling pad sites, access roads, and freshwater and produced-water holding ponds, both on private and public lands. After some significant pollution incidents, the state increased regulation of unconventional gas development, making Pennsylvania's regulations some of the strictest in the country. In some areas, such as the Tioga State Forest, gas well development occurred on lands previously mined for coal and suffering from acid coal mine drainage. Southwestern Oil and Gas partnered with the Tioga County Concerned Citizens Committee to install limestone channels in the Tioga State Forest to mitigate acidification caused by historical deep coal mining activities, restoring parts of the Tioga River and Fall River to trout-bearing status. Requiring environmental restoration or mitigation for natural gas development permits does not negate the risks of development, but it can reduce the overall burden on the environment.

On the other side of the country, California, no stranger to water scarcity, is breaking new ground with its Sustainable Groundwater Management Act, established in 2014. Where the federal CWA is essentially silent on groundwater resources, California's state action created local Groundwater Sustainability Agencies throughout the state that are charged with avoiding "significant and unreasonable" reductions in quality and quantity of the state's groundwater resources. In addition to exploring the role of markets and big data as tools for improving groundwater management, a coalition of southern California actors is exploring groundwater net metering schemes, where a framework similar to that successfully utilized in the residential solar photovoltaic sector would reward landowners for efforts they undertake to recharge groundwater aquifers. Agricultural management is also largely state and local. Compliance with State Pollutant Discharge System Permits for point-source pollution and implementation of best agricultural practices to minimize non-point source pollution reduces negative impacts on water resources from food production.

Research has shown that avoiding over-application and choosing appropriate timing of fertilizers, as well as maintaining vegetated riparian buffers around streams, can significantly reduce agricultural nutrient loading. Sediment traps in drainage ditches, which allow for the pooling of water and settling of suspended sediment during heavy precipitation events, can also be effective. County-level Soil and Water Conservation Districts provide critical educational programs for farmers and individualized technical assistance in creating manure management plans and runoff reduction strategies.

In the Finger Lakes region of New York, the tourist industry is built on clean lakes, bucolic scenery, and more recently, wineries, breweries, and local foods. While farms benefit from the climate-moderating effect of the lakes, they typically do not draw water from the lakes (relying on groundwater wells instead) and are not impacted by the water quality in the lakes the way lakefront owners are. Harmful algal blooms threaten the health of the lakes, and mitigation will likely depend on significantly increasing the participation of upland farmers in the management of the watershed as a whole, and the willingness of lakefront property owners and municipalities dependent on the lakes to contribute financially too expensive measures to reduce nutrient loading from farms. The conundrum created by temporal, geographical, and economic separation between nonpoint source pollution and its most salient impacts challenges to water resource management throughout the USA, from watersheds as small as a Finger Lake to those as large as the Chesapeake Bay. Though legislation aimed at addressing nutrient runoff and HABs issue in New York has not yet been framed specifically using the language of the nexus, successful mitigation strategies will require mobilization of a variety of actors—state and local officials, farmers, lakefront property owners, academics, and extension agents—able to recognize and address the problem as a food, energy, and water problem simultaneously.

Watershed managers can help coordinate planning throughout the watershed, engaging critical stakeholders including farmers, landowners, industries, and municipalities. These professionals can spearhead diverse initiatives such as reducing stormwater flows, updating municipal codes to promote watershed-friendly practices, and seeking funding for development and implementation of formal watershed management plans. Particularly in areas where there is no dominant municipal user in a watershed, money, and incentives to promote the reduction in pollution, watershed managers can be critical to improving water quality and mitigating negative impacts of energy and food production activities.

8.3.2 Energy

Thermoelectric power plants fueled by coal, natural gas, biomass, or uranium provide roughly 80% of US electricity. As Peer and Sanders point out, many new thermal power plants rely on groundwater for cooling, a situation that risks stressing underground aquifers. To the extent, those new power plants fueled by natural gas instead of coal, the cooling water required for each kilowatt-hour (kWh) of electricity generated will be lower. Unfortunately, much of the new thermoelectric power plant capacity is being developed in states whose freshwater resources are already under stress, like Texas and California.

Although demand for cooling water may be peaking, efficiencies in water use could be offset by increased demand for water in **carbon capture and sequestration** (CCS) technologies. This would be especially true if a price or tax on carbon emissions pushed plant operators to implement CCS rather than close coal-fired plants. Moreover, as power plants rely increasingly on groundwater resources for cooling water, state regulators in dry states may look to the Safe Drinking Water Act for grounds to limit those withdrawals. Scarcer water supplies and tighter policies may well spur continued advances in alternative cooling strategies for thermoelectric power plants in the coming decades (Peer and Sanders 2018, p. 620).

Vermont's Farm to Plate Strategic Plan, for instance, addresses the role of agricultural land in providing opportunities for increasing energy efficiency and providing space for renewable energy installations. Such space may turn out to be a vital ingredient to the success of the state's renewable energy future. Vermont's comprehensive energy plan calls for 90% renewable energy across all sectors by 2050, not including the resurrection of the storied Vermont Yankee nuclear power plant, which was shut down in 2014.

8.3.3 *Food*

Models for more comprehensive food regulation include several statewide initiatives which, while focused on strengthening the economic viability of the farming industry and reducing health care costs related to a poor-quality diet, include references to energy–agriculture or water–agriculture linkages.

New York's Farm to School program, in which 43% of New York school districts participated in 2016, promotes the use of farm produce by local school districts, potentially reducing the carbon impact of food distribution. However, the NYS Department of Agriculture and Markets states the goals of the legislation are to "strengthen local agriculture, improve student health, and increase local food systems awareness," without mention of reducing energy consumption or impacts on water resources. Next door, the Massachusetts Local Food Action Plan places a similar priority on increasing "production, sales, and consumption of Massachusetts-grown foods" and creating agriculture-related jobs, but also acknowledges the importance of protecting water resources and the environment more broadly.

8.4 Human Rights as a Guiding Principle?

In Sect. 6.2.4, we noted that the notion of human rights with respect to food, energy, and water has begun to emerge in international discussion and, more concretely, through treaties under the United Nations system. Some countries have included one or more of the FEW rights (especially food, secondarily water) in their constitutions or national laws, and many governmental programs attempt to implement one or more of the FEW rights for all citizens or all inhabitants. In the USA, access to FEW resources at the most basic level is not assured. But a growing number of federal, state, and local programs attest to the increasingly strong political drive to support affordable access to these necessities.

Both federal and state programs implement a limited entitlement for food, although adversarial politics can lead to threats to reduce or eliminate funding for these programs. While there is no federal assistance program for routine water access in the USA, both FEMA and state programs provide bottled water universally in emergencies, and some states (e.g., California) have begun to develop programs that guarantee residential water to all residents.

The Federally funded, state-administered Low Income Energy Assistance Program defrays costs of heating and cooling for low income families, and some states and cities. (e.g., Pennsylvania and New York) have their own programs that provide financial assistance with utility bills, and on-bill financing for energy-saving upgrades. These are not always the result of government largesse, however; one such program in New York, for instance, depends on the generosity of customers voluntarily adding a dollar or two to their bill payments in order to assist others who are unable to cover their own energy expenses. Finally, nonprofit organizations, such as the Salvation Army and religious institutions, have long provided funds to low-income individuals for basic commodities such as food and utilities on a request or application basis, filling gaps left by government programs.

While a universal human right to FEW resources remains elusive for now, growing awareness in some academic and policy circles points to the problems that arise in the *absence* of such rights, and may help pave the way to a more hopeful and stable future. One example is the growing prevalence of international conferences and publications centered on energy poverty. A concept hardly used a decade ago, energy poverty has become a common theme among social scientists concerned with energy and energy access. Similarly, the relatively new notion of food deserts, at least as applied in countries that otherwise enjoy relatively high degrees of development (as measured by GDP, for example), such as the USA, suggests a growing awareness among scholars, activists, and decision-makers that not everyone has similar access to affordable, nutritious food. The nearly universal outrage over the Flint, Michigan water crisis and nationwide concern of harmful algal blooms portend changes in the way we think about water management.

It is telling, perhaps—damning, even—that these three concepts reflecting governments' failure to guarantee a human right to FEW resources are gaining political traction now that it is clear they apply in more-developed countries, not just in some remote "Third World," where they have long existed. Clearly, national and international governing structures need to make giant strides toward improving human well-being around the world, including access to reliable and affordable FEW resources.

Key Points

- In the USA, most laws that address food, energy, or water are directed at protecting or promoting the development of one resource or the other.
- US laws generally do not target the interactions among critical consumable resources like water, energy, and food.
- In some cases, laws devoted to energy development or conservation, optimizing food production and safety, or efficient use and conservation of water require intentional considerations of the impact on the other nexus resources. However current legal structures requiring a given law regulating energy, or food, or water to explicitly analyze and address the implications for the other resources are ad hoc, rather than systematic.
- At the federal level, the National Environmental Policy Act (NEPA) provides a mechanism for consideration of the environmental and social externalities of federal actions, and state analogs to NEPA similarly constrain state actions.
- Municipalities may implement local land-use laws that have significant impacts on food, water or energy—such as zoning laws and "Right to Farm" laws.
- State and local governments can play a significant role in raising awareness and educating the public about the FEW nexus, thereby increasing support for local laws that recognize connections among FEW resources.

Discussion Points and Exercises

1. Give three important examples of the food–energy–water nexus discussed in this chapter. Next, write one or two paragraphs describing a facility in or near your hometown that highlights the food–water–energy nexus.
2. Briefly explain the link between thermal discharges and water quality. Then use the Internet to identify four to five plant or animal species that may be particularly affected by thermal discharges.
3. Provide three examples of mechanisms designed to reduce negative nexus impacts resulting from the production or use of one or more of the "three great consumables" at the local or regional level. Briefly explain each mechanism works.
4. Identify four key areas where food policy implicates the FEW nexus and give one or two concrete examples for each area. Then suggest at least two areas where food policy could reduce energy or water intensity in food production.
5. Why does the development of biofuels represent an especially important area where the FEW nexus should be considered?
6. Use the Internet to find data on two biofuels crops in the USA, including total acreage planted, and irrigation water required. Then graph those data using a spreadsheet program like Excel or Google Sheets. Be sure to provide a meaningful title for your graph, and to include units on the graph's axes.
7. Plan a meal you can make for yourself that includes at a minimum a vegetable, a protein, and a grain or starch. Make a list of the ingredients. Go to the local supermarket and find all the ingredients you would need (you do not actually have to purchase them, just look at them.) Using the labels, determine as accurately as you can where each ingredient was grown or processed. Then, use a mapping program like Google Maps or Google Earth to make a map of where each ingredient originated.
8. Using your results from Question 7, calculate the distance between the origin of each ingredient (as best you can figure) and your home or campus. Add up the mileage for all the ingredients, then do a rough calculation to estimate the petroleum used to truck or ship your meal. Refer to a reliable source such as the Energy Information Administration (www.eia.gov) for data on fuel consumption of various vehicles (train, truck, and ship).
9. Much of the USA's long-distance shipment of food grains occurs by rail, which is far more efficient than a truck. CSX, one of the country's major rail shippers, claims it can move one ton of freight 500 miles using only one gallon of diesel fuel. Do some "back-of-the-envelope" calculations to compare that per-ton efficiency to the efficiency of an 18-wheeler truck carrying 15 tons of freight. Use EIA or a similarly reliable source to determine fuel consumption (mpg) for the trucks (you will likely find this data called "average fleet fuel economy" or similar. Be sure to calculate your calculated value has the same units so that it can be compared directly to the train's efficiency.
10. How much virtual water does the USA export annually? Virtual water is a concept designed to account for the water needed to produce certain goods, usually foods. A tomato grown in a Canadian hothouse and sold in California,

for instance, means that no water was used in water-scarce California to produce that tomato, while all the benefits of consuming it (including its embodied water) accrue to someone in California rather than Canada. Use government data to ascertain how many soybeans the US exports in a year, and how much water each unit of soybeans takes to produce (soybeans' "water intensity"). Use a spreadsheet program to chart the change in soybean exports and their associated "virtual water" amounts over the past 10 years.

11. Using what you understand about the interdependence among food, energy, and water, create a poster or a comic strip that will communicate these connections to people in your community. You may use images from the Internet (with proper citation), or you might wish to sketch them by hand.
12. Using the same two biofuels crops you used in Question 3, research the prior land use on the majority of the acreage is now growing those biofuels. What consequences might change the land use to cropping biofuels have for food and water resources in those areas? Remember, all farmland is not created equal; whereas some areas might require significant inputs of water and fertilizer, others might be far more fertile naturally. Therefore, be sure to specify your assumptions about where each croup was grown and the type of farmland most prevalent there.
13. The debate over who should regulate natural resources is intimately tied to the struggle between states' rights and federal authority. For each resource—Food, Water, and Energy—discuss at least two benefits of allowing states to regulate the development of the resource and two benefits of allowing the federal government to regulate the development of the resource. If you can, find some examples from recent public debates over oil and gas development, water use, or food production to illustrate your points.
14. The debate over hydraulic fracturing and, to a lesser extent, gas pipelines has been quite heated in states such as Colorado, New York, Pennsylvania, and West Virginia. Some have characterized it as a "rural versus urban" debate, where urban dwellers were largely against leasing rural lands to gas drilling and rural people, particularly farmers living on very slim margins, were in favor of being allowed to lease their lands to oil and gas companies. Pick one of the four states listed above and research the food, energy, and water implications of allowing the development of unconventional shale gas through hydraulic fracturing in that state. Use peer-reviewed or government agency sources. Write a 300-word summary of your findings.

References

Ackerman, F., & Fisher, J. (2013). Is there a water-energy nexus in electricity generation? Long-term scenarios for the western United States. *Energy Policy, 59*, 235–241. https://doi.org/10.1016/j.enpol.2013.03.027.

Berkman, M. (2015). The electricity-water nexus: Is a crisis imminent? *Water Policy, 17*(6), 1163–1175. https://doi.org/10.2166/wp.2015.060.

Beyranevand, L. J., & Leib, E. M. B. (2017). Making the case for a national food strategy in the United States. *Food and Drug Law Journal, 72*, 225–261.

Black, M. (2016). *The atlas of water: Mapping the world's most critical resource* (3rd ed.). Oakland, CA: University of California Press.

Buzby, J. C., & Wells, H. F. (2007). Meeting Fruit and Vegetable Dietary Recommendations. US Department of Agriculture, Accessed July 13. https://www.ers.usda.gov/amber-waves/2007/april/meeting-fruit-and-vegetable-dietary-recommendations/.

California Energy Commission. (n.d.). Water-Energy Nexus. Retrieved from http://www.energy.ca.gov/research/iaw/water.html

Craig, R. K., & Roberts, A. (2015). When will governments regulate nonpoint source pollution? A comparative perspective. *Boston College Environmental Affairs Law Review, 42*(1), 1–64.

Federal Register. (2016). *Renewable fuel standard program: Standards for 2017 and biomass-based diesel volume for 2018*. (Vol. 81 No. 238, p 89746). Retrieved from https://www.gpo.gov/fdsys/pkg/FR-2016-12-12/pdf/2016-28879.pdf

Government Accountability Office. (2016). "Government Efficiency and Effectiveness: Opportunities to Reduce Fragmentation, Overlap, and Duplication and Achieve Other Financial Benefits (GAO-16-579T)." Last Modified April 13, 2016, Accessed March 20. https://www.gao.gov/products/GAO-16-579T

Hall, N. D. (2007). Transboundary pollution: Harmonizing international and domestic law. *University of Michigan Journal of Law Reform, 40*(4), 681–746.

Hamilton, N. D. (2013). Harvesting the law: Reflections on thirty years of agricultural legislation. *Creighton Law Review, 46*(4), 563–590.

Maupin, M. A., Kenny, J. F., Hutson, S. S., Lovelace, J. K., Barber, N. L., & Linsey, K. S. (2014). *Estimated use of water in the United States in 2010: US geological survey circular 1405*. https://pubs.usgs.gov/circ/1405/pdf/circ1405.pdf

Paerl, H. W., Gardner, W. S., Havens, K. E., Joyner, A. R., McCarthy, M. J., Newell, S. E., et al. (2016). Mitigating cyanobacterial harmful algal blooms in aquatic ecosystems impacted by climate change and anthropogenic nutrients. *Harmful Algae, 54*, 213–222. https://doi.org/10.1016/j.hal.2015.09.009.

Peer, R. A. M., & Sanders, K. T. (2018). The water consequences of a transitioning U.S. power sector. *Applied Energy, 210*, 613–622. https://doi.org/10.1016/j.apenergy.2017.08.021.

Pollans, M. J. (2016). Drinking water protection and agricultural exceptionalism. *Ohio State Law Journal, 77*(6), 1195–1258.

Saundry, P. (2019). Review of the United States energy system in transition. *Energy, Sustainability and Society, 9*(1).

Stokes, J. R., & Horvath, A. (2009). Energy and air emission effects of water supply. *Environmental Science & Technology, 43*(8), 2680–2687. https://doi.org/10.1021/es801802h.

Further Reading

Ackerman, F., & Fisher, J. (2013). Is there a water-energy nexus in electricity generation? Long-term scenarios for the western USA. *Energy Policy, 59*, 235–241. https://doi.org/10.1016/j.enpol.2013.03.027.

Berkman, M. (2015). The electricity-water nexus: Is a crisis imminent? *Water Policy, 17*(6), 1163–1175. https://doi.org/10.2166/wp.2015.060.

Beyranevand, L. J., & Leib, E. M. B. (2017). Making the case for a national food strategy in the USA. *Food and Drug Law Journal, 72*, 225–261.

Government Accountability Office Energy–Water Nexus. (n.d.). Retrieved from https://www.gao.gov/key_issues/energy_water_nexus/issue_summary#t=1

Peer, R. A. M., & Sanders, K. T. (2018). The water consequences of a transitioning U.S. power sector. *Applied Energy, 210*, 613–622. https://doi.org/10.1016/j.apenergy.2017.08.021.

U.S. Geological Survey. (n.d.). *How much water do we use?* https://owi.usgs.gov/vizlab/water-use/.

Wise, T. A., & Cole, E. (2015). *Mandating food insecurity: The global impacts of rising biofuel mandates and targets. Working paper no. 15-01*. Global Development and Environment Institute.

Chapter 9
Ecosystems and Ecosystem Services

Nathanial Matthews, Wei Zhang, Andrew Reid Bell, and Lara Treemore-Spears

9.1 Introduction

This chapter identifies core principles of ecology and environmental science applicable to the nexus and discusses the tightly coupled **socio-ecological systems** that establish ecosystems services (ES) within the nexus. The chapter begins by outlining key ecological principles and their interdependencies, demonstrating how ecosystem services are provided related to food, energy, and water. It then illustrates how ecosystems and ecosystem services are valued with respect to the nexus. Next, it outlines the relationship of ecosystem services to the Sustainable Development Goals (SDGs) (see Sect. 3.7) before concluding with a case study examining the importance of ecosystem services in erosion control services and **conservation agriculture**.

As is outlined in other Chapters, population growth (Sect. 3.3), a degraded natural resource base, and climate change (Chap. 11) have intensified, and are projected

N. Matthews (✉)
Global Resilience Partnership, Nairobi, Kenya
e-mail: nathanial.matthews@kcl.ac.uk

W. Zhang
Environment and Production Technology Division, International Food Policy Research Institute, Washington, DC, USA
e-mail: w.zhang@cgiar.org

A. R. Bell
Department of Environmental Studies, New York University, New York, NY, USA
e-mail: andrew.reid.bell@nyu.edu

L. Treemore-Spears
Health Urban Waters, College of Engineering, Wayne State University, Detroit, MI, USA
e-mail: treemorespears@wayne.edu

P. Saundry, B. L. Ruddell (eds.), *The Food-Energy-Water Nexus*, AESS Interdisciplinary Environmental Studies and Sciences Series,
https://doi.org/10.1007/978-3-030-29914-9_9

to further intensify pressure on the world's food, energy, and water systems and the environment (Ringler et al. 2016).

Ecosystems underpin the functionality and health of the food–energy–water nexus. Although the environment and ecosystems are not explicit within many definitions of the nexus, without functioning ecosystems and healthy stock and flow of ecosystem services our global water and food systems and our energy provision would be immeasurably impacted.

The United Nations Convention on Biological Diversity (CBD 1992) defines an **ecosystem** as "a dynamic complex of plant, animal and microorganism communities and their non-living environment interacting as a functional unit." For the purpose of this chapter we use Walker et al. (2006) definition of ecosystem services as "the combined actions of the species in an ecosystem that perform functions of value to society." This definition highlights links between ecosystems and people with an emphasis on the system that is key to the delivery of services.

Ecosystem services are the benefits that humankind derives from ecosystems, and can be usefully grouped into four broad categories:

- **provisioning**, such as the production of food and water;
- **regulating**, such as the control of climate and disease;
- **supporting**, such as nutrient cycles and crop pollination; and
- **cultural**, such as spiritual and recreational benefits.

In addition to these classifications, Walker et al. (2006) include a classification for **habitat** services that comprise the maintenance of the life cycles of individual species and their genetic diversity, through habitat quality and quantity, e.g., natural vegetation type, structure and distribution that supports a sufficiently diverse genepool to sustain a species; adequate reproductive rates for an animal species to be able to be used as food; habitat which serves to support pollinator or pest control agents that are beneficial to agricultural systems. Habitats at the nexus primarily protect and maintain stocks, flows and interrelationships of biologically diverse plant and animal species, which underpin and ensure the resilience of many of the provisioning, regulating and cultural services provided by ecosystems.

There are many nexus interlinkages, dependencies, and trade-offs within ecosystem services. Food, water and some types of energy themselves can be both provisioning and supporting services depending on how they are used.

Despite the importance of ecosystem services to our health and well-being and economies, they are often significantly undervalued in our current economic, political, and social systems.

Freshwater ecosystems, for example, including lakes, ponds, rivers, streams, springs, and wetlands are important habitats for diverse species and provide provisioning, supporting, regulating and **cultural ecosystem services** that underpin the health, livelihoods, and well-being of billions of people (Matthews 2016). Despite their importance and their links to food, energy, and water, freshwater ecosystems are some of the most heavily altered ecosystems on earth, and these alterations have had significant negative impacts on their health and functionality (**ecosystem health** and **ecosystem integrity**) (Carpenter et al. 2011).

In our food systems, ecosystems services are vital to the functioning of our entire agricultural system. Food supply, fresh water and clean air are all examples of provisioning services. Soil **biota** and soil organic matter (SOM), nutrient cycling, and crop pollination are key supporting services that play critical roles in driving **ecological processes** that lead to ecosystem goods and services, upon which human civilization is predicated (Lavelle 1997). Soil health also plays an important role in regard to mitigating climate change and combatting land degradation for agricultural production. Regulating services control water and air quality, climate and pest populations, while cultural ecosystem services can be seen in the exchange and development of seeds, spiritual values, and food culture.

In our waterways, freshwater ecosystems including lakes, ponds, rivers, streams, springs, and wetlands are home to approximately 126,000 species. In addition to being an important home for **biodiversity**, these aquatic ecosystems provide provisioning, supporting, regulating and cultural ecosystem services that underpin the health, livelihoods, and well-being of billions of people (Aylward et al. 2005). Food supply chains, for example, account for 92% of the world's freshwater consumption and about 20% of the energy (Allan and Matthews 2016). Healthy functioning freshwater ecosystems not only underpin our food systems and are deeply connected to our energy systems they also provide us with the majority of our drinking water. Freshwater is a provisioning service providing water for domestic use, power generation, irrigation, and transportation. Inland waterways, including rivers, wetlands, and lakes also depend on the hydrological cycle and freshwater. These water bodies provide recreational and cultural values and also support inland fisheries, which are a critical livelihood and nutrition source for millions of people.

Some energy systems are also dependent on other ecosystem services. Approximately 10% of the world's energy consumption is supplied by ecosystem services such as fuelwood and plant material, and this rises to up to 40% of the energy consumption in developing countries (World Energy Council 2016). The food–energy–water nexus and ecosystem services are deeply interlinked, but they are currently under significant stress.

The shift to the **Anthropocene**, the current geological age during which human activity has been and continues to be the dominant influence on climate and the environment, has resulted in a rising population, changing food habits, increased consumption and waste, and climate change that have all increased the threats to and demands on ecosystems and ecosystem services (Vörösmarty et al. 2010).

9.1.1 Ecological Concepts that Frame the Nexus

The complexity of food, energy, and water systems as discussed in Chap. 2—and the complex interactions of ecological processes with social systems—makes it challenging to provide a brief summary of the underlying environmental processes that are the most applicable to the nexus. The purpose of discussing the processes themselves in Sect. 9.2 therefore will be to highlight the areas of natural science in which it may be necessary to have a working knowledge to understand a nexus problem, and for which it may require mastery and new research to solve. This discussion of ecosystem concepts and mechanisms in Sect. 9.2 is organized into the following categories:

- Hydrological and biogeochemical cycles
- Energy flow
- Land and soil
- Biota

Core natural science concepts within these overarching topics that will be discussed include variability, **disturbance**, storage, **weathering**, **resilience**, **ecosystem diversity**, **carrying capacity**, **adaptation**, feedback loops, and trophic, pathogenic, and toxicant interactions. There are of course many other important concepts and interactions to understand in ecological systems but the intention here is to build somewhat upon the basics learned in introductory natural sciences courses toward nexus scientific understanding.

Because the water cycle is discussed at length in Sect. 2.5, and Chap. 11 addresses climate change, this chapter focuses more attention on other dynamics. To the extent that natural ecological systems can be viewed as part of **natural infrastructure**, the concepts outlined in Chap. 10 also may be adapted for understanding ecosystems, particularly **interdependency**, **cascading failures**, **robustness**, **buffering**, and others.

9.1.2 Socio-ecological Systems: A Framing for Ecosystems in the Nexus?

Socio-ecological systems are an approach or framing that starts with the premise that humans and nature are deeply integrated and that they should be understood as one integrated system. As has been outlined above, this is evidenced by the fact that all ecosystems are shaped by people, and all people need ecosystems. This is especially true in our highly globalized and heavily populated world.

Ecosystems are prototypical **complex adaptive systems** (see Sect. 2.2) in that they are composed of many diverse entities that interact within networks. Complex adaptive systems demonstrate specific characteristics that are important for framing and understanding the criticality of ecosystems at the nexus and for finding solutions that best manage nexus interactions and cross-sectoral trade-offs for decision makers. These include:

- **Sudden transitions and tipping-points**. Complex systems show non-linear dynamics. They may suddenly move from a high degree of stability to instability.
- **Limited predictability**. The unpredictable behavior of complex adaptive systems makes them difficult to predict, and therefore, history cannot be a predictor of future events and results in a great deal of uncertainty.

- **Large events**. Relatively small changes may lead to large effects. This is the case if a complex system is close to a tipping point
- **Self-organization**. Complex adaptive systems operate without central control. However, they are often characterized by a certain order. They, as it were, organize themselves from the bottom-up.

Using socio-ecological systems to frame the FEW nexus helps us to understand the deep interactions between human and natural systems. For example, the loss of supporting services in food due to climate change in Syria (60% farm loss and 80% livestock abandonment) has been identified as a key destabilizing factor that initiated the war and caused mass migration into Europe. Although these interdependencies and trade-offs were not clear before, they demonstrate the role ecosystems and the nexus play within complex adaptive systems.

Continuing to undervalue our ecosystems and ignore the complex interactions, trade-offs, and interdependencies across the nexus will amplify risks to human development (Chap. 3) and ultimately undermine prosperity. Section 9.3 outlines how ecosystems and ecosystems services can be valued and undervalued within the nexus.

9.2 Ecosystem Concepts and Mechanisms Relating to the Nexus

9.2.1 Hydrological and Biogeochemical Cycles

9.2.1.1 Water Cycle and Atmospheric Processes

Despite the global connectedness of the water cycle as discussed in Sect. 2.5, substantial variability in hydrologic and atmospheric conditions exist in space and time across the earth's surface, associated with factors such as landform, climate, elevation, soil, and season. Because of the dependency of life on the presence water, the resulting variety in the amount of water and its chemical characteristics—based on its interactions with the surrounding geology and **biogeochemical cycles**—provides the conditions for the formation of many different ecosystems and associated **ecological niches** to support the organisms that are associated with those ecosystems.

As discussed in Chap. 11, rapid climate change due to anthropogenic influence is a problem for organisms that have evolved slowly over time to be adapted to a specific ecosystem because they may be less able to change on the timeframe required to cope with conditions that quickly become wetter or drier on a long-term basis.

To the extent that food systems depend on organisms that rely for some of their support upon neighboring ecosystems that are associated with a particular moisture regime—such as pollinator insects that cause flowering plants to produce seed or fruit and parasitoids that control crop pest insects—the water cycle has both direct and indirect effects on food production. Changes in wind velocity and timing due to climate change may also affect seed production for wind-pollinated plants that produce cereal grains.

9.2.1.2 Biogeochemical Cycles

Biogeochemical cycles in ecosystems influence the availability of the nutrients organisms require to live, grow, and reproduce. In nexus studies the phosphorus, nitrogen, carbon, and sulfur cycles are most commonly of interest; other chemical elements are important in smaller amounts dependent on the ecosystem. These cycles are driven by both biotic and abiotic factors.

Nutrients in ecosystems cycle repeatedly move through various states and forms, with the potential to return to the same or other consumers in the food web. For example, the carbon exhaled by animals as CO_2 may be stored in plant leaves that are consumed by those animals; or the plant leaves may fall off and decompose, making the component nutrients available for uptake by the roots of that plant or other organisms. The biological unavailability of atmospheric nitrogen (N_2) to plants and animals illustrates the importance of this biological decomposition activity in the nitrogen cycle; it is the activity of microorganisms that break down organic matter in the soil that provides much of the nitrogen (nitrate NO_3^- and ammonium NH_4^+) available to plant and animal life. As more research occurs into the role of microbial processes in biogeochemical cycles, we are finding that soil microbes also influence storage of other nutrients such as carbon and phosphorus that are applicable for agricultural crop growth and sustainability.

The sulfur cycle is of interest in nexus studies because sulfur is an essential element for the formation of plant and animal proteins and it is also a by-product of fossil fuel consumption, in addition to being produced naturally, such as in volcanic eruptions. The atmospheric reaction of sulfur emissions (as sulfur dioxide SO_2 and hydrogen sulfide H_2S) with air and water may result in acid deposition (as sulfuric acid, the primary component of acid rain H_2SO_4) that pollutes land and water, in some cases making inland lakes devoid of fish and other aquatic life. Conversely, sulfur deposited on land through the biogeochemical cycle (as sulfate SO_4^{2-}) may support agricultural production, when it occurs at less-than-toxic concentrations. As with nitrate, sulfate does not bind readily to mineral soil that typically is composed of negatively charged anions; this means that positively charged particles in soils such as organic materials are essential for nutrient retention.

9.2.1.3 Energy Flow

Primary energy originates from the sun and flows through ecosystems, with various ecosystem components storing energy. For example: through photosynthesis, plants convert sunlight and nutrients into sugars and complex molecules that are stored as plant tissues which may then be consumed at higher trophic levels (such as by an aphid that is a pest on food crops; which may then be consumed by a lacewing insect larvae that is a predator and control on insect pests; which may then be consumed by a bird). Energy flow is often described as "one-way"—as opposed to cyclical nutrient recapture—because energy either changes form or is converted to heat such that it is not re-consumed by its previous consumers.

Energy flow and storage efficiency in an ecosystem depends on the living organisms and other components that are present; only a fraction of the energy stored at one trophic level is typically transferred to the next higher level. For example, organisms such as birds and mammals are inefficient relative to insects at storing energy, because they must spend much of the energy they consume on maintaining their body temperature. This principle can be utilized to increase human food supply sustainability, for example, by substituting insect protein for other animal protein.

The flow of energy into living tissues that then die and decompose can concentrate the release of energy into the decomposition process itself—or into energy stored as fossil fuels formed through geologic pressure exerted on decaying plant and animal matter.

9.2.2 *Land and Soil*

9.2.2.1 Formation and Weathering Processes

Soil is a layer of relatively loose minerals that have been **weathered** from the earth's crust and have undergone a variety of other physical, chemical and biological processes over a period of time making them suitable to support plant growth. Geological variations across the earth's crust contribute to different chemical composition in the associated soil that is formed through surficial processes such as the hydrologic cycle, **erosion**, and **sedimentation**. Weathering of rock via either mechanical or chemical processes results in the loosening or transformation of materials that may then be eroded, transported, and deposited in other locations and in other forms, and may make raw minerals available for use by organisms—these are the parent materials from which soils are then formed through biological and other processes.

Soil texture is important to an understanding of the agricultural potential of land, and a region's human and animal **carrying capacity**. Typically described by the relative amounts of sand, silt, and clay particles, soil texture originates from the rock the soil develops from. For example, granite contains sand-sized particles of quartz—which do not readily weather—and feldspar which weathers into fine-grained clay particles. The resulting clays are significant for supporting life because their layer-like structure with many negatively charged anionic adsorption sites provides a structure that can retain essential cations of calcium, magnesium, potassium, iron, and aluminum, while the sand in a soil originating from weathered granite supports a texture that allows the soil to drain. The erosion, transport, and deposition of weathered granite bedrock from the Canadian Shield to the Midwestern USA by glaciers 10,000–25,000 years ago is one of the reasons there is a high concentration of fertile well-drained soils there.

In contrast to recently glaciated soils, highly weathered surface soils—which may have been in place for hundreds of thousands of years—typically have lost many nutrients and their structure and chemistry have changed due to the weathering process. These changes follow predictable chemical patterns based on parent materials and climate, with the oldest soils containing only relatively resistant primary minerals like quartz, silicate clays, and aluminum and iron oxide clays—the latter of which may be observed as a pronounced reddish soil color. In warm climates with high rainfall such as the southeastern USA the weathering-induced loss of nutrient adsorption sites on the soil's clay particles may allow substantially more fertilizer to be lost from the root zone than would occur on a Midwestern US soil (the associated nitrogen loss to groundwater supplies can have significant human health effects). In warm dry climates like the southwestern USA where evaporation often exceeds precipitation, areas with a high groundwater table may experience the concentration of salts and calcium carbonate in a concrete-like layer at the soil surface that discourages plant growth. Both the history of the soil's formation and the current climate therefore have considerable influence on the vegetation that can be supported on the land.

Landscapes containing older soils may also contain newer soil formations due to the activity of volcanoes, wind, and flooding, resulting in specific characteristics depending on landscape features, geologic characteristics, climate, and the amount of time soil formation processes have been at work. For example, volcanic ash is a unique material in its light weight and ability to hold water and be compacted; eruptions may deposit it widely across the landscape or concentrate it thickly in lower areas, changing the growing characteristics in those areas. Silt, clay and fine sand particles also may be transported long distances by wind and the resulting soil, termed **loess**, typically has high potential for fertility under the right climatic conditions due to the nutrient-holding capacity associated with its fine particles. Transport of fine particles via the sedimentation process is one of the reasons for the richness of soils formed in the **alluvium** of river floodplains.

The biogeochemical cycle and other soil formation processes work together to form different soil structures—also called **pedoliths** or peds—over geologic timeframes, starting with unconsolidated deposits or parent materials. In the case of loess, its chemical composition results in particulate alignment that allow the soil to cleave off in near-vertical bluffs. More typically, agriculturally productive soil contains small blocky peds that allow for good aeration and drainage of plant roots and ample nutrient retention. In contrast an overly cultivated soil may have had its peds broken down to a size that negatively affects aeration, drainage and plant growth, through the mechanical mixing and compaction of particles, chemical modification, and loss of soil microorganisms due to plowing and irrigation.

9.2.2.2 Soil–Biological Interactions

The biologically active portion of a soil profile is generally to the depth of plant roots—which may be only a few centimeters for lawn grass or several meters for prairie plants in the Midwestern USA. Climate and its temperature and precipitation therefore influence soil formation because of both the associated weathering and the vegetation types that will grow. Moderate conditions—not too cold or hot and not too dry or wet—tend to create the most biologically productive soils.

Topographic position in the landscape, in which gravity works on both materials and water to collect in low areas, may result in concentrated areas of organic materials that have fallen or been transported to those areas as well as soil that is saturated by water. The presence of water slows the rate at which those organic materials decompose both because of the lower temperatures typically present in water and also because of the relatively anaerobic saturated conditions which result in less efficient decomposition processes. In warmer climates biological activity is increased, increasing the rate at which organic material decomposes; hence the lower organic content of tropical soils.

Microorganisms that occupy the soil such as bacteria play an essential role not only in decomposition of matter into its component parts, making nutrients available to other organisms, but also in symbiotic interactions with plants and other microorganisms such as fungi to enhance nutrient utilization, increase carbon storage and other functions just becoming known to science.

Anthropogenic processes alter the way in which natural processes function, by changing soil structure and reducing soil organic matter. Under moderate climatic conditions with natural grassland or forest ecosystem processes at work, it may take 500 years to generate a 2.5-cm thick layer of topsoil rich in organic matter and nutrients. An unvegetated cultivated field during a heavy rain—because it lacks the plant roots and natural pedolithic structure that would otherwise stabilize and retain the soil—can lose that much soil in a couple of hours.

9.2.2.3 The Land–Water Interface

Water comes in contact with soil not only from precipitation and the global water cycle but also from surface water such as rivers, streams, and lakes, and groundwater that emerges naturally from springs and via human pumping. The underlying geology influences both groundwater and surface water chemistry, which may supply helpful nutrients such as calcium, magnesium, and iron, and also elements toxic to many organisms at high concentrations such as arsenic and sulfur.

Because groundwater may emerge as natural springs in surface water bodies, both its chemistry and temperature influence the organisms that can occupy those ecosystems. In particular, aquatic organisms such as certain insect and fish species that require higher dissolved oxygen content to survive may only be found in rivers fed by cold water springs because cold water is capable of holding more oxygen than warm water. The presence of particular benthic macroinvertebrate species is often interpreted as an indicator of water quality because an invertebrate species assemblage indicates prolonged conditions over a period of time better than several water quality measurements would.

The biogeochemistry and physical characteristics of land and soil interacting together with water in all forms (e.g., rainfall, surface-water runoff, groundwater) have profound impacts on ecosystem characteristics and functions, and the biota those ecosystems can support. Human land management to supply food, energy, and water particularly influences these dynamics. While a full discussion of human impacts on ecosystems at the land–water interface is beyond the scope of this book, Sect. 9.4 provides a case study for erosion control that illustrates ecosystem conservation practices at the nexus.

9.2.2.4 Habitat Connectivity

The connectivity of habitats and surface waters across the landscape can significantly contribute to the functioning of ecosystems by maintaining the flow of resources needed by all organisms to thrive. Conversely, habitat fragmentation can isolate species from food and water sources necessary for their long-term survival, particularly in the face of climate change that is causing plant species to change their distribution to cooler and wetter locations than they have historically occurred. Even in patchy habitat, connection of patches by corridors can allow gene exchange while permitting herbivores to follow their host plants according to the appropriate environmental conditions, and predators to follow herbivores. A species is therefore more resilient against the loss of a single habitat patch when all habitat remains connected by corridors. Natural processes such as disturbance—due to factors such as wind and wildfire—and river incision during seasonal high-volume rain events can also cause habitat fragmentation by separating stream beds from their floodplains or removing habitat patches, such as in a tornado.

Landscapes that are resilient to unpredictable events preserve:

- a robust set of natural processes,
- buffers to lessen the impact of disturbance on ecosystems, and
- connectivity that considers a variety of interdependencies and likely disturbance regimes.

9.2.3 Biota

9.2.3.1 Vegetation

Native plant species that have occupied earth for millions of years and cultivated agricultural plant species that have been around for thousands of years bear many similarities due to their common phylogenetic origins. Domesticated food plants exhibit several key differences from the bulk of the vegetation observed in the natural environment, however, largely owing to the reasons these species were domesticated in the first place: relatively large seed or fruit size, and high productivity. Annual cereals like wheat produce large amounts of big nutrient-rich seeds because the plants that sprout from the seed must survive wherever the seed falls and complete the entire reproductive cycle in a single growing season—making these seed characteristics competitively advantageous for the plant.

Before wheat was domesticated, the plant also benefited from having much of its seed drop quickly upon ripening, hiding it from depredation by animals, and preventing it from getting too dry for germination by protecting it in the soil. During the domestication of wheat, therefore, humans selected for individual plants that held more seed on the plant, because that seed was more easily harvested. This selection process resulted in those and other related genetic characteristics being carried forward to the next generations of domestic wheat. The suite of characteristics that have been bred into domestic plants has thereby made the plants more desirable to humans in a variety of ways, but not always in ways that conserve water and energy resources during their growth, harvest and processing. As a result of this dynamic, many domesticated annual plants would not thrive without human cultivation because the characteristics that have been bred into them are not adaptive for survival of the plant in the absence of human intervention.

In contrast to the dominance of annual plant species in conventional agriculture, natural ecosystems are more characteristically dominated by perennials including trees, shrubs, vines, perennial flowering herbaceous plants and perennial grasses. While annual plants sometimes play a significant role in natural ecosystems—particularly in climates that provide a competitive advantage to plants that can go dormant in seed form during dry summers—annuals in nature more typically rely on regular disturbance to be able to compete with their relatively well-rooted, long-lived perennial neighbors.

Fruit- and nut-producing trees, shrubs and vines are the primary counter-example to annual plants in agriculture. The longer lives of these plants are inherently coupled with a longer establishment phase before they can yield food. This tends to average the effects of climate, natural moisture regime, soil and other factors that influence vegetation growth. While these types of vegetation are sometimes grown through intensive human intervention that sustains them in locations where they otherwise would not grow (e.g., citrus trees in areas where frost occurs), generally these plants grow more sustainably and with less human intervention than annual crops. A variety of agroforestry practices have been developed that conserve use of energy and water and couple the production of tree crops with annual or perennial plants or with livestock grazing. These practices can reduce the amount of nitrogen fertilizer that is lost from the soil—and that potentially impacts human groundwater wells and health—because the deeper and more extensive root systems of trees can uptake excess nitrogen that annual plants are unable to use before it runs off.

The development of new perennial food crops is a growing area of research as discussed in Chap. 21, because of their potential for conserving energy, water, and natural ecosystem functions. Conventional perennial plants used for food or livestock forage, such as potatoes and alfalfa, are replanted either every year or every few years to keep yields up at a commercially viable level and reduce likelihood of losses due to disease or pests. This keeps external inputs high and limits the ecosystem benefits that could be provided by a purely perennial crop. Grains and nitrogen-fixing plant species such as legumes—which require less external nitrogen fertilizer inputs because they are able to use atmospheric nitrogen or N_2 more directly—are the types of plants most frequently pursued in conceptual alternatives to resource-intensive agriculture. Better understanding is needed about potential candidate species and the selection process that could be used for their adaptation to widespread cultivation.

9.2.3.2 Animals

Animal-plant interactions play important roles in ecosystems and agriculture, both beneficial and detrimental. The primary roles animals play in plant survival, reproduction and fertilization are as pests, herbivores, pollinators, and decomposers; secondary roles are played as control-agents for pest species, and in the formation of soil structure that can help water to be retained rather than being lost from the soil. In some cases, highly specific interactions between a host plant and its associated pollinator species require both the animal and the plant in order for both to survive and reproduce.

Animals themselves may be agents of both ecosystem disturbance and maintenance. For example, large migratory ungulates on grassland, such as buffalo in the USA and elephants in Africa, were long thought to be contributors to over-grazing and loss of grassland productivity. As more research has been done on the complex ecosystem dynamics driven by animals, the need for maintaining spatial heterogeneity and placing limits on human landscape modifications such as fenced boundaries and artificial water sources have emerged as more important factors that prevent animals from adhering to natural limits on carrying capacity.

9.2.3.3 Microorganisms

Although animals such as insects may play similar roles to microorganisms in below-ground ecosystems, bacteria and fungi are the most powerful and relatively unseen mediators of these ecosystem processes. As discussed in relation to biogeochemical cycles, soil structure, and soil–biological interactions, these microorganisms play essential roles that include retaining water, cycling nutrients, and supporting plant growth. Without fungi, we would have neither the decomposition required to break down nutrients into more available forms to support plant growth, nor the creation of important food products using yeast (a unicellular fungus).

On the less beneficial side, bacteria and fungi also represent some of the most economically important diseases of agricultural plants and animals and destroyers of agricultural products. Additionally, to the extent that fuel wood is attacked by fungi, energy availability is also at risk from the aggressive ability of fungi to decompose both living and dead plant tissue. These dynamics, however, have unacceptable impacts primarily when other stressors put ecosystems out of balance.

9.3 Valuing Ecosystems and Ecosystem Services

9.3.1 Why Do We Need to Evaluate the Value of Ecosystem Services?

When values of ecosystem services are evaluated and incorporated into decision-making, we can make more informed decisions about managing the **natural capital** (including biodiversity and ecosystem services) that underpins human well-being and livelihoods. Specifically, we can use the values to choose desired management practices, technologies, and policies to maximize social welfare, minimize unintended costs, or balance various societal objectives, including economic development, health, environmental protection, and conservation. This is especially important for decision-making regarding the nexus, where there are often many "hidden" interdependencies, trade-offs, and synergies that need to be measured and evaluated.

For example, valuation studies reveal trade-offs between management options (e.g., between the use of agrochemical inputs and regulating/**supporting ecosystem services** that maintain agricultural productivity) and farming systems (e.g., conventional vs. organic). Economic valuation of ecosystem services through **ecosystem accounting** can also highlight the potential appeal of changes in agricultural management that deliver enhanced ecosystem services—specifically those supporting and **regulating ecosystem services** that lack markets.

Equally important is to evaluate and account for the impacts of varied agricultural systems on ecosystem services. Such an assessment can help illustrate the long-term sustainability of production systems that supply nutritious food in required quantities without negatively impacting on the environment and human health. Sukhdev et al. (2016) demonstrate the challenges of this within the nexus, stating that current metrics for agricultural performance do not recognize or account for the hidden costs and benefits of the whole system and that food metrics must be urgently overhauled, or the Sustainable Development Goals will never be achieved.

An important caveat is that the economic values are only useful to the extent that they can capture the revealed or stated market and non-market benefits of ecosystem services with monetary metrics. It is important to recognize that monetarization is not feasible for many important contributions of ecosystem services to society (social equity, distributional effects, **intrinsic value**, etc.) that need to be considered when weighing options with respect to management of ecosystem services (or **ecosystem management**).

It is also critical to acknowledge that the diversity of values of nature and its contributions to people's good quality of life are associated with different cultural and institutional contexts and are hard to compare on the same yardstick. This makes it necessary to expand the way society recognizes the diversity of values that need to be promoted in decision-making at the nexus and to embrace pluralistic valuation approaches as adopted by organizations such as the Intergovernmental Platform on Biodiversity and Ecosystem Services (IPBES).

9.3.2 *How Can It Be Done?*

Beginning two decades ago, scientists took on the ambitious task of characterizing broad categories of ecosystem services that are fundamentally important to humans (see discussion in Swinton and Zhang 2005). Some have gone farther: Costanza et al. (2016) not only characterized the Earth's major ecosystem services but also estimated their "aggregate annual monetary value." That particular attempt to put a price sticker on the planet's entire ecosystem services has been justly criticized by economists for violating microeconomic principles of diminishing marginal utility, budget constraints, and comparison of most feasible alternatives (e.g., Pearce 1998; Bockstael et al. 2000; Daily et al. 2000). However, the continuing, frequent citation of Costanza et al.'s article highlights the need felt by many to link monetary values to ecosystem services and to better understand ecosystem trade-offs within the nexus.

Depending on how consumers and producers experience an ecosystem service, there are many different methods to estimate its value (Freeman 2003; Shiferaw et al. 2005). The methods used for agricultural ecosystem services focus on values that people obtain from the use of the services. For example:

1. **revealed preference methods** capture the values of individuals as revealed by how the act/spend money in existing markets; and
2. **factor input valuation methods** estimate the value of an ecosystem service which can substitute for an existing marketed input, or when the service contributes to measurable marketed output, the economic value of changes in the level of the service can readily be inferred using information from the related input and/or crop markets. For example, the value of biological nutrient cycling can be estimated by its fertilizer **replacement value** or **replacement cost** in cereal-legume systems (Bundy et al. 1993). This refers to the concept of **Substitutability**.

   Under factor input valuation methods, **production input** (the value of inputs required to production), **profitability trade-off**, and **stated preference methods** estimate the value of changes to the status quo, such as changing current farmer cropping systems to include ecologically recommended practices.

So far economic research examining the economic value of the natural enemy contribution to crop profitability has been limited, with the exception of Zhang and Swinton (2009, 2012). Their approach to measuring the economic value of the natural enemy complex was based on model-predicted densities of the pest and its predators and a dynamic optimal pest control model that explicitly incorporates the economic value of natural enemy survival when making profit-maximizing decisions on insecticide applications.

Other attempts at placing economic values on pest regulation ecosystem services, such as Losey and Vaughan (2006) and Pimentel et al. (1997), estimated the total cost of averted pest damage due to all pest control practices and then attributed a fraction of the total pest control benefit to natural enemies. While these aggregate values provide snapshots of the possible magnitude of the benefit from natural pest control, they ignore the local context and are also subject to the critiques of not being economically credible.

Some modeling tools such as those discussed in Chap. 15 also provide a quantitative means to valuing ecosystems and ecosystem services in the context of FEW systems analysis. Typically, these models include ecosystem components (e.g., water, land, vegetation, biota) as well as economics and financial analysis capabilities, so they can be used to quantify impacts to ecosystems from FEW activities as well as valuing the services provided to FEW systems.

9.3.3 Ecosystems and the Nexus Within the Sustainable Development Goals (SDGs)

In Sect. 3.7, we reviewed the 2015 UN Sustainable Development Goals (SDGs) and their and their associated targets, which agreed to by all 193 UN member countries and represent a major step toward better integration of ecosystem services into the sustainable development efforts of meeting water, food and energy targets. The SDGs requires careful balancing between the environmental, social, and economic dimensions of specific development challenges.

Unfortunately, it is oftentimes much too easy to address one component of the food, energy, and water nexus in any specific situation and let the other two fall by the wayside. For example, mitigating climate change through biofuel production in any given country may compete with an agenda to end hunger and achieve food security: both endeavors might be competing for the same land and water resources.

The logical priority for many countries may be to choose the social benefits of food production over the environmental benefits of green energy, unless long-term **costs and benefits**, both tangible and intangible, are accounted for. Although a mechanism for measuring how successfully countries achieve the multiple goals remains a challenge, the acknowledgment that all goals should the achieved simultaneously represents a major step forward relative to the siloed sectoral approach in the previous global development agenda.

Achieving the SDG targets may require even more difficult decisions regarding trade-offs at the Nexus compared to earlier development goal schemes. The scale of the SDGs and the context in which they are being implemented are very different from that of the Millennium Development Goals (MDGs). The MDGs consisted of just 8 goals and 18 targets, whereas the Sustainable Development Goals have 17 goals and 169 targets (see Sects. 3.5 and 3.7).

The world is also a much more connected and integrated place in 2016 than it was in 2000. Global financial markets, supply chains, people, and the climate are more connected than ever (Allan and Matthews 2016). This interconnectivity presents huge opportunities, but also significant systemic risks, as evidenced by the impact of weather shocks and financial crises on 2008 food price spikes.

To begin to address deeply interconnected challenges such as these, we must gain an understanding of how best to fit all the SDG puzzle pieces together in a logical manner. According to a review of the targets and goals by the International Council for Science (ICSU ISSC 2015), all SDGs benefit to some degree from ecosystem protection, restoration, and sustainable use. In order to realize the ambitions embodied in the Sustainable Development Goals, it will be essential to manage ecosystems to protect nature and improve the supply of, and equitable access to, environmental benefits while protecting the species and functions present within them (DeClerck et al. 2016).

The importance of integrating ecosystem services into the effort toward sustainable development has been highlighted in several recent studies. Costanza and colleagues emphasize the need for aggregate metrics of human and ecosystem well-being to replace growth in the gross domestic product as the primary development goal for nations. They indicate that there is compelling new research relating ecosystem services and natural and social capital to human well-being, which can be integrated with the Genuine Progress Indicator to produce an expanded version that connects more directly with the SDGs.

Considering interlinkages between Sustainable Development Goals and the nexus, and the urgency to achieve SDGs simultaneously, the Landscapes for People, Food and Nature Initiative proposes integrated landscape management as a fundamental means of implementation of the SDGs and to manage trade-offs across the nexus. Integrated landscape management offers an action-oriented means to achieve multiple Sustainable Development Goal targets simultaneously at local and subnational levels.

Wood and DeClerck (2015) elaborate on the importance of concentrating on strengthening specific ecosystem-service-related targets in the poverty reduction, food security, human health, and water SGDs, as each of these depend heavily on ecosystems in order to be sustainably achieved. They suggest that interventions can include: managing in-field biodiversity to enhance resilience to climate-related impacts (SDG1) and combat malnutrition (SDG2); planning landscape configuration to maximize multiple agro-ecosystem services such as pest control and pollination to improve food security (SDG2) and reduce disease risk (SDG3); or implementing riparian and field margin buffers to secure clean water (SDG6).

Managing for ecosystem services, which link human well-being to conservation outcomes, is essential for achieving the SDGs. To synthesize knowledge articulating where and how ecosystem services contribute to meeting Sustainable Development Goals, Wood and colleagues conducted an expert survey on evidence for the contribution and importance of ecosystem services to SDG targets. Experts judged ecosystem services could make contributions to achieving at least 37 targets across 10 Sustainable Development Goals, of which contributions were considered of high importance for attaining 32 targets. Food and water provisioning, habitat and biodiversity maintenance, and erosion control services contributed to the largest number of SDG targets. The results highlight that proper management of these services offers opportunities for synergistic outcomes across multiple SDGs.

Our understanding of the interaction across the SDGs and within the nexus is increasing, but our ability to manage them remains limited. As addressed in Chaps. 17 and 20, engaging diverse scholars and practitioners can facilitate practical and sustainable solutions to FEW nexus issues.

9.4 Case Study: Erosion Control Services and Conservation Agriculture (CA)

Erosion control services provide an excellent focus to examine the challenges and opportunities for ecosystem services provision in the nexus. Vegetation (such as trees, shrubs, grasses, crops, or anything else with roots growing in soil) helps to stabilize soil in place, maintain soil structure and porosity, and break up the energy of falling precipitation or flooding. In doing so, it helps to manage the rate at which soil is carried away in surface waters as sediment.

Topsoil forms very slowly—on the order of a few centimeters per century—and where vegetation has been removed and topsoil exposed, it can be vulnerable to being washed away much faster than it is produced. In agricultural settings, this can damage livelihoods at several scales: at the farm scale, as the farmer's capacity to earn a living from productive land is stripped away; and at the landscape scale, as sediment is carried into surface streams, rivers, and reservoirs, damaging aquatic habitats and the livelihoods of those who fish (food) and filling reservoirs whose functions would otherwise include flood regulation, irrigation storage (water), or hydropower production (energy).

Meeting demands for food without eroding soil has been a challenge for millennia; the earliest known writings on soil management date back to the Roman era, with similar struggles being repeated across Europe, North America, South America, and particularly in agricultural landscapes with developing economies. In every context the trade-offs are the same: farmers till and plow their soils to speed the incorporation of fertilizer and organic matter, easing the task of planting and in the best of cases, increasing productivity over the coming season. In doing so, however, they break up root structures and leave loose soils exposed precisely as rains begin to fall, a perfect formula to strip the topsoil away and erode the long-term productivity of the land.

Where land is abundant, people have historically moved on from each exhausted plot to new land, as Romans growing crops in ancient Italy, and Southern plantation owners growing cotton in the USA each did. When we have reached the limits of easily available land, our best answer to these problems has remained the same for millennia as well—minimal soil disturbance and a cover crop to preserve soil integrity, and an intercropping or rotation of nitrogen-fixing legumes to help maintain soil fertility.

Today, this set of practices is commonly known as **conservation agriculture (CA)**, promoted in various forms in agricultural areas across the world. In settings where goals extend beyond erosion control to the regulatory and habitat services provided by forests, the principles of CA may be tied with those of agroforestry, in what has been called "Conservation Agriculture with Trees" (CAWT). The request on land users may also be simply to refrain from cropping entirely.

However beneficial CA or CAWT might be for the landscape, encouraging farmers to practice them can be difficult as "[Actions] that are optimal for farmers are not necessarily consistent with their societies' interests" (Montgomery 2007; p.237).

In particular, though practicing CA can bring benefits to farmers in the mid- to long-term, it brings more costs and risks in the short term, as weeding efforts increase, crop residues spread out on plots are no longer available for burning or as forage for animals, and hardpans of long-tilled, compacted soils get waterlogged and damage crops. Under such conditions, an incentive is often necessary for the short term to encourage farmers to try; where preserving or restoring ecosystem services requires farmers to cultivate less (or no) land, the need to provide an incentive is even clearer and may be needed in perpetuity.

Such incentives may take the form of subsidies or other support from government agencies (Pannell et al. 2006), though these programs often have limited life spans, and it is common to see farmers stop the encouraged practice once incentives stop flowing (Andersson and D'Souza 2014). Ideally, a party or parties are willing to pay farmers to keep up their practices indefinitely, and where this is the case (and that willingness can be channeled effectively), there is potential for **payments for ecosystem services (PES)** program.

It is not the aim of this chapter to provide a detailed examination of payment for ecosystem services. However, there are a few key metrics of what kind of agricultural systems might be good candidates:

First and foremost, agricultural systems where changes in the practices of one group of resource users would make an economic difference in the lives of some other group of individuals or firms; to the extent that those affected would be willing to pay for the change; where those being paid are among the poorer members of society; PES has the greatest potential to contribute toward both ecosystem services provision and poverty alleviation.

Second, agricultural systems where payments from buyers can be easily pooled, and where payments to providers can be easily distributed, so that transaction costs do not overwhelm the potential gains from trade.

In the case of erosion control services, smallholder fishers benefiting from improved aquatic habitats might not be able to pay much for the service, nor would it be easy to pool any payments they might be able to make; on the other hand, beneficiaries of electricity from a hydropower provider (a direct beneficiary of erosion control services) have their payments pooled through the billing process (Bell et al. 2016); to the extent that farmers are accessible as groups through cooperatives or via local extension offices, the transaction costs on the provider side are minimized as well. To this end, in landscapes where hydropower is a large part of electricity provision, such as the Shire River Basin in Malawi (Ward et al. 2015), paying to keep the house cool and the lights on may play a key part in the modern-day solution to keeping soils on farms and intact.

9.5 Conclusion

Ecosystem services are fundamental to the functioning of the food–energy–water nexus. In truth, our supplies of freshwater, our foods such as grains, vegetables, fish and meats, and much of our electricity come to us only through regulatory and **provisioning ecosystem services** on land and in the water. How we consume and manage these ecosystem services at the nexus has both trade-offs and synergies. In providing water, food, and energy to our growing global population, there are increasingly significant environmental impacts.

Despite the critical importance of ecosystem services within the nexus, the environmental trade-offs of our food, energy, and water provisions are often overlooked or undervalued. Economic valuation of ecosystem services can play an important role in helping to manage trade-offs, but economic valuations do not adequately capture the many important contributions of ecosystem services to society, and hence they do not provide a comprehensive picture for decision-makers.

A comprehensive evaluation framework that accounts for both monetary and non-monetary benefits throughout the value chain is needed (see the TEEB Agriculture and Food study for an effort in addressing this).

The pathway to achieving the Sustainable Development Goals provides a critical space for the sustainable management of the food, energy, and water nexus in relation to ecosystem services. Many of the SDGs, however, have trade-offs and synergies between goals and this leaves decision-makers with difficult decisions as to how trade-offs will be distributed.

There are many areas where the importance of ecosystem services within the nexus is brought into sharp focus. Erosion control services and conservation agriculture, for example, provide a useful space to examine the challenges and opportunities for ecosystem services provision within the nexus. The case study of conservation agriculture reminds us that nexus opportunities and challenges may emerge at different time scales, for example, practicing Conservation Agriculture can bring benefits to farmers in the mid- to long-term, but potentially more costs and risks in the short term. Payments for ecosystem services may provide a useful tool in managing trade-offs and interconnections across food, energy, and water.

In a world of increasing uncertainty punctuated by rapid population growth, degraded natural resource bases, and climate change all intensifying pressure on the world's food, energy, and water systems, there is more need now than ever before to better understand the interlinkages and properly manage the trade-offs of water, energy, and food on our ecosystems.

Key Points

- Ecosystem services are fundamental to the functioning of the food, energy, and water nexus.
- There are increasingly significant environmental impacts in providing food, energy, and water to our growing global population.
- How we consume and manage ecosystem services within the nexus has both trade-offs and synergies.
- How trade-offs are managed by decision makers is complex and involves political, social, economic and environmental considerations.
- Economic valuation of ecosystem services can play an important role in helping to manage trade-offs, but economic valuations do not adequately capture the many important contributions of ecosystem services to society.
- Payments for ecosystem services may also provide a useful tool in managing trade-offs and interconnections across food, energy, and water.

Discussion Points and Exercises

1. Identify five recent developments on the valuation of ecosystems.
2. Identify the local/state/national agencies with responsibility for ecosystem protection.
3. Identify the nonprofits, NGOs, and advocacy groups whose mission it is to protect ecosystem services.
4. Identify the most prominent/authoritative report on ecosystem services.
5. Identify the biggest three barriers to the protection of ecosystem services in your country.
6. Discuss how to best manage trade-offs for ecosystem services in the context of a development project like large-scale intensive agricultural production or a large hydropower dam.
7. Discuss the implications of the failure of different ecosystem services (provisioning, supporting, regulating, and cultural) on society, politics, and economics in your country.
8. Discuss the role that ecosystem services play in your daily lives.
9. Using the WWF Living Planet Report, identify what you think is the most important ecosystem service.

References

Allan, T., & Matthews, N. (2016). The water, energy and food nexus and ecosystems: The political economy of food non-food supply chains. In F. Dodds & J. Bartram (Eds.), *The water, food, energy and climate nexus: Challenges and an agenda for action* (pp. 78–89). Routledge - Earthscan: Oxon.

Andersson, J. A. & D'Souza, S., 2014. From adoption claims to understanding farmers and contexts: A literature review of Conservation Agriculture (CA) adoption among smallholder farmers in southern Africa. *Agriculture, Ecosystems & Environment*, 187, pp.116–132. Retrieved January 5, 2015, from http://linkinghub.elsevier.com/retrieve/pii/S0167880913002685.

Aylward, B., Bandyopadhyay, J., Belausteguigotia, J. C., Borkey, P., Cassar, A. Z., Meadors, L., Saade, L., Siebentritt, M., Stein, R., Tognetti, S., & Tortajada, C. (2005). Freshwater ecosystem services. *Ecosystems and Human Well-being: Policy Responses, 3*, 213–256.

Bell, A., Matthews, N., & Zhang, W. (2016). Opportunities for improved promotion of ecosystem services in agriculture under the Water-Energy-Food Nexus. *Journal of Environmental Studies and Sciences*. https://doi.org/10.1007/s13412-016-0366-9.

Bockstael, N. E., Freeman, A. M., Kopp, R. J., Portney, P. R. & Smith, V. K. (2000). On measuring economic values for nature.

Bundy, L. G., Andraski, T. W., & Wolkowski, R. P. (1993). Nitrogen credits in soybean-corn crop sequences on three soils. *Agronomy Journal, 85*, 1061–1067.

Carpenter, S. R., Stanley, E. H., & Vander Zanden, M. J. (2011). State of the world's freshwater ecosystems: Physical, chemical, and biological changes. *Annual Review of Environment and Resources, 36*, 75–99.

CBD (Convention on Biological Diversity). (1992). CBD convention text. Retrieved March 15, 2017, from http://www.cbd.int/convention/text

Costanza, R., Fioramonti, L., & Kubiszewski, I. (2016). The UN sustainable development goals and the dynamics of well-being. *Frontiers in Ecology and the Environment, 14*, 59. https://doi.org/10.1002/fee.1231.

Daily, G. C., Söderqvist, T., Aniyar, S., Arrow, K., Dasgupta, P., Ehrlich, P. R., Folke, C., Jansson, A., Jansson, B. O., Kautsky, N., & Levin, S. (2000). The value of nature and the nature of value. *Science, 289*(5478), 395–396.

DeClerck, F. A. J., Jones, S., Attwood, S., Bossio, D., Girvetz, E., Chaplin-Kramer, B., Enfors, E., Fremier, A., Gordon, L. J., Kizito, F., Lopez Noriega, I., Matthews, N., McCartney, M., Meacham, M., Noble, A., Quintero, M., Remans, R., Soppe, R., Willemen, L., Wood, S., & Zhang, W. (2016). Agricultural ecosystems and their services: The vanguard of sustainability? *Current Opinion in Environmental Sustainability, 23*(December 2016), 92–99. https://doi.org/10.1016/j.cosust.2016.11.016.

Freeman, A. M. (2003). *The measurement of environmental and resource values: Theory and methods* (2nd ed.). Washington DC: Resources for the Future.

ICSU ISSC. (2015). Review of the sustainable development goals: The science perspective. Paris: International Council for Science (ICSU). Retrieved March 23, 2016, from http://www.icsu.org/publications/reports-and-reviews/review-of-targets-for-the-sustainable-development-goals-the-science-perspective-2015/SDG-Report.pdf

Lavelle, P. (1997). Faunal activities and soil processes: Adaptive strategies that determine ecosystem function. *Advances in Ecological Research, 27*, 93–132.

Losey, J. E. & Vaughan, M. (2006). The economic value of ecological services provided by insects. *Bioscience, 56*(4), 311–323.

Matthews, N. (2016). People and freshwater ecosystems: Pressures, responses, and resilience. *Aquatic Procedia, 6*, 99–105.

Montgomery, D. R. (2007). *Dirt: The erosion of civilizations*. Berkeley, CA: University of California Press.

Pannell, D. J., et al. (2006). Understanding and promoting adoption of conservation practices by rural landholders. *Australian Journal of Experimental Agriculture, 46*(11), 1407. Retrieved from http://www.publish.csiro.au/?paper=EA05037.

Pearce, D. (1998). Cost benefit analysis and environmental policy. *Oxford review of economic policy, 14*(4), 84–100.

Pimentel, D., Wilson, C., McCullum, C., Huang, R., Dwen, P., Flack, J., Tran, Q., Saltman, T., & Cliff, B. (1997). Economic and environmental benefits of biodiversity. *BioScience, 47*(11), 747–757.

Ringler, C., Willenbockel, D., Perez, N., Rosegrant, M., Zhu, T., & Matthews, N. (2016). Global linkages among energy, food, and water: An economic assessment. *Journal of Environmental Studies and Sciences, 6*(1), 161–171.

Shiferaw, B., Freeman, H. A., & Navrud, S. (2005). Valuation methods and approaches for assessing natural resource management impacts. In B. Shiferaw, H. A. Freeman, & S. M. Swinton (Eds.), *Natural resource management in agriculture: Methods for assessing economic and environmental impacts* (pp. 19–51). Wallingford: CABI Publishing.
Sukhdev, P., May, P., & Müller, A. (2016). Fix food metrics. *Nature, 540*, 33–34.
Swinton, S. M., & Zhang, W. (2005). *Rethinking ecosystem services from an intermediate product perspective*. Selected paper, the annual meeting of American Agricultural Economics Association, Providence, Rhode Island, July 24–27, 2005.
Vörösmarty, C. J., McIntyre, P. B., Gessner, M. O., Dudgeon, D., Prusevich, A., Green, P., Glidden, S., Bunn, S. E., Sullivan, C. A., Reidy Liermann, C., & Davies, P. M. (2010). Global threats to human water security and river biodiversity. *Nature, 467*, 555–561.
Walker, B., Salt, D., & Reid, W. (2006). *Resilience thinking: Sustaining people and ecosystems in a changing world*. Washington, DC: Island Press.
Ward, P. S., Bell, A. R., & Parkhurst, G. M. (2015). *Heterogeneous preferences and the effects of incentives in promoting conservation agriculture in Malawi, Washington, DC*. Retrieved from http://ebrary.ifpri.org/cdm/ref/collection/p15738coll2/id/129169
Wood, S. L., & DeClerck, F. (2015). Ecosystems and human well-being in the Sustainable Development Goals. *Frontiers in Ecology and the Environment, 13*, 123. https://doi.org/10.1890/1540-9295-13.3.123.
Wood, S. L., Jones, S. K., Johnson, J. A., Brauman, K. A., Chaplin-Kramer, R., Fremier, A., Girvetz, E., Gordon, L. J., Kappel, C. V., Mandle, L. & Mulligan, M. (2018). Distilling the role of ecosystem services in the Sustainable Development Goals. *Ecosystem services, 29*, 70–82.
World Energy Council. (2016). *World energy resources*. Retrieved September 26, 2018, from https://www.worldenergy.org/wp-content/uploads/2016/10/World-Energy-Resources-Full-report-2016.10.03.pdf
Zhang, W., & Swinton, S. M. (2009). Incorporating natural enemies in an economic threshold for dynamically optimal pest management. *Ecological Modeling, 220*, 1315–1324.
Zhang, W., & Swinton, S. M. (2012). Optimal control of soybean aphid in the presence of natural enemies and the implied value of their ecosystem services ecological economics. *Journal of Environmental Management, 96*(1), 7–16.

Further Reading

CGIAR Research Program on Water, Land, and Ecosystems (WLE). (2014). *Ecosystem services and resilience framework* (p. 46). Colombo: International Water Management Institute (IWMI). CGIAR Research Program on Water, Land, and Ecosystems (WLE). https://doi.org/10.5337/2014.229 Retrieved from file:///C:/Users/Nathanial%20Matthews/Downloads/Ecosystem%20Services%20and%20Resilience%20Framework-revised.pdf.
Millennium Ecosystem Assessment. (2005). Retrieved from http://www.millenniumassessment.org/en/index.html
The Intergovernmental Science-Policy Platform on Biodiversity and Ecosystem Services (IPBES). (2018). Retrieved from https://www.ipbes.net/library
WWF. (2016). *Living planet report*. Retrieved from http://awsassets.panda.org/downloads/lpr_living_planet_report_2016.pdf
World Energy Council. (2016). *World Energy Resources*. Retrieved September 26, 2018, from https://www.worldenergy.org/wp-content/uploads/2016/10/World-Energy-Resources-Full-report-2016.10.03.pdf

Chapter 10
Infrastructure

Benjamin L. Ruddell, Hongkai Gao, Okan Pala, Richard Rushforth, and John Sabo

10.1 Introduction: What Is Infrastructure?

Infrastructures handle high-volume goods and services that require heavily capitalized, large-scale, durable, reliable, shared, interdependent, and specialized systems. **Infrastructure** facilitates social, economic, and environmental functions by achieving a high degree of efficiency at a low **marginal cost** to produce, transport, distribute, quality-control, and allocate high-volume goods and services. Infrastructure development usually requires large, long-term investments and substantial consideration of risk, change, and **vulnerability to extreme events** during the design phase. Water and wastewater service provision, stormwater management, bulk freight transport, bulk storage of food, energy, and water, waste management, distribution warehouses, internet address registries, fiber optic lines and switches, electrical power service, heavy manufacturing, roads, law enforcement and security, education, financial services and regulations, property rights, health services, and government are examples of goods and services that tend to be infrastructure-heavy.

B. L. Ruddell (✉) · R. Rushforth
School of Informatics, Computing, and Cyber Systems, Northern Arizona University, Flagstaff, AZ, USA
e-mail: Benjamin.Ruddell@nau.edu; Richard.Rushforth@nau.edu

H. Gao · J. Sabo
Wrigley Global Institute of Sustainability, School of Life Sciences, Arizona State University, Tempe, AZ, USA
e-mail: hongkai.gao@asu.edu; John.L.Sabo@asu.edu

O. Pala
Center for Geospatial Analytics, College of Natural Resourses, North Carolina State University, Raleigh, NC, USA
e-mail: opala@ncsu.edu

P. Saundry, B. L. Ruddell (eds.), *The Food-Energy-Water Nexus*, AESS Interdisciplinary Environmental Studies and Sciences Series,
https://doi.org/10.1007/978-3-030-29914-9_10

Supply Chains are sequences and steps during the production and delivery of goods and services that are heavily infrastructure-dependent. This is especially true for FEW commodity supply chains.

Modern infrastructure is a spectacular success story. It is essential to understand why modern infrastructure is so successful, so that we can understand why it is so essential to preserve and adapt that infrastructure. Think about the incredible feats of productivity, reliability, and efficiency that are made possible by modern infrastructure—for example, in the USA. Consider the following:

- A freight train can move a ton of freight a thousand miles on a gallon of fuel.
- Massive ships carry the products of whole nations across oceans to customers on the other side of the world with such efficiency that they are cost competitive with goods produced in-country.
- The natural, grey (dams, pumps, canals), and soft (treaties) infrastructures of the Colorado River bring high quality water from the snowcapped peaks of Wyoming a thousand miles southwest (and then hundreds of miles back uphill!) to the cities of Los Angeles, San Diego, Phoenix, and Tucson, at a cost and reliability equal to local water sources even after 20 years of crippling drought.
- Electrical power almost never goes out, and is shockingly affordable for such a high-quality and diversely useful energy source.
- We can send messages, and even complicated media and digital datasets, around the world without error at the speed of light.
- During the Great Plains Drought of 2012 and the Great California Drought of 2016 that drastically cut water availability for crops in the world's two greatest agricultural centers, nobody starved in the USA owing to global agricultural trade and massive grain stockpiles.
- Most of us can afford to drive our cars and trucks wherever we want without a thought to the cost of the fuel.
- Passenger automobile safety improves nearly every year.
- Food and water reliability and safety is so high that your risk of illness from what you eat is negligible—whereas for most human generations, starvation and waterborne illness were routine causes of death. Packages are delivered rapidly, reliably, and for pennies (or dollars) per parcel.
- There is more than enough food, energy, water, shelter, and housing for everyone (although access and equity are imperfect).
- Law enforcement can apprehend criminals across multiple countries and continents, and violent crime is rare (and national security threats vanishingly rare).
- Our cumulative taxes are generally far less than half our income, and provide historically excellent (albeit inconsistent) educational, security, health, environmental, law enforcement, recreational, regulatory, and other comprehensive government services.

10.1.1 *What Isn't Infrastructure?*

Small-scale and **distributed infrastructures** do exist, but when we discuss infrastructure we generally emphasize the largest scales of size, cost, longevity, dependency, complication, and complexity. If a system could be removed and replaced quickly, affordably, and without a great deal of disruption to the function of the greater socioeconomic system's function, it is probably not infrastructure. For example, if a local fast food restaurant were replaced with a salon (or even if all the fast food restaurants were replaced), this would probably not bankrupt your city, and would have only a modest impact on the lifestyle and function of the surrounding residents and businesses. The roads, power lines, and water pipes would need to change much. However, if the roads, power lines, telecommunications, or water systems serving the city were removed, the converse is not true: your neighborhood restaurant and all other residential and commercial activities would be out of service.

There are hard and soft FEW infrastructures.

Examples of **hard infrastructure** include pipes and water mains; power lines and plants; refineries; dams, canals, and aqueducts; roads; railways; ports, border crossings, and ports of entry; rich agricultural soil and cultivated acreage; heavy factories; warehouses, storage tanks, and silos; landfills; and natural resources, such as rivers, aquifers, and petrol basins.

Examples of **soft infrastructures** include laws and courts, regulatory agencies, government institutions, social institutions, cultures, private individuals, families, companies, trade agreements, borders, and markets.

Hard infrastructure is engineered, owned, managed, and have a significant physical footprint that is difficult to change. Soft infrastructure is behavioral, social, complex, flexible, and often difficult to observe to control or manage (let alone engineer). Soft infrastructures are no less locked in than hard infrastructures because human cultures and institutions are often as resistant to change as concrete and asphalt. While both hard and soft infrastructure are important to FEW system functioning, but soft infrastructure is fundamental to FEW system functioning because it determines the rules and purposes for hard infrastructure. Both hard and soft infrastructure are **values-laden** as they **lock-in** and reinforce the human values that guided their creation. Economies of scale and cost efficiency are two clear values embodied by modern, large-scale hard infrastructure. **Socio-ecological-technical systems** comprise both hard and soft infrastructure systems. **Knowledge systems** are social systems for thinking, remembering, and communicating, and are an important and often forgotten category of soft infrastructure.

Green infrastructure is a class of hard infrastructure that emphasizes the intentional creation or augmentation of (usually engineered) natural ecological systems to provide goods and services. Green infrastructure blurs the line between technology, policy, and ecology. Artificial wetlands are an example of green infrastructure, they provide both stormwater management and water quality services.

Natural infrastructure is a class of hard infrastructure that provides goods and services to people without any need for investment or management. Rivers are an excellent example of natural infrastructure that provide all kinds of water, water quality, waste management, and transportation services to people.

Grey infrastructure is a class of hard infrastructure that uses primarily artificial, engineered, and nonbiological materials and systems. Roads, pipes, power lines, and sewers are good examples of grey infrastructure.

The capital expense of infrastructure, in combination with the difficulty to exclude people from using infrastructure once it is built, naturally lends itself to socialized, heavily regulated ownership and governance structures. Typically, hard infrastructure systems are owned and managed by soft infrastructure systems, such as government agencies or large publicly regulated private enterprises. Infrastructures are often multipurpose and public-use, or are publicly owned with fee-based uses, such as with water systems, airports, seaports, and roads and bridges.

As was noted in our discussion of criteria in Chapter 1, FEW goods and services are among the lifelines of cities and of civilization, so they need to be extremely affordable and reliable for all people (Criterion 1 and 4, see Sect. 1.3.1). FEW infrastructures are therefore by definition critical infrastructures (Criterion 2), whether or not they are officially "protected" by the government (Chap. 14).

Lock-in is a key property of both hard and soft infrastructure. Once infrastructure is built, human civilization tends to incrementally build around it and **lock-in** to specific configurations that are prohibitively expensive to reverse because the economy, ecology, and society have become heavily dependent on the infrastructure. Locked-in infrastructure is inflexible and unadaptable. For example, most cities have evolved around major rivers and at natural cross-roads, because the natural infrastructure of rivers has historically provided nearly cost-free waterborne transportation, sanitation, and water supply. FEW infrastructures are a geographical foundation for human cities and civilization. If the river dries up, will you replace the river or move the city? Either option is staggeringly expensive and fraught with political, social, economic, and environmental considerations. Lock-in is not entirely negative. Societies and especially business decisions require some degree of certainty and commitment in order to invest with confidence. Infrastructure lock-in ensures some level of commitment about the future, and facilitates long term investment as a result.

Interdependency is an infrastructure property wherein each type of infrastructure is heavily dependent on one or several other infrastructures. Each interdependency creates vulnerability to **cascading failures**, where failure in one infrastructure causes failures in other infrastructures. For example, most layers of the modern FEW system are dependent on the electrical power grid, and even brief disruptions to electricity supply will disable water systems, financial and transaction systems, governance systems, transportation systems, and food distribution systems. Interdependency also means that the location of one infrastructure depends on another. For instance, steel mills depend on huge quantities of ore and electrical power, and are therefore located near ports, coast lines, and large (often nuclear) power plants. Railroads need to link to ports and cities, as well as to power plants, refineries, and large mines and sources of other raw material. A change in regulations will force cascading changes in other layers of the system. If you change one infrastructure layer, you change many, due to interdependency! This is lock-in squared.

Many practitioners view water security and water infrastructure as the key component to the FEW nexus, and the focal point that will bring together the global challenges that the world economy will face in the coming decades (Hoff 2011; Rasul and Sharma 2016). The higher risks of water resources are caused by the abrupt climate change, the more frequent extreme weather, and the rapid growth of water demand in both developing and developed countries. Agriculture is a major user of ground and surface water globally, especially in the USA, and accounts for approximately 70 percent of global consumptive water use. Therefore, safeguarding the water is essential to meet food demand in a nonstationary world. Additionally, in the energy sector, water is used for cooling during thermoelectric power generation (e.g., coal, nuclear, natural gas), 98 percent of power supply depends on the availability of water in an increasingly water-stressed world. Maintaining water security in a nonstationary world poses a huge challenge not only for water systems, but also maintaining FEW systems.

The electrical power grid is another infrastructure of huge concern. The power grid is the world's most complex machine, and provides the indispensable electrical energy to operate critical systems like lighting, telecommunications, controls, pumps, traffic systems, and payments. The power grid is vulnerable to heat waves, solar storms, targeted physical and cyber-attacks, extreme weather, and increasingly rapid technological change. It is not clear whether the power grid as it currently exists can survive the advent of massive renewable energy, battery, and electrical vehicle deployment. The power grid's extreme cost and importance are both an adaptive asset and a liability; an asset because there is a great deal of money and governance capacity focused on the problem; a liability because of the extreme consequences of failure.

Why are FEW infrastructures such a problem in the twenty-first century? The basic reason is the rapid rate of change and unpredictability of the modern world. Because of lock-in, it is difficult to adapt infrastructure to changing climate, demographics, culture, and technology—at least as infrastructure has been traditionally conceived. But in the rapid changing world, increasing the resilience of FEW infrastructure to increasing population, extreme natural events (e.g., flood and drought) and social instability (e.g., terrorism) is not an option—it is a must (Milly et al. 2008)! FEW infrastructure systems must be smarter, not just stronger, and more resilient to withstand a diverse set of emerging risks. Successfully managing these risks is crucial for leaders globally, bringing the need for infrastructure resilience to the fore.

How will we adapt our infrastructure to the rapid growth and change of the modern world? Some think that the whole idea of infrastructure needs to be reimagined to reduce or eliminate lock-in and avoid stranded capital. **Stranded capital** is where large amounts of capital investments in infrastructure (or other systems) are rendered useless and valueless due to changing regulation, economics, climate, etc.

This capital investment is irrecoverable—climate change creates a huge stranded capital risk for oil rigs and refineries, ports, large dams, and irrigation districts. As infrastructure is a long-term investment, small mistakes in planning may result in large impacts in a rapidly changing, unpredictable future.

We cannot afford to make infrastructure mistakes—financially or otherwise. Most of US infrastructure, and much global infrastructure, was built during a different era: burgeoning economic growth, low public debt, and large government surpluses. Rebuilding infrastructure during lean fiscal times is an entirely different proposition. An ultimate cause of FEW infrastructure collapse in the twenty-first century has been, and will continue to be, fiscal exhaustion. In the usual cycle soft socioeconomic infrastructures fail first, followed by fiscal exhaustion of the proximate infrastructure-governing institutions, followed by the slow neglect of hard infrastructures, followed by "sudden" collapse of hard infrastructures. Our twenty-first century infrastructure investments need to be smarter, more resilient, more cost-effective, more modest, and more fiscally sound.

So, how can we build infrastructure resilience?

Eliminating lock-in is hard to imagine, but lock-in could be reduced by shifting to **distributed infrastructure** that facilitates services on a smaller, more local scale; increasing the diversity of the system; shortening lead times to make changes; and reducing the risk of large losses and stranded capital. Nonetheless, distributed small-scale infrastructures tend to have higher marginal costs, which advantages traditional infrastructure during long-term infrastructure planning. Local governance is a form of distributed soft infrastructure. Is distributed infrastructure still "infrastructure"? Yes and no. Yes because it serves the same functions. No because it lacks the systemic lock-in property and it mitigates many of the systemic problems associated with large scale infrastructure. If we can provide the services of infrastructure, without the problems, this is a good thing—regardless of whether we still call this "infrastructure."

Improving our foresight by investing in science and information is also a good strategy to improve our resilience to future change. However, foresight is emphatically not 20/20, and history is full of examples of catastrophic failures to anticipate change… and we should not make the mistake of imagining we are exceptional in this regard. The world is full of Black Swans, Knowable Unknowns, and Knightian Unknowns. The **Black Swan** is an unforeseeable and consequential event that has never happened before, like the emergence of the atomic bomb during World War 2, the founding of Islam, or the creation of social media and the internet. **Knowable Unknowns** are important decision factors that we include in our calculations and that we could in principle estimate or observe, but which have not yet been estimated or observed to an adequate precision (perhaps due to cost or lacking knowledge), like the exact volume of water remaining in California's freshwater aquifers or the likelihood that the Earth will be struck by a large meteor tomorrow. **Knightian Unknowns** are important decision factors that we know we cannot know, or that we do not even know exist, like (from the perspective of a medieval plague doctor) the existence of microbes as a key factor in medical pathology.

Decoupling is a strategy to reduce infrastructure vulnerability to cascading failures by buffering or reducing the interdependency property. For instance, backup generators and islanding are tactics to remove vulnerability to power grid failure by selectively decoupling a facility or community from the power grid. However, decoupling reduces efficiencies and erodes the effectiveness of the infrastructure.

Anticipatory Adaptation is a resilience strategy that involves monitoring and anticipating changes that will affect the infrastructure, followed by an attempt to change the infrastructure in time to compensate for the changes. This is a sound strategy, but in the case of large FEW infrastructures we may not be able to see far enough into the future, with sufficient accuracy and lead time, in order to make the necessary changes.

Robustness is a strategy that chooses infrastructures that perform well under a very wide range of possible futures, reducing the need to precisely predict the future. Robust investments are low-risk investments, but they may come at the cost of some cost and efficiency in the short term. Robustness can be engineered into infrastructure, but has historically been ignored in favor of efficiency; this has proven to be a mistake.

Insurance, and especially reinsurance, attempts to price-in the cost of change and uncertainty about the future to investment decisions. Insurance is an efficient mechanism for pooling risk, but it fails to handle systematic risks that affect almost everyone or that impact very large scales. This is a problem for infrastructure insurance, because infrastructure tends to focus on the systematic and the large scale.

Unfortunately, the current infrastructure in the USA is a continued dire need of overhaul. On March of 2017, ASCE released its Infrastructure Report Card with D+, reflecting the significant backlog of needs facing the USA's infrastructure writ large. Aging and underperforming FEW infrastructures become a drag on the national economy, especially for aging hydraulic infrastructures (dams, levees) and transportation system (aviation, roads, bridge, rail, transit, ports, and inland waterways). The infrastructure here is functional, but in many cases outdated, inefficient, under-sized, and beginning to fail. This should be viewed as an opportunity to adapt to the future—and it is an opportunity we cannot afford to waste.

10.2 Infrastructure Supply Chains Form a Multilayer Network

Coupled infrastructures and supply chains form a network—a very complicated network with many nodes (i.e., producers, consumers, storage), many layers (i.e., multitype or multiplex), representing multiple modes of transportation (i.e., pipe, road, rail, ship, water, wire), with many different FEW commodities sharing that network. These are Process Networks, not simple networks (see Sect. 2.2). In order to manage these coupled infrastructure systems, we need to understand the spatial structure of the network, in addition to the process details of each infrastructure system. Important process details include: capacity, utilization of capacity, marginal cost of use, capital and maintenance cost, reliability, vulnerability-to, and resilience-to, along with the details of the couplings between network layers.

Capacity is the maximum rate at which goods and services could be produced, consumed, or transported using the infrastructure. For instance, a power transmission line may have a capacity of 100 kW, and a canal might have a capacity of 1 million gallons/h.

Utilization is the percentage, fraction, or simple rate at which the infrastructure is employed to produce, consume, or transport goods and services. Utilization cannot exceed 100% of the capacity. Infrastructures often experience chaotic failures when utilization approaches capacity, because the stress involved in near-capacity operations tends to trigger many unpredictable problems. Power grids are known to become unreliable when utilization exceeds 85–90% of capacity, so power grid regulations in the USA typically require operation below 85% of capacity.

Marginal cost to use the capacity is the cost of processing the next unit of a good or service using the infrastructure, not including capital and maintenance cost. For instance, a thousand gallons of potable water in the USA costs roughly 1 dollar (2018 USD). This marginal cost might include maintenance, operational, and energy costs, but does not include the **capital cost** or finance charges required to build the infrastructure. Water customers in the USA typically pay close to the marginal cost for water, whereas capital costs and water prices are heavily subsidized.

Capital cost is the cost to build an infrastructure, not including maintenance and operational costs. Capital costs tend to be large, and in many cases are heavily subsidized using publicly funded grants and low-cost loans.

Reliability is the percentage or fraction of the time that an infrastructure is adequately functioning. Actual reliability and designed (predicted) reliability are two different concepts. Often the actual reliability is lower than the designed (predicted) reliability). Most infrastructures in the USA are designed to keep reliability well above 99% after considering all known hazards such as earthquakes, storms, heat waves, floods, and accidents. Reliability is a common form of regulatory quality requirement imposed on infrastructure operators by their governors. Reliability is a measure of safety and quality for infrastructure.

Vulnerability-to some hazard is the degree of damage, downtime, cost, degradation, or loss of life that will occur if a specific magnitude and duration of hazard occurs against a specific infrastructure. The longer and more severe the hazard, and the weaker the infrastructure's design standards, the more vulnerable the infrastructure is to the hazard.

Resilience-to some hazard is the infrastructure's capacity to quickly and efficiently adapt to and recover from a damaging event. Resilience usually involves change, because if a failure occurred in the system, the system often needs to be redesigned. Resilience of hard infrastructure is aided by its redundancy, diversity, (in)vulnerability, and by the effectiveness of the governing soft infrastructures.

Couplings between layers are points of contact between different types of infrastructure—like where the power grid runs water pumps, or where a utility's staff act to maintain or restore a water pump. These couplings are key functional relationships and key factors for reliability, vulnerability, and resilience. Wherever a coupling exists, it is possible to control for cascading failures and to manage vulnerability.

FEW infrastructure systems are inherently circular because every producer of outputs is also a consumer of inputs to some extent. This is especially true for electrical energy and water supplies, and for road-based transportation services, which are used by practically everyone in the system. Water and energy (especially electricity) are major and non-substitutable inputs to almost every production process and infrastructure. Ordinary people, or consumers, are part of this "circular economy" because as people consume food, energy, and water, they produce labor that produces these goods.

The FEW system is connected to every other human and natural system on Earth; it needs to be understood as the "Food–energy–water-everything" system (FEW^e). Some of the more important connections of FEW infrastructure to everything else are to climate, weather, hydrology, finance, manufacturing, human health, security and defense, terrestrial ecology, marine ecology, and microbiology.

Circularity can be measured; it is the degree to which an economy, or flows of goods, services, or information, cycle and feed back to the original source; for example, if half of a community's food is produced locally, the circularity of that food economy is 0.5.

Circularity exists within each of the individual "F," "E," and "W" sectors. Oil, natural gas, coal, and uranium are "primary" energy commodities: they are produced by directly extracting them from the Earth and can directly provide heating energy. When these primary energy commodities are used as inputs to fuel electrical power plants, the resulting electrical power is a "secondary" energy source. Similarly, raw water extracted from rivers, aquifers, and oceans is a primary commodity, but treated potable water may be a secondary commodity if it undergoes significant treatment, and treated wastewater is always a secondary commodity. Wild seafood and wild game along with naturally sourced fresh produce are primary food resources, but processed foods and meats and industrially produced produce tend to be secondary food sources. The Earth's biosphere is a producer, transporter, and storage provider for primary FEW natural resources (especially water); people merely distribute and consume these commodities, reaping the free benefits of Earth's massive natural infrastructure. Secondary commodities are more expensive than primary commodities, and have much higher life cycle impacts because of the infrastructure costs and the efficiencies lost in the conversion of coal to electricity, grain to beef, or river water to tap water.

Electricity is the main energy input to all steps of the water infrastructure, and roughly 4 percent of US electricity is used to operate the water infrastructure, that is, moving and treating water and wastewater by public and private entities (Copeland and Carter 2017). However, regional differences can be significant. In California, the State Water Project (SWP) pumps water almost 2000 **ft** over the Tehachapi Mountains. The SWP is the largest single user of energy in California. It consumes an average of 5 billion kWh/year, accounting for about 2–3 percent of all electricity consumed in California. Overall, as much as 19% of the state's electricity consumption is for pumping, treating, collecting, and discharging water and wastewater (Copeland and Carter 2017). Water is heavy: extracting it from rivers and streams, pumping it from aquifers, and conveying it over hills and into storage facilities is a highly energy-intensive process.

The food system provides another interesting example of connections between FEW infrastructure and services. Most of the bulk grains produced in the USA are used as livestock feed or as inputs to biofuel processes, rather than being eaten by people (Pimentel and Pimentel 2003). US livestock directly consume more than 7 times as much grain as the entire American population. The grains fed to U.S. livestock are sufficient to feed about 840 million people following a plant-based diet (Pimentel and Pimentel 2003). Additionally, about 15% of the total global energy consumption was derived from biomass in 2007, which threatens the food security particularly in developing countries (Uhlenbrook 2007). Food production and transportation consumes about 10 percent of energy in the USA, and up to 30 percent of global energy consumption (FAO 2011). Food production accounts for 70 percent of global freshwater use. As food demand increases with population growth, it will require both more water and more energy. In the USA, most of the water consumption is applied as irrigation water to grow crops and cooling thermoelectric power (Maupin et al. 2010). Water supply infrastructure, food production transportation and storage infrastructure, and fuel infrastructure are closely coupled.

Because FEW systems are interconnected and circular, there are infinite degrees of dependency between the FEW system components and infrastructures. This is not a simple, linear, efficient system defined entirely by inputs and outputs; rather, it is a complicated, complex system with many redundant processes and pathways that balance productivity, sustainability, risk, and resilience in the presence of randomness and uncertainty. An individual infrastructure component may be engineered as a simple linear system (e.g., a canal or roadway), but the assemblage of all the FEW[e] components are not a simple linear system. Complexity and redundancy is the key to the system's resilience, but it also obscures the systems-level structure and function and introduces the potential for unexpected outcomes.

Given the interdependency, circularity, complexity, and randomness involved in the world's FEW systems, and socio-ecological systems more generally, cascading failures are an outcome that should be expected for FEW infrastructures. For example, cascading failures can occur when the electrical power grid failures quickly—within seconds—cause food infrastructure failures via refrigeration and transportation failures, and water infrastructure failures via pump and water treatment failures. Water and food failures can propagate to the energy system as well, albeit on the timescale of days and months rather than seconds.

Traditional systems engineering strategies against cascading infrastructure failures are redundancy, buffering, storage, and decoupling. **Redundancy** underpins resilience and reliability by allowing us to switch sources when one input is disrupted. A backup power generator provides redundancy **Buffering** delays the propagation of effects from one part of the network to another, increasing our ability to wait out problems and slowing their impacts. Buffering can include demand modification, for example water restrictions during drought. **Storage** serves as both a buffer and a redundancy because it allows an input to be completely turned off without affecting a process—at least until the storage is exhausted. A reservoir provides is a buffer that provides storage. Adding storage to FEW systems has made them more reliable and resilient, and can also improve economic efficiency by allowing FEW producers and consumers to ride out peaks and valleys in commodity pricing. But storage is not enough to prevent failures or cascades. Decoupling is a strategy to make remove dependency on a network during an emergency. For example, a house is completely separated from the power grid and runs on its own solar energy is decoupled.

We also need to mind the soft infrastructure, and make sure that we have the human systems in place to anticipate, redesign, adapt, repair, and recover from disruptions in the FEW system. These strategies are implemented at interconnection points between infrastructure layers.

Because a large fraction of the inputs to FEW systems are ecosystem services, FEW systems are unusually dependent upon the stability and healthy function of the climate, environment, and natural ecosystems. The production, transportation, and storage of clean freshwater are mostly provided by the Earth's water cycle. The water cycle takes huge amounts of solar energy as an input to distill and transport ocean water to the continents, where rain falls into watersheds, soaks into storage in aquifers, and then rivers that reliably release and transport stored groundwater year-round to our large farms and cities (Falkenmark 1977). If the climate changes and disrupts this water cycle, people will need to build and power water infrastructures capable of desalinating and transporting a literal river of water from the ocean to the continental inland—an expensive proposition. Other examples abound; farms depend heavily on growing climate and soil quality for cost effective production, and electrical energy is increasingly produced by solar, wind, and hydroelectric resources. This **natural infrastructure** has an important role in the FEW system, so changes to natural infrastructure, ecosystems, and climate have demonstrable cascading effects throughout the whole FEW system.

10.3 Infrastructure Supports the FEW Supply Chain

Food, Energy and Water supply chains are supported by infrastructure. FEW commodities tend to be massive and voluminous, and therefore require substantial infrastructure (Criterion 2). Inflow/Outflow or Import/Export are implied in every step of the supply chain. At each step in the supply chain there is a point of **origin** of the flow, a point of **destination** for the flow, a **route** traveled by the flow, and a **mode of transportation** for the commodity flow. The mode of transportation is always a type of infrastructure, such as a truck operating on a road system. The route follows a spatial path, consumes time, and crosses a distance. The points of origin and destination exist in space and time, and these represent human producers and consumers, storage facilities, and also natural systems like aquifers that can produce ecosystem goods and services. Most supply chains involve a variety of **intermediate producers**

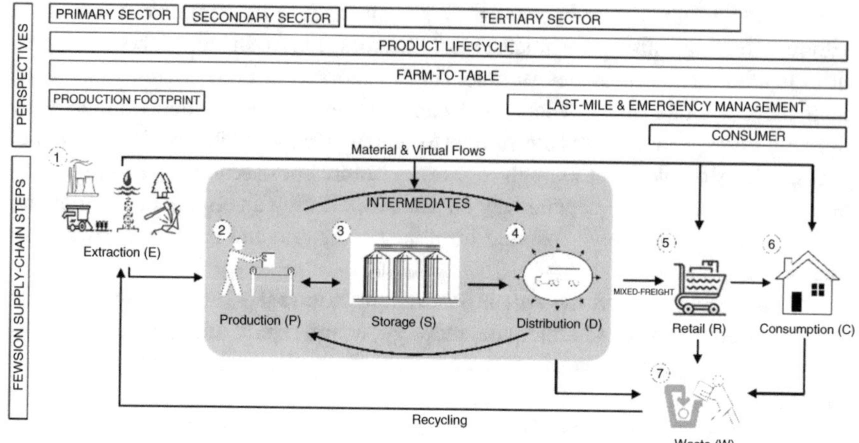

Fig. 10.1 A generalized supply chain is shown above that shows the different life cycle stages of a produce and its components. First, a product is extracted (E) from raw materials, where it is then transformed through a production step (P), moves through Storage (S) and Distribution (D), after which it is eventually brought to market (R), consumed (C), and discarded (W). Not every commodity contains all these steps, sometimes just an individual link, e.g., P → S, but all commodities go through a similar seven-stage process of varying degrees of complexity. Used with permission of the FEWSION project

that add value along the supply chain, for instance when a producer of raw crude oil delivers that oil to a refinery to produce derivative products like gasoline; these intermediates tend to be "business to business" transactions.

It is possible to describe supply chains and infrastructure very generally as a sequence of seven typical supply chain steps, each of which utilizes infrastructure as appropriate to the type of commodity involved. Figure 10.1 summarizes these seven typical supply chain steps.

1. Extraction (from the environment, at the source, primary industry only): E
2. Production (manufacturing, processing, value-add, business to business): P
3. Storage (mass stockpiling, hubs, reserves): S
4. Distribution (transportation, warehousing, last mile distribution): D
5. Retail (last mile delivery): R
6. Consumption (by human end users, not companies or producers): C
7. Waste (from all sources, both producers and consumers, including recycling and reuse): W

What follows is a graphical summary of commodities, supply chains, and infrastructures for each of the major components of the FEW infrastructure and supply chain. We make reference to FEWSION commodity codes (https://fewsion.us) which contain SCTG commodity codes to explicitly link these diagrams to commonly used meso-scale commodity flow data for the USA. We will refer to the FEWSION commodity codes as "FC" in the remainder of this chapter.

Depending on your perspective within the greater FEW system, you might focus on one part of the supply chain and infrastructure. Emergency Managers tend to focus on the "last mile" connecting warehouses and distribution hubs to consumers, because this tends to be the source of risk during emergencies such as in the aftermath of storms. Consumers and also political decisions tend to focus on their relationship with retail servicers and deliveries, which are the average person's only substantial point of contact with the supply chain. Life Cycle analysts and sustainability experts try to consider the impacts and waste streams generated along the full supply chain, for instance in the farm-to-table production and delivery of food. Each step in the FEW supply chain involves both hard and soft infrastructure. These diagrams emphasize hard infrastructure, but each step in the chain also involves soft infrastructures of governance and ownership.

The **primary sector** is the set of natural resource extraction industries that perform fishing, farming, mining, and drilling operations; these industries tend to be very infrastructure-intensive and create heavy footprints (Criterion 2). Extraction industries make use of bulk storage stockpile infrastructure.

The **secondary sector** is the manufacturing value-added industry, which is also relatively infrastructure-intensive and footprint-heavy, albeit not to the extent of the primary sector. Manufacturing makes heavy use of warehouse infrastructure.

The **tertiary sector** is the service sector, which tends to add a lot of economic value, tends to be relatively urban and lightweight on its infrastructure requirements. Services tend to be produced and consumed over short distances and do not use much infrastructure besides commercial real estate buildings and roads, with the exception of information, banking, and consulting services that tend to make heavy use of communication and computing infrastructure and personal transportation systems (i.e. airlines).

10.3.1 Food

There is a difference between agricultural products and "food" for human consumption. Most agricultural products do not become human food, at least not directly or without a lot of value-added manufacturing. It is necessary to distinguish between agricultural products and "food." The term "agri-food supply chains" (ASC) is used to describe the systems that bring agricultural and horticultural products from production to consumption (Aramyan et al. 2006). Even though ASCs are similar to other supply chains, factors such as food quality and safety, weather variability, limited shelf life, and demand and price variability makes it complex, therefore challenging to manage (Ahumada and Villalobos 2009).

Agricultural products are produced at the farms, ranches, hatcheries, fish farms, fruit and tree nut farms, with the exception of wild-caught fish. Most of the agricultural products have a farm-to-table connection where extraction is directly connected to the retail or consumption stage. The conventional supply line follows storage and processing stages. Livestock and poultry are first fattened by using stored animal feed obtained from farms and then slaughtered, portioned and packaged. Grains, nuts, fruits, and vegetables are stored in grain silos or cold storage before they are sent to food processing facilities where they are cleaned, prepared into various products, and then packaged. All the fish and seafood products are processed and packaged in the fish processing facilities to be ready for consumption. Note that "Animal Farms/Ranches/Hatcheries (E)" stage has a direct path to the "Slaughter Houses/Processing Plant (P)" for products that does not need to go into feedlots. All the products are transported to regional warehouse for distribution to restaurants, grocery stores, and food banks to connect with the consumer.

In Fig. 10.2 we combine various commodities that are interrelated or similar in their supply chain steps. All agricultural products (FC: 1100000) as well as all food products (FC: 1110000) are represented. Agricultural products include live animals and fish (FC: 1100001), cereal grains (FC: 1100002), other agricultural products (FC: 1100003) such as vegetables, fruits, flowers, sugar beets, sugar canes, unprocessed coffee, etc., as well as animal feed (FC: 1100004) such as cereal straw, husks, forage products, raw hides, dog food, cat food, etc. Food products include meat and seafood (FC: 1111005), milled grain products (FC: 1112006), and other food stuffs (FC: 1113007), such as milk, cream, cheese curd, vegetable oils, soups, broths, fruit and vegetable juices, and more.

In extraction stages waste is generated in the form of air pollution, water pollution and animal waste which falls under FEWSION waste categories "Chemical and Industrial" (FC: 5570041), "Green Organic" waste (FC: 5572041) and "non-potable reclaimed water" (FC: 3342999). Waste from feedlots and processing plants also fall into those categories along with the waste that goes into landfills "Municipal

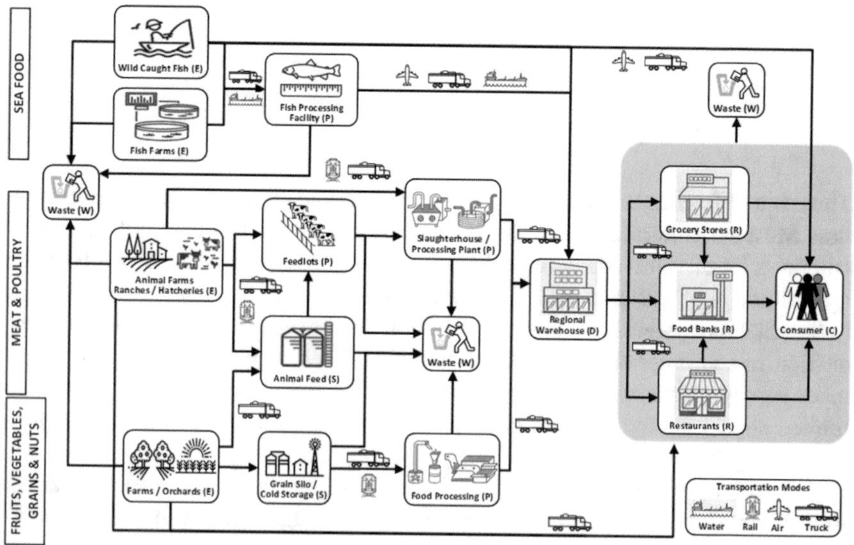

Fig. 10.2 The supply chain and infrastructure for food; agricultural, cereal grain, meat, poultry, and seafood products. Used with permission of the FEWSION project

Solid Waste—Landfill" (FC: 5568041). On the customer side, spoiled and unused food items along with discarded packaging are sent to landfills (FC: 5568041) and recycled (Municipal Solid Waste—Recycling FC: 5568041). In addition, wastewater is generated by consumption of the food items (FC: 3331999).

The meat supply chain starts with livestock like cows, pigs, and chickens, which are brought by truck and rail to facilities that slaughter and pack the meat. Livestock are the main input, and processed meat products are the secondary food source output. These facilities may resemble modern factories, or may be traditional butcher shops. If a butcher shop, this facility combines production, transportation, storage, and distribution in one operation. Water and electricity are used in these facilities. Seafood may or may not be considered "meat" depending on the culture and industry language, but seafood processes are similar. The traditional supply of live animals is through grazing and foraging in grasslands or by catching wild fish. The industrial feedlot system utilizes bulk grains like corn, and also hay, as the major inputs rather than relying on wild food sources for the animals. The majority of bulk grains go to feed livestock.

Packaged meat is transported by refrigerated truck from the packing facility across public roadways to logistical warehouses, and then distributed to groceries and restaurants. Packaged meat is produced in large quantities for mass international export by a limited number of global regions, and many countries import meat products from those major regions. Those meat exporting regions tend to coincide with large surpluses of bulk grain production. Packaged meat is stored using commercial refrigeration at various points along the transportation chain, including at the packing facility and at the warehouse. Meat has a short shelf life measured in days to months (refrigerated) depending on the product, so refrigeration is an essential modern technology. Refrigeration requires large amounts of electricity and is moderately expensive. One of the most important roles of animals in the food supply chain is as storage of calories. Field crops are seasonal, but animals can be kept alive until needed and then slaughtered. As a result, one of the key storage reservoirs in the food system is in livestock, before the animals enter the meat infrastructure. A major trigger of famine in developing countries is that livestock can die *en masse* from drought or lack of feed, and this suddenly eliminates a major stored supply of food that is normally available.

Bulk grain agricultural products are foundational to the modern civilization's food supply chain. Bulk grain production e.g., for corn, soybean, or wheat, is an industrial scale process involving significant capital placed in improvements and maintenance and cultivation of entire regions and large fractions of the Earth's surface. The most suitable land for grain production is well-drained but flat, with rich organic soil and moderate temperatures and abundant rainfall during the growing season. The US–Canadian Great Plains are arguably the most advantaged major rain-fed grain-producing region in the world, but southern Brazil, parts of Australia, parts of southern Africa, and much of Europe are also excellent grain regions.

Industrial grain production requires many industrial inputs. Fertilizer is used to maintain soil phosphorous and nitrogen, and this fertilizer is an energy-intensive or mined product shipped from regional and global sources. Herbicides and pesticides are used in large quantities to ensure yield and quality; residues of these chemicals can make their way into the grain supply. Seed is used in small quantities; today's seed is often genetically modified in the USA to enhance drought and pest resistance and improve yield. Most bulk grain is rain-fed, as in the Corn Belt of the central USA, but some bulk grain is irrigated. Expensive heavy equipment is used on farms, including tractors, cultivators, combines, and trucks that are capable of rapidly covering thousands of acres of land; this heavy equipment is often leased or shared among farms so its capacity can be more fully utilized.

Bulk grain is produced in such large masses and volumes that it is transported by rail, barge, and ship more often than by truck, after trucks bring the grain from the farm to railheads. Bulk grain is stored in silos that keep the grain dry, free from pests, and well ventilated; these silos are normally located at collection hubs along railways where local farmers transfer their grain to wholesalers. Bulk grain can be stored for years if it is kept dry and kept safe from pests. Rain-fed grain supply is regional, seasonal and unpredictable, with grain harvests occurring once or perhaps twice per year. Irrigated grain operations are much more reliable and are very different because they occur mostly in dry, hot regions with several growing seasons per year resulting in much higher yields. Supply is concentrated in a number of major exporting regions around the world that produce massive surpluses of bulk grain. If a major producing region is compromised by drought or other weather problems, yields can drop by half or more in that region for the year. Industrial farming techniques have reduced crop failures and improved yields in bad years, reducing this problem.

Although grain production is concentrated in a handful of major exporting regions and is regionally unreliable, demand for grain exists wherever people live and is nonstop, because the grain feeds biofuel and livestock operations (and human flour production) that have relatively constant demands. The modern global market for bulk grain has ensured that supplies are usually reliable on the international market, because all exporting regions are not hit simultaneously by problems such as drought. However, people who do not participate in global markets can suffer severe shortages and famines during local failure of their grain harvest.

Most food in modern countries comes from factories, not farms. Non-meat packaged foods have more in common with factory-manufactured products than traditional foodstuffs. A wide range of produce, packaged meat, grains, and food-grade chemicals are inputs to these factory processes, along with a lot of electricity and water, and natural gas for cooking. Food science is a branch of chemistry that optimizes products for flavor, shelf life, nutritional content, and cost, and that manages food quality and safety processes. Packaging technology is an essential part of food processing, because it allows processed foods to be shipped and stored for periods of months to years without refrigeration. Packaged food is a secondary food source—much like electricity is a secondary energy source.

Packaged food is shipped by ordinary truck on public roadways, from food processing facilities to logistical warehouses, and then by truck to groceries and restaurants. Unlike meat and dairy products, most packaged foods are shelf-stable and require no refrigeration. Packaged food can be easily shipped internationally and

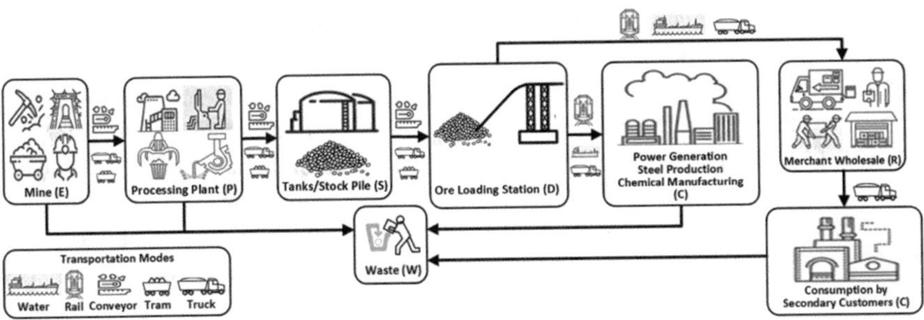

Fig. 10.3 The supply chain and infrastructure for energy (coal) and mined products. Used with permission of the FEWSION project

across jurisdictional boundaries for specialty foods, but most of the system is national in scale. Packaged food is a major part of the modern food supply, especially in the USA. Because it is relatively easy to store and can accept inputs from global sources, the production and transportation of packaged food helps to alleviate seasonal mismatches between local crop production and the constant demand by people for food.

Soft infrastructure in the food sector involves **consumers**, **owners**, **financiers**, and **governance** (Fig. 10.3).

In the USA the food infrastructure is privately owned by a combination of small and large businesses, known as "agribusiness." In the developing world the food infrastructure tends to be locally owned, or even owned by the consumers themselves (e.g., subsistence farming, family farming). Finance tends to be private. Consumers are ultimately people, but may also be value-added food processing and manufacturing operations or even energy companies (in the case of grain ethanol). Governance over the food system ranges from local market control to national regulatory agencies like the Food and Drug Administration (FDA) in the USA. The FDA enforces grading labels and quality standards that distinguish grades such as organic, prime, and "natural," and requires labeling of ingredients and nutritional values. Most countries require some kind of random testing for food quality and safety; the primary concern of this testing is the prevention of contamination of food products by infectious diseases. Unlike bulk grain or most raw FEW commodities, packaged food and meat tends to be non-subsidized and is usually sold at market prices. Bulk grains and some dairy products tend to be price-controlled or subsidized as part of national food security policies.

Regulated and subsidized markets serve to prevent agribusiness from collapsing during droughts and periods of low prices, and stabilize the supply of bulk grain in what would otherwise be a highly risky and cyclical business. This type of regulation and subsidy is politically popular in countries with a large traditional farming population, and also ensures that grain is normally overproduced. This helps ensure food security. Bulk grain trade serves as a large substitutionary mechanism for local water, as pointed out by Allan (1998) as the "virtual water trade." There are many countries in arid regions that feed their populations and animals largely with imported grain, dramatically reducing their need for water to irrigate crops. Notably, Saudi Arabia and other Arabian Gulf nations import a large fraction of their grain from overseas including the USA. Other countries like Iran have chosen to attempt self-sufficiency in bulk grains despite an adverse climate and supply situation, because they do not trust western grain suppliers. Bulk grain has been overproduced and exported by nations like the USA as a geopolitical strategy both to ensure food self-sufficiency, and to allow for food aid to nations suffering famine, and to create political leverage through foreign food dependencies on the USA. Food, like energy and water, can be a tool of power.

Quite a few academic papers detail the extent of this field by providing literature reviews summarizing the work done in various aspects of food supply chain systems (Chandirasekaran 2017; Cunningham 2001; Liljestrand and Fredriksson 2015; Shukla and Jharkharia 2013; Handayati et al. 2015). Another resource providing essential information by detailing the conceptual model of US food supply chain system is recently published by the National Research Council (2015). Other sources used for this section are Nature Conservancy (2016), Septani et al. (2016), and Van der Vorst et al. (2007).

10.3.2 Energy: Coal (and Mining Products)

The ore is extracted from surface and underground mines. In the case of coal, majority of the ore comes from surface mining operations. Extracted ore is then transferred to processing plants where it undergoes both physical and chemical transformation. This process involves, crushing, removing impurities and producing different ore for different purposes. Extraction to loading station transfers are mostly achieved through conveyor belts, trams or trucks. Processed ore is stored in stock piles or storage tanks and shipped through ore loading stations. Freight trains transport 70% of the coal in the USA, and barges and ships are used when applicable. Trucks are not frequently used for coal transportation.

Processed ore is shipped to various industrial facilities that are used for manufacturing, production and power generation (coal) to be consumed. It also is shipped to wholesale merchants to be sold to secondary customers. Statistics from 2017 show that 92% of the coal is used for power generation, 7.2% is used by other industrial customers and only less than 2% is consumed by commercial and residential customers (US EIA 2017). In extraction, processing and consumption stages waste (FC: 5565000) is generated as well as wastewater (FC: 3330000) and reclaimed water (FC: 3340000). Transfer to retail and consumer stages are achieved through trains, trucks and tankers (from overseas sources).

In FEWSION **database** mining products are listed under industrial sector (FC 4400000) with five sub categories such as stone, natural sand, gravel, non-metallic minerals and metallic minerals (FCs: 4401010—4401014). However, fossil fuels such as coal (FC: 2211015) and coal n.e.c. (FC: 2216019) that are also illustrated in this supply-chain diagram have their own FEWSION categories under fossil fuels sub sector of energy sector.

Soft infrastructure associated with coal and mining involves **consumers**, **owners**, **financiers**, and **governance**.

Coal and mining products are almost entirely consumed by other producers in the energy and manufacturing sectors, and are often subject to significant materials-manufacturing and refining before delivery further down the supply chain. The mining infrastructure is owned by large private corporations which are often internationally managed, and is usually financed by private or sovereign-wealth concerns. Mining operations are owned by many concerns around the world, but they are particularly important and emphasized in developing nations. Mining is often extremely damaging to the environment (especially to freshwater ecosystems), and is often very hazardous to workers, and is therefore subject to heavy regulatory oversight in most countries—but this oversight is frequently found insufficient to prevent environmental and human damage, even in the USA. Cleanup of toxic materials at abandoned or financially failed mining operations is a long-term environmental and economic problem in many places.

10.3.3 Energy: Natural Gas

Natural Gas (FC: 2215196) is extracted from onshore and offshore wells through conventional drilling and hydraulic fracturing operations. From the wellhead, raw natural gas it is pumped to a processing plant to be cleaned by separating impurities, fluids, and various hydrocarbons. Water, oil, and sulfur are some of the waste products (FC: 5565000) of this process. Processing plants produce "pipeline quality" dry natural gas, also known as methane and when needed stored in underground storage. Dry natural gas reaches the customers in two ways. If it is going to an overseas market, it goes through a liquid natural gas (LNG) plant to be liquefied and loaded onto tanker ships at an LNG shipping terminal. Upon arrival, it would go through the regasification process and sent to consumers through a market hub. If it is not going to an overseas market, it goes through a market hub to be priced and traded. The market hubs are located at the intersection of major pipeline systems (i.e., Henry Hub in Louisiana, which is the principal hub in the USA) and serve local distribution companies, power plants, and industrial manufacturing facilities. Distribution companies mainly serve industrial, commercial, business, and residential consumers through distribution pipeline systems that they build and maintain. However, depending on location they also may serve power plants and industrial manufacturing facilities (Fig. 10.4).

Processing plant produces another product called natural gas liquids (NGL). This product is sent to a fractionator where they are broken down to their base components, such as ethane, propane, t and butane (Purity NGLs). Purity NGLs are stored at a NCG storage facility and sold to petrochemical plants to be turned into chemical products that industrial customers need such as ethylene, propylene, butadiene, and aromatics.

The extraction and processing stages generate wastewater (FC: 3330000) as well as non-potable reclaimed water (FC: 3342999). In the extraction stage, water is

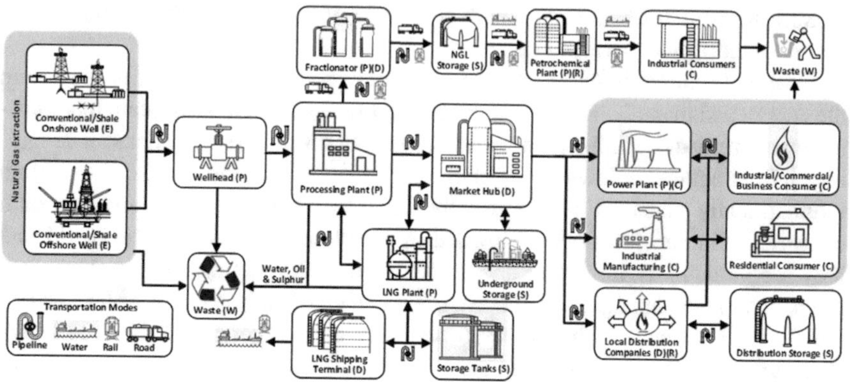

Fig. 10.4 Supply chain and infrastructure used for natural gas. Used with permission of the FEWSION project

mostly used for sludge preparation and cooling. There is also chemical waste from the drilling, hydraulic fracturing and refining processes, but there is no FEWSION code for chemical waste. Most of the waste at the consumption side is from emissions caused by the consumption of the natural gas and its by-products.

Natural gas, like oil, exists in naturally formed basins under the ground where it has been stored for geological timescales. Natural gas reservoirs occur in many parts of the world and in most countries, but some regions lack them entirely. The USA is rich in natural gas, but it is concentrated in Southern California, the Rocky Mountains, the Southern Great Plains, the Gulf of Mexico, Michigan, and Appalachia. Unlike oil, which often requires capital-intensive and long-term drilling operations that can only be undertaken by major corporations, natural gas is relatively easy to produce using modern hydraulic fracturing techniques and drilling, so there are currently a large number of small wells. Natural gas is produced by drilling into these reservoirs.

Natural gas is transported by being routed directly from the wellhead into a collection network where many small pipelines link individual wells to major transmission pipelines. Natural gas is shipped internationally in liquid form using special freighters and terminals. In cities, natural gas is distributed using small low-pressure lines that run along most streets and into most buildings—much like electricity.

Natural gas storage is centralized at major terminals located at ports and along pipelines, normally in large compressed tanks that store natural gas in its compressed liquid form. Demand for Natural Gas correlates with winter heating uses, morning and evening hot water heating and cooking uses, and late afternoon peak power generation. Storage, supply, and demand management are not as problematic in natural gas infrastructures as in electrical power infrastructures, but supply problems do sometimes occur during extreme winter weather.

The soft infrastructure of natural gas involves **consumers**, **owners**, **financiers**, and **governance**.

Natural gas is a major input to many chemical processes, and is also a very cost-effective and clean-burning fuel for cooking and heating in homes and businesses. Importantly, much of the natural gas in the USA is now used for natural gas turbine power plants, owing to the relatively low cost, low water use, high ramp rates, low capital costs, and low air pollution of natural gas turbines compared with other energy sources. Because natural gas is relatively easy to produce using small wells, there are a large number of small private gas producers mixed with the major corporations in the production market. The transportation and storage system in the USA is privately owned, in a model for pipeline ownership that resembles the private railroad model. Natural gas prices have fallen dramatically in the last 10 years due to the introduction of hydraulic fracturing technologies, so natural gas is among the least expensive energy sources at the moment. Natural gas production, like oil production, frequently occurs under public license on publicly owned lands, and is therefore subject to significant public oversight in the USA. Eminent Domain is sometimes applied for the development of pipelines in the USA. Natural gas is traded internationally, and Europe is heavily dependent on Russia for natural gas. This creates vulnerability in Europe to Russian pressure in the energy supply chain. The USA, another major producer, has historically banned natural gas exports since 1975 for reasons including the promotion of energy self-sufficiency; these policies have recently been relaxed, however. Pollution of water and air is a major regulatory issue for natural gas production, with controversy around hydraulic fracturing water pollution and methane leakage growing as current issues. Natural gas use generates less greenhouse gas than coal or oil products but is still a significant source of CO_2 from combustion, and pipeline leaks are a problematic source of methane emissions.

10.3.4 Energy: Liquid Petroleum Fuels, Fuel Oil, and Gasoline

Crude OIL is extracted from onshore and offshore wells through conventional drilling and hydraulic fracturing operations. It is produced in 32 US states and in US coastal waters. Approximately 65% of total US crude oil production came from five states: Texas (38%), North Dakota (11%), Alaska (5%) California (5%), and New Mexico (5%). A global approach shows that about 100 countries produce crude oil, however 48% of the world's total crude oil production comes from five countries: Russia (13%), Saudi Arabia (13%), USA (12%), Iraq (6%), and Iran (5%) (US EIA 2017). From the wellhead, it is pumped out to be transferred to tankers, pipelines, rail carts, and trucks to be transported to refineries and short-term storage facilities.

The refined product is also stored in short-term storage facilities, which helps regulate the fluctuations in supply and demand and supplies product for the U.S. **Strategic Petroleum Reserve** (SPR). The USA has 615 million barrels of storage capacity at short-term storages near refineries and 950 million barrels of storage capacity at long term storage terminals that make up the SPR. An additional 89 million barrels can be stored in the extensive pipeline system (America's Petroleum

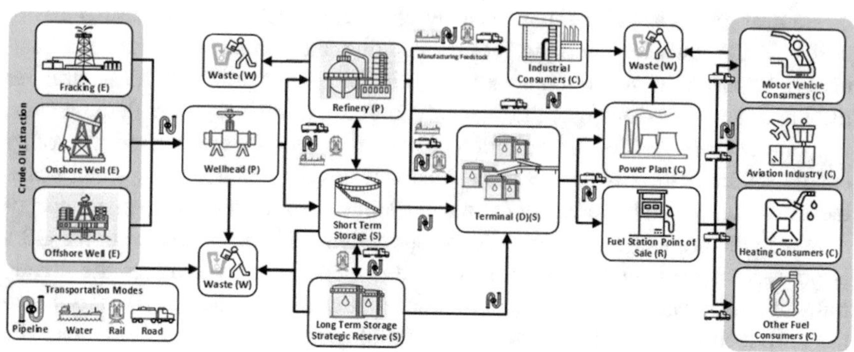

Fig. 10.5 The infrastructure and supply chain for petroleum derived liquid fuels such as fuel oil, gasoline, kerosene, diesel, and crude oil. [Sources: (An et al. 2011; Lima et al. 2016; America's Petroleum Institute; American Fuel and Petrochemical Manufacturers (AFPM), and U.S. Energy Information Administration)]

Institute 2019). The refinery as well as storage facilities connects to a distribution terminal that provides gasoline, diesel, heating oil, propane, and jet fuel to appropriate customers through a fuel station point of sale. The refineries provide oil and its by-products to industrial facilities such as plastic, organic chemical, and pharmaceutical manufacturers, in addition to the power plants for power generation (Fig. 10.5) (AFPM 2018).

Many stages of the petroleum, fuel oil, and gasoline supply chain contribute to the waste generation (FC: 5565000). The extraction and processing stages generate wastewater (FC: 3330000) as well as non-potable reclaimed water (FC: 3342999). In the extraction stage, water is mostly used for oil well sludge preparation and cooling of the generators as well as other mechanical parts. There is also chemical waste from the drilling, hydraulic fracturing and refining processes. Accidental spills or spoiled products in SPR generate waste at storage and distribution/transportation stages. Most of the waste at the consumption side is from emissions caused by the consumption of this commodity.

Gasoline, along with diesel, kerosene, and naphtha, and also natural gas are the outputs of a refining process that distills various grades of crude oil (petroleum). A modest amount of toxic residues are also a result of the distillation process. Besides crude oil, refineries use electricity and water, and sometimes natural gas as inputs. Refineries tend to be located along oil pipelines and railroads that transport large volumes of crude oil. Refineries in the USA are concentrated in oil-producing regions like the Gulf Coast but are also scattered across the map to serve many local markets. Crude oil basins are widely distributed on Earth and occur in most countries, but "cheap oil" that is economically recoverable and profitable occurs in only a few locations; historically, in the USA and Gulf of Mexico area, Middle East, and Russia. More recently, high crude oil prices have motivated high-technology and capital-intensive drilling operations that tap undersea basins, and shade-based and sand-based oil extraction. These newer crude oil sources are more expensive and environmentally risky.

Oil and other petroleum-derived liquid fuels are very easy to transport and distribute. Crude oil is an energy-dense and stable liquid that is economical to transport long distances using freighters and pipelines—or when necessary, at a higher cost using railcars. Refined fuels like gasoline are also economical to transport in this fashion, but the economics favor moving crude oil and then refining the fuel close to the target market. Refined fuels are transported in bulk with a mixture of modes including rail, pipeline, and tanker trucks running on public roads.

Refined fuels like gasoline are economical to store, and oil companies commonly keep months of crude oil or gasoline on hand at refineries and transportation hubs in large steel storage tanks. Refineries are capital intensive to build and they operate at a fixed rate (like coal and nuclear power plants), so storage is used to balance out seasonal changes in supply, demand, and price of oil and refined fuels. Countries like the USA maintain large strategic reserves of crude oil that can supply months or years of national demand, for emergency purposes. Because gasoline is less stable than crude oil and degrades over a span of months, there is a preference for stockpiling crude oil in bulk and maintaining smaller reserves of gasoline to keep the age of the fuel low. Gasoline in the USA is usually distributed to filling stations using tanker trucks running short distances over public roadways, to bring the gasoline from refineries and pipelines to retail sales locations in cities.

Soft infrastructure for petroleum-based fuels involves **consumers**, **owners**, **financiers**, and **regulators**.

Gasoline is mostly consumed by private citizens for the operation of personal vehicles. There is a preference for diesel over gasoline for heavy vehicles like trucks and construction equipment, and for military applications, because diesel is more energy-dense and is more shelf-stable than gasoline. Kerosene is preferred for aviation fuels because it is the most energy-dense, and weight is critical for aircraft. The liquid fuel infrastructure is privately owned and operated, or owned and operated by state-owned companies. In the USA the system is privately owned by a large number of large and medium sized companies and is largely market-driven. Internationally, many major petroleum operations are state-owned. The liquid fuel infrastructure tends to be privately capitalized except in state-owned circumstances, and operates for profit at each stage of the system (crude oil production, transportation, refining, distribution, retail sales). In the USA, fuel prices are largely unregulated and follow market dynamics, but the government has occasionally applied pressure to suppress price spikes during emergencies.

Price fluctuations for gasoline in the USA are driven mostly by constraints on refinery capacity, but also by crude oil prices. Many oil-producing countries subsidize gasoline and liquid fuels for their own residents, selling the fuel far below market price. Internationally, many of the countries that export a large amount of oil have state-owned oil companies that operate to fund the government, to build citizens' support for the government through fuel subsidies and handouts, and as tools of diplomacy and economic warfare, rather than as for-profit private concerns. Because oil is an international market for a critical FEW resource, governance is a fundamentally political issue. Because a large fraction of crude oil production and transportation crosses national boundaries, crude oil has frequently been used as a tool of diplomacy and economic warfare in geopolitics. Many countries prioritize domestic crude oil production and storage to reduce their strategic vulnerability to international pressure applied through the oil infrastructure. In the USA, domestic energy policy since 1980 has emphasized domestic oil production and fuel efficiency largely for the purpose of reducing dependency on foreign crude oil—a policy that has been very successful, as the USA is now mostly independent of foreign supply and no longer very vulnerable to 1970s-style oil embargoes.

Eminent domain is sometimes applied for the siting of oil pipelines, and this has been extremely controversial in some cases in the USA, for example the recent Dakota Access pipeline. In the USA, Environmental regulations and social concerns have made oil exploration controversial, especially on sensitive public lands like ANWR and in offshore applications. The Exxon Valdez spill and BP Deepwater Horizon spill were major political and environmental events in US history, and resulted in major losses to those companies. However, oil production interests have usually prevailed in the long term in the regulatory debate.

Greenhouse gas emissions from the burning of fuels are a major issue of policy concern. Although tailpipe emissions of greenhouse gas emissions have improved in recent decades along with vehicle fuel efficiency (gasoline and diesel), tailpipe emissions of greenhouse gasses continue to rise rapidly for air transportation which burns large amounts of jet fuel (kerosene). Additionally, air quality impacts from tailpipe emissions of vehicles are a major health concern in many major cities worldwide, because these emissions contribute to NO_X and SO_X concentrations and dangerous smog. Catalytic converters and scrubber technologies are important infrastructure for the mitigation of this air pollution.

A major policy concern facing financiers of the petroleum industry is the issue of "stranded capital." Capital can be stranded when a poor investment decision is made. In the current instance, hundreds of billions of dollars are invested by private companies in oil field infrastructure that might become stranded if greenhouse gas regulations reduce demand for petroleum fuels.

10.3.5 Energy: Electrical Power

There are two main categories of power generation as shown on the very left of the diagram above. These are conventional and renewable power generation. Incoming commodities that are used for power generation include coal, natural gas, oil, nuclear rods, and water. Renewable technologies harness energy directly from environment without needing to have specific commodity inputs. However, renewable energy generation is not necessarily waste free. The materials and facilities used for power generation, such as photovoltaic panels and wind turbines, have to be replaced periodically. In addition, there is some wastewater generated through cleaning, cooling and steam generation. In 2016 34% or the power generated through technologies using natural gas, 30% using coal, 20% using nuclear materials, 6% hydroelectric, 6% using wind, 2% using biomass, 1% using solar, and 1% using oil (Fig. 10.6) (US EIA 2017).

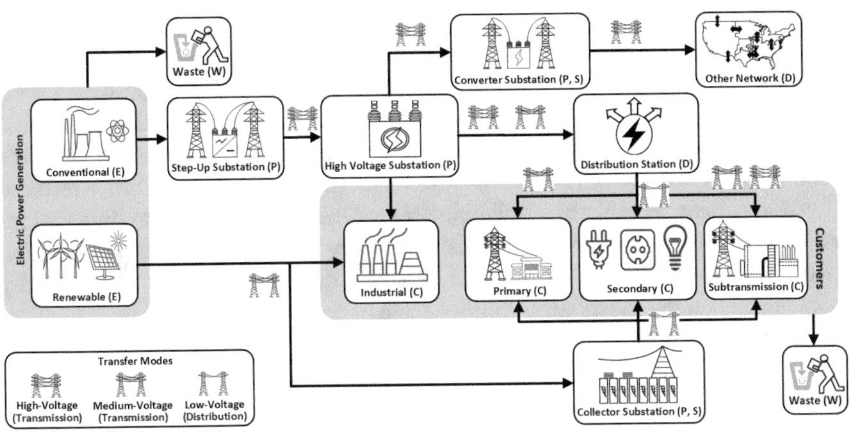

Fig. 10.6 The supply chain and infrastructure for electrical power. Used with permission of the FEWSION project. [Sources: (US DOE, US EPA, Glover et al. 2012)]

Power generated by conventional technologies converted into high-voltage power through a "step-up" substation and maintained in the transmission network through high-voltage substations. This power serves industrial customers (138–230 kV) who maintains their own substations for voltage adjustment and through a converter substation transferred to other transmission networks such as neighboring regional or neighboring country network. This power also serves subtransmission (26–69 kV), primary (4–13 kV), and secondary (120–240 V) customers through a distribution substation after the voltage is reduced for appropriate customer.

Power generated through renewable energy technologies is transferred to industrial customers by means of medium-voltage transmission lines. This also serves subtransmission, primary, and secondary customers through a collector substation.

FEWSION only contains data on where it is generated (E) and where it is consumed (C), so an electricity flow in the FEWSION database is from Extraction directly to Consumption (E → C) and identified with FEWSION code (FC) 2200000. While the FEWSION database presents a simplified reality of electricity flows, the above diagram shows the complexity involved in balancing electricity generation with electricity consumption on the electric grid to make the E → C flow reliable and cost efficient. Waste is generated at the point of generation (E), e.g., coal ash, emissions, spent nuclear waste (FC: 5565000), wastewater (FC: 3330000), reclaimed water (FC: 3340000), along the way through transmission and distribution loss, and at the point of consumption (C) through heat loss.

Electricity is produced using several energy sources. Hydroelectric power uses a river's potential energy to turn a turbine-linked generator, either in dam-reservoir or run-of-river implementations. It depends on river flows as an input. Thermoelectric geothermal power uses the heat from the Earth's molten interior to make steam or otherwise turn a turbine-linked generator. Thermoelectric power usually uses Fossil Fuel, making steam by burning primary fuels like coal, natural gas, oil to turn a turbine-linked generator, and emitting large amounts of CO_2 and air pollutants in the process. Can require a large amount of water for cooling, unless "dry cooling" systems are installed—and these cost more money. Thermoelectric solar power uses the sun's radiation and heating energy to make steam or otherwise turn a turbine-linked generator. It can require some water for washing solar panels and for steam cycles. Wind power uses a simple air turbine to turn a generator when the wind blows, and uses no water. Tidal power uses tidal ocean flows to turn a turbine-linked generator. Photovoltaic solar power uses the sun's radiation to directly create a voltage using silicon chips. This is the only production source that is often distributed in small units on rooftops; other production mechanisms are usually implemented at the much larger "utility scale."

Electricity is transported by the Power Grid. High-voltage AC lines are used to transport electrical power from large power plants to cities. Sometimes these lines are hundreds of miles long, and incur significant voltage losses in transmission. High-voltage transmission lines are vulnerable to damage by storms and fires. Because these lines are expensive and difficult to site, power production facilities are usually built along existing transmission lines. Power Grids are among humanity's most complex machines, and they involve countless power plants, transmission systems, transformers, substations, distribution systems, and organizations. A power grid can be any size, but they tend to be regional in scope and tend to loosely obey national boundaries. It is economical to transport electrical power within a power grid, and harder to transport it between power grids. Failures can cascade across a Power Grid's transmission lines but stop at a grid's boundaries; historical cascading failures in the USA have tended to affect one or two balancing regions within a grid, but not the whole grid. The USA and Canada share two grids, the Western Interconnection and the Eastern Interconnection. The Western Interconnection also serves a small part of northwestern Mexico. Texas maintains its own power grid that stops at the state's borders but can selectively interconnect using transformers. Electricity is distributed using substations and mid-voltage distribution circuits that are normally less than a mile long.

Electrical energy has historically been very difficult and expensive to store, which requires producers to be extremely agile and "ramp" their production rates rapidly to match supply with demand at every moment. Failures to match supply with demand during peak demand periods is a common cause of failure in these systems, because when insufficient or excessive voltage is available at any point in the system, breakers trip to shut off power and protect the power grid from damaging itself. **"Baseload" power plants** using coal and nuclear technologies are difficult to ramp, and are normally operated at a constant rate. But demand for power goes up and down by a factor of 2–3 in most cities, peaking seasonally in the summer and daily in the late afternoon and early evening, so the gap between the peaks and valleys of demand must be filled with a combination of storage, demand response, and ramping power production. Grid-scale battery technologies and thermal storage technologies are beginning to become cost-effective. Electric cars are increasingly storing power in their batteries. An excellent technology both for storage and for ramping production is the "pump and store" hydroelectric facility that pumps water uphill to store energy and/or releases water and high-demand times to generate energy. Hydroelectric facilities and natural gas turbines are able to rapidly ramp production to match demand peaks. "Demand Response" is another substitute for storage, and it involves programs that reduce power demand at peak times to keep supply and demand for power in balance.

The soft infrastructure of the power grid involves **consumers**, **regulators**, **owners**, and **financiers**.

Consumers are mostly buildings, including factories, homes, and businesses, with a major demand also by the water infrastructure sector. However, increasingly, electric cars are consuming a lot of electrical power. In the USA, public and private utilities own the entire electrical energy infrastructure in most cases. Merchant power plants are owned separately from the transmission and distribution systems, but many power plants are owned by the distribution utility in vertically integrated arrangements. Private utilities bring their own capital, but public utilities are capitalized with various models. Electrical power (i.e., peak demand charges, Watts) is priced separately from electrical energy (i.e., **Watt-hours**). Because storage is expensive for electricity, many utilities charge more for on-peak power and energy during high-demand periods. Prices are usually regulated and are relatively stable, increasing in small increments each year. In the USA these utilities are usually overseen by publicly elected bodies and executive branch agencies that approve investment decisions and pricing and set reliability standards. The U.S. EPA enforces regulations designed to protect endangered species on land and in rivers, affecting large hydroelectric and solar projects, and limits air pollution especially from fossil fuel powered plants, including recently CO_2 emissions. In the USA, "Balancing Regions" are subdomains of a given power grid that are managed by a single "Balancing Authority." This authority is not the same as the utilities and power plant operators. This authority has the responsibility to make sure that supply and demand for power are balanced at all times and places within the Balancing Region, and that reliability and quality of power meets the standards established by regulations.

10.3.6 Water

In the FEWSION database, water is classified into five subcategories: "fresh water" (FC: 3300000), "potable water" (FC: 3310000), "saline water" (FC: 3320000), "wastewater" (FC: 3330000), and "reclaimed water" (FC: 3340000). The water supply chain diagram above represents all but saline water. Wastewater and reclaimed water are also referred in other supply chain explained in this chapter. Water is pumped out of a river or a lake through a pump station to be sent to a water treatment station to be processes through filtering and chlorination. Processed water is sent to industrial, business and residential consumers through mainlines where pumps are employed when it is not possible to use gravity to distribution and when water needs to be stored in water towers. Water also is extracted from underground aquifers through wells to be used for water/food production process or industrial applications such as cooling. Water is returned to the environment after it goes through certain processes depending on the use. Water used in agriculture goes through riparian vegetation to be filtered before it returns to the reservoir. Water used in mining and other industrial applications such as animal growing and slaughtering operations, is collected in a pond to let the particulate matter settle before it returned to the environment through agricultural watering or evaporation. Water used by residential and business consumers is processed at a wastewater treatment plant before it is returned to the environment.

Huge volumes of raw freshwater are exclusively provided by the Earth's water cycle that distills and transports the water from oceans to continents, delivering it as rain and snow, especially in humid regions and in high-elevation locations within arid regions. Freshwater extracted from rivers, lakes, or aquifers is the main input to the water production system. Uniquely for water infrastructure, the output is largely the same as the input, although there may be some losses and quality transformations in the process. Electricity is the other major input, to run pumps. Raw freshwater is mostly produced for irrigation of crops. Water quality testing is usually not conducted for raw water, but this is changing because of the threat of bioaccumulative environmental contaminants that could concentrate in irrigation-fed crops and be passed along to people via the food supply chain.

Raw freshwater is produced in higher volumes and masses than any other commodity, so transportation requires highly specialized infrastructure; in this case, concrete-lined canals are the solution of choice, although large pipelines and tunnels are sometimes used. Gravity-driven flow is preferred, but sometimes large electrically powered pumping stations are needed when canals lift water, such as the Central Arizona Project in the USA which lifts Colorado River water thousands of vertical feet over hundreds of miles of canals. Canals suffer losses from evaporation and leakage, but these losses are usually small, under a few percent of total flows.

Raw freshwater for irrigation is used almost exclusively in the growing season. Crops demand water every few days, but storage is not required because canal systems are designed to constantly switch deliveries between distribution systems to maintain constant flows. Canal systems are often coupled with large storage reservoirs on rivers, and the dams of those reservoirs are capable of storing the river's flow during flood seasons and releasing that water to canal systems during the growing season. This storage strategy allows growing season demands to be satisfied with flood season flows, maximizing the utilization of river flows across the water year. A typical water year starts at the end of the summer growing season, when the water balance in the system is at its annual minimum (Fig. 10.7).

A few raw water systems are coupled with mass storage systems that use raw water transfers to maintain local water balances in areas where water consumption exceeds the local hydrologic system's sustainable rate of supply. For example, the Central Arizona Project in the USA pumps excess Colorado River water into the aquifers under Phoenix utilizing a legally defined water banking infrastructure to keep those heavily utilized aquifers sustainably recharged. The owners of the water are then entitled to withdraw it from the aquifer later, for example if it is needed during a drought emergency. Snowpack and glaciers on high mountains is an important natural storage infrastructure that effectively provides the same services as

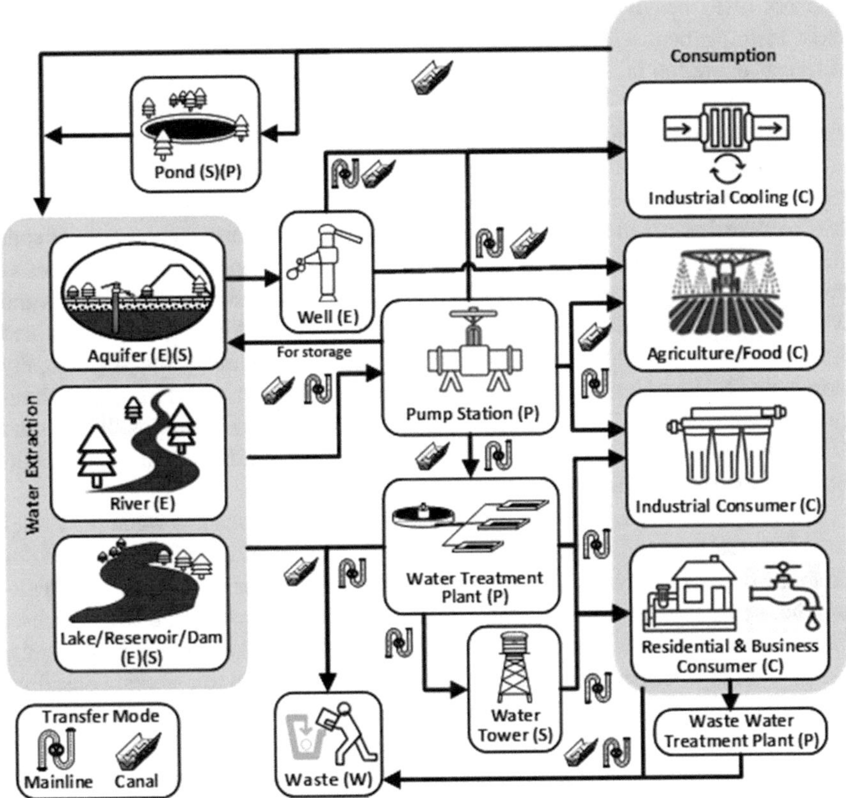

Fig. 10.7 Supply chain and infrastructure for water, including raw, potable, and waste water. Used with permission by the FEWSION project

large dams, by collecting winter season precipitation and slowly releasing it to rivers during the growing season. Global warming trends are especially strong in mountainous regions, and this warming effectively reduces the ability of mountains to provide raw freshwater storage infrastructure services.

The soft infrastructure for raw water is comprised of **consumption**, **ownership**, **finance**, and **governance**.

Raw water is mostly consumed by irrigated agriculture, which is capable of using raw freshwater as a direct input to growing operations. Some raw freshwater is used by cities and factories that do not make use of municipal water supplies. For example, the south-north transfer in China will provide a large fraction of the raw freshwater needed by northern Chinese megacities, by moving water hundreds of miles from the humid southern Yangtze river to the more arid North China Plain. Paper mills, steel mills, and power plants are common examples of industries that need a high volume of non-potable quality freshwater, and that normally utilize raw water as a result. Water canals and dams are usually publicly owned and financed and are often heavily subsidized. Raw water conveyance infrastructures usually exist to implement a legal agreement on water use in a shared river basin. Their physical operation is usually governed by a single jurisdiction such as a state or county, but the water that they convey is frequently subject to laws and treaties between that jurisdiction and other parties that share the freshwater source.

Large raw freshwater infrastructures are a tool of national political power, and are frequently exploited by more powerful political actors to the detriment of the less powerful actors in shared river basins. These infrastructures are sources of conflict (or cooperation) between nation-states. Some of the more notable examples of river basins in recent tension are the Colorado River that is shared by US States and Mexico, the Nile River that originates in the nations of Africa's Rift Valley and Ethiopia but is politically dominated by Egypt, the Jordan River that is politically dominated by Israel but is fed by Lebanon, Syria, and Jordan, and the Mekong River that drains southern China through Myanmar, Laos, Thailand, Cambodia, and Vietnam. By contrast, the Danube River in Europe drains a large number of neighboring countries, but has been shared successfully.

Potable water and wastewater treatment infrastructures take raw water (or wastewater) as the primary input, utilizing a large amount of electricity to pump water from the canal, river, lake, brackish, sewer, ocean, or aquifer source, and then treating that water to meet drinking water quality or other applicable standards. Surface water intakes and wells feed pipes, which bring raw water to water treatment plants. Water treatment ranges from the simple addition of chlorine to elaborate desalination techniques. Chlorination is very cost effective, but desalination of ocean and brackish water requires heavy capital investment and also a large amount of electrical or thermal energy. Desalination usually involves distillation, which requires a lot of thermal heating energy, or filtration, which requires a lot of electrical energy.

Advanced disinfection techniques are becoming more common, including UV and ozone treatment. Because of the threats of environmental hazards like chemical or mining spills, and terrorism, along with emerging contaminants like pharmaceuticals, the use of water quality testing and sensing techniques is becoming more common. Treated water quality poses more of a long-term threat to environmental quality than short-term security threat, however, because the large volumes of water in these systems dilute contaminants rapidly and chlorine residuals quickly degrade infectious agents, making municipal potable water systems resistant (albeit not immune) to these threats.

A major topic of discussion in cities is the use of treated (or reclaimed) wastewater as the input to potable water systems. This is called direct potable reuse for the case where wastewater is directly piped into potable water systems, or indirect potable reuse, in the case where wastewater is discharged to a river or aquifer before being reused for potable supply. Potable water is usually produced by municipal water providers, but is sometimes "self-supplied" by industries that operate their own supplies. Industries tend to employ customized treatment processes that match their processes, and often exceed drinking water quality standards.

Wastewaters originate from all human economic processes, including agriculture, industry, power generation, and urban uses. Agricultural wastewaters tend to be unregulated and untreated, and are considered "nonpoint" sources of water pollution. Power plants that use water for cooling tend to produce only thermal pollution, which is not treated. Industrial and urban water uses produce what we usually think of as wastewater—that is, sewage. Sewage is water mixed with human and industrial waste; common waste components include nutrients like nitrogen and phosphorous, along with salt, acid, various household and industrial chemicals, trash & solids, and trace amounts of minerals, metals, and pharmaceuticals.

Many technologies are used for wastewater treatment, depending on the quality of the sewage and the requirements for quality of the treated wastewater. Wastewater plants use a large amount of electricity to operate these technologies. A typical municipal sewage plant will use a linear sequence of:

1. screening solids;
2. coarse grid removal,
3. sedimentation to remove fine solids;
4. aeration to oxidize and degrade organic matter and contamination;
5. clarification; and
6. final disinfection and polishing.

The polishing step can use high technology processes like UV treatment, ozonation, and reverse osmosis to bring the treated wastewater back up to potable water quality standards if the wastewater will be reused. Solid sludge is produced in large quantities and needs to be disposed of; often it is taken to landfills, or treated and used as fertilizer for agriculture owing to its high nitrogen and phosphorous content. Use of sewage sludge carries many risks of contamination and must be undertaken with care. Many industrial facilities employ specialized and hazardous chemicals and employ specialized treatment processes to remove those chemicals. Many industrial facilities then recycle that wastewater to reduce potable water requirements and avoid the need to obtain wastewater discharge permits. Industries that operate in cities frequently pre-treat wastewater before discharging it into municipal sewers.

Combined sewer overflow (CSO) sewers are a major infrastructure problem in world cities. Until late in the twentieth century, wastewater treatment in cities was uncommon, and wastewater was discharged untreated into rivers. Storm drains in Paris, London, and New York for hundreds of years doubled as sewers, because both flows went to the same place—the river. It is difficult to retrofit wastewater treatment plants onto CSO's, because wastewater and stormwater are mixed. Where it is impossible to build separate wastewater collection sewers, infrastructure megaprojects like the tunnels under Chicago are needed to collect, store, and then treat the combined overflow.

Reclaimed wastewater's purple pipes are primarily used to supply irrigation water to golf courses and public parks. The aquatic environment is the primary receiver of wastewater, and many ecosystems have come to depend on it—including the so-called "accidental wetlands" that occur at the outlets of wastewater plants. Reclaimed wastewater is frequently injected into aquifers to maintain their water balance, as in the water banks that operate in Southern Arizona.

Potable water is transported locally and not over long distances, for multiple reasons. It is difficult to maintain constant system pressures across long distances. Water quality degrades over time and distance. Most importantly, the institutions that operate the potable water infrastructure rarely cross municipal or property boundaries. Potable water is critical infrastructure because it is essential for human life and for nearly all economic activities. This means reliability of supply is extremely important. Demand for potable water in cities varies strongly by time of day and time of year. Peaks in residential demand occurring at the start and end of the workday as people wash and cook, and peaks in commercial demand occur during the workday, with low demands at night. Because potable water is used for landscaping and outdoor water use in cities, demand also peaks during the warm season and growing season when plants need more watering. If a fire starts, the system must deliver extremely high rates of supply for 1 or 2 h. In order to match supply with demand at all times of day and year, and to ensure reliable supplies during temporary failures of electrical power, pumps, or unavailable or contaminated source water supplies, potable water systems incorporate storage. Usually there is between 24 h and 5 days of storage.

Because water supply uses a large fraction of electrical power demands (USGS), a best practice is to produce potable water and treat wastewater at times of day when electricity is inexpensive, and use stored potable water at times of day when electricity is expensive.

Potable water is easy to store in large steel tanks. Elevated storage is a best practice, because it can maintain system pressures in the absence of electrical power or functioning pumps, and it is provided by placing storage on hills or in water towers.

Potable water distribution works on zones that distribute treated water from one or more water treatment facilities to one or more users. A zone operates at a fixed positive pressure that is high enough to provide strong flow rates. A zone is a grid or network of small pipes that are connected together in a series of loops and junctions. Positive system pressures result in leakage, which is usually less than 5 percent in a well-maintained system, but can be 30–50 percent in the oldest systems. The distribution system's design is usually limited by the dual purpose of firefighting, which has much higher peak demands, rather than by residential and business uses. Wastewater distribution is usually run by gravity (downhill) where possible, using traditional sewer designs.

The soft infrastructure of potable and waste water systems is comprised of **consumption**, **ownership**, **finance**, and **governance**.

Potable water is used for drinking, sanitation, outdoor landscaping, firefighting, construction, social & lifestyle, and most industrial processes, in residences and businesses, especially in cities. Potable water is an essential and non-substitutable input into nearly all economic processes, and especially for human life and residential uses.

Cities cannot exist without high quality potable water supplies and wastewater services, and this fact has held constant for the entirety of human civilization. Potable water systems are owned by water utilities, which are publicly regulated entities that may be publicly or privately owned. Often water utilities are owned by municipalities in the USA. The municipal utilities of larger communities frequently serve their small neighbors as well. Ownership is heavily fragmented, with many very small to medium sized local providers in a region or metropolitan area. This fragmentation has prevented some water utilities from developing the economies of scale and the level of sophistication and staffing needed to address twenty-first century challenges.

Finance for potable water and wastewater systems is usually capitalized using bonds issued by utilities, but in the USA it is common for federal and state grants to subsidize major construction projects. Potable water and wastewater utilities tend to sell water and treated wastewater at a price equal to their marginal unit cost, meaning that they charge roughly the unit cost of operating and maintaining their pumps, treatment plants, and distribution networks, assuming that raw water is priced at zero. Some utilities charge a flat connection fee as a part of their fee structure or instead of unit pricing.

Sometimes "tier" pricing is used charging residential users a higher price for water above a threshold representing a basic "human right" to water (see Sect. 6.2.4), so that a basic quantity of water for drinking and washing is provided at a subsidized and low price, but extra water for outdoor and lifestyle uses costs more. This encourages water conservation and provides a progressive pricing model. Most utilities in the USA do not have the authority to set water prices, and water prices are set below financially sustainable levels. Finance for potable water systems is a major problem, especially as water infrastructures age and need more maintenance and replacement, and as raw water rights become more expensive. In the USA, the EPA mandates strict standards for quality and reliability of potable water systems, although enforcement varies. Potable water systems, whether publicly owned or privately owned, tend to be small and local and are governed at that scale.

Wastewater discharge into the natural environment is heavily regulated, in the USA by the EPA through the Clean Water Act and the National Pollutant Discharge Elimination System (NPDES). Maximum concentrations of pollutants and maximum temperatures are one mechanism. Another mechanism is the establishment of Total Maximum Daily Loads (TMDLs) that establish environmental quality minimums in the receiving body of water. In the USA, regulations have successfully reduced discharges from point sources (i.e., from pipes), and a majority of water pollution in many areas is now from nonpoint sources like runoff from farms and city streets. Nonpoint pollution sources have proven much more resistant to regulation and control than point sources. So-called "emerging contaminants" such as pharmaceuticals are not covered by existing water quality regulations, but are of concern and may be regulated in the future.

Key Points

- FEW systems are infrastructure-heavy because of the large volume and low cost of FEW commodities.
- Infrastructure is an expensive long-term investment in marginal efficiency.
- Infrastructure creates lock-in.

- Infrastructure must be designed for robustness and resilience, because it is very vulnerable to stranded capital and extreme events.
- FEW Infrastructure forms a multilayer network that involves interdependency and cascading failure risk, especially from the electrical power grid and raw water supply to other system components.
- Buffers are a key feature to ensure the reliability of FEW systems.
- Each layer of the FEW infrastructure handles a specialized commodity, and involves a unique supply chain structure.
- FEW supply chains share some common features, including seven standard steps.

Discussion Points and Exercises

1. Read the ASCE Infrastructure Report Card. Note the parts of the report card that pertain to FEW commodity supply chains and/or infrastructures. (a) Do you agree with the reasons for ASCE's generally low ratings of the US infrastructure for FEW? (b) What are the specific consequences to the US FEW system, of poorly maintained or inefficient infrastructure? (c) What fraction of infrastructure types mentioned by ASCE are FEW-related?
2. Define and discuss the differences between Vulnerability, Risk, Robustness, and Resilience as it pertains to infrastructure
3. Define and discuss the differences between Hard, Soft, Grey, Green, and Natural infrastructure.
4. What is the essential difference between "infrastructure," specifically, as compared with any type of system that produces or moves a good or service?
5. Is lock-in a problem or a benefit of infrastructure? Give an example where it is a problem and an example where it is a benefit.
6. Give an example of storage in each major infrastructure category discussed in this chapter.
7. Identify the most effective, efficient, or commonly used mode of transportation for each major commodity category discussed in this chapter.
8. Discuss: what is a supply chain, and how can it be conceptually modeled? (What are the key components of a supply chain?)
9. Is it possible to design infrastructure that will probably not experience a severe failure over the next 100 years? 1000 years? Why or why not?
10. Discuss stranded capital, interdependency, lock-in, and buffers; (a) Is it possible to avoid stranding capital in infrastructure investments, creating interdependency, generating lock-in, or using large buffers, and if so, how? (b) Would your strategy compromise reliability, robustness, or efficiency of the system? (c) What is the best trade-off between these multiple objectives?
11. Discuss: what is the most unique feature of each FEW commodity's supply chain, and why?
12. Discuss: what careers and professions are responsible for FEW infrastructures?
13. Identify at least one important agency or law that governs the FEW infrastructure in your location, at the local scale (not federal/national scale).

References

An, H., Wilhelm, W. E., & Searcy, S. W. (2011). *Biofuel and petroleum-based fuel supply chain research: A literature review* (Vol. 35).

Allan, J.A. (1998) Virtual Water A Strategic Resource Global Solutions to Regional Deficits. Groundwater, 36, 545 546.

Aramyan, L., et al. (2006). Performance indicators in agri-food production chains. In J. H. Wijnands & C. Ondersteijn (Eds.), *Quantifying the agri-food supply chain* (pp. 49–66). Dordrecht: Springer.

Ahumada, O., & Villalobos, J. R. (2009). Application of planning models in the agri-food supply chain: A review. *European Journal of Operational Research, 196*, 1–20.

AFPM–America's Fuel and Petroleum Manufacturers. (2018). *America's fuel and petrochemical supply chain*. https://digital.afpm.org/infographic/americas-fuel-and-petrochemical-supply-chain/

American Petroleum Institute (API). (2019, January). *Energy: Understanding our oil supply chain*. https://www.api.org/~/media/Files/Policy/Safety/API-Oil-Supply-Chain.pdf.

Chandirasekaran, G. (2017). Agri-food supply chain management: Literature review. *Intelligent Information Management, 9*, 68–96.

Cunningham, D. C. (2001). The distribution and extent of agrifood chain management research in the public domain. *Supply Chain Management: An International Journal, 6*(5), 212–215.

Copeland, C. & Carter, N. (2017) Energy-Water Nexus: The Water Sector's Energy Use. Congressional Research Service.

Falkenmark, M. (1977). Water and mankind—A complex system of mutual interaction. *Ambio, 6*(1), 3–9.

FAO. (2011). Global food losses and food waste – Extent, causes and prevention. Rome

Glover, J. D., Sarma, M. S., & Overbye, T. (2012). *Power system analysis & design, SI Version*. Hampshire: Cengage Learning.

Handayati, Y., Simatupang, T. M., & Perdana, T. (2015). Agri-food supply chain coordination: The state-of-the-art and recent developments. *Logistics Research, 8*(1), 5.

Hoff, H. (2011). Understanding the Nexus. Background Paper for the Bonn 2011 Conference: The Water, Energy and Food Security Nexus. Stockholm Environment Institute (SEI), Stockholm.

Liljestrand, K., & Fredriksson, A. (2015). Capturing food logistics: A literature review and research agenda. *International Journal of Logistics Research and Applications, 18*(1), 16–34.

Lima, C., Relvas, S., & Barbosa-Póvoa, A. P. F. D. (2016). Downstream oil supply chain management: A critical review and future directions. *Computers & Chemical Engineering, 92*, 78–92.

National Research Council. (2015). *A framework for assessing effects of the food system*. Washington: National Academies Press.

Maupin, M. A., Kenny, J. F., Hutson, S. S., Lovelace, J. K., Barber, N. L., and Linsey, K. S. (2014). Estimated use of water in the United States in 2010: U.S. Geological Survey Circular 1405, 56 p., http://dx.doi.org/10.3133/cir1405.

Rasul, G. & Sharma, B. (2016.) The nexus approach to water–energy–food security: an option for adaptation to climate change. *Climate Policy, 16*(6), 682–702. https://doi.org/10.1080/14693062.2015.1029865

Septiani, W., Marimin, M., Herdiyeni, Y., & Haditjaroko, L. (2016). Method and approach mapping for agri-food supply chain risk management: A literature review. *International Journal of Supply Chain Management, 5*, 51–64.

Shukla, M., & Jharkharia, S. (2013). Agri-fresh produce supply chain management: A state-of-the-art literature review. *International Journal of Operations and Production Management, 33*(2), 114–158.

The Nature Conservancy. (2016). U.S. Beef Supply Chain: Opportunities in fresh water, wildlife habitat, and greenhouse gas reduction. Available from http://www.nature.org/ourinitiatives/urgentissues/landconservation/globalagriculture/sustainability-and-the-us-beef-supply-chain.xml

Pimentel, D and Pimentel, M. (2003). Sustainability of meat-based and plant-based diets and the environment, *The American Journal of Clinical Nutrition, 78*(3), 660S-663S. https://doi.org/10.1093/ajcn/78.3.660S

US EIA – U.S. Energy Information Administration. (2017). *Oil: Crude and petroleum products explained.* Retrieved January, 2019, from https://www.eia.gov/energyexplained/index.php?page=oil_

Uhlenbrook, S. (2007). Biofuel and water cycle dynamics: what are the related challenges for hydrological processes research? *Hydrological Processes, 21*(26).

van der Vorst Jack, G. A. J., da Silva Carlos, A., & Trienekens, J. H. (2007). *Agro industrial supply chain management*. Roma: FAO.

Further Reading

Allenby, B., & Fink, J. (2005). Toward inherently secure and resilient societies. *Science, 309*(5737), 1034–1036.

American Petroleum Institute. (n.d.). *Energy infrastructure*. Retrieved February 15, 2019, from https://energyinfrastructure.org/

ASCE Infrastructure Report Card. (n.d.). https://www.infrastructurereportcard.org/

Bakke, G. (2017). *The grid: The fraying wires between Americans and our energy future*. New York: Bloomsbury. ISBN-10: 1632865688.

Brian, H. (2005). *Infrastructure: The book of everything for the industrial landscape*. New York: W. W. Norton & Co.

Brooks, N. (2003). *Vulnerability, risk and adaptation: A conceptual framework.* Tyndall centre for climate change research working paper 38.38: 1–16. Energy-Water.

Cash, D. W., et al. (2003). Knowledge systems for sustainable development. *Proceedings of the National Academy of Sciences, 100*(14), 8086–8091.

Chester, M. V., & Allenby, B. (2018). Toward adaptive infrastructure: Flexibility and agility in a non-stationarity age. *Sustainable and Resilient Infrastructure, 3*, 1–19.

Committee on a Framework for Assessing the Health. (2015, June 17). Environmental, and social effects of the food system; Food and Nutrition Board; Board on Agriculture and Natural Resources; Institute of Medicine; National Research Council. In M. C. Nesheim, M. Oria, & P. T. Yih (Eds.), *A framework for assessing effects of the food system.* Washington, DC: National Academies Press.

Devold, H. (2013). *Oil and gas production handbook: An introduction to oil and gas production*. Lulu.com.

DOE. (n.d.). *United States electricity industry primer, Office of electricity delivery and energy reliability, U.S. Department of Energy, DOE/OE-0017*. Retrieved July 2015, from https://www.energy.gov/sites/prod/files/2015/12/f28/united-states-electricity-industry-primer.pdf

EIA. (n.d.). *Energy explained, U.S. Energy Information Administration*. Retrieved February 15, 2019, from https://www.eia.gov/energyexplained/

EPA Water Infrastructure. (n.d.). https://www3.epa.gov/region9/waterinfrastructure/waterenergy.html

EPA. (n.d.). *Sustainable water infrastructure*. Retrieved February 15, 2019, from https://www.epa.gov/sustainable-water-infrastructure

Giordano, T. (2012). Adaptive planning for climate resilient long-lived infrastructures. *Utilities Policy, 23*, 80–89.

Markolf, S. A., Chester, M. V., Eisenberg, D. A., Iwaniec, D. M., Davidson, C. I., Zimmerman, R., et al. (2018). Interdependent infrastructure as linked social, ecological, and technological systems (SETSs) to address lock-in and enhance resilience. *Earth's Future, 6*(12), 1638–1659.

Milly, P. C. D., et al. (2008). Stationarity is dead: Whither water management? *Science, 319*(5863), 573–574.

O'Rourke, T. D. (2007). Critical infrastructure, interdependencies, and resilience. *BRIDGE-Washington-National Academy of Engineering, 37*(1), 22.

Taleb, N. N. (2007). *The black swan: The impact of the highly improbable* (Vol. 2). Random house, ASIN: B00139XTG4.

USGCRP. (2018). Impacts, risks, and adaptation in the United States. In D. R. Reidmiller, C. W. Avery, D. R. Easterling, K. E. Kunkel, K. L. M. Lewis, T. K. Maycock, & B. C. Stewart (Eds.), *Fourth national climate assessment, Volume II* (p. 1515). Washington, DC: U.S. Global Change Research Program. https://doi.org/10.7930/NCA4.2018.

Vugrin, E. D., et al. (2010). *A framework for assessing the resilience of infrastructure and economic systems. Sustainable and resilient critical infrastructure systems* (pp. 77–116). Berlin: Springer.

Chapter 11
Climate Change

Peter Saundry

11.1 Introduction

Climate patterns (operating over years-to-centuries) and weather events (occurring over minutes-to-months), significantly impact all aspects of food, energy, and water (FEW) systems—from production to conversion processes, to infrastructure requirements, to demand. Recognition of climatic conditions is essential in all FEW systems that are not within controlled environments.

The purpose of this chapter is to frame FEW systems in the context of both natural climate conditions and variations, and in the context of anthropogenic climate change. In this text, we will generally use the word "variable" and "variability," to describe natural fluctuations, oscillations, cycles, and changes in the earth's climate, while generally using the word "change" for human impact on the earth climate.

We will explore how anthropogenic climate change and efforts to mitigate it are expected to impact FEW systems, individually and collectively. We will briefly look at climate modeling and the lessons for integrating food, energy, and water systems. While climate considerations are usually explicit in FEW system models, they are also often included as implicit assumptions in models about crop choices, seasonal changes in energy demand, or water flows. Many of the topics introduced in this Chapter will be addressed in greater depth in later chapters.

A full exploration of climate change is beyond the scope of this book. For a detailed summary of the extensive physical scientific basis for climate change, the reader is referred to volume 1, the Fifth Assessment Report (AR5) of the Intergovernmental Panel on Climate Change (IPCC 2013, 2014a, b). This chapter, especially Sect. 11.4, includes many passages quoted from AR5. To avoid repetition

P. Saundry (✉)
Energy Policy and Climate, Advanced Academic Programs, Krieger School of Arts and Sciences, Johns Hopkins University, Washington, DC, USA
e-mail: psaundr1@jhu.edu

P. Saundry, B. L. Ruddell (eds.), *The Food-Energy-Water Nexus*, AESS Interdisciplinary Environmental Studies and Sciences Series,
https://doi.org/10.1007/978-3-030-29914-9_11

of the same citation, all quotes not otherwise attributed, are attributable to (IPCC 2013, 2014a, b).

Even without consideration of human-induced changes to the climate, many elements of FEW systems in many locations are unsustainable for non-climatic reasons, such as the degradation of local ecosystem functions. Thus, mitigating or adapting to climate change is only part of developing sustainable FEW systems that meet the needs of human societies.

11.2 Background

There are many well-known natural cycles in the earth's climate system. There is also significant natural variability in weather from year-to-year, including the occurrence of extreme events such as droughts, storms, heatwaves, and extremely cold conditions.

Without considering anthropogenic climate change, it would be possible to model FEW systems based on long-term observations of climate conditions and the factors that influence them, including seasonal cycles, variations, and the frequency of extreme events. However, anthropogenic climate change requires recognition of changes (some nonlinear) so that the future will not look like the past. Modeling future climates means combining an appreciation for both natural climate patterns and their variations and the impacts of human-induced changes to the earth's climate system.

Anthropogenic emissions of greenhouse gases (GHGs) exist alongside natural sources and sinks of greenhouse gases intensifying the earth's natural greenhouse effect. Emissions of carbon dioxide and methane are the most significant anthropogenic sources of GHGs, about 72%, and 19% respectively in 2016 (Janssens-Maenhout et al. 2017). **Both the level of GHGs present in the earth's atmosphere and the rate of increase are unprecedented over the period of human history**.

Anthropogenic emissions of GHGs are changing the earth's climate in many ways, including the general warming of the atmosphere. The earth's lower atmosphere is about 0.8 °C (1.4 °F) warmer since 1880. While average global temperature is a useful and widely cited metric of climate change, it is only one aspect of the climate and in most practical situations, not the most useful (see Chap. 13). **Anthropogenic climate change results in a wide variety of effects and impacts which manifest themselves in different ways in different locations**.

The United Nations Framework Convention on Climate Change (UNFCCC) has a stated goal to "stabilize greenhouse gas concentrations in the atmosphere at a level that would prevent dangerous anthropogenic interference with the climate system." In the 2015 Paris Agreement, the nations of the world adopted a goal of "keeping a global temperature rise this century well below 2 °C above pre-industrial levels and to pursue efforts to limit the temperature increase even further to 1.5 °C."

Some effects of climate change, such as sea level rise and the average number of days of temperature extremes, have long-term incremental impacts on all aspects of FEW systems, from production to utilization to population movements. The ability of systems and communities to adapt to such changes will vary. For example, sea level rise will result in more frequent, or even permanent inundation of low-level coastal areas, especially in areas that are unable to build (or cannot afford to build) sea walls or other protections. Such incremental changes add stress to food, energy, and water systems, which can be adapted to with greater or lesser success depending on the local geographic and economic circumstances.

Other effects of climate change, such as more frequent and intense extreme weather events will have very short-term as well as long-term impacts. For example, rapid and difficult population movements are likely to be "triggered" by extreme events are rather than incremental changes. A community already vulnerable to flooding or severe storms may experience a single event resulting in population displacement. Extreme weather events such as Hurricane Katrina (2005) and Hurricane Irma and Hurricane Maria (2017) resulted in both large-scale breakdowns in FEW systems and population movements in one of the wealthiest nations on earth, underscoring the challenges to building resilient communities. Chapter 4 will explore this subject in greater detail.

The combination of slow incremental changes and extreme events within the range experienced already in human history can combine to create impacts more significant than previously experienced. A storm surge atop higher sea level is more likely to overtop seawalls illustrates how the likelihood that extreme events will be even more impactful on FEW systems and human societies in the coming decades.

Total anthropogenic GHG emissions and anthropogenic GHG emissions per capita vary significantly from country to country for many reasons. However, the use of fossil fuels for energy and changes in land use for agriculture and forestry are the two largest anthropogenic sources of greenhouse gases. In 2010, the burning of fossil fuels for energy accounted for approximately two-thirds of global anthropogenic GHG emissions, or 32 (±2.7) of 49 (±4.5) gigatonnes of carbon dioxide equivalent ($GtCO_2eq$.) (IPCC 2014b). Agriculture, deforestation, and other land use changes accounted for approximately 24% or 10–12 $GtCO_2eq$. Here, emission shares are based on their impact or **global warming potential** over a period of 100-years.

Therefore, energy and food systems, and to a lesser extent, water systems, are significant contributors to anthropogenic emissions of GHGs. Efforts to mitigate climate change will necessarily require significant changes to FEW systems. Because FEW systems cause climate change and are impacted by it, it is appropriate to think of climate change as mediating the interactions between FEW systems on a global scale.

11.3 Communicating Risk, Probability, and Scientific Confidence

Two significant challenges in communicating science, particularly climate change science, to lay audiences are the difficulty that many have in understanding: (1) scientific uncertainty; and (2) probability-based risk. The challenges in communicating these concepts manifest themselves concerning FEW systems and climate change.

Communicating scientific confidence in a manner accessible to a lay audience is possible for some issues. For example, a non-scientist can understand the ability to have high confidence in a long-term (decadal) climate projections while having very low confidence in relatively near-term (weeks) weather forecasts by recognizing the common experience of differences between average conditions in winter and summer. However, communicating about more nuanced scientific issues.

The Intergovernmental Panel on Climate Change has developed a useful and more systematic approach to address these two communication challenges, which has broad applicability and utility is communicating scientific confidence and risk to non-scientific audiences. In the word of the IPCC, "[c]onfidence increases towards the top-right corner as suggested by the increasing strength of shading. Generally, the evidence is most robust when there are multiple, consistent independent lines of high-quality evidence."

Table 11.1 shows the IPCC's depiction of statements about scientific evidence and agreement and how they are related to confidence in the scientific community. Thus, high scientific confidence is a result of robust evidence and high agreement between sources. Conversely, where there is limited evidence and low agreement, there is low scientific confidence. These phrases appear frequently below.

Table 11.1 A depiction of evidence and agreement statements and their relationship to confidence developed by the Intergovernmental Panel on Climate Change

Agreement ↑	Evidence (type, amount, quality, consistency) →		
	High Agreement Limited Evidence	High Agreement Medium Evidence	High Agreement Robust Evidence
	Medium Agreement Limited Evidence	Medium Agreement Medium Evidence	Medium Agreement Robust Evidence
	Low Agreement Limited Evidence	Low Agreement Medium Evidence	Low Agreement Robust Evidence

Confidence ↑
High Confidence
Medium Confidence
Low Confidence

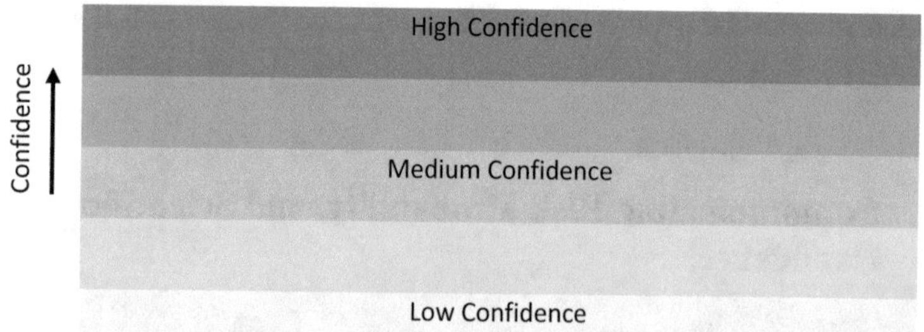

"Confidence increases towards the top-right corner as suggested by the increasing strength of shading. Generally, the evidence is most robust when there are multiple, consistent independent lines of high-quality evidence." (Source IPCC 2013)

Table 11.2 Likelihood terms associated with outcome probability developed by the Intergovernmental Panel on Climate Change as used in the Fifth Assessment Report (Source: IPCC 2013, 2014a)

Term	Likelihood and outcome
Virtually certain	99–100% probability
Very likely	90–100% probability
Likely	66–100% probability
About as likely as not	33–66% probability
Unlikely	0–30% probability
Very unlikely	0–10% probability
Exceptionally unlikely	0–1% probability

Table 11.2 shows the IPCC's use of specific terms associated with probability (likelihood) of a particular outcome. This probability, when related to adverse outcomes, is synonymous with risk. Again, terms such as "virtually certain," "likely," and "unlikely" are generally used here with reference to their definition in Table 11.2.

Climate change adds additional stress on FEW systems, increasing the overall vulnerability to disruptions of a region and its instability. As a result, climate change increases the likelihood of events with adverse impacts. Thus, climate change is a "**risk multiplier**." This framing of climate change risk does not distinguish between increasing existing risks and new risks but views climate change in terms of overall risk.

11.4 Impacts

FEW systems have been developed in the context of their regional climates and conditions, such as natural resources, ecosystem services, population centers and demands, latitude, topography, and infrastructure. Inflows and outflows of FEW commodities are defined by local demands and external influences such as economics and politics. Therefore, any consideration of climate change impacts must recognize local conditions and the geographic scale of those conditions. For example, water flows are usually considered on a scale of watersheds, while resource consumption is considered on the scales related to population centers and political regions. Weather events and climate change also have regional scales that strongly influence how they impact FEW systems.

In Chap. 2, we noted how each separate system (food, energy, and water) draws upon outputs from the other two and provides inputs to them, creating "first-order" and higher-order relationships. Thus, in addition to recognizing long-term impacts from incremental climate changes and short-term impacts from extreme weather events, it is also possible to look at likely climate change impacts in terms of the following:

1. Separate food, energy, and water systems.
2. Direct, or "first-order" interactions of one system on another, such as the demands on water by energy systems.
3. Higher-order "cascading" impacts that affect the relationship between all three sectors.

Here, we review climate change impacts framed in this manner.

11.4.1 Water

11.4.1.1 Climate Change and the Water Cycle

> Freshwater-related risks of climate change increase significantly with increasing greenhouse gas (GHG) concentrations (robust evidence, high agreement)

The general warming of the earth associated with increases in atmospheric concentrations of GHGs will result in higher evaporation of surface water and more water vapor in the earth's atmosphere. This intensifies the earth's natural water cycle impacting precipitation (rain, snow, hail, and sleet) and the supply of freshwater water resources. Intensifying the earth's natural water cycle will also mean more intense storms resulting in increased flooding risks, and, at the other end of the spectrum, some areas will see more extended periods between precipitation result in more frequent drought conditions.

> So far there are no widespread observations of changes in flood magnitude and frequency due to anthropogenic climate change, but projections imply variations in the frequency of floods (limited evidence, medium agreement). Flood hazards are projected to increase in parts of South, Southeast, and Northeast Asia; tropical Africa; and South America (limited evidence, medium agreement). Since the mid-20th century, socioeconomic losses from flooding have increased mainly due to greater exposure and vulnerability (high confidence). Global flood risk will increase in the future partly due to climate change (*limited evidence, medium agreement*).

> Once precipitation falls, its movement can also be affected by climate change. Higher temperatures usually result in less snowfall and more rainfall in winters and rapid thawing of ice and snow in springs leading to higher rates of runoff and streamflow earlier in the year and lower rate later. 'In regions with snowfall, climate change has altered observed streamflow seasonality, and increasing alterations due to climate change are projected.' (robust evidence, high agreement).

Further,

> because nearly all glaciers are too large for equilibrium with the present climate, there is a committed water resources change during much of the 21st century, and changes beyond the committed change are expected due to continued warming; in glacier-fed rivers, total meltwater yields from stored glacier ice will increase in many regions during the next decades but decrease thereafter (robust evidence, high agreement). Continued loss of glacier ice implies a shift of peak discharge from summer to spring, except in monsoonal catchments, and possibly a reduction of summer flows in the downstream parts of glacierized catchments.

Water storage will change in various ways in response to changes in precipitation and movement.

> "Climate change is projected to reduce renewable surface water and groundwater resources significantly in most dry subtropical regions (robust evidence, high agreement). In contrast, water resources are projected to increase at high latitudes. Proportional changes are typically one to three times greater for runoff than for precipitation." Further, "for each degree of global warming, approximately 7% of the global population is projected to be exposed to a decrease of renewable water resources of at least 20% (multi-model mean)" Such changes will have impacts on water storage infrastructure such as dams and reservoirs.

Where total precipitation drops,

> this will intensify competition among agriculture, ecosystems, settlements, industry, and energy production, affecting regional water, energy, and food security (limited evidence, medium to high agreement).

Three types of droughts should be recognized, each describing a significant and sustained drop in the presence of water:

1. **Meteorological drought** referring to precipitation.
2. **Hydrological drought** referring to streams, lakes, reservoirs, and groundwater.
3. **Agricultural drought** referring to soil moisture.

The relationship between the three is strong but complex.

Precipitation and water flows are strongly influenced by local features such as topography and land cover. Therefore, while it is possible to speak with high confidence about changes in precipitation averages over large areas, it is much more difficult to speak about precise changes in specific locations.

In addition to impacts of climate change on freshwater systems, there are also impacts on marine (saltwater) systems. The most significant are sea level rise and changes to marine ecosystems, including fisheries.

Effective responses to water use challenges, be they rooted in natural variations in the water cycle, or the impacts climate change are often rooted in **Integrated Water Resource Management**, which coordinates different water uses with other factors like food and energy systems, ecosystems, and socioeconomic goals.

11.4.1.2 Climate Change Impacts on Water for Human Consumption

Climate change will have an array of impacts on human consumption of water; impacts that will depend significantly on local and regional geography, including:

- 'Higher ambient temperatures, which reduce snow and ice volumes and increase the evaporation rate from lakes, reservoirs, and aquifers. These changes decrease natural storage of water, and hence, unless precipitation increases, its availability. Moreover, higher ambient temperatures increase water demand, and with it the competition for the resource (medium to high agreement, limited evidence).
- Shifts in the timing of river flows and possible more frequent or intense droughts, which increase the need for artificial water storage.
- Higher water temperatures, which encourage algal blooms and increase risks from cyanotoxins and natural organic matter in water sources, requiring additional or new treatment of drinking water (high agreement, medium evidence). On the positive side, biological water and wastewater treatment are more efficient when the water is warmer.
- Possibly drier conditions, which increase pollutant concentrations. This is a concern especially for groundwater sources that are already of low quality, even when pollution is natural as in India and Bangladesh, North and Latin America and Africa; here arsenic, iron, manganese, and fluorides are often a problem.
- Increased storm runoff, which increases loads of pathogens, nutrients, and suspended sediment.
- Sea level rise, which increases the salinity of coastal aquifers, in particular where groundwater recharge is also expected to decrease."

A warmer climate will increase water demand in most cases. Impacts will be more significant where water is already scarce, infrastructure poor, and populations are increasing rapidly. However, there are significant opportunities to increase the efficiency of water use and adopting practices such as water reuse that will reduce adverse impacts on human populations. Also, a variety of climate impacts on water infrastructure (Chap. 10) are likely including natural water capture and storage by ecosystems (Chap. 9), extreme events damaging reservoirs, dams, aqueducts, drinking water treatment plants, pipes, and wastewater treatment plants.

11.4.1.3 Climate Change Impacts on Water for Food

Agricultural productivity will be impacted both directly by climate change (see Sect. 11.4.2 below), through impacts on first order interactions of food-systems with water, and through impacts on higherorder system interactions.

Changes in the earth's hydrologic cycle will impact the supply of water to crops from rainfall and soil moisture. Such changes will change the demand for water via irrigation. Demand for irrigation is expected to increase in many areas such as the USA, Europe, and parts of Asia, while some areas may see reduced demand for irrigation, such as South Asia and parts of China. The availability of water for irrigation can mitigate the drop in agricultural productivity associated with agricultural droughts during meteorological droughts. However, water availability for irrigation depends on water storage in natural locations like aquifers or reservoirs and on infrastructure and energy to move water from storage sites to fields.

As already noted, climate change will intensify the earth's hydrologic cycle and create greater variability in precipitation in most areas. The impact of this on food production will depend on many factors, including the level of damage from intense rainfall, soil characteristics, infrastructure, the likelihood of flood, and actions to balance water supply and demand. Drying will be particularly pronounced in the subtropics, the portion of the globe between the wetter tropical regions and mid-latitude regions (typically 23° to 35° latitude). This will result in significant impacts on irrigation water supply also effects on groundwater depletion in such locations as northern Mexico, southern USA, northern and southern Africa, northern India, and larger areas of China and Australia.

Most intense rainfall may lead to more soil erosion, reduced soil depth, and degradation of fertility. This may further impact water quality for other uses.

Of course, greater efficiency in agricultural water use through better management and technological innovation can reap benefits separate from climate change impacts.

11.4.1.4 Climate Change Impacts of Water for Energy

The world's energy systems are currently dominated by fuels that utilize large volumes of water in production and use. For example, fossil fuels, which provided 81% of the world's Total Primary Energy (TPE) in 2015 (IEA 2017), have considerable water demands.

Production of **oil and natural gas** (53% world TPE in 2015), requires significant water usage. In particular, the techniques of enhanced oil recovery and hydraulic fracturing are water-intensive. These techniques are only viable in locations where sufficient water is available. Refining of petroleum and processing of natural gas also require considerable water and are built with water available being one of several important factors. **Coal** production (28% of world TPE in 2015) uses water for dust suppression and cleaning, among other uses.

Thermoelectric power plants, which provided over 77% of the world's electricity in 2015 (Coal: 39%; natural gas: 23%; oil: 4%; and nuclear: 11%) use significant water for cooling. For example, an estimated 45% of total water withdrawals in the USA in 2010 were used for thermoelectric power plants; 38% of total freshwater withdrawals. Although nearly all of the water withdrawal is returned after use, availability of water is a significant factor in siting thermoelectric power plants.

Consequently, of the production and transformation of fossil fuels will be impacted by changes in water availability or greater competition for water for other uses because of climate change raises significant challenges, exacerbated during times of drought.

Hydropower, which provided over 16% of the world's electricity in 2015 and 2.5% of TPE, is inherently susceptible to variations in total water resources and seasonal fluctuations related to rainfall cycles. Drought conditions in the western USA reduced hydroelectricity generation each year from 2011 (7.8% of US generation) to 2015 (6.1%) until a strong "**El Niño**" warming phase in the Pacific Ocean brought heavy rains in 2016–2017 resulting in 2 years of increased hydroelectric power generation (7.5% in 2017). Areas with greater total precipitation may see an increase in hydropower if the additional precipitation can be captured. However, higher average temperatures will also result in high evaporation of water stored in reservoirs behind hydroelectric dams.

Seasonal rainfall in North America results in hydropower generation peaks in late spring and minimums in early fall. In areas where water is stored in the snowpack, a warmer climate will result in quicker melting and move generation peaks forward. Hydropower in different parts of the world will be affected according to local conditions and climate change impacts.

About 10–15% of **bioenergy crops** currently being utilized (corn, soy, sugar cane, and rapeseed) typically have irrigation demands. In addition, there is water usage in the conversion process from feedstock to biofuels.

The importance of local topography on climate effects will be particularly crucial to the knock-on impacts of water on energy. This will be important both through the availability of water resources and through impacts on facilities located on the coastline and at risk of flooding related to sea level rise.

There is also a cross-cutting impact of climate change that should be recognized first—the potential for climate change to forcing population movements through slowly building stresses or extreme weather events.

In all of the areas, potential increases in competition for water will raise important issues of water rights (see Sect. 5.2.3, Chap. 8, and Sect. 20.3).

11.4.2 Food

In the preceding section, we have addressed the knock-on impacts on food production from climate impacts on the water cycle. Here we will look at other impacts of food production and food systems. While climate impacts on food security are most significant in terms of production, there are many other climate impacts of the food system. It is important to remember that, as noted in Sect. 3.2, food security depends on a wide range of factors unrelated to climate, including demographics and diet.

11.4.2.1 Climate Change Impacts on Food Production

Most food production does not occur in controlled climates (e.g., greenhouses). As such, crops are selected with recognition of seasonal variations in temperature, sunlight, rainfall, and other climatic conditions. Other local factors such as soils, terrain, pests, availability of irrigation, other inputs, and socioeconomic consideration also impact choices of a crop. Some changes will be regional in scale, such as a warming climate making the best location for some crops farther north, causing some pests to move into or out of a region. Other changes may be very local such as a localized decrease in soil moisture or availability of irrigation water. As the climate changes, decisions about the most appropriate crop will be impacted.

Increased atmospheric carbon dioxide (CO_2) generally stimulates plant growth because it increases photosynthesis. However, the size of the effect is crop-dependent. There is a more significant effect on some crops such as wheat, rice, cotton, soybean, sugar beets, and potatoes than on other crops such as corn, sorghum, and sugarcane. The size of the effect is also impacted by the influence of water, nutrients, temperature, weeds, and parasites.

However, CO_2 emissions are usually accompanied by other pollutants (e.g., hydrocarbons and nitrogen oxide emitted from vehicles and power plants), which interact with ultraviolet radiation in the lower atmosphere to create ozone (O_3). In addition to being a human health hazard, elevated levels of ozone reduce plant photosynthesis and retard plant growth. Unsurprising, where steps have been taken to reduce air pollution in wealthier nations, ozone impacts on crops are less than in rapidly developing nations where pollution reduction technologies are deployed to a lesser degree and local air pollution a significant problem.

> Elevated O_3 since preindustrial times has very likely suppressed global production of major crops compared to what they would have been without O_3 increases, with estimated losses of roughly 10% for wheat and soybean and 3 to 5% for maize and rice. Impacts are most severe over India and China but are also evident for soybean and maize in the USA.

Food crops and farm animals are sensitive to **weather extremes** such as high and low temperatures, rainfall, hail, flooding, and wind. Impacts depend on the particular crop or animal and the intensity and duration of the extreme. Local geographic factors are, as always, important.

Climate change will also impact food production through changes in **agricultural pests**. Pests include weeds, fungi, beetles, moths, mice and rats, birds, and many other living organisms, as well as diseases that they impart to plants and animals. Changes in climate will impact each pest differently, including geographic shifts, their survivability through dormant periods, their growth and impact during other times, and the vulnerability of plants and animals to their effects. Agricultural use of pesticides can limit some impacts and will likely need to be used in different ways under different climates in different locations. However, some parasites are not controllable with pesticides.

Overall,

> the effects of climate change on crop and terrestrial food production are evident in several regions of the world (high confidence). Negative impacts of climate trends have been more common than positive ones. Positive trends are evident in some high latitude regions (high confidence)… periods of rapid food and cereal price increases following climate extremes in key producing regions, indicating a sensitivity of current markets to climate extremes, among other factors.
>
> For the major crops (wheat, rice, and maize) in tropical and temperate regions, climate change without adaptation will negatively impact production for local temperature increases of 2 °C or more above late-20th-century levels, although individual locations may benefit (medium confidence). Projected impacts vary across crops and regions and adaptation scenarios, with about 10% of projections for the period 2030–2049 showing yield gains of more than 10% and about 10% of projections showing yield losses of more than 25%, compared to the late 20th century. After 2050, the risk of more severe impacts increases. Regional [studies] … show crop production to be consistently and negatively affected by climate change in the future in low-latitude countries, while climate change may have positive or negative effects in northern latitudes (high confidence). Climate change will increase progressively the inter-annual variability of crop yields in many regions (medium confidence).

While the preceding has focused on land-based agriculture, climate change is also significant for food from marine and freshwater fisheries. Warming of oceans results is a reduction of fish catches in lower (warmer) latitudes, and likely expansion of fish catches in higher (colder) latitudes. Ocean acidification caused by the absorption of excess CO_2 in the atmosphere is also impacting fisheries. Of course, such changes are occurring against an ongoing trend in unsustainable fishing at all latitudes.

One aspect of food production frequently overlooked is the nutritional content of food, which is also impacted by climatic conditions. Such impacts are like other climate impacts, specific to the crop and the attribute (temperature extremes, CO_2 concentration, etc.)

Finally, the likelihood of more intense and frequent weather events (storms, floods, droughts, etc.) means that, in the absence of other changes to agriculture, it is likely that there will be greater variability in food production as the climate warms threatening the stability of food supplies.

11.4.2.2 Climate Change Impacts on Land for Water

Land can be set aside or managed for ecosystem functions such as for water capture and storage based upon existing local precipitation patterns and ecosystem services. Changes in precipitation will necessitate changes in land requirements for water, which may include new lands utilized for water capture and storage diverted from agricultural use.

11.4.2.3 Climate Change Impacts on Land for Energy

Renewable energy sources such as wind and solar can be land-intensive, and impactful if not managed in tandem with of purposes. Since these sources of energy are still emerging, and the ideal location for such wind and solar are primarily based on wind strength and solar radiation (isolation) reaching a given area, climate change will likely not have a decisive effect on future wind and solar but will influence decision making through impacts on wind and cloud cover. The significant expansion of on-shore wind and solar will bring challenging land use trade-offs.

Biofuels accounted for approximately 4% of the world's road transportation fuel in 2016 (IEA 2017). Some advocate significant expansion of biofuel use as a mitigation strategy for climate change. Using existing biofuel feedstocks such as sugar, corn, and soybean for such an expansion of biofuels would quire significant diversion of land for that purpose. Moreover, climate changes may alter locations where biofuel feedstocks might be optimally grown and the best locations for biofuel processing facilities. However, the development of advanced biofuels using non-food crops would likely reduce competition with food crops for food cropland.

As noted in Chap. 3, traditional forms of biomass—wood, charcoal, leaves, agricultural residue, animal/human waste, and urban waste—are significant energy sources, especially for rural populations and in many developing countries. Therefore, climate change impacts of the plant growth and agriculture will have a significant impact on the energy sources of those who depend on traditional biomass for energy.

11.4.3 Energy

Climate change will affect existing energy systems primarily through direct impacts on infrastructure and demand and indirectly through impacts on water. Mitigating emissions of greenhouse gases will require near wholesale decarbonization of a global energy system that obtained 85% of its primary energy from nonrenewable hydrocarbons.

11.4.3.1 Climate Change and Energy Production and Movement

Energy production has traditionally been based upon the location of natural resources, reflecting the local availability of hydrocarbons and suitable water flows for hydropower. The recent rise of renewables reflects the same pattern with wind patterns and solar insolation (the amount of solar energy incident on a surface), both being products of climate. Energy infrastructure reflects the level and range of connectivity required to produce energy resources in a given location and move them to demand centers. Energy production infrastructure, like all infrastructure, is sensitive to a wide range of local conditions, including:

- Offshore oil and gas production and offshore wind generation are susceptible to storms. Such facilities are routinely shut down and evacuated in advance of major storms.
- Coastal and island facilities are vulnerable to tropical cyclones (e.g., hurricanes and typhoons), storm surges, and sea level rise. In 2017, Hurricane Maria devastated much of the electric grid of Puerto Rico, leaving nearly all of its over three million inhabitants without electricity for weeks and a significant part of its population without electricity for months.
- Away from the coast, low-lying power plants, refineries, rail line, electric grid facilities, and other facilities are susceptible to flooding following extreme rainfall events, as well as strong wind events, wildfires, and other climate-impacted occurrences.
- Heatwaves can damage rail lines, which are the primary mode of transport for coal. For example, temperatures over 35 °C can cause buckling of rails known as "sun kinks."
- Extreme low-temperature events like the 2014 "polar vortex" event in the USA, can cause power plants to become non-operational.

There are many examples of each of these types of impact in all parts of the world.

As has been noted already, increased warming of the earth's atmosphere causes an intensification of the earth's water cycle resulting in more frequent and more extreme weather events. Thus, the impacts listed above are projected to become more frequent as a result of climate change.

Impacts of climate change on energy systems through impacts on water are noted above and include influences on the production of oil and natural gas production, cooling for thermoelectric power plants, water flows for hydropower, and irrigation for bioenergy crops.

Impacts of climate change on energy systems through impacts on land are also noted above and include influences on land use for energy production, as well as for traditional and modern forms of biomass.

11.4.3.2 Climate Change and Energy Demand

Warmer temperatures and more extremes in temperature will result in more energy demand for cooling and less for warming. Thus, energy demand will, in most locations, result in a net increase in energy use, if other factors such as efficiency and demand management do not change.

In mid- and low-latitude regions, electricity demand peaks in summer months, particularly on hot afternoons. Because current electricity systems have little capacity to store energy, real-time demand and generation must be in balance. Therefore, electric power generation capacity is built to meet peak demand with some margin of safety. Thus, many electric power plants operate only for brief periods of extreme demand during (usually) hot periods or (at higher latitudes) cold periods. Such peak demand services represent one of the significant inefficiencies of electric systems.

More extreme heat events will necessitate additional peak generation capacity to meet higher peak demands.

In regions with high energy insecurity (see Sect. 3.2), the lack of energy services will result in higher adverse impacts during heat waves such as heat stroke and other health effects.

11.4.3.3 Climate Change Impacts on Energy for Water and Food

In Chap. 1, the use of energy for water and food were noted. The energy requirements for the pumping and movement of water, for treating water before and after use, and for the heating and cooling of water are all significant. Energy demands for food production, processing, movement and refrigeration, and storage are also significant. Changes in water and food systems, therefore, change the demand energy use through a variety of paths. Conversely, changes in energy systems can change the availability and cost of energy for water- and food-related activities. Making some more or less viable and economical.

11.4.4 Cascading Impacts

The prior three sections have recognized how climate change impacts on one sector have "knock-on" impacts on other sectors. Where those knock-on impacts are powerful, climate change impacts can "cascade" through FEW systems.

Cascades typically begin with a disturbance that either damages FEW resources (e.g., soil, water supplies, or energy sources) or infrastructure required to energy food, energy, or water security. Extreme climate or weather events can sometimes damage more than one resource or infrastructure. Short-term or long-term actions to address problems in one sector then cause additional stress or new problems in other sectors.

Case studies throughout this book provide examples of cascading effects. Here we will explore two examples of cascading impacts, which are likely to happen more frequently and with greater intensity in the future because of anthropogenic climate change.

11.4.4.1 Drought in California, USA, 2011–2017

From 2012 to 2016, the state of California experienced severe drought, as part of an extended drought episode that spread across most of the contiguous USA and parts of Mexico (Heim 2017). Low snow and rainfall reduced water flows directly, and the water stored in the snowpack of the Sierra Nevada Mountains, which yield runoff and stream flows through traditionally dry periods. The drought impacted each of the food, energy, and water sectors of the state in ways that were compounded by cascading influences. A strong El Nino event finally broke the drought in late 2015 and early 2016.

The severity of the drought peaked in 2014 and 2015, during which the California Department of Resources reduced water use through a series of mandatory restrictions, impacting all users in all sectors. When state allocations of surface water were reduced, farmers increased the pumping of groundwater increasing energy use and groundwater depletion.

California's 77,500 farms constitute about 1.2% of US farmlands, produce over $50 billion/year of crops, including over a third of the vegetables and two-thirds of the fruits and nuts grown in the USA (Pathak et al. 2018). About 90% of California's crops are irrigated. The agricultural sector had direct losses estimated at over $2 billion in 2014.

Approximately 21% of in-state electricity generation in California comes from hydropower. In 2011 and 2017, the years before and after the drought, California generated 43 TWh of net electric energy from hydropower. In 2014, the net generation was 16 TWh, a 62% decline. Net hydropower was just 14 TWh in 2015.

Increased soil aridity, tree death, and groundwater pumping created long-term issues such as the increased risk of wildfires, reduced soil productivity, and mudslides associated with heavy rain.

11.4.4.2 Coastal Bangladesh

The low-lying coast of Bangladesh is subject to frequent flooding from annual monsoon rains, tropical cyclones, and tidal surges, often resulting in substantial population displacements (Penning-Rowsell 2013). For example, in 2007, Cyclone Sidr killed over 3400, and severely damaged coastal mangrove forests (the Sundarbans) and agriculture, exacerbating food insecurity. Anthropogenic climate change will intensify the sources of flooding and damage, as well as the loss of land to sea level rise.

Pumping of groundwater, sea level rise, and shrimp farms on the coast (which require brackish water) are causing salinity intrusion, which reduces drinking water supplies, degrades agricultural soils, and damages other coastal fisheries (Nuruzzaman et al. 2014). The importance of fisheries is reflected by the fact that 56% of the animal protein intake in Bangladesh comes from fish (FAO 2018). Fifteen percent of the Bangladesh population were considered undernourished in 2016, approximately 24.5 million people (FAO).

A quarter of Bangladesh's population did not have direct access to electricity in 2016 (World Bank 2017). A more substantial part of the population relies on traditional biomass for cooking (just 18% of the population had access to clean cooking in 2016), heating or other energy services. Thirty-five percent of Bangladesh's primary energy comes from renewable sources, nearly all of it traditional biomass. Access to biomass is stressed by flooding, storms, salinity intrusion, and clearing of additional land for agriculture (although agricultural wastes are a source of biomass energy). Conversely, intensive use of biomass degrades food and water systems.

Anthropogenic climate change is not the only stressor on the coast of Bangladesh, rising population and demand for food, energy, and water is also a significant stressor. However, extreme events can trigger cascading impacts throughout FEW systems.

11.5 Climate Mitigation

The effects of anthropogenic climate change are already apparent, and adverse effects are being experienced in specific locations (IPCC 2014a). Analysis of the drivers of anthropogenic climate change along with the vulnerability of large parts of humanity to adverse impacts of anthropogenic climate change has led the global community to adopt adopted a goal of "keeping a global temperature rise this century well below 2 °C above pre-industrial levels and to pursue efforts to limit the temperature increase even further to 1.5 °C."

In 2015, as part of the Paris Agreement, nearly every country in the world submitted a Nationally Determined Contribution (NDC) listing steps that they are willing to take voluntarily to reduce their future contribution to anthropogenic climate change. NDCs are not legally binding and therefore are an example of "soft" international law. There is a wide range of ambitions contained in NDCs which are not enforceable or, at the time of writing, reported consistently. Even if all NDCs adopted at or following the 2015 Paris Agreement were fully implemented, the projected rise in temperature would be significantly higher than 2 °C.

Mitigating anthropogenic climate change requires a wide range of human interventions, including significant alterations to energy and food systems.

Climate change mitigation policies exist everywhere alongside other policies seeking to achieve other aims such as poverty reduction and prosperity; food, energy, and water security; health; education; social justice; and many other aspirations. How climate change mitigation policies and other policies manifest themselves and affect outcomes are specific to the context of every location.

IPPC's Fifth Assessment Report (AR5) makes clear that

> deep cuts in emissions will require a diverse portfolio of policies, institutions, and technologies as well as changes in human behaviour and consumption patterns (high evidence; high agreement). There are many different development trajectories capable of substantially mitigating emissions; the ability to meet those trajectories will be constrained if particular technologies are removed from consideration. It is virtually certain that the most appropriate policies will vary by sector and country, suggesting the need for flexibility rather than a singular set of policy tools. In most countries, the actors that are relevant to controlling emissions aren't just national governments. Many diverse actors participate in climate policy from the local to the global levels—including a wide array of nongovernmental organizations representing different environmental, social, business and other interests. (robust evidence, medium agreement)

Further,

> policies to mitigate emissions are extremely complex and arise in the context of many different forms of uncertainty. While there has been much public attention to uncertainties in

> the underlying science of climate change … profound uncertainties arise in the socioeconomic factors [related to mitigation]. Those uncertainties include the development and deployment of technologies, prices for major primary energy sources, average rates of economic growth and the distribution of benefits and costs within societies, emission patterns, and a wide array of institutional factors such as whether and how countries cooperate effectively at the international level.

Effective mitigation policies rest of perception of risk and social benefits; recognition of uncertainty; an understanding of costs and benefits; the economic, social, and cultural conditions of the community targeted; and an ability to integrate all critical issues, including food, energy, and water systems. Such policies are more likely to achieve their aims without unintended or unforeseen side effects.

Central to mitigation efforts on a global scale are the issues of sustainable development (SD) explored in Chap. 3.

> First, the climate threat constrains possible development paths, and sufficiently disruptive climate change could preclude any prospect for a sustainable future (medium evidence, high agreement). Thus, a stable climate is one component of SD.
>
> Second, there are synergies and trade-offs between climate responses and broader SD goals, because some climate responses generate co-benefits for human and economic development, while others can have adverse side effects and generate risks (robust evidence, high agreement).

11.5.1 Energy

In 2010, the burning of fossil fuels accounted for 69% of global anthropogenic emissions of carbon dioxide (IPCC 2014b). Nearly half of these emissions (approximately 35% of total anthropogenic GHG emissions) are related to the extraction, movement, conversion, storage of energy to end-users. This includes the electric power sector, where the use of coal is the largest source of GHG emissions. The transportation sector, dominated by the use of petroleum-based fuels, accounted for approximately 23% of total anthropogenic GHG emissions.

One common approach to viewing GHG emissions from energy use is the Kaya Identity:

$$GHG\,\text{Emissions} = \text{Population} \times \frac{GDP}{\text{Population}} \times \frac{\text{Energy}}{GDP} \times \frac{GHG\,\text{Emissions}}{\text{Energy}} \tag{11.1}$$

where GDP is Gross Domestic Product.

The Kaya Identity is based in on approach to GHG emissions that emphasizes energy use, and in particular, energy use for economic activity. The Kaya Identity is therefore limited in its ability to explore non-energy sources and sinks of GHGs and energy emission unrelated to economic activity. With these significant limitations in mind, the Kaya Identity does bring into focus the importance of:

1. Population and population growth
2. Average economic prosperity per capita

3. Aggregate energy intensity
4. GHG intensity of energy

The first two terms of the Kaya Identity bring into focus issues central to human development that were explored in Chap. 3.

The third term of the Kaya Identity, the aggregate energy intensity is a commonly cited metric for energy efficiency. However, aggregate energy intensity encompasses all uses of energy regardless of their relationship to GDP. For example, residential energy use and personal driving do not measurably add to GDP but are significant components of the energy intensity numerator. Further, certain types of energy use are susceptible to "offshoring" where a product or service consumes energy in another country before being imported and contributing to the economy. Finally, economies change their structure over time in ways they affect energy use but do not reflect changes in energy efficiency. Decomposing energy use reduces some problems associated with the use of aggregate energy intensity into components that can be studied separately for insights into differences in the use of energy in different parts of an economy before being re-aggregated.

The fourth term in the Kaya identity, the carbon intensity of energy, is again a metric that aggregates all forms of energy. Decomposition of energy sources provides a clearer understanding of the climate impacts of different energy sources.

The utility of metrics like energy and carbon intensity, and broad formulations like the Kaya Identity, are highest when applied narrowly with carefully chosen boundaries, and with discrete (i.e., separate) understandings of each component within an area of study. These issues with be explored further in Chap. 12.

The first three terms of the Kaya Identity also highlight the demand side of energy use, while the final term highlights the supply side as it related to climate change. Reductions in energy use through conservation and greater efficiency complement efforts to decarbonize energy supply. The energy-related demands of food and water systems are significant and highlight the win-win solutions that can arise from nexus thinking.

For example, during the California drought described above, water efficiency policies enacted on utilities and end-users resulted in significant energy savings and reductions in GHG emissions (Spang et al. 2018).

In a world of over seven billion people seeking to live materially prosperous lives, reduced consumption alone will not reduce GHG emissions sufficiently to avoid major adverse impacts from climate change. IPPC's Fifth Assessment Report (AR5) makes clear that

> The stabilization of GHG concentrations at low levels requires a fundamental transformation of the energy supply system, including the long-term substitution of unabated fossil fuel conversion technologies by low-GHG alternatives (robust evidence, high agreement).
>
> Concentrations of CO_2 in the atmosphere can only be stabilized if global (net) CO_2 emissions peak and decline toward zero in the long term. Improving the energy efficiencies of fossil power plants and/or the shift from coal to gas will not by itself be sufficient to achieve this. Low-GHG energy supply technologies are found to be necessary if this goal is to be achieved.

As already noted, nearly half of energy-related emissions and approximately 35% of total anthropogenic GHG emissions are related to the extraction, movement, conversion, storage of energy to end-users—the supply side of energy systems. Fortunately,

> Multiple options exist to reduce energy supply sector GHG emissions (robust evidence, high agreement). These include energy efficiency improvements and fugitive emission reductions in fuel extraction as well as in energy conversion, transmission, and distribution systems; fossil fuel switching; and low-GHG energy supply technologies such as renewable energy (RE), nuclear power, and carbon dioxide capture and storage (CCS).

Challenges to reducing and decarbonizing energy use also exist in the transportation, industry, commercial, residential, and agricultural sectors. Each sector has distinct characteristics that shape possible pathways to decarbonization.

The commercial and residential sectors, for example, account for about one-third of energy use and 19% of GHG emissions. Energy use in these sectors is primarily related to buildings and the energy services delivered inside them. Electricity provides the largest share of energy used in buildings for heating and cooling space, heating of water, refrigeration of food, lighting, and the powering of a host of machines and devises. Thus, on the demand side, there are many pathways to increase the efficiency of obtaining energy services. On the supply side, since electricity can be produced from many energy sources, there are also additional pathways with allowing for competition and trade-offs between technologies, especially renewable (low- and non-carbon) energy sources.

Renewable energy sources include a wide array of technologies, including hydropower, solid biomass, liquid biofuels, biogas (methane from renewable sources), wind, solar, and geothermal technologies.

> Infrastructure and integration challenges vary by RE technology and the characteristics of the existing background energy system (medium evidence, medium agreement). Operating experience and studies of medium to high penetrations of RE indicate that these issues can be managed with various technical and institutional tools. As RE penetrations increase, such issues are more challenging, must be carefully considered in energy supply planning and operations to ensure reliable energy supply, and may result in higher costs.

In contrast, over 90% of energy use in the transportation sector is from petroleum-based fuels. Therefore, the pathways to the decarbonization of transportation are significantly fewer than in buildings. Demand-side strategies include vehicle efficiency, alternatives to travel, alternative modes of transportation. Supply-side strategies include the use of biofuels and natural gas, and emerging electric vehicles. Electric vehicles represent the most significant alternative transportation energy option because it opens up the various pathways and options associated with the electricity sector.

Various policies that advance low carbon energy sources have different attributes that make them more or less effective in different contexts.

> Greenhouse gas emission trading and GHG taxes have been enacted to address the market externalities associated with GHG emissions (high evidence, high agreement). In the longer term, GHG pricing can support the adoption of low-GHG energy technologies due to the resulting fuel- and technology-dependent markup in marginal costs. Technology policies (e.g., feed-in tariffs, quotas, and tendering/bidding) have proven successful in increasing the share of RE technologies (medium evidence, medium agreement).

> The success of energy policies depends on capacity building, the removal of financial barriers, the development of a solid legal framework, and sufficient regulatory stability (robust evidence, high agreement). Property rights, contract enforcement, and emissions accounting are essential for the successful implementation of climate policies in the energy supply sector.

11.5.2 Food systems

In 2010, agriculture, deforestation, and other land use changes accounted for 24% of global anthropogenic GHG emissions (IPCC 2014b). Agricultural emissions of GHGs come from many sources, including how soils are managed, the digestive system of livestock (enteric fermentation), manure management, cultivation of waterlogged crops like rice, urea fertilizers, and the burning of agricultural residues.

Because soils and plants can be a sink for GHGs, net emissions can be reduced by both lowering emissions and carbon sequestration in soils and biomass. Thus, mitigation strategies on the supply-side of food systems include the following:

- Reducing emissions from croplands and livestock (principally methane and nitrous oxide).
- Minimizing carbon losses from, and maximizing carbon sequestration in, biota and soils. This can be achieved through better agricultural practices such as **no-till cropping** and the maximizing tree growth through reduced deforestation and increased **afforestation** and **reforestation**. (This approach can be built upon outside of food systems by using renewable products such as wood instead of steel or concrete in construction. In a similar vein, some forms of biomass use for energy can reduce GHG emissions.)
- Reducing energy inputs to agriculture. Direct energy inputs to agriculture include energy for machinery, irrigation, and food preservation. Indirect energy inputs include those used to produce fertilizers, pesticides, equipment, and other inputs. GHG emissions related to indirect energy use are usually counted as emissions from another sector, such as industry.
- Increasing productivity (yield) of crops, livestock, and fish for the same or reduced energy inputs (direct and indirect). This can be accomplished through crop varieties with higher yield or lower input demands, more efficient agricultural/aquaculture practices, new productive uses of farm waste, and reduced loses in the supply chain.

Demand-side mitigation strategies include:

- Reducing food waste (which also reduces GHG emissions from landfills).
- Modifying diets toward food choices that have lower associated GHG emissions. Related to this, but separate from food systems, is the use of agriculture/aquaculture products to displace products with higher GHG emissions such as construction material and fuels.

11.5.3 Water

Water systems have both direct and indirect GHG emissions. Direct GHG emissions include those associated with wastewater and the decomposition of its biological matter. When treated in a wastewater facility, biogas can be captured and utilized as a renewable form of energy. However, approximately 47% of wastewater produced in the domestic and manufacturing sectors is not treated (IPCC 2014b). Further, water systems of farms are used to manage animal waste, resulting additional emissions in the agricultural sector. As noted above, the cultivation of waterlogged crops like rice also leads to emissions.

The use of anaerobic digesters to produce and utilize biogas and biosolids from wastewater is a well-known mitigation strategy. In agricultural settings, this strategy merges food, energy, and water systems. However, technological and economic challenges face the wider use of such systems.

Indirect emissions associated with water systems include that associated with the use of energy throughout water systems (see Sect. 2.5). More efficient use of water and other demand-side mitigation strategies reduce the GHG footprint of water use.

11.5.4 Integrated Mitigation

Climate change mitigation requires significant changes to FEW systems. Some mitigation strategies are single sector in their focus, such as shifting away from coal generation of electricity, improved management of soils, and capturing biogas from wastewater. However, nearly all strategies have **knock-on effects** on other sectors. Moving away from coal reduces the water required for cooling power plants (even when switch to natural gas); managing soils for carbon storage is usually accompanied by a reduction in energy-intensive inputs; and capturing biogas from wastewater creates a renewable form of natural gas for energy. The example mentioned above of water efficiency requirements implemented in response to drought in California, reducing energy use, and thus, GHG emissions is illustrative of suck knock-on benefits.

Such knock-on benefits are most significant where interactions between FEW systems are strongest (e.g., agriculture, cities); FEW systems compete for ecosystem services (e.g., from Chap. 9), infrastructure (e.g., from Chap. 10), and economic resources (e.g., from Chap. 5); or conflicts between stakeholders in different aspects of FEW systems exist (e.g., from Chaps. 18–20).

In these three classes of situations, mitigation strategies that integrate FEW systems from the outset achieve a more significant impact.

11.6 Modeling

The modeling of FEW systems is addressed in depth in Chap. 15. Here we explore some fundamentals of climate modeling, which have led to the integration of FEW systems with climate.

Climate models simulate the main components of the climate system. These include the atmosphere, but also incoming solar radiation; absorbed and reemitted radiation; the water cycle and other geochemical cycles; the oceans, the cryosphere, and the biosphere; and interactions between these and other components.

Climate models at the global scale are called **General Circulation Models** (GCMs) and represent the climate by a three-dimensional grid composed of cells defined by vertical layers and horizontal elements. The number and thickness of layers above and below the surface of the earth are defined within each GCM according to its scientific objective. Sometimes larger cells are preferred for ease of computation while sometimes smaller cells are preferred to allow more careful consideration of small-scale factors.

Under the Millennium Development Goals (Sect.3.5), GHG emissions were recognized under Goal 7, "to ensure environmental sustainability." This led to many efforts to assess linkages between climate, land, energy, and water (CLEW) in the context of sustainable development. A climate, land, energy, and water (CLEW) modeling framework was developed and launched in 2009 by several agencies within the United Nations system (see Sect. 6.2.2). CLEW models aimed to map important interactions between food, energy, and water systems in the context of climate change. Further:

> In addition to incorporating such a mapping and quantification of key relationships, the tool should be designed for use in the following applications.
>
> - Decision making: A well-formulated integrated CLEW tool would help decision and policymakers assess options in terms of their likely effects on the broad CLEW system. The tool should be able to transparently evaluate the trade-offs reflected in different options.
> - Policy assessments: Given limited resources, it is important for policymakers to ensure that policies are as cost-effective as possible. If a single policy can achieve multiple objectives, it may advance development more than policies focused separately on single objectives. A CLEW tool should, therefore, provide a more complete, multisystem policy assessment.
> - Facilitating policy harmonization and integration: there are instances of very contradictory policies, for example, electricity subsidies that accelerate aquifer depletion, which in turn lead to greater electricity use and subsidy requirements. A CLEW tool would help harmonize potentially conflicting policies.
> - Technology assessments: Some technology options can affect multiple resources; for example, nuclear power could reduce GHG emissions and reduce the exposure to volatile fossil fuel markets. Although it would use water for cooling, nuclear power can generate electricity for seawater desalination. As with policies, a CLEW tool should allow a more inclusive assessment of technological options.
> - Scenario development: Another goal is to elaborate consistently on scenarios of possible socio-economic development trajectories with the purpose of identifying future development opportunities as well as understanding the implications of different policies.

> This is important for understanding whether current development is sustainable, and for exploring possible alternative development scenarios and the kinds of technology improvements that might significantly change development trajectories.
>
> (IAEA 2009).

The CLEW tool launched significant modeling work on country-level FEW integration studies in the context of climate change (Bazilian 2011) which continues (UN DESA 2018).

Key Points

- Anthropogenic emissions of greenhouse gases are intensifying the earth's natural greenhouse effect with significant impacting of all aspects of human societies, including FEW systems in a manner that is largely negative and necessitates adaptation to the consequences on new climate regimes.
- Communication of scientific certainty and risk is difficult. Many lessons can be learned from the approach developed by the Intergovernmental Panel on Climate Change (IPCC) about communicating these critical concepts.
- Mitigation of climate change falls heavily on energy systems because approximately two-thirds of anthropogenic GHG emissions are the result of burning fossil fuels for energy.
- Mitigation of climate change also falls of food systems because nearly a quarter of anthropogenic GHG emissions are the result of agriculture, deforestation, and other land use changes.
- Climate change is profoundly impacting water systems, which also contribute to GHG emissions when used to absorb and move biological waste.
- While many climate mitigation strategies are focused on a single sector, especially energy and agriculture, integrate FEW systems can be far more effective where the three systems have inherently strong interactions; compete for ecosystem services, infrastructure, or economic resources; or result in conflicts between communities.
- Advances in climate modeling over recent decades have resulted in the integration of FEW systems in many ways.

Discussion Points and Exercises

1. Describe three direct, or "first-order" interactions and three higher order "cascading" impacts of human demographics on food, energy, and water systems.
2. Describe three direct, or "first-order" interactions and three higher order "cascading" impacts of international or national governance on food, energy, and water systems.
3. Describe three direct, or "first-order" interactions and three higher order "cascading" impacts of ecosystem degradation on food, energy, and water systems.
4. Describe three direct, or "first-order" interactions and three higher order "cascading" impacts of hard infrastructure on food, energy, and water systems.
5. Use the Kaya Identity as a model for similar equations applicable to externalities (one for each) related to (a) food, (b) energy, and (c) water. Describe the strength and weaknesses or each new identity.

6. Explore how a Kaya Identity relationship might be applied to an **integrated FEW system**. Describe its strength and weaknesses.
7. Describe how mitigation of ecosystem degradation might apply to food, energy, and water systems.
8. Describe how embedded energy (see Sect. 7.4) applies to climate change impacts and possible mitigation strategy. How might embedded greenhouse emissions be applied to international climate change treaties? What are the strengths and weakness of such an approach.

References

Bazilian, M. (2011). Considering the energy, water and food nexus: Towards an integrated modelling approach. *Energy Policy, 39*, 7896–7906.

FAO (2018) The State of Agricultural Commodity Markets 2018. Agricultural trade, climate change and food security. Food and Agriculture Organization of the United Nations Rome. https://www.fao.org/3/I9542EN/i9542en.pdf

Heim, R. (2017). Comparison of the early twenty-first century drought in the United States to the 1930s and 1950s drought episodes. *Bulletin of the American Meteorological Society, 98*(12), 2579–2592.

IAEA (International Atomic Energy Agency). (2009). Annex VI: Seeking sustainable climate land energy and water (CLEW) strategies. In *Nuclear technology review*. Vienna, Austria: International Atomic Energy Agency.

International Energy Agency. (2017). *Tracking progress: Transport biofuels*. Retrieved from https://www.iea.org/etp/tracking2017/transportbiofuels/.

Janssens-Maenhout, G., et al. (2017). *Fossil CO2 & GHG emissions of all world countries*. JRC Science for Policy Report, European Commission. Retrieved from http://edgar.jrc.ec.europa.eu/booklet2017/CO2_and_GHG_emissions_of_all_world_countries_booklet_online.pdf.

Nuruzzaman, A., et al. (2014). Causes of salinity intrusion in coastal belt of Bangladesh. *International Journal of Plant Research, 4*(4A), 8–13.

Pathak, T. B., et al. (2018). Climate change trends and impacts on California agriculture: A detailed review. *Agronomy, 8*(3), 25.

Penning-Rowsell, E. C., Sultana, P., & Thompson, P. M. (2013). The 'last resort'? Population movement in response to climate-related hazards in Bangladesh. *Environmental Science & Policy, 27*(S1), S44–S59.

Spang, E. S., Holguin, A. J., & Loge, F. J. (2018). The estimated impact of California's urban water conservation mandate on electricity consumption and greenhouse gas emissions. *Environmental Research Letters, 13*(1).

World Bank. (2017). *Global tracking framework 2017*. Retrieved from http://www.worldbank.org/en/topic/energy/publication/global-tracking-framework-2017.

Further Reading

Blunden, J., Arndt, D. S., & Hartfield, G. (2018). State of the climate in 2017. *Bulletin of the American Meteorological Society, 99*(8), Si–S332. https://doi.org/10.1175/2018BAMSStateoftheClimate.1.

Hayhoe, K., et al. (2004). Emissions pathways, climate change, and impacts on California. *Proceedings of the National Academy of Sciences of the United States of America, 101*(34), 12422–12427.

Howitt, R. E., et al. (2014). *Economic analysis of the 2014 drought for California agriculture*. Davis, CA: Center for Watershed Sciences, University of California.

Huq, N., et al. (2015). Climate change impacts in agricultural communities in rural areas of coastal Bangladesh: A tale of many stories. *Sustainability, 7*(7), 8437–8460.

International Energy Agency. (2018). *Key world energy statistics*. Retrieved from http://www.iea.org/publications/freepublications/publication/KeyWorld2017.pdf.

IPCC. (2013). *Climate change: The physical science basis*. Contribution of Working Group I to the Fifth Assessment Report of the Intergovernmental Panel on Climate Change. Retrieved from http://www.ipcc.ch/report/ar5/wg1/.

IPCC. (2014a). *Climate change 2014: Impacts, adaptation, and vulnerability*. Contribution of Working Group II to the Fifth Assessment Report of the Intergovernmental Panel on Climate Change. Retrieved from http://www.ipcc.ch/report/ar5/wg2/.

IPCC. (2014b). *Climate change 2014: Mitigation of climate change*. Contribution of Working Group III to the Fifth Assessment Report of the Intergovernmental Panel on Climate Change. Retrieved from http://www.ipcc.ch/report/ar5/wg3/.

Liu, Q. (2016). Interlinking climate change with water-energy-food nexus and related ecosystem processes in California case studies. *Ecological Processes, 5*, 14.

Maupin, M. A., et al. (2014). *Estimated use of water in the United States in 2010*. U.S. Geological Survey Circular 1405 (56 p). Retrieved from https://pubs.usgs.gov/circ/1405/.

McMahon, J. E., & Price, S. K. (2011). *Water and energy interactions*. Lawrence Berkeley National Laboratory. Retrieved from https://escholarship.org/uc/item/5pr6r5h6#page-1.

National Research Council. (1979). *Carbon dioxide and climate: A scientific assessment*. Washington, DC: The National Academies Press. https://doi.org/10.17226/12181.

O'Riodan, J., & Sandford, R. W. (2015). *The climate nexus: Water, food, energy and biodiversity in a changing world*. Calgary: Rocky Mountain Books.

Skaggs, R., et al. (2012). *Climate and energy-water-land system interactions*. Richland, WA: Pacific Northwest National Laboratory, The U.S. Department of Energy.

Tasbirul Islam, M., et al. (2014). Current energy scenario and future prospect of renewable energy in Bangladesh. *Renewable and Sustainable Energy Reviews, 39*, 1074–1088.

United Nations Department of Economic and Social Affairs. (2018). *Global climate, land, energy & water strategies*. Retrieved from https://unite.un.org/sites/unite.un.org/files/app-global-clews-v-1-0/landingpage.html.

USGCRP. (2017). *Climate science special report: Fourth National Climate Assessment*, Volume I (470 pp). D. J. Wuebbles, D. W. Fahey, K. A. Hibbard, D. J. Dokken, B. C. Stewart, T. K. Maycock (Series Eds.). Washington, DC: U.S. Global Change Research Program. https://doi.org/10.7930/J0J964J6. Retrieved from https://www.globalchange.gov/nca4.

USGCRP. (2018). *Impacts, risks, and adaptation in the United States: Fourth National Climate Assessment*, Volume II (1515 pp). D. R. Reidmiller, C. W. Avery, D. R. Easterling, K. E. Kunkel, K. L. M. Lewis, T. K. Maycock, B. C. Stewart (Series Eds.). Washington, DC: U.S. Global Change Research Program. https://doi.org/10.7930/NCA4.2018. Retrieved from https://www.globalchange.gov/nca4.

Part II
Tools

Chapter 12
Questions and Scales

Michael Carbajales-Dale, Emre Eftelioglu, Carey W. King, Fernando R. Miralles-Wilhelm, Benjamin L. Ruddell, Peter Saundry, and Shashi Shekhar

12.1 Introduction to Questions and Scales

In Chap. 1, we noted that criteria that lead us to integrate FEW systems are human-centric, and the ultimate purpose of studying the nexus of FEW systems is to guide human decision-making. Integrated FEW systems are coupled natural-human systems, and the scholarship of FEW is fundamentally connected to ecological function and human use. Thus, Nexus research is an applied science.

M. Carbajales-Dale
Environmental Engineering & Earth Sciences, Clemson University, Clemson, SC, USA
e-mail: madale@clemson.edu

E. Eftelioglu
Engineering and Data Sciences, Cargill Inc., Wayzata, MN, USA
e-mail: emre_eftelioglu@cargill.com

C. W. King
Energy Institute, The University of Texas at Austin, Austin, TX, USA
e-mail: careyking@mail.utexas.edu

F. R. Miralles-Wilhelm
Earth System Science Interdisciplinary Center and Global Water Team, University of Maryland and The Nature Conservancy, College Park, MD, USA
e-mail: fwilhelm@umd.edu

B. L. Ruddell
School of Informatics, Computing, and Cyber Systems, Northern Arizona University, Flagstaff, AZ, USA
e-mail: Benjamin.Ruddell@nau.edu

P. Saundry, B. L. Ruddell (eds.), *The Food-Energy-Water Nexus*, AESS Interdisciplinary Environmental Studies and Sciences Series,
https://doi.org/10.1007/978-3-030-29914-9_12

In Sect. 3.2, we recognized that food, energy, and water security require these resources to be available, accessible, utilizable, stable and reliable, sustainable, and resilient. Achieving and maintaining FEW security entails human decision-making related to every aspect of each resource and the complex interrelationships between them (Chap. 4).

In this chapter, we will build upon the key issues raised in Sect. 1.5 regarding decision-making contexts and noted that there is tremendous variation in geographic factors and time consideration. Geography defines the environmental conditions and the natural resources initially present within a system. Geopolitics and international governance (Chap. 6) and trade (Chap. 7) connect the physical, biological, and human resources of one subsystem to the greater regional and/or global system, and thereby connect each local system to many other local systems. Connections express themselves quickly but not instantaneously, so time lags and especially political and economic decision timescales are usually a significant factor. Decision-making occurs within a context of the culture, social, legal customs operating at the scale of the decision-makers who have varied economic, technological, economic, and other factors acting upon them (Chap. 4).

We will begin by exploring the relationship between science and the type of human-centric challenges confronted in the nexus of FEW systems. We will then explore the wide range of scales in space and time which arise in FEW nexus studies. These scales are rooted in factors related to decision-making; natural, political, and cultural geography; ecological functioning; engineering and infrastructure; scientific practice and capabilities; economics; and other considerations such as social structure, politics, culture, demographics, and human aspirations. Some of these factors have been explored in earlier chapters, and others will be explored in later chapters.

In addition to reviewing the questions and scales at which we need to measure, collect data, model, and carry out significant computation work on FEW systems. This chapter sets up a deep exploration of these topics in Chaps. 13–16. In Chap. 17, we will return to the integration of these areas science and connect them to decision-making at the Nexus. We will explore of communities of science are require for effective research and communities of practice are required for the effective application of Nexus science to problem-solving in the real world.

P. Saundry (✉)
Energy Policy and Climate, Advanced Academic Programs, Krieger School of Arts and Sciences, Johns Hopkins University, Washington, DC, USA
e-mail: psaundr1@jhu.edu

S. Shekhar
Computer Science & Engineering, College of Science & Engineering, University of Minnesota, Minneapolis, MN, USA
e-mail: shekhar@umn.edu

12.2 Connecting Practical Questions with Scientific Capabilities

12.2.1 *Pasteur's Quadrant and Use-Inspired Science*

In his book, Pasteur's Quadrant, Donald Stokes frames scientific research according to two considerations; the quest for fundamental understanding and considerations of applied use by society. This leads to a quadrant graph (Fig. 12.1).

Research that is aimed at a fundamental understanding of how the universe works but with little or no utility to society is termed **basic research**. Research that is aimed at an applied societal use with little or no consideration for advancing the fundamental understanding of nature is termed **applied research**. Research that both advances fundamental understanding and is useful to society in the relatively near-term is termed **use-inspired research**. The lower-left quadrant represents science that is not of substantial applied or basic value and is to be avoided. The upper-right quadrant of the graph is termed Pasteur's quadrant after the nineteenth-century scientist whose work was both fundamental and use-inspired. We should all aspire to do use-inspired research, but that is challenging to achieve.

For example, research on the biochemical reactions underpinning photosynthesis of a flower that is not used for any practical purpose is an example of basic research. However, such basic research might become important for a practical application to FEW systems in the coming decades.

Research on lowering the manufacturing costs of a wind turbine is an example of pure applied research. Such research may not advance the fundamental understanding of the world, but it is clearly important to energy systems. Because wind turbines make no operational demands on water as thermo-electric power plants do, there are implications for water systems for lowering the cost of wind turbines. Because wind turbines do not make water available for irrigation as dual-purpose

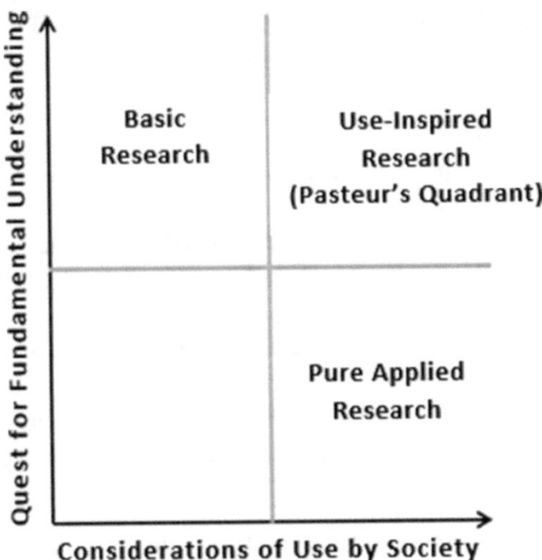

Fig. 12.1 Pasteur's Quadrant. (Source: Stokes (1997))

hydroelectric dams do, there are implications for food systems for lowering the cost of wind turbines.

Research on how societies address the trade-offs between water use for energy, agriculture, industry, human health and sanitation, and environmental uses is an example of use-inspired research. The fundamental understanding of why and how human societies make decisions about the different uses of water is a fundamental challenge in social science basic research. It is also a topic that has clear applications to societies striving to develop sustainable water policies at the nexus of food, energy, and water systems.

Integrating FEW systems to develop near-term solutions requires basic, applied, and also use-inspired research. This textbook is primarily focused on scholarship in Pasteur's quadrant while recognizing the importance of long-term basic research and near-term applied research.

FEW system nexus studies are an example of use-inspired science. The traditional approach to analysis has been defined by a decision-maker who approaches the problem from the perspective of a narrowly defined, single subsystem, which is that decision-makers' primary concern. For example, if a water perspective is adopted, then food and energy systems are users of the resource; from a food perspective energy and water are inputs; from an energy perspective, water as well as bioresources (e.g., biomass in the form of energy crops) are generally an input or resource requirement, and food and water are both generally the outputs. Food and water supply, as well as wastewater treatment, require significant amounts of energy. Of course, areas such as food-as-fuels (i.e., biofuels) tend to blur these descriptions due to additional impacts associated with land-use, land-use change, and use of the available biomass resource. However, such tradition single-sector perspectives inherently underemphasize the fact that these are three highly coupled systems.

Although this primary perspective bias may not be directly addressed through improved modeling tools, one role that such tools can play is to provide information to sectoral-based policymakers on the implications of potential choices for other sectors. These implications are usually easy to translate as food, energy, and water systems have many characteristics in common.

- All currently leave billions of people without access (quantity and/or quality of access).
- All have rapidly growing global demand.
- All are constrained by natural resources that are economically recoverable on Earth.
- All are "global goods" and involve international trade and geopolitics.
- All have strongly differing regional availability and temporal variations in supply and demand.
- All have strong connections to climate change and the environment.
- All are critical for peoples' security.
- All operate in heavily regulated markets, although regulation is quite varied and diverse between market sectors and between societies.
- All require the explicit identification and treatment of risks.

This is all very complicated, highly interdisciplinary, multiscalar, and highly cross-cutting. No single scientist or decision-maker is able to master the full picture.

As a result, gains from FEWS nexus science are increasingly achieved when decision-makers gain access to scientifically valid analytical tools (data/models) that cover the range of applications from planning (larger scales) to on-the-ground implementation (smaller scales). More effective implementation of FEW system models is achieved when these tools are coproduced with end-users, integrating stakeholders in the food, energy, and water sectors and the broader community to be served by the application of such tools (e.g., ecosystem services, health, other economic activities) in framing critical questions for research and helping plan for implementation. Such factors are addressed further in Chaps. 17 and 21.

As noted in Chaps. 4 and 5, economic analysis and modeling are necessary when assessing FEW system trade-offs. For instance, water and energy are crucial inputs to the production of goods and services. Tightening constraints may introduce the potential for reductions in production activities. Increasing water demand and scarcity has the potential to increase market prices for food, energy, and water and to lead to redistributions of these increasingly scarce resources. Economics governs most FEW decisions and is an essential science for these systems models.

In the case of water, increasing scarcity in one area is likely to result in part in the increased purchase of food products from another area. When this occurs, significant structural adjustment can occur and needs to be managed with sensitivity in order to ensure that overall economic activity and employment is not reduced in the short term, or, in extreme cases, result in food insecurity, migration, and conflict. Actual outcomes will depend on the capacity of a community to adjust; rates of technological progress in the development of water efficiency in energy and food production; knowledge provision; and institutional, governance, and planning arrangements to facilitate efficient investment and synergies in water and energy planning.

Ultimately, the feasibility of transitioning to a FEW system approach in practice will depend not simply on the technology and policy characteristics, but also on the economic and climate/environmental impacts, the manner in which they interface with other technologies, physical and institutional infrastructures, and a range of societal norms.

12.2.2 Physical Questions and Values Questions

Because FEW systems are coupled natural-human systems, a recognition of both physical-science-based and human-values-based questions (as developed in Chap. 4) is essential. This distinction is sometimes called physical (or natural) science and social science. It is essential to recognize the subjective human values involved in many FEW nexus questions and delineate them from objective scientific aspects of FEW nexus scholarship. Example human-values-based questions are:

- How can FEW outputs (goods, services) most efficiently be produced and distributed?
- What outputs should be maximized? Food, energy, or water separately or together?
- Where should inputs come from?

- What alternative technologies and processes could do better?
- What are the economic costs and prices of inputs and outputs? Shadow prices?
- What environmental values are considered? Whose environmental values are considered?
- What is the impact on human health, in both the near and long term?
- How are cultural/social values and impacts identified and incorporated?
- How are diverse inputs and outputs weighted and/or balanced?
- Is there transparency and equity in making these decisions?
- Are the human outcomes of the decisions just? According to whom?
- What human interests benefit and what human interests are harmed?
- How can the harmed express disagreement or receive compensation?
- How well does the soft-infrastructure (governance) process weigh these considerations?
- How does the governance process need to be changed to better weigh these considerations?

There are several types of questions we can ask in FEW systems. An incomplete list includes Descriptive, Correlative, Causality, Efficacy, and Management questions. Each of these questions can be addressed using metrics, data, computing, and models. Example physical-science-based questions are as follows:

- **Descriptive questions**: production levels, emissions, inputs and outputs, impacts.
- **Correlation/Relational questions**: relative importance of energy sources, levels of food, energy, and water insecurity in different communities, the relationship between income/education levels/ethnicity and food, energy, and water insecurity, health impact assessments (e.g., epidemiology).
- **Causality questions**: what is the causal relationship between access to treated water and disease? What is the effect of efficiency policies on energy use? What is the effect of food assistance programs (e.g., the Supplemental Nutritional Assistance Program, or SNAP) on food security?
- **Adaptive management questions**: what changes in management might optimize desired outcomes?

12.2.3 *Decision Science and Questions at Scale*

Decision science also depends on questions of scale, both geographic and temporal. For example, at what scale is a question framed for research and examination? For example, trying to address flooding impacts for one small community may be futile if the entire watershed is subject to flooding. Alternatively, an examination of the US Renewable Fuel Standard (RFS) alone would miss the global implications of this policy. Temporal scales also matter. As noted in Sect. 4.3, decision-makers may also be responding to whether an event is a "press" event that occurs gradually and over time or a "pulse" event that happens quickly. The catastrophic wild fires that occurred in California in both 2017 and 2018 are examples of pulse events but also signify a perhaps longer-term shift in regional dynamics through changes over time in press events.

12.3 Scales of Questions and Decisions

Spatial scales relate to biophysical phenomena and processes, environmental conditions, the biophysical resources present within a system, and the flows of goods and services, financial resources, and people. Temporal scales related to these phenomena have a very wide range from daily (diurnal, e.g., insolation) and seasonal/annual (e.g., precipitation, to longer-term climatic-, drought-, fire-, economic-, demographic-, and even ice-age cycles).

Using average numbers, such as the average amount of precipitation in a given year, may mask temporal scales that matter. As noted in Chap. 11, one impact of climate change is increasingly heavier precipitation and more drought. While average amounts of precipitation in a given year may be increasing, lack of precipitation during key points of a growing season may hinder crop production. In other words, a precipitation cycle that is drought, drought, then flood is on average fine but is entirely problematic for plant growth.

The human dimensions of FEW systems are shaped by cultural/religious and economic motivations, governmental laws, and public and private policies also operating at a range of spatial scales defined by the extent of a governmental or economic structure (town, city, county, state/province, country, and multinational); an economic or trade or migration system; or a cultural/demographic common denominator. Political and economic and demographic timescales are usually a significant factor in influencing FEWS decisions. For example, election cycles can be an important driver.

In fact, these anthropocentric scales tend to be the primary and indispensable scales in FEW systems, a fact that physical scientists and engineers often forget. This forgetfulness leads to overly narrow definitions of the problems and questions and cripples both the scientific accuracy and applied utility of the resulting science and engineering. We cannot attain Pasteur's quadrant without getting the scales right. Many human systems serve explicitly to overcome and/or bridge physical **process scales** toward the purpose of allowing people and societies to transcend the small-scale physical constraints and variations. Vice versa, politicians, and business people often forget that their options are ultimately constrained by large-scale physical processes that are hard to see from their perspective. This forgetfulness of physical reality leads to sustainability problems and ineffective policy.

In FEW systems, it is, therefore, essential for all human agents to consider scale, and to pay some attention to the "other" scales outside one's primary scale of focus. We need to know what system components are important to consider, and what is acceptable to omit, in any given circumstance; scale is often an essential tool with which to make these judgments. In a system, our success as scientists and policy-makers always depends on an accurate awareness of scale.

We have already noted that complex systems (see Sect. 2.2) are frequently composed of diverse subsystems operating at different scales. Unsurprisingly, one of the key challenges of any nexus study, is integrating components that operate at different spatial and temporal and social scales. Different physical components in a nexus study may be studied at different scales for scientific reasons. For example, in a study of the food system, data might be collected at a scale of plants, fields, farms, and/or crop-growing regions. Other components of the food system might be oriented around a geographic feature such as a river, a watershed, an aquifer, an ecoregion, or a market area or trade zone. Against such diverse geographies, we need well-reasoned and explicitly argued scientific reasons for the choice of scale to study.

A researcher must choose scale of a study for an integrated FEW system based on the human and physical processes that are most important, not based on the convenience of methods, data, or tools that are provided by one scientific discipline participating in the research. However, one must, of course, utilize methods, data, and tools that function at the appropriate human and physical scales. This often means that we need to build new methods, data, and tools to study FEW systems, or, at least, that we must translate those methods, data, and tools from their original domain into the FEW system framing and scale where we are now working. The meso-scale is often an appropriate compromise on scale that allows us to bring these elements together.

One scale gap that is commonly experienced in FEW nexus studies is between the scales associated with external stressors (e.g., climate) and those associated with socioeconomic impacts (e.g., regional droughts, reduced or increased agricultural production, energy shortages). Another common scale gap is between shocks and stresses (or presses and pulses, for the ecologist); shocks are fast, and sharp transient dynamics and stresses are slow and persistent dynamics. Consider also the scale gap between individual action—which can be very agile and predictable—and community response—which can be very slow and unpredictable.

We also noted that complex systems often have hierarchies of scale, importance, and control (see Sect. 2.2). Despite their heterogeneity, interdependency, and emergent properties, some subsystems exert more control than others, and some scales are more important than others. Complex systems have distributed control, but there are centers and hubs of control. In FEW systems, some of the centers include climate, weather, natural resource availability, national governance, technology, international conflict and politics, cities, cultural consumption preferences, financial investment, infrastructure, and large corporate business.

The appropriate spatial and temporal scales must, therefore, balance the requirements and resources of science, the spatial and temporal scales of the structures and processes being studied, the existing and possible data, and the needs of the end-users of the results and decision-makers. Key to the appropriate choice of scale is the phenomenon of **emergence**, where specific characteristics of the entire system "emerge" from the interconnections between the fine-scale parts of the system (again, see Sect. 2.2). A poor choice of scale can obscure emergent properties, while a good choice of scale can accurately highlight emergent properties. For instance, Chap. 14 describes the importance of the meso-scale as a "Goldilocks" scale where urban–rural differences and infrastructure effects become apparent, but where we also tend to have substantial data and tools available to study FEWS. By contrast, the national scale is too coarse to reveal process, and the **establishment scale** is too fine for the data and tools we have available.

Because of the difference in geographic and temporal scales, meso/macro scale metrics are often more focused on societal goals and outcomes than those at the establishment level. Social issues are often focused on the availability, exchanges, and distribution of resources and outputs from all facilities (e.g., all farms, power plants, and rivers), whereas the owner or operator of a single **establishment** at a single address only has direct control over local operations.

The spatial extent of the impact of a single establishment is compounded by the global nature of supply chains; a single establishment can have regional impacts (e.g., air emissions or water consumption) or even global impacts (e.g., the establishment could use steel from another country).

Another major distinction between the establishment and meso/macro scales is the extent of the temporal boundary used when analyzing the system. At the establishment level, the temporal boundary is defined by the lifetime of the establishment. For example, the analysis of a power plant starts with the extraction of materials used to manufacture the plant and ends with decommissioning and subsequent disposal of these materials. At the meso/macro scale, there are no well-defined beginning and end times (e.g., the steel-production industry has been around for over a century). The extent of the temporal boundary is defined somewhat arbitrarily by the analyst, which for planning purposes is often defined to be a year; for example, the annual water consumption by all power plants within a basin.

12.3.1 Scales and Decision-Makers

As noted in Sect. 2.6.3, in practical situations, the unifying identity of a system is typically the question (or problem) that is being addressed: "how do we manage a system in a certain context and in which food, energy, and water are critical interacting components, to achieve desirable outcomes?" The key word and key phrase in such a question are "context" and "desirable outcomes." Answering the question will require defining issues of spatial and temporal scale (defining its boundaries and external factors) metrics, data, modeling, and computing.

12.3.2 Spatial Scales

Throughout the sciences, it is common to utilize micro-, meso-, and macro-scale approaches. While the meaning of these terms is somewhat different in various scientific fields, there is significant commonality.

Micro-scale approaches look at a system in terms of individual elements. These can be biophysical elements operating under fundamental laws of nature or anthropogenic elements such as consumers, firms, families, workers, or public utilities operating under social forces such as laws, economics, or cultural factors.

Chapter 14 identifies two very common micro-scales at which data is available for FEWS, establishment, and process.

In physics, for example, a micro-scale approach might deal with individual atoms or molecules operating under the fundamental forces of nature, such as electromagnetism and gravity. In contrast, a macro-scale approach in physics looks at large numbers of atoms or molecules operating under laws, such as those of thermodynamics, that describe their averaged emergent properties such as temperature, volume, and pressure. Meso-scale physics typically deals with intermediate systems where micro-scale features are still relevant, but macro-scale approaches are useful in describing the overall system's equilibrium state.

In meteorology, micro-scale studies explore very localized phenomena such as small-scale weather events like individual local storms. This includes understanding flows of heat, water, and gases between adjacent parts of the atmosphere or with local parts of the biosphere or hydrosphere on a scale typically less than 1 km, and approaching the 1–10 m scale where human and ecological microclimates exist on the landscape. In contrast, macro-scale meteorology is usually referred to as "synoptic-scale" meteorology, explores systems on the scale of 1000 km or more where factors like the earth's rotation and related "cyclonic" phenomenon like the Coriolis force and factors like greenhouse gas concentration are important. Intermediate meso-scale meteorology addresses phenomena such as topography, the atmospheric boundary layer, clouds, coastlines, and the cold and warm "fronts" common to weather maps, which give rise to weather conditions over 1–100s of kilometers.

12.3.2.1 Micro-scale

Broadly speaking, **micro-scale** FEW systems are those where the components are largely irreducible and have well-defined conditions and modes of interactions. Micro-scale systems are sometimes referred to as "fine-scale" because of the fundamental and granular consideration of elements. Emergent phenomena (see Sect. 2.2) emerge from micro-scale processes and interactions, but these phenomena become visible only at meso-scales.

Micro-scale **food systems** include the following:

- Individual gardens, fields, farms, and facilities.
- Individual plants, soils, mineral flows, hydrology, local climate conditions, fertilizers, pests and their management, and other local factors.
- Hydroponic and aquaponic systems.
- Food processing facilities that process, convert, and manufacture food products.
- Food storage and distribution.
- Individuals and families who are laborers.
- Individual and household consumers of food.

Micro-scale **energy systems** include the following:

- Energy production sites and firms (e.g., an oil well or biofuel feedstock growth site).
- Energy conversion sites and firms (e.g., a power plant or a refinery or an electric utility company).
- Individual devices, processes, and machines.
- Buildings or industrial facilities like a factory with a well-defined operation.
- Residences and buildings and the individuals who manage their energy systems.
- Government agencies that regulate energy firms and public utilities.
- Individuals who provide labor in energy systems and their labor unions.

Micro-scale **water systems** include the following:

- Hydroponic systems where water acts as a medium for transporting nutrients to plants.
- Water treatment and wastewater treatment facilities.
- Hydroelectric and thermoelectric power plant water systems.
- A variety of industrial facilities in which water flows have a critical function.
- Irrigation systems which can range from a single field to a farm.
- Pumps and canals.
- Water-using processes and machinery.
- Water providing utilities.
- Water transfer, storage, and wholesale infrastructures.
- Individuals and households who are micro-scale producers (e.g., wells, rain barrels) and consumers of water.
- Commercial and industrial buildings.

12.3.2.2 Meso-scale

Meso-scale systems include larger systems where the precise conditions and interactions of irreducible components are important, but averaging is also utilized. This includes systems where large-scale economic and political forces are important such as in cities, counties, and peri-urban areas, which act as hubs of processing, transit, consumption, and, sometimes, production (often referred to a **Network Level** perspective) (see Sect. 1.4.2). Meso-scale systems can also include landscapes, watersheds, and foodsheds. Examples include the following:

Meso-scale **food systems**:

- Landscape systems encompassing many farms or agricultural communities.
- "Foodsheds" that serve a particular population or city.
- City regulations placed on food supply chain quality or sustainability.
- Wholesale, warehouse, and retail supply chains and storage facilities for a particular regional food product or location (recognizing that the elements of a supply chain are micro-scale elements).
- Bulk grain storage.
- Labor organizations of workers in food systems.
- Advocacy groups for farmers, food industries, specific commodities, and consumers.

Meso-scale **energy systems**:

- Human communities like cities, including all of the energy generation, importation, storage, transmission, conversion and consumption within the boundaries of the community.
- Particular energy resources, fuels, and energy products (e.g., electricity systems).
- City regulations placed on energy sourcing, quality, greenhouse gas emissions.
- Major electrical power plant, fuel refinery, bulk fuel storage, and oil and gas field infrastructure.
- Labor organizations of workers in energy systems.
- Government agencies that regulate energy systems.

Meso-scale **water systems**:

- Irrigation systems for agricultural districts.
- Human communities like cities and municipalities, including all of the water production and collection, importation, storage, transmission, distribution, consumption, disposal, and post-disposal treatment within the boundaries of the community.
- City water resource planning including infrastructure and also ownership and management of water resources, water rights, and large water facilities.
- Groundwater systems which drive the evolution of aquifers or the movement of pollutants.
- Aquatic ecosystems where water quantity, quality, and movement impact an essential natural resource.
- Large dams and hydropower facilities.
- Most "water transfers."
- Watersheds, water basins, drainage basins and catchment areas where the water flows in a given area go to a common outlet such as a reservoir or a bay.
- State and regional water policy advocacy groups, along with some national governance.

12.3.2.3 Macro-scale

In **Macro-scale** systems, the average state and the fluctuations around averages and trends in average are dominant considerations, rather than the individual granular details. This includes large economic and political systems, international trading systems, and global systems.

Macro-scale **food system**:

- Large agribusiness food systems which handle a set of crop and animal supply chains serving multiple cities, regions, and/or nations.
- Bulk agricultural products and food product distribution infrastructure.
- National governance of market and nonmarket trade and exchange of food.
- Agricultural and food technology.

- Binational trade and investment in food production.
- International treaties and aid policies for food.
- Global migratory flows of labor in food systems.

Macro-scale **energy systems**:

- National power grids and pipeline networks.
- International energy cartels.
- Energy technology.
- National energy policies (e.g., self-sufficiency).
- International and interregional flows of fuel commodities.

Macro-scale **water systems**:

- Large river systems especially international rivers.
- Water transfers between regions, climate zones, and continental divides.
- Bulk virtual (or embedded) water trading between nations.
- The earth's natural hydrologic cycle that delivers water from oceans to continents as precipitation.
- Water technology.
- National water law and water policy.
- International water policy and aid advocacy groups.

12.3.3 Temporal Scales

Temporal scales are conceptually similar to spatial scales. However, temporal scales tend to be named differently. The most common temporal scales for FEW system processes are Interval, Daily, Monthly, Seasonal, Annual, and Decadal. (Climate, as noted in Chap. 11, is often viewed on scales of hundreds or thousands of years). Models and Data are often available at only one or two of these scales. Process, Micro, Meso, and Macro concepts apply equally to temporal and spatial scales, but the meaning for temporal scales varies with context. Meso often refers to monthly or seasonal data in the temporal context, whereas micro would correspond to interval timescales, and macro to annual or decadal timescales.

The **Interval timescale** is for high-frequency processes, and this kind of data is only collected by digital "smart" data loggers and meters. The power grid is the FEW process that operates natively at interval timescales of seconds or even fractions of seconds. Power grid failures that last more than a few seconds tend to trigger blackouts because these systems operate at nearly the speed of light. Another FEW system at the interval timescale is a building's water use; as toilets are flushed and appliances are operated, the building's water use changes dramatically. Water use spikes in cities at the start and end of every workday. People eat roughly three times per day, and the preparation of that food happens at equivalent intervals. Finally, weather changes at timescales of minutes to hours.

The Daily timescale captures the main periodic rhythm of life on planet Earth: the rising and setting of the sun. Most processes of weather, life, society, and production, both human and natural, change from day to night; production and consumption of FEW goods change dramatically from day to night, for instance when energy use spikes at night in the winter due to lighting and heating usage, or when water use in cities drops nearly to zero at night. Human activities of all kinds change on weekends and holidays, or even during breaks in big events like sports tournaments.

The Monthly and Seasonal timescales are closely related, and Monthly timescale tends to be a human unit of reference for seasonal patterns. Seasons are the other main periodic rhythm of life on planet Earth. The climate varies with seasons, bringing warm, cold, wet, dry, productive, and dormant cycles to our ecosystems. We use a lot more water when it is seasonally warm, and a lot more energy when it is cold (or extremely hot). The water cycle produces a lot of extra water during rainy and wet seasons, and this water is stored in reservoirs to be used by people during warm and dry seasons. The foundation of our food supply chain is grain and vegetable production, and this has traditionally occurred only when the weather is sufficiently warm and wet (e.g., the growing season); these crops must be stored for use the rest of the year following the harvest season. Seasons differ dramatically from place to place, and this is a major reason that humanity's FEW systems and FEW cultures differ from region to region.

The annual timescale is the most natural human accounting timescale for FEW systems because it averages and totals across all of the seasonal and finer-scale events that occur during a year. We normally think in terms of calendar years, but some professionals use custom years; hydrologists use "water years, " which start at the end of the dry and warm season when water storage is lowest. For example, the State of California measures its water year starting on October 1st of each year. Government and private accountants and census-takers are often mandated to perform annual data collection. There are few natural or human processes besides accounting that specifically operate at the annual timescale.

The decadal timescale is important for consideration of long-term trends such as growth or decline of populations and economies, technological change, and climate change. FEW system sustainability and planning problems often exist at decadal timescales.

12.4 Metrics, Data, Models, Computing, and Decisions

There is a clear relationship between metrics, data, mathematical modeling, and computational demands. Any measurable datum and variable within a model may be used as a metric for decision making and to drive action. As depicted in Fig. 12.2, there is a pathway through data collection, model development, understanding, and action with a subsequent feedback loop in the opposite direction. This feedback loop is used to update and confirm that micro-scale physical measurements are consistent with meso and macro-scale emergent system descriptions.

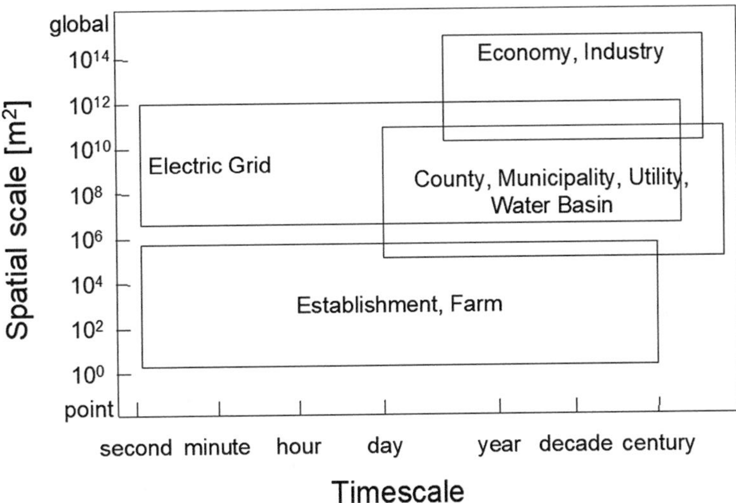

Fig. 12.2 Metrics can characterize various systems or subsystems across spatial (e.g., geographic extent) and temporal dimensions, each of which spans several orders of magnitude

12.4.1 Metrics

We value what we measure, and we measure what we value. Metrics represent both data from measuring the physical world and data representing the values of stakeholders within society. Thus, metrics can be represented with units that are a mix of both concepts, such as the quantity of water consumed (physical unit) relative to the gross domestic product (an emergent, values laden economic accounting metric).

We use metrics to inform stakeholders on how to interpret and use models as well as to verify that models accurately reflect and use data measured from the physical world. Thus, there is tremendous utility in well-defined and consistent metrics that all (or most) stakeholders agree to use for decision-making and modeling (Fig. 12.3).

12.4.2 Data

Data are collected to answer questions and must be collected specifically to address the research questions. However, there are also questions we can ask about data. What is the quality and structure of the data? Are the data accessible? At what scale and **resolution** are the data available?

After one carefully identifies the scale at which a FEWS question or problem must be studied and managed, one quickly discovers that complete systemic data is not fully available at that scale. This is arguably the most severe practical barrier to the study of FEWS.

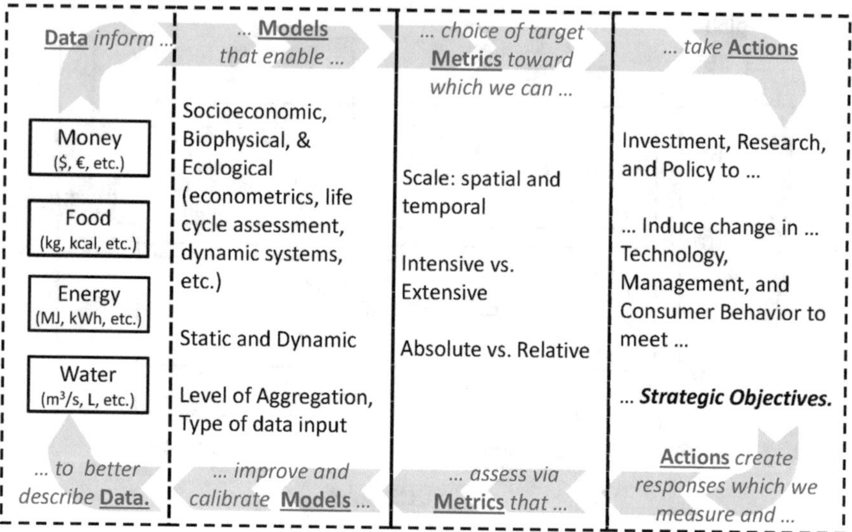

Fig. 12.3 Continuous feedback between data, models, metrics, and actions that seek to reach the strategic objectives of various FEW stakeholders

Chapter 14 provides a detailed treatment of the issues surrounding data availability and data management for FEWS. Data is generated both by observations and models. This data has attributes of structure, quality, scale, and resolution. Resolution pertains to the level of detail with which the data describes the spatial, temporal, and process-level variations in the system, and is subtly distinct from scale in concept.

The data management life cycle is important to master in order to ensure that other researchers and decision-makers can access the data that you generate. Data and privacy ethics are the major concern when collecting FEWS data and must be carefully attended to; in many cases, it is ethics rather than other factors, that limit the availability of micro-scale data for FEWS work. Depending on what scale your work requires, you should be aware of some common problems with data availability at that scale; Chapter 14 reviews these problems along with the other issues summarized above.

12.4.3 Models

An integrated approach to modeling of FEW systems that is able to address the research questions and stakeholder needs will need not only to couple modeling tools for food, energy, and water systems but connect to decision science, data integration, and **visual analytics**. The logic of such an integrated approach is depicted in Fig. 12.4.

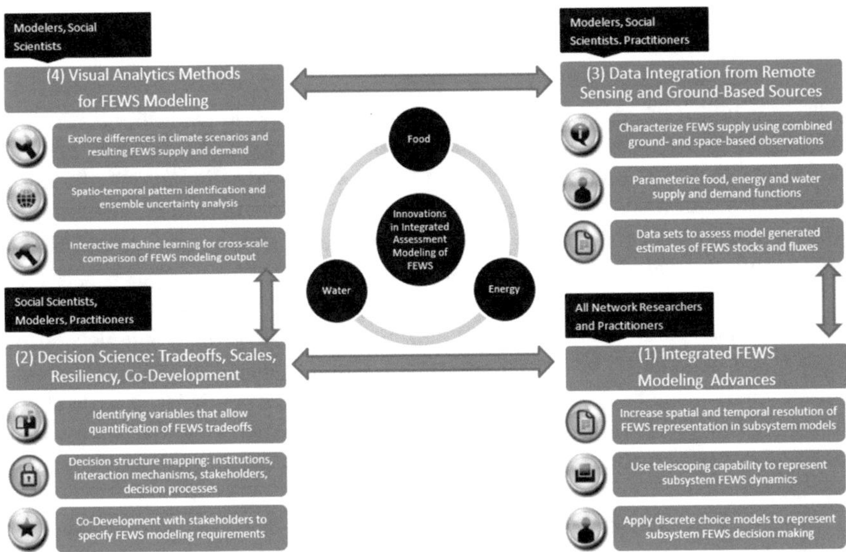

Fig. 12.4 Logical diagram for the integration of advances in FEW system modeling

In Chap. 1, we reviewed seven different framings of the nexus of FEW systems that allow the analysis of FEW systems in a coupled fashion. In Chap. 2, we showed how a system science approach allows recognition of internal (endogenous) interactions and feedbacks among the food, energy, and water systems, while other variables are included to recognize external (exogenous) drivers on the system.

Planning and development challenges at the nexus of FEW systems are likely to involve other factors such as land-use, urbanization, demographics, and environmental protection. A number of data and modeling platforms have been developed to support the assessment of FEW sector developments under different economic and environmental policy conditions, and to support integrated resource development in the different sectors. Typically, however, these data and modeling tools are designed for different purposes, and linkages between the food, energy, and water sector development are limited. Moreover, the level of technical detail and complexity in the models can preclude their application for upstream sector strategy development, a crucial analytical need in development planning.

In Chap. 2, we also noted that bottom-up and top-down approaches can be envisioned in developing an approach for integrated modeling of FEW systems, particularly with respect to the needed economic analysis. For instance, the output from the different food, energy, and water system models can be incorporated into an economic model that enables us to look at different policy options. This first approach involves bottom-up model coordination that draws on and extends existing system models. A second, top-down approach, can be focused on extending economy-wide **computable general equilibrium (CGE) models** to account for food, energy, and water as both factors of production and consumption goods, including aggregation of model output to CGE sectors and regions.

A hybrid FEW system model development approach could focus on a more flexible modeling framework that combines the strengths of the bottom-up and top-down approaches. This would recognize the distinctive dynamics of FEW systems and their interactions, and overcome a number of inherent limitations of linking individual sector models or the CGE framework to explore the nexus. These limitations are likely to become increasingly problematic for policy making (and associated analysis) as interactions and pressures increase.

In Chap. 2, we explored FEW systems as complex systems characterized by heterogeneous parts with complex interactions that make them interdependent, coevolving, and subject to distributed control. Unsurprisingly, modeling FEW systems is challenging because interactions are complex and nonlinear, operate at different time and spatial scales, and are characterized by dynamic interactions between physical systems and the institutional and social systems that interact with and manage them.

We also noted that the characteristics of the whole system emerge from the interactions between the components of the system giving rise to stability and changes that are often hard to predict and sensitive to initial conditions. Thus, integrated models of FEW systems can produce insights and discoveries that do not emerge from research on food or energy or water systems alone; the synergy among these components provides pathways to produce new knowledge and practical applications to solve the challenges of FEW sustainability and security.

All framings have a shared scientific challenge in integrating and providing coherence to the nexus—disparate spatial and temporal scales that are inherent to physical, biological, social/behavioral processes within an integrated FEW system. The diversity of scales also applies to interactions and feedbacks between components. The scales associated with external stressors (e.g., climate) may be quite different from those of impacts (e.g., regional droughts, impacts on agricultural production, energy shortages).

Continuing challenges for integrated modeling in this emerging field are identifying and addressing shared needs of FEW systems stakeholders, facilitating tailored analyses over different geographical regions, computing in varying spatial and temporal scales, improving system efficiencies, and addressing FEW system vulnerabilities and resilience to human and natural stressors. Chapter 15 expands on this topic.

12.4.4 Computing

Computer science is a set of scientific tools and techniques that have revolutionized how science is being conducted. For the last decade, one of the main reasons for the snowball effect in scientific breakthroughs was the use of computational tools and techniques that made large and complex computations possible.

Although it is often underappreciated, computer science is one of the driving forces of FEW research. From the collection of data using sensors and tools to the conversion of these data to machine understandable code, as well as restructuring and indexing the data for efficient storage and queries, computer science plays a big role in FEW Nexus research. In addition, the manipulation of this data such as merge, join, sort, aggregate, and analytics on these such as machine learning and statistical analysis are made available by computer science. For example, the raw analog temperature data collected from a wireless sensor is converted to bits, these are stored in a short-term memory on the device, transferred through computer network by encapsulating in a specific network protocol, checked for the correctness of the transfer by the computer algorithms on the receiver side, then transferred to a database table by specific tags/column names. Once at the database table, this data can be indexed by the location, time, and so on to make faster retrieval and queries possible. Similarly, once queried, machine learning algorithms can provide statistical insights as well as predictions (Fig. 12.5).

Similar in many domains, for specific purposes such as FEW nexus data collected from a variety of sources, different computer science approaches were developed through time to improve efficiency and scalability when faced with large specialized datasets. As an example, from the FEW nexus domain, due to the location-aware nature of remote sensing imagery, storage, manipulation, and analytics on these data is different from classical photographs taken by a handheld device.

Spatial datasets impose complex interactions in space and time, as well as correlations and relationships between locations. Tobler's first law of geography states that "Everything is related to everything else, but near things are more related than distant things." Thus, those specific properties of spatial data should be taken into account when performing data science tasks. First, spatial big data exhibits spatial autocorrelation effects. In other words, we cannot assume that nearby samples are statistically independent. Thus, data analysis techniques that ignore spatial autocorrelation may perform poorly, such as low prediction accuracy. Second, spatial inter-

Scientific Methods

		Computation
Causal& Mechanistic Models	**Theory** Causal Model Building Mechanistic Model Building	**Numerical Solutions** Simulations (known as 3rd Leg of Science) Example: GCM, SWAT, etc.
Data-driven Empirical Models	**Controlled Experiments** Experimental Design, Statistical Analysis - Hypothesis Testing and Ranking	**4th Paradigm** Data Intensive Computing, Big Data, Data Science - Data driven hypothesis generation

Fig. 12.5 Scientific methods and the role of computing

actions are not isotropic and vary across directions. Third, spatial dependency exists in multiple spatial scales. Finally, spatial big data exhibits heterogeneity, that is, identical feature values may correspond to distinct class labels in different regions. Thus, learned predictive models may perform poorly in many local regions. Depending on the scale, these effects may get exasperated. Moreover, scale may affect the computational performances of algorithms that were used for FEW research. Therefore, researchers are encouraged to find a correct balance between performance and accuracy, depending on needs.

Key Points

- FEWS questions tend to be inherently multiscalar, but there are key scales at which data, models, metrics, and processes are more readily available.
- Pasteur's quadrant for use-inspired research is aspirational for FEWS practitioners and scholars.
- Scales can be important to determine in framing, researching, and implementing key FEW challenges and solutions.
- Spatial scales include establishment, process, micro, meso, and macro.
- Temporal scales include interval, daily, monthly/seasonal, annual, and decadal, although scales of thousands to millions of years to arise in climate modeling.
- Scale takes on a unique meaning for metrics, data, models, and computing.
- Metrics, data, models, and computing each answer important questions for FEWS work.

Discussion Points and Exercises

1. What are the characteristics of a well-defined question or study in a FEWS context?
2. Identify a FEWS question that you are very interested in.

   (a) What is the best scale of space and time at which to study that question, considering what you know about the physical and human processes involved in this part of the system?
   (b) What is the best scale of space and time at which to study that question, based on what you know about the availability of data and tools?
   (c) Do the answers to (a) and (b) above match, and if not what can you do to close the gap?

3. If you are doing research work on FEWS, identify which part of Pasteur's quadrant that work most nearly belongs to, and explain why. What could you do to move your work into Pasteur's quadrant?
4. Do you agree with our characterization and nomenclature for the primary scales in a FEW system? Is something missing?
5. What questions are not being asked about FEWS right now?
6. Consider and discuss whether you think basic or applied research is more important and/or more attractive to you, personally—and why. Does your discipline, occupation, or program provide a different answer to this question from your personal answer?

Further Reading

Asian Development Bank. (2013). *Thinking about water differently: Managing the water, energy and food nexus*. Philippines: Asian Development Bank.

Boyd, D., & Crawford, K. (2012). Critical questions for big data: Provocations for a cultural, technological, and scholarly phenomenon. *Information, Communication & Society, 15*(5), 662–679.

Cambridge Econometrics (CE) and Sustainable Europe Research Institute (SERI). (2010). *A scoping study on the macroeconomic view of sustainability*. Final report for the European Commission, DG Environment.

Flandrin, P. (1998). *Time-frequency/time-scale analysis* (Vol. 10). San Diego: Academic Press.

Herod, A. (2003). Scale: The local and the global. In *Key concepts in geography* (Vol. 229, p. 234). Thousand Oaks, CA: Sage.

Liverman, D. (2004). Who governs, at what scale and at what price? Geography, environmental governance, and the commodification of nature. *Annals of the Association of American Geographers, 94*(4), 734–738.

Marston, S. A., Jones, J. P., III, & Woodward, K. (2005). Human geography without scale. *Transactions of the Institute of British Geographers, 30*(4), 416–432.

McCann, E. J. (2003). Framing space and time in the city: Urban policy and the politics of spatial and temporal scale. *Journal of Urban Affairs, 25*(2), 159–178.

McLaughlin, D., & Kinzelbach, W. (2015). Food security and sustainable resource management. *Water Resources Research, 51*, 4966–4985. https://doi.org/10.1002/2015WR017053.

Paasi, A. (2004). Place and region: Looking through the prism of scale. *Progress in Human Geography, 28*(4), 536–546.

Preston, B. L., King, A. W., Ernst, K. M., Absar, S. M., Nair, S. S., & Parish, E. S. (2014). Scale and the representation of human agency in the modeling of agroecosystems. *Current Opinion in Environmental Sustainability, 14*, 239–249.

Schneider, D. C. (2001). The rise of the concept of scale in ecology: The concept of scale is evolving from verbal expression to quantitative expression. *AIBS Bulletin, 51*(7), 545–553.

Stokes, D. (1997). *Pasteur's quadrant: Basic science and technological innovation*. Washington, DC: Brookings Institution Press.

Voinov, A., & Cardwell, H. (2009). The energy-water nexus: Why should we care? *Journal of Contemporary Water Research & Education, 143*, 17–29.

Zaheer, S., Albert, S., & Zaheer, A. (1999). Time scales and organizational theory. *Academy of Management Review, 24*(4), 725–741.

Chapter 13
Metrics

Michael Carbajales-Dale and Carey W. King

13.1 Introduction

To mitigate the consequences of FEW interdependences and to guide policy action, decision-makers and stakeholders can benefit from using clearly developed indicators and performance metrics.

This chapter presents a high-level framework to categorize FEW metrics; demonstrate how different metrics might be favored over others; and explain how metrics and models are used to inform and direct actions.

Decision-making and planning are not only about numbers but also honest, clear language to communicate the science and data correctly.

See Chaps. 4 and 20 for discussion of how people adapt and collaborate in the face of FEW stresses. Chapter 14 discusses the various types of data and their challenges that are the core inputs for metrics. Chapter 15 discusses interdependent modeling which simulates outcomes that can be calibrated and compared to historical data and metrics, as well as establishes future possibilities for meeting target goals defined by metrics via stakeholder engagement.

M. Carbajales-Dale
Environmental Engineering & Earth Sciences, Clemson University, Clemson, SC, USA
e-mail: madale@clemson.edu

C. W. King (✉)
Energy Institute, The University of Texas at Austin, Austin, TX, USA
e-mail: careyking@mail.utexas.edu

P. Saundry, B. L. Ruddell (eds.), *The Food-Energy-Water Nexus*, AESS Interdisciplinary Environmental Studies and Sciences Series,
https://doi.org/10.1007/978-3-030-29914-9_13

13.1.1 The Importance of Metrics

> We try to measure what we value. We come to value what we measure.
>
> —Donella Meadows (Meadows 1998)

Metrics are useful to measure what we value and facilitate effective stakeholder communication, engagement, and decision-making around FEW activities, regulations, and targets. As suggested by the quote above, there are two primary intents of metrics:

> First, metrics attempt to capture what society values.
>
> Second, society is itself molded by the ongoing effort to bestowing the measured quantities with greater value.

There is, therefore, a purposeful dimension to this act of measurement: a hope to mold society in a specific way, in an attempt to better align societal actions to specific values. An example familiar to most people is the use of atmospheric carbon dioxide concentrations as a metric by which to discuss and guide action on climate change mitigation (Chap. 11). Well-defined metrics are crucial for the ability of stakeholders and decision-makers to sift through competing arguments for and against different FEW nexus policies. However, different stakeholders might emphasize one set of metrics over another to focus attention on what they deem most important. Figure 12.3 shows how metrics enable conversations regarding how data we measure translate to values of different stakeholders. Metrics are more reliable when more stakeholders agree that they accurately summarize both data and values, and vice versa.

In the economic sector, gross domestic product (GDP) is a widely used metric of economic development and a significant driver of economic policies. However, GDP is deficient as a measure of other important societal concerns such as quality of life and environmental conditions (see Sect. 3.4). As a result, excessive use of GDP can result in policies which increase GDP but have detrimental effects on other areas (Heun et al. 2015). In recent decades, efforts have been made to extend the assessment of economic activities beyond the simple accounting of market prices (see Chap. 5).

Certain computational frameworks help quantify societal energy and material investments and environmental impacts associated with the provision of goods and services. These computational methods help translate data from one domain (e.g., water) to another (e.g., energy), and can enable the definition of metrics useful in both domains. Examples of such computational frameworks include **Net Energy Analysis** (NEA) (Slesser 1974) and **Life Cycle Assessment** (LCA) (ISO 2006) (see Sect. 13.2.1) which seek to comprehensively quantify the net impact of all activities and impacts of a process. The State of California for example has incorporated LCA into evaluating its Low Carbon Fuel Standards. Such frameworks can be implemented at both the establishment (micro) and meso/macro scales.

Similar extensions to industry- and economy-wide accounting frameworks have been developed in fields like **material flow analysis** (also called substance flow analysis) which look at the flows and stocks of certain materials/substances in a system.

The United Nations **System of Integrated Environment-Economic Accounts** (SEEA) is a "framework that integrates economic and environmental data to provide a more comprehensive and multipurpose view of the interrelationships between the economy and the environment and the stocks and changes in stocks of environmental assets, as they bring benefits to humanity" (UN 1993). SEEA uses market exchange information (i.e., financial intersectoral payments) to infer biophysical flows of material and energy. Since the publication of the SEEA, many countries and states have developed their own versions. However, there is currently no biophysical metric analogous to GDP, although candidates include *domestic processed output* and *total domestic output*—both measures of the total mass of material flowing through an economy (Matthews et al. 2000).

The **ecological footprint** has become a popular indicator of the environmental impact of economic activities at multiple levels by parsing the environmental impacts of a person, community, or activity in terms of the natural land area required to service the activity (Wackernagel & Rees 1998). The global human population is currently living beyond the planet's available land area (Rees & Wackernagel 2013).

Other methods account all economic expenditures positively (i.e., as income), to GDP, even when that activity constitute negative societal costs such as the cleanup of polluted sites. Such spending should be subtracted from GDP (Daly 2013).

The **Genuine Progress Indicator** (GPI) measures both the benefits and costs of a region's economic activity to determine when further growth becomes uneconomic (Lawn 2003). Studies indicate that for many industrialized countries, GPI is stagnant or even declining despite increases in GDP (Kubiszewski et al. 2013).

As was noted in Chap. 12, there are several considerations involved in metric development including spatial and temporal issues as well as quantity and quality issues.

13.1.2 Metrics and Models

The choice of model used to characterize the system will influence what type of metrics are used and eventually any action that is undertaken. This also occurs in the opposite direction where the type of metric used will determine what sort of data is collected and subsequently what sorts of models can be built.

Box 13.1 Thermoelectric Power Plants

Power plants can be made more efficient at establishment scale. Drivers for improving efficiency are from wholesale markets (global scale) and regulations (establishment and global scale). Operation of a plant is dictated by global scale interdependence (e.g., wholesale market, optimal least-cost dispatch). Water and food impacts are from the price of electricity and direct water use by thermal plants. Ultimate feedback to improve establishment operations comes from its operation within a global context (e.g., if it does not operate enough, it needs to improve or retire).

In the design and operation of a thermoelectric power plant, fuel efficiency (and thus water intensity) is an overarching concern. Process engineering models optimize the operation of the plant to maximize efficiency, and thereby minimize cost. Due to past improvements in fuel efficiency of thermal power plants, increased gains come at significant diminished returns. Most of the reductions in the water intensity of thermal power generation have come from new designs, such as combined cycle power plants, as well as use of non-thermal technologies such as wind power and solar photovoltaics.

However, a significant consideration in determining the actual operation of the plant is the marginal cost of electricity as dispatched by the independent system operator (ISO), who is unconcerned with plant efficiency improvements, that is, plant characteristics are fixed for short-term dispatch.

The metric of interest (in a wholesale electric market) is the market price of electricity relative to the marginal cost for the power plant, irrespective of plant efficiency.

13.1.3 Metrics and Data

Data availability can constrain what models have been developed, which may then indirectly impact future data collection (hence availability) through the mechanism depicted in Fig. 12.3 by limiting the models and metrics that can be developed in the future. Current data collection methods and practices entail constraints on what metrics are derived, used, and accepted. Metrics only based upon modeling might not be as readily accepted as those that can be calculated using measured data (e.g., streamflow).

Consistent state, regional, and national-level data are usually not collected at greater than annual frequency. This is, however, highly dependent on the type of data. For example, in the USA, there is currently no single agency charged with collecting nation-wide data on water withdrawal and consumption, which is generally only available from the U.S. Geological Survey at 5-year intervals. There is now an effort to integrate various water models through a new National Water Model, which went online in 2016. Food (agriculture) and energy (with their own federal departments) have better data availability, often on monthly timescales.

In the electric power sector of the USA, Independent System Operators (ISOs) operate in many parts of the USA to coordinate, control and monitor the operation of the electrical power system. ISOs have a need to monitor electric grids at time scales of seconds to minutes, and some (such as California's CAISO, and Texas' ERCOT) make this information publicly available at those resolutions. County data are sometimes available, but data at finer spatial detail (e.g., zip code or block group) is generally not available due to disclosure or **data protection** concerns. Further, the Energy Information Administration makes a plethora of data available via its website that describe flows of primary energy and energy carriers such as electricity.

13.1.4 Metrics and Computing

Power and availability of computing resources can also constrain which metrics and models can be used. Sometimes computing resources might constrain the requisite complexity with which a model can represent the real-world system, requiring a more simplified model or spatial or temporal resolution may need to be sacrificed.

For example, attempts to create regionally specific climate models (requiring more detailed spatial resolution) were postponed until sufficiently powerful (and low cost) computing resources were available. Some high-resolution and stochastic phenomena, such as cloud cover and rainfall, might never be *modeled or measured* at the desired resolution; thus, modeling results and data will always be approximations. Whatever the limitations, care must be taken when extrapolating (or interpolating) results beyond the resolution at which they were generated.

13.2 Methodological Frameworks

In this section, we develop a methodological framework for establishing a suite of FEW metrics based on system science principles and life cycle assessment (LCA). This will allow us to develop clear both the ***scale***, or scope (see Sect. 13.2.2), and ***taxonomy*** (see Sect. 13.2.3) of any given metric.

13.2.1 Life Cycle Assessment

Life cycle assessment (LCA) is a methodological framework for assessing the environmental impacts associated with a product, service or activity. Within LCA, an analysis is typically restricted to the product or establishment level using a process-based approach, built "bottom-up" from engineering models of individual unit processes, for example, a coal-fired burner. We might extend this approach to a regional level by aggregating several facilities.

An additional, and often complementary, approach is based upon environmental extensions to **economic input-output (IO) tables**, which allow a comprehensive snapshot of the whole economy in a particular year. A further distinction is made between attributional and consequential analyses.

An **attributional analysis** is a descriptive or comparative analysis of how the world currently is, that looks at the impact of the product system under the assumption that the background economic system remains constant (e.g., the environmental impact of one power plant using carbon capture and storage (CCS) technology).

In contrast, a **consequential analysis** looks at how the world *could be* by determining the impact of production under changes to the broader economy, for example under a scenario that all coal-fired electricity use CCS. In this case, we would to dramatically scale up CCS technology, requiring significant structural changes within the economy.

13.2.2 Metric Scale

Given the identification of metrics at the establishment level (e.g., from LCA), two questions arise: are these metrics appropriate at the meso/macro scale, and how should we translate metrics between the establishment and meso/macro scales? In its simplest form, one can aggregate each establishment-level average or marginal impact, such as greenhouse gas (GHG) emissions per gallon of treated water, by the total production of water from each establishment. In this case, any error or uncertainty at the establishment level would be transmitted to the meso/macro scale.

The computational structure of LCA provides a useful framework for defining metrics pertinent to the FEW nexus. Within this structure, we can define two matrices:

A: A square $n \times n$ matrix (**A**) depicting the technological network of interacting processes and facilities, with n columns representing processes and n rows representing products or services being exchanged.

B: An $n \times m$ matrix (**B**) again with n columns representing processes but with m rows representing interactions occurring outside the set of modeled processes—in other words, within larger-scale economic, social, or environmental domains.

Assuming a vastly simplified technological network composed of only three processes (food, energy, and water production) generating three highly aggregated products (food, energy, and water), as depicted in Fig. 13.1, we can quantify metrics defined by the economy and broader social and environmental contexts.

Having defined the distinct scales of analysis as establishment and meso/macro and having identified that LCA enables tracking of the multiple interactions between the three key elements of food, energy, and water, we now turn to other considerations regarding different categories of metrics.

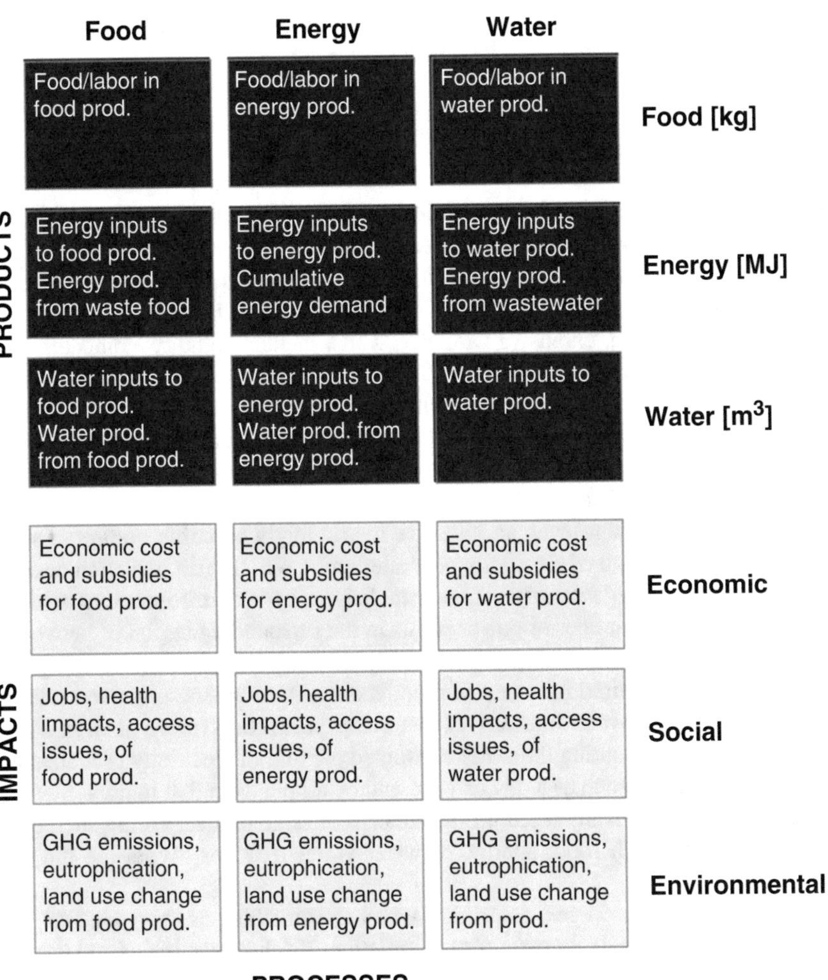

Fig. 13.1 Metrics can be defined using the structural framework of LCA (Heijungs & Suh 2002). Food, energy, and water production (columns) require inputs and output products (green rows) as well as producing economic, social, and environmental impacts (pink rows). The combination of several production processes forms a network (green matrix) with associated environmental impacts (pink matrix)

13.2.3 *Metric Taxonomy*

As stated previously, the purpose of metrics is to guide action in the hopes of more closely aligning activity with strategic objectives (see Fig. 13.1). For instance, one of the first steps taken to manage water in integrated water resources management is an accounting of current water budgeting to set target water budgets for each establishment or network that operates in a water basin.

Two dimensions can help define the purpose of an existing or proposed metric: (A) *intensive* vs. *extensive* and (B) *absolute* vs. *relative*.

13.2.3.1 Intensive vs. Extensive

In thermodynamics, **intensive** properties are those whose value is independent of the amount of material present. Examples are temperature, pressure, and mass density. The density of water is the same (1000 kg/m^3 at standard temperature and pressure) whether you have filled a bathtub or one of the Great Lakes.

On the other hand, **extensive** variables, which include volume, momentum, and mass, change depending on the quantity of material present. For example, the mass of water in your bathtub, measured as density multiplied by the water's volume, is much lower than the mass of the water in Lake Michigan, despite having the same density.

We draw a parallel from thermodynamics to describe **technological metrics** and **socioeconomic metrics**.

At the establishment level, an intensive metric might be **cubic meters** of water consumed *per* kilowatt of electricity generated [m^3/kWh]. In fact, we often describe such metrics as *water intensity*. At the establishment scale, this intensive metric is independent of the number of power plants in the surrounding region or network.

The inputs into the LCA framework in Fig. 13.1 are **intensive metrics** (e.g., the amount of input required for a unit of product output). At the meso/macro scale, we might calculate the water intensity of electricity production (again as m^3/kWh) in each region, by accounting the water consumed per unit of electricity produced in a certain time period, such as a day or year, and by all plants in that region.

Extensive metrics are based on the concept of scale or size, asking such questions as, "How much food, energy, or water is desired?" or "What are the total inputs needed?"

Intensive metrics do not determine whether something is "too big" or "not enough," as they are largely metrics of the relationships between inputs and outputs. In contrast, extensive metrics help understand 'how much' *relative to* what is known about total availability or limiting quantity (see Sect. 13.2.3.2 below).

Examples of extensive metrics include the total water consumption by a thermal power plant over its lifetime or within a given year (establishment-level), and total water consumption for electricity produced for all power plants within a given region in a given year (meso/macro scale).

Those familiar with life cycle assessment will notice that at the establishment level the difference between intensive and extensive metrics is a change in the functional unit (e.g., what is our reference flow).

Equation (13.1) represents a simple way of remembering the relationship between extensive and intensive metrics, where q and y are extensive metrics with-distinct units, and $\frac{q}{y}$ is an intensive metric that relates them. Typically q is an input and y is an output.

$$q = \frac{q}{y} y \qquad (13.1)$$

The computational structure of LCA can be used to aggregate inputs and impacts of a process, or a mix of processes that are part of a network, at various time scales depending upon how each process is defined (Heijungs & Suh 2002). The columns of matrix **A** can represent a set of unit processes within an establishment to assess establishment metrics, or they can represent establishments that are interacting within a network. Further, it is computationally trivial to combine both concepts into a single matrix model.

For example, consider a power plant in a water basin such that there is interest in how much water consumption is allocated to the power plant from within the water basin it resides. Consider two LCA processes: one for constructing the power plant (a "one-time" occurrence of construction) and an additional process (among possible others) for operating the power plant for an output of 1 kWh ("continuous" material flows during operation). Constructing the power plant is one functional unit, and operating the power plant to produce 1 kWh is another functional unit. There is water consumption associated with constructing the power plant, say 100 million gallons (Mgal), an extensive metric (and "absolute," see next section). Thus, before the power plant has begun operation because it has yet to produce electricity output, its kWh water intensity, an intensive metric, is infinite (i.e., 100 Mgal divided by 0 kWh). Further, this water consumption might have occurred in another water basin far away.

Consider that during operation the plant consumes 0.5 gal of water for each kWh generated, an intensive metric that largely describes the water requirements to cool the steam cycle of the power plant, and that it generates 20 million kWh each year.

It thus consumes 10 Mgal/year (an extensive real-time metric) during operation in the water basin in which it resides.

If the power plant operates for 10 years, its total water consumption is 100 Mgal + 10 Mgal/year × 10 years = 200 Mgal, and it generates 200 million kWh, for 10-year water intensity of 200 Mgal/200 million kWh = 1 gal/kWh. Half of the water was for construction and the other half for operation.

If the power plant operates for 40 years, the average water intensity is 100 Mgal + 10 Mgal/year × 40 years = 500 Mgal, and it generates 800 million kWh, for a water intensity of 500 Mgal/800 million kWh = 0.625 gal/kWh. Here one-fifth of the water was for construction (the same 100 Mgal), and four-fifths was for operation.

Thus, the longer the assumed operational lifetime of the power plant, the closer the water intensity of electricity (gal/kWh) of the entire power plant life cycle becomes to the operational water intensity. The distinction between the temporal flows of both inputs and outputs is crucial such that lifetime LCA metrics are not confused with real-time LCA metrics.

This example demonstrates three points:

1. Environmental impacts can be local, regional/global, or both.
2. Full life cycle intensive metrics approach the real-time (operational) intensive metrics the longer the establishment operates (or alternatively, the life cycle and real-time metrics converge by the end of the life cycle).
3. How the definition of the functional unit changes when representing an intensive versus extensive metric.

13.2.3.2 Absolute vs. Relative

The second dimension we use for defining metrics is *absolute* vs. *relative*. **Absolute metrics** have units of some sort (cubic meters, kilowatt-hours, etc.), whereas **relative metrics** are dimensionless, essentially a rephrasing of absolute metrics relative to some normalization factor.

At the establishment level, absolute metrics can be converted to a relative metric by scaling to an industry standard (e.g., best practice) or a target performance metric (e.g., Carnot efficiency). For example, the Corporate Average Fuel Economy fuel efficiency and emissions standards for company-wide vehicle sales determine whether a company's vehicles produce emissions at levels below that of the target standard as well as its competitors.

Box 13.2 Boundaries Are Not Always Political

We are often used to thinking of "boundaries" in terms of political boundaries between countries or between other governmental jurisdictions (e.g., counties, states). However, within the environmental sustainability literature, there is a concept of 'planetary boundaries,' or sustainable limits defined by scientific research and translated via government or industry policy (Rockström, J., et al. 2009, Heijungs et al. 2014). Ordinarily, LCA studies produce intensive characterizations of products or services that do not address whether the product or service fits within sustainable consumption patterns. The imposition of these so-called planetary boundaries enables the determination of a sustainable pattern of consumption for the products and services under analysis.

The term "planetary" is also somewhat misleading since in some cases the spatial region would be more restrictive, depending on the impact being analyzed. This is most clearly demonstrated in the case of water management, where the spatial region of interest would be the extent of the watershed where the processes and activities.

Limits are most appropriately defined at the meso/macro scale as an absolute metric (e.g., water supply, availability, or precipitation within the watershed). One way to achieve the imposition of the limit is using the distance-to-target approach, by which the weighting factor of the impact is scaled as one approaches the limit. For example, to determine stress on a resource, we can scale by a factor f:

$$f = \frac{D}{A} \tag{13.2}$$

where A and D are availability and demand, respectively, of food, water, energy, or another resource of interest (Boulay et al. 2015).

An example water stress indicator is the amount of water available in a region (e.g., water in a basin as typical rainfall, flow rates, etc.) relative to water demand (e.g., total evapotranspiration for crops and irrigated lawns, evaporation from thermoelectric power plant cooling). Therefore, the impact of water use increases as the limit is approached, that is, as $A \rightarrow D$.

It is the knowledge of how close a system is to reaching one or more boundaries that provides feedbacks for prompting operational changes or new actions. Chapter 19 discusses three case studies of large watersheds. These case studies demonstrate how each basin is experiencing effects from one or more specific limits, or boundaries, that influence both the withdrawal and consumption of water from the basin as a source as well as the discharge of pollutants into the basin as a sink. Other non-water limits, such as deforestation goals, also affect water infrastructure and use, such as via hydropower dams in the Amazon basin.

Thus, the use of relative metrics is crucial in discussing the management and distribution of limited resources. A pure focus on absolute metrics avoids discussion of the distribution of resources because there is no conceptual limit on the absolute quantity of that resource (i.e., there is no need to discuss how to distribute an infinite quantity).

Ultimately, the definition of establishment-level performance targets raises issues regarding the distribution of resources, impacts, and accountability within the larger context of the network or region being analyzed. This distribution of access and accountability is often the focus of decision-making and planning and often makes consensus difficult, particularly in the context of historic access rights, such as those to water. This now prompts a discussion of how metrics—both absolute and relative, both intensive and extensive—fit into the context of informing action.

13.2.3.3 Summary of Intensive–Extensive and Absolute–Relative Metric Combinations

Given the two categories of metrics from A and B, there are effectively four possible combinations of metrics at the establishment or meso/macro scale.

Table 13.1 shows examples of each combination of metrics in the context of the FEW nexus.

Absolute–intensive metrics are those that have the form of "unit q/unit y." Generally, but not always the numerator (unit q) quantifies an input to a process, and the denominator (unit y) quantifies an output from the process (as depicted in Fig. 13.1 and Eq. (13.1)), but an output/input framework can also be used, as in efficiency metrics. In the context of biophysical units within FEW systems, there can be nine combinations of metrics (energy–water, water–food, food–energy, etc.). There can also be socioeconomic metrics with non-FEW units, such as financial cost. An example of an absolute–intensive metric is the electricity input per unit of treated water [kWh/m^3] produced by a reverse osmosis desalination plant.

An absolute–intensive metric can be dimensionless, and **energy return on investment (EROI)**, which has units of energy output per unit of energy invested, is an example. These types of metrics help understand how a technology or management practice might impact an input-output relationship: What type of water

Table 13.1 A two-by-two matrix comparing the different types of metrics between the absolute–relative and intensive–extensive dimensions

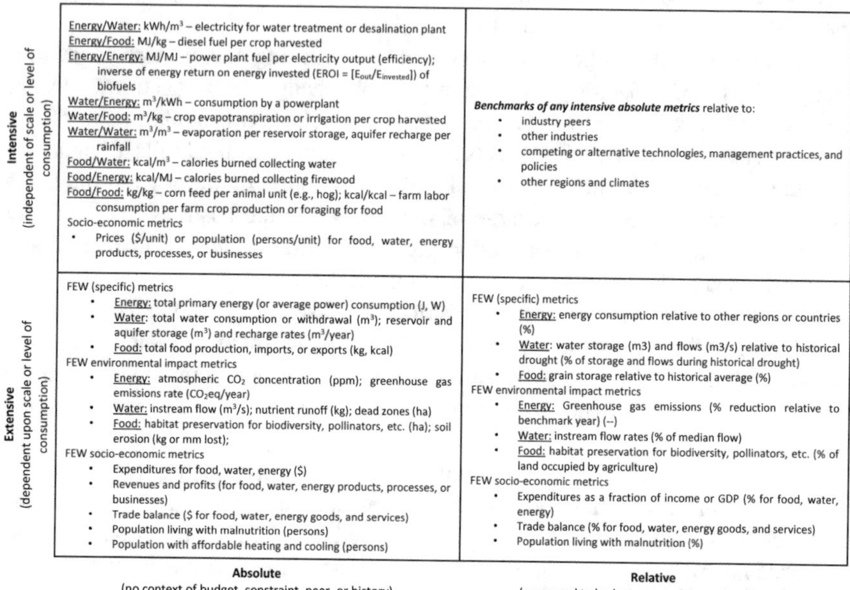

<table>
<tr><th></th><th>Absolute
(no context of budget, constraint, peer, or history)</th><th>Relative
(compared to budget, constraint, peer, or history)</th></tr>
<tr><td>Intensive
(independent of scale or level of consumption)</td><td>Energy/Water: kWh/m³ – electricity for water treatment or desalination plant
Energy/Food: MJ/kg – diesel fuel per crop harvested
Energy/Energy: MJ/MJ – power plant fuel per electricity output (efficiency); inverse of energy return on energy invested (EROI = [E_{out}/$E_{invested}$]) of biofuels
Water/Energy: m³/kWh – consumption by a powerplant
Water/Food: m³/kg – crop evapotranspiration or irrigation per crop harvested
Water/Water: m³/m³ – evaporation per reservoir storage, aquifer recharge per rainfall
Food/Water: kcal/m³ – calories burned collecting water
Food/Energy: kcal/MJ – calories burned collecting firewood
Food/Food: kg/kg – corn feed per animal unit (e.g., hog); kcal/kcal – farm labor consumption per farm crop production or foraging for food
Socio-economic metrics
• Prices ($/unit) or population (persons/unit) for food, water, energy products, processes, or businesses</td><td>Benchmarks of any intensive absolute metrics relative to:
• industry peers
• other industries
• competing or alternative technologies, management practices, and policies
• other regions and climates</td></tr>
<tr><td>Extensive
(dependent upon scale or level of consumption)</td><td>FEW (specific) metrics
• Energy: total primary energy (or average power) consumption (J, W)
• Water: total water consumption or withdrawal (m³); reservoir and aquifer storage (m³) and recharge rates (m³/year)
• Food: total food production, imports, or exports (kg, kcal)
FEW environmental impact metrics
• Energy: atmospheric CO_2 concentration (ppm); greenhouse gas emissions rate (CO_2eq/year)
• Water: instream flow (m³/s); nutrient runoff (kg); dead zones (ha)
• Food: habitat preservation for biodiversity, pollinators, etc. (ha); soil erosion (kg or mm lost);
FEW socio-economic metrics
• Expenditures for food, water, energy ($)
• Revenues and profits (for food, water, energy products, processes, or businesses)
• Trade balance ($ for food, water, energy goods, and services)
• Population living with malnutrition (persons)
• Population with affordable heating and cooling (persons)</td><td>FEW (specific) metrics
• Energy: energy consumption relative to other regions or countries (%)
• Water: water storage (m3) and flows (m3/s) relative to historical drought (% of storage and flows during historical drought)
• Food: grain storage relative to historical average (%)
FEW environmental impact metrics
• Energy: Greenhouse gas emissions (% reduction relative to benchmark year) (--)
• Water: instream flow rates (% of median flow)
• Food: habitat preservation for biodiversity, pollinators, etc. (% of land occupied by agriculture)
FEW socio-economic metrics
• Expenditures as a fraction of income or GDP (% for food, water, energy)
• Trade balance (% for food, water, energy goods, and services)
• Population living with malnutrition (%)</td></tr>
</table>

treatment system requires the least amount of electricity? What type of crop requires the least amount of irrigated water?

Relative–intensive metrics can be viewed as *benchmarking* absolute–intensive metrics: for example, the water intensity of a power plant [m³/kWh] compared to an industry best practice. For example, this metric could be expressed as some percentage relative to the average value for an absolute–intensive metric (e.g., 30% greater than average) or as a percentage rank relative to peers (e.g., a value better than 75% of all peer facilities). These types of metrics help understand how a technology or management practice compares to alternatives, and they help answer questions such as the following: Is my establishment one of the most resource-efficient? Am I using best practices?

A practical way to view an ***absolute–extensive*** metric is as an absolute–intensive metric multiplied by the total quantity of output (e.g., amount of unit *y* in the denominator, see Eq. (13.1)). Thus, absolute–extensive metrics consider the total quantity of inputs (money, water, food, energy, etc.) needed. For example, a city with a population of 100,000 targeting a per capita rate of water consumption of 200 L/person/day would need 20 million liters of water per day. Questions not addressed for absolute–extensive metrics include: Is 20 million liters of water per day (20 ML/day) available in a drought? Are other cities consuming less water?

The translation of absolute–extensive metrics to ***relative–extensive*** metrics is largely the process of including the concept of a target or limiting level of con-

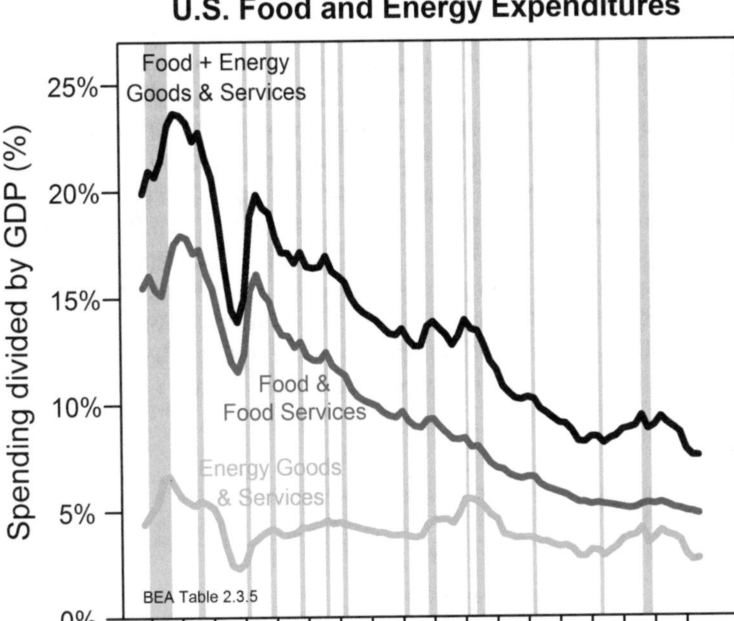

Fig. 13.2 The percentage of US consumer expenditures (1929–2017) for food and energy declined for over 60 years until 2002, after which the percentage has been approximately constant. (Source (BEA, 2018))

sumption, for example by comparing it to historical situations (e.g., drought), estimates of resource size (e.g., rainfall, water storage, fossil reserves, grain storage), or values of peers.

For the example of the city targeting 20 ML/day, it is critical for stakeholders in water planning to know the relative change (e.g., % relative to baseline year) of both total city-wide and per capita consumption, and whether each has increased or decreased over time.

Further, describing the water consumption of a city or farm as a fraction of the total consumption in a river basin is a common way to describe water allocation. A city, state, or country might decrease its per capita consumption of food, energy, or water over time even when its total consumption of each might increase due to increased population.

In addition, the proportion of income, total spending, or GDP spent on food, energy, and water services represent examples of economic FEW relative–extensive metrics, as depicted for the USA in Fig. 13.2. By looking at a time series like this one, it is possible to address the implication of changing historical trends, such as the no-longer-declining share of expenditures for food.

13.2.3.4 Example Metrics for a Thermal Power Plant

Water consumption is required for the cooling of coal, nuclear, natural gas, and biomass-fired power plants that use steam cycles. We use this energy–water example to demonstrate an example of each scale and taxonomic class of metric combination for the same technology (see Table 13.2). At the establishment level (e.g., the power plant), the volume of water consumed or withdrawn per unit of electricity generated [m^3/kWh] is an absolute–intensive metric, as is the cost [\$/m^3] of obtaining and delivering water to the power plant. This cost can dictate the type of cooling system one might design for the power plant—for example, for a new power plant a cost of >~1\$/m^3 could incentivize the use of a dry cooling tower versus a wet cooling tower (King 2014). In principle, a high cost of water would be representative of a region in which a high percentage of available water (a relative–extensive metric) is already allocated to all uses.

Table 13.2 Example (energy–water) metrics for each taxonomic class, at both the establishment and meso/macro scale level, for a thermal power plant

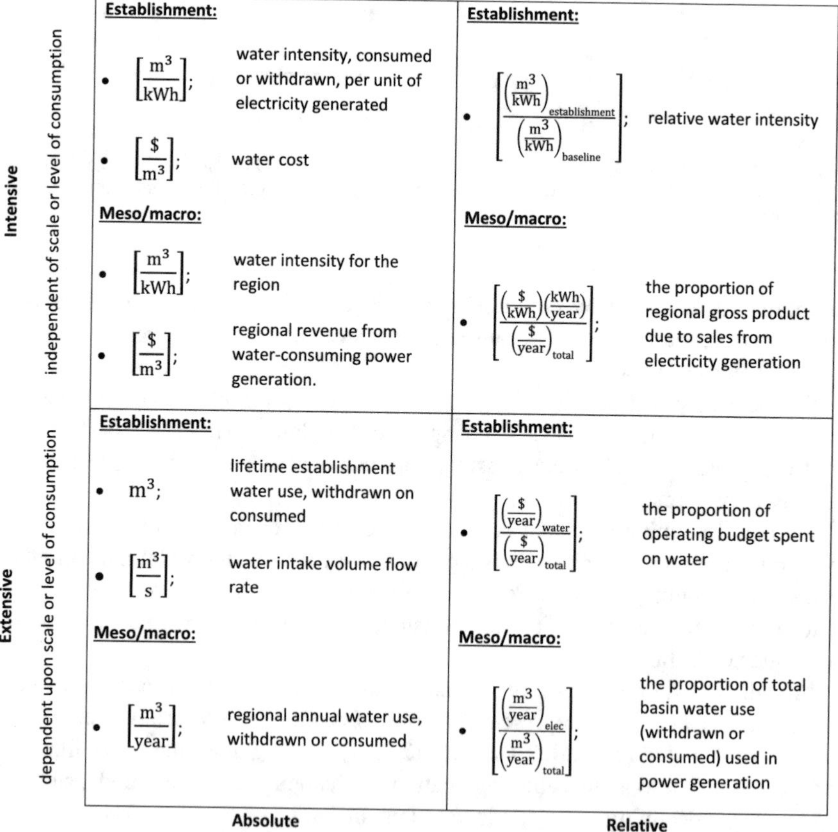

	Absolute no context of budget, constraint, peer, or history	Relative compared to budget, constraint, peer, or history
Intensive independent of scale or level of consumption	**Establishment:** • $\left[\frac{m^3}{kWh}\right]$; water intensity, consumed or withdrawn, per unit of electricity generated • $\left[\frac{\$}{m^3}\right]$; water cost **Meso/macro:** • $\left[\frac{m^3}{kWh}\right]$; water intensity for the region • $\left[\frac{\$}{m^3}\right]$; regional revenue from water-consuming power generation.	**Establishment:** • $\left[\frac{\left(\frac{m^3}{kWh}\right)_{establishment}}{\left(\frac{m^3}{kWh}\right)_{baseline}}\right]$; relative water intensity **Meso/macro:** • $\left[\frac{\left(\frac{\$}{kWh}\right)\left(\frac{kWh}{year}\right)}{\left(\frac{\$}{year}\right)_{total}}\right]$; the proportion of regional gross product due to sales from electricity generation
Extensive dependent upon scale or level of consumption	**Establishment:** • m^3; lifetime establishment water use, withdrawn on consumed • $\left[\frac{m^3}{s}\right]$; water intake volume flow rate **Meso/macro:** • $\left[\frac{m^3}{year}\right]$; regional annual water use, withdrawn or consumed	**Establishment:** • $\left[\frac{\left(\frac{\$}{year}\right)_{water}}{\left(\frac{\$}{year}\right)_{total}}\right]$; the proportion of operating budget spent on water **Meso/macro:** • $\left[\frac{\left(\frac{m^3}{year}\right)_{elec}}{\left(\frac{m^3}{year}\right)_{total}}\right]$; the proportion of total basin water use (withdrawn or consumed) used in power generation

Environmental advocates looking to reduce impacts to marine ecosystems often compare absolute–intensive metrics across many types of power plant designs, thereby forming a relative–intensive ranking of which type of technology consumes or withdraws the least amount of water (Averyt et al. 2011, King 2014).

The total annual consumption by a power plant (m^3/year, calculated as m^3/kWh × kWh/year) is an absolute–extensive metric for its water input. This measurement, calculated for a single power plant (establishment scale) or all power plants in a region or water basin (meso/macro scale), is relevant for ensuring legal water access, or water rights, and is critical for water planning processes (King et al. 2008).

In contrast, the proportion of all power plant operating costs associated with water provision and the cooling system is a relative–extensive metric. For example, the U.S. Environmental Protection Agency (EPA) has set a benchmark water intake velocity (measured in feet per second, or ft/s) for new power plants, effectively driving the inclusion of wet or dry cooling towers in newly designed thermal plants.[1]

This absolute–intensive metric of water intake velocity (input) divided by power output (e.g., MW of power) is coupled with a location-specific relative–extensive metric ("proportion of mean annual flow of a freshwater or stream") and, in effect, limits the range of technological solutions. Because environmental impacts from water intake such as impingement and entrainment of aquatic organisms are specific to the establishment and the site, applying a single threshold water intake velocity to all power plants (residing in basins with varying levels of water availability) is applying an establishment-level absolute–extensive metric but to every establishment at the large scale (Nuclear Energy Institute 2008).

Because of the difference in geographic and temporal scales, meso/macro-scale metrics are often more focused on societal goals and outcomes than those at the establishment scale.

Again, considering power plants, water consumption for power generation in a region such as a state or water basin could serve as an absolute–intensive [m^3/kWh] or absolute–extensive [m^3/year] metric. However, the *fractional* contribution to a regional metric, such as the percentage of annual water consumption associated with power generation, is a relative–extensive metric.

For metrics such as the amount of water withdrawal and consumption for power generation associated with a region (e.g., m^3/year) regional or national political entities tend to summarize these numbers. An example is the USA Geological Survey acting as the government agency providing sector-level annual water withdrawal and consumption estimates for each state and the nation overall (Solley et al. 2009, Maupin et al. 2014, Diehl & Harris 2014).

[1] See EPA 40 CFR Parts 122 and 125 (http://www.gpo.gov/fdsys/pkg/FR-2014-08-15/pdf/2014-12164.pdf): "First, the intake flow of the cooling water intake structure is restricted, at a minimum, to a level commensurate with that which could be attained by use of a closed cycle, recirculating cooling system. Second, the design through-screen intake velocity is restricted to 0.5 fps (foot per second). Third, the total quantity of intake is restricted to a proportion of the mean annual flow of a freshwater river or stream, or to a level necessary to maintain the natural thermal stratification or turnover patterns (where present) of a lake or reservoir except in cases where the disruption is beneficial, or to a percentage of the tidal excursions of a tidal river or estuary."

Lastly, the percentage of water consumption or withdrawal associated with thermal power plant operations in comparison with total regional water availability or human appropriation is categorized as a relative–extensive metric.

Continuing this discussion of power plant water use at the meso/macro scale, another example of the use of metrics to achieve two different strategic objectives is exemplified by the differential use of two relative–extensive metrics. Environmental advocates tend to focus on the fraction of total water withdrawal[2] in the USA for thermal power plant cooling (~45%), while electric utilities and their industry organizations tend to emphasize the proportion of total water consumption[3] for thermal power plant operations (~3–4%) (Solley et al. 2009, Maupin et al. 2014, Diehl & Harris 2014). One percentage seems "large" while the other seems "small." This example shows that terminology referring to metrics with different physical and mathematical assumptions is important. This is one reason why one should not use the word "use" to refer to water but instead specifically use words such as "consume" or "withdraw."

13.3 Case Study Texas Drought of 2011

13.3.1 Background

As an example of the FEW nexus and the exacerbation of the interdependencies therein, we summarize the after-effects from the 2011 drought in Texas. From late 2010 and throughout most of 2011, Texas received the lowest amount of rainfall over 12 months ever recorded (Nielsen-Gammon 2011). Further, that year was also the hottest on record, with many parts of the state experiencing a record number of 100 + °F days. Consequently, water flow in some rivers eventually became so low that it was insufficient for all water users to receive the surface water legally allotted to them, and subsequent decisions impacted the food and energy sectors (Fig. 13.3).

Water law in Texas is similar to that much of the western USA, with surface water rights based upon the concept of prior appropriation (Texas Water Development Board. n.d.). Surface water rights granted under the prior appropriation doctrine have the following characteristics:

- An assigned priority date, which determines the holder's priority for available water. Regardless of the priority date, whenever there is less water than is needed to satisfy all water rights in a basin, each appropriated right is subordinate to domestic and livestock users for the use of the available water.

[2] Water *withdrawal* refers to water that is removed from the local environment for human purposes without consideration of whether that water is returned to the local environment in liquid form.

[3] Water *consumption* refers to the difference between water withdrawal from the local environment and the quantity of water that is returned to the local environment in liquid form. Generally, water consumption is due to evaporation and evapotranspiration or its embodiment in some product (e.g., food).

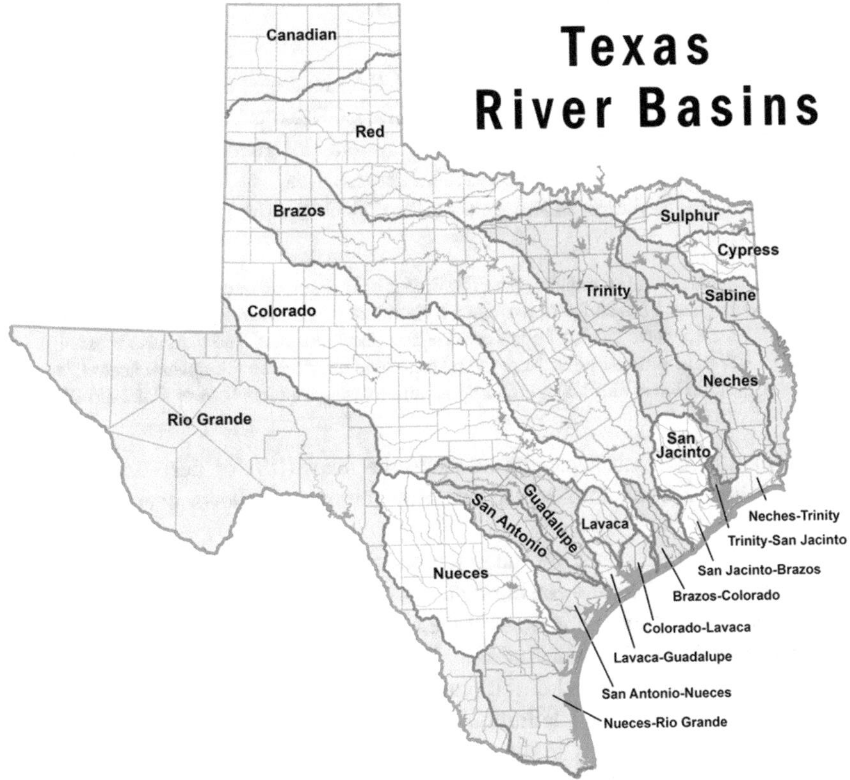

Fig. 13.3 Like most of Texas during a severe drought in 2011, the water availability within the adjacent Brazos and Colorado River basins of Texas were severely impacted, but existing management produced different levels of cooperation within the basins (Texas Commission on Environmental Quality 2018a)

- A specified volume of water that the holder may take or use within each year and a diversion rate if water diversion occurs. Access to this volume of water is subject to varying degrees of reliability depending upon the availability of water and the priority date of the holder.
- An ability to impound water (e.g., to store it in a reservoir above a dam), to divert water (e.g., to pump it from the stream), or both.

The **Texas Commission on Environmental Quality (TCEQ)** is the state agency that issues surface water rights. In case of drought, senior water rights holders can ask the TCEQ to "call" younger water rights, effectively cutting off junior rights holders from withdrawing water and thereby giving prioritized access to senior rights holders. The next two sections contrast responses to this issue from two adjacent river basins in Texas—the Colorado and the Brazos.

13.3.2 Colorado River (Texas) Basin

The **Lower Colorado River Authority (LCRA)** manages the allocation of water within the Colorado River Basin in Texas. The LCRA operates the Highland lakes (Lakes Buchanan and Travis) as a single system designed to supply water and address varied interests within the LCRA's service area.

Regarding their **Water Management Plan (WMP)**, the Lower Colorado River Authority states,

> during drought, the plan requires the curtailment or cutback of interruptible stored water from the Highland Lakes for downstream agriculture so that firm water supplies will be available for the basic needs of cities, businesses, and industries. The plan also prescribes how LCRA must provide water from the lakes to help meet the environmental needs of the lower Colorado River and Matagorda Bay at various trigger points (Lower Colorado River Authority (n.d).

The metrics here are water inflows into the lakes (**ft**^{3}/s or **acre-ft**/month) and environmental inflows into Matagorda Bay (cumulative flows over 1-, 3-, and 6-month spans).

Due to the 2011 drought, for the first time in its history, the LCRA had to invoke the WMP to deny Colorado Basin rice farmers their allocations of irrigation water, thus impacting the food component of the FEW nexus.

The Lower Colorado River Authority Water Management Plan has been a working document since its initial development in the late 1980s, with a summary of a few of its metrics, rules, and definitions that allow it to effectively manage the water basin and prescribe actions in the case of future droughts described here (Lower Colorado River Authority 2014).[4]

1. The WMP sets forth **criteria** for declaring a **Drought Worse than Drought of Record (DWDR)** and includes a calculation of a metric known as the "**Combined Firm Yield**" of Lakes Buchanan and Travis, which is the amount of water that can be supplied annually from lakes Buchanan and Travis through repeat circumstances from the Drought of Record.
   (a) **Firm yield** is defined as the amount of water that a reservoir could have produced annually if it had been in place during the worst drought of record. For the two main LCRA reservoirs (Lake Travis and Lake Buchanan), the firm yield is a storage quantity of water currently set at 434,154 acre-ft/year determined using hydrologic simulation from the TCEQ Water Availability Model (Texas Commission on Environmental Quality 2018b). The Chap. 19 case study of the Great Lakes and St. Lawrence watershed shows that even for a watershed with extremely large storages and flows of water, stakeholders

[4] A full description is beyond the scope of this paper. The summary here is from the latest WMP, and the differences from WMP active during the drought of 2011 are not material for this discussion.

find it important to forge strong agreements to limit large scale water diversions from sources within the basin.

(b) Declaring and canceling a DWDR occurs when an ongoing drought has a real likelihood of becoming a new Drought of Record. A DWDR declaration would trigger action to cut off interruptible stored water and implement mandatory pro rata curtailment of Firm Water demands. The LCRA requirements for declaring a DWDR are given by the following metrics:

- **Drought duration** of at least 24 consecutive months.
- **Drought intensity** greater than that of the Drought of Record, as measured by inflows into Lakes Buchanan and Travis.
- **Combined storage** in lakes Buchanan and Travis of less than 600,000 acre-ft. (Note that the LCRA uses the value of 600,000 acre-ft as a safety factor to account for the uncertainty inherent in the hydrologic modeling of firm yield, helping to avoid the triggering of a DWDR in the event of a repeat of historic hydrology.)

2. The WMP sets the "**Evaluation Dates**" of March 1 and July 1 when the LCRA determines the water supply condition, the interruptible stored water available for agriculture, and effective environmental flow criteria.
3. The WMP allows the LCRA to stop releasing interruptible stored water for agricultural operations, even in the middle of a crop season, if reservoir storage drops below certain predetermined levels.

Thus, the Lower Colorado River Authority Water Management Plan uses a combination of different metrics that are most relevant for meso/macro-scale management: absolute–extensive and relative–extensive metrics.

Of the metrics mentioned above, relevant absolute–extensive metrics include the following:

- Firm Yield (acre-ft/year).
- Water Inflows into the lakes (ft^3/s or acre-ft/month).
- Environmental Inflows into Matagorda Bay (cumulative flows over 1-, 3-, and 6-month spans).

However, it is probable that the reference of drought plans to the Drought of Record is the most critical metric driving not only the LCRA WMP but also the WMP of the entire state.

For Texas, the drought of record occurred during the 1950s, with water planning and water curtailment operations ultimately based upon the relative–extensive metric of how similar current conditions are to the conditions that would come about from a repeat of this drought of record.

In the event of a drought triggering the curtailment of interruptible water rights, primarily irrigation, the plan also indicates pro rata, or proportional, cuts in water access based on the total quantity of surface water rights, another relative–extensive metric to govern water allocation.

13.3.3 Brazos River Basin

In contrast to the Colorado River, there was no basin-wide management of water allocation for the Brazos River. Arguably, as a result of the 2011 drought, the Brazos River basin in Texas became the center of a legal battle regarding the water allocation and the right of the State of Texas, via the Texas Commission on Environmental Quality, to have some discretion in interpreting state surface water law (Malewitz 2015).

In 2011, a senior water rights holder asked the TCEQ to call water rights on the Brazos River.

As the TCEQ went down the prioritized list of water rights for the basin, it eventually came to water rights for some cities and thermoelectric power plants, putting them in a difficult position. Cutting water access to power plants could alienate elected officials and their power-consuming constituents, thus reducing the capacity for electricity supply and in turn causing increased electricity prices during the hottest summer on record.

Consequently, the TCEQ chose not to cut off legal water access to power plants or cities in 2011, citing public health, safety, and welfare concerns; however, they did call water rights for farmers, continuing curtailment of water for senior irrigation rights into 2012, but not for power generation (Texas Commission on Environmental Quality, 2013).

As a result, the Texas Farm Bureau sued TCEQ, claiming the TCEQ's decision criteria in this instance cannot replace the prior appropriation doctrine for long-term governance. Over the ensuing few years, the Farm Bureau won initial legal rulings, along with two appeals.

Eventually, the TCEQ appealed to the Texas Supreme Court. In February 2016, the Court resolved the legal issue by choosing not to consider the appeal, thus leaving in place lower court rulings that sided with the farmers and against the TCEQ (Malewitz 2016).

> Effectively, the justices left in place a lower court's ruling that said Texas cannot give special treatment to cities or power generators over more "senior" water rights holders on parched rivers—even if the state declares it necessary to protect the 'public health, safety, and welfare.'

13.3.4 Use of Metrics to Improve Drought Management

Plans are often created in preparation for their execution in time-sensitive situations in which decision-making is needed, like emergencies. In the Texas case study previously discussed, the surface water rights priority doctrine serves as such a plan during a case of drought; however, the plan is only valuable if it can practically be enacted during such a time of constraint.

The TCEQ used words such as "public health" and "safety" to discuss why they could not cut off the water supply to thermal power generators. However, how much water and electricity are truly needed to enable health and security? ***To make these words actionable, they must be translated into quantifiable metrics.***

For example, according to the World Health Organization (WHO), every person should have access to at least 50 L/day [L/person/day] of water that is safe, or clean for washing, cooking, and drinking; affordable, costing less than 3% of household income; and accessible, or less than 1 km and/or 30 min away (United Nations n.d.). In contrast, in the Texas State Water Plan, the Texas Water Development Board estimates that Texas municipal water demand for all purposes in 2010 was approximately 650 L/person/day.

We can convert these absolute–intensive metrics of per capita water use into a relative–intensive metric to indicate that the average Texan has 13 times the WHO standard for freshwater access. This metric could be used to discuss safe and healthy levels of water consumption during drought. Furthermore, the maintaining proper pressure in the municipal water supply for fighting fires is a critical safety issue and is a reason for larger per capita water availability.

In turn, a certain amount of electricity per person is needed to be healthy and safe. A minimum amount of electricity (less than 2% of the total in Texas of over 430 TWh/year) is required to run water and wastewater services, with additional resources necessary for hospitals, police stations, traffic lights, and other city services (Stillwell et al. 2010). Texas' 2014 retail sales of electricity were 379 TWh (10^{12} Wh), comprised of 37% residential, 36% commercial, and 26% industrial sales. These can be used as starting metrics to quantify desired power demand and allocation during drought.

13.4 Conclusion

The central issue of the FEW nexus is the interaction of multiple dimensions: economic, social, and environmental. To be *comprehensive*, the more metrics, the better; however, to be *comprehensible*, fewer metrics are preferable.

A range of metrics using a range of units is needed to understand FEW interdependencies, and multiple considerations across a range of dimensions feed into our framework for defining FEW metrics. Life cycle assessment (LCA) is one valuable framework for calculating and defining metrics in order to characterize FEW systems. The inputs to LCA can come from both data and models.

In considering metrics by which we can improve decision-making, there are a number of important considerations including *scale* and *taxonomic class*. The metric spatial and temporal *scales* at which activity takes place, ranges in space from establishment-level to the meso/macro (or meso/macro) scale within a specific region composed of many interacting facilities while ranging in time from concerns of seconds to decades. The metric *taxonomic class* defines it across two dimensions (intensive vs. extensive and absolute vs. relative) and how each dimensional combination can be used to reference how the metric relates to decision-making.

However, the metrics are most useful if we can both calculate them from observed data and simulate them in models that give us insight into future impacts, options, and interactions with feedbacks and constraints. Future research should focus on modeling methods that can translate across the various categories of metrics we have outlined here. This is important is because FEW constraints are primarily expressed by extensive metrics, both biophysical and socioeconomic, but many decisions and investments are made based upon establishment-level intensive improvements.

We hope that the framework outlined here might provide a useful lens by which to categorize metrics and inform future discussion and planning efforts. Proper metrics can be used to expand the impact categories (economic, social, environmental) to include impacts on other community capitals, including political, cultural, built, and human. See Chap. 3 to consider how various metrics might relate the indicators identified for the UN Sustainable Development Goals.

If data or models identify critical thresholds or leverage points, then these "boundaries" can be used to create relative metrics that inform resilience of a system of interest. For example, there seems to be a strong connection between the percentage of GDP spending on energy commodities and economic recession. We might already be reaching a fundamental constraint on current FEW interdependencies at the overall economic (meso/macro) scale. Therefore, understanding FEW constraints is becoming more important as many long-term trends now indicate diminishing returns (King 2015).

Key Points

- Metrics provide an important bridge between data, models, policy, and ultimately behavior.
- Our choice of metric can constrain which data we collect, how our models are framed, and thus our ability to understand the world.
- There is a trade-off between *comprehensiveness* (more metrics) and *comprehensibility* (fewer metrics).
- The chapter presents a taxonomy of metric dimensions across which metrics can vary: intensive vs. extensive and absolute vs. relative.
- FEW metrics can be applied at different system spatial and temporal scales.
- Different stakeholders can prefer different metrics and system scales depending upon their strategic objectives.
- Life cycle assessment provides a useful framework within which to understand interactions among food–energy–water systems and thus guide our choice of metrics.

Discussion Points and Exercises

1. Describe why the identification of metrics is an important step in action toward building a more sustainable food–energy–water nexus.
2. Identify a historical example within the FEW nexus which follows the data → models → metrics → actions process outlined in Fig. 12.3. Was this example successful in your view (i.e., the example might or might not have produced an action that you agree with)?
3. Describe how data (un)availability might impact our choice of metrics (with reference to examples) and determine an approach to mitigate this problem.

4. Discuss some limitations with the use of life cycle assessment for identifying metrics of interest.
5. Identify examples where spatial and temporal scales impact the specification of appropriate metrics.
6. Apply the metric taxonomy framework as demonstrated in Table 13.1 to fill out the table below with metrics that would be appropriate for the cases below. What similarities do you see in metrics between the different cases?
   (a) Food—a dairy farm and regional dairy distribution network.
   (b) A bio-refinery and regional supply network.
   (c) A wastewater treatment establishment and regional network.

Intensive	Energy/water:	Energy/water:
	Energy/food:	Energy/food:
	Energy/energy:	Energy/energy:
	Water/energy:	Water/energy:
	Water/food:	Water/food:
	Water/water:	Water/water:
	Food/water:	Food/water:
	Food/energy:	Food/energy:
	Food/food:	Food/food:
	Socio-economic metrics:	Socio-economic metrics:
Extensive	FEW (specific) metrics	FEW (specific) metrics
	Energy:	Energy:
	Water:	Water:
	Food:	Food:
	FEW environmental impact metrics	FEW environmental impact metrics
	Energy:	Energy:
	Water:	Water:
	Food:	Food:
	FEW socio-economic metrics	FEW socio-economic metrics
	Energy:	Energy:
	Water:	Water:
	Food:	Food:
	Absolute	Relative

7. For one of the cases of Question 6, choose one FEW metric out of each quadrant and specify the category of stakeholder(s), or an example stakeholder, that might want to use each metric.
8. For the case study of the 2011 Texas drought and the Lower Colorado River Authority, what do you feel were two critical aspects of water management during the drought? State why each aspect was important.
9. Discuss strengths and weaknesses of the framework outlined in this chapter for identifying food–energy–water nexus metrics of interest.

References

Averyt, K., Fisher, J., Huber-Lee, A., Lewis, A., Macknick, J., Madden, N., Rogers, J., & Tellinghuisen, S. (2011). *Freshwater use by U.S. power plants: Electricity's thirst for a precious resource. A report of the energy and water in a warming world initiative*. Cambridge, MA: Union of Concerned Scientists.

BEA. (2018). *Table 2.3.5. Personal consumption expenditures by major type of product*. National Income and Product Accounts, U.S. Bureau of Economic Analysis. Retrieved from https://apps.bea.gov/iTable/iTable.cfm?reqid=19&step=2#reqid=19&step=2&isuri=1&1921=survey.

Boulay, A., Bare, J., De Camillis, C., Döll, P., Gassert, F., Gerten, D., Humbert, S., Inaba, A., Itsubo, N., Lemoine, Y., Margni, M., Motoshita, M., Núñez, M., Pastor, A., Ridoutt, B., Schencker, U., Shirakawa, N., Vionnet, S., Worbe, S., Yoshikawa, S., & Pfister, S. (2015). Consensus building on the development of a stress-based indicator for LCA-based impact assessment of water consumption: Outcome of the expert workshops. *The International Journal of Life Cycle Assessment, 20*(5), 577–583.

Daly, D. (2013). A further critique of growth economics. *Ecological Economics, 88*, 20–24.

Diehl, T. H., & Harris, M. A. (2014). *Withdrawal and consumption of water by thermoelectric power plants in the United States, 2010*. U.S. Geological Survey Scientific Investigations Report 2014–5184 (p. 28), https://doi.org/10.3133/sir20145184.

Heijungs, R., & Suh, S. (2002). *The computational structure of life cycle assessment, eco-efficiency in industry and science* (Vol. 11). Dordrecht, Netherlands: Springer.

Heijungs, R., de Koning, A., & Guinée, J. B. (2014). Maximizing affluence within the planetary boundaries. *International Journal of Life Cycle Assessment, 19*(6), 1331–1335.

Heun, M. K., Carbajales-Dale, M. K., & Haney, B. R. (2015). *Beyond GDP: National accounting in the age of resource depletion* (Lecture notes in energy) (Vol. 26). Cham: Springer International Publishing.

ISO. (2006). *ISO 14040:2006 Environmental management—Life cycle assessment—Principles and framework*. International Organization for Standardization. Retrieved from https://www.iso.org/standard/37456.html.

King, C. W. (Ed.). (2014). *Thermal power plant cooling: Context and engineering*. New York: ASME.

King, C. W. (2015). The rising cost of resources and global indicators of change. *American Scientist, 103*(6), 410. Retrieved from https://www.americanscientist.org/article/the-rising-cost-of-resources-and-global-indicators-of-change.

King, C. W., Duncan, I. J., & Webber, M. E. (2008). *Water demand projections for power generation in Texas*. Report prepared for the Texas Water Development Board contract no. 0704830756. Retrieved from http://www.twdb.state.tx.us/wrpi/data/socio/est/Final pwr.pdf.

Kubiszewski, I., Costanza, R., Franco, C., Lawn, P., Talberth, I., Jackson, T., & Aylmer, C. (2013). Beyond GDP: Measuring and achieving global genuine progress. *Ecological Economics, 93*(C), 57–68.

Lawn, P. A. (2003). A theoretical foundation to support the Index of Sustainable Economic Welfare (ISEW), Genuine Progress Indicator (GPI), and other related indexes. *Ecological Economics, 44*(1), 105–118.

Lower Colorado River Authority. (2014). Exhibit A: Technical papers. In *Proposed revised water management plan, with appendices*. Technical report. Retrieved October 19, 2018, from https://www.lcra.org/water/water-supply/water-management-plan-for-lower-colorado-river-basin/Documents/2014-wmp-application-files/AppendixA_TechPapers_2014Oct_FINAL.pdf.

Lower Colorado River Authority. (n.d.). *Water management plan*. Retrieved October 19, 2018, from https://www.lcra.org/water/water-supply/water-management-plan-for-lower-colorado-river-basin/Pages/default.aspx.

Malewitz, J. (2015). *Water ruling cuts state's power in droughts*. Texas Tribune. Retrieved October 19, 2018, from https://www.texastribune.org/2015/04/02/huge-water-ruling-court-sides-ranchers/.

Malewitz, J. (2016). *Major water case, win for ranchers is loss for cities*. The Texas Tribune. Retrieved October 19, 2018, from https://www.texastribune.org/2016/02/19/major-water-case-win-farmers-loss-cities/.

Matthews, E., Amann, C., Bringezu, S., Fischer-Kowalski, M., Hüttler, W., Kleijn, R., Moriguchi, Y., Ottke, C., Rodenburg, E., Rogich, D., et al. (2000). *The weight of Nations: Material outflows from industrial economies*. Washington, DC: World Resources Institute. Retrieved October 19, 2018, from http://pdf.wri.org/weight_of_nations.pdf.

Maupin, M. A., Kenny, J. F., Hutson, S. S., Lovelace, J. K., Barber, N. L., & Linsey, K. S. (2014). *Estimated use of water in the United States in 2010*. U.S. Geological Survey, Circular 1405. Retrieved October 19, 2018, from https://pubs.usgs.gov/circ/1405/.

Meadows, D. (1998). *Indicators and information systems for sustainable development*. Report to the Balton Group, The Sustainability Institute, Hartland Four Corners, Vermont.

Nielsen-Gammon, J. W. (2011). *The 2011 Texas drought, a briefing packet for the Texas legislature, October 31, 2011*. A report of the Office of the State Climatologist.

Nuclear Energy Institute. (2008). *NEI fact sheet: Water consumption at nuclear power plants*. Retrieved October 19, 2018, from http://neinuclearnotes.blogspot.com/2008/03/nei-fact-sheet-on-water-consumption-at.html.

Rees, W. E., & Wackernagel, M. (2013). The shoe fits, but the footprint is larger than earth. *PLoS Biology, 11*(11), e1001701.

Rockström, J., et al. (2009). A safe operating space for humanity. *Nature, 461*(7263), 472.

Slesser, M. (1974). *Energy Analysis Workshop on Methodology and Conventions: 25th–30th August, 1974, Guldsmedshyttan, Sweden*. Stockholm: IFIAS.

Solley, W. B., Pierce, R. R., & Perlman, H. A. (2009). *Estimated use of water in the United States in 1995*. U.S. Geological Survey Circular 1200. Retrieved October 19, 2018, from https://pubs.er.usgs.gov/publication/cir1200.

Stillwell, A. S., King, C. W., Webber, M. E., Duncan, I. J., & Hardberger, A. (2010). The energy-water nexus in Texas. *Ecology and Society, 16*(1), 2. Retrieved October 19, 2018, from http://www.ecologyandsociety.org/vol16/iss1/art2/.

Texas Commission on Environmental Quality. (2013). *Suspensions of permitted state surface water diversions in the Brazos River Basin*. Retrieved October 19, 2018, from https://www.tceq.texas.gov/assets/public/response/drought/water-right-letters/07-02-13Brazos-domestic.pdf.

Texas Commission on Environmental Quality. (2018a). *Texas river basins maps*. Retrieved October 19, 2018, from https://www.tceq.texas.gov/agency/data/maps.html.

Texas Commission on Environmental Quality. (2018b). *Water availability models*. Texas Commission on Environmental Quality. Retrieved October 19, 2018, from https://www.tceq.texas.gov/permitting/water_rights/wr_technical-resources/wam.html.

Texas Water Development Board. (n.d.). *A Texans guide to water and water rights marketing*. Retrieved October 19, 2018, from http://www.twdb.texas.gov/publications/reports/infosheets/doc/WaterRightsMarketingBrochure.pdf.

United Nations. (1993). *Integrated environmental and economic accounting: Interim version* (Studies in methods) (Vol. 61). New York: United Nations.

United Nations. (n.d.). *The human right to water and sanitation*. Media Brief. Retrieved October 19, 2018, from http://www.un.org/waterforlifedecade/human_right_to_water.shtml.

Wackernagel, M., & Rees, W. (1998). *Our ecological footprint: Reducing human impact on the earth* (Vol. 9). Gabriola Island, BC: New Society Publishers.

Further Reading

Daly, H. (1999). *Uneconomic growth and the built environment: In theory and in fact. Reshaping the built environment: Ecology, ethics and economics*. Washington, DC: Island Press.

Foley, J. A., DeFries, R., Asner, G. P., et al. (2005). Global consequences of land use. *Science, 309*(5734), 570–574.

Hoekstra, A. Y., & Chapagain, A. K. (2007). Water footprints of nations: Water use by people as a function of their consumption pattern. *Water Resources Management, 21*(1), 35–48.

Hoekstra, A. Y., Chapagain, A. K., Aldaya, M. M., & Mekonnen, M. M. (2011). *The water footprint assessment manual: Setting the global standard*. Enschede, The Netherlands: Water Footprint Network.

King, C. W., & Webber, M. E. (2008). Water intensity of transportation. *Environmental Science & Technology, 42*(21), 7866. https://doi.org/10.1021/es800367m.

Kounina, A., et al. (2013). Review of methods addressing freshwater use in life cycle inventory and impact assessment. *The International Journal of Life Cycle Assessment, 18*(3), 707–721.

Chapter 14
Data

Benjamin L. Ruddell

14.1 Introduction: Framing Data Between Metrics and Models

Data provides the practical means by which we conduct scientific investigation of metrics (Chap. 13) that facilitate informed operational decision-making, and of the construction and parameterization of predictive models (Chap. 15). Especially today in the computer age where computing power is abundant, adequate data now tends to be the limiting factor on the quality of our estimation, modeling, understanding, decision-making, and prediction. Fortunately, there is a lot of data available about FEW systems. But, this data is not easy to locate, access, or utilize. This data is often privacy-restricted and privileged. This data is patchy with surprisingly large gaps for critical layers and scales of FEW systems. The lack of seamless, high-quality, synthetic datasets describing FEW systems across sectors and scales is one of the major practical barriers to FEW systems work at the present time. Most data sets were collected to answer a single question about a single layer, process, and scale in the FEW system, but systems science and systems management requires data that interoperates across layers and scales.

In the abstract sense **data** are simply facts and assumptions, and in the digital era, these facts and assumptions are usually (but not always) quantitative and numeric in nature. Data may be observations gathered from the real world, in which case the data are empirical. However, empirical observations still imply both a conceptual model (a question) and an observational model (an apparatus). Data are increasingly, these days, the outputs of complicated computer models; these model output data are predictions (or post-dictions) that assimilate empirical observations and

B. L. Ruddell (✉)
School of Informatics, Computing, and Cyber Systems, Northern Arizona University, Flagstaff, AZ, USA
e-mail: Benjamin.Ruddell@nau.edu

P. Saundry, B. L. Ruddell (eds.), *The Food-Energy-Water Nexus*, AESS Interdisciplinary Environmental Studies and Sciences Series,
https://doi.org/10.1007/978-3-030-29914-9_14

inputs along with theoretical constraints and predictions. If we could afford and obtain high-quality observations of all aspects of the FEW system (including the future of the system…), we would not have any practical use for models. However, models are useful measures to fill in the gaps in our observations, predict the future (which is currently impossible to observe!), and economize when our theory provides an adequate shortcut to avoid expensive empirical work. Models are also useful for the development and testing of hypothesis and theory using the scientific method, in which case empirical data are compared with modeled data. **Assimilation** is a modeling technique that blends theoretical model estimates with observations to produce an optimally accurate dataset. Data provide **information** which is the answer to a question; to the extent that data answer interesting questions, data and information are one and the same. Information theory concerns the measurement of the quality of these questions, answers, and hypotheses.

As reviewed in other chapters, there are many different types of questions we want to answer about FEW systems for purposes of science, management, and modeling. Each of these questions requires data, either modeled or observed. The most common types of data we need to answer questions about a FEW system at any scale include:

- Flows and supply chains of commodities, goods, services, and finances.
- Routes of transportation including Origin and Destination.
- Physical and Legal availability.
- Financial constraints, Price, and Cost.
- Infrastructure Capacity and Utilization.
- Supply and Demand.
- Environmental impacts or dependencies.
- Peak vs. average rates and totals.
- Mode of transportation.
- Storage, available and utilized.
- Inputs and Outputs.
- Environmental and Ecological quality.
- Natural resource availability and stocks.
- Boundaries and Governance.
- Regulatory and Legal constraints.
- Social objectives and performance metrics.
- Product performance metrics.
- Performance benchmarks.
- Life cycle effects and footprints.
- Teleconnections.
- Dynamics of Perturbation and Response, Shock and Stress.
- Risk, vulnerability, sustainability, resilience, and exposure.

These FEW systems data may originate from models—or from empirical methods such as sensors, surveys, or inventories. **Sensors** are used to make large volumes of observations and measurements automatically. Examples of sensors for FEW systems include smart meters for water, gas, or electric service, RFID chips placed on products for supply chain monitoring, remote sensors on satellites or UAVs, streamflow and water quality gauges, smart-agriculture measurements of soil moisture, and so on. **Surveys** are census methods administered by researchers for collecting usage, production, or transportation data. Census methods are generally employed at the establishment scale (space) and annual to decadal scale (time) using an affordable statistical sampling technique, with data released in aggregated form to preserve privacy. **Inventory data** is collected by businesses for their own internal purposes. Inventories account for how much of a product is available at an establishment, on order, or en route. Inventory data is a central component of private sector **supply chain management**. Inventory data tends to be private and may be considered **trade secrets**. These three bottom-up data types are critical for understanding the "last mile" of the FEW supply chain and the flows of FEW commodities in general, along with emergency management.

FEW systems data has several attributes, including but not limited to structure, quality, spatial resolution, temporal resolution, process resolution, and scale. Data management best practices are essential both for users and originators of data at every stage of the **data life cycle**. Important privacy and ethics principles compromise the accessibility of bottom-up data and compromise the utility of top-down data resources. Major strengths and weaknesses in data availability and accessibility are reviewed for application domains of the FEW system.

This chapter does not attempt to identify all current FEW system data sources (an impossible task!) but does employ numerous examples. Most of these examples are drawn from the USA. Each chapter of this book surveys exemplary datasets that are germane to its specific concepts and applications, so the reader should look to the chapters for guidance on where to find especially useful data for each topic. The chapter's end matter surveys some particularly useful datasets for the FEW system, and the reader should inspect these sources for further education.

14.2 Data Attributes: Structure, Quality, Scale and Resolution

14.2.1 Data Structure (and Type)

FEW systems are coupled natural-human systems that may often be types of socio-ecological systems. The native conceptual model for FEW systems is usually a multitype (or multiplex, Baggio et al. 2016) network with many qualitatively different types of agents, behaviors, processes, boundaries, stocks, and flows. The special type of mathematics that characterizes this system is a graph or network—specifically, the process network (Ruddell and Kumar 2009). This type of graph features nodes that represent many types of natural or human processes or agents in the system; these processes and agents act on the FEW system network by transforming inputs into outputs, producing and consuming goods and services, and causing transportation and movement between nodes in the system. Nodes and edges on the graph are located in space and time. The edges in the graph are the "couplings" between agents, and these edges represent directional (or bidirectional) and flows of mass, energy, information, goods, and services. These stocks and flows change over time. This FEW system process network structure is a close relation to the environmentally extended and multi-regional input–output data models (I/O, Miller and Blair 2009). However, the FEW system is more complicated than I/O, because it considers stocks, storage, environmental impacts, and boundaries in the system in addition to the characterization of the system's nodes and their inputs and outputs.

Inputs and Outputs are the flows of goods and services in and out of a process. These inputs and outputs may include raw materials, finished products and services, and wastes. Input–output coefficients describe the efficiency of a process and the connectedness between input and output stocks and flows. **Stocks** are accounted for quantities of material, goods, currency, and so on that exist at a specific space-time location. Stocks may be human (e.g., bank account, gasoline tank) or natural (e.g., aquifer, coal seam), and are often synonymous with storage or buffers in the system. **Flows** are accounted quantities of material, good, currency, and so on that move from one space-time location to another—that is, from an origin to a destination, from a producer to a consumer, and so on. Flows may be human (e.g., currency payment, oil by pipeline) or natural (e.g., water in a river). Natural flows are often considered ecosystem services. Flows may be valuable, as in the instance of goods and services or currencies, but they may also be undesirable, such as flows of wastes and pollutants. Information also flows. **Origins and Destinations** identify the space-time locations where a good or service was transported (or flowed) from, and to, respectively. **Route** identifies the precise mode of transport, pathway, and waypoints followed by a flow that transports a good or service. **Agents** are the natural or human entities who control stocks, and who produce, consume, originate, terminate, and take inputs and outputs concerning flows; natural processes may be characterized as agents with predictable behavior.

Ontologies name and describe the categories of objects in a dataset and the relationships between these types. Subtypes of ontologies include controlled vocabularies or codesheets which are lists of categories and objects that are allowable in a dataset, and data models (or object–entity relationship diagrams) detail the formats and relationships between different objects in a dataset (e.g. Ruddell and Kumar 2006).

An ontology is a conceptualization of categories that captures their qualitative types and describes the relationships between different types. Establishments and processes (or aggregations thereof) need to be named using controlled vocabularies. "Codes" are a very simple type of controlled vocabulary. For example, the North American Industry Classification System (NAICS, OMB 2017) is a coding system that categorizes different types of business establishments according to the general type of goods and services they produce. NAICS was created for the U.S. Census Bureau in partnership with Canadian and Mexican authorities to track business and tax statistics in North America—but not for FEW systems research. The U.S. Department of Agriculture maintains a system of codings for agricultural production associated with the National Agricultural Statistics Service (NASS 2015). The U.S. Geological Survey maintains a system of codings for water use in its Water Census of the USA (Maupin et al. 2014). The U.S. Energy Information administration similarly classifies energy production and use (EIA 2017). There are many other examples of data codings that apply to a single dataset and layer of the FEW system—but they do not form complete FEW systems ontologies. The FEWSION project has undertaken a relatively complete but not particularly detailed ontology for the FEW system (FEWSION, n.d.; FEWSION Codesheet 1.0, 2019).

Codings created for disparate purposes often do not line up one-to-one between different layers of the FEW system network. These datasets are "heterogeneous," meaning that they are not a part of a consistent and coherent ontology. As a result, a common research task in FEW systems is the development of FEW system ontologies that relate one dataset to another. Unfortunately, at this early stage of FEW systems research there has not been sufficient progress on the creating of standardized ontologies including the relational data models, controlled vocabularies, and crosswalks, and as a result, most of this work currently involves custom and application-specific bilateral crosswalks. There is currently a pressing need for the development of standard FEW systems ontologies.

In the absence of standard FEW systems ontologies, there are a number of problems that need to be solved in order to pair up two datasets. When more than two datasets need to be linked, it is necessary to either construct a complete *relational data model* or to choose a single *lowest common denominator* (LCD) dataset to which all other datasets will be related in the ontology. For instance, if several datasets and models have different resolution, one can standardize them at the coarsest resolution of space, time, and category. This is roughly the approach taken by the FEWSION project in its selection of the county, month, and FEWSION/SCTG+ category codes (FEWSION Codesheet 1.0, 2019; BTS, n.d.); this resolution is more or less compatible with all of the available data comprising the systems network description. Building a robust and widely encompassing FEW systems ontology is a significant undertaking that has not yet been completed, although various input–output and commodity flow modeling teams have made strides especially at the coarser resolutions. The LCD approach is a simpler but also less flexible solution. The steps involved in the more common LCD approach commonly include, but are not limited to:

1. Identifying a formal coding and controlled vocabulary for each dataset.
2. Establishing a crosswalk between each pair of controlled vocabularies.
3. Developing aggregation or disaggregation factors to relate datasets with mismatching resolution.

Data **Format** is the most common colloquial shorthand for a data structure in FEW science and management work. Most communities have de-facto best practices for data structure encoded in their preferred data formats. For instance, many climate models now use a specific kind of NetCDF implementation of the HDF5 format to capture the voluminous and multi-type gridded data that is produced by these models. This practice is not mandated by any regulatory body, but it has been adopted by the community and is, therefore, a de-facto standard. There are not de-facto standards yet for the young FEW systems community or related fields of footprinting, sustainability analysis, life cycle analysis, agent-based modeling, and network analysis. The reader is directed to the various repositories listed later in this chapter, and in the end matter, for examples of de-facto data structure standards.

A common practice is the use of "flat" CSV text files and spreadsheets for smaller volumes of data, with an accompanying "header" that lists the names and units of each column of data, and a separate documentation file that defines terms. Often FEW data text files are pseudo-spatial in that they are cross-coded and joined to spatial locations such as rivers, counties, states, etc. that can be mapped with a Geographic Information System. Remote sensing datasets for FEW topics have their own well-developed format standards that are not unique to the FEW systems domain of applications.

Learning to make use of these formats and de-facto standards is one of the core outcomes and skillsets resulting from a research apprenticeship or early career training experience within a specific domain of application. There are no general rules—only best practices within your application domain. Learn to carefully identify which **community of practice** you will be working within and interoperating with, and to ask experienced experts in that community the right questions about what is the best, easiest, and most well-tested set of practices for data structures and formats. It is inadvisable to become an early adopter who tries to introduce new technologies, standards, and data structures to a community before you become an expert on the older ways. A little bit of due diligence at the early stage of your work will save you a great deal of time and yield much more rapid productivity. Data format and structure is like a language; none is right, all have their place and intended use, and it is most important that everyone is "on the same page."

14.2.2 Data Quality

The major issues in FEW systems **data quality** are validity, completeness, precision (and accuracy), resolution, and provenance. We need sufficient data quality to do effective research and make effective operational decisions. Quality is entirely relative to the intended use.

1. **Validity** is the idea that data demonstrably corresponds to reality. Validity must be rigorously ensured by using formal methods, carefully applied, in data collection. Because most FEW systems datasets are derived from statistical sampling and surveys, it is important that those survey samples are unbiased and have a large enough sample size. The design of the survey itself must be rigorous and studied, with verifiable performance.

2. **Completeness** may be straightforward to ascertain if there are empty fields and gaps in the dataset. If we are lucky, the creators of the data left incomplete data blank, rather than employing gap-filling techniques. Gap filling, especially if done without disclosure, can seriously compromise data quality by misrepresenting completeness and compromising validity.
3. **Precision (and accuracy)**. The precision of valid data tends to be well-characterized so that we know how close the numbers are to reality. When mixing data of varying or unknown precision, it is important to accurately describe the precision of the resulting derived data products or conclusions. Accuracy and precision are closely related concepts via statistics.
4. **Resolution**. The resolution is the level of spatial, temporal, or categorical detail at which a dataset is aggregated; for instance, the "energy" category involves natural gas, gasoline, oil, electricity, etc., and "monthly" is a finer temporal resolution than "annual." Resolution and scale are distinct but sometimes closely correlated concepts; scale refers to the size of a process whereas resolution refers to the detail of the data. If resolution is not finer than scale, there is a fundamental problem with the validity of the data (but this is common).
5. **Provenance** is the lineage and origin of the data. **Auditability** is the gold standard for provenance. However, this standard is rarely achieved for scientific research data because this requires that the quality and provenance of data can be verified by tracking it upstream through the data life cycle to its source, following a chain-of-custody; this auditability is achievable if each step in the data chain (or supply chain) maintains records about its internal processes and also its first-degree connections both upstream and downstream in the system. Do we know where the data came from, what methods were used to process it, and what the known problems are? Valid data tends to have solid provenance.

14.2.3 Data Scale and Resolution

Data Scale and Resolution are closely related concepts. Scale normally refers to the dominant size, speed, generalizability, or frequency of a real-world process. Resolution is the analogy to scale in the data domain. The two will sometimes be used interchangeably in this text. There are some scales and resolutions of data and process that are most relevant to FEW systems, and these are reviewed below. There are four primary spatial resolutions or scales of data in FEW systems: Macro, Meso, Establishment, and Process. These spatial scales allow us to describe agents, processes, and nodes in the FEW system's network structure at some level of spatial aggregation. All scales are useful for answering the same types of questions, but the coarser data is less actionable, and the finer data is more expensive, less available, and carries privacy and security implications. Transportation processes exist at all scales; these processes are associated with the edges and flows between nodes in the FEW system's network graph structure.

14.2.3.1 Spatial Scale and Resolution

The macro-scale involves data aggregation at spatial scales ranging from regions, states, large river basins, or nations, up to the planetary boundary. No greater scale currently exists while humanity is confined on "spaceship earth." At the macro scale, there are few coherent bottom-up data sources, but top-down data sources are both widely available and usually adequate to approximately describe the system. The annual or decadal timescale often corresponds with Macro scale data. The macro scale is also the most common size for the "Macro" scale where averages are taken across extremely broad categories of space, time, and category (Chap. 9). Macro resolution data tends to be minimally useful for FEW systems research, as it tends to capture only the broadest gradients in economic development level or climate type between nations.

Validity and Provenance tend to be good for truly global data (e.g., remote sensing) because data sources and methods are transparent and standardized, but precision tends to be very poor. It is not usually possible to relate global, national, or regional numbers to specific agents on the ground that can make decisions or shape policy based on the information. There are not currently global decision-makers in the FEW space, although there are national agents that operate to a limited extent in the FEW spaces concerning geopolitics and national security. Macro data is actionable for answering questions about regional stresses, regional interdependencies, geopolitical **hotspots**, and planetary footprints and sustainability boundaries. Macro data is inadequate for most FEW systems questions because the FEW system operates primarily at much finer "human" scales. There are no major challenges for data privacy, security, or availability for Macro data. Macro scale data is easy to work with conceptually and computationally and makes nice maps for communication. When working with macro data be aware of nonstandard and extremely incomplete bottom-up data that is "pretending" to be standardized and seamless top-down data; some United Nations information sources, along with for instance the USGS Water Census, fit this type.

The major discontinuity in the top-to-bottom continuum of FEW systems data is between the top-down macro-scale sources and the bottom-up meso-scale sources. It is easy to associate the two by downscaling macro- to meso-scales, but they cannot be expected to show similar results because they come from dramatically different source data. Macro-data is rarely validated by bottom-up comparisons or aggregated from bottom-up data.

The meso-scale is where top-down and bottom-up methods usually meet and is the finest scale at which aggregated census-style data is aggregated up from establishment-level data to avoid the need to invoke the use of private, PII, or PCII data. Meso is one of the two primary spatial scales at which FEW systems function (community scale and process scale; Lant et al. 2019); this scale involves data aggregation at spatial scales of neighborhoods, blocks, municipalities, watersheds, irrigation districts, or counties; seasonal or monthly timescales often correspond with meso-scale data. Meso data captures sub-national gradients in the economy and climate, differences between suburbs and urban cores, and differences between urban and rural economies.

In the USA, the USDA provides meso-scale survey data of uniformly high quality for crop production using standard national collection techniques with documented provenance, but the USGS provides meso-scale data for water use that is aggregated from nonstandard local and state sources with quality and provenance that varies dramatically from state to state. A good example of meso-scale data is the county-level agricultural production dataset provided by the U.S. National Agricultural Statistics Service (NASS 2015), which breaks out annual food production at the level of individual commodities like corn, or soybeans, beef, or the Commodity Flow Survey that provides source and destination commodity transportation information between cities and counties in the USA (BTS 2012).

Aggregation methods at the meso-scale (like all aggregation) are often nonstandard and opaque, and disaggregated source data is by definition usually unavailable, so meso-scale data cannot be quality controlled or audited if this is the case. Establishing the provenance of meso-scale data, for this reason, is hard. Meso-scale data accuracy and precision can be very good in one nation, region, or city, but very poor in another. Additionally, the largest and most essential establishments (i.e., the "fat tail" participants) are often dropped from meso-scale aggregations to preserve business privacy, as in the case where one large company is responsible for the vast majority of energy production in a local area. This is a significant hazard in meso-scale data products and can result in under-reporting of the most essential establishments in the FEW enterprise.

Some of the most important social, economic, and government interests in FEW operate at the meso-scale, so meso-scale information is highly actionable especially for local policy and infrastructure purposes. A large fraction, and probably the majority, of the information content of bottom-up FEW data, is preserved at the meso-scale. Meso-scale data is actionable for identifying infrastructure-scale dependencies, stresses, and vulnerabilities, for managing city-scale FEW systems, for linking cities with their rural FEW support systems, for water resource management, for mapping ecosystem-level hotspots in streams and terrestrial systems, for making coarse-level sourcing and siting decisions, and for assessing vulnerabilities at the scale of most human and natural disasters. Individual businesses are not resolved, but clusters of FEW businesses are resolved at the meso-scale.

Meso-scale data is more computationally challenging to work with, but is still relatively feasible given modern computing resources and software tools. Because of the plurality of spatial units that encode meso-scale data, mismatched spatial domains create significant problems for systems analysis.

The **Establishment** scale is often called the "address" scale or "customer" scale of an individual building, residence, or facility. The establishment spatial scale may correspond to any timescale of data. The Establishment scale is intrinsically private, PCII, and/or PII categorized unless an exception is granted (e.g., for a public establishment that is not critical infrastructure). Establishments are often the resolution at which data is reported due to the survey methods (mailed to specific addresses) and sensors (e.g., water or gas meters) that are employed, but establishments tend to represent an agglomeration of processes under a single ownership and management, rather than an irreducible process. Nevertheless, the establishment spatial scale is often the closest we can get in practice to the "micro" scale for analysis and research methods purposes (Chap. 9).

Scales that refer to a networked entity than a spatial unit are the **Firm**, which is a single economic entity that operates at multiple establishments, and the Enterprise, which is a broader arrangement that may include multiple Firms and Enterprises. A common confounding factor for establishment data is that firms tend to aggregate data at the level of the Firm rather than collecting it for individual Establishments. Firm-level data may be incorrectly coded as establishment-level data or vice versa.

Examples and terms for the establishment scale include address, individual, business, facility, point, process, warehouse, treatment plant, dock, terminal, grain elevator, or reservoir. Establishment scale data is of private origin and is collected utilizing surveys, transactions, permits, tax records, and increasingly digital surveillance and tracking—usually by the government or large corporations. A good example of establishment scale data is the **private data** on US businesses that are provided via carefully restricted access by the U.S. Census Bureau's Federal Statistical Research Data Centers.

Establishment scale data is highly variable in availability and quality, depending on who is collecting the data, the purpose of the data collection, and whether there are government requirements and standards for data collection. One type of data will be highly accurate, and 100% accessible, as in the USA for property tax records, and another type of data will be practically nonexistent, as for most individuals' and companies' FEW consumptions and purchasing data. This data is not originally or primarily collected for FEW research purposes, so some information that is critical for FEW systems analysis can be completely deficient, and other data is abundant at the establishment level.

Establishment scale data is highly actionable and valuable for research because the FEW system's function is an aggregation of Billions of microeconomic and microenvironmental decisions that are often made at the scale of a household or business facility. Individual data can change individual decisions and empower direct and immediate solutions by providing actionable information about FEW system quality, reliability, sourcing, etc. Establishment scale data is "big data" and poses daunting challenges for data science and computational analysis.

The **Process** scale involves the creation, transformation, or transportation of a specific product, good, or service, and implicates the inputs and outputs of the process. The process scale most nearly corresponds to the "micro" scale that is irreducible (Chap. 9). Examples and terms for the process scale are a product, service, machine, generating unit, farm field, stage, step, module, and value-add. Process scale data may be associated with a firm or enterprise rather than an establishment, or, alternatively may be associated with the establishment where the process is carried out. Examples of processes include no-till corn growing, power generation with a natural gas turbine, a dishwasher, lawn watering, bathroom operation, paper milling, and so on. This data is usually private and may be considered Trade Secrets but is generally not categorized by law as PCII or PII. This data is essential for the determination of product-level performance metrics and Key Performance Indicators (Chap. 12).

Establishment scale data is not detailed enough for some important applications. What we really need for FEW systems research is process scale data. The process scale is the natural scale for calculations of efficiency, inputs, and outputs because processes by definition (as opposed to establishments) transform inputs into outputs. Most establishments host dozens or even thousands of FEW processes including human, technological, and environmental processes. Working at the establishment scale confounds these processes, and prevents us from precisely linking establishment-level inputs with the multiple products and services that are the outputs of a typical establishment. For example, many electrical power plants have several different types of generators using different processes, in a single facility, and refineries and farms likewise produce multiple outputs using multiple processes at the same establishment. Human processes tend to be associated with establishments in n:1 cardinality relationships and to be smaller than establishments. Processes can be of any size and are not usually associated with establishments.

The most actionable process information concerns major businesses and utilities and infrastructure processes because these are stakeholders in the FEW system that concentrate a large volume of FEW decisions in a small number of establishments. Fortunately, these large operators are among the most sophisticated agents in the FEW system, and they are often interested in FEW systems research. Industry and utility consortia like The Sustainability Consortium are forming around data sharing, and these efforts are among the most promising developments for establishment scale primary data collection and sharing. Notably, product sustainability labeling efforts are process-level efforts.

The process-scale data is a frontier of data science and sensor technology that links to industrial engineering, microeconomics, business, behavioral, social science, policy, and systems management. The internet of things, combined with new business models and regulatory requirements, will move toward process-level data collection in FEW systems. Process scale data shares data privacy and quality and **computational complexity** issues with establishment scale data. Process scale data may be simpler and more precise than establishment data because processes' inputs and outputs tend to be more sharply defined—especially for the outputs that are products—than for establishments. Many businesses can clearly identify the inputs and output of a process or a product, and also for the business as a whole including all its establishments, but do not collect the same data at the level of individual facilities.

The methods for obtaining data across these scales and resolutions tend to broadly fit two classes: top-down and bottom-up. **Top Down** methods involve estimation or approximate measurement from afar (often, from space!) and intrinsically enforce a "mass balance" so that all is included in the aggregate measure. Typically, top-down methods are very limited in their resolution, validity is difficult to confirm, and the validity of top-down data degrades further via disaggregation error as resolution grows finer, but privacy and completeness are not problems. **Bottom Up** methods involve direct measurement at the process scale, and subsequent aggregation and gap filling of many process observations to estimate "mass balance" at larger scales; typically, bottom-up methods are expensive, require cross-organizational coordination and communication, have large gaps in completeness, and create privacy problems, but validity, precision, and resolution are not problems (Fig. 14.1).

There are four primary temporal resolutions for FEW system data, each of which corresponds to major temporal scales of FEW system dynamics: Decadal, Annual, Seasonal, and Hourly. Coarser time resolution data is useful for understanding long-term structural trends in the FEW system but is not as useful for operational decision-making or process modeling, and it lags the present by many years—too

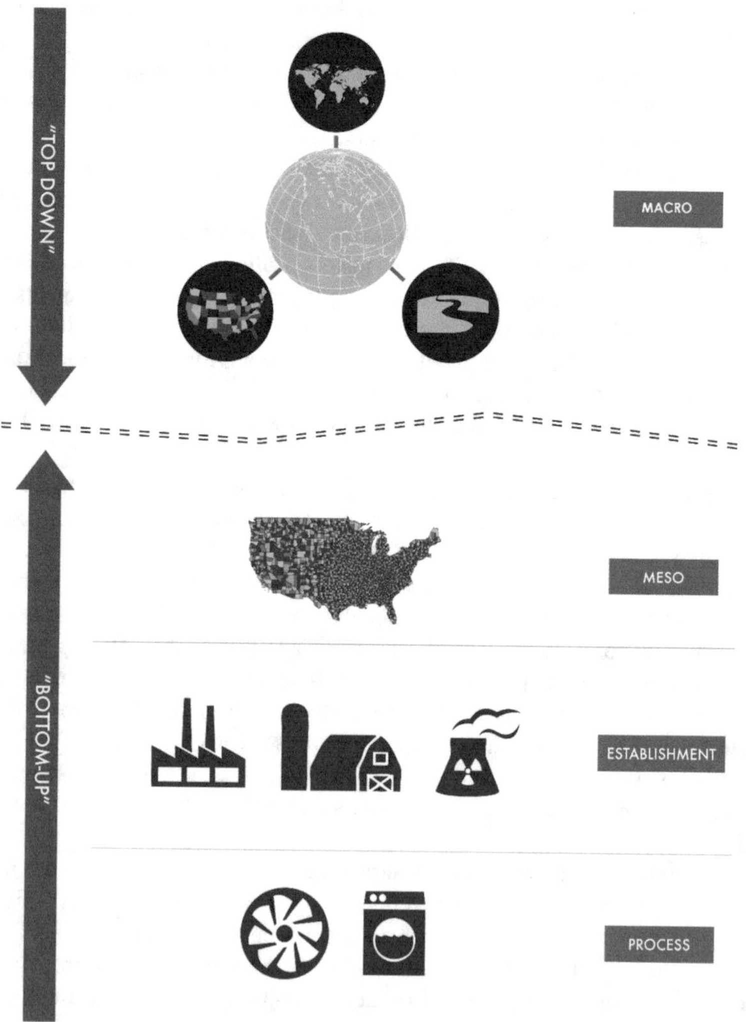

Fig. 14.1 Illustration of the usual relationship between Macro, Meso, Establishment, and Process scales, and the top-down and bottom-up methods, in a FEW system

long to use for operational decision-making. The coarsest possible FEW systems data is Paleo data, which describes ancient patterns in FEW systems before modern data became available. Long-term climate models are similarly coarse, with resolution on an (arguably) decades-to-centuries timescale.

Fine time resolution data can sometimes be used as "early warning" of problems and for operational decision-making. Fine dynamics also allow scientists to establish cause and effect in FEW systems robustly and to build accurate models (as opposed to simply observing long-term trends). However, fine temporal resolution data below the Annual scale is relatively rare and unavailable in FEW systems, even for establishments and processes that are identified at a fine spatial resolution. The higher-frequency and near-real-time sensor stream data needed for operations pose serious software engineering challenges.

The decadal timescale is most common for census data publication. Decadal timescale FEW system process dynamics include for instance changes in technology, economic and population growth, demographics, changing culture and consumption patterns, collapse and conflict, natural capital accounts and natural resource sustainability, infrastructure development, changing trade patterns, changing legal environments, climate, strategic reserves and storage systems, elections, and policy. Annual timescale dynamics are often aggregated up to decadal scales for publication and reporting.

The annual timescale is the most common establishment scale survey data reporting resolution. However, these annual data tend to represent processes and dynamics that are either slower or faster. Annual data tend to be coarse aggregations of hourly and seasonal processes and patterns, summarizing total outcomes but losing most of the process dynamics arising from finer scales. Annual data also tend to reflect decadal trends but contain more noise than the decadal data with respect to those trends. Downscaling annual data reports to the more appropriate seasonal scale, or obtaining seasonal data, is a fundamental problem for FEW systems research and operations.

Seasonal timescale data is rarely collected empirically but is a common resolution for model output. This seasonal timescale corresponds to everything from days to months but is most commonly 1 month. Seasonal processes in FEW systems include Weather norms and averages impacting agricultural production, transportation, and energy demand, drought, high or low water levels, the water year with low storage after summer and high storage in spring, water demand for agriculture during warm seasons, annual harvests of grain and crops, urban water demand patterns (peaking in summer for irrigation), urban energy demand patterns (peaking in winter for heating and in summer for cooling), liquid fuel demand cycles (higher in summer due to more travel), storage system filling and discharge, streamflow, renewable energy production, water stress, water quality (especially temperature), and utility pricing.

Hourly timescale data refers to scales from seconds to days, but nominally 1 h—because increments of 1 h resolve the key diurnal patterns of the human life cycle and business activities impacting the FEW system but also because weather and ecological functions change dramatically between day and night. Urban electrical demands are higher in the evening due to lighting demands and higher mid-afternoon in summer due to air conditioning demands; these "peak" demand periods determine the generation capacity needed in the electrical power system and require that the generators can "ramp" power production rapidly up and down to match demand peaks. Urban water demands peak morning and evening corresponding to washing and cooking times. Cities store water; they produce it at night when electrical power is less expensive, and demand is lower and consuming it during the day. Cities need to handle fires and power outages without losing potable water supply and use elevated storage to maintain system pressure during these events. Power outages usually last hours and at most days, because of storms and other random damage to the system. Solar and wind renewable energy production has a strong diurnal pattern, with solar peaking late morning and wind peaking at times determined by geography. Liquid fuels are burned mostly during the day by vehicles in city traffic, and this is when most air pollution is generated in cities as a result. "Demand management" in the water and electrical utility space focuses primarily on reducing peak rates of demand for electricity and water in these systems. When water and energy demands are "on peak" at high-usage times of day, utility prices tend to be higher. People eat and wash at specific times of day, and this drives diurnal water and energy demands in cities, in addition to creating strong "food peaking periods" at three times during the day (breakfast, lunch, supper).

Process Scale and Resolution

There are various levels of detail with which we can describe processes and encode our knowledge as data—that is, multiple levels of Process Resolution. Process Classification is an exercise in Ontology. There is a need to establish Process Classifications and Ontologies for FEW systems so that processes and products can be benchmarked and described in relation to the greater system's geometry.

For example, the U.S. Geological Survey's National Water Census describes the nation's water using processes at a coarse resolution, using broad categories such as "agriculture," "public supply," "industrial," and "thermoelectric" to describe wide swathes of processes. At the other extreme, Process Classification Frameworks (PCFs) provide highly precise and detailed descriptions of industry-specific processes, from the perspective of industrial engineering and quality control applications (e.g., APQC 2017). PCFs do not exist yet for many FEW-related topics.

Two of the most common economic classifications for FEW systems in the USA are the Standard Classification of Transported Goods (SCTG) Codes used by the U.S. Department of Transportation to code freight flows, and the North American Industry Classification System (NAICS) used for industry type and product coding; NAICS can be ontologically mapped (cross-walked) to the Standard Industrial Classification (SIC) which is an older and US-focused ontology. The FEWSION project has undertaken to develop a new ontology for the food, energy, and water system. The FEWSION codes build on the SCTG and NAICS framework while reorganizing some categories of particular interest to FEW researchers, and also adding categories like electrical energy, potable water, and wastewater.

14.3 FAIR Data Management and the Data Life Cycle

14.3.1 The Data Life Cycle

The **Data Life Cycle** involves experimental and observational design and approval, data collection, quality control, **metadata** description, data curation with a **repository**, discovery (if necessary) of the data resource, integration of the data with other data, and analysis of the data to answer questions. Sound data management practices are essential to enable the full completion (and repetition) of the data life cycle. Data grows more valuable as it proceeds along its life cycle, and even more so with reuse and repeated integration. You should make sure to implement sound data management practices with any data that you collect so that it can benefit yourself and others. (Re)using a scientific workflow or utilizing a mature data repository's workflow systems for data processing can help a great deal.

Unfortunately, we are still very early in the information age, and we have little successful experience with the implementation of the data life cycle to date. Common struggles include critically flawed metadata, use of nonstandard data formats, the lack of use of a formal and well documented ontology or controlled vocabulary, archival on websites or outside standard repositories where findability and accessibility are poor, poor and irrecoverable initial quality control on data and metadata, and lack of issuance of a unique identifier (DOI) for data (Ruddell, 2006). Most critically, the "half-life" problem rapidly erodes our data resources as poor data management processes result in datasets and/or critical metadata or the tools to access those data vanishing steadily over time. The half-life of research data has been estimated at under 10 years; this is an indefensible tragedy, given the expense and value of data in a digital age (Ruddell et al., 2014).

14.3.2 FAIR Data Management

The **FAIR Principles** for scientific data management and stewardship were published in 2016, and provide guidelines for improving findability, accessibility, interoperability, and reuse of data. Findability emphasizes the use of a globally unique identifier for the dataset, use of that identifier within a rich metadata file that is easy for both humans and computers to use, and the metadata is registered with an appropriate searchable index service.

Accessibility emphasizes that data and metadata are retrievable using a standard and open and free communication protocol, and that metadata accessibility persists even when data are not accessible. Interoperability emphasizes that metadata and data:

- Use a formal and widely used language or format.
- Use controlled vocabularies that follow FAIR principles.
- Appropriately reference other FAIR metadata and data.

Reusability emphasizes that metadata and data are richly described, including a clear data usage license, detailed provenance, and the use of domain-specific community standards.

The reader will have an easier time both in publishing and accessing FEW systems data by working with vendors and repositories that implement FAIR principles (e.g. Wilkinson et al., 2016). Implementation decisions concerning these FAIR principles for data publication tend to focus on the choice of a repository for your data. The Coalition for Publishing Data in the Earth and Space Sciences (COPDESS) provides some detailed advice on best practices for implementing the FAIR principles for your research project or other data product project—including the selection of a repository, data and metadata format, compliance with your funding agency's data publication requirements, and so on.[1]

The Library is the ancient institution that specializes in managing the data life cycle. Libraries still (mostly) specialize in data as the printed word, and are the classic archetype of a repository for data.

A **Repository** is generalized library service that curates data and that ideally provides FAIR data management services. Libraries, journals, web portals, search services, and agency data services may be examples of repositories, although it is a rare repository that provides all components of FAIR data management services that you might require. For instance, it is relatively easy in 2018 to access and catalog data using search services, but it is still challenging to find a high-quality repository that is willing to permanently curate your datasets at no charge, and it is rare to find a repository that will donate consulting and expert services to help you reformat, metadata-markup, and package your data properly to comply with FAIR principles.

Different types of repositories and data entities provide different services and serve different roles. If you are working or studying at a university, your library's role is to assist you not only with search and access to information but also with the curation of data from your research. However, many university libraries lack the specialized capacity to do this. Journals are rapidly transitioning from publishers of the written word into full-service knowledge and data curation services and repositories, but some are further along this path than others as of 2018. In many cases, journals require scholars to curate their data elsewhere. If you work for a state or federal agency, in most cases your agency has its own funding and services to curate and provide access to data—but only to those data originated by that specific agency. Private consultancies and data services may provide excellent data products, but these tend to exist behind an access paywall and tend to be targeted to provide narrow information to specific clients with deep pockets—not at the typical researcher. Data consortia provide community standards for data and metadata format and content and may provide indexing and search services, but these consortia rarely provide the curation services themselves. Search services (e.g., Google) have grown sophisticated and broadly skilled in recent decades, but often fail to provide

[1] The Coalition for Publishing Data in the Earth and Space Sciences (COPDESS) provides some detailed advice on best practices for implementing the FAIR principles for your research project or other data product project.

hits on "niche" data products that are the most relevant for your application because these services are general-purpose for a broad audience.

Metadata are "data about data" which describe a dataset's authorship, contents, format, resolution, timing, provenance, sources, methods, a license of use, globally unique identification, etc. Metadata is essential for FAIR data management. Metadata is an essential component of FEW systems data because valid uses of data require a rigorous treatment of quality. The most widely accepted metadata standard for geospatial datasets is the ISO 19136 standard, which is **XML**-encoded and machine-readable by GIS systems. However, when integrating multiple datasets, we routinely find that some of them are missing metadata or that metadata is noncompliant (Ruddell et al. 2014). This can make it difficult to ascertain the origin (i.e., provenance), precision, or methods for quality control on the source datasets. However, a longstanding best practice is to bury the details of the dataset's construction in the original report or published paper that made use of the data—that is, a "publication of record," so this is a good place to look for details on data quality. Regardless of whether the source metadata is intact, it is essential to poduce high-quality metadata to accompany any integrated FEW systems datasets that are published.

Beginning with the Dublin Core workshops, the concept of "core" metadata is now popularized (Dublin Core Metadata Initiative, n.d.). Some of the more important or "core" metadata elements are defined by the metadata standard, but they typically include:

- Title
- Author(s)
- Date of Original Publication (not date the metadata was created)
- Abstract, which usually describes provenance, methods, precision, and other details
- File Format
- Data model or Ontology utilized (which may be implied in some file formats)
- Variable definition headers and Variable units
- Language (e.g., English or "en")
- Conditions of use, License, Permissions, and Security
- Publication of record (for citation purposes)
- GIS projection and datum
- Temporal resolution
- Spatial resolution
- Temporal coverage or bounds
- Spatial coverage or bounds
- Keyword(s) from controlled vocabularies and ontologies
- Identity of the controlled vocabulary (may be implied by the data model or ontology)

ISO is the International Organization for Standardization (or Organisation Internationale de Normalization, in the French), which maintains many essential metadata standards including the widely used ISO19136 geospatial metadata standard and the XML-based Geography Markup Language (GML) that is important for describing much FEW systems data. **XML** is the Extensible Markup Language which is a generic standard for structuring metadata (and less commonly, data) in a fashion that is both machine-readable and human-readable, and also easily auditable against a community standard (e.g., ISO metadata standard). **Federation** is a strategy for linking multiple repositories to enable sharing and search of diverse databases while allowing the originators of the data to maintain control and ownership of data if they desire.

14.3.2.1 Examples of Open Data Repositories for FEW (Mostly US-Focused, as of 2018)

1. Environmental Data Initiative
2. CUAHSI Water Data Center
3. Commodity Flow Survey
4. FHWA Freight Analysis Framework
5. BTS Waterborne Freight
6. USGS Water Use in the USA
7. USDA NASS QuickStats
8. USDA Census of Agriculture
9. Energy Information Administration
10. US Census Bureau Economic Census
11. US Department of Transportation Bureau of Transportation Statistics (BTS)
12. United Nations FAOSTAT
13. BLS Labor Stats
14. BEA Input-Output Accounts Data
15. U.S. Census Bureau Foreign and Port Trade Data
16. EPA's Environmental I/O Model USEEIO
17. EIA Electric System Operating Data
18. USBR Reclamation Water Information System (RWIS)
19. UN/FAO Aquastat
20. WTO Trade and Tariff Data
21. US CIA World Factbook
22. USAID Famine Early Warning Systems Network (FEWS NET)
23. NREL Open Energy Information (OpenEI)
24. North American Cooperation on Energy Information (NACEI)
25. Vulcan Project North American GHG Emissions
26. PRISM Model
27. Multi-Resolution Land Characteristics Consortium (NLCD etc.)
28. USDA NRCS Geospatial Data
29. USGS National Map
30. Google Public Data Directory
31. Google Earth Engine Data Catalog
32. NASA Global Imagery Browse Service (GIBS)

33. NASA VIIRS/Night-Lights
34. NASA GRACE Water Storage Anomalies v1
35. USDA NASS Food Cold Storage Reserves
36. WRI Aqueduct
37. US Data.gov
38. US Library of Congress

14.4 Privacy and Ethics Principles for FEW Data

FAIR data management principles are intended primarily for application to public data, so they must be carefully translated and qualified (but not discarded) for application to private and **sensitive data** that are only ethically made accessible to a specific group of qualified users. If data = information = knowledge = power, and if FEW systems can be manipulated to gain power over people (as they often are!), then data concerning people's critical FEW infrastructures and lifelines is both powerful and dangerous in the wrong hands. There is a great deal of very clear risk, whereas the benefits of releasing this data are often unclear—or are very clearly not worth the risk. As a result, in the USA and most other countries the government and census data repositories containing FEW systems data err on the side of caution by releasing only heavily aggregated and redacted FEW systems data, or none at all, and regulations often prevent private companies and nongovernmental organizations from releasing this data. In many other cases, private entities elect not to release private data unless they are strongly incentivized to do so. This is an ethically defensible, practical, and precautionary choice, but a deeply frustrating one for FEW systems researchers. Because privacy and ethics are such central considerations in FEW systems data accessibility and research, the student or researcher must have a grasp of the logic behind these obstacles in order to successfully navigate the metaphorical data landscape.

Data Ethics concerns the determination of right and wrong data practices, especially for sensitive and private data. The conversation on Data Ethics concerns the ownership of data, transparency over how it is to be used, consent to how it will be used, privacy and control over privacy of some data, value of the data and knowledge of data transactions, and accessibility of specific types of datasets by specific types of people or algorithms. Important questions include, for instance, "Who owns the data on my food, energy, and water use?" "Can I control who knows my food, energy, and water use and sources?" "If I tell you my food, energy, and water use, are you allowed to tell others without asking me?" and "Do I have a right to know how you are using my food, energy, and water data?" Imagine a future where data science is capable of measuring and predicting almost anything about a person's needs, choices, and behavior. Should we use that data-driven foresight to prevent people from making harmful choices, or should we allow people to exercise their own free will for better or worse? And who, if anyone, should possess that foresight? These are critical questions for the twenty-first century—and for the twenty-first century's critical food, energy, and water systems.

If there are potential sensitivities or risks involved in the disclosure of data, that data is **sensitive**. Most FEW systems data is sensitive, along with (for instance) a person's medical history (e.g., HIPAA), financial status, student records (e.g., FERPA), web searches, and locations of travel. If data is covered by an individual's or organization's right to privacy, this data is **private**. Privacy is most commonly granted by custom and law for information about which nobody but the individual or organization concerned has a legitimate need to know. **Trade Secrets** are private data that a business elects to hold private to avoid providing an unfair advantage to competitors; many businesses consider their supply chains including FEW usage and production to be trade secrets.

Data Protection is the law, and/or means of enforcement of the law, protecting private or sensitive data and exercising the right to data privacy. You cannot achieve Data Ethics or privacy without effective data protection practices. Data protection principles include minimization, accuracy control, expiry, legitimate purpose, transparency (accountability), integrity (effectiveness), and confidentiality. Privacy International identifies seven key data protection principles (quoted below; Privacy International, n.d.).

14.4.1 Privacy International Data Protection Principles

- Fair, Lawful and Transparent: The processing of personal data should be lawful and fair and done in a transparent manner.
- Purpose Limitation: Personal data should be processed for a specified, explicit, and legitimate purpose, stated at the point of collection, and further processing (remains) compatible with this purpose.
- Minimization: The processing of personal data should be adequate, relevant and limited to the necessity of the purpose for which it is being processed.
- Accuracy: Personal data that is processed should be accurate, complete and measures should be taken to ensure it is up to date.
- Storage Limitation: Personal data should only be retained for the period of time that is necessary for the purposes for which it was processed.
- Integrity and Confidentiality: Appropriate measures must be taken to ensure the security of data and systems, and to protect personal data from loss, unauthorized access, destruction, use, modification, or disclosure.
- Accountability: Those that process personal data must be accountable for demonstrating compliance with the above principles, their obligations, and facilitate and fulfill the exercise of these rights.

It is easy to see that many modern businesses violate these data protection principles routinely, for instance by failing to limit the purpose of customer data's use to the originally intended purpose, failure to be transparent about how data is being used or sold, or by failing to protect the confidentiality of the data from hackers, or by minimizing data collection to that which is strictly necessary.

However, consider the counterargument—especially from a government's point of view—that it is difficult to know exactly how data could be legitimately used. Many types of data are only useful when combined with other data that may not yet exist. The value of these data grows over time. For example, by "joining" independently collected energy use, water use, demographic, building-type, and process level data, it might be possible for us to precisely understand how to help make a given business, household, or process more efficient and sustainable. What if this exact type of combined use of the data was not anticipated at the time of its collection several decades earlier, and the person or business about whom the data was originally collected no longer exists to seek consent? Is it ethical to make this use of the data, or not?

Need-to-Know is a precautionary data access principle that minimizes access to sensitive data to only those persons with a legitimate purpose. This access minimization, along with the screening and identification of persons accessing data, provides a basic level of data protection along with deterrence of and accountability for unethical or illegal employment of the data. **Security through Obscurity** is a tactic, often unintentional or implicit, of protecting data by minimizing its findability, accessibility, and interoperability. Many government agencies practice security through obscurity, as do many private individuals. Government agencies do this primarily as a cost savings measure and precautionary measure because they do not have a funded mandate to provide FAIR data management services that also protect data properly.

Need-to-know is integral to many data protection processes, most notably classified and categorized data access. Classified data is only available to highly qualified and exhaustively screened individuals like military personnel or high-ranking officials. In the USA, Protected Critical Infrastructure Information (**PCII**) is a government-enforced national security categorization (not classification) originating after the September 11th, 2001 attacks. Under this categorization, some food energy or water infrastructure data is available only on a need-to-know basis. Research may not qualify as a need-to-know, depending on its utility and on who is doing the research. PCII data specifically includes the precise "target list" locations of key pumps, transformers, storage depots, pipelines, or transmission lines for food, energy, and water.

The bedrock and central Data Ethics principle is that the benefits of collection and use of private or sensitive data must outweigh the risks from the perspective of the object of the data collection (that is, the people and ecosystems involved). This principle will be familiar to any university researcher who has undergone the Institutional Review Board (IRB) process to scrutinize their data collection methods. In other words, the privacy of data must be balanced against the legitimate benefit of the use of the data—that is, the **utility** of the data. One cannot usually optimize both utility and privacy, because they are in tension in most cases. For instance, water efficiency researchers could derive significant utility by accessing individual customers' water and energy use data, but this release could be sensitive and risky for some individual customers and so most US States grant a right of privacy to the customer's billing data (but allow release of aggregated deidentified billing data, which minimizes risk and still allows substantial utility). **Data utility** underpins need-to-know.

A closely corollary data ethics principle is that of **Informed Consent**, which is written explicit consent granted by a private party to allow public release of private data or its use for a specifically defined purpose. Informed consent requires that the private party is reasonably capable of being informed and of granting consent based on a description of the intended data use and its risks and benefits and that the consenting party is under no compulsion to do so. As a result, informed consent cannot be granted in most cases by prisoners, citizens of autocracies and dictatorships, children, and so on. Informed consent guarantees that the object of the study understands and agrees with the data ethics of those doing the data collection. Ethical researchers are required to obtain informed consent before collecting new human subjects (objects) data in most cases. Private businesses usually insist, by the habit of legal counsel, that their customers grant informed consent to the use of all private customer data as a precondition of doing business. As a result, large private businesses like McDonalds, Wal-Mart, or Aldi possess a great deal of precise customer-level data about individuals' food consumption, and this data is effectively the property of the private business instead of the originating individual. Public energy and water utilities have historically neglected to obtain informed consent, and are therefore unable to release utility customer data.

Personally Identifiable Information (**PII**) is a distinct data protection categorization identifying some types of an individual person's data as being explicitly covered by data privacy and data protection law. In the USA, HIPAA (health data) and FERPA (student data) laws, along with utility customer billing data privacy laws, are good examples of PII categorization. **Utility customer data** concerns an individual's potable water, sewerage, natural gas, heating oil, or electricity usage. **Safe Harbor** is a legal data protection against release of specifically named PII data, and safe harbor principles apply unless a qualified **expert determination** is made that a narrowly defined scope of data release does not create a significant risk of identification and/or harm to the person(s) involved in this specific context and application of the data based on the data protection methods employed. Expert Determination usually requires careful de-identification of PII data to remove the risk that an individual's identity will be disclosed publicly, while simultaneously maintaining the utility of the data for the specifically proposed application.

If you want to access private data such as a customer's or business's food consumption or water or energy usage, you need to ask the private party for informed content and release of that data. If you want to access categorized public data such as health records, student records, precise infrastructure data, or utility customer billing data for energy or water usage, you will need to apply and satisfy the authorities' ethical and legal concerns regarding your need-to-know and your expert determination that your methods of data collection and use favorably balance risk against benefit to the objects of the data (and then additionally you must overcome the security through obscurity). If you want to access classified data, you need to enter the national security sector… and then you must still demonstrate need-to-know. In the USA, many National Laboratories have the necessary security credentials and facilities to work with PCII and classified data, and are the primary venues for research using these data. Also in the USA, the U.S. Census Bureau's Research Data Centers are secure facilities for research using PII data under proper data protection protocols. **Differential Privacy** is a database technology that allows users to query only those fields and joins that are permitted based on their credentials; this preserves the full potential utility of the data while tailoring privacy to suit the user's application. Differential privacy is a promising future tool to enable research, but it is not yet broadly implemented for FEW systems data repositories.

Aggregation is a common technique for de-identifying PII or PCII or other private or sensitive data by reporting the space-time location, stock, flow, etc. for a group instead of an individual. The coarser the resolution of the aggregation, the greater the privacy and lower the utility and quality of the data. For utility customer data, the Rule of Fifteen is a common best practice governing legally minimal aggregation. The **Rule of Fifteen** is a common legal threshold in US State law specifying that utility customer data and other PII data must be de-identified through aggregation into groups of not less than 15 individuals, any one of which comprises not more than 15% of the group's total usage volume. This rule is a crude approximation of the census bureaus' careful statistical practice of aggregating their PII survey data at a resolution of census blocks, tracts, municipalities, or counties (etc.) to preserve the privacy of individuals and companies' data. The Rule of Fifteen is easy to apply to residential utility customers, of which a city will have many thousands, but it difficult to apply to industrial and business customers, of which a city may have only one or a few of a given type. An especially problematic type data quality problem—incompleteness is created by census bureaus when especially large companies are dropped from datasets for the sake of anonymization and privacy. Because the "fat tail" of the distribution of users—the extremely large producers and consumers—are responsible for a large fraction of the FEW system function, they cannot be dropped without dramatically damaging the completeness and validity of the data. Thus aggregation is not an adequate solution to the privacy and sensitivity problem, because it fails for some of the most important and valuable use cases involving key industries or address-level analysis of energy and water usage.

The opposite technique, **Disaggregation**, attempts to reverse aggregation and achieve finer resolution using assumptions which trade reduced validity for increased precision/accuracy. Disaggregation is fundamentally a modeling method that cannot create new information and is therefore only marginally useful—and should be used only with great caution to avoid misrepresentation. It is far better to go get valid data at the resolution required, if possible than to estimate those data using disaggregation. If this is not possible, one should ask themselves whether it is ethical in this application to make inaccurate estimates of data that are being intentionally held private for reasons of sensitivity. If the real data are sensitive, then surely the estimated and inaccurate data are equally sensitive and also potentially misleading.

14.5 Domain-Specific Strengths and Gaps in FEW Systems Data

Every place and time in the world has a unique FEW system, and the data needs and accessibility are every bit as diverse as the systems they represent. In the developing world, formal FEW systems data may not be available from governments and companies and may need to be collected as needed directly at the source. In authoritarian countries, FEW systems data may not be accessible because it is classified as a state secret. In liberal countries, some FEW systems data is readily available from government statistical bureaus in the aggregate, but detailed data about establishments and individuals may be fiercely protected by privacy law and critical infrastructure security restrictions. In all cases, there are critical gaps in FEW systems data coverage. Often these are due to privacy and security concerns, but sometimes these are due to the difficulty or expense of collecting or modeling a domain of data, or due to a lack of capacity on the part of the government.

Major application domains for FEW system data include census, infrastructure management, business supply chain intelligence, product tracking, life cycle analysis and footprinting, last mile and emergency management, input-output, stock and storage management, trade, commodity flow, thresholds regulations and carrying capacities, waste management, vulnerability, products, and customer level data tracking.

It is impossible to survey all the available datasets or gaps in data for FEW systems. However, the reader is directed to the FEWSION project for a USA oriented list of relevant datasets, and to the National Science Foundation's INFEWS program funded projects list for an overview of projects that are aiming to address specific high priority gaps. Below we list some particularly notable data assets and gaps in US FEW systems data.

14.5.1 Strengths

- Global model estimates (e.g., GCAM, climate models)
- Global remote sensing of agricultural yield and land use
- Global scale estimates of groundwater depletion (e.g., GRACE)
- Meso-scale data, for example, at the county scale
- Annual and Decadal data in general
- Global scale input–output and life cycle models
- Freight flow data for voluminous and massive commodities at the meso-scale
- Selected strategic reserve storage (e.g., petroleum reserves, cold storage of food)
- Things we can see from space, like nighttime lights for energy use, or land use for agricultural production
- Large urban and industrial water users and wastewater dischargers, at the meso-scale

- Energy production and generation processes
- Production data in general, at the meso-scale
- PCII and secure Military and Homeland Security intelligence on critical facilities
- Land availability and suitability, for example, for agriculture

14.5.2 Gaps

- Establishment data of all kinds
- Process data of all kinds, especially for factory processes and agribusiness
- Fine-resolution water use
- Water consumption, wastewater, and return flows
- Water availability
- Spatially explicit supply chain
- Product or process level performance benchmarks
- Storage data is patchy; water storage, in particular, is missing
- Electrical power grid function and utilization
- Transaction data in general
- Monthly data in general
- Hourly data in general
- Consumption data in general at all scales
- Legal data in general (e.g., water rights)
- Spatially or temporally explicit meso-scale input–output and life cycle
- Stocks and flows, with exceptions noted above
- Valid and verifiable meso-scale and global scale data using bottom-up methods
- Publicly available corporate supply chains
- Things we cannot see from space, like power plants, food processing facilities
- Differential privacy methods to efficiently work with PII and PCII data
- Standardized "reference" datasets describing the integrated system as a whole
- "Last Mile," supply chain data, linking to retail and distribution facilities for emergency response
- "Real-time" data without long lead times
- Linkages from the meso-scale to the establishment scale

Key Points

- Data comes from both observations and models and is useful to generate metrics.
- Data quality has several aspects including but not limited to validity, completeness, precision, resolution, and provenance; this quality is relative to the end use.
- Data management best practices must be followed throughout the Data Life Cycle (e.g., FAIR (meta)data principles); this implies a central role for data repositories, data sharing consortia, standards bodies, and search services.

- Rules for privacy, protection, ethics, and also laws governing individually identifiable and critical infrastructure data are overriding concerns for FEW systems data.
- Some common data aggregation and access principles explain the type of FEW systems data that is commonly accessible (and nonaccessible) in the USA.
- There are common data scales at which FEW systems are described, and these data scales do not match all problem scales; high-quality FEW systems data is accessible for some geographies, scales, and application domains, and not for others.
- FEW systems data cover a myriad of highly specialized public and private applications, and these are voluminous, complex, and diverse with respect to data structure and standard, as well as the repositories that handle each application.

Discussion Points and Exercises

1. Many types of data are only useful when combined with other data that may not yet exist. The value of these data grows over time. For example, by "joining" independently collected energy use, water use, demographic, building-type, and process level data, it might be possible for us to precisely understand how to help make a given business, household, or process more efficient and sustainable. What if this exact type of combined use of the data was not anticipated at the time of its collection several decades earlier, and the person or business about whom the data was originally collected no longer exists to seek consent? Is it ethical to make this use of the data, or not?
2. Have you ever had your privacy violated? How did it feel?
3. Is there a specific type of food, energy, or water data about yourself or your business that you would not be comfortable sharing with the world? Why or why not? And, at what level of aggregation of that data with others' data would you become comfortable with the release? Why?
4. Identify an example of data "security through obscurity" in the FEW data space, and distinguish that example clearly from a similar case where data protection was intentionally applied to restrict access to data.
5. Is there a specific type of FEW data that you need for your work and have been unable to access? Why was that—data protection practices, law, privacy, or poor application of FAIR principles?
6. Explain the four major FAIR data management principles.
7. Choose a data repository with which you are familiar and evaluate it against both the FAIR data management principles and data ethics principles.
8. Identify an appropriate repository for a dataset that you would like to publish, following the COPDESS and/or Stanford Library guidelines. Attempt to follow that repository's processes to publish your data.
9. Identify the steps in the Data Life Cycle, in your own words.
10. Have you ever been unable to find a dataset? Why and what was the specific problem?
11. Describe how data aggregation works and precisely identify the difference between adequate and inadequate levels of data aggregation for a private or sensitive FEW dataset with which you are familiar.

12. Who owns the data on your food, energy, and water use?
13. How can you control who knows your food, energy, and water use and sources?
14. If you tell someone else your food, energy, and water use, are they allowed to tell a third party without asking you?
15. Do you have a right to know how your data are being used?
16. Should we use data-driven foresight to prevent people from making harmful choices, or should we allow people to exercise their own free will for better or worse? And, who if anyone should possess that foresight?
17. Name the most important and widely used metadata standard for a community of practice with which you are familiar.
18. Name the most important and widely used data standard for a community of practice with which you are familiar.
19. Is Mark Zuckerberg right—is privacy dead? Should we let it die? Why, or why not? Does anything need to change?
20. Do you disagree with any of Privacy International's data protection principles? Why or why not, and in what specific instance would you agree or disagree?
21. If your town or country went to war tomorrow with a dangerous enemy, what parts of your food, energy, and water data would you not want that enemy to have access to?
22. What is the difference between data that is generated by a model, as opposed to data that is generated by empirical observations?
23. Identify an end use of FEW data. For instance: developing your personal water footprint including both direct and indirect uses of water. Determine what would be adequate data quality for that end use, and attend to each aspect of data quality described in this chapter. Would any of the data quality requirements change if you had to defend that number in a court of law, or during a tax audit?
24. What is an ontology? Give an example of a FEW data space.

References

APQC. (2017). *American productivity and quality center process classification framework*. Retrieved April 28, 2017, from https://www.apqc.org/industry-specific-process-classification-frameworks.

Baggio, J. A., et al. (2016). Multiplex social ecological network analysis reveals how social changes affect community robustness more than resource depletion. *Proceedings of the National Academy of Sciences of the United States of America, 113*(48), 13708–13713.

BTS. (2012). *Commodity flow survey*. U.S. Department of Transportation Bureau of Transportation Statistics. Retrieved April 28, 2017, from https://www.rita.dot.gov/bts/sites/rita.dot.gov.bts/files/publications/commodity_flow_survey/index.html.

BTS. (n.d.). *SCTG Codes*. Retrieved December 4, 2018, from https://www.bts.gov/archive/publications/commodity_flow_survey/classification.

Dublin Core Metadata Initiative. (n.d.). Retrieved December 4, 2018, from http://dublincore.org.

EIA. (2017). *Annual energy outlook 2017*. Washington, DC: U.S. Energy Information Administration.

FEWSION. (n.d.). Retrieved December 4, 2018, from https://fewsion.us.

FEWSION Codesheet 1.0. (2019). Retrieved December 4, 2018, from https://fewsion.us/data.

Lant, C., et al. (2019). The U.S. food–energy–water system: A blueprint to fill the meso-scale gap for science and decision-making. *Ambio, 48*, 251–263.

Maupin, M. A., Kenny, J. F., Hutson, S. S., Lovelace, J. K., Barber, N. L., & Linsey, K. S. 2014. *Estimated use of water in the United States in 2010* (56p.). U.S. Geological Survey Circular 1405. https://doi.org/10.3133/cir1405

Miller, R. E., & Blair, P. D. (2009). *Input-output analysis: Foundations and extensions*. Cambridge: Cambridge University Press.

NASS. (2015). *Agricultural statistics 2015*. Washington DC: U.S. Department of Agriculture National Agricultural Statistics Service (NASS), U.S. Government Printing Office.

OMB. (2017). *North American industry classification system United States 2017 (NAICS)*. Executive Office of the President Office of Management and Budget (OMB). Retrieved April 28, 2017, from https://www.census.gov/eos/www/naics/2017NAICS/2017_NAICS_Manual.pdf.

Privacy International. (n.d.). *A guide for policy engagement on data protection. Part 3: Data protection principles*. Retrieved December 4, 2018, from https://privacyinternational.org/sites/default/files/2018-09/Part%203%20-%20Data%20Protection%20Principles.pdf.

Ruddell, B. L. (2006). Scientific metadata: Back to basics. In *ICHE Conference 2006*. Philadelphia, PA: Drexel University.

Ruddell, B. L., & Kumar, P. (2006). Hydrologic data models, Chapter 5. In P. Kumar (Ed.), *Hydroinformatics: Data integrative approaches in computation, analysis, and modeling*. Boca Raton, FL: CRC Press.

Ruddell, B. L., & Kumar, P. (2009). Ecohydrologic process networks: 1. Identification. *Water Resources Research, 45*, W03419. https://doi.org/10.1029/2008WR007279.

Ruddell, B. L., Zaslavsky, I., Valentine, D., Beran, B., Piasecki, M., Fu, Q., & Kumar, P. (2014). Sustainable long term scientific data publication: Lessons learned from a prototype Observatory Information System for the Illinois River basin. *Environmental Modelling & Software, 54*, 73–87. https://doi.org/10.1016/j.envsoft.2013.12.015.

Wilkinson, M. D., et al. (2016). The FAIR guiding principles for scientific data management and stewardship. *Scientific Data, 3*, 160018. https://www.go-fair.org/fair-principles/.

Further Reading

Best Practices for Data Management. (n.d.). *DataOne*. Retrieved December 4, 2018, from https://www.dataone.org/best-practices.

Boyd, D., & Crawford, K. (2012). Critical questions for big data: Provocations for a cultural, technological, and scholarly phenomenon. *Information, Communication & Society, 15*(5), 662–679.

COPDESS. (n.d.). *Enabling FAIR data FAQs*. Retrieved December 4, 2018, from http://www.copdess.org/enabling-fair-data-project/enabling-fair-data-faqs/.

Data Best Practices. (n.d.). *Stanford libraries*. Retrieved December 4, 2018, from https://library.stanford.edu/research/data-management-services/data-best-practices.

Data Life Cycle. (n.d.). *DataOne*. Retrieved December 4, 2018, from https://www.dataone.org/data-life-cycle.

DHS. (n.d.). *Protected Critical Infrastructure Information (PCII) Program*. Retrieved December 4, 2018, from https://www.dhs.gov/pcii-program.

Dwork, C. (2008). Differential privacy: A survey of results. In *International Conference on Theory and Applications of Models of Computation*. Berlin: Springer.

Floridi, L., & Taddeo, M. (2016). What is data ethics? *Philosophical Transactions of the Royal Society A, 374*, 20160360.

HHS. (2003). *HIPAA privacy rule for research*. Retrieved April 24, 2018, from https://www.hhs.gov/sites/default/files/ocr/privacy/hipaa/understanding/special/research/research.pdf?language=es.

HHS. (2012). *Guidance regarding methods for de-identification of protected health information in accordance with the Health Insurance Portability and Accountability Act (HIPAA) Privacy Rule, November 26, 2012*. Retrieved May 1, 2018, from https://www.hhs.gov/sites/default/files/ocr/privacy/hipaa/understanding/coveredentities/De-identification/hhs_deid_guidance.pdf.

ICPSR. (2018). *Recommended informed consent language for data sharing*. Institute for Social Research at the University of Michigan. Retrieved May 1, 2018, from https://www.icpsr.umich.edu/icpsrweb/content/datamanagement/confidentiality/conf-language.html.

Knaflic, C. N. *Storytelling with data: A data visualization guide for business professionals*. Hoboken, NJ: Wiley 978-1119002253.

McKinney, W. (2017). *Python for data analysis: Data wrangling with Pandas, NumPy and IPython* (2nd ed.). Newton, MA: O'Reilly Media 978-1491957660 (or current edition).

O'Neil, C., & Schutt, R. (2013). *Doing data science: Straight talk from the frontline*. Newton, MA: O'Reilly Media 978-1449358655.

Press, G. (2013, May 28). A very short history of data science. *Forbes*. Retrieved December 4, 2018, from https://www.forbes.com/sites/gilpress/2013/05/28/a-very-short-history-of-data-science/#2fbb886355cf.

Retrieved May 1, 2018, from http://www.wpuda.org/assets/Energydocs/model%20data%20privacy%20guideline%20for%20large%20utilities%2009%2008%2016.pdf.

Richards, N. M., & King, J. H. (2014). Big data ethics. *Wake Forest Law Review, 49*, 393.

State of Washington. (n.d.). *Data privacy guidelines for large utilities*. Washington, DC: State of Washington.

Towards Data Science. (n.d.). Retrieved December 4, 2018, from https://towardsdatascience.com/.

Chapter 15
Modeling

Fernando R. Miralles-Wilhelm

15.1 Introduction

Managing FEW systems requires modeling tools to understand the merits of different decisions, policies, and investments given potential future constraints and the wider social, environmental, and economic contexts in which these are made. This chapter reviews integrated modeling tools used to support the analysis of FEW systems; especially those used for integrated planning; and the identification and evaluation of trade-offs and synergies.

Integrated FEW system models include representations of **Coupled Natural-Human Systems** (e.g., the energy system, agriculture and land use, water supply and use, the economy, and the climate). Through this integration, these models allow for exploration of FEW system interactions, and the interactions between these systems and other key external forces such as climate change, socioeconomic and technological change, and policy interventions. There is a clear relationship between FEW system modeling and the metrics reflecting interactions.

While several modeling frameworks are described in Sect. 11.3, only a small number of models and projects have been actually implemented in practice. Ongoing research and applications of FEW system modeling consist of development of principles, algorithms, data requirements and model formulations for understanding and evaluating the potential of implementing FEW system nexus approaches within a systems perspective (Chap. 2). Outputs and products of these efforts are quantitative tools that focus on FEW system planning in order to identify primary opportunities

F. R. Miralles-Wilhelm (✉)
Earth System Science Interdisciplinary Center and Global Water Team,
University of Maryland and The Nature Conservancy, College Park, MD, USA
e-mail: fwilhelm@umd.edu

P. Saundry, B. L. Ruddell (eds.), *The Food-Energy-Water Nexus*, AESS Interdisciplinary Environmental Studies and Sciences Series,
https://doi.org/10.1007/978-3-030-29914-9_15

and constraints to FEW system development, indicating priorities for more detailed analysis as well as providing a characterization of alternative system configurations that meet integrated FEW objectives.

15.2 Overview of Existing FEW System Modeling Approaches

A number of modeling platforms have been developed to support the assessment of food, energy, and water systems under different economic and environmental conditions and to support resource development policies. For instance, water models include consideration of water utilization for hydroelectricity expansion versus other uses, while some energy and food/agricultural modeling tools include calculations of water requirements for different technology investments.

Typically, however, such models are designed for different purposes, and linkages between the systems are limited. Moreover, the level of technical detail and complexity in the models can preclude their application for planning and strategy management, a crucial need in the integrated analysis of FEW systems.

Recent reviews of existing integrated resource assessment and modeling literature focused on FEW systems have shown that the analysis of individual systems (such as energy or water systems) is undertaken routinely but is often focused only on a single resource or has often been applied on an aggregated scale for use at regional or global levels and, typically, over long time periods.

Likewise, the analytical tools used to support decision-making are equally fragmented. For instance, examples of existing tools used for energy system analysis include the MESSAGE, MARKAL, and LEAP models (Sect. 2.4). A commonly used model for water system planning is the Water Evaluation and Planning (**WEAP**) system, and for water scarcity and food security planning, the Global Policy Dialogue Model (**PODIUM**) is well established.

However, these and other models, in one way or another, lack the data and methodological components required to conduct an integrated analysis of FEW systems towards decision-making, policy development, and other contexts, especially where these may be needed at a country/state/local scale. Generally, they focus on one resource and ignore the interconnections with other resources, have overly simplified spatial representations, are grand policy "research" rather than short-term applied "policy" or "decision support" models, or analyze scenarios which are too long term for practical use by decision-makers.

The sections below describe **sector-centric models** as well as efforts in the integrated modeling of FEW systems. This sampling of modeling tools is not meant to be exhaustive, but rather highlight some representative examples of modeling approaches to the analysis of FEW systems that have been documented in the literature. FEW system modeling is an area of active research and development, so many of the tools described in the chapter are the subject of continuous improvements and applications.

15.2.1 Food-Centric Approaches to FEW System Modeling

Food production models, particularly in the agricultural sector (crop models), have been the subject of extensive research, development, and applications worldwide. In addition to this, because of the connections between climate and agricultural systems, there is also a variety of agricultural models that have been linked or coupled to climate models. A few representative food-centric models are described here, highlighting their applications to FEW systems nexus issues.

DSSAT and GOSSYM are two of the most widely used dynamic crop growth models for corn, soy, wheat, cotton, and other major crops, and their formulation, development, and application have been well.

Both models simulate water, carbon, and nitrogen processes in plant root zones. They predict crop growth (with detailed plant chemistry, morphogenesis, and phenology) and soil responses to environmental stresses, primarily from heat, water, carbon, and nutrients.

Both models enable parallel computing and have been coupled with a regional **Climate-Weather Research and Forecasting** (**CWRF**) models, and tested for credibly simulating cotton and corn yields over the Cotton and Corn Belts in the USA. CWRF is the climate extension of the Weather Research and Forecasting (WRF) model. CWRF simulates surface radiation and terrestrial hydrology and has significantly improved regional precipitation skill over NOAA seasonal forecasts and over NCAR and another GCMs' climate simulations. CWRF resolves the synoptic and meso-scale processes governing regional climate anomalies and changes essential to crop production.

FASOMGHG **(Forest and Agricultural Sector Optimization Model-GreenHouse Gases version)** simulates the allocation of land over time to competing crops (food, feed, fuel, fiber), livestock, forestry, and urban activities plus the impacts of changing land allocation and production practices.

FASOMGHG outputs the effects on commodity markets and the environment, as well as the welfare and market impacts of policies that influence land allocation and alter production activities within these sectors.

FASOMGHG covers the major agricultural activities across the continental USA and represents agricultural production, processing and markets, aquifer water withdrawal, renewable fuels production, and land use. It captures biophysical and economic processes determining the technical, economic, and environmental implications of bioenergy production, climate change, and policy intervention.

FASOMGHG has been used to address a wide variety of scenarios relevant to FEW systems nexus issues. These include how climate change and bioenergy expansion influence land use, crop mix, input usage, land values, livestock and commodity production/prices, energy and fertilizer use, exports, greenhouse gas fluxes, and environmental emissions. It has also been used to evaluate responses to carbon programs, adaptation strategies to climate change, land use erosion related rules, population growth, and food demand, farm program provisions, and export promotions plus many other analyses at regional and national levels.

BioCro is a model for perennial biomass feedstocks, including Miscanthus, switchgrass, and willow. It captures both biochemical and biophysical mechanisms of carbon assimilation, plant growth, and water movement through the soil and into the atmosphere. It is capable of predicting the impacts of interannual variability in drought, temperature, and their timing.

The BioCro model provides a common structure for all bioenergy crops, simulating the mechanisms by which plants respond to rising [CO_2] and climate change, including water and carbon fluxes, and so avoiding any confounding of species comparison.

Box 15.1 Case Study: DSSAT/CCWRF
A coupled food–climate model (**DSSAT/CCWRF**) simulates strong crop–climate feedbacks with important consequences on regional climate, hydrology, and yields (Fig. 15.1). In the Midwestern USA, corn–climate interactions decrease temperature (~2 °C) and increase precipitation (~1 mm/day). They also notably affect remote regions through meso-scale circulation, causing warmer temperatures in the Southwest-Mexico and more rainfall in the Cotton Belt (Southeast), with similar magnitudes. This demonstrates the need to incorporate full crop–climate coupling, which has teleconnected effects beyond where crops are grown.

15.2.2 Energy-Centric Approaches to FEW System Modeling

Conventional energy planning is primarily concerned with siting and cost requirements for energy generation in the context of transmitting the produced energy to population centers. Except for hydropower-dominated systems, the availability of

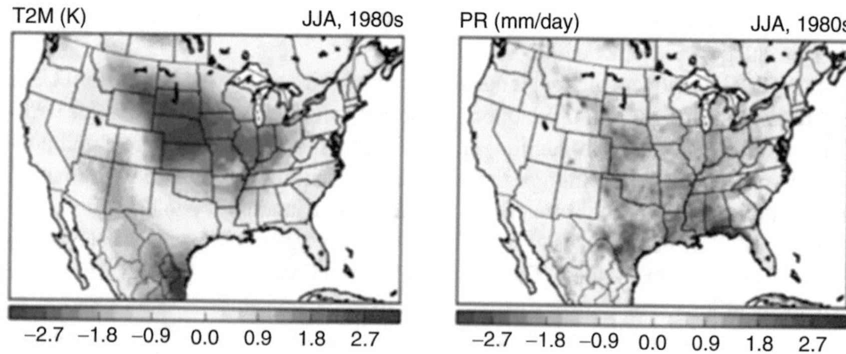

Fig. 15.1 Simulated summer mean differences in surface (2 m) temperature and precipitation with and without incorporating the crop–climate interactions and feedbacks from corn growth over the Corn Belt in the Midwest USA (Xu et al. in prep)

water supply necessary for power generation at the upstream planning stage is typically assumed to exist and is often not considered to be a limiting factor in operations although it is accepted that potential constraints will be an important factor.

The consumptive use of water necessary for the generation of energy production required by water infrastructure is not considered dynamically within models. In these situations, there is an inherent multiplier on both energy and water demands that may be overlooked when employing the traditional approach to modeling and analysis. While this effect may be quite marginal in regions with ample supplies of both water and energy, it could become a central cross-sector constraint in regions with resource scarcity and will require accurate evaluation and analysis.

Energy sector models have advanced substantially over the past decades, and these can also incorporate estimates of water demand for energy production through simple coefficients of water utilization per unit of output. A wide range of models is available, from fairly basic electricity capacity expansion models to very detailed electricity network models to economy-wide general equilibrium models with representations of various types of energy supply and demand. However, the energy models do not address total water availability, and its dynamic nature or (economic as well as volumetric) trade-offs among water uses.

In some advanced models, water availability and variability is taken into account as it affects hydropower production and with that other supply options to the system. The linkages of such water availability and variability with other sectors are usually handled by incorporating exogenous constraints or parameters in the energy models (e.g., minimum environmental or navigation outflows, quotas for irrigations, among others).

***LEAP* (Long-range Energy Alternatives Planning)** is a widely used energy systems model used for energy policy analysis and climate change mitigation assessment developed at the Stockholm Environment Institute. LEAP is an integrated, scenario-based modeling tool that can be used to track energy consumption, production and resource extraction in all sectors of an economy. It is typically used to account for both energy sector and non-energy sector greenhouse gas (GHG) emission sources and sinks. LEAP has been used together with WEAP to analyze trade-offs in the energy–water nexus.

***TIMES* (The Integrated MARKAL-EFOM System)** model generator was developed as part of the IEA-ETSAP (International Energy Agency Energy Technology Systems Analysis Program), an international community which uses long-term energy scenarios to conduct in-depth energy and environmental analyses. The TIMES model combines two different components to model energy systems: a technical engineering approach (energy technologies) and an economic (least cost optimization) approach. TIMES uses linear-programming to produce a least-cost energy system, optimized according to a number of user constraints, over medium to long-term time horizons. TIMES is used for the exploration of possible energy futures based on contrasted scenarios.

Box 15.2 Case Study: Energy–Water Nexus Modeling in South Africa

This case study application (Rodriguez et al. 2017) focuses on incorporating a representation of water supply and infrastructure costs into an energy systems model (TIMES) to better reflect the interdependent nature of the energy–water nexus in South Africa; the water supply challenges facing the energy system are therefore of primary interest.

The water-smart energy systems model (SATIM-W) embeds the various water supply options in a least-cost optimization procedure for the energy system, so that the cost of water is captured as energy and water sector investments are planned, and any changes in these investments due to implementing this nexus approach can be quantified, as compared with taking a siloed view of the two planning areas.

The results of this case study (e.g., Figs. 15.2 and 15.3) demonstrate the process and type of tools that can be employed to examine the energy–water nexus in a national level planning context and the insights that can be gained from water-smart energy planning.

A number of relevant energy–water policy scenarios in South Africa were explored, and the results show that specific energy sector policies can have significant implication for both new investments in water supply infrastructure and in some cases can lead to stranded energy and water investments, reinforcing the importance of planning in these sectors through a nexus approach.

A key finding of the study is that a national-level energy systems optimization model can be readily regionalized in terms of energy resource supply and power plant locations, and the regional costs and limitations for water supply can be incorporated into the energy model to create a water-smart energy sector planning tool.

This work has demonstrated the importance and value of employing an enhanced modeling tool to better assess the energy–water nexus challenges. Recommendations for further development of the SATIM-W model and its wider application for energy and water planning have resulted in additional areas of improvement of the model to further expand the coverage and insights that can be obtained.

15.2.3 Water-Centric Approaches to FEW System Modeling

Conventional water systems planning is primarily concerned with supporting the development of water resources and manage the distribution of water in time and space in order to allocate the water supplied by various sources to meet a specific set of objectives or demands. Most water allocation modeling assumes that there are always adequate energy supplies available to facilitate the diversion, pumping, and treatment of water. Few, if any, of the water allocation models, quantify the imposed energy consumption associated with different water demands. This approach does not adequately reflect the dynamic interplay between energy and water, especially

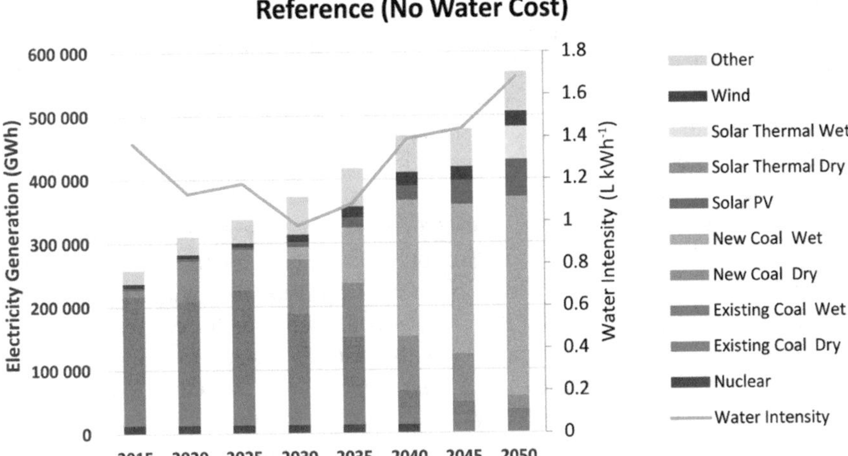

Fig. 15.2 Reference (no water cost) electricity generation by type (with water intensity)

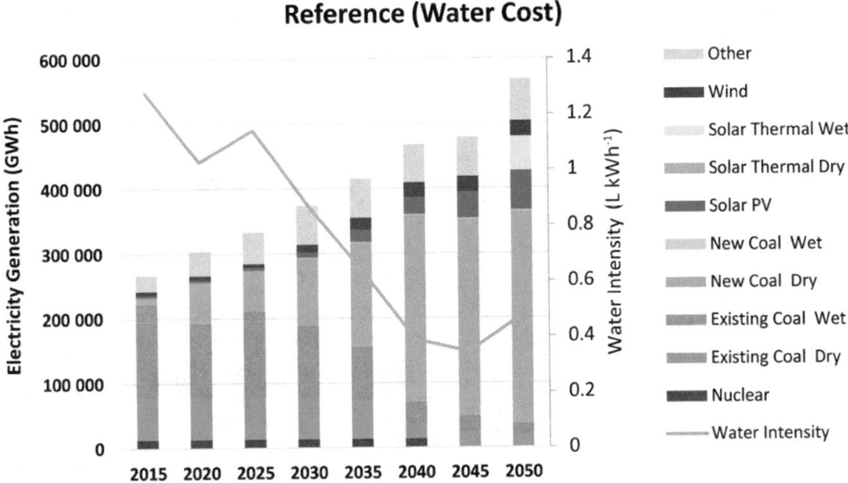

Fig. 15.3 Electricity generation and water use intensity simulated for South Africa using a water-smart TIMES model (SATIM-W) by type (Rodriguez et al. 2017)

when considering the large energy demands that may be incurred as a result of transporting (pumping) and treating water to meet a particular end use.

Water system models range from physical hydrology modeling approaches, focused on simulating the cycling of water between the atmosphere, land surface, and subsurface, to water systems models that incorporate anthropogenic demands for allocation of water to satisfy a set of objectives and constraints.

A representative model in the hydrology space is ***SWAT*** **(Soil and Water Assessment Tool)**, which has been continuously developed to simulate streamflow, in-stream water quality/yield/supply, surface runoff, groundwater recharge, nutrients, bacteria, pathogens and sediments across a wide range of watershed scales and environmental conditions, and to assess the impacts of climate change, land use, agricultural production, wetland drainage, and water management. SWAT has been coupled with CWRF to capture crop–hydrology–climate interactions for distributed modeling over the continental USA, incorporating surface–subsurface watershed processes, most pollutant sources, agricultural practices, and human interventions.

Water systems models such as ***WEAP* (Water Evaluation and Planning)** focus on the simulation of natural hydrological processes (e.g., evapotranspiration, runoff, and infiltration) to enable assessment of the availability of water within a catchment, integrating anthropogenic activities, such as water demands for energy generation and food production, superimposed on the natural system to influence water resources and their allocation (i.e., consumptive and non-consumptive water demands) and enable evaluation of the impact of human water use. WEAP can be used to simulate tailored scenarios specified by the model user.

Similar modeling approaches to water systems model development can be found in the system dynamics literature. Such approaches combine physical processes based on water balance principles, with water demands from diverse human activities. Examples in this line of work are WaterSIM, spatial system dynamics modeling of water resources systems, and the global water availability model. A useful synthesis review of system dynamics modeling approaches is provided by Mirchi et al. (2012).

Box 15.3 Case Study: Physical Impacts of Climate Change on Water Resources

This water systems modeling study (Miralles-Wilhelm et al. 2017) presents an investigation of the impacts of climate change on water resources throughout the world, and specific effects on water-dependent sectors of the economy such as urban, energy, and agriculture. The impacts are projected to vary regionally (Fig. 15.4) and are likely to include changes in average hydroclimate patterns (precipitation, surface runoff, and streamflow), as well as increases in the probability of extreme events.

Prudent management of water resources will be pivotal in addressing the climate challenge—both for adapting to the effects of climate change as well as for meeting global greenhouse gas mitigation goals. The precise consequences of climate change on the hydrological cycle are uncertain, which makes adaptation especially challenging.

Uncertainty regarding impacts is partly a consequence of the limitations of climate models; despite improvements in climate science, the Global Circulation Models developed to project climate futures generate a wide range of projections that often disagree on both the direction and magnitude of precipitation changes. Furthermore, these models do not have the precision required for planning and managing water resources. In addition to this, changes in the hydrological cycle imply that future water systems may not resemble the past (non-stationarity), so historic trends as used in engineering designs, no longer serve as a reliable guide for assessing and managing future risks.

Global Water Scarcity Index (WSI)

2050

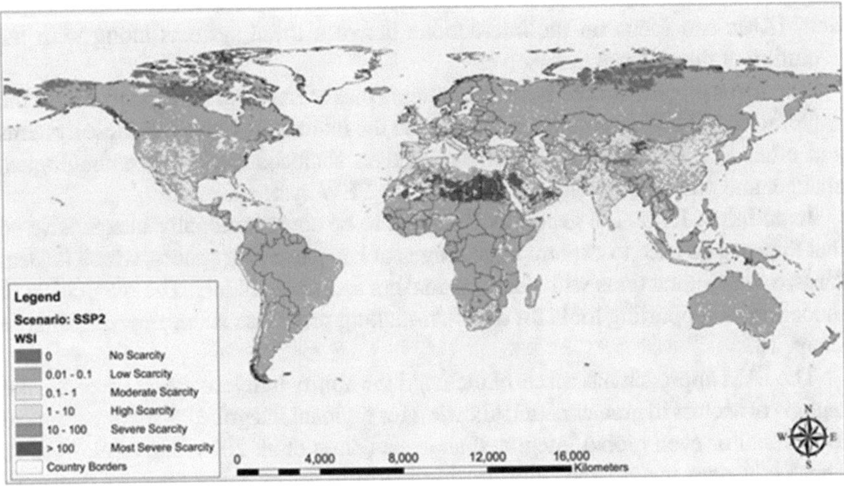

Fig. 15.4 Water Scarcity Index globally simulated using the global water availability model (Hejazi et al. 2014a, b)

15.2.4 Integrated FEW System Modeling Approaches

In response to the need for an integrated analytical approach across sectors, a new generation of FEW system modeling tools and approaches are being developed that include representations of food, energy, and water systems. For example, **Integrated Assessment Models (IAMs)** have been enhanced over the last decade to include interacting representations of FEW systems.

The IAM approach has recently included the ability to telescope-in on particular regions or sectors in greater detail, allowing for regional integrated assessment linked to national or even global integrated analysis (Zhou et al. 2014; Liu et al. 2015a). These advances make IAMs a promising platform for bringing together decision-makers across sectors and levels of decision-making. Other FEW systems modeling approaches are focused on enhancing and coupling individual food, energy, and water system models to analyze the integrated FEW systems. In such model architectures, the initial state of the land/water use system, and a set of relevant exogenous forcings drive climate, agricultural, and hydro system models, to predict fine-scale climate, crop, and water responses. The agronomic responses, through scale-aware coupling, feed bio-hydro-economic system models, including agricultural to assess the consequences on resources management and economics at farm, regional and national scales. Some examples of these integrated modeling tools are described in this chapter.

Integrated modeling approaches to FEW systems such as Integrated Assessment Models (IAMs) can couple representations of multiple human and natural systems (e.g., the energy system, agriculture and land use, water supply and use, the economy, and climate) into a single computational platform. By achieving this integration, IAMs can focus on the interactions between these systems along with the evolution of the systems themselves.

The focus on interactions between systems makes IAMs an ideal platform for the exploration of FEW system interactions, and the interactions between these systems and other key forcers such as climate change, socioeconomic and technological change, and policy interventions in any of the FEW systems.

In addition, IAMs are generally designed to be computationally inexpensive so that they can be used to explore large ranges of key parameter spaces, which is ideal for two-way interactions with decision-makers and stakeholders. The integration of modeling as supporting tools for decision-making processes is developed further in Chap. 16.

The IAM approach has recently included the ability to telescope-in on particular regions or sectors in greater detail, allowing for regional integrated assessment linked to national or even global integrated analysis (Zhou et al. 2014; Liu et al. 2015a). These advances make IAMs a promising platform for bringing together decision-makers across sectors and levels of decision-making. Other FEW systems modeling approaches are focused on enhancing and coupling individual food, energy, and water system models to analyze the integrated FEW systems. In such model architectures, the initial state of the land/water use system, and a set of relevant exogenous forcings drive climate, agricultural and hydro system models, to predict fine-scale climate, crop, and water responses. The agronomic responses, through scale-aware coupling, feed bio-hydro-economic system models, including agricultural to assess the consequences on resources management and economics at farm, regional and national scales. Some examples of these integrated modeling tools are described in this chapter.

As recognition of the importance of FEW systems has grown, the focus of IAMs has shifted from studying the interactions of a limited number of systems (energy–economy–climate) at the global level to consideration of multiple interactions (energy–water–land–economy–climate) and at greater temporal, spatial, and process resolution.

Recent research has explored the effects of climate on water systems and their connections to energy, agriculture, and climate and bioenergy and agricultural systems.

Box 15.4 Case Study: Modeling the FEW System Nexus in the Middle East and North Africa (MENA)

The objective of this work is to develop and illustrate an analytical framework that can help formulate integrated (nexus) approaches for FEW system activities in the MENA region. This work places focus on the following:

Box 15.4 (continued)

1. An analysis of the current status of water resources in the region using an Integrated Assessment Model (**GCAM: Global Change Assessment Model**).
2. A scenario analysis focused on water scarcity and potential impacts on the energy and food sectors in the region.
3. Recommendations for further analysis that can inform policy making at the national level, and contribute to ongoing efforts towards integrated planning at the regional level.

Two illustrative examples of water resources management scenarios were proposed to better understand the implications of different water management approaches on water scarcity, energy generation and food production in the MENA region.

The ***Unlimited Water*** **scenario** assumes unlimited water resources where all sectors within the economy can achieve all their water demands with no water constraints. This serves as a benchmarking scenario and to quantify the projected changes in the water and agricultural sectors under no water constraints.

The ***Limited Water*** **scenario** focuses on constraining the water demands to the available water resources (renewable surface and groundwater, nonrenewable groundwater resources, desalinated water) within each river basin. In this scenario, two options for limiting water resources at the basin level were simulated. The first is to account for the increase in the cost of nonrenewable groundwater as the groundwater aquifers get depleted over time but with no limitation on the total amount of extracted water, as discussed in Kim et al. (2016). The second is using a methodology to estimate the amount economically available groundwater and constructing marginal cost resources curves.

The *LimitedWater* scenario also incorporates adaptation measures to be deployed as a means to mitigate the water scarcity problem. More specifically, the expansion of desalination and more efficient irrigation technologies are included as adaptation measures. This is done to shed light on the level of necessary adaptation to close the water gap in the region and the associated investment costs that are associated with those measures. Also, by comparing this scenario to the UnlimitedWater scenario, one can estimate the economic impacts associated with limitations in water on the economy of the region. The impacts of constraining water translate into impacts on water use for agriculture.

Figures 15.5 and 15.6 compare the *UnlimitedWater* and *LimitedWater* scenarios with respect to agricultural production across a number of crops. Noteworthy large reductions in production occur in Saudi Arabia (almost threefold reduction when limiting water) and Yemen (approximately 60% overall reduction) as a result of constraining water demand.

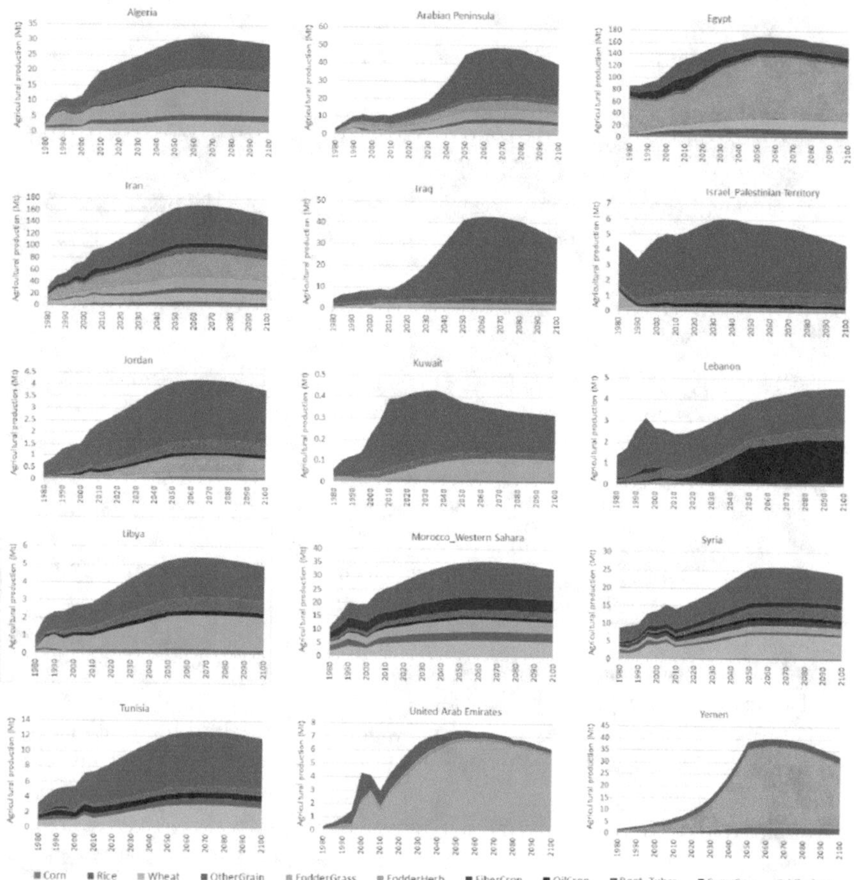

Fig. 15.5 Total agricultural production per crop type in the MENA region (in Mt), UnlimitedWater scenario (water withdrawals from surface and groundwater sources are not constrained)

15.3 Some Key Research Questions in FEW System Modeling

The interconnectedness of FEW systems necessitates the use of a coupled model framework capable of representing complex feedbacks that may shift or vary as critical thresholds are reached. The framework must be able to capture the systems' response to changing environmental conditions, which will be determined in part by feedbacks from the FEW systems themselves. Here we describe some of the critical pairwise interactions.

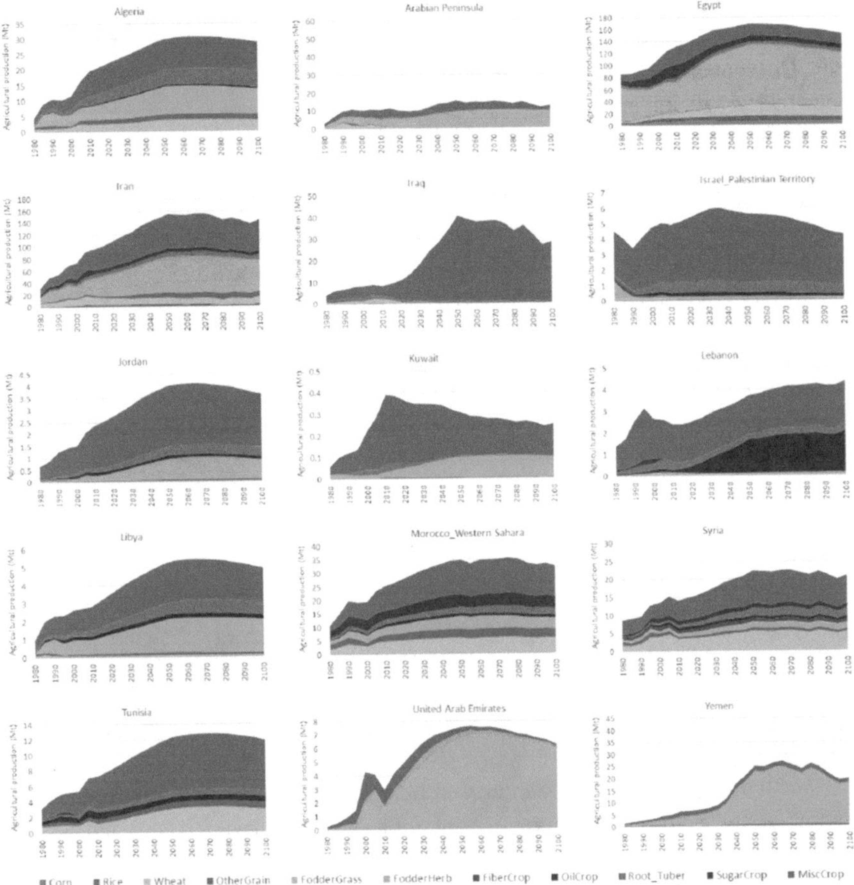

Fig. 15.6 Total agricultural production per crop type in the MENA region (in Mt), LimitedWater scenario (water withdrawals from surface and groundwater sources are constrained to prevent irreversible depletion of these sources)

15.3.1 Food–Energy

In the USA, legal mandates for biofuels (see Sects. 5.2.1 and 8.2.1) have resulted in land use change that has outpaced agricultural and biofuel policies. For instance, the conversion of almost 40% of the US corn crop to energy production has shifted land and water resources from food to energy and contributed to high crop prices. These bioenergy-influenced economic and resource pressures have led to riskier growing practices in many regions, with reduced crop rotation, increasing tendencies towards monoculture, and more crops being planted on erosion-prone or ecologically sensitive land.

Such changing agricultural practices could have far-reaching impacts for water quantity and quality, cascading into long-term effects on crop yield and food security. Consequently, modeling of FEW systems should represent changing bioenergy land use and its impact on crop distribution and yield.

Growing bioenergy on marginal land in the form of perennial vegetation may alleviate food-versus-fuel trade-off concerns while reducing the carbon debt and indirect land-use costs associated with food-based biofuels. However, such expansion could have regional and widespread effects on climate, land, and water, which must be assessed using an advanced coupled model system.

Thus, FEW system models should address key questions such as: Will large-scale deployment of the new generation of heat-resistant, drought-tolerant, or higher-reflectivity crop hybrids deliver high yields, increase water use efficiency, or reduce canopy temperatures sufficiently to meet future food demand under available water supply while maintaining soil and water quality and offsetting greenhouse gas warming? If not, what will the deficiencies be?

15.3.2 Energy–Water

Increasing irrigation for crop production would surge energy demand for water conveyance, deeper pumping, and desalination, so FEW system models account for the energy cost of different irrigation scenarios. Additionally, bioenergy expansion may have significant impacts on the water cycle, depending on the regional climate and hydrologic conditions.

Fertilizing and harvesting biomass on marginal land may substantially increase sediment and nutrient discharge, exacerbating already stressed inland and coastal waters. Furthermore, the relative water consumption of perennial versus annual crops differs strongly depending both on crop type and location.

Therefore, FEW system models must simulate various potential biomass crops' effects on local water cycles, accounting for regional climate and other impacts, to answer key questions such as: Where and which bioenergy crops can be produced sustainably (both economically and environmentally) without stressing water supply, soil quality, or food production, and how will they affect regional climate and carbon footprint?

15.3.3 Water–Food

Water scarcity, variability, and uncertainty are already threatening US agriculture resilience, with increasing vulnerability to drought. Agricultural productivity depends on optimal temperature and reliable water supply, and agronomic advances have not reduced crop yield sensitivity to drought (Hatfield et al. 2014).

As food demand continues to increase, rainfall alone will not support the needed yield increases, and greater strain will be placed on water resources. Shifts in crop type, whether to meet demand or in response to changing regional climates, will change evapotranspiration, infiltration, and runoff rates, impacting water availability and regional climate. Crop extensification into marginal land will increase the extent of these impacts.

FEW system models should represent how water availability will drive crop choice and distribution, and the feedback effects these have on water quantity and quality. Key questions are: How will the water cycle (along with the coupled carbon and nitrogen cycles) be altered by changes in agricultural land use? How will this coupled system respond to progressive climate change and feedback on regional water supply?

15.3.4 Food–Energy–Water–Climate

The above intersystem connections must be considered in the context of the changing climate, which both affects and is affected by FEW systems and their interactions, and FEW system models must, therefore, represent not only the current systems but also their evolution under changing climate conditions.

Climate change impacts on FEW systems are already visible and projected to increase. Rising temperatures are altering traditional growing zones, shifting food and bioenergy crops into nontraditional, formerly unsuitable marginal land, with new impacts on both of these environments.

Climate change alters the physiological and structural properties of terrestrial ecosystems, perturbing their soil–plant–atmosphere interactions. Severe weather events such as heat waves, droughts and floods are projected to increase in frequency and/or intensity, with significant consequences for short- and long-term food and bioenergy production. For example, critical vegetation losses due to drought can lead to drastic increases in soil erosion and upland sediments impounding flood control reservoirs. They may remarkably increase evapotranspiration, amplifying catchment water storage anomalies.

Reduced yield due to environmental change may drive sharp price increases, with potential effects on future crop choice and land use, and this should be captured by economic component(s) in FEW system models.

The impacts of FEW systems on climate are equally important. For instance, agricultural management practices (e.g., irrigation, fertilization, tillage, crop rotation) can significantly alter the land surface properties controlling energy and water fluxes.

These feedbacks may alter regional-local climate over seasons to decades, thus modifying the biophysical environment in which crops grow. Likewise, irrigation, which changes terrestrial hydrology, may cause much larger evapotranspiration than its induced local plus recycled remote rainfall, resulting in a net water loss; this alters surface water and energy budgets. The resultant near-surface cooling tends to suppress moist convection and reduce rainfall.

By contrast, moistening may enhance or suppress local convection when the antecedent soil is relatively wet or extremely dry, causing precipitation to increase over irrigated areas during normal and pluvial years but decrease during drought years. Irrigation may also induce meso-scale circulations that recycle the evapotranspiration water for more downwind precipitation and streamflow, or reduce atmospheric moisture convergence for overall less regional rainfall; these impacts may differ substantially by region through land–atmosphere feedbacks.

Attempts to mitigate or respond to changing climate conditions may have far-reaching, often unrecognized consequences that feedback into the climate and cause further changes. It is therefore insufficient to look at projected climate uncoupled from regional FEW systems.

Recent efforts to model effects of agricultural landscapes and irrigation practices on regional climate and vice versa have not included system feedbacks, dynamic crop phenology (especially growing stages susceptible to climate stresses), and water management practices.

FEW system models should account for these interactions by coupling regional climate, agroecosystem, watershed, and groundwater models to address key questions such as: At what spatiotemporal scales will weather extremes, climate anomalies, or climate changes significantly impact the resilience of regional FEW systems, and, conversely, at what scales can they be notably affected by land use changes resulting from adaptation and mitigation strategies?

Key Points

- This chapter presents the state-of-the-art and challenges ahead in the development of integrated modeling tools to support the analysis of food, energy, and water systems (FEWS). A focus is placed on interdisciplinary mathematical modeling toward integrated planning and identification/evaluation of trade-offs and synergies in developing such systems.
- A fundamental challenge in the management of FEW systems is that, although these systems are tightly linked, the decisions about them are often made independently by decision-makers focusing on each system in isolation or with only minimal consideration of the other systems. Similarly, modeling in support of these decisions is also frequently focused on single systems with only minimal consideration of the other systems.
- A number of modeling platforms have been developed to support the assessment of food, energy, and water systems under different economic and environmental conditions and to support resource development policies. Typically, however, such models are designed for different purposes, and linkages between the F-E-W systems are limited.
- Continued challenges for integrated modeling in this emerging field are identifying and address shared needs of FEW systems stakeholders, facilitating tailored analyses over different geographical regions, computing in varying spatial and temporal scales, improving system efficiencies and addressing FEW system vulnerabilities and resilience to human and natural stressors.

Discussion Points and Exercises

1. Discuss some of the limitations of existing FEW system modeling approaches.
2. Highlight a few differences between sector-centric and integrated FEW systems modeling approaches.

3. Discuss how climate has been incorporated into FEW system modeling. How would you propose improved ways of doing this?
4. Discuss how remote sensing and earth observation data (see Chap. 14) can be incorporated to improve FEW system modeling methods.
5. Pick an example of a FEW system metric from Chap. 13 and develop a conceptual approach to quantify this metric through FEW system modeling.

References

Food-Centric Approaches to FEWS Modeling

Adams, D., Alig, R., McCarl, B. A., Murray, B. C., Bair, L., Depro, B., Latta, G., Lee, H.-C., Schneider, U. A., & Callaway, J. (2005). *FASOMGHG conceptual structure, and specification: Documentation*. Unpublished paper, Texas A&M University, Department of Agricultural Economics, College Station, TX.

Baker, D. N., Lambert, J. R., & McKinion, J. M. (1983). *GOSSYM: A simulator of cotton crop growth and yield* (134p). Technical Bulletin, South Carolina Agricultural Experiment Station.

Baker, D. N., & Landivar, J. A. (1990). The simulation of plant development in GOSSYM. In T. Hodges (Ed.), *Predicting crop phenology* (pp. 153–170). Boston, MA: CRC Press.

Beach, R. H., & McCarl, B. A. (2010). *U.S. Agricultural and Forestry Impacts of the Energy Independence and Security Act: FASOM Results and Model Description* (p. 178p). Research Triangle Park, NC: RTI International.

Boote, K. J., Jones, J. W., Hoogenboom, G., & Pickering, N. B. (1998). The CROPGRO model for grain legumes. In G. Tsuji, G. Hoogenboom, & P. Thornton (Eds.), *Understanding options for agricultural production* (Vol. 7, pp. 99–128). Berlin: Springer.

Chen, L., Liang, X.-Z., DeWitt, D., Samel, A. N., & Wang, J. X. L. (2015). Simulation of seasonal U.S. precipitation and temperature by the nested CWRF-ECHAM system. *Climate Dynamics, 46*(3), 1–18. https://doi.org/10.1007/s00382-015-2619-9.

Hodges, H. F., Whisler, F. D., Bridges, S. M., Reddy, K. R., & McKinion, J. M. (1997). Simulation in crop management—GOSSYM/COMAX. In R. M. Peart & R. B. Curry (Eds.), *Agricultural systems modeling and simulation* (pp. 235–282). New York: Marcel Dekker Inc.

Hoogenboom, G., Jones, J., & Boote, K. (1992). Modeling growth, development, and yield of grain legumes using SOYGRO, PNUTGRO, and BEANGRO: A review. *Transactions of the ASAE (USA), 35*(6), 2043–2056.

Jones, C. A., Kiniry, J. R., & Dyke, P. (1986). *CERES-Maize: A simulation model of maize growth and development*. College Station, TX: Texas A&M University Press.

Jones, J. W., Hoogenboom, G., Porter, C. H., Boote, K. J., Batchelor, W. D., Hunt, L., Wilkens, P. W., Singh, U., Gijsman, A. J., & Ritchie, J. T. (2003). The DSSAT cropping system model. *European Journal of Agronomy, 18*(3), 235–265.

Kiniry, J. R., Williams, J. R., Vanderlip, R. L., Atwood, J. D., Reicosky, D. C., Mulliken, J., Cox, W. J., Mascagni, H. J., Hollinger, S. E., & Wiebold, W. J. (1997). Evaluation of two maize models for nine U.S. locations. *Agronomy Journal, 89*(3), 421–426.

Larsen, S., Jaiswal, D., Bentsen, N. S., Wang, D., & Long, S. P. (2015). Comparing predicted yield and yield stability of willow and Miscanthus across Denmark. *Global Change Biology Bioenergy*. https://doi.org/10.1111/gcbb.12318.

Lee, H. (2002). *The dynamic role for carbon sequestration by the U.S. agricultural and forest sectors in greenhouse gas emission mitigation*. Unpublished Ph.D. Thesis, Texas A&M University.

Liang, X. Z., Xu, M., Gao, W., Reddy, K. R., Kunkel, K., Schmoldt, D. L., & Samel, A. N. (2012a). Physical modeling of U.S. cotton yields and climate stresses during 1979 to 2005. *Agronomy Journal, 104*(3), 675–683.

Liang, X. Z., Xu, M., Gao, W., Reddy, K. R., Kunkel, K., Schmoldt, D. L., & Samel, A. N. (2012b). A distributed cotton growth model developed from GOSSYM and its parameter determination. *Agronomy Journal, 104*(3), 661–674.

Liang, X.-Z., Xu, M., Yuan, X., Ling, T., Choi, H. I., Zhang, F., Chen, L., Liu, S., Su, S., & Qiao, F. (2012c). Regional climate-weather research and forecasting model. *Bulletin of the American Meteorological Society, 93*(9), 1363–1387.

Liang, X.-Z., & Zhang, F. (2013). The cloud–aerosol–radiation (CAR) ensemble modeling system. *Atmospheric Chemistry and Physics, 13*(16), 8335–8364.

Liu, S., Wang, J. X. L., Liang, X.-Z., Morris, V., & Fine, S. S. (2015). A hybrid approach to improving the skills of seasonal climate outlook at the regional scale. *Climate Dynamics, 46*(1–2), 483–494. https://doi.org/10.1007/s00382-015-2594-1.

McKinion, J., Baker, D., Whisler, F., & Lambert, J. (1989). Application of the GOSSYM/COMAX system to cotton crop management. *Agricultural Systems, 31*(1), 55–65.

Miguez, F. E., Zhu, X. G., Humphries, S., Bollero, G. A., & Long, S. P. (2009). A semimechanistic model predicting the growth and production of the bioenergy crop Miscanthus × giganteus: Description, parameterization and validation. *Global Change Biology Bioenergy, 1*, 282–296.

Miguez, F. E., Maughan, M., Bollero, G. A., & Long, S. P. (2012). Modeling spatial and dynamic variation in growth, yield, and yield stability of the bioenergy crops Miscanthus × giganteus and Panicum virgatum across the conterminous United States. *Global Change Biology Bioenergy, 4*, 509–520.

Pang, X., Letey, J., & Wu, L. (1997). Yield and nitrogen uptake prediction by CERES-Maize model under semiarid conditions. *Soil Science Society of America Journal, 61*(1), 254–256.

Reddy, K. R., Hodges, H. F., & McKinion, J. M. (1997). Modeling temperature effects on cotton internode and leaf growth. *Crop Science, 37*(2), 503–509.

Reddy, K. R., Kakani, V. G., McKinion, J. M., & Baker, D. N. (2002). Applications of a cotton simulation model, GOSSYM, for crop management, economic and policy decisions. In L. R. Ahuja, L. Ma, & T. A. Howell (Eds.), *Agricultural system models in field research and technology transfer*. Boston, MA: CRC Press.

Skamarock, W., Klemp, J., Dudhia, J., Gill, D., Barker, D., Duda, M. G., Huang, X.-Y., Wang, W., & Powers, J.G. (2008). A description of the advanced research WRF Version 3. *NCAR Technical Note*. NCAR/TN–475+ STR (113p).

Tsvetsinskaya, E. A., Mearns, L. O., & Easterling, W. E. (2001). Investigating the effect of seasonal plant growth and development in three-dimensional atmospheric simulations. Part I: Simulation of surface fluxes over the growing season. *Journal of Climate, 14*(5), 692–709.

Wang, D., Jaiswal, D., Lebauer, D. S., Wertin, T. M., Bollero, G. A., Leakey, A. D. B., & Long, S. P. (2015). A physiological and biophysical model of coppice willow (Salix spp.) production yields for the contiguous USA in current and future climate scenarios. *Plant, Cell and Environment, 38*(9), 1850–1865. https://doi.org/10.1111/pce.12556.

Yuan, X., & Liang, X. Z. (2011). Improving cold season precipitation prediction by the nested CWRF-CFS system. *Geophysical Research Letters, 38*(2). https://doi.org/10.1029/2010GL046104.

Energy-Centric Approaches to FEWS Modeling

Dale, L., Karali, N., Millstein, D., Carnall, M., Vicuña, S., Borchers, N., Bustos, E., O'Hagan, J., Purkey, D., Heaps, C., Sieber, J., Collins, W., & Sohn, M. (2015). An integrated assessment of water-energy and climate change in Sacramento, California: How strong is the nexus? *Climatic Change, 132*, 223–235.

Hunt, P., Miller, J., Ducey, T., Lang, M., Szogi, A., & McCarty, G. (2014). Denitrification in soils of hydrologically restored wetlands relative to natural and converted wetlands in the Mid-Atlantic coastal plain of the USA. *Ecological Engineering, 71*, 438–447.

Loulou, R., Goldstein, G., & Noble, K. (2004). *Documentation for the MARKAL Family of Models*. ETSAP.
Loulou, R., Remne, U., Kanudia, A., Lehtila, A., & Goldstein, G. (2005). *Documentation for the TIMES Model—Part I* (pp. 1–78). ETSAP.
Mirchi, A., Madani, K., Watkins, D., & Ahmad, S. (2012). Synthesis of system dynamics tools for holistic conceptualization of water resources problems. *Water Resources Management, 26*, 2421–2442.
Rodriguez, D., Delgado, A., Bazilian, M., Ahjum, F., Cullis, J., Delaquil, P., Goldstein, G., Liden, R., Merven, B., Miralles-Wilhelm, F., Sohns, A., Stone, A., & Toman, M. (2017). Water Contrains South Africa's Energy Future: A Case Study on Integrated Energy-Water Nexus Modeling and Analysis, *International Journal of Engineering Science, 6*(10), 1–25.
Yates, D., & Miller, K. (2013). Integrated decision support for energy/water planning in California and the southwest. *International Journal of Climate Change: Impacts and Responses, 4*(1), 49–63.

Water-Centric Approaches to FEWS Modeling

Ahmad, S., & Simonovic, S. (2004). Spatial system dynamics: New approach for simulation of water resources systems. *Journal of Computing in Civil Engineering ASCE, 18*(4), 331–340.
Arnold, J. G., Srinivasan, R., Muttiah, R. S., & Williams, J. R. (1998). Large area hydrologic modeling and assessment Part I: Model development. *Journal of the American Water Resources Association, 34*(1), 73–89.
Comín, F. A., Sorando, R., Darwiche-Criado, N., García, M., & Masip, A. (2014). A protocol to prioritize wetland restoration and creation for water quality improvement in agricultural watersheds. *Ecological Engineering, 66*, 10–18.
Daloğlu, I., Cho, K. H., & Scavia, D. (2012). Evaluating causes of trends in long-term dissolved reactive phosphorus loads to Lake Erie. *Environmental Science & Technology, 46*(19), 10660–10666.
Denver, J., Ator, S., Lang, M., Fisher, T., Gustafson, A., Fox, R., Clune, J., & McCarty, G. (2014). Nitrate fate and transport through current and former depressional wetlands in an agricultural landscape, Choptank Watershed, Maryland, United States. *Journal of Soil and Water Conservation, 69*(1), 1–16.
Ducey, T., Miller, J., Lang, M., Szogi, A., Hunt, P., Fenstermacher, D., Rabenhorst, M., & McCarty, G. (2015). Soil physicochemical conditions, denitrification rates, and abundance in North Carolina coastal plain restored wetlands. *Journal of Environmental Quality, 44*(3), 1011–1022.
Ficklin, D. L., Luo, Y., Stewart, I. T., & Maurer, E. P. (2012). Development and application of a hydroclimatological stream temperature model within the Soil and Water Assessment Tool. *Water Resources Research, 48*(1). https://doi.org/10.1029/2011WR011256.
Ficklin, D. L., Stewart, I. T., & Maurer, E. P. (2013). Effects of climate change on stream temperature, dissolved oxygen, and sediment concentration in the Sierra Nevada in California. *Water Resources Research, 49*(5), 2765–2782.
Garg, K. K., Bharati, L., Gaur, A., George, B., Acharya, S., Jella, K., & Narasimhan, B. (2012). Spatial mapping of agricultural water productivity using the swat model in Upper Bhima catchment, India. *Irrigation and Drainage, 61*(1), 60–79.
Gassman, P. W., Reyes, M. R., Green, C. H., & Arnold, J. G. (2007). The soil and water assessment tool: Historical development, applications, and future research directions. *Transactions of the ASABE, 50*(4), 1211–1250.
Gober, P., Wentz, E., Lant, T., Tschudi, M., & Kirkwood, C. (2011). WaterSim: A simulation model for urban water planning in Phoenix, Arizona, USA. *Environment and Planning B: Urban Analytics and City Science, 38*(2), 197–215.

Hejazi, M., Edmonds, J., Clarke, L., Kyle, P., Davies, E., Chaturvedi, V., Wise, M., Patel, P., Eom, J., Calvin, K., Moss, R., & Kim, S. (2014a). Long-term global water projections using six socioeconomic scenarios in an integrated assessment modeling framework. *Technological Forecasting and Social Change, 81*, 205–226. https://doi.org/10.1016/j.techfore.2013.05.006.

Hejazi, M. I., Edmonds, J., Clarke, L., Kyle, P., Davies, E., Chaturvedi, V., Wise, M., Patel, P., Eom, J., & Calvin, K. (2014b). Integrated assessment of global water scarcity over the 21st century under multiple climate change mitigation policies. *Hydrology and Earth System Sciences, 18*(8), 2859–2883. https://doi.org/10.5194/hess-18-2859-2014.

Hejazi, M. I., Voisin, N., Liu, L., Bramer, L., Fortin, D., Huang, M., Hathaway, J., Kyle, P., Leung, L. R., Li, H.-Y., Liu, Y., Patel, P., Pulsipher, T., Rice, J. S., Tesfa, T. K., Vernon, C. R., & Zhou, Y. (2015). 21st Century U.S. emissions mitigation increases water stress more than the climate change it is mitigating. *Proceedings of the National Academy of Sciences, 112*(34), 10635–10640.

Hively, W. D., Hapeman, C. J., McConnell, L. L., Fisher, T. R., Rice, C. P., McCarty, G. W., Sadeghi, A. M., Whitall, D. R., Downey, P. M., & de Guzmán, G. T. N. (2011). Relating nutrient and herbicide fate with landscape features and characteristics of 15 subwatersheds in the Choptank River watershed. *Science of the Total Environment, 409*(19), 3866–3878.

Jang, J.-H., Jung, K.-W., & Yoon, C. G. (2012). Modification of SWAT model for simulation of organic matter in Korean watersheds. *Water Science & Technology, 66*(11), 2355–2362.

Khoi, D. N., & Suetsugi, T. (2014). Impact of climate and land-use changes on the hydrological processes and sediment yield—A case study for the Be River Catchment, Vietnam. *Hydrological Sciences Journal, 59*(5), 1095–1108.

Lang, M., McCarty, G., Oesterling, R., & Yeo, I.-Y. (2013). Topographic metrics for improved mapping of forested wetlands. *Wetlands, 33*(1), 141–155.

Luo, Y., Ficklin, D. L., Liu, X., & Zhang, M. (2013). Assessment of climate change impacts on hydrology and water quality with a watershed modeling approach. *Science of the Total Environment, 450*, 72–82.

Michalak, A. M., Anderson, E. J., Beletsky, D., Boland, S., Bosch, N. S., Bridgeman, T. B., Chaffin, J. D., Cho, K., Confesor, R., & Daloğlu, I. (2013). Record-setting algal bloom in Lake Erie caused by agricultural and meteorological trends consistent with expected future conditions. *Proceedings of the National Academy of Sciences, 110*(16), 6448–6452.

Miralles-Wilhelm, F., Clarke, L., Hejazi, M., Kim, S., Gustafson, K., Muñoz-Castillo, R., & Graham, N. (2017). *Physical impacts of climate change on water resources*. World Bank Discussion Paper.

Neitsch, S. L., Arnold, J. G., Kiniry, J. R., & Williams, J. R. (2011). *Soil and Water Assessment Tool theoretical documentation: Version 2009* (618p). Texas Water Resources Institute Technical Report 406. Texas Water Resources Institute.

Palao, L. K. M., Dorado, M. M., Anit, K. P. A., & Lasco, R. D. (2013). Using the Soil and Water Assessment Tool (SWAT) to assess material transfer in the Layawan watershed, Mindanao, Philippines and its implications on payment for ecosystem services. *Journal of Sustainable Development, 6*(6), 73–88.

SEI (Stockholm Environment Institute). (2001). *WEAP: Water evaluation and planning system—User guide*. Boston, MA: SEI.

Sieber, J., Yates, D., Purkey, D., & Huber Lee, A. (2004). WEAP: A demand, priority and preference driven water planning model: Part 1: Model characteristics. *Water International*. https://doi.org/10.1080/02508060508691893.

White, M., Santhi, C., Kannan, N., Arnold, J., Harmel, D., Norfleet, L., Allen, P., DiLuzio, M., Wang, X., & Atwood, J. (2014). Nutrient delivery from the Mississippi River to the Gulf of Mexico and effects of cropland conservation. *Journal of Soil and Water Conservation, 69*(1), 26–40.

Yepsen, M., Baldwin, A. H., Whigham, D. F., McFarland, E., LaForgia, M., & Lang, M. (2014). Agricultural wetland restorations on the USA Atlantic coastal plain achieve diverse native wetland plant communities but differ from natural wetlands. *Agriculture, Ecosystems & Environment, 197*, 11–20.

Zhanxue, Z., Broersma, K., & Mazumder, A. (2012). Impacts of land use, fertilizer and manure application on the stream nutrient loadings in the Salmon River watershed, south-central British Columbia, Canada. *Journal of Environmental Protection, 3*, 809–822.

Integrated FEWS Modeling Approaches

Blanc, E., Strzepek, K., Schlosser, A., Jacoby, H., Gueneau, A., Fant, C., Rausch, S., & Reilly, J. (2014). Modeling U.S. water resources under climate change. *Earth's Future, 2*(4), 197–224. https://doi.org/10.1002/2013EF000214.

Chaturvedi, V., Hejazi, M., Edmonds, J., Clarke, L., Kyle, P., Davies, E., & Wise, M. (2015). Climate mitigation policy implications for global irrigation water demand. *Mitigation and Adaptation Strategies for Global Change., 20*(3), 389–407. https://doi.org/10.1007/s11027-013-9497-4.

Crona, B. I., & Parker, J. N. (2011). Network determinants of knowledge utilization preliminary lessons from a boundary organization. *Science Communication, 33*(4), 448–471.

Fawcett, A., Iyer, G. C., Clarke, L. E., Edmonds, J. A., Hultman, N. E., McJeon, H. C., Rogelj, J., Schuler, R., Alsalam, J., Asrar, G. R., Creason, J., Jeong, M., McFarland, J., Mundra, A., & Shi, W. (2015). Can Paris pledges avert severe climate change? *Science, 350*(6265), 1168–1169. https://doi.org/10.1126/science.aad5761.

Hibbard, K. A., & Janetos, A. C. (2013). The regional nature of global challenges: A need and strategy for integrated regional modeling. *Climatic Change, 118*(3–4), 565–577.

Kim, S. H., Hejazi, M. I., Liu, L., Calvin, K. V., Clarke, L. E., Edmonds, J., Kyle, P., Patel, P. L., & Wise, M. A. (2016). Balancing global water availability and use at basin scale in an integrated assessment model. *Climatic Change*. https://doi.org/10.1007/s10584-016-1604-6.

Kraucunas, I., et al. (2015). Investigating the nexus of climate, energy, water, and land at decision-relevant scales: The Platform for Regional Integrated Modeling and Analysis (PRIMA). *Climatic Change, 129*(3–4), 573–588.

Liu, L., Hejazi, M., Patel, P., Kyle, P., Davies, E., Zhou, Y., Clarke, L., & Edmonds, J. (2015a). Water demands for electricity generation in the U.S.: Modeling different scenarios for the water–energy nexus. *Technological Forecasting and Social Change, 94*, 318–334. https://doi.org/10.1016/j.techfore.2014.11.004.

Schlosser, A. P., Kicklighter, D., Dutkiewicz, S., Reilly, J., Wang, C., Felzer, B., Melillo, J. M., & Jacoby, H. D. (2009). Probabilistic forecast for twenty-first-century climate based on uncertainties in emissions (without policy) and climate parameters. *Journal of Climate, 22*(19), 5175–5204. https://doi.org/10.1175/2009JCLI2863.1.

Strachan, N., Fais, B., & Daly, H. (2016). Reinventing the energy modelling–policy interface. *Nature Energy, 1*, 16012.

Zhou, Y., Clarke, L., Eom, J., Kyle, P., Patel, P., Kim, S. H., Dirks, J., Jensen, E., Liu, Y., Rice, J., Schmidt, L., & Seiple, T. (2014). Modeling the effect of climate change on U.S. state-level buildings energy demands in an integrated assessment framework. *Applied Energy, 113*, 1077–1088. https://doi.org/10.1016/j.apenergy.2013.08.034.

Food-Energy (Bioenergy)

Anderson, C. J., Anex, R. P., Arritt, R. W., Gelder, B. K., Khanal, S., Herzmann, D. E., & Gassman, P. W. (2013). Regional climate impacts of a biofuels policy projection. *Geophysical Research Letters, 40*(6), 1217–1222.

Barton, B., & Clark, S. E. (2014). *Water & climate risks facing U.S. corn production: How companies and investors can cultivate sustainability* (p. 71pp). Austin, TX: CERES.

Gelfand, I., Sahajpal, R., Zhang, X., Izaurralde, R. C., Gross, K. L., & Robertson, G. P. (2013). Sustainable bioenergy production from marginal lands in the U.S. Midwest. *Nature, 493*(7433), 514–517.

Hamilton, S. K., Hussain, M. Z., Bhardwaj, A. K., Basso, B., & Robertson, G. P. (2015a). Comparative water use by maize, perennial crops, restored prairie, and poplar trees in the U.S. Midwest. *Environmental Research Letters, 10*(6), 064015.

Johnston, C. A. (2014). Agricultural expansion: Land use shell game in the U.S. Northern Plains. *Landscape Ecology, 29*(1), 81–95.

Lark, T. J., Salmon, J. M., & Gibbs, H. K. (2015). Cropland expansion outpaces agricultural and biofuel policies in the United States. *Environmental Research Letters, 10*(4), 044003.

Le, P. V., Kumar, P., & Drewry, D. T. (2011). Implications for the hydrologic cycle under climate change due to the expansion of bioenergy crops in the Midwestern United States. *Proceedings of the National Academy of Sciences of the United States of America, 108*(37), 15085–15090.

Mladenoff, D. J., Sahajpal, R., Johnson, C. P., & Rothstein, D. E. (2016). Recent land use change to agriculture in the U.S. lake states: Impacts on cellulosic biomass potential and natural lands. *PLoS One, 11*(2), e0148566. https://doi.org/10.1371/journal.pone.0148566.

Nonhebel, S. (2005). Renewable energy and food supply: Will there be enough land? *Renewable and Sustainable Energy Reviews, 9*, 191–201.

Plourde, J. D., Pijanowski, B. C., & Pekin, B. K. (2013). Evidence for increased monoculture cropping in the Central United States. *Agriculture, Ecosystems & Environment, 165*(2013), 50–59.

Qin, Z., Zhuang, Q., & Cai, X. (2015). Bioenergy crop productivity and potential climate change mitigation from marginal lands in the United States: An ecosystem modeling perspective. *GCB Bioenergy, 7*, 1211–1221.

Robertson, G. P., Hamilton, S. K., Del Grosso, S. J., & Parton, W. J. (2011). The biogeochemistry of bioenergy landscapes: Carbon, nitrogen, and water considerations. *Ecological Applications, 21*, 1055–1067.

Tilman, D., Hill, J., & Lehman, C. (2006). Carbon-negative biofuels from low-input high-diversity grassland biomass. *Science, 314*, 1598–1600.

Werling, B. P., Dickson, T. L., Isaacs, R., Gaines, H., Gratton, C., Gross, K. L., Liere, H., Malmstrom, C. M., Meehan, T. D., Ruan, L., & Robertson, B. A. (2014). Perennial grasslands enhance biodiversity and multiple ecosystem services in bioenergy landscapes. *Proceedings of the National Academy of Sciences of the United States of America, 111*, 1652–1657.

Energy-Water

Environmental Protection Agency. (2013). *National rivers and streams assessment, 2008–2009 results*. Retrieved from http://www.epa.gov/national-aquatic-resource-surveys/nrsa.

Hamilton, S. H., ElSawah, S., Guillaume, J. H., Jakeman, A. J., & Pierce, S. A. (2015b). Integrated assessment and modelling: Overview and synthesis of salient dimensions. *Environmental Modelling & Software, 64*, 215–229.

Schornagel, J., Niele, F., Worrell, E., & Boggermann, M. (2012). Water accounting for (agro) industrial operations and its application to energy pathways. *Resources, Conservation and Recycling., 61*(2012), 1–15.

World Bank. (2013). *Thirsty energy, water papers*. Water Partnership Program.

World Energy Council. (2010). *Water for energy*. London, UK: World Energy Council.

World Water Assessment Program (WWAP). (2012). *The United Nations World Water Development Report 4*. Paris: UNESCO.

Yimam, Y. T., Ochsner, T. E., Kakani, V. G., & Warren, J. G. (2014). Soil water dynamics and evapotranspiration under annual and perennial bioenergy crops. *Soil Science Society of America Journal, 78*(5), 1584–1593.

Climate-FEWS

Alter, R. E., Fan, Y., Lintner, B. R., & Weaver, C. P. (2015). Observational evidence that Great Plains irrigation has enhanced summer precipitation intensity and totals in the Midwestern US. *Journal of Hydrometeorology, 16*(4), 1717–1735.

Boucher, O., Myhre, G., & Myhre, A. (2004). Direct human influence of irrigation on atmospheric water vapour and climate. *Climate Dynamics, 22*(6–7), 597–603.

DeAngelis, A., Dominguez, F., Fan, Y., Robock, A., Kustu, M. D., & Robinson, D. (2010). Evidence of enhanced precipitation due to irrigation over the Great Plains of the United States. *Journal of Geophysical Research-Atmospheres, 115*. https://doi.org/10.1029/2010JD013892.

Dunbar, J. A., Allen, P. M., & Bennett, S. J. (2010). Effect of multiyear drought on upland sediment yield and subsequent impacts on flood control reservoir storage. *Water Resources Research, 46*, W05526.

Field, C. B., Lobell, D. B., Peters, H. A., & Chiariello, N. R. (2007). Feedbacks of terrestrial ecosystems to climate change. *Annual Review of Environment and Resources, 32*, 1–29.

Godfray, H. C. J., Beddington, J. R., Crute, I. R., Haddad, L., Lawrence, D., Muir, J. F., Pretty, J., Robinson, S., Thomas, S. M., & Toulmin, C. (2010). Food security: The challenge of feeding 9 billion people. *Science, 327*(5967), 812–818.

Harding, K. J., & Snyder, P. K. (2012a). Modeling the atmospheric response to irrigation in the great plains. Part II: The precipitation of irrigated water and changes in precipitation recycling. *Journal of Hydrometeorology, 13*(6), 1687–1703.

Harding, K. J., & Snyder, P. K. (2012b). Modeling the atmospheric response to irrigation in the great plains. Part I: General impacts on precipitation and the energy budget. *Journal of Hydrometeorology, 13*(6), 1667–1686.

Hatfield, J., Takle, G., Grotjahn, R., Holden, P., Izaurralde, R. C., Mader, T., Marshall, E., & Liverman, D. (2014). Chapter 6: Agriculture. In J. M. Melillo, T. C. Richmond, & G. W. Yohe (Eds.), *Climate change impacts in the United States: The third National Climate Assessment* (pp. 150–174). Washington, DC: U.S. Global Change Research Program, 150–174.

Im, E. S., Marcella, M. P., & Eltahir, E. A. B. (2014). Impact of potential large-scale irrigation on the West African monsoon and its dependence on location of irrigated area. *Journal of Climate, 27*(3), 994–1009.

IPCC. (2013). Climate change 2013: The physical science basis. In T. F. Stocker, D. Qin, G.-K. Plattner, M. Tignor, S. K. Allen, J. Boschung, A. Nauels, Y. Xia, V. Bex, & P. M. Midgley (Eds.), *Contribution of Working Group I to the Fifth Assessment Report of the Intergovernmental Panel on Climate Change* (p. 1535pp). Cambridge, UK: Cambridge University Press.

Karlen, D. L., Lal, R., Follett, R. F., Kimble, J. M., Hatfield, J. L., Miranowski, J. M., Cambardella, C. A., Manale, A., Anex, R. P., & Rice, C. W. (2009). Crop residues: The rest of the story. *Environmental Science & Technology, 43*(21), 8011–8015.

Kustu, M. D., Fan, Y., & Rodell, M. (2011). Possible link between irrigation in the U.S. High Plains and increased summer streamflow in the Midwest. *Water Resources Research, 47*(3). https://doi.org/10.1029/2010WR010046.

Lo, M. H., & Famiglietti, J. S. (2013). Irrigation in California's Central Valley strengthens the southwestern U.S. water cycle. *Geophysical Research Letters, 40*(2), 301–306.

Lobell, D., Bala, G., Bonfils, C., & Duffy, P. (2006). Potential bias of model projected greenhouse warming in irrigated regions. *Geophysical Research Letters, 33*(13), 2–5.

Lobell, D., Bala, G., Mirin, A., Phillips, T., Maxwell, R., & Rotman, D. (2009). Regional differences in the influence of irrigation on climate. *Journal of Climate, 22*(8), 2248–2255.

Lobell, D. B., Roberts, M. J., Schlenker, W., Braun, N., Little, B. B., Rejesus, R. M., & Hammer, G. L. (2014). Greater sensitivity to drought accompanies maize yield increase in the U.S. Midwest. *Science, 344*(6183), 516–519.

McGuire, V. L. (2009). *Water-level changes in the High Plains aquifer, predevelopment to 2007, 2005–06, and 2006–07* (9pp). U.S. Geological Survey Scientific Investigations Report 2009–5019. Retrieved from http://pubs.usgs.gov/sir/2009/5019/.

Melillo, J. M., Richmond, T. C., & Yohe, G. W. (Eds.). (2014). *Climate Change Impacts in the United States: The Third National Climate Assessment* (p. 841pp). Washington, DC: U.S. Global Change Research Program.

Ort, D. R., & Long, S. P. (2014). Limits on yields in the corn belt. *Science, 344*, 483–484.

Osborne, T., Slingo, J., Lawrence, D., & Wheeler, T. (2009). Examining the interaction of growing crops with local climate using a coupled crop-climate model. *Journal of Climate, 22*(6), 1393–1411.

Ozdogan, M., Rodell, M., Beaudoing, H. K., & Toll, D. L. (2010). Simulating the effects of irrigation over the United States in a land surface model based on satellite-derived agricultural data. *Journal of Hydrometeorology, 11*(1), 171–184.

Porter, J. R., & Semenov, M. A. (2005). Crop responses to climatic variation. *Philosophical Transactions of the Royal Society B: Biological Sciences, 360*(1463), 2021–2035.

Pryor, S. C., Scavia, D., Downer, C., Gaden, M., Iverson, L., Nordstrom, R., Patz, J., & Robertson, G. P. (2014). Chapter 18: Midwest. In J. M. Melillo, T. C. Richmond, & G. W. Yohe (Eds.), *Climate change impacts in the United States: The third National Climate Assessment* (pp. 418–440). Washington, DC: U.S. Global Change Research Program.

Qian, Y., Huang, M., Yang, B., & Berg, L. K. (2013). A modeling study of irrigation effects on surface fluxes and land–air–cloud interactions in the Southern Great Plains. *Journal of Hydrometeorology, 14*(3), 700–721.

Roberts, M., Long, S., Tieszen, L., & Beadle, C. (1993). Measurement of plant biomass and net primary production of herbaceous vegetation. In *Photosynthesis and production in a changing environment* (pp. 1–21). Berlin: Springer.

Seager, R., Ting, M., Held, I., Kushnir, Y., Lu, H., Vecchi, G., Huang, H.-P., Harnik, N., Leetmaa, A., Lau, N.-C., Li, C., Velez, J., & Naik, N. (2007). Model projections of an imminent transition to a more arid climate in southern North America. *Science, 316*, 1181–1184.

Shafer, M., Ojima, D., Antle, J. M., Kluck, D., McPherson, R. A., Petersen, S., Scanlon, B., & Sherman, K. (2014a). Chapter 19: Great plains. In J. M. Melillo, T. C. Richmond, & G. W. Yohe (Eds.), *Climate change impacts in the United States: The third National Climate Assessment* (pp. 441–461). Washington, DC: U.S. Global Change Research Program.

Shafer, M., Ojima, D., Antle, J. M., Kluck, D., McPherson, R. A., Petersen, S., Scanlon, B., & Sherman, K. (2014b). Chapter 19: Great plains. In J. M. Melillo, T. C. Richmond, & G. W. Yohe (Eds.), *Climate change impacts in the United States: The third National Climate Assessment* (pp. 441–461). Washington, DC: U.S. Global Change Research Program.

Sollenberger, L. E., Srygley, R., Stöckle, C., Takle, E. S., Timlin, D., White, J. W., Winfree, R., Wright-Morton, L., & Ziska, L. H. (2010). *Climate change and agriculture in the United States: Effects and adaptation* (196pp). USDA, Technical Bulletin 1935.

Sorooshian, S., AghaKouchak, A., & Li, J. (2014). Influence of irrigation on land hydrological processes over California. *Journal of Geophysical Research: Atmospheres, 119*(23), 13137–13152.

Teuling, A. J., Van Loon, A. F., Seneviratne, S. I., Lehner, I., Aubinet, M., Heinesch, B., Bernhofer, C., Grunwald, T., Prasse, H., & Spank, U. (2013). Evapotranspiration amplifies European summer drought. *Geophysical Research Letters, 40*(10), 2071–2075.

Tilman, D., Balzer, C., Hill, J., & Befort, B. L. (2011). Global food demand and the sustainable intensification of agriculture. *Proceedings of the National Academy of Sciences of the United States of America, 108*(50), 20260–20264.

Tuinenburg, O. A., Hutjes, R. W. A., Stacke, T., Wiltshire, A., & Lucas-Picher, P. (2014). Effects of irrigation in India on the atmospheric water budget. *Journal of Hydrometeorology, 15*(3), 1028–1050.

Vose, R. S., Applequist, S., Bourassa, M. A., Pryor, S. C., Barthelmie, R. J., Blanton, B., Bromirski, P. D., Brooks, H. E., DeGaetano, A. T., Dole, R. M., & Easterling, D. R. (2014). Monitoring

and understanding changes in extremes: Extratropical storms, winds, and waves. *Bulletin of the American Meteorological Society, 95*(3), 377–386.
Walthall, C. L., Hatfield, J., Backlund, P., Lengnick, L., Marshall, E., Walsh, M., Adkins, S., Aillery, M., Ainsworth, E. A., Ammann, C., Anderson, C. J., Bartomeus, I., Baumgard, L. H., Booker, F., Bradley, B., Blumenthal, D. M., Bunce, J., Burkey, K., Dabney, S. M., Delgado, J. A., Dukes, J., Funk, A., Garrett, K., Glenn, M., Grantz, D. A., Goodrich, D., Hu, S., Izaurralde, R.C., Jones, R. A. C., Kim, S.-H., Leaky, A. D. B., Lewers, K., Mader, T. L., McClung, A., Morgan, J., Muth, D. J., Nearing, M., Oosterhuis, D. M., Ort, D., Parmesan, C., Pettigrew, W. T., Polley, W., Rader, R., Rice, C., Rivington, M., Rosskopf, E., Salas, W. A., Sollenberger, L. E., Srygley, R., Stöckle, C., Takle, E. S., Timlin, D., White, J. W., Winfree, R., Wright-Morton, L., & Ziska, L. H. (2010). *Climate change and agriculture in the United States: Effects and adaptation* (196pp). USDA, Technical Bulletin 1935.
Wei, J., Dirmeyer, P. A., Wisser, D., Bosilovich, M. G., & Mocko, D. M. (2013). Where does the irrigation water go? An estimate of the contribution of irrigation to precipitation using MERRA. *Journal of Hydrometeorology, 14*(1), 275–289.
Wuebbles, D. J., Meehl, G., Hayhoe, K., Karl, T. R., Kunkel, K., Santer, B., Wehner, M., Colle, B., Fischer, E. M., Fu, R., Goodman, A., Janssen, E., Kharin, V., Lee, H., Li, W., Long, L. N., Olsen, S., Seth, A., Sheffield, J., Tao, Z., & Sun, L. (2014). CMIP5 climate model analyses: Climate extremes in the United States. *Bulletin of the American. Meteorological Society*. https://doi.org/10.1175/BAMS-D-12-00172.1.

Integrated Assessment Models

Edmonds, J., & Reilly, J. (1983). A long-term global energy—Economic model of carbon dioxide release from fossil fuel use. *Energy Economics, 5*(2), 74–88. https://doi.org/10.1016/0140-9883(83)90014-2.
Clarke, L., Jiang, K., Akimoto, K., Babiker, M., Blanford, G., Fisher-Vanden, K., Hourcade, J.-C., Krey, V., Kriegler, E., Löschel, A., McCollum, D., Paltsev, S., Paltsev, S., Rose, Shukla, P. R., Tavoni, M., van der Zwaan, B. C. C., & van Vuuren, D. P. (2014). Assessing transformation pathways. In O. Edenhofer, R. Pichs-Madruga, Y. Sokona, E. Farahani, S. Kadner, K. Seyboth, A. Adler, I. Baum, S. Brunner, P. Eickemeier, B. Kriemann, J. Savolainen, S. Schlömer, C. von Stechow, T. Zwickel, & J. C. Minx (Eds.), *Climate change 2014: Mitigation of climate change. Contribution of Working Group III to the Fifth Assessment Report of the Intergovernmental Panel on Climate Change*. Cambridge, UK: Cambridge University Press.
Davies, E. G. R., Kyle, P., & Edmonds, J. A. (2013). An integrated assessment of global and regional water demands for electricity generation to 2095. *Advances in Water Resources, 52*, 296–313. https://doi.org/10.1016/j.advwatres.2012.11.020.
Edmonds, J., & Reilly, J. (1983b). Global energy and CO2 to the year 2050. *The Energy Journal*, 21–47.
Kyle, P., Müller, C., Calvin, K., & Thomson, A. (2014). Meeting the radiative forcing targets of the representative concentration pathways in a world with agricultural climate impacts. *Earth's Future, 2*(2), 83–98. https://doi.org/10.1002/2013EF000199.
Reilly, J., Stone, P. H., Forest, C. E., Webster, M. D., Jacoby, H. D., & Prinn, R. G. (2001). Uncertainty and climate change assessments. *Science, 293*, 430–433.
Reilly, J., Paltsev, S., Strzepek, K., Selin, N. E., Cai, Y., Nam, K.-M., Monier, E., Dutkiewicz, S., Scott, J., Webster, M., & Sokolov, A. (2013). Valuing climate impacts in integrated assessment models: The MIT IGSM. *Climatic Change, 117*(3), 561–573. https://doi.org/10.1007/s10584-012-0635-x.
Schlosser, C. A., Strzepek, K., Gao, X., Fant, C., Blanc, É., Paltsev, S., Jacoby, H., Reilly, J., & Gueneau, A. (2014). The future of global water stress: An integrated assessment. *Earth's Future, 2*(8), 341–361. https://doi.org/10.1002/2014EF000238.

Scott, M. J., Daly, D. S., Hejazi, M. I., Kyle, G. P., Liu, L., McJeon, H. C., Mundra, A., Patel, P. L., Rice, J. S., & Voisin, N. (2016). Sensitivity of future U.S. Water shortages to socioeconomic and climate drivers: a case study in Georgia using an integrated human-earth system modeling framework, *Climatic Change 136*:233–246.

Sokolov, A.P., Stone, P. H., Forest, C. E., Prinn, R., Sarofim, M. C., Webster, M., Paltsev, S., Schlosser, C. A., Kicklighter, D., Dutkiewicz, S. Reilly, J., Wang, C., Felzer, B., & Jacoby, H. D. (2009). Probabilistic Forecast for 21st Century Climate Based on Uncertainties in Emissions (without Policy) and Climate Parameters, Report No. 169, MIT Joint Program on the Science and Policy of Global Change, Feb 2009.

Voisin, N., Liu, L., Hejazi, M., Tesfa, T., Li, H., Huang, M., Liu, Y., & Leung, L. R. (2013). One-way coupling of an integrated assessment model and a water resources model: Evaluation and implications of future changes over the U.S. Midwest. *Hydrology and Earth System Sciences, 17*(11), 4555–4575. https://doi.org/10.5194/hess-17-4555-2013.

Von Lampe, M., Willenbockel, D., Ahammad, H., Blanc, E., Cai, Y., Calvin, K., Fujimori, S., Hasegawa, T., Havlik, P., Heyhoe, E., Kyle, P., Lotze-Campen, H., Mason d'Croz, D., Nelson, G. C., Sands, R. D., Schmitz, C., Tabeau, A., Valin, H., van der Mensbrugghe, D., & van Meijl, H. (2014). Why do global long-term scenarios for agriculture differ? An overview of the AgMIP global economic model intercomparison. *Agricultural Economics, 45*(1), 3–20. https://doi.org/10.1111/agec.12086.

Webster, M. D., Babiker, M., Mayer, M., Reilly, J. M., Harnisch, J., Hyman, R., Sarofim, M. C., & Wang, C. (2002). Uncertainty in emissions projections for climate models. *Atmospheric Environment, 36*(22), 3659–3670. https://doi.org/10.1016/S1352-2310(02)00245-5.

Weyant, J. P., de la Chesnaye, F. C., & Blanford, G. J. (2006). Overview of EMF-21: Multigas mitigation and climate policy. *The Energy Journal, 27*, 1–32.

Weyant, J. P. (2004). Special issue-EMF 19 Alternative technology strategies for climate change policy. *Energy Economics, 26*, 501–755.

Wigley, T. M. L., Richels, R., & Edmonds, J. A. (1996). Economic and environmental choices in the stabilization of atmospheric CO2 concentrations. *Nature, 379*(6562), 240–243. https://doi.org/10.1038/379240a0.

Further Reading

Graham, N. T., et al. (2018). Water sector assumptions for the shared socioeconomic pathways in an integrated modeling framework. *Water Resources Research, 54*, 6423–6440 Retrieved from https://agupubs.onlinelibrary.wiley.com/doi/10.1029/2018WR023452.

Motesharrei, S., et al. (2014). Human and nature dynamics (HANDY): Modeling inequality and use of resources in the collapse or sustainability of societies. *Ecological Economics, 101*, 90–102 Retrieved from https://www.sciencedirect.com/science/article/pii/S0921800914000615.

Motesharrei, S., et al. (2016). Modeling sustainability: Population, inequality, consumption, and bidirectional coupling of the earth and human systems. *National Science Review, 3*(4), 470–494 Retrieved from https://academic.oup.com/nsr/article/3/4/470/2669331.

Muñoz-Castillo, R., et al. (2017). Uncovering the green, blue, and grey water footprint and virtual water of biofuel production in brazil: A nexus perspective. *Sustainability, 9*(11), 2049 Retrieved from https://www.mdpi.com/2071-1050/9/11/2049.

National Science Foundation (NSF). (2014). *Food, energy and water transformative research opportunities in the mathematical and physical sciences*.

Perrone, D., & Hornberger, G. (2014). Water, food, and energy security: Scrambling for resources or solutions? *WIREs Water, 1*, 49–68. https://doi.org/10.1002/wat2.1004.

Chapter 16
Computing

Emre Eftelioglu and Shashi Shekhar

16.1 Introduction

This chapter reviews how computer science and technologies are contributing to FEW systems research. First, some background on computer science is provided to familiarize readers with basic concepts. Then, it provides an overview of computer science as a scientific tool as well as FEW Nexus specific challenges, followed by several case studies that are presented to illustrate in greater detail the role computing in nexus-related endeavors.

16.2 What Is Computer Science?

In a nutshell, computer science can be defined as the study of information. It uses science, engineering, and mathematics to understand information and create an efficient way to collect, store, and use it. Some activities of computer science are primarily science, for example, experimental algorithms, computational science, and experimental computer science. However, it also has some pure engineering roots, such as design, development, software engineering as well as computer engineering. Moreover, it uses mathematics where the computational complexity,

---

E. Eftelioglu (✉)
Engineering and Data Sciences, Cargill Inc., Wayzata, MN, USA
e-mail: emre_eftelioglu@cargill.com

S. Shekhar
Computer Science & Engineering, College of Science & Engineering,
University of Minnesota, Minneapolis, MN, USA
e-mail: shekhar@umn.edu

P. Saundry, B. L. Ruddell (eds.), *The Food-Energy-Water Nexus*, AESS Interdisciplinary Environmental Studies and Sciences Series,
https://doi.org/10.1007/978-3-030-29914-9_16

mathematical software and numerical analysis play a big role. Therefore, despite the misconception of many, computer science is not only about programming.

Thus, the next subsections will introduce some basics of computer science to help readers understand the role of computer science in FEW nexus research.

16.2.1 What Is a Computer Program?

A computer program is a set of machine-understandable instructions which aims to perform specific tasks on a computer. Computer programs consist of two parts, namely, algorithms and data structures. Algorithms are ordered step-by-step instructions that aim to solve a problem. The idea of an algorithm does not necessarily require a computer (Hopcroft et al. 2001). For example, in mathematics, an algorithm may be the strategy to calculate the derivatives of a function. To solve the same problem, different algorithms may be used. However, algorithms may use strategies to provide efficiency, to scale up to large datasets and to provide outputs in a reasonable time.

16.2.2 Computational Complexity Theory

The efficiency of a computer algorithm is often analyzed through a set of mathematical approaches which are considered under computational complexity theory (Papadimitriou 1994). The theory formalizes this intuition, by introducing mathematical models of computation to study these problems and quantifying the amount of resources needed to solve them, such as time and storage. In addition, it also provides a qualitative viewpoint where the question of whether a problem is algorithmically decidable or not. In other words, it tries to reply whether there exists an algorithm to solve the problem. For example, quadratic integer programming is undecidable (Wolsey 1998).

At the end, the theory gives insight about what can (decidable) and cannot be computed (non-decidable) as well as the resources needed to do so. For the decidable problems, there are time complexity classes defined to introduce the hardness of providing the decisions. These decidable problems are yes/no problems which are tried to be answered by the algorithm. For example, in graph theory "Is the graph connected?" is such a question to be answered by an algorithm.

(a) *Time complexity*: In order to quantify the complexity classes, time complexity, which is denoted by Big-O, is used. For example, suppose we are given an input with a size of "n," if the problem can be solved by a square function of the input, i.e., n^2, then the time complexity is shown with $O(n^2)$.

Class P problems: P is the class of problems that can be decided in polynomial time, i.e., those for which the running time is $O(n^k)$, for $k \in N$. For example, to find the largest element of an unsorted list of numbers, an algorithm must look at all the numbers in the list and keep a record of the largest number it has seen so far. The algorithm must look at each entry only once. The algorithm's execution time is thus directly proportional to the number of elements it is handling. Although many problems are class P, there are harder problems.

Class NP problems: NP stands for "nondeterministic polynomial time." These kinds of problems are the ones that cannot be solved in polynomial time unless there are some "hints" given. In other words, these problems have polynomial time verifiers. Deciding is hard for such problems, but verifying a given example is easier. For example, finding prime factors of a large number is an NP problem. Verifying a solution requires some multiplications but solving the problem is much harder since it requires systematically trying out all the candidates.

NP-hardness: These are the problems, which most people believe cannot be solved in polynomial time, even though nobody can prove a super-polynomial lower bound. Intuitively, this is like saying that if we could solve one particular NP-hard problem quickly, then we could quickly solve any problem whose solution is easy to understand, using the solution to that one special problem as a subroutine (Karpinski & Kannan 2004). NP-hard problems are at least as hard as any problem in NP. For example, the number of colors needed to color a map such that the neighboring regions do not have the same color is considered to be an NP-Hard problem to solve.

NP-completeness: NP-Complete problems are the hardest problems to solve among the NP problems. For those, no polynomial time algorithm has yet been discovered, nor has anybody yet been able to prove that no polynomial-time algorithm exists for any of them. The interesting part is, if any one of the NP complete problems can be solved in polynomial time, then all of them can be solved. An example of an NP-Complete problem is energy generation and scheduling. Suppose there are hundreds of energy generators. Scheduling the turn on and shut down operations of these generators by the changing variables (demand, etc.) is proved to be NP-Complete.

(b) *Space (Memory) complexity*: Just like the fact that the computational (time) complexity measures the amount of computing resources an algorithm uses; memory complexity defines the memory space that an algorithm uses. For the space complexity, several aspects need to be considered. These are the amount of memory for the code of the algorithm, the input data, the output data and the space for the intermediate steps of the algorithm (workspace for the algorithm execution). Although these were equally important in history, today the most important one is the amount of memory needed for the intermediate steps of the algorithm.

From a practical perspective, algorithms and data structures are not the only concerns for efficient program design. An efficient algorithm may perform differently on different platforms due to the use of different development platforms. To understand the performance variability between platforms, one may need to understand the types of programming languages. Next, we will introduce a brief introduction to programming languages.

Complexity of Problems in FEW Nexus: Without knowing the specifics of the problems, it is hard to come to a general conclusion about the computational complexity of FEW nexus problems. However, even when the computational complexity is not an issue, the datasets in FEW nexus are large, and the complex relationships and dependencies between different FEW datasets may cause the problem to be computationally intensive. Especially, when the datasets are large, the space (memory) complexity will play an important role in determining how to approach the problem. As a rule of thumb, the building blocks of the problem should be analyzed to see whether the user needs a high-performance computing platform to handle the computations. In addition, the user may use an approximation instead of an exact solution to reduce the computational cost.

16.2.3 Programming Languages

Modern programming languages consist of a set of vocabulary and rules that instruct a computer to perform some tasks. They are specialized to ease the programming and implementation of algorithms for the programmers (Krishnamurthi 2012). In other words, it translates the human-readable instructions to machine-readable ones. Although there are many ways to classify them, a simple classification will be high-level and low-level representing the abstraction level of the language (Martin-Löf 1982).

Low-level languages are often considered faster than higher level languages because the overhead of translation between different abstraction levels reduces the speed of high-level languages. For example, machine code can be considered as the lowest level that can be processed directly by a computer without a need for translation or interpretation (Krishnamurthi 2012). On the downside, low-level language codes are extremely hard to understand and debug causing them to be not as prevalent around the scientific community. Moreover, lower level programming languages require a programmer to keep track of stacks and memory locations that is often hard and error-prone. Thus, many of the programming languages that the reader may encounter fall under the high-level programming language category.

High-level programming languages can also roughly be divided into two categories. Compiled programming languages, as the name suggests, are needed to be translated into machine-readable code before running. Once compiled, the program can run directly on the target machine without any intervention (Friedman et al. 2001; Mitchell 1996). C and C++ are examples of compiled programming languages. Interpreted programming languages such as Python, are compiled on the fly, meaning that they are not directly run by the target machine, but another program reads and executes the original source code. Interpreted programming codes are first converted to an intermediate code, then processed by another interpreter that converts this code to machine instructions (Krishnamurthi 2012).

Interpreted programming languages have the advantage of being more portable and easier to code and implement. For example, reading a file in C++ may require 15 lines of code, whereas in Python, the same job can be done with one line of instruction. In addition, due to its ease of use, automatic memory management, and platform independence, even people without a programming background can perform data analytics tasks without dealing with the details of machine architecture and performance issues which makes them better suited for the scientific community (Millman & Aivazis 2011). However, the biggest drawback of interpreted languages is that they often execute slower compared to compiled languages. This is because the compiled languages check the code for syntax and type errors, translate the code to machine interpretation at the compile stage and convert it to a platform-specific binary executable before executing them. Thus, the compiled languages are faster since the source code is already converted into native platform specific code and the code can be directly executed without any more translation. On the other hand, the interpreter of interpreted languages will run through the code line by line and execute each command by checking for inconsistencies on the fly. Moreover, the translation to machine code will happen at the execution time making them slower. However, due to the new technologies such as just-in-time compilation (Akeret et al. 2015) and parallelization approaches (Rubinsteyn et al. 2012), interpreted languages are closing the gap of the performance issues.

Programming Languages for FEW Nexus Research: As opposed to software engineering, in data analytics, several aspects should be considered when selecting a programming language for data analytics. First, some programming languages have a faster learning curve than others. Particularly for scientists from other scientific disciplines than computer science, learning the language quickly and creating prototypes and doing analysis without considering the underlying data structure and memory management is important. Thus, such languages reduce the time for scientists to analyze their data instead of learning to manage the language structure. Second, due to their popularity, some languages have great learning tools as well as online communities for users to ask questions and get their answers quicker. Also, there are many tutorials that can be leveraged by users' own problems that are widely available for free. Finally, some languages have a wide variety of ready-to-use packages that can simply be downloaded and used.

Two such languages are R Python (n.d.) and Python R Web Site (n.d.). R is a scripting language that is widely used by statisticians and especially the data manipulation and statistics tools in R are strong. Since R comes with its own **Integrated Development Environment (IDE)**, users can also leverage the benefits of code completion and syntax highlighting when writing their code. When working with geospatial data and doing geostatistical analysis, R provides lots of packages that can be used.

Python is another language that can be used for FEW Nexus research. Python is more flexible since it supports a better/faster execution times, more efficient algorithms and the availability of parallelization approaches. In addition, since it is a programming language, it has a bigger community online, and there are thousands of packages that are specialized in almost all fields in the scientific community. As of

2018, most users of Python are using package management tools (e.g., Anaconda n.d.) to keep track of which packages are used in their application as well as help ease the portability of their code. Also, Python has an option of using web browser-based IDEs such as Jupyter notebooks which are getting increasingly popular due to their code completion, portability, and presentation (web-based tutorials, etc.) capabilities.

Finally, it is worth noting that the scientific community is increasingly doing their developments in the cloud because of the machine-independent development capabilities as well as data management/storage options provided by these tools. For example, **Amazon Web Services** (n.d.) provide data management by its "Buckets (e.g., S3)" and its web-based development environment (e.g., AWS Sagemaker) with ready to use data analytics tools embedded.

16.3 Challenges in Computer Science

As defined above, some computational problems are extremely hard to solve. In addition, those may require large storage space to solve the problem. To make things worse, "Big Data," which is often defined by four V's, namely, volume, variety, velocity, and value, adds an additional challenge to the current approaches. Volume refers to the increasing amount of data that are collected from different sources. Variety refers to the different types of data that are collected from many sensors and velocity refers to the speed of collection. Finally, the Value terms emphasize the analytic application of the data and its potential associated value.

Most data related to FEW is Big Spatial Data with more challenging characteristics such as autocorrelation between samples, non-isotropy across dimensions, and heterogeneity.

Agricultural datasets are great examples of big data (Bronson & Knezevic 2016). These datasets which are often collected from multiple sources, including the soil sensors in the field, remote sensing imagery from satellites as well as unmanned aerial vehicles (UAV), water and energy consumption datasets (Eftelioglu et al. 2016b) for each individual field and meteorological datasets represents how the agricultural data represents the variety aspect of the big data (Eftelioglu et al. 2016a). In addition, these datasets have huge volumes caused by the hundreds or thousands of variables with high resolution over a large geographical area, multiplied by the finer temporal frequency of data collection. Finally, due to the advancements on cheap sensors and improved remote sensing capabilities thanks to the UAVs, the data velocity is much faster than before. So far, this trend seems to continue to grow in the future.

To address the challenges of hard computational problems, as well as the issues, arise with big data, several approaches are used.

First, from a hardware perspective, supercomputers were introduced to computer science. These computers often have many CPUs or **GPUs** that are specialized in doing a high number of parallel computations of smaller floating point or matrix operations. For the most part, supercomputing (sometimes referred to as high-performance computing or HPC) is used to model simulations or perform brute force calculations. Supercomputers are research tools. Historically (in the 1970s) this was weather forecasting and aerodynamics (planes, spacecraft, cars), then nuclear weapons simulations and radiation shielding modeling (the height of the Cold War in the 1980s), and in recent years the emphasis has been on cracking decryption and molecular dynamics modeling. For example, for weather forecasting (Webster 2012), it is the sheer amount of data that needs to be processed. There are hundreds of thousands of weather stations around the world, each one recording dozens of variables (wind speed/direction, temperature, humidity, pollen count). All this data needs to be compared to last night's data, last week's data, and historical data, all of which must be fetched from huge memory and storage banks. Generally, most of these calculations can occur at the same time, and so massive parallelism means that weather forecasts can be produced in seconds or minutes, rather than hours (Baillie et al. 1997).

Second, these computers are backed by distributed file systems which aim to divide the data into multiple storage units, compute and save intermediate steps in the disk, then merge the outputs into a single result. To understand why such a strategy is used, one needs to understand the performances of these units. CPUs or GPUs are extremely fast components that can handle trillions of floating-point operations in a second. However, the data, which is used as the input, resides in the disks that are thousands of times slower than the CPU and GPU units.

Since a computer system is as fast as its slowest component, historically, the disks were the bottlenecks for operations that require large input or intermediate data. These distributed file system approaches aim to address the speed issue by dividing the data into smaller chunks, using them individually by each CPU/GPU and thus reducing the cost of read/writes into disk (Howard et al. 1988). For example, Hadoop Map-Reduce framework uses this strategy (Borthakur 2007). The term Map-Reduce refers to two separate and distinct tasks that Hadoop programs perform. The first is the map job, which takes a set of data and converts it into another set of data, where individual elements are broken down into tuples (key/value pairs). The reduce job takes the output from a map as input and combines those data tuples into a smaller set of tuples and thus giving the results.

Due to the advances in cheap and abundant memory space, another recent approach is using memory instead of disk to perform such distributed data processing. Although this second approach, namely, (Apache Spark 2015), is not completely replacing the Hadoop (it still uses Hadoop Distributed File System), it uses in-memory processing of data on top of Hadoop's Map-Reduce framework and thus improves the performance since memory read/writes are thousands of times faster than disk read/writes (Intel 2013; Jain & Somni 2013; Reed & Dongarra 2015). Therefore, Spark's in-memory processing delivers near real-time analytics for data machine learning, sensors, data monitoring and analytics which is particularly important to address the Velocity issues of Big Data. For example, agricultural datasets, which include soil properties, humidity, and so on, that are collected from the field sensors in real time can be analyzed near real-time by using such an approach.

Finally, it is worth mentioning the current state-of-the-art approaches that leverage these high-performance capabilities. One particular and often heard term is Deep Neural Networks. **Neural networks** that mimic the information flow in a brain are increasingly getting attention from the computer science research community. Originally proposed in the late twentieth century, research on neural networks stagnated due to the limitation of the hardware at that time. The increased computational power by GPUs with thousands of cores as well as in memory matrix manipulations, re-started the neural network research again. Especially in the computer vision domain, Deep Neural Networks (i.e., deep learning) proved to be promising.

Some recent applications of deep learning in food–energy–water research includes identification of objects (trees, farm fields, etc.) from remote sensing imagery, identification of healthy/unhealthy animals from the camera feed in farms as well as automated detection of ripe fruits from trees. This research is evolving, and in the future, it is expected to have more use cases for GPU usage for deep learning purposes. Readers are encouraged to read on these tools but beware that these tools are often a black box that affects their interpretability.

16.4 Computer Science as a Scientific Tool

The use of computer science in scientific methods is twofold:

Causal and mechanistic models often start with model building and theories. Then these models are numerically solved and backed by simulations. Many such models require a complicated process of solving differential equations that are hard to solve by hand and paper. In addition, often the simulations are hard without the use of a computer. For example, SWAT and GCM techniques require heavy use of supercomputers.

Data-driven/empirical models involve controlled experiments and statistical analysis. Often, these methods start with hypothesis creation followed by hypothesis testing as well as hypothesis ranking. Since such approaches are hard to prove by mathematical theories (due to uncertainties, complications, etc.), they often rely on a large amount of data in hand (fourth Paradigm of Science). Since these big datasets require extensive analysis, computational tools that use distributed data analysis (e.g., big data tools such as Hadoop and Spark) are required to perform those. For example, the ozone hole over Antarctica was detected by such an approach.

16.4.1 Nexus Data from Computer Science Perspective

In order to understand the role of computer science in food, energy, and water research, one may need to understand the different types of data that are related to nexus research. Since the details of the importance of Data is explained in Chap. 14, in this section, we will provide a view of how different types of data affects the computer science approach to be used in nexus research.

From a computer science perspective, data is a collection of objects that are associated with attributes and their values. These attributes can be nominal, ordinal, interval or ratio. Nominal attributes represent different categories of data. For example, the type of cars (e.g., hatchback, SUV, sedan) can be a nominal attribute. Ordinal attribute refers to quantities with some order. For example, the grades of a class can be ordinal. Interval attribute is like the ordinal attribute, but the intervals are equally split. For example, the temperature readings from a sensor can have interval attribute in **Celsius** or **Fahrenheit** scale. Finally, ratio attribute is an interval data with a natural zero point. For example, temperature readings in Kelvin scale are ratio datasets because they have an absolute zero value and the steps of each degree have a same degree of magnitude. Those attributes depending on the application domain can also be discrete or continuous; the former has only finite number or countably infinite set of values whereas the latter has real numbers as attribute values.

Datasets can also be divided into categories depending on their type (Srivastava et al. 2005). Record datasets include data matrices, document data and transaction datasets. Graph datasets, where each data item is considered as nodes and their relationships are considered edges between nodes, include molecular structures, social network datasets (e.g., friendship relation), or websites with the links between each other. Finally, ordered datasets include spatial, temporal, and spatio-temporal data and sequential data.

Datasets that are used in computer science tasks comes from a variety of sources including texts, videos, and photographs from social networks, web, and so on and numerical data sets that are collected from sensors and/or generated by mathematical models. These collected datasets can be used in a variety of tasks in computer science from data mining to machine learning and artificial intelligence. For example, a set of photos can be used to train a deep learning tool (Schmidhuber 2015), that can further be later used to identify the objects in new input photos. Similarly, using text mining from social network data streams (e.g., tweets) may help identify the implicit connections between people. Since nexus datasets are often associated with a location (e.g., longitude, latitude) as well as temporal information, the datasets that will be described in this section will only cover spatial and spatiotemporal datasets. However, interested readers may read the referenced sources in this chapter to understand better the types of datasets that are used in other fields.

16.4.2 Spatial and Spatiotemporal Datasets

Spatial data represents the location, size, and shape of an object on planet Earth such as a building, lake, mountain or township. Spatial data may also include attributes that provide more information about the entity that is being represented. If those datasets include temporal information associated with them, they are considered as spatiotemporal datasets.

These datasets can roughly be divided into three categories: Raster datasets are the representation of the world as a surface divided into a regular grid of cells (e.g., pixels). Raster models are useful for storing data that varies continuously, as in an aerial photograph, a satellite image, a surface of chemical concentrations, or an elevation surface. For example, remote sensing imagery collected from UAVs can be considered as raster data, and the pixel values may represent elevation above sea level, chemical concentrations, or rainfall, etc. Vector datasets are the representation of the world using points, lines, and polygons. Vector models are useful for storing data that has discrete boundaries (**discrete data**), such as country borders, land parcels, and streets. For example, in land cover allocation problem, vector data such as polygons may be used to identify the different types of lands. Graph datasets, which are often considered together with vector data, are graphs with vectors and edges associated with spatial properties. A road network or electric grid in a city can be considered as such type of data.

Due to the different nature of these datasets, the models, as well as data structures that are used for these, are different. For example, raster datasets where each pixel (a square grid cell) represents information associated with a specific location can be treated as a multidimensional image. The data structures for these datasets are thus matrices where (x, y) coordinates represent a square area and the additional dimensions represent the sensor readings. Thanks to the autocorrelation effect causing similar values at nearby locations, efficient image compression, and indexing algorithms are used to handle these.

On the other hand, vector datasets are mostly mathematical equations that define the geometric shapes which are relatively smaller datasets. However, depending on the resolution, the details of each polygon may be represented in a more complicated manner, causing the size and rendering of these to increase by finer resolution. As a simple example, a pentagon in finer resolution may be represented as a rectangle in a coarser resolution. To handle the rendering issues for different resolutions and provide a data generalization, algorithms for vector simplification (e.g., Douglas–Peucker algorithm) were developed. However, it is worth noting that the algorithms for vector data analytics are complex due to the inherent topological relationships between features.

Graphs are studied for a long time due to their use in a variety of application domains such as electronic circuit design, social network analysis, spatial analytics, fluid dynamics, etc. Spatial network graphs have a variety of differences compared to other graph datasets due to the varying planarity (e.g., bridges on roads, etc.), temporal dimension, multi-constrained node structures (e.g., lights and stop signs on roads) as well as their sparsity. Many data structures that were developed for graph datasets are already used for spatial network graphs, but more specialized tools and algorithms are also present such as Lagrangian shortest path framework (Gunturi et al. 2011) for temporally aware routing (Evans et al. 2014), and shortest path/travel time approaches using (Dijkstra 1959), and A$*$ algorithms.

16.5 Challenges and Opportunities for Computer Science

The FEW Nexus poses tough challenges as well as exciting opportunities for transformative research in computer science. The topics discussed in this section are inspired by the contributions of participants at the INFEWS Data Science Workshop held in 2015.

Although spatial computing scientists expect to tackle issues on many fronts in the coming years, this section will concentrate on challenges relevant to the FEW Nexus. We begin by discussing the challenges of coupling and fusing multiple model types (Webber 2015) for the FEW Nexus, followed by a review of key issues that arise from spatiotemporal modeling. Next is a discussion about how methodologies for guiding stakeholders to reach consensus on FEW issues. A proposal is made to develop a life cycle thinking methodology for understanding the FEW Nexus. The final section presents novel computing research questions that address issues of data uncertainty, incompleteness, and bias in spatiotemporal Nexus data.

16.5.1 Coupling and Fusing Multiple Models

As described above, *causal and mechanistic models* often start with model building and theories. Moreover, such models are critical to understand and evaluate the potential of implementing FEW nexus approaches. For example, the climate model that coupled ocean and atmosphere models created a breakthrough in El Nino prediction. Similarly, models are beneficial for understanding the role of "forces" such as climate change, socioeconomic and technological change, and policy interventions. These examples show that models serve to understand and quantify the risks. This subsection presents the key issues, science questions, and challenges that arise when modeling for the FEW Nexus. Note that in Chap. 15, a variety of models are discussed through a sector-centric categorization. In this chapter, we used another taxonomy of models where these are categorized as causal and mechanistic models (e.g., process models) and data-driven models.

(a) *Mechanistic and causal models versus data-driven models.*

Traditionally, FEW models have been based on mechanistic process models (e.g., see Sect. 1.4). For example, a model of passive hyperspectral sensing might consider incident solar radiation, re-radiation at various wavelengths, attenuation and refraction by the atmosphere, and noise in a satellite-borne CCD ("charge-coupled device"). A model of hydrology might encompass precipitation, solar radiation, and wind inputs, canopy capture, through-fall, surface runoff, soil infiltration, evapotranspiration (Elshorbagy et al. 2010; Solomatine 2005), and groundwater storage and release into streams (Fatichi et al. 2016). Each of the individual processes involved could, in turn, be based on appropriate mechanisms (plant physiology, percolation rates, etc.).

The key advantage of such models is their mechanistic basis, which enhances their utility to explain system responses to prevailing and counterfactual state conditions without the need of collecting a complete set of actual data (which sometimes is impossible to collect). Since these models are based on known scientific laws, they are typically valid over a wide range of input states and action choices. However, they can be computationally expensive (e.g., requiring the solution of partial differential equation systems in high dimensions) and may require approximations (e.g., linearization) to achieve computational tractability. These approximations may reduce their accuracy.

To handle accuracy problems and reduce the number of approximations, supercomputers which have high computing power are widely used for such tasks. Finally, it should be noted that these models are mostly coded in low-level languages (e.g., FORTRAN) due to their use of past several decades and the overhead of higher-level languages (e.g., object-oriented languages) makes them unattractive. Also, often, these models do not implement multiple optimization strategies since the models are changing so frequently that hand optimization of any particular model version is not useful.

Data-driven models are increasingly attractive to the scientists due to the cheap and abundant data and the availability of technologies that can store, manage and process such large datasets. These models are capable of capturing "training examples" of ground truth that can be used to develop empirical relations between the state variables of the system without explicit descriptions of the underlying physical processes. One example is the identification of farm fields with known crops and recording of the corresponding satellite observations under a variety of observing conditions and then learning a model that predicts the crops on the ground from the CCD signals. Similarly, a data-driven model of hydrology would directly map inputs (precipitation, solar radiation, and wind) to outputs (evapotranspiration, soil moisture, stream flows).

The key advantage of data-driven models is that in a data-rich world, they can provide more accurate predictions. Non-parametric, nonlinear models can adapt to the complexity of the process and yet permit very fast computation of predictions. Because data-driven models are based on statistical methods, their uncertainty can be easily quantified. The primary drawback of data-driven models is that their range of validity is restricted to inputs that are like those observed during model fitting. Without a mechanistic foundation, the models have difficulty extrapolating beyond the state conditions that are described by the "training" data.

FEW systems are rarely stationary because of such factors as climate change, disturbances, and complex couplings with other systems. This threatens to take data-driven models outside their "validity horizon" (the region where their predictions are valid). Such data-driven approaches require a large number of datasets, and High-Performance Computing (HPC) infrastructures to avail the processing. Computational tools that use distributed data structures and analysis (e.g., big data tools such as Hadoop and Spark) are better suited to improve the I/O (input/output) cost of spatial queries (e.g., retrieving a set of farm polygons within a given spatial range) for such models. Users of data-driven approaches often use higher level languages (e.g., Python, R) that already have the appropriate scientific tools (coded by someone else) for their tasks as well as due to their simplicity of use and re-usability of the code for sharing in the scientific community.

(b) *Coupling multiple models*: The essential nature of the FEW Nexus requires developing models that represent human and natural systems (e.g., energy, agriculture, and land use, water supply and use, economy, and climate). Furthermore, these models may be mechanistic models, data-driven models, or some combination of the two. Many FEW models reflect the coupling of empirical and model-based approaches, often incorporating optimization methods to assist in identifying the level of process complexity that is consistent with the observations (Guillibert 2015; Mohtar & Daher 2012; NSF 2015).

Coupled models raise issues of end-to-end testing, debugging, evolution, uncertainty quantification, and validation. For these tasks, good visualization tools [45] are essential. Two related important questions are how to enable modelers and end users to understand the essential effects of the coupling of multiple models and how does the coupling alter the behavior of the individual models and how does this affect their validity.

(c) *Variable temporal and spatial scale*: Both the data and the models in the FEW Nexus typically involve multiple spatial and temporal scales. For example, an individual farmer deploying precision agriculture may need a model of soil moisture and nutrients at the 1-m scale at a weekly timescale. At the same time, the operator of a system of hydroelectric dams may only need precipitation estimates at the scale of an entire watershed and yet need to model the hydrograph at 10-min intervals.

The data collected to create and validate such models is similarly characterized by a wide range of spatial and temporal scales Zhang (2010). Human population, an important input to economic models, is measured in the USA every 10 years and released at the level of census blocks. The MODIS satellite instrument scans the entire planet every day at scales of 250 m, 500 m, and 1 km cells, whereas the LANDSAT 7 instrument produces data at 16–60 m and a temporal resolution of 16 days (Allen et al. 2007; Norman et al. 2003; Steering Committee for NASA Technology Roadmaps 2016).

Differences of scale raise two important issues: First, when creating or evaluating models, if the spatiotemporal scale at which the model operates is different from the spatiotemporal scale at which the data are observed, then we must find a way to bridge that gap. Second, when coupling multiple models, if the models operate at multiple spatiotemporal scales, then we must find a way to link them properly. Currently, this linking is done using interpolations (Goovaerts et al. 2010), or estimations which cause results to be inaccurate or cause missing information at the finer scale.

(d) *Spatio-temporal modeling*: Understanding the FEW nexus from a computer science perspective requires developing effective and efficient spatiotemporal models. Observational data on FEW are spatiotemporal in nature. Remote sensing technologies continuously monitor the land surface of the earth and collect valuable data about water bodies, agricultural fields, and so on. For groundwater observation, the U.S. Geological Survey is creating a national network of gauge stations to develop a water census for both quality and quantity.

Over the last several decades, the spatial, spectral, and temporal resolutions of FEW datasets have improved greatly through the deployment of improved remote sensing and ground sensing systems. These improvements have led to the collection of synoptic scale data and enabled a variety of new FEW applications. However, these improvements also pose novel computing questions and challenges due to the unique nature of spatial and spatiotemporal data (Shekhar et al. 2011a; Shekhar et al. 2015a; Shekhar et al. 2011b). Spatial and spatiotemporal FEW data have unique characteristics that can be summarized as follows:

Spatiotemporal autocorrelation: Per the first law of geography (Tobler 1970), everything is related to everything else, but nearby things are more related than distant things. What this law tells us is that spatial data are not independently distributed, and autocorrelation exists (Shekhar et al. 2015b). For spatiotemporal data, temporal autocorrelation exists in addition to spatial autocorrelation. For example, in remote sensing images, nearby pixels are often in the same thematic class (e.g., forest or water), and consecutive snapshots of the same locations are often very similar.

Spatial heterogeneity: Spatial heterogeneity describes the fact that samples often do not follow an identical distribution in the entire space due to varying geographic features. Thus, a global model for the entire space fails to capture the varying relationships between features and the target variable in different regions. For example, the spectral characteristics of remote sensing images on surface water in tropical areas probably look different from those in temperate areas.

Non-Euclidean space, anisotropy: Current spatiotemporal modeling techniques often assume Euclidean space with isotropic and symmetric spatial neighborhoods. However, in many real-world applications, data are collected in a network space (e.g., river networks, road networks) (Isaak et al. 2014; Oliver et al. 2014a; 2014b). One of the main challenges of spatiotemporal data modeling in a network space is to account for the unique network structure (e.g., one-ways, connectivity, left-turns). Such network structure often violates the common assumption that spatial dependency is isotropic in Euclidean space, and thus requires asymmetric neighborhoods with directionality. Recently, some new research has been conducted in spatial network statistics and computer/data science methods, e.g., network K-function and network spatial autocorrelation, network point cluster analysis and clumping method, network point density estimation, network spatial interpolation (**Kriging**). Nevertheless, most techniques apply only to Euclidean space. Future research is needed in this area.

Temporal non-stationarity: The statistical distribution of FEW data is not stationary, as assumed by many traditional models; it changes over time. The stationarity assumption means that the model created in the training phase of

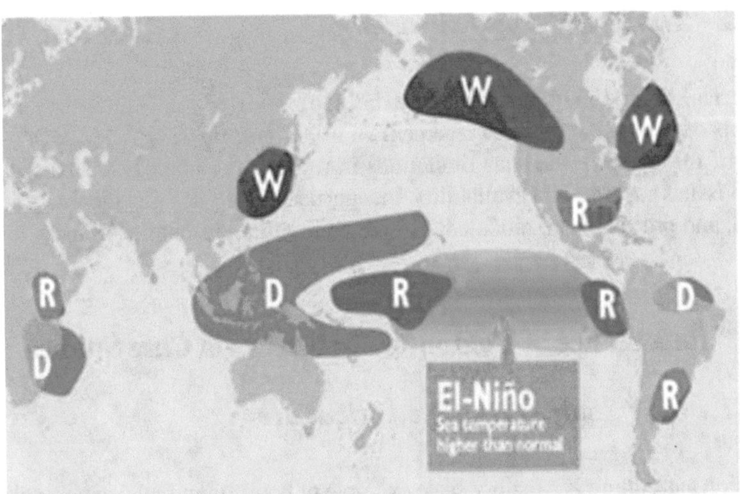

Fig. 16.1 Worldwide climatic impacts of warm El Nino events (Chen et al. 2004) during the northern hemisphere winter, D indicates drought, R indicates unusually high rainfall and W indicates abnormally warm periods (Food and Agriculture Organization n.d.)

the tool is assumed to be consistent over time. In other words, if the conditions change over time, the generated model may not reflect the actual case. For example, a model that is created for climate conditions of a specific agricultural field may not reflect the present conditions of the field due to climate change and therefore may cause incorrect future projections. Thus, there is a need for tools that will adapt the model to changes over time.

Spatiotemporal coupling and teleconnection: Spatiotemporal coupling patterns include spatiotemporal (mixed drove) co-occurrences (for unordered patterns) (Shekhar & Huang 2001), spatiotemporal cascades (for partially ordered patterns) (Mohan et al. 2010), and spatiotemporal sequential patterns (for totally ordered patterns). A spatiotemporal teleconnection is a pattern of significantly positive or negative temporal correlations between spatial time series data at a great distance. Figure 16.1 shows an example of spatiotemporal teleconnection patterns in climate data.

Multiple scales, resolutions, and sensors: FEW data are collected by many different types of sensors, with various scales and spectral, spatial and temporal resolutions. The current state of the art often reduces the resolution of data to make them compliant with each other. Such reductions cause loss of information that could be had at finer resolutions.

Large volume and velocity: The underlying data volume (terabytes to petabytes) and velocity (gigabytes to terabytes per day) of these applications is very large. NASA alone collects petabytes of earth images every year. Climate model simulations (Gazzola et al. 2007), or projection data about the variables of the earth are increasing even faster. Such volume and velocity exceed the capability of traditional spatiotemporal computational platforms.

16.5.2 Numerical Approximation Models

Numerical approximation models can be applied to describe the climate-driven impacts of changes in land management on water quality and carbon sequestration. Each of these approaches has limitations in representing the FEW nexus systems due to issues such as data availability, incongruence in scales at which data are collected, and processes are modeled, and the ability to represent uncertainty.

16.6 Importance of Computing in Nexus via Case Studies

16.6.1 Case Study: Precision Agriculture

Precision agriculture is a facility scale example of how computing can help scientists from a FEW Nexus perspective. It is an agricultural decision support system whose tools allow site-specific (i.e., location-aware) use of water, pesticides, and fertilizers by measuring crop health, soil nutrients, and moisture. It enables farmers to improve yields, reduce unnecessary applications of fertilizers and pesticides, preserve natural resources, and contend with impending weather events (Dixon & McCann 1997; McBratney et al. 2005; Press, The National Academies 1997).

Apart from the benefit of improving yields, precision agriculture improves the efficient use of energy and water as well. In agricultural practices, one of the implicit costs is the cost of energy. From the application of fertilizer and seeds to pumping irrigational water, energy is widely used for agriculture. Moreover, the indirect costs of energy use in fertilizer production and desalination of irrigational water also increase the energy cost.

With precision agriculture, water run-offs are prevented, and water is used only on the required locations in the field depending on the soil moisture levels measured by soil sensors. Therefore, by using precision agriculture for food production, not only the food aspect of FEW nexus is improved but also energy and water are used more efficiently.

Precision agriculture uses a wide variety of computing tools to achieve its success. As explained earlier, spatial data is collected from a variety of sources including soil sensors installed in the ground, tractors that collect soil samples from a variety of locations in the field, and unmanned aerial vehicles (UAV) and satellites that collect remote sensing imagery. For example, Normalized Difference Vegetation Index (NDVI) imagery, which determines the greenness of plants, is widely used in the field by precision agriculture applications in large farms. Using these collected remote sensing data, the locations where fertilizer and irrigations are needed are determined. Finally, variable rate applicator devices apply fertilizer and water just as needed.

The key technology that allows these data to be merged is their location information, which, as noted earlier, is made available by Global Navigation Satellite Systems (GNSS) (Davis et al. 2012; GNSS 2011).

Once these datasets are collected, they are stored in (spatial) databases, which allow location-based queries about soil properties, plant properties, farm management practices, and yield. For example, queries such as "where is the location in the field with high moisture and low fertilizer?" can be easily answered using spatial databases.

Data mining tools are widely used for precision agriculture tasks. For example, hotspot analysis in spatial data mining is used to determine the hotspots of locations where the soil has high nutrient or moisture levels to prevent applying more fertilizer or irrigation to these locations.

Similarly, co-location pattern mining tasks are used to determine which combination of fertilizer and water makes the crops healthier and in which location of the field. In addition, spatial data mining tasks allow predicting future yield from a field. These yield projections help farmers make a crop type decision for the next harvesting season. They also give more insights about the causes of low or high yields, best crop type for a specific type of soil and location in a field, etc.

Similarly, path planning and navigation systems promote efficient traversing of farms without unnecessary soil compaction, (spatial) statistical analysis tools help delineate management zones in large farms, and spatial decision support systems help optimize yield while preserving energy and water use by site-specific irrigation selection.

Computerized map visualizations are also integral to precision agriculture practices since they allow farmers to understand inter and intra-field variability of key agricultural components, e.g., crop health, nutrients, and moisture, etc. (Fig. 16.2)

Visualizations of these components along with future projections serve as a decision support system to help precision agriculture farmers decide where to

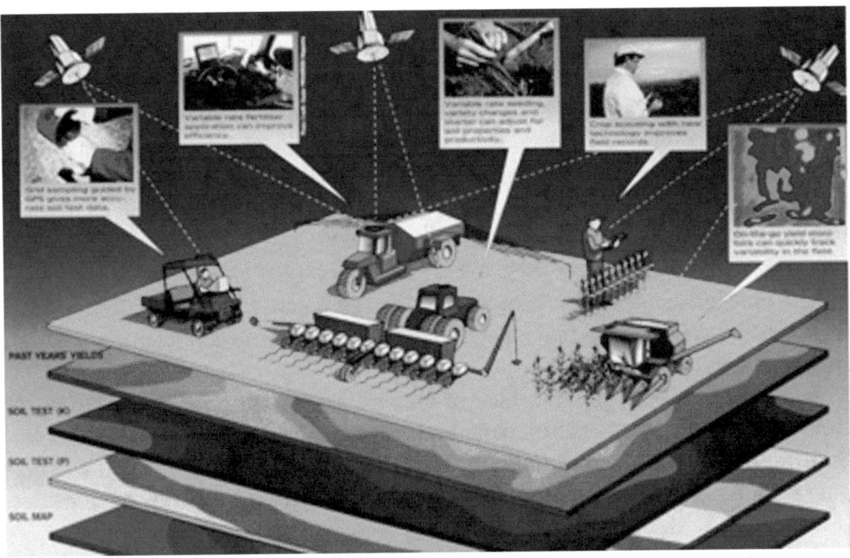

Fig. 16.2 Illustration of precision agriculture (Plant et al. 2000)

use different kinds of seeds or what yields to expect in different parts of their fields in different fields. An illustration of such visualizations can be seen in Sect. 17.6.2.

Finally, it should be noted that precision agriculture is highly dependent on the decision support component of the data cycle. Farmers rarely deal with, nor are they interested in the details of data collection, management, mining, etc. but they do make decisions using projections of future events. Their primary concern is support for decisions using projections of not only the projected amount of yield but also the possible alternative types of crops for the next year.

The spatial computing techniques outlined here are also applicable to many FEW problems, including but not limited to the efficient and sustainable use of resources by precision agriculture practices, better resource management by virtual water trading (Oki & Kanae 2004), supply chain relocation (Allan 2003; Melo et al. 2009; Min & Zhou 2002), etc.

16.6.2 Case Study: Geodesign for Landscape Modeling

Collaborative geodesign (Minnesota 2016) aims to involve stakeholders in landscape design using a tool which interactively lets them test and receive feedback for different design decisions. The benefit of this approach is that the stakeholders can immediately see the impact of their design decisions on biophysical and social indicators. Collaborative geodesign is a step forward to initiate discussions among stakeholders and domain scientists. Thus, the actual parties (i.e., stakeholders) that are affected by the decisions have the opportunity to communicate their concerns with the scientific community as well as the policymakers to make better and more realistic design decisions.

Collaborative Geodesign is a network scale (see Sect. 12.3) example of how computing tools can be used to improve food and water resources without posing a negative impact on stakeholders. Basically, it explores a computational framework that brings stakeholders from a wide range of societal sectors together to tackle the landscape redesign challenge (Fi. 16.3b). Stakeholders will divide the study area into partitions and assign management practices to the partitions to improve multiple ecosystem service objectives (e.g., water quality, soil quality, revenue).

The final land allocation is achieved through an iterative process of design, evaluation, and comparison to find a better balance among these conflicting ecosystem service objectives.

In the design step, the interface of Geodesign system provides a variety of supporting data layers to help stakeholders to make design decisions (Fig. 16.3a). Stakeholders work together to delineate partitions on the interactive touch display and choose the best management practice for each partition (Slotterback et al. 2016).

In the evaluation step, the Geodesign system involves state-of-the-art soil and water physics modeling tool (Gassman et al. 2007; Nelson et al. 2009) to generate

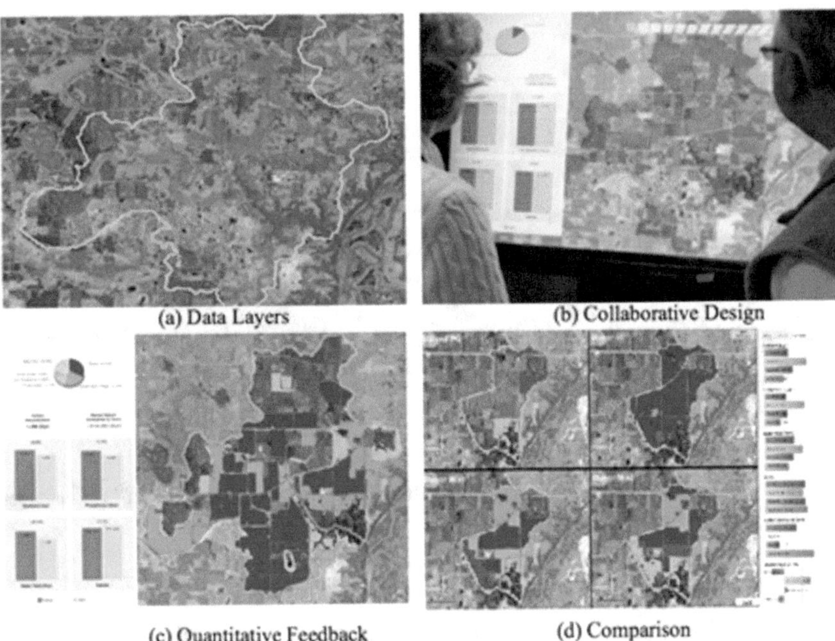

Fig. 16.3 Collaborative geodesign

quantitative feedback on the multiple objectives (Fig. 16.6). Stakeholders send their land allocation designs to the system and refine them based on the feedback created.

Finally, in the comparison step, stakeholders explore different design strategies which lead to different final land allocation designs. The Geodesign system allows stakeholders to compare multiple designs based on mapping and quantitative feedback (Fig. 16.3d).

Collaborative Geodesign is a social-learning process among stakeholders, who may represent a wide range of social sectors and intend to emphasize different ecosystem service objectives. For example, farmers tend to care more about food production, while increasing food production may lead to water pollution issues and conflict with the goals of water conservationists.

A case study of collaborative Geodesign was conducted at the Seven Mile Creek watershed, Minnesota. The watershed drains directly to the Minnesota River, which is a tributary of the Mississippi River. Although the watershed has an average slope of less than 2%, its flat upland transitions quickly into a ravine-zone before draining into the Minnesota River.

The land allocation solutions created by stakeholders are shown in Fig. 16.4 (F, G, H, I), where the x-axis denotes total amount of investment and y-axis denotes the percent of sediment reduction in water. Compared to the baseline scenario, stakeholder designs are still conservative and do not significantly improve water quality compared to the Pareto frontier achieved by linear programming (A, B, C, D, E) (Junger et al. 2010).

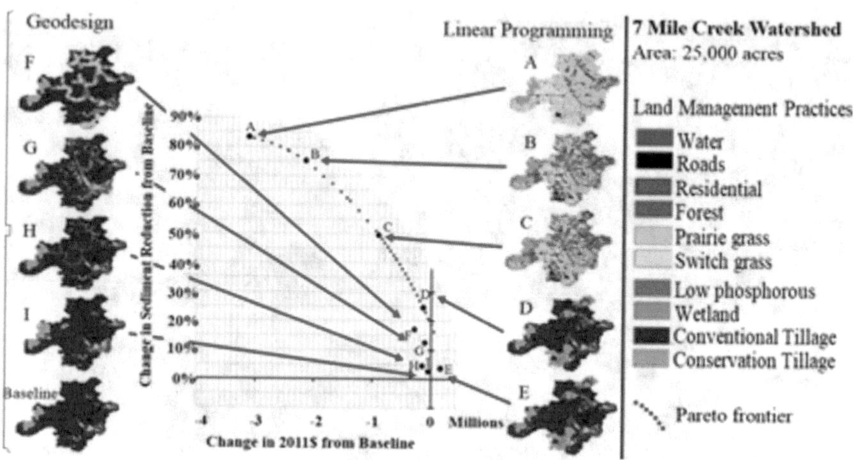

Fig. 16.4 Preliminary results on Geodesign and linear programming. Examples of land management practices are shown in the legend on the right

Although linear programming indeed can maximize water quality under a given budget limit, the mapping of solutions shows a large amount of land fragmentation (e.g., tiny patches of management practices), which are not practical for farm equipment operations.

Ongoing research has targeted reducing land fragmentation in linear programming results (e.g., spatially constrained Geodesign Optimization) (Xie et al. 2017).

The goal of innovating optimization techniques for land allocation is not to completely replace Collaborative Geodesign, but to help stakeholders identify new opportunities in this complex land allocation challenge.

Collaborative Geodesign is an important practice towards improving the sustainability of the FEW nexus through land allocation. It encourages stakeholder collaboration among multiple social sectors and enforces social learning to help reach consensus on a better-balanced production of food, biofuel, and clean water. As an emerging field, Geodesign is an active research field that encourages innovative data science approaches (e.g., optimization, machine learning) to assist stakeholders in identifying new opportunities in land allocation.

16.6.3 *Case Study: The Global Agricultural Monitoring (GEOGLAM)*

GEOGLAM (2016) is the Group on Earth Observations Global Agricultural Monitoring Initiative. It is a global scale project launched by the Group of Twenty (G20) Agriculture Ministers in 2011. The participating countries produce more than

80% of global crop production and have a substantially strong impact on international food markets. GEOGLAM aims to improve agricultural monitoring at national, regional, and global scales using state-of-the-art remote sensing technologies. Ultimately, the goal is improving food production and food security.

Data collection: GEOGLAM starts with global data collection using integrated satellite imageries and ground-based measurements. Earth observations from a variety of facilitates are coordinated and preprocessed to prepare reliable data sources for global crop monitoring.

Crop monitoring: The initiative monitors primarily four crops, namely, corn, wheat, rice, and soybeans. GEOGLAM offers capacity for crop classification (e.g., growing conditions, status), crop yield prediction as well as an early warning for countries at risk for food insecurity. The monitoring tasks are not executed using a single model but based on a set of existing models and tools developed nationally or internationally. Existing models developed at a local region may contain methods and parameters tuned to local climates, weather conditions, etc. which make them unsuitable for a different geographic region. Thus, for more accurate results, the models are implemented at different scales to account for spatial heterogeneity and integrated together to enable monitoring at a global scale.

Interactive tools: GEOGLAM provides tools with interactive interfaces for global crop monitoring. Data generated by the system can be visualized either on a map (Fig. 16.5) or through a table-based view. Spatially, users can navigate and select a set of regions on the map interface and generate reports as needed (e.g., recent crop condition). Temporally, users can visualize and generate reports for different years and make comparisons. GEOGLAM also publishes a monthly bulletin on global crop conditions with a variety of maps and charts (Fig. 16.6).

GEOGLAM provides accurate, timely, and sustained global crop monitoring and yield forecasts. More importantly, it offers a framework and platform for global collaboration among countries through data and model sharing. It is a success story that, no less importantly, also highlights the potential for computer science to help secure our planet's food, energy, and water resources.

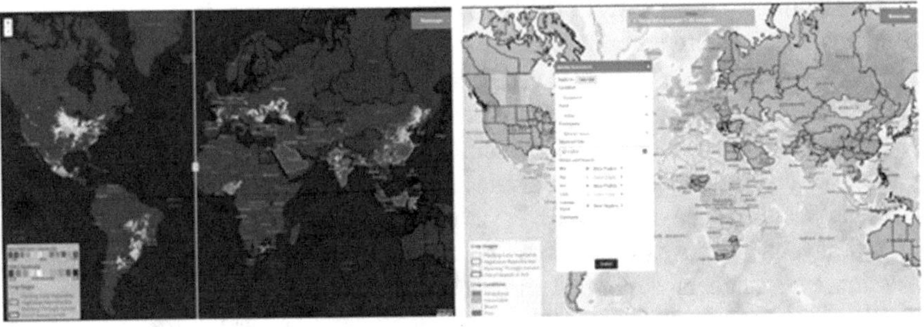

Fig. 16.5 Interactive interfaces of crop monitoring provided by GEOGLAM

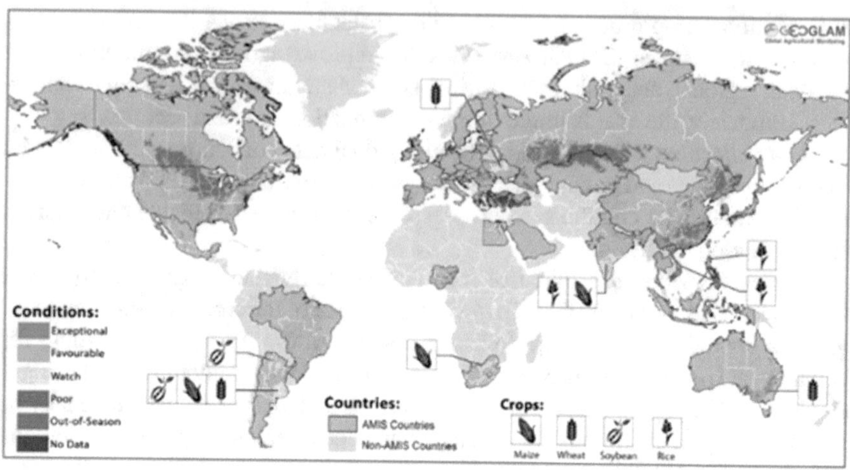

Fig. 16.6 Synthesis map on crop conditions in GEOGLAM monthly report

16.7 Summary

The research on Food, Energy and Water systems is highly dependent on computational tools and services. To understand the role of computations in FEW Nexus research, this chapter provided basic background information on computer science.

Key Points

- Computer science is not only software development. It involves experimental algorithms, computational science, and experimental computer science. However, it also has some pure engineering roots, such as design, development, software engineering as well as computer engineering.
- Algorithms have different computational costs associated. This cost governs how to solve the problems. Solutions can either be a heuristic, an approximation or an exact for a problem. In addition, FEW datasets are large and sophisticated. Therefore, the tools and algorithms should be selected depending on the data size.
- It is important to select a programming language that has good scalability and easy learning pace. The availability of ready-to-use packages is also an important factor since different communities evolved around each of these. Two particularly important languages that can be leveraged by the FEW Nexus researchers are Python and R. These languages have large online resources that can be leveraged to find answers for most common problems.
- The increasing amount of data creates a bottleneck for traditional computational tools. These data are often referred to as Big Data, represented by 3Vs, i.e., Volume, Velocity, and Variety. To overcome the associated issues with big data, computer science researchers developed distributed architectures. Similarly, cloud-based, GPU based and parallelized computational tools are available for users.

- FEW nexus datasets are spatial and spatiotemporal in nature, and the key issues such as autocorrelation, anisotropy, teleconnections, and heterogeneity should be considered when working on these datasets. Specialized tools for spatial and spatiotemporal datasets are topics of ongoing research and readers are encouraged to follow up the news regarding these.
- Stakeholders and decision makers need tools to interactively participate in the decision-making process to have a joint solution for the FEW nexus related problems. Similarly, life cycle thinking is important for a successful FEW Nexus research. Data availability, validity, and computational challenges are key concerns.
- Visualizations of FEW nexus data along with future projections can serve as a decision support system to decision makers and stakeholders with less knowledge about the underlying interconnected components. Thus, useful and intuitive visualizations should be developed.

Discussion Points and Exercises

1. Discuss the differences between time and space complexity of computational problems.
2. Which programming languages are better suited for FEW Nexus researchers? Why?
3. What are the three V's of Big Data? Give examples of Big Datasets from FEW Nexus.
4. Among scientific methods, Data-Driven Empirical Models are getting much interest in the data science community nowadays. What is the cause of this?
5. One particularly important component of FEW Nexus datasets is their spatial and spatiotemporal nature. What properties of spatial and spatiotemporal datasets create new challenges in addition to the challenges of Big Data?
6. What are the key advantages of using mechanistic models and data-driven models? How can these be fused together to leverage their advantages?
7. What are the useful tools that can help guide the stakeholders build consensus? Why are visualizations important for this particular task?
8. What are the concerns for Life Cycle Assessment from a computer science perspective? How can these concerns be overcome?
9. Precision agriculture is a good example of using computational tools for improving efficiencies as well as productivity. Which computing tools are used in precision agriculture? Are there any emerging technologies that can be added to this list?
10. What is collaborative geo-design? What other components can be added to the geo-design concept? Discuss the similarities and dissimilarities between geo-design and life cycle assessment.
11. What data are available at the GEOGLAM website. How one can use these datasets for the FEW Nexus research? Are there any other datasets that would be useful for crop monitoring?

References

General Computational Science

Akeret, J., Gamper, L., Amara, A., & Refregier, A. (2015). HOPE: A Python just-in-time compiler for astrophysical computations. *Astronomy and Computing, 10*, 1–8. https://doi.org/10.1016/j.ascom.2014.12.001.

Amazon Web Services Web Site. (n.d.). https://aws.amazon.com/.

Anaconda Web Site. (n.d.). https://www.anaconda.com/.

Friedman, D., Wand, M., & Haynes, C. (2001). Essentials of programming languages. *Journal of Functional Programming, 13*, 829–831. https://doi.org/10.1017/S0956796803254873.

Hopcroft, J. E., Motwani, R., & Ullman, J. D. (2001). Introduction to automata theory, languages, and computation, 2nd Edition. *ACM SIGACT News, 32*, 60–65. https://doi.org/10.1145/568438.568455.

Karpinski, M., & Kannan, R. (2004). Approximation algorithms for NP-hard problems. *Oberwolfach Reports, 1*, 1461–1540. https://doi.org/10.4171/OWR/2004/28.

Krishnamurthi, S. (2012). Programming languages: Application and interpretation, 1–207. http://scholar.google.com/scholar?hl=en&btnG=Search&q=intitle:Programming+languages:+Application+and+interpretation#0.

Martin-Löf, P. (1982). Constructive mathematics and computer programming. *Studies in Logic and the Foundations of Mathematics, 104*(C), 153–175. https://doi.org/10.1016/S0049-237X(09)70189-2.

Millman, K. J., & Aivazis, M. (2011). Python for scientists and engineers. *Computing in Science and Engineering*. https://doi.org/10.1109/MCSE.2011.36.

Mitchell, J. C. (1996). Foundations for programming languages. *Computers & Mathematics with Applications, 32*(10), 130. https://doi.org/10.1016/S0898-1221(96)90016-4.

Papadimitriou, C. H. (1994). Computational complexity. *Computational Complexity*. https://doi.org/10.1006/jcom.1995.1011.

Python. (n.d.). https://www.python.org/.

R Web Site. (n.d.). https://cran.r-project.org/.

Rubinsteyn, A., Hielscher, E., Weinman, N., & Shasha, D. (2012). Parakeet: a just-in-time parallel accelerator for {Python}. *USENIX Conf. on Hot Topics in Parallelism, 14*.

Wolsey, L. A. (1998). Integer programming. *Mathematical Programming, 98*(1–3), 1–2. https://doi.org/10.1186/1471-2105-11-S11-S4.

Challenges

Apache Spark. (2015). Apache Spark™ – Lightning-fast cluster computing. Spark.Apache.Org. 2015. http://spark.apache.org/.

Baillie, C., Michalakes, J., & Skalin, R. (1997). Regional weather modeling on parallel computers. *Parallel Computing, 23*(14), 2135–2142. https://doi.org/10.1016/S0167-8191(97)00104-X.

Borthakur, D. (2007). The Hadoop distributed file system: Architecture and design, Hadoop project website. *Hadoop Project Website, 11*(2007), 1–14.

Bronson, K., & Knezevic, I. (2016). Big data in food and agriculture. *Big Data & Society, 3*(1), 1–5. https://doi.org/10.1177/2053951716648174.

Eftelioglu, E., Jiang, Z., Ali, R., & Shekhar, S. (2016a). Spatial computing perspective on food energy and water nexus. *Journal of Environmental Studies and Sciences, 6*(1), 62–76. https://doi.org/10.1007/s13412-016-0372-y.

Fatichi, S., Vivoni, E. R., Ogden, F. L., Ivanov, V. Y., Mirus, B., Gochis, D., Downer, C. W., et al. (2016). An overview of current applications, challenges, and future trends in distributed process-based models in hydrology. *Journal of Hydrology*. https://doi.org/10.1016/j.jhydrol.2016.03.026.

Howard, J. H., Kazar, M. L., Menees, S. G., Nichols, D. A., Satyanarayanan, M., Sidebotham, R. N., & West, M. J. (1988). Scale and performance in a distributed file system. *ACM Transactions on Computer Systems, 6*(1), 51–81. https://doi.org/10.1145/35037.35059.

Intel. (2013). Intel distribution for **Apache Hadoop** Software.

Jain, T., & Somni, R.. (2013). Intel distribution for Apache Hadoop, *4*(3), 530–532. http://www.intel.com/content/dam/www/public/us/en/documents/articles/intel-distribution-for-apache-hadoop-product-brief.pdf.

LeCun, Y., Bengio, Y., & Hinton, G. (2015). Deep learning. *Nature, 521*(7553), 436.

Reed, D. A., & Dongarra, J. (2015). Exascale computing and big data. *Communications of the ACM, 58*(7), 56–68. https://doi.org/10.1145/2699414.

Shwartz-Ziv, R., & Tishby, N.. (2017). Opening the black box of deep neural networks via information. ArXiv Preprint ArXiv:1703.00810.

Webster, P.. (2012). Climate change simulation: NASA's weather supercomputer. CSC World Spring. http://www.csc.com/cscworld/publications/81769/81773-supercomputing_the_climate_nasa_s_big_data_mission.

Computer Science Tools

Dijkstra, E. W. (1959). A note on two problems in connexion with graphs. *Numerische Mathematik, 1*(1), 269–271. https://doi.org/10.1007/BF01386390.

Douglas, D. H., & Peucker, T. K. (1973). Algorithms for the reduction of the number of points required to represent a digitized line or its caricature. *Cartographica: The International Journal for Geographic Information and Geovisualization, 10*(2), 112–122.

Evans, C., Gartrell, A., Gomez, L., Mouyebe, M., Oxley, D., Popuri, S. K., Neerchal, N. K., & Mehta, A. (2014). Improving the computational efficiency of downscaling GCM data for use in SWAT.

Frew, J. E., & Dozier, J. (2012). Environmental informatics. *Annual Review of Environment and Resources, 37*, 449–472.

Gunturi, Venkata, M. V., Nunes, E., Yang, K. S., & Shekhar, S. (2011). A critical-time-point approach to all-start-time lagrangian shortest paths: A summary of results. *Lecture Notes in Computer Science (Including Subseries Lecture Notes in Artificial Intelligence and Lecture Notes in Bioinformatics), 6849*, 74–91. https://doi.org/10.1007/978-3-642-22922-0_6.

Schmidhuber, J. (2015). Deep learning in neural networks: An overview. *Neural Networks*. https://doi.org/10.1016/j.neunet.2014.09.003.

Shekhar, S., Evans, M. R., Kang, J. M., & Mohan, P. (2011a). Identifying patterns in spatial information: A survey of methods. *Wiley Interdisciplinary Reviews: Data Mining and Knowledge Discovery, 1*(3), 193–214. https://doi.org/10.1002/widm.25.

Shekhar, S., Feiner, S. K., & Aref, W. G. (2015a). Spatial computing. *Communications of the ACM, 59*(1), 72–81. https://doi.org/10.1145/2756547.

Srivastava, J., Desikan, P., & Kumar, V. (2005). Foundations and Advances in Data Mining. *Foundations and advances in data mining*, 275–307. https://doi.org/10.1007/b104039.

Tobler, W. R. (1970). A computer movie simulating urban growth in the detroit region. *Economic Geography, 46*, 234. https://doi.org/10.2307/143141.

Challenges and Opportunities

Abadi, L. S. K., Shamsai, A., & Goharnejad, H. (2014). An analysis of the sustainability of basin water resources using vensim model. *KSCE Journal of Civil Engineering, 19*(6), 1941–1949. https://doi.org/10.1007/s12205-014-0570-7.

Allen, M., Rodrigo, M. F., Vargas, O., Graham, E. A., Swenson, W., Hamilton, M., Taggart, M., & Harmon, T. C. (2007). Soil sensor technology: Life within a pixel. *BioScience, 57*(10), 859. https://doi.org/10.1641/B571008.

Brander, M., Tipper, R., Hutchison, C., & Davis, G. (2008). Consequential and attributional approaches to LCA: A guide to policy makers with specific reference to greenhouse gas LCA of biofuels. *Econometrica Press, 44*, 1–14. http://onlinelibrary.wiley.com/doi/10.1002/cbdv.200490137/abstract%5Cnhttp://www.globalbioenergy.org/uploads/media/0804_Ecometrica_-_Consequential_and_attributional_approaches_to_LCA.pdf%5Cnhttp://d3u3pjcknor73l.cloudfront.net/assets/media/pdf/approachest.

Chang, K.-T. (2006). *Introduction to geographic information systems*. Boston: McGraw-Hill Higher Education.

Chen, D., Cane, M. A., Kaplan, A., Zebiak, S. E., & Huang, D. (2004). Predictability of El Niño over the past 148 years. *Nature, 428*(6984), 733–736. https://doi.org/10.1038/nature02439.

Cheng, R., Emrich, T., Kriegel, H. P., Mamoulis, N., Renz, M., Trajcevski, G., & Züfle, A. (2014). Managing uncertainty in spatial and spatio-temporal data. In *Proceedings – International conference on data engineering* (pp. 1302–1305). https://doi.org/10.1109/ICDE.2014.6816766.

Costanza, R., & Voinov, A. (2001). Modeling ecological and economic systems with STELLA: Part III. *Ecological Modelling, 143*(1–2), 1–7. https://doi.org/10.1016/S0304-3800(01)00358-1.

Eftelioglu, E., Li, Y., Tang, X., Shekhar, S., Kang, J. M., & Farah, C. (2016b). Mining network hotspots with holes: A summary of results. *Lecture Notes in Computer Science (Including Subseries Lecture Notes in Artificial Intelligence and Lecture Notes in Bioinformatics), 9927*, 51–67. https://doi.org/10.1007/978-3-319-45738-3_4.

Elshorbagy, A., Corzo, G., Srinivasulu, S., & Solomatine, D. P. (2010). Experimental investigation of the predictive capabilities of data driven modeling techniques in hydrology – Part 2: Application. *Hydrology and Earth System Sciences, 14*(10), 1943–1961. https://doi.org/10.5194/hess-14-1943-2010.

ESRI, Redlands. (2011). *ArcGIS desktop: Release 10*. CA: Environmental Systems Research Institute.

European Commission. (2015). *Eurostat database, Luxemburg*. European Commission 2008.

Food and Agriculture Organization. (n.d.). *Farmers brace for extreme weather conditions as El Nino effect hits Latin America and Australia*. Retrieved October 1, 2017, from http://www.fao.org/NEWS/1997/970904-e.htm.

Frischknecht, R., & Rebitzer, G. (2005). The ecoinvent database system: A comprehensive web-based LCA database. *Journal of Cleaner Production, 13*(13–14), 1337–1343.

Gazzola, V., van der Worp, H., Mulder, T., Wicker, B., Rizzolatti, G., & Keysers, C. (2007). Aplasics born without hands mirror the goal of hand actions with their feet. *Current Biology, 17*(14), 1235–1240. https://doi.org/10.1016/j.cub.2007.06.045.

Gonzalez-Dugo, M. P., Neale, C. M. U., Mateos, L., Kustas, W. P., & Li, F. (2006). Comparison of remote sensing-based energy balance methods for estimating crop evapotranspiration. *Remote Sensing*, 63590Z. https://doi.org/10.1117/12.690056.

Goovaerts, P., Vellidis, P., & Ortiz, B. V. (2010). Geostatistical modeling of the spatial variability and risk areas of southern root-knot nematodes in relation to soil properties. *Geoderma, 156*, 243–252. https://doi.org/10.1016/j.geoderma.2010.02.024.Geostatistical.

Graedel, T. E., Allenby, B. R., & Comrie, P. R. (1995). Matrix approaches to abridged life cycle assessment. *Environmental Science & Technology, 29*(3), 134A–139A.

Greilich, S., Glasmacher, U. A., & Wagner, G. A. (2002). Spatially resolved detection of luminescence: A unique tool for archaeochronometry. *Naturwissenschaften, 89*(8), 371–375. https://doi.org/10.1007/s00114-002-0341-z.

Guillibert, P. (2015). *FEW nexus resource platform*. http://www.water-energy-food.org/.

Harris, J. C. (2010). Clark University vicennial conference on psychology and pedagogy. *Archives of General Psychiatry, 67*(3), 218. https://doi.org/10.1001/archgenpsychiatry.2010.16.

Heijungs, R., & de Koning, A. (2011). Computational challenges in huge LCA and EEIOA systems. In *Innovations in sharing environmental observations and information*. Aachen: Shaker Verlag.

Isaak, D. J., Peterson, E. E., Ver Hoef, J. M., Wenger, S. J., Falke, J. A., Torgersen, C. E., et al. (2014). Applications of Spatial Statistical Network Models to Stream Data. *Wiley Interdisciplinary Reviews: Water, 1*(3), 277–294. https://doi.org/10.1002/wat2.1023.

Marcus, P.. (n.d.). *Business leaders pitch water sharing between urban, rural communities*. The Durango Herald.

Miao, Y., Mulla, D. J., Randall, G. W., Vetsch, J. A., & Vintila, R. (2009). Combining chlorophyll meter readings and high spatial resolution remote sensing images for in-season site-specific nitrogen management of corn. *Precision Agriculture, 10*(1), 45–62. https://doi.org/10.1007/s11119-008-9091-z.

Miller, H. J., & Han, J. (2001). *Geographic data mining and knowledge discovery. Knowledge creation diffusion utilization*. Boca Raton, FL: CRC Press. http://books.google.com/books?id=1ipxxOt_79sC.

Mohan, P., Shekhar, S., Shine, J. A., & Rogers, J. P. (2010). Cascading spatio-temporal pattern discovery: A summary of results. *Sdm, 24*(11), 327–338. http://epubs.siam.org/doi/abs/10.1137/1.9781611972801.29%5Cnhttp://www.it.uniovi.es/opnet/IADIS2004_english.pdf.

Mohtar, R. H., & Daher, B. (2012). *Water, energy, and food: The ultimate nexus. Encyclopedia of agricultural, food, and biological engineering*. London: Taylor & Francis. http://wefnexustool.org/docs/water,%20energy,%20and%20food_the%20ultimate%20nexus%20(mohtar,%20daher,%202012).pdf.

None. (2016). *Identifying interdisciplinary data science approaches and challenges to enhance understanding of interactions of food systems with energy and water systems*. http://www.spatial.cs.umn.edu/few/few_report_draft.pdf.

Norman, J. M., Anderson, M. C., Kustas, W. P., French, A. N., Mecikalski, J., Torn, R., Diak, G. R., Schmugge, T. J., & Tanner, B. C. W. (2003). Remote sensing of surface energy fluxes at 10^1-m pixel resolutions. *Water Resources Research, 39*(8). https://doi.org/10.1029/2002WR001775.

NSF. (2015). *Dear colleague letter: SEES: Interactions of food systems with water and energy systems*.

Okabe, A., & Sugihara, K. (2012). *Spatial analysis along networks: Statistical and computational methods (statistics in practice)*. New York: John Wiley & Sons.

Oliver, D., S. Shekhar, J.M. Kang, R. Laubscher, V. Carlan, and A. Bannur. (2014a). A K-main routes approach to spatial network activity summarization. *IEEE Transactions on Knowledge and Data Engineering, 26*, 1464–1478. IEEE. https://doi.org/10.1109/TKDE.2013.135

Oliver, D., Shekhar, S., Zhou, X., Eftelioglu, E., Evans, M. R., Zhuang, Q., Kang, J. M., Laubscher, R., & Farah, C. (2014b). Significant route discovery: A summary of results. *Lecture Notes in Computer Science (Including Subseries Lecture Notes in Artificial Intelligence and Lecture Notes in Bioinformatics), 8728*, 284–300.

Rebitzer, G., Ekvall, T., Frischknecht, R., Hunkeler, D., Norris, G., Rydberg, T., Schmidt, W., Suh, S., Weidema, B., & Pennington, D. W. (2004). Life cycle assessment: Framework, goal and scope definition, inventory analysis, and applications. *Environment International, 30*(5), 701–720. https://doi.org/10.1016/j.envint.2003.11.005.

Roisin, B. R. (2008). *Life-Cycle Assessment (LCA)*. http://engineering.dartmouth.edu/~cushman/courses/engs171/LCA.pdf%5Cnpapers2://publication/uuid/86C92A41-44D7-46B3-9092-43176DA8B84B.

Sala, S., Reale, F., Cristobal-Garcia, J., Marelli, L., & Pant, R. (2016). Life cycle assessment for the impact assessment of policies. *EUR 28380 EN*.

Shekhar, S., Evans, M. R., Kang, J. M., & Mohan, P. (2011b). Identifying patterns in spatial information: A survey of methods. *Wiley Interdisciplinary Reviews: Data Mining and Knowledge Discovery, 1*(3), 193–214. https://doi.org/10.1002/widm.25.

Shekhar, S., & Huang, Y. (2001). Discovering spatial co-location patterns: A summary of results. In *Advances in spatial and temporal databases. SSTD 2001* (pp. 236–256). Berlin, Heidelberg: Springer. https://doi.org/10.1007/3-540-47724-1_13.

Shekhar, S., Jiang, Z., Ali, R., Eftelioglu, E., Tang, X., Gunturi, V., & Zhou, X. (2015b). Spatiotemporal data mining: A computational perspective. *ISPRS International Journal of Geo-Information, 4*(4), 2306–2338. https://doi.org/10.3390/ijgi4042306.

Slotterback, C. S., Runck, B., Pitt, D. G., Kne, L., Jordan, N. R., Mulla, D. J., Zerger, C., & Reichenbach, M. (2016). Collaborative geodesign to advance multifunctional landscapes. *Landscape and Urban Planning, 156*, 71–80. https://doi.org/10.1016/j.landurbplan.2016.05.011.

Solomatine, D. P. (2005). Encyclopedia of Hydrological Sciences. *Encyclopedia of hydrological sciences, 14*. https://doi.org/10.1002/0470848944.

Sonnemann, G., Vigon, B., Broadbent, C., Curran, M. A., Finkbeiner, M., Frischknecht, R., Inaba, A., et al. (2011). Process on "Global guidance for LCA databases". *The International Journal of Life Cycle Assessment, 16*(1), 95–97.

Steering Committee for NASA Technology Roadmaps. (2016). *NASA Space Technology roadmaps and priorities revisited. Nap 23582*. https://doi.org/10.17226/23582.

Swedish SPINE Database. (n.d.).

Trajcevski, G. (2011). Uncertainty in spatial trajectories. In *Computing with spatial trajectories* (pp. 63–107). New York: Springer. https://doi.org/10.1007/978-1-4614-1629-6_3.

Uddin, M. S., Armstrong, P. R., & Zhang, N. (2006). Accuracy of grain moisture content prediction using temperature and relative humidity sensors. *Applied Engineering in Agriculture, 22*(2), 267–273.

Umer, M., Kulik, L., & Tanin, E. (2010). Spatial interpolation in wireless sensor networks: Localized algorithms for variogram modeling and kriging. *GeoInformatica, 14*(1), 101–134. https://doi.org/10.1007/s10707-009-0078-3.

USGS. (2011). *USGS groundwater watch*. http://groundwaterwatch.usgs.gov/AWLSites.asp?S=302416087505501&ncd=.

USGS and Christiansen, G. (1984). U.S. Geological Survey circular. *Circular*. https://pubs.er.usgs.gov/publication/cir930N.

Vaisman, A., & Zimányi, E. (2014). *Data warehouse systems. Information and software technology* (Vol. 51). Berlin, Heidelberg: Springer. https://doi.org/10.1007/978-3-642-54655-6.

Webber, M. (2015). Energy, water and food problems must be solved together. *Scientific American, 312*(2), 5.

Wilensky, U. (2012). *NetLogo home page*. Evanston, IL: Northwestern University. http://ccl.northwestern.edu/netlogo/index.shtml.

Zhang, J. (2010). Multi-source remote sensing data fusion: Status and trends. *International Journal of Image and Data Fusion, 1*(1), 5–24. https://doi.org/10.1080/19479830903561035.

Importance of Computing

Allan, J. A. (2003). Virtual water – The water, food, and trade nexus. Useful concept or misleading metaphor? *Water International, 28*(1), 106–113. https://doi.org/10.1080/02508060.2003.9724812.

Davis, L. A., Enge, P. K., & Gao, G. X. (2012). *Global navigation satellite systems: Report of a joint workshop of the National Academy of Engineering and the Chinese Academy of Engineering*. Washington, DC: The National Academy Press. ISBN: 978-0-309-22275-4.

Dixon, J., & McCann, M. (1997). *Precision agriculture in the 21st century. Geospatial and information technologies in crop management*. Washington, DC: National Academies Press. https://doi.org/10.17226/5491.

Gassman, P. W., Philip, W., Manuel, M. R., Reyes, R., Colleen, C. H., Green, H., & Arnold, J. G. (2007). The soil and water assessment tool: Historical development, applications, and future research directions. *Transactions of the ASAE, 50*(4), 1211–1250.

GEOGLAM. (2016). *Crop monitor: A geoglam initiative*. http://www.geoglam-crop-monitor.org/pages/currentreport.php?id=201603&type=SY&interface=AMIS.

GNSS. (2011). *GNSS systems reports*.

Junger, M., Naddef, D., Pulleyblank, W. R., Rinaldi, G., Liebling, T. M., Nemhauser, G. L., Reinelt, G., & Wolsey, L. A. (2010). *50 years of integer programming 1958–2008: From the early years to the state-of-the-art*. Berlin, Heidelberg: Springer. https://doi.org/10.1007/978-3-540-68279-0.

McBratney, A., Whelan, B., Ancev, T., & Bouma, J. (2005). Future directions of precision agriculture. *Precision Agriculture, 6*(1), 7–23. https://doi.org/10.1007/s11119-005-0681-8.

Melo, M. T., Nickel, S., & Saldanha-da-Gama, F. (2009). Facility location and supply chain management – A review. *European Journal of Operational Research, 196*(2), 401–412. https://doi.org/10.1016/j.ejor.2008.05.007.

Min, H., & Zhou, G. (2002). Supply chain modeling: Past, present and future. *Computers and Industrial Engineering, 43*(1–2), 231–249. https://doi.org/10.1016/S0360-8352(02)00066-9.

Minnesota. (2016). *Buffer law amendments*. http://www.bwsr.state.mn.us/buffers.

Nelson, E., Mendoza, G., Regetz, J., Polasky, S., Tallis, H., Richard Cameron, D., Chan, K. M. A., et al. (2009). Modeling multiple ecosystem services, biodiversity conservation, commodity production, and tradeoffs at landscape scales. *Frontiers in Ecology and the Environment, 7*(1), 4–11. https://doi.org/10.1890/080023.

Oki, T., & Kanae, S. (2004). Virtual water trade and world water resources. *Water Science and Technology, 49*(7), 203–209. http://www.waterfootprint.org/Reports/Report12.pdf.

Plant, R. E., Pettygrove, G. S., & Reinert, W. R. (2000). Precision agriculture can increase profits and limit environmental impacts. *California Agriculture, 54*(4), 66–71. https://doi.org/10.3733/ca.v054n04p66.

Press, The National Academies. (1997). *Precision agriculture in the 21st century: Geospatial and information technologies in crop management*. Washington, DC: The National Academies Press. %5Curl%7Bwww.nap.edu/openbook.php?record_id=5491%7D%5Cnhttp://books.google.fr/books?hl=fr&lr=&id=HHErAAAAYAAJ&oi=fnd&pg=PA1&dq=searcy(1994+)+engineering+systems+for+site-specific+management&ots=F3EZlW-WZk&sig=8k3--9Yr9bqWlN82RwN7_CU63Ak.

Xie, Y., Yang, K., Shekhar, S., Dalzell, B., & Mulla, D. (2017). *Spatially constrained geodesign optimization (GOP) for improving agricultural watershed sustainability*.

Further Reading

Cormen, T., C. Leiserson, R. Rivest, C. Stein. 2009. Introduction to algorithms, 3rd edn. Contemporary Sociology, 25. https://doi.org/10.2307/2077150.

Denning, P. J. (2005). Is computer science science? *Communications of the ACM, 48*(4), 27. https://doi.org/10.1145/1053291.1053309.

Gustafson, M. (2014). Big data and agriculture. *Agri Marketing*. https://doi.org/10.1002/ajh.23643.

McKinsey & Company. 2011. Big data: The next frontier for innovation, competition, and productivity. McKinsey Global Institute. no. June. 156. https://doi.org/10.1080/01443610903114527.

Mitchell, T. M. (1997). *Machine learning*. New York: McGraw-Hill Education.

Ng, A. (2017). Machine learning yearning. http://www. mlyearning.org/(96).

Pang-Ning, T., M. Steinbach, and V. Kumar. 2006. Introduction to data mining. Library of Congress. https://doi.org/10.1016/0022-4405(81)90007-8.

Ramakrishnan, R., & Gehrke, J. (2002). *Database management systems* (3rd ed.). New York: McGraw-Hill.

Shekar, S., & Chawla, S. (2003). *Spatial databases: A tour. spatial databases: A tour*. Upper Saddle River, NJ: Prentice Hall. https://doi.org/papers3://publication/uuid/E626604D-1F38-4DCD-878A-C1E029178A87.

Wing, J. M. (2006). Computational thinking. *Communications of the ACM, 49*(3), 33. https://doi.org/10.1145/1118178.1118215.

Chapter 17
Applying Science to Practice

Emre Eftelioglu, Fernando R. Miralles-Wilhelm, Rabi Mohtar, Benjamin L. Ruddell, Peter Saundry, and Shashi Shekhar

17.1 Introduction

Throughout this book, we have noted the interdisciplinary nature of research at the nexus research: engaging the full range of physical, life, social, and engineering sciences, and integrating them through sophisticated models. FEW Nexus research, therefore, requires scientists with diverse disciplinary training to work together.

E. Eftelioglu
Engineering and Data Sciences, Cargill Inc., Wayzata, MN, USA
e-mail: emre_eftelioglu@cargill.com

F. R. Miralles-Wilhelm
Department of Atmospheric and Ocean Science, College of Computer, Mathematical, and Natural Sciences, University of Maryland (and The Nature Conservancy), College Park, MD, USA
e-mail: fwilhelm@umd.edu

R. Mohtar
Faculty of Agricultural and Food Sciences, American University of Beirut, Beirut, Lebanon

Departments of Civil Engineering and of Biological and Agricultural Engineering, WEF Nexus Initiative, Texas A&M University, College Station, TX, USA
e-mail: mohtar@aub.edu.lb

B. L. Ruddell
School of Informatics, Computing, and Cyber Systems, Northern Arizona University, Flagstaff, AZ, USA
e-mail: Benjamin.Ruddell@nau.edu

P. Saundry (✉)
Energy Policy and Climate, Advanced Academic Programs, Krieger School of Arts and Sciences, Johns Hopkins University, Washington, DC, USA
e-mail: psaundr1@jhu.edu

S. Shekhar
Computer Science & Engineering, College of Science & Engineering, University of Minnesota, Minneapolis, MN, USA
e-mail: shekhar@umn.edu

P. Saundry, B. L. Ruddell (eds.), *The Food-Energy-Water Nexus*, AESS Interdisciplinary Environmental Studies and Sciences Series,
https://doi.org/10.1007/978-3-030-29914-9_17

We have also noted that the objective of Nexus research is to support decision-making ranging from the level of an individual facility to the global Sustainable Development Goals. It is, therefore, critical to precisely understand the stakeholder's decisions and questions before producing analysis and data. A degree of uncertainty can be tolerated, but science is not at all useful at all to decision-makers if the wrong questions, even slightly wrong or misframed questions. This is perhaps the single most important lesson in the application of science to practice. Thus, FEW Nexus research interfaces with non-scientific stakeholders in critical ways.

One definition of the word *practice* is "the actual application or use of an idea, belief, or method, as opposed to theories relating to it" (Oxford Dictionaries). Thus, we can think of the FEW Nexus practice as the application of Nexus research to real-world problems. Both scientists and non-scientists involved in such work can be termed "practitioners."

As explored in Chap. 2, the Nexus is a complex interlinked system. Changes in one aspect of food, energy, or water management can create unintended consequences in another aspect due to feedback loops among actors, resources, and the socio-economic and environmental processes operating in each one of these realms.

Knowledge enhances action that can move beyond analysis toward implementation. Knowledge should increase validity, accountability, and responsibility for solution options (Lang et al. 2012). Applied FEW Nexus projects, therefore, requires the integration of diverse skill sets, expertise, and experience to address the complex problems surrounding such issues as equitable access to food, water, and energy. Scientists and nonscientific stakeholders are required to work together to frame questions, connect research to application, and ensure effective implementation and monitoring.

The term *stakeholders* include individuals and organizations in the science, business, sociological, and political sectors: all are essential to interdisciplinary or transdisciplinary research and its effective application. The spread of the community skills and interest and the spectrum they cover from knowledge to action defines these communities and differentiates them from professional disciplinary societies. Multi-stakeholder participation is an essential element for addressing sustainability challenges and developing solutions palatable to civil society (Kates et al. 2001).

The human decision-making process is shaped by the goals and capacities of various stakeholders and by the technological capabilities each of these has. A central challenge at the Nexus is helping multiple stakeholder communities negotiate the inevitable trade-offs that arise when changes in policy, planning, or management are introduced to the interlinked systems.

There is rarely a linear flow of science to decision-making. Decision-makers utilize science to a lesser or greater degree but are influenced by their own experiences and values, as well as the objectives of any community or constituency that they represent, before making choices and judgments. Thus, the relationship between science and decisions is a complex one. Where the products of science are more aligned with the processes and needs of decision-makers, they are more influential. Where science is more effectively communicated to and understood by decision-makers, it is more influential. How risk and uncertainty are understood and perceived can strongly influence the impact of science. Thus, communication is a critical part of the application of science to practice (see Sect. 21.3.7).

Stakeholders and decision-makers often have difficulty making sense of complexity, of the inherent uncertainty associated with it, and of its implications for decision-making. Tracing the impacts of interactions through a complex system over time is challenging for most people. These impacts and the possible interventions to address them present a range of trade-offs for which optimal solutions are elusive. Stakeholders' attitudes and behaviors towards the Nexus can change over time as new information becomes accessible to them, and with evolving beliefs and values. All of this is compounded by questions of scale (Chap. 12), both geographic and temporal.

Stakeholders and decision-makers need better tools for generating, communicating, and visualizing Nexus scenarios, including their impacts, likelihood, and uncertainty. Having the participation of multiple stakeholders in decision-making processes and providing tools that are co-developed with the concerns and needs of the stakeholders from the beginning is very important to making robust decisions and taking steps towards successful decisions.

There is a growing body of research dedicated to developing the modeling and data tools and interfaces that encourage active participation and not just information transfer and consumption. Such tools help stakeholders jointly make sense of complex problems and together design the solutions upon which they can all agree. Such tools build on existing planning support tools, which have tended to focus on complicated modeling tools that stakeholders, especially stakeholders without appropriate scientific training, have difficulty engaging with. For example, the use of "**serious games**" allows the integration of complex models into interfaces that stakeholders can use to test scenarios; see more on this in Chap. 20.

This chapter discusses how stakeholders can come together as Communities of Practice to utilize tools that enhance their ability to make decisions by maximizing areas of agreement and minimizing areas of conflict. Four case studies are utilized to illustrate such applications. Further examples are described in Chaps. 18 (Cities), and 19 (Watersheds), and the application of such tools in conflict management is developed further in Chap. 20. In Chap. 21, we will conclude this book but returning to the challenges and opportunities of applying science to practice at the nexus.

17.2 Applying Decision Science to Practice

Decision science was originally a study of the management of systems and of operations research; it has since taken on a broader scope of application to sustainability and policy (Simon 1960; Roy 1993; Kraucunas et al. 2015). More on decision science is in Sect. 4.3.

As was noted in Chaps. 4 and 12, decision science brings a wide array of tools to decision-making at the FEW nexus. Applying metrics, data, modeling, and cutting-edge computing to questions aligned with the needs of stakeholders can dramatically improve our understanding of linked socio-ecological-technological systems and the management of food, energy, and water systems.

To address these interrelated challenges, processes that are decision-driven and participatory have been co-developed and implemented, which make research more responsive to decision-making needs in real case applications (i.e., FEW system planning, policy, management). This decision science perspective requires linking modeling with decision-making through a recursive, iterative process of co-production and adaptive management.

It is increasingly common for model developers to consider multiple stakeholders' perspectives in the design process in an effort to enhance the relevance and usability of models and decision support systems. For instance, research has examined how modelers themselves frame key concepts such as sustainability in the design process, how multiple stakeholder groups interpret models and visualizations, the factors affecting the adoption of models for decision-making purposes, and decision-maker perceptions of how uncertainty is treated and visualized in models and visualizations. Such research reflects a growing awareness of the importance of understanding the way that individual actors and social groups frame issues and how their perspectives play out through social processes in participatory, integrated resources management. This approach was developed, studied, and refined through sustained engagement between scientists and resource management stakeholders; the process involved data collection through interviews, focus groups, questionnaires, participant observations, and document analysis. See Sect. 11.3 for a description of the methodology developed by the Intergovernmental Panel on Climate Change for communicating risk, probability, and scientific uncertainty.

The co-development process is central to the successful application of Nexus science to real-world decision-making (e.g., Lubchenco 1998). Prior research has demonstrated empirically that the degree to which scientific and technical knowledge is successfully applied to decision-making depends on three criteria:

- salience: significance and impact;
- credibility: trusted by key stakeholders; and
- legitimacy: recognized as conforming to the rules of high-quality science such as reproducibility and peer review.

The degree to which these criteria are maximized is determined by the level of joint participation by multiple scientific disciplines and stakeholders, accountability to scientific and policy concerns, and production of models, decision support systems, and other "boundary objects" that serve as points of articulation to mediate the interactions between scientists and decision-makers.

Two-way, iterative engagement between producers and users of scientific information is also key to building trust and to better understanding the needs of policy and what scientists can provide to assist policy formulation. It has also been empirically demonstrated that collaborative processes that facilitate social learning and the co-production of knowledge can help overcome concerns about credibility, saliency, legitimacy of research, and related boundary objects. First, perceived credibility and legitimacy can be enhanced as individual concerns are heard, and collective decisions made to improve the research and information. Second, opportunities for privately or confidentially expressing opinions can be important when group settings can inhibit the sharing of controversial viewpoints. Third, collaborative modeling enhances cooperative decision-making, as compared to individual or personal displays. See also Sect. 20.3.4 on **collaborative governance**.

Through an actively managed process, FEW system collaborators can cooperatively negotiate scale, level, and specificity of the high-resolution integrated assessments to ensure that analyses address questions that are scientifically important and decision-relevant. The challenge lies in explaining the decision-space of uncertainty in an accessible, transparent manner, and without presenting all possible permutations and potential ramifications of the uncertainty itself. When framed in a context consistent with sustainable resource management and planning objectives, and within timelines consistent with decision processes, modeling can help assuage stakeholder concerns about the usefulness of FEW system modeling for decision-making and ensure the outputs can be used to help make decisions.

17.3 Tools for Decision-Making

Working with stakeholders to evaluate alternative scenarios in the FEW Nexus requires collaboration between stakeholder facilitators, planners, social scientists, geospatial/biophysical scientists, data scientists, and communication professionals. To be effective, the process of generating computational tools and visualization support for stakeholder decision-making should include stakeholders in the design process of these tools and in a manner that goes beyond surveying their questions and goals. Non-scientist stakeholders should also have a say in the assumptions, rules, processes, and outputs represented throughout the decision-making process. Such input can help guide how models are built and how results are conveyed.

In general, scientists transform data into metrics using different kinds of models. Existing models, however, do not easily connect from local to regional to global scales: the processes and data represented vary at these different scales. Finer-scale with coarser-scale models need to be connected to help transform data and assumptions into performance indicators across wider spatial and temporal scales. Rather than coupling models, cross-scale connection could be achieved through ranges or ensembles of local to global models, or process-based to data-driven models. This will help stakeholders, and decision-makers understand how gains at one scale might translate into costs at a different scale, putting them in a stronger position to collectively design meaningful landscape change. Pattern recognition and classification tools could help generalize models that evaluate alternative land, water, and energy management scenario impacts on various performance indicators and at different spatial or temporal scales. In addition, running scenarios through models may be a helpful way to test assumptions, try on future outcomes, and see what results.

Figure 12.4 illustrated the logic of an integrated approach to modeling of FEW systems that can address the research questions and stakeholder needs, must both couple modeling tools for food, energy, and water systems, and also connect to decision science, data integration, and visual analytics.

Computational tools can create platforms for conducting conversations, group design, planning exercises, and for transforming data into performance indicators.

Data scientists use a wide range of tools for data cleaning, processing, and visualization; many of these have a rather steep learning curve. FEW nexus evaluations and solutions, tiered interface-based platforms that allow for easier data manipulation are needed to enable participation of a wider range of stakeholders. Stakeholders can be quite frustrated if the model is too much of a "black box."

At the same time, data and computational scientists are also stakeholders in the FEW nexus participation processes. Effective, scientifically derived solutions must be tested in practice and accepted by the larger community: stakeholders need to be mutually aware of their needs, desires, and limitations when developing participatory tools and approaches.

The following tools can facilitate the participation and meaningful interaction of stakeholders at all steps in the design and decision-making process.

17.3.1 Data Integration via Remote Sensing and Ground-Based Sources

Water availability is a first-order control over crop yield and energy production. Integrated FEW system modeling approaches are fundamentally constrained by the hydrologic cycle and the conservation of mass. The term "water availability" is defined here as an aggregate measure of surface soil moisture, root zone soil moisture, groundwater, precipitation, surface water impoundments, and runoff; as such, these dynamics represent water that could be utilized for supply for diverse activities such as food production or energy generation.

Integrated Assessment Modeling (IAM) of water systems can use inputs from weather forecast models or reanalysis products to generate spatially and temporally detailed water availability maps crucial for predicting the coupled FEW systems output potential. This will require research to characterize water supply using observations from a comprehensive array of ground- and space-based observations of the Earth's hydrologic system.

The integrated data platform will provide a more complete and accurate characterization of the available water supply than is possible through a separate analysis of each data source. Further, when coupled with spatiotemporal measurements of agriculture food and energy production, this data platform will support the analysis of the dynamics between water availability, energy generation, and agricultural production.

The synthesis of diverse hydro-climate, agronomic, energy, and economic data measurement sources into a unified platform can yield a unique and valuable tool for use in diagnosing and predicting water sustainability, agricultural resilience, and energy security. In addition, incorporating legal or governance drivers into biophysical models may also help identify key stressors and opportunities.

Sources of hydrologic system measurements include national, state, and local governmental agencies that quantify available groundwater, soil moisture, surface water impoundments, and snow that span a range of spatial and temporal scales. Measurements of hydrologic fluxes and atmospheric state variables can also be employed to characterize the movement of water within the system and across system boundaries, as well as the forces that drive FEW system demands. Collectively, this information describes the physical conditions that largely control crop growth, dictate irrigation demand, inform crop management decisions, define thermoelectric energy requirements, and constrain biofuel production rates.

The ground-based observations used in the characterization of water availability across the world are available through the World Meteorological Organization (https://www.wmo.int) and the Automated Surface Observing System (http://www.nws.noaa.gov/asos/). In the absence of groundwater measurements, satellite-based gravimetric measurements from the Gravity Recovery and Climate Experiment (GRACE) can be used. Preliminary research using the GRACE satellites indicates a far more severe decline in global groundwater budgets than has been previously estimated.

Preliminary results demonstrate a high degree of agreement with ground-based observations (i.e., correlation greater than 0.88) and suggest GRACE measurements contain information related to groundwater storage changes. A suite of reanalysis products (e.g., MERRA-Land, MEERA-2, Land Information System, NLDAS2) provide additional information about groundwater and surface water storage. In addition to ground-based observations, satellite-based soil moisture and snow depth retrievals from passive microwave sensors are available from 1979 onwards.

Figure 17.1 provides an example of such a synthesis of data sources. The 9-year time series plot for a location in northern India (Uttar Pradesh) shows the dominant

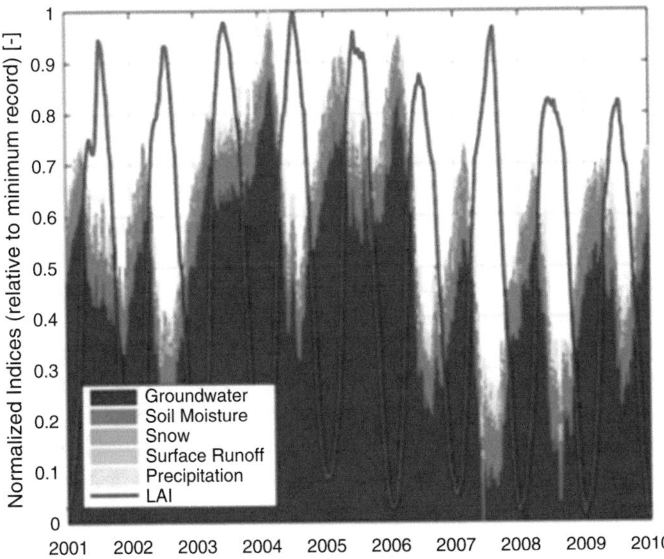

Fig. 17.1 Terrestrial water availability in Uttar Pradesh, India, from 1 January 2001 through 1 January 2010. Estimates include groundwater, soil moisture, snow, surface runoff, and precipitation flux. The red line represents the leaf area index (LAI) and serves as a proxy for vegetation dynamics. Additional hydrologic variables not shown include air temperature and evapotranspiration

components of the terrestrial water cycle, including groundwater, soil moisture, snow, surface runoff, and precipitation. A normalized index is employed to succinctly display the relative magnitude of each component of the terrestrial water cycle.

This relatively simple example, when coupled to agricultural demand, thermoelectric water demand, and biofuel production, provides a holistic record of freshwater availability that can be linked and subsequently used as input to a FEW system model across a range of spatial and temporal scales. Further, historical observational data sets enable the evaluation of model-generated estimates of terrestrial water availability. By ensuring that model estimates of water availability agree with historical observations, it provides more confidence in model projections of future water availability.

17.3.2 Integrated Assessment Modeling of FEW Systems

Over the three decades since its inception, IAMs have made fundamental contributions to our understanding of human system dynamics and human-Earth system interactions, providing critical insights into how energy technologies, land-use practices, and other human activities interact with one another and influence greenhouse gas emissions and land-cover change. Of importance in this context, IAMs are a primary means by which researchers at global and national levels have explored possible pathways to climate stabilization or the achievement of global temperature goals.

A range of challenges remain, however, in the development and use of integrated tools such as IAMs for FEW system decision-making across scales.

First, because integrated modeling requires representations of all three FEW systems, and across regional scales, the temporal and spatial resolution may frequently be sacrificed in the name of integration. Yet, carefully chosen focus areas of higher-resolution can substantially enhance the usefulness of FEW system modeling for decision-making. Major scientific questions remain about when, where, and how to create this higher resolution; when it matters and when it does not.

FEW systems interact at multiple scales from local to global. Although IAMs are increasingly proving effective at exploring these interactions at global and regional scales and over multi-year time periods, they are not designed for, nor applied at smaller, local scales, which may be of interest to many stakeholders and decision-makers. Important scientific questions remain about how to effectively transfer the methods applied to understand larger-scale phenomena at more local levels while retaining linkages to larger-scale phenomena such as agricultural or energy trade.

In most cases, exploration of issues at finer regional scales has involved soft-coupling IAMs with regional sectoral models such as regional hydrology models and power systems models. However, these approaches raise important computational and software challenges and have not taken full advantage of the integrated character of IAMs. In contrast, the low computational intensity of IAMs makes them ideal for the exploration of a large number of scenarios. This is particularly valuable for two-way interactions and co-development with decision-makers.

In response to the need for regional or local integrated assessment, recent research has focused on including variable spatial resolution, or a "telescoping" ability. This is a comparable effort to the capability for multiple scales in Earth System models. The capability has already been used to disaggregate the US energy and economy systems into 50 states, and a province-level version of an IAM has recently been created for China.

However, this variable-spatial-resolution capability has not yet been applied to explore sub-state level FEW system dynamics. Place-based research can be used as a means to take on several critical challenges that present themselves in applying IAMs to address sub-state resolution of FEW systems (e.g., major urban centers, sub-basin representation and major water demand sectors, grid-level agricultural representations).

One important modeling challenge in this regard is the representation of hydrology. IAMs typically use a gridded global hydrologic model with a fixed spatial resolution and allocate water demands among the competing water users in river basins globally. This scale (in the order of 0.1–0.5°) is large enough to limit the effects of between-basin water flows, except for major water transfer projects.

At finer spatial scales, it will be critical to represent the linkages among sub-basins, i.e., upstream-to-downstream dynamics, which implies major advances in a reduced form hydrology modeling approach that has thus far not needed to include these water transfers. This telescoping over a much smaller scale than traditionally applied in IAMs will also dictate investigating the scalability of the current representation of the hydrologic processes (e.g., runoff generation, river) to finer spatial scales. Future efforts in FEW system modeling using IAMs can achieve this by incorporating interbasin water flow capabilities within the hydrology model or switch to a finer resolution model (e.g., SWAT).

Second, the successful development and use of modeling tools for decision-making require a two-way process in which decision-makers are co-developers with the modeling teams. Questions about when and where higher resolution is important or beneficial are dependent critically on the specific decisions and decision processes to be supported by modeling. This co-development, thus, requires the creation and strengthening of social networks that include decision-makers and analysts.

This challenge involves the representation of human decision-making, which applies across the FEW system nexus. In the case of IAMs, economic decisions are made using a discrete choice formulation (comparing two or more discrete alternatives, e.g., McFadden 1980). At sufficiently large spatial scales, there exists sufficient information to calibrate these discrete choice functions. However, at finer spatial scales, information about historical choices may be insufficient to calibrate the tools, particularly under the assumption of major new technology options or price changes.

A working hypothesis is that these discrete choice models can be used for finer-scale representations, but that doing so will require innovative methods that utilize a wider set of calibration data from across a wider range of regions, potentially using clustering algorithms to identify comparable regions for analysis.

Both of these challenges raise a final fundamental challenge, which is how to create meaningful sub-state IAMs in the absence of sufficient data. Research on data integration (the lower right quadrant in Fig. 12.4), will help to bridge data gaps, but many gaps can be expected to remain, particularly at local scales and in developing countries that may not have the same tradition of data collection and management as the USA. For this reason, an important scientific challenge and focus of FEW systems modeling research will be creating multiple meaningful scenarios that effectively span the parameter uncertainty space in the context of limited data. In addition to linkages to data research, this will necessarily also rely on the co-development of parameter sets with stakeholders (2) and visual analytics methods (3) as part of that co-development process.

Third, and related, the development teams for models and for the use of models in decision-making must consist not only of modelers, but also of *social scientists* that understand how different actors frame the problems and potential solutions and how scientists, modelers, and decision-makers interpret key concepts such as uncertainty; *data development experts* that can create usable information from often incomplete sources and characterize the uncertainty in this information; *visualization and other communication experts* to bridge the divide between modelers and decision-makers.

These challenges are particularly acute in developing countries with less data and modeling infrastructure. Overcoming these challenges will be critical if the world is to address FEW system needs while simultaneously making regional/local decisions that fulfill other related needs such as sustainable development goals (Sachs 2012) and climate change commitments.

17.3.3 Visual Analytics Methods

Understanding future FEW system conditions and their impacts is based, in part, on the ability to model the key drivers and the underlying processes of complex interactions between FEW systems and human systems. Various food, energy, water, climate, and socioeconomic scenarios can reveal different projections of future FEW system conditions and their cascading impacts. It is important for analysts and policymakers to design strategies (e.g., for sustainability purposes) based on different scenario results.

These scenarios can be modeled by the interaction between ensembles of FEW system models with varied model parameters. This assumes that those scenarios can approximately describe such interactions between natural and human systems given limited observations, imperfect assumptions, and finite model choice. Thus, it becomes necessary to explore differences in modeled FEW system scenarios and the resulting variability in supply and demand. Most visual analytics tools for modeling focus on both spatiotemporal pattern identification and uncertainty analysis of an ensemble. For pattern identification and hypothesis generation purposes, visual analytics tools tend to implement a single model for understanding its spatiotemporal characteristics. Inter-comparison of model simulations is helpful in providing insight into the uncertainty of an ensemble analysis because it can help understand the impact of model structures on the outputs.

Other research has begun exploring the reliability and quality of climate projections through intercomparison of ensemble model simulations with varied model parameters. Such intercomparisons can help researchers gain insight into how model structures and varied parameters might impact model outputs and the way in which different models develop across scales and evolve. However, little research has explored the impact of spatial variation and scale on similarity analysis of modeled scenarios. Intermodel comparisons have also been used to address questions of nutrient imbalanced in the Chesapeake Bay region; these multimodel comparisons have resulted in significant changes in policy for the Chesapeake Bay Program.

Visual analytics of FEW systems modeling efforts can expand on the state-of-the-art by developing novel geovisual analytics methods that will enable users to explore the impact of spatial variations and scales for climate scenario comparisons. Such work is critical at the intersection of FEW systems: tools need to allow for regionally integrated assessment linked to nationally or even globally integrated analysis and in terms of the interactions between ensembles of models of natural and human systems with varied model parameters. In the Chesapeake Bay, a publicly available program called BAY-CAST now allows anyone to work with the actual models used by the policy analysis with the Chesapeake Bay Program.

Moving forward, an important goal in FEW systems modeling is to develop visual analytics methods for cross-scale comparison of outputs from IAMs, with a focus on the impact of spatial and temporal scales and variations on FEW system scenarios. One major challenge is exploring the impacts of these scenarios over different spatial scales. Methods for scenario comparison have traditionally explored using clustering methods to group scenarios with similar outputs. Clusters can then be compared for commonalities across input parameters, and changes to sensitivity explored. However, a common issue in data classification is that items near a classification boundary are often mislabeled. As such, the goal is to augment automated spatial classification schemes by utilizing interactive machine learning as part of the cluster creation step.

Recently, the exploration of multidimensional similarity and interactive machine learning has become a major focus in the visual analytics community: this approach combines ideas from previous investigations and will focus on developing interactive brushing techniques for interactive classification relabeling.

Little work has been done in interactive machine learning with respect to utilizing geographic projections as an underlying representation of data distributions. Instead, most work has focused on dimension reduction techniques and steering in the reduced space (e.g., Choo et al. 2010). Key challenges here include the impact of classification reassignment on distances in multivariate space and visually translating those distances to distances in physical or geographical space.

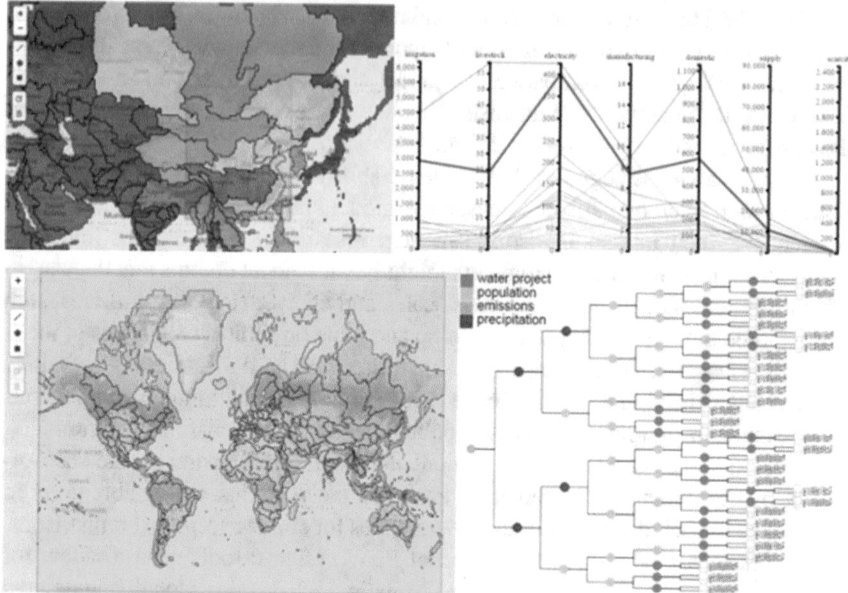

Fig. 17.2 Preliminary work clustering ensemble outputs from an IAM in a visual analytics framework. By identifying similar spatiotemporal feature structures between runs, parameter sensitivity can be explored and communicated in terms of risk-based trade-offs

Along with the interactive clustering, enabling sensitivity analysis can be done by creating a metric on how elements change groupings from one classification scheme to another. This will allow analysts to determine the sensitivity of elements to changes in the different distance functions used by various classification algorithms. This sensitivity can be used to assign certainty values to cluster groupings, which could then be visualized or used as variables in regression models and other advanced analyses.

Figure 17.2 illustrates preliminary work in coupling IAM ensembles with a visual analytics platform. New methods are needed for ensemble grouping and presentation, methods that capture user interactions, and develop visualizations that can identify groups of FEW system model runs with similar spatiotemporal properties at varying levels of resolution.

17.3.3.1 Mapping FEWS (ArcMap, Open GIS)

Domain-based scientists use the GIS software ArcMap for spatial data processing at scales ranging from farms to cities to watersheds. Open GIS tools (e.g., QGIS) as well as web-based tools (Kepler.GL, etc.) that enable users from different backgrounds to interact with spatial and spatiotemporal datasets are available and have become competitive with commercial GIS platforms. Agent-based, cellular automata, and network modeling platforms such as RePast or NetLogo are used to represent human behavior and attitudes in response to alternative scenarios for land, water, and energy management, and for the process-based environmental mechanisms underlying such behaviors. One of the larger advantages of commercial software is that visualization and mapping functions may be streamlined and easily accessible.

17.3.3.2 Systems Dynamics Modeling Software

Systems dynamics modeling software, such as Stella and Vensim, allow users to visualize (via causal loop diagrams) and simulate (via system dynamics models) stocks and flows of matter and energy through feedbacks in an interconnected system. Life Cycle Analysis is used to understand the fuller impacts of consumption and production choices through the entire process of extraction, production, consumption, and disposal.

17.3.4 Online Platforms

An essential class of tools are online platforms that allow researchers and practitioners to store, share, and exchange data, models, software, and other tools, and to engage with each other in developing ideas, strategies, and solutions.

R Shiny is a library to develop interactive charts, data visualizations, and applications to be hosted on the web using the R language. The R Shiny library makes it easy to build online environments for exploring data and analysis. Tableau is another powerful commercial online visual interface platform.

An online platform can provide various ways to represent multiple interests and outcomes, helping to mediate stakeholder negotiations. A data collection and integration tool can help gather information and study the economic impact of farmers selling water rights on agriculture, the energy industry, urban growth, and stock markets. Roleplaying and integrated biophysical-economic modeling could help better understand trade-offs, promoting consensus.

The International Energy Agency, Energy Information Administration, and several other agencies have produced powerful web-based infographic tools that provide public access to a shared basis in quantitative knowledge about FEWS. FEW-View and HydroViz are academically produced, online visualization environments for FEWS.

17.3.5 Immersive Decision Environments

Making important decisions calls for a carefully crafted environment in which to consider and make the decision. Most countries have deliberative Chambers of Congress or Parliaments, in which decisions can be considered carefully and facilitated using precise rules of order. Modern technology has made it possible to innovate the decision environment through the use of interactive simulations, virtual reality, and data-augmented discussions. This strategy goes by names like: "Serious Gaming" (Sušnik et al. 2018), "Decision Theater" (White et al. 2010), and "Immersive Analytics" (Chandler et al. 2015), among others; see also Sect. 20.3.3.

However, this strategy is more easily described than implemented. It takes a great deal of investment in software, physical infrastructure, human resources, data, and modeling, along with extensive research into the stakeholder's identity and decision processes, to create an effective, immersive decision environment. It is essential that the decision-maker trusts the environment's representation of reality. This may explain why few of these environments have actually been constructed in practice, and why those environments that have been constructed tend to be single-purpose, highly focused environments intended for specific decisions. General-purpose decision environments are beyond the current state of the art.

17.4 Decision-Making Under Uncertainty

Imperfect spatiotemporal data from diverse sources related to food, energy, and water systems creates uncertainty in data science approaches to FEW data analytics and decision making. Modeling and efficient management of uncertain spatiotemporal data from the perspective of its broad impacts and applicability in the FEW Nexus is a challenging task. The main reason behind this is that uncertainty can come from various factors, including source data, generalization error of cross-validation in supervised learning, the sensitivity of modeling choices for unsupervised learning, and a tendency to "false positives" due to multiple selections. This chapter concludes with a brief overview of current issues and solutions in the management of uncertain spatiotemporal data and their transformative potential in Nexus research.

One of the main motivations for investigating the efficient management of uncertainty in the context of agriculture and its impact on the consumption of other resources—most notably, water and energy—is that it is simply a fact of life in many different facets.

Most data-gathering devices, whether sensors for measuring different physical phenomena (possibly organized in a network), imaging devices (e.g., satellite-based) for remote sensing, surveys, citizen science, or other, are inherently imprecise. There is always a limit on the accuracy of the measurements, and the accuracy can vary based on the existing conditions. For example, at higher temperatures, there may be a higher error in measuring humidity; clouds may occlude the view of the satellite, etc. The impact is especially amplified when certain types of actual measurements must be used for indirectly estimating other values.

In practice, regardless of the underlying technology, data samples are obtained in discrete time-instants and at discrete geo-locations. However, the data are used to approximate a continuous behavior of the underlying phenomena: implying that one must resort to interpolation using models and take this into account in storage and querying techniques. The physical coverage of a spatial range of interest may not be uniform due to physical constraints (inaccessibility of terrain) or other operational constraints. Worse, sensing may not provide sufficient density/coverage throughout the entire region of interest. In such scenarios, one is compelled to resort to techniques for estimating the values of the physical phenomena throughout the geospatial domains, relying on techniques such as *Kriging*. The fidelity of the source data and models of data must match the decision-maker's need for information: imprecise matches introduce uncertainty.

The relevant data for FEWS science are heterogeneous: data obtained from different sensing devices vary in type and format, and in ontology (Chap. 14). For instance, the water-resources data of the Food and Agriculture Organization (AQUASTAT) summarized by continents and the major region is an Excel file; food data (FASTAT) is maintained in a different software package. Similarly, the USGS-produced National Land Cover Data (NLCD) uses a 21-class classification scheme, whereas the USDA-generated Crop Data Layer (CDL) has more than 100 classes. Not only do the data come from multiple sources at different scales (see Chap. 12), but the data are also subject to additional (implicit) uncertainty due to the different levels of granularity of its storage in (spatial and/or spatiotemporal) data warehouses. Heterogeneous data not collected by a single agent or for a single purpose cannot be combined without introducing various uncertainties. Finding a significant variable that allows linkages between various models can be a critical challenge.

Thus, uncertainty and risk are part of all FEWS Nexus decisions: as such, it must be explicitly addressed when making decisions or providing decision support. It is important to know how much uncertainty is too much to make the decision and to know whether the metrics, data, models, computing, and tools available can achieve a tolerable uncertainty. It may be better to postpone decision making in the presence of too much uncertainty, rather than making a serious mistake based on inadequate information. However, paralysis by analysis can also be its own form of risk: waiting to make a decision until everything is absolutely certain can also lead to problematic outcomes.

17.5 Communities of Science and Practice

Bringing about the tools described above requires an interdisciplinary community of scientists with the diverse skills needed to develop the Big Data, tools, and digital platforms. These tools require knowledge sharing and communication with nonscientific stakeholders and decision-makers to put the knowledge into practice.

Community of Science and Practice is a common concept in social sciences, private and public organizations, governments, education, and international development. Multi-stakeholder participation is an essential element for addressing sustainability challenges and necessary for developing solutions palatable to civil society (Kates et al. 2001).

The term stakeholders includes academia, research, business, sociological, and political sectors: all are essential to interdisciplinary or transdisciplinary research. The inclusion of diverse skillsets, expertise, and experience help in addressing complex problems such as equitable access to food, water, and energy. Knowledge enhances action that can move beyond analysis toward implementation; knowledge should increase validity, accountability, and responsibility for solution options (Lang et al. 2012). Essentially the spread of the community skills and interest and the spectrum they cover from knowledge to action defines these communities and differentiates them from disciplinary professional societies.

A Community of Science and Practice (CoSiP) is a group of individuals sharing an interest in and knowledge of a craft or profession, and who can offer a means to learn about and share knowledge and practices regarding a specific set of issues, in our case, the FEW Nexus. Wenger (2011) defined the *Community of Practice* as people who share a domain of interest and learn to improve action regarding that interest through regular interaction: sharing and generating new knowledge (Wenger and Snyder 2000). The activities of a Community of Practice include solving problems through brainstorming, seeking or sharing information, experiences, coordination, and synergies, providing feedback to ideas, and mapping knowledge and knowledge gaps (Wenger 2011).

Such a Community also contributes to integrating personal growth and empowerment with the broader goals of the community (Li et al. 2009). For example, the WEF Nexus CoSiP is developing a global science-policy platform that will allow the sharing of data, knowledge, and best practices through identifying the data gaps and developing a common accounting framework. Actively involving stakeholders in this process can be critical to a policy-relevant outcome. FEW Nexus Communities of Practice are immature at this time but emerging and evolving.

One existing example of such a working community is the Sustainable Water–energy–food Nexus Working Group of Water Future. This group is part of the Sustainable Water Future Programme (Water Future) of Future Earth and provides a global platform facilitating international scientific collaboration to drive solutions to the world's water problems. The objectives of the Working Group include drawing on expertise from researchers with a wide range of scientific backgrounds and representing different regions of study to bring together the ongoing initiatives and research groups to advance and exchange cutting edge "Nexus" research, apply it in pilot regions in close cooperation with stakeholders, and disseminate findings to the scientific community and decision-makers (Water Future 2017). Numerous other examples exist, some of which are described in Mohtar and Lawford (2016).

The regional and thematic nature of FEW nexus issues attests to the value in developing these communities at local and national levels focusing on regional specific Nexus Hotspots (Mohtar and Daher 2016). *Hotspots* are critical thematic topics or threatened locations for which the nexus approach has the potential for strong societal impact. Guiding principles for building and maintaining a thriving scientific community of practice are presented. These include identification of a shared conceptual vision that is clear in purpose and provides for the evolution and growth of the Community, as well as for both internal and external perspectives and multiple levels of participation by its members.

Yet the commonalities of particular local 'hotspots' attest to the value of sharing knowledge, experience, and best practices at the global level (Mohtar and Lawford 2016). A global platform will provide for these leverages:

1. Enables and empowers the primary resources management debate;
2. Bridges sound science with science-based policy-making;
3. Offers the multiscale stakeholder community the benefits associated with effective, equitable distribution and management of WEF resources.
4. Provides the platform for trade-offs analysis; these trade-offs are integral to effective decision making: the current allocative model has failed to correctly consider them, and this fact will increasingly lead to conflict unless corrected. Understanding and addressing the trade-offs will allow the development of new business models that leverage technological innovations, social and political awareness, and ultimately will provide knowledge-based clarity for financial leverage.

These four leverages will allow the community to define the roles of the private and **public sectors** and civil society. We will return to communities of science and practice in Sect. 21.3.1.

17.6 Case Studies

17.6.1 Case Study 1: Collaborative Geo-Design

Another example of novel participation enabling approaches has supported the improvement of management practices in rural landscapes to produce food and renewable energy while attaining better water quality, and sequestering greenhouse gases that lead to climate change. Stakeholders used landscape designs involving coupled biophysical and economic models to evaluate how alternative water and landscape management scenarios could improve the provision of ecosystem services in comparison with baseline practices.

Results from this modeling were then placed in a geospatial database, linked to an ArcMap server and a touchscreen. Stakeholders used the touchscreen to identify locations where they would like to place alternative management practices on the landscape. Their landscape geo-designs were then used to determine the impacts of their decisions on the production of food, renewable energy, water quality, carbon sequestration, and terrestrial bird habitat. Results were also easily visualized as a set of ecosystem indicators meaningful to stakeholders.

17.6.2 Case Study 2: Visualized Simulations

In a third example, stakeholders and researchers participated in developing simulations, a set of tangible and mobile interfaces, and facilitation guidelines to help participants jointly explore the problem space and design solutions to water shortages and flooding.

The simulations were process-based, explicitly showing land-cover and water flow in a simple, yet reasonably realistic, way so that stakeholders could inspect the models and examine and adjust their assumptions. Participants created scenarios on a tangible interface (a map and wooden tokens) to try out management alternatives via simulations that provided immediate feedback on a range of outcomes. Each participant held a mobile interface to which all the simulation outcomes were downloaded, saved, and then sorted by the preferences of everyone (property damage, infrastructure investment, area flooded, runoff to downstream neighbors, etc.). Participants could further inspect each scenario and select their favorites, per their values, and then discuss the trade-offs across stakeholder perspectives.

Assisted by a facilitated exploration and synthesis process, this participatory modeling setup allowed stakeholders to learn about the impacts of land and water use on hydrology, and with this understanding, design and try out several scenarios, negotiate across costs and benefits for each scenario, and move towards concrete solutions that helped them address their multiple concerns.

17.6.3 Case Study 3: Colorado River Controversy

A final example illustrates how decisions in one area of the FEW nexus affect the other areas, thereby necessitating tools for impact evaluation and stakeholder negotiation. Water rights refer to the right to use water from various sources, such as rivers, streams, or groundwater. Water rights laws and management differ from country to country and from state to state in the USA (see Chaps. 7 and 8). In the western USA, for example, water rights can be sold or mortgaged like other property. Water rights owners, such as farmers or landowners, can sell their water rights to municipalities and energy providers, who may wish to purchase water rights from farmers for expanding urban and commercial needs. Coupled with the inevitable changes in climate patterns and possibly diminishing river flows, significant shifts in water usage and allocation will affect food production, urbanization, and energy growth.

An ongoing controversy surrounding the Colorado River and fears of agricultural "dry-up" due to the demands of cities and businesses is an example of how such shifts in water patterns can negatively affect agriculture and divide the community. Even if the consequences of water rights transfers are not negative, they can still affect markets and sustainability rates. In Sect. 13.3, the Texas case study shows that the Colorado River problem was triggered by drought in Texas and illustrates the interconnectedness of the water sector and the avalanche effect on multiple stakeholders from different sectors.

Key Points

- Applying science to practice involves stakeholders, decisions, decision tools, and communities of practice.
- An integrated approach to modeling of FEW systems that is able to address the research questions and stakeholder needs will need not only to couple modeling tools for food, energy, and water systems but connect to decision science, data integration, and visual analytics.
- In response to the need for an integrated analytical approach across sectors, a new generation of FEW system modeling tools and approaches are being developed that include representations of food, energy, and water systems. For example, Integrated Assessment Models (IAMs) have been enhanced over the last decade to include interacting representations of FEW systems.
- Uncertainty is a key part of most FEWS decisions and must be explicitly addressed.
- Communities of practice, decision tools, and platforms for FEWS are evolving, but still immature.
- Successful case studies are a good way to learn about how to apply science to practice.

Discussion Points and Exercises

1. Find a good example of a data integration system that includes remote sensing data and the FEW system, and present it. What do you like about it? What are its limitations?
2. Read about integrated assessment models (IAM's). Is this something we need for FEW systems? Why or why not? What are your concerns, if any, about the use of IAM's?
3. Find a good example of visual analytics that includes the FEW system, and present it. What do you like about it? What are its limitations?
4. Find a good example of an online platform or collaborator that is used in your area of work. Try to use it. Do you think you would benefit from using this kind of tool in your work? Why or why not?
5. Find a good example of an immersive decision environment (if you can!). Try it out. Do you think the immersive decision environment is useful? Why or why not?
6. What is the most important kind of uncertainty preventing good decisions, in your area of work? Is the uncertainty human, or is it found in the data and predictions?
7. Identify two or three communities of practice in your area of work. Join one, and present on the options, service opportunities, activities, resources, and relationships that this community provides.
8. Come up with your own case study (imitating Sect. 17.6).

References

Visual Analytics

Choo, J., Lee, H., Kihm, J., & Park, H. (2010). iVisClassifier: An interactive visual analytics system for classification based on supervised dimension reduction. In *IEEE symposium on visual analytics science and technology* (pp. 27–34). Salt Lake City, UT: IEEE.

Immersive Environments

Chandler, T., et al. (2015, September). Immersive analytics. In *2015 Big Data Visual Analytics (BDVA)* (pp. 1–8). IEEE.

Sušnik, J., et al. (2018). Multi-stakeholder development of a serious game to explore the water-energy-food-land-climate nexus: The SIM4NEXUS approach. *Water, 10*(2), 139. https://doi.org/10.3390/w10020139.

White, D. D., Wutich, A., Larson, K. L., Gober, P., Lant, T., & Senneville, C. (2010). Credibility, salience, and legitimacy of boundary objects: Water managers' assessment of a simulation model in an immersive decision theater. *Science and Public Policy, 37*(3), 219–232.

Communities of Practice

Kraucunas, I., et al. (2015). Investigating the nexus of climate, energy, water, and land at decision-relevant scales: The Platform for Regional Integrated Modeling and Analysis (PRIMA). *Climatic Change, 129*(3–4), 573–588.

Li, L. C., et al. (2009). Use of communities of practice in business and health care sectors: a systematic review. *Implementation Science, 4*, 27. https://doi.org/10.1186/1748-5908-4-27.

McFadden, D. (1980). Econometric models for probabilistic choice among products. *The Journal of Business, 53*(S3), S13. https://doi.org/10.1086/296093.

Mohtar, R. H., & Daher, B. (2016). Water-energy-food nexus framework for facilitating multi-stakeholder dialogue. *Water International*. https://doi.org/10.1080/02508060.2016.1149759.

Mohtar, R. H., & Lawford, R. (2016). Present and future of the water-energy-food nexus and the role of the community of practice. *Journal of Environmental Studies and Sciences, 6*, 192. https://doi.org/10.1007/s13412-016-0378-5.

Roy, B. (1993). Decision science or decision-aid science? *European Journal of Operational Research, 66*(2), 184–203.

Sachs, J. D. (2012). From millennium development goals to sustainable development goals. *The Lancet, 379*(9832), 2206–2211.

Water Future Website. (2017). Retrieved June 2017 from http://water-future.org/working_groups/sustainable-w-e-f-nexus-working-group/.

Wenger, E.. (2011). *Communities of practice: A brief introduction.* University of Oregon Libraries. A part of the STEP Leadership Workshop Collection. Retrieved June 2017 from http://hdl.handle.net/1794/11736.

Wenger, E. C., & Snyder, W. M.. (2000). *Communities of practice: The organizational frontier.* Harvard Business Review. Jan–Feb 2000 issue (pp. 139–145). Retrieved June 2017 from https://hbr.org/2000/01/communities-of-practice-the-organizational-frontier.

Further Reading

Andrienko, G., Andrienko, N., Mladenov, M., Mock, M., & Politz, C. (2012). Identifying place histories from activity traces with an eye to parameter impact. *IEEE Transactions on Visualization and Computer Graphics, 18*(5), 675–688.

Beer, C., Reichstein, M., Tomelleri, E., Ciais, P., Jung, M., Carvalhais, N., Rodenbeck, C., Arain, M. A., Baldocchi, D., Bonan, G. B., Bondeau, A., Cescatti, A., Lasslop, G., Lindroth, A., Lomas, M., Luyssaert, S., Margolis, H., Oleson, K. W., Roupsard, O., Veenendaal, E., Viovy, N., Williams, C., Woodward, F. I., & Papale, D. (2010). Terrestrial gross carbon dioxide uptake: Global distribution and covariation with climate. *Science, 329*, 834–838.

Bizikova, L., Roy, D., Swanson, D., Venema, H. D., & McCandless, M. (2013). *The water-energy-food security nexus: Towards a practical planning and decision-support framework for landscape investment and risk management*. Winnipeg, Manitoba: International Institute for Sustainable Development.

Borowski, I., & Hare, M. (2007). Exploring the gap between water managers and researchers: Difficulties of model-based tools to support practical water management. *Water Resources Manage, 21*(7), 1049–1074.

Bosilovich, M. G., Akella, S., Coy, L., Cullather, R., Draper, C., Gelaro, R., & Kovach, R. (2015). *MERRA-2: Initial evaluation of the climate*. Technical Report Series on Global Modeling and Data Assimilation, NASA/TM–2015-104606/Vol. 43.

Brown, E. T., Liu, J., Brodley, C. E., & Chang, R. (2012). Dis-function: Learning distance functions interactively. In *IEEE conference on visual analytics science and technology* (pp. 83–92). Seattle, WA: IEEE.

Burch, S., Sheppard, S. R. J., Shaw, A., & Flanders, D. (2010). Planning for climate change in a flood-prone community: Municipal barriers to policy action and the use of visualizations as decision-support tools. *Journal of Flood Risk Management, 3*(2), 126–139.

Cash, D. W. (2001). "In order to aid in diffusing useful and practical information": Agricultural extension and boundary organizations. *Science, Technology & Human Values, 26*(4), 431–453.

Chen, K., & Liu, L. (2006). iVIBRATE: Interactive visualization based framework for clustering large datasets. *ACM Transactions on Information Systems, 24*(2), 245–294.

Clark, W. C., Tomich, T. P., van Noordwijk, M., Guston, D., Catacutan, D., Dickson, N. M., & McNie, E. (2011). Boundary work for sustainable development: Natural resource management at the Consultative Group on International Agricultural Research (CGIAR). *Proceedings of the National Academy of Sciences of the United States of America*. https://doi.org/10.1073/pnas.0900231108.

Cliburn, D. C., Feddema, J. J., Miller, J. R., & Slocum, T. A. (2002). Design and evaluation of a decision support system in a water balance application. *Computers & Graphics, 26*(6), 931–949.

Cosgrove, B. A., Lohmann, D., Mitchell, K. E., Houser, P. R., Wood, E. F., Schaake, J. C., & Robock, A. (2003). Real-time and retrospective forcing in the North American Land Data Assimilation System (NLDAS) project. *Journal of Geophysical Research, 108*(D22). https://doi.org/10.1029/2002JD003118.

Cutts, B. B., White, D. D., & Kinzig, A. P. (2011). Participatory geographic information systems for the co-production of science and policy in an emerging boundary organization. *Environmental Science & Policy, 14*(8), 977–985.

Dilling, L., & Lemos, M. C. (2011). Creating usable science: Opportunities and constraints for climate knowledge use and their implications for science policy. *Global Environmental Change, 21*(2), 680–689.

Fujimura, J. H. (1992). Crafting science: Standardized packages, boundary objects, and "translation.". In A. Pickering (Ed.), *Science as practice and culture*. University of Chicago Press. pp. 168–211. http://www.philsci.univ.kiev.ua/UKR/courses/asp/asp-lit/Andrew%20Pickering%20(Ed.)-Science%20as%20Practice%20and%20Culture-University%20of%20Chicago%20Press%20(1992)%20(1).pdf

Furrer, R., Knutti, R., Sain, S. R., Nychka, D. W., & Meehl, G. A. (2007). Spatial patterns of probabilistic temperature change projections from a multivariate Bayesian analysis. *Geophysical Research Letters, 34*(6).

Guo, D., Chen, J., MacEachren, A. M., & Liao, K. (2006). A visualization system for space-time and multivariate patterns (VIS-STAMP). *IEEE Transactions on Visualization and Computer Graphics, 12*(6), 1461–1474.

Harris, G. (2002). *Energy, water, and food scenarios*. Best Partners.

Hellegers, P., & Zilberman, D. (2008). Interactions between water, energy, food and environment: Evolving perspectives and policy issues. *Water Policy, 10*(S1), 1–10.

Hu, Q., Johnston, E., & Hemphill, L. (2012). Fostering cooperative community behavior with IT tools: The influence of a designed deliberative space on efforts to address collective challenges. *The Journal of Community Informatics, 9*(1).

Jin, H., & Guo, D. (2009). Understanding climate change patterns with multivariate geovisualization. In *ICDM workshops* (pp. 217–222).

Isendahl, N., Dewulf, A., Brugnach, M., Francois, G., Moellenkamp, S., & Pahl-Wostl, C. (2009). Assessing framing of uncertainties in water management practice. *Water Resources Management, 23*(15), 3191–3205.

Isendahl, N., Dewulf, A., & Pahl-Wostl, C. (2010). Making framing of uncertainty in water management practice explicit by using a participant-structured approach. *Journal of Environmental Management, 91*(4), 844–851.

Khan, S., & Hanjra, M. A. (2009). Footprints of water and energy inputs in food production – Global perspectives. *Food Policy, 34*, 130–140.

Kates, R. W., et al. (2001). Sustainability science. *Science, 292*(5517), 641–642. https://doi.org/10.1126/science.1059386.

Kehrer, J., & Hauser, H. (2013). Visualization and visual analysis of multifaceted scientific data: A survey. *IEEE Transactions on Visualization and Computer Graphics, 19*(3), 495–513.

Kelly, R. E. (2009). The AMSR-E snow depth algorithm: Description and initial results. *Journal of the Remote Sensing Society of Japan, 29*(1), 307–317.

Knutti, R. (2008). Hotter or not? Should we believe model predictions of future climate change? *Significance, 5*(4), 159–162.

Knutti, R., Furrer, R., Tebaldi, C., Cermak, J., & Meehl, G. (2010). Challenges in combining projections from multiple climate models. *Journal of Climate, 23*(10), 2739–2758.

Knutti, R., Masson, D., & Gettelman, A. (2013). Climate model genealogy: Generation CMIP5 and how we got there. *Geophysical Research Letters, 40*(6), 1194–1199.

Kraucunas, I., et al. (2015). Investigating the nexus of climate, energy, water, and land at decision-relevant scales: The Platform for Regional Integrated Modeling and Analysis (PRIMA). *Climatic Change, 129*(3–4), 573–588.

Kumar, S. V., Peters-Lidard, C. D., Tian, Y., Houser, P. R., Geiger, J., Olden, S., & Lighty, L. (2006). Land information system – An interoperable framework for high resolution land surface modeling. *Environmental Modelling and Software, 21*, 1402–1415.

Ladstädter, F., Steiner, A. K., Lackner, B. C., Pirscher, B., Kirchengast, G., Kehrer, J., Hauser, H., Muigg, P., & Doleisch, H. (2010). Exploration of climate data using interactive visualization. *Journal of Atmospheric and Oceanic Technology, 27*(4), 667–679.

Lang, D. J., et al. (2012). Transdisciplinary research in sustainability science: practice, principles, and challenges. *Sustainability Science, 7*(Suppl 1), 25. https://doi.org/10.1007/s11625-011-0149-x.

Lubchenco, J. (1998). Entering the century of the environment: A new social contract for science. *Science, 279*, 491–497.

Masson, D., & Knutti, R. (2011). Spatial-scale dependence of climate model performance in the CMIP3 ensemble. *Journal of Climate, 24*(11), 2680–2692.

Mitchell, K. E., Lohmann, D., Houser, P. R., Wood, E. F., Schaake, J. C., Robock, A., & Cosgrove, B. A. (2004). The multi-institution North American Land Data Assimilation System (NLDAS): Utilizing multiple GCIP products and partners in a continental distributed

hydrological modeling systems. *Journal of Geophysical Research, 109*(D07S90). https://doi.org/10.1029/2003JD003823.
Miralles-Wilhelm, F. (2016). Development and application of integrative modeling tools in support of food-energy-water nexus planning—A research agenda. *Journal of Environmental Studies and Sciences*. https://doi.org/10.1007/s13412-016-0361-1.
Mohtar, R. H. (2017). Making the leap into the nexus: Changing nexus. Presentation at the Watershed conference, Vatican City, 22 Mar 2017.
Mohtar, R. H., & Daher, B. (2016). Water–energy–food nexus framework for facilitating multi-stakeholder dialogue. *Water International*. https://doi.org/10.1080/02508060.2016.1149759.
Mushtaq, S., Maraseni, T. N., Maroulis, J., & Hafeez, M. (2009). Energy and water tradeoffs in enhancing food security: A selective international assessment. *Energy Policy, 37*, 3635–3644.
Owe, M., de Jeu, R., & Walker, J. P. (2001). A methodology for surface soil moisture and vegetation optical depth retrieval using the microwave polarization difference index. *IEEE Transactions on Geoscience and Remote, 39*(8), 1643–1654.
Pahl-Wostl, C. (2009). A conceptual framework for analysing adaptive capacity and multi-level learning processes in resource governance regimes. *Global Environmental Change, 19*, 354–365.
Pelekis, N., Andrienko, G., Andrienko, N., Kopanakis, I., Marketos, G., & Theodoridis, Y. (2012). Visually exploring movement data via similarity-based analysis. *Journal of Intelligent Information Systems, 38*(2), 343–391.
Potter, K., Wilson, A., Bremer, P.-T., Williams, D., Doutriaux, C., Pascucci, V., & Johnson, C. (2009a). Visualization of uncertainty and ensemble data: Exploration of climate modeling and weather forecast data with integrated visus-cdat systems. *Journal of Physics: Conference Series, 180*, 012089, IOP Publishing.
Potter, K., Wilson, A., Bremer, P.-T., Williams, D., Doutriaux, C., Pascucci, V., & Johnson, C. (2009b). Ensemble-vis: A framework for the statistical visualization of ensemble data. In *IEEE international conference on data mining workshops* (pp. 233–240). Miami, FL: IEEE.
Rienecker, M. M., Suarez, M. J., Gelaro, R., Todling, R., Bacmeister, J., Liu, E., & Bosilovich, M. G. (2011). MERRA – NASA's modern-era retrospective analysis for research and applications. *Journal of Climate, 24*, 3624–3648.
Roy, B. (1993). Decision science or decision-aid science? *European Journal of Operational Research, 66*(2), 184–203.
Sanyal, J., Zhang, S., Dyer, J., Mercer, A., Amburn, P., & Moorhead, R. (2010). Noodles: A tool for visualization of numerical weather model ensemble uncertainty. *IEEE Transactions on Visualization and Computer Graphics, 16*(6), 1421–1430.
Scott, C. A., Pierce, S. A., Pasqualetti, M. J., Jones, A. L., Montz, B. E., & Hoover, J. H. (2011). Policy and institutional dimensions of the water–energy nexus. *Energy Policy, 39*(10), 6622–6630.
Serrat-Capdevila, A., Browning-Aiken, A., Lansey, K., Finan, T., & Valdes, J. B. (2009). Increasing social–ecological resilience by placing science at the decision table: The role of the San Pedro Basin (Arizona) decision-support system model. *Ecology and Society, 14*(1), 37.
Serrat-Capdevila, A., Valdes, J. B., & Gupta, H. V. (2011). Decision support systems in water resources planning and management: Stakeholder participation and the sustainable path to science-based decision making. In C. Jao (Ed.), *Efficient decision support systems—Practice and challenges from current to future* (pp. 423–440). Rijeka, Croatia: InTech.
Simon, H. A. (1960). *The new science of management decision*. New York: HarperCollins.
Smith, D. W., Welch, M., Bennett, K. E., Padgham, J., & Mohtar, R. (2017). Building a WEF Nexus Community of Practice (NCoP). *Current Sustainable/Renewable Energy Reports, 4*(3), 168–172. Springer.. https://doi.org/10.1007/s40518-017-0080-6.
Star, S. L., & Griesemer, J. R. (1989). Institutional ecology, 'translations' and boundary objects: Amateurs and professionals in Berkeley's Museum of Vertebrate Zoology, 1907–1939. *Social Studies of Science, 19*(3), 387–420.
Tapley, B. D., Bettadpur, S., Ries, J. C., Thompson, P. F., & Watkins, M. M. (2004). GRACE measurements of mass variability in the earth system. *Science, 305*, 503–505.

Tebaldi, C., & Knutti, R. (2007). The use of the multi-model ensemble in probabilistic climate predictions. *Philosophical Transactions of the Royal Society of London A: Mathematical, Physical and Engineering Sciences, 365*(1857), 2053–2075.

Tebaldi, C., Smith, R. L., Nychka, D., & Mearns, L. O. (2005). Quantifying uncertainty in projections of regional climate change: A Bayesian approach the analysis of multimodel ensembles. *Journal of Climate, 18*(10), 1524–1540.

White, D. D. (2013). Framing water sustainability in an environmental decision support system. *Society & Natural Resources, 26*(11), 1365–1373. https://doi.org/10.1080/08941920.2013.788401.

White, D. D., Wutich, A. Y., Larson, K. L., & Lant, T. (2015). Water management decision-makers' evaluations of uncertainty in a decision support system: The case of WaterSim in the decision theater. *Journal of Environmental Planning and Management, 58*(4), 616–630.

Williams, D. (2014). Visualization and Analysis Tools for Ultrascale Climate Data. Eos, Transactions American Geophysical Union. 95. 10.1002/2014EO420002 https://eos.org/science-updates/visualization-analysis-tools-ultrascale-climate-data

Wutich, A., Lant, T., White, D. D., Larson, K. L., & Gartin, M. (2010). Comparing focus group and individual responses on sensitive topics: a study of water decision-makers in a desert city. *Field Methods, 22*(1), 88–110.

Yokohata, T., Annan, J. D., Hargreaves, J. C., Jackson, C. S., Tobis, M., Webb, M., & Collins, M. (2012). Reliability of multi-model and structurally different single-model ensembles. *Climate Dynamics, 39*, 599–616.

Part III
Applications

Chapter 18
Cities at the Nexus

Craig Harris, Natalie Lyon, Carol Miller, Kami Pothukuchi, Lara Treemore-Spears, and Yongli Zhang

18.1 Introduction

The definition of what constitutes a *city* differs substantially worldwide, with some countries defining an urban area as a locality with at least 200 inhabitants and others, such as the United Nations, defining a city as a locality with more than 50,000 inhabitants. These and other differences between places designated as cities—such as in the level of service and quality of urban infrastructures such as roads, electricity, sanitation, drinking water, communication, health, security, and education—make global generalizations about the food–energy–water Nexus in cities nearly

C. Harris
Department of Sociology, School of Social Science, Michigan State University, East Lansing, MI, USA
e-mail: harrisc@msu.edu

N. Lyon · K. Pothukuchi
Department of Urban Studies & Planning, College of Liberal Arts & Sciences, Wayne State University, Detroit, MI, USA
e-mail: natalie.lyon@wayne.edu; k.pothukuchi@wayne.edu

C. Miller
Civil and Environmental Engineering, College of Engineering, Wayne State University, Detroit, MI, USA
e-mail: ab1421@wayne.edu

L. Treemore-Spears (✉)
Health Urban Waters, College of Engineering, Wayne State University, Detroit, MI, USA
e-mail: treemorespears@wayne.edu

Y. Zhang
Civil and Environmental Engineering, College of Engineering, Wayne State University, Detroit, MI, USA
e-mail: zhangyl@wayne.edu

P. Saundry, B. L. Ruddell (eds.), *The Food-Energy-Water Nexus*, AESS Interdisciplinary Environmental Studies and Sciences Series,
https://doi.org/10.1007/978-3-030-29914-9_18

impossible. Nevertheless, it is certainly the case that in all cities, food, and energy and water are characterized by the five criteria of criticality, infrastructure, footprint, affordability, and policy (Sect. 1.3).

It is important, however, to distinguish cities (*urban* areas) as being different from *rural* areas that have lower population density and smaller population aggregates and that usually are characterized by significant primary production (farming, fishing, forestry, mining) and tourism. Areas immediately surrounding a city, which have mixed urban and rural characteristics are termed *peri-urban*.

As discussed in this chapter, the dynamics and drivers in urban systems that are largely physically separated from the agricultural resources needed to support them are inherently different from those in rural systems. The purpose of this chapter is, therefore, to discuss the FEW Nexus in cities—and their associated non-agricultural population centers and metropolitan areas—and to provide examples of Nexus innovation in several specific cities, rather than to draw general conclusions about cities as a whole.

18.2 Context: The Sociopolitical-Economic Drivers

In this section, we will discuss the sociopolitical-economic drivers of the FEW nexus in cities. Our approach will emphasize that drivers of the nexus are situated in multiple sectors of the *socio-ecological system* (see Sect. 9.1.1) and at multiple interacting levels of social organization (international and national, state and regional, local, facility, household, and individual). Acknowledging that the drivers of the FEW nexus depend on the overall structure, organization, and culture of the society, a neoliberal society like the USA will exhibit different drivers from those in a communist society like Cuba. In essence, the drivers of the FEW nexus form a multi-level, multi-sectoral system (see Sect. 2.2) wherein the forces interact continuously with feedback to produce particular nexus outcomes. Those outcomes may be characteristic of a society as a whole or of particular scales such as cities or industries, depending on the drivers.

In analyzing the drivers of the FEW nexus, we take a *social organizational* or *macrostructural approach* that views the nexus as being driven by the three types of institutions:

1. *private sector* institutions;
2. the state and the *public sector* institutions; and
3. social impact organizations or **civil society** (also known as nongovernmental organizations).

These three fundamental types of institutions, in turn, generate systems of science and technology, regimes of demography, and the knowledge, attitudes, and practices of human behavior, with feedback loops, as human actions influence the paths of these institutions.

18.2.1 *The Private Sector and the Economy in Cities*

In almost all societies in the world today, the private sector is one of the main drivers of the FEW nexus. The *private sector* can be conceptualized as the portion of the economy that is not under direct government control, generally regarded as being composed of individuals and companies, and characterized by the intention of producing a profit. As a state-controlled economy, Cuba is the main counter-example of strong private sector involvement, and as a result, the FEW nexus in Havana is very distinctive. In this chapter, any example from a state-driven economy will be clearly specified.

In most contemporary societies discussed in this chapter, the physical (and to some extent intellectual) means of production are privately owned; the social relations of production result from private transactions in a labor market; inputs to the production of goods and services are bought and sold in private transactions; and the goods and services produced are sold in private transactions. In some highly industrial and post-industrial societies, the government is a significant supplier of inputs and a significant purchaser of goods and services, but the majority of the economy is private.

Finally, whether a society is characterized by classical **capitalism**, welfare state liberalism, or neoliberalism, the driving impetus of the owners and managers of extractive, industrial, commercial, and financial capital is to make monetary profits and accumulate monetary wealth.

18.2.1.1 Food Systems

Food systems (see Sect. 2.3) in cities are dominated by national and transnational food corporations. This means that decisions about food production (what, how, and where) and distribution are heavily influenced by the profit motive and are made by actors often spatially and socially distant from the city. It also means that food systems are characterized by very heavy physical and social infrastructures (Criterion 2).

In cities where decisions about where to locate are made exclusively by food wholesalers, and grocery retailers are motivated only by profit, urban areas with a high concentration of low-income individuals may have limited access to affordable and nutritious fresh foods. Such areas are often called *food deserts* and are characteristic of many US cities. They are important because they violate the criteria of criticality and affordability (Criteria 1 and 3). In some cases, research shows that urban areas are shunned by grocery retailers despite market strength due to challenges posed by lack of suitably sized parcels, permitting, a more diverse inventory relative to suburban standards, and the perception and reality of crime.

Further, many of the connections between FEW components occur distant from the city under consideration. The resulting FEW Nexus interactions may occur across an urban-rural continuum with product life cycle costs that include substantial transportation inputs, carbon footprint, and greenhouse gas emissions, giving rise to the concept of "**food miles**." While the miles food travels is not a complete reflection of its sustainability, estimates of the water or energy footprint of a city must take the hidden life cycle costs of its food supply into account, including transportation mechanisms within the city for people to access their food, how and when the food is grown, and the energy used for its preparation. Indeed, this is part of the reason that the food footprint of cities is so heavy.

Although the most impactful drivers in the private sector have been large scale, smaller-scale social actors like households, food hubs, urban farmers, local entrepreneurs, and farming cooperatives also play a role. Household gardening and urban farming may be part of a social empowerment movement or may instead be a matter of economic necessity that may have implications for social equity and stigma. Small scale private sector actors at the FEW nexus include, for example, local entrepreneurs implementing new technologies of vertical farming and high production greenhouses utilizing water-saving green infrastructure and renewables such as solar energy.

18.2.1.2 Water Systems

In many capitalist societies, there has been some trend toward water and wastewater service privatization, especially in the global South. Critics of water privatization claim that the profit motive will lead to quality meeting only minimum legal standards and pricing that excludes a significant portion of the population (the affordability criterion). Proponents of water privatization claim that the profit motive drives efficiencies and reduction of water and energy losses. These claims of critics and proponents apply both to drinking water supply and wastewater services.

From the perspective of the private sector, part of the challenge faced in determining water pricing is the great differentials in quantities of water use among customers. The household in a small apartment consumes relatively little water, the majority of which may enter the wastewater treatment system; the suburban home with a large yard consumes a moderate amount of water, some of which is lost to evaporation and evapotranspiration from the lawn and some of which may run off into storm drains requiring wastewater treatment; and the urban factory may consume a large amount of water, with varying degrees of reprocessing, recapture, and reuse.

A private water supply utility, like any for-profit company, wants to maximize its accumulation of wealth, so it seeks to sell as much water as possible at profitable rates. To sell to the farm or factory, the water must have a relatively low unit price, slightly above marginal cost. This means that the small residential user may be required to pay a higher unit price to cover the total cost of supply not satisfied by the prices paid by larger users. In some states, such as Michigan, the state Constitution and case law require user fees to be proportionate to the cost of providing the service, subjecting water utilities to additional scrutiny when calculating costs and setting rates (the policy criterion).[1]

[1] See Michigan Supreme Court *Bolt v. City of Lansing*, 459 Mich. 152 (1998) for a distinction between user fees and taxes requiring voter approval under the Headlee Amendment to the Michigan Constitution (Mich Const 1963, art 9, sections 25 and 31).

Private corporations may also be able to leverage their substantial political clout to negotiate inexpensive deals. For example, the state of Michigan permits Nestle Corporation to pump water at the rate of 400 gallons per minute or 576,000 a day, from a well near the town of Evart in the western part of the state for its Ice Mountain label. This transaction cost Nestle a one-time application fee of $5000 and an annual reporting fee of $200, and raises critical questions about whether water should be treated as a commodity, a commercial product, or a human right, and the cost to local communities and society as a whole due to ground water declines and the ubiquitous plastic waste that result (Fleming 2018).

18.2.1.3 Energy Systems

Although some cities in liberal capitalist and neoliberal societies have public electricity providers and distributors, primary forms of energy are provided and distributed by the private sector. This includes coal, crude oil, and natural gas for residential, commercial, and industrial uses. Although these primary forms of energy—along with secondary energy such as electricity—may be produced and distributed by local and regional firms, their transmission, pricing and availability are determined to a significant extent by national and transnational corporations and organizations (the infrastructure criterion) within the context of extensive public regulation (such as in the case of the US electric power grid) or subsidization (in the case of US oil and gas exploration)—the policy criterion. Purely private, rather than public, electricity production and distribution more typically occur in higher-income economies that can support public-debt-driven private investment.

As with water, one of the main concerns with energy supply is the affordability criterion. In temperate and arctic cities, people need the ability to heat their homes (the criticality criterion). In some states, such as New York and Massachusetts, tenants have an enforceable legal right to heat (the policy criterion). In most cities, people require transportation to get to their place of employment and to procure food and health services. One of the drivers for the creation of *food deserts* is the distance that must be traveled to reach a broadline (high-volume, low-cost) purveyor of groceries. In the absence of public transportation options, market pricing may make it challenging or impossible for some people to purchase the energy required to travel the distance to meet these basic human needs.

The private sector also significantly determines the demand for energy. Industrial factories, large office buildings, and mass transportation companies can determine to a significant extent both the amount of energy and the kinds of energy that they consume, based on their own private investments in physical plant and operational infrastructure. As more sustainable business models are adopted, quantities of energy consumed are decreased, and the supply shifts from non-renewables to renewables.

18.2.2 *The Public Sector and Governance in Cities*

In this section, we focus on the ways in which the public sector and the governance process can drive actions at the FEW nexus.

A city sits at the intersection of multiple governments. These governments differ in their scale, spatial purview, and substantive responsibility. Thus, the public sector that influences the FEW urban nexus includes, in addition to the city government itself, international and national governments, state or provincial governments, county and regional governments, and borough or ward governments, as well as school districts, water districts, and sometimes energy utility districts, air pollution control districts and others. For example, Chicago has over 1000 applicable governmental bodies, and the New York metropolitan area has over 1400 at various scales. These governments, their agents, and elected officials may work together to varying degrees; politics and the legal system may come into play where conflicts in ordinances, laws, administrative rules, policies, and their interpretation arise. See Chapter 20 for discussion of Human Conflicts and resolution strategies at the FEW nexus. Furthermore, the authority of public or private interests may be in flux at any given point in time, as demonstrated by the movement toward remunicipalization of water and energy systems in various instances (returning private services to municipal authority—the policy criterion), treating these resources as a commons rather than a commodity.

Understanding the public sector as a driver of the urban FEW nexus is further complicated by the plethora of overarching governmental and economic systems present throughout the world. For example, in **pluralist governmental systems**, no set of interests in the society has sufficient power to determine government action to a complete extent, and the organizations and processes of government decide what is in the public good and act accordingly. In comparison, the *statist system of government* assumes that the state itself has interests and acts in ways that conserve and enhance its power over material, human and monetary resources, and over the legitimate use of force. Depending on the economic system in place, government may act to benefit the interests of one or more specific capitals (extractive, agricultural, industrial, commercial, financial)—and/or to benefit the long-term interests of **capitalism** as a mode of socio-economic organization—consolidating wealth to a greater degree in private control (such as in the USA) or public control (such as in China's state capitalism of the twenty-first century); in contrast, purely communist economic systems in which there is officially no private wealth are increasingly rare (such as Cuba and North Korea). This section will discuss the FEW Nexus in pluralist and statist governmental systems that function within a capitalist private sector, as discussed previously in this chapter.

18.2.2.1 Food Systems

The roles of municipal governments in the agriculture/food sector have increased in recent years. Traditionally charged with ensuring the safety of food in food service and retail establishments, city governments have now become more active in food policy. These activities include:

- the establishment of **food policy councils** (see 18.4.1 for a case study example);
- the establishment and operation of market facilities for local farmers and small-scale food manufacturers and vendors;
- the adoption of regulations to permit small-scale crop and animal production in residential areas;
- the adoption of municipal curbside compost pickup (as seen in Austin, Boston, Chicago, and Victoria); and
- declaring food to be a right of citizenship (as seen in Belo Horizonte, Brazil).

Some cities also use their own purchasing activities to favor fair trade foods and foods with other types of ethical certification. Some cities make public land available for household or small-scale commercial food production, and/or adopt policies that encourage the use of vacant land for those purposes (the affordability criterion). See 18.4.2 for an illustrative case study of Detroit, MI.

Schools, hospitals, and prisons can be either public or private enterprises, but all have been influenced by local, state, and federal policies that foster farm to institution linkages. Schools can contribute to the awareness of the local food system by developing school gardens.

Local and regional governments increasingly incorporate food production, distribution, and consumption, and the management of food system wastes such as through composting in urban planning activities, especially with discussions of productive landscape infrastructure and ecology.

Even in cities that encourage innovative food-related policies and practices, challenges still exist. For example:

1. Water-related costs may negatively impact the feasibility of raising urban food (the criticality and affordability criteria) such as costs of obtaining water from the municipally treated drinking water system, treating stormwater run-off from farms and gardens in cities that have a stormwater utility fee, and adhering to stringent building codes, which can hinder the widespread use and cost-savings of rainwater capture, storage, and reuse devices.
2. Labor related-costs may also be higher in cities than in rural areas due to the increased cost of living (which may be in part an artifact of the high per-capita cost of providing water, energy and other infrastructure in the relatively vacant urban centers that have the land capacity to support gardening).
3. Policies for tax assessment may prevent small vacant properties from being combined to result in the lower per-area rates offered to other industries, and urban zoning and land use policies may conflict with food production (for example, restricting vegetation height or animals such as chickens, or allowing building height to be increased in a way that casts too much shade on adjacent gardens).
4. The public process for updating policies may be obscure, cumbersome, or controlled by players who are unaware of or do not support food–energy–water innovation.

These financial, political, and administrative dynamics create pressure, even in high-vacancy cities without commensurate pressure for development, to treat urban gardens as temporary or interim uses only, rather than as permanent ones.

18.2.2.2 Water Systems

As noted above, municipal and regional governments historically have been very active in the provision of water and wastewater services. Water agencies are required by state and federal governments to maintain certain minimum levels of drinking water quality for human health, monitoring water contaminants, and reporting results to consumers and citizens publicly. Wastewater effluent similarly must meet minimum contamination standards, and in some cases, its volume is also publicly regulated. Public water agencies have, in some cases, been successfully pressured to ensure that water and wastewater services are available even to those households that lack sufficient income to pay for the services; this is sometimes framed as a discussion of water as a human right (the policy criterion).

Like the private sector, public water-providing agencies must consider the allocation of scarce and costly water supplies among residential uses (basic human needs), agricultural uses (local food production), industrial uses (providing employment and tax revenue), and aesthetic uses (lawns and golf courses), which may be particularly challenging in arid climates with limited water resources. Legislative and legal drivers may be instituted at the state and regional levels addressing rate structures to recapture costs in the public interest that may not be capturable via market forces. The cost of providing water and wastewater services is particularly driven by the significant amounts of energy required for its distribution (the footprint criterion), and many efforts are underway to reduce these demands. Some coastal cities are investigating the possibility of the desalination of seawater, which may require less energy than pumping groundwater or transporting distant surface water (such as in San Diego County, California). It takes energy to move water from the source to the city, and it takes energy to treat the water supply, distribution system, and wastewater prior to final discharge to surface waters (the infrastructure criterion). Another driver in older cities is the age of the water infrastructure, which may contribute to leaks and water losses that increase costs both related to energy use and system maintenance.

Public wastewater effluent water quality standards are based in the USA on designated public uses of rivers, lakes, and groundwater for drinking water and human contact and fishing. In older US cities, agencies have been particularly challenged to reduce **combined sewer overflows (CSOs)**—a common engineering practice of the twentieth century in which sewage and storm wastewater flows were combined—when large precipitation events overload the wastewater processing facility, raw sewage is released into the receiving body of water, as discussed in Chap. 10, Infrastructure. Public innovation to reduce the storage and energy costs of stormwater management, contributing to CSO reduction, includes the use of large-scale green stormwater infrastructure (such as porous pavement and bioswales), and when these and other practices are designed as discussed in Sect. 18.3.3, they can also recapture water supply for use in food production. Some agencies are also recovering the nutrients that remain after the processing of wastewater, producing biogas for energy production by composting, or selling the digested, pasteurized and pelletized biosolids as fertilizer to urban and peri-urban farms and home gardeners, although the possible presence of pathogens and low-level persistent pollutants such as PCBs, dioxin or heavy metals may limit land-application of treated sewage sludge in some areas and depend on its treatment.

While innovative individually implemented solutions may not be public in scope, they often require substantial public discourse and policy revisions to be able to implement. Putting treated household wastewater directly back into the household water supply system (through **greywater systems** or **pee-cycling**) reduces the energy required both for water supply and treatment, and dry toilets (either urine-diverting or composting) can somewhat reduce pressure on water supply while reducing the need for wastewater treatment and supporting the reuse of human excreta for agriculture. These household- and facility-scale solutions demonstrate that although the density of human population and industry in cities concentrates related water quality impacts there, supportive public policy and education to promote these solutions have the potential to support greater food, energy, and water security.

18.2.2.3 Energy Systems

Municipal governments that are energy providers may be able to improve upon the public benefits that private energy providers supply in that they can coordinate the provision of energy with land use planning and other municipal services (e.g., water and wastewater, transportation). One example of this is the creation of heating districts where the generation of electricity is used to produce hot water (as in Detroit, see Sect. 18.5.2) for residential and commercial heating or for industrial uses or for hydroponic and aquaponic agriculture.

Other cities are piloting electricity generation from the water flowing through their water infrastructures (as in Portland, see Sect. 18.5.1). Some cities have begun installing wind generation and solar collection capacity on public land to feed into the local electric grid, and some coastal cities are exploring the possibilities of tidal power.

In addition, some cities have adopted policies mandating that all energy used by the city and its agencies come from renewable sources (i.e., hydroelectric, wind, solar). In some cases, public hydroelectric dams have had the unanticipated impact of creating fishing opportunities that provided food for segments of the local population, and in other instances, public investments to mitigate fisheries damage due to private dams have been necessary (criticality and affordability criteria).

One of the important dimensions of smart cities is the reduction of energy use for transportation (to see how one Brazilian city has achieved this, refer to Sect. 18.5.3 on Curitiba). City policies increasing densification in order to reduce energy consumption can help urban areas transition toward sustainability, with intentional ecovillages designed for efficiency of the production and consumption of energy, water, and food (as in Tianjin, refer to Sect. 18.5.4).

18.2.3 *Civil Society and Cities*

Except in totalitarian societies, civil society exerts important influences on collective social good at the FEW Nexus, both overall and specifically in urban areas. In civil society organizations, we include many of what have been called nongovernmental organizations (NGOs) and faith-based organizations. We choose the interchangeable use of the terms *Social Impact Organization* or *Civil Society Organization* for several reasons, detailed here because they exemplify terminology challenges posed by interdisciplinary work:

1. To avoid ambiguity in the negative definition of the term "nongovernmental organizations"; each organization that is not governmental—business, social media, and 501(c)(3) designated "non-profit" organization—exists for a particular purpose, not for an unspecified "non" purpose.
2. Civil Society Organization has a problematic abbreviation (CSO) that also stands for Combined Sewer Overflow in the field of wastewater treatment, which would be difficult to use in a publication relating to water; the abbreviation for Civil Society Actor or CSA also stands for Community Supported Agriculture (a membership farm common in urban markets).

18.2.3.1 Food Systems

Civil society or social impact organizations have been leaders in creating alternative forms of food security for persons who are food-challenged (the criticality and affordability criteria). These organizations provide soup kitchens, food banks, civic food networks, and double-up bucks (a program in which government assistance to low-income individuals for food purchase is doubled by farmers when used for fresh produce). They may advocate for local food in general—and for policies that support and foster urban agriculture, such as favorable regulation, public food markets, and vacant land reuse. At the facility scale, civil society organizations have created and supported incubators for new small-scale food processing and manufacturing firms and vendors, and organized food waste recycling from grocery stores and food service establishments to deliver it to soup kitchens and food banks for consumption (i.e., new infrastructure). At the meso-scale, civil society or social impact organizations have provided new and existing farmers and food processors/manufacturers with information on how to accomplish their objectives with less water and less energy.

18.2.3.2 Water Systems

At the macro-scale, social impact organizations have been strong advocates for water availability and water quality, in some cases attempting to advance policies that recognize a right to water. They have advocated with water and wastewater service providers for pricing structures that provide affordability to income-challenged households and for policies that avert shutting off service when bills are not paid. In many states, they have advocated for the reform of public service regulation and limitations on clientelism. They have also advocated for the development of lakes and rivers within the city's boundaries for recreation and tourism.

18.2.3.3 Energy Systems

Civil society organizations have been very active in advocating for greater efficiency (energy conservation) in residential and commercial uses. They have advocated with municipal and private energy providers for pricing structures that provide affordability to low-income households, and for policies that avert shutting off service when bills are not paid. They have advocated for reduced use of coal in local electricity plants to reduce air pollution and greenhouse gases, and for the transition to low carbon energy sources (a reduced footprint). They have advocated for more public transportation and for the reconfiguration of urban areas to save energy. They have advocated for more generation of wind and solar energy on municipal land, and for revisions in zoning regulations and building codes that would foster greater adoption of residential solar collection.

Policies favoring **degrowth**, or the overall downscaling of production and consumption to reduce the environmental impact of human activities (i.e., the footprint), have been promoted because, among other things, they would reduce the energy used to landfill municipal solid waste. Many of the concepts in the degrowth movement have the potential to address the FEW Nexus in an integrated way, as do organizations that promote sustainability.

18.2.4 Global Climate Change

At least in the short and medium-term, one of the strongest potential drivers of the FEW nexus in urban areas is global climate change and its associated socio-ecological impacts. (Refer to Chap. 11 for a focus on Climate Change nexus impacts broadly). Global climate change is expected to result in higher mean air, land, and water temperatures that will exacerbate the urban heat island effect. Climate change will include greater intra-annual and interannual variation in temperature and precipitation, and more frequent and severe hurricanes and typhoons. Cities vary in their planned response and policies relating to climate change, with coastal cities generally further ahead in their planning than inland cities and those of the Great Lakes region. However, engineering stormwater treatment systems for larger storms is a current topic of discussion even in inland cities, as is increasing the security of water supplies to adapt to decreasing surface water sources.

One of the impacts of climate change that is already being seen is more frequent and severe flooding of low-lying coastal and riverine cities. Municipal governments are constructing barriers to control flooding and altering land use and land cover patterns to modify, mitigate, and adapt to flooding and coastal erosion. Urban planning social impact organizations have developed a coastal management framework called a "coastal hazard wheel" that may also be used by governments to provide a consistent approach to multi-hazard-assessments at the local, regional and national levels; identification of hazard management options for a specific coastline; and standardized language for communicating coastal information.

When the increasingly large and frequent storms associated with climate change overwhelm **combined sewer systems** of older cities such as Detroit, MI, consequences may be federally imposed through the National Pollutant Discharge Elimination System (NPDES) permit that applies to large urban wastewater treatment systems (see Sect. 8.2.3). In the case of Detroit, a financial driver for green infrastructure construction was created through the NPDES permit, which requires the water utility to spend a certain amount on building these alternative water management systems over the life of the permit (see Sect. 18.5.2).

Thus, cities face a fourfold challenge:

1. chart a course for sustainable development;
2. adapt to the emerging impacts of global climate change
3. develop their resilience or ability to recover from the extreme events that characterize global climate change; and
4. mitigate the ways in which they contribute to global climate change (reduce their footprints).

While the urban FEW Nexus is driven by climate change, fossil fuel consumption associated with urban food, energy, and water supply is also one of the drivers of anthropogenic global climate change itself. All of the things that the private sector, public sector, and civil society do to reduce carbon emissions and increase carbon capture contribute to mitigating the impacts of global climate change. And vice versa: all of the things that the three sectors do to increase carbon emissions and reduce carbon sequestration contribute to exacerbating the impacts of global climate change.

18.2.5 Knowledge, Attitudes, and Practices

In this section, we will focus on individuals, households, and informal groups.

While corporations, governments, and civil society organizations are composed of individuals, these individual members act as agents of, and within the structure of, a particular organization. In this section, we focus on the knowledge, beliefs, and attitudes of individuals, and how those attributes interact with resources and constraints to produce behavior affecting the FEW Nexus in cities. For this chapter, we do not elaborate a detailed model of beliefs, attitudes, and practices. We fully acknowledge that the beliefs and attitudes of individuals are highly influenced by the actions of organizations in the private, public, and civil sectors and that the behavior of individuals is both potentiated and constrained by the resources made available to them by economic, political and social structures. In this section, we emphasize the agency of individuals and households in pursuing the interests they have chosen based on their particular mix of ideology, knowledge, and resources.

18.2.5.1 Food Systems

In the decisions they make, individuals may try to consume foods that are produced locally and that have required fewer energy and water inputs (reducing their footprint). They also may encourage retailers to provide such foods. In general, buyers are limited in their ability to know the energy or water footprint of a particular food because of the lack of related standardized data or labeling, although European countries are starting to require this. Indeed, it is not universally true that a meal kit or ready-to-eat food is more energy or water-intensive than food prepared from scratch at home.

Individuals may choose to grow some of their own food in a backyard or community garden (criticality and affordability) or to grow food for exchange or sale in a non-profit farming cooperative or a city market. Farmers and buyers may join together in community-supported agriculture. Grocery store managers and food service operators may decide to participate in food waste reduction, reuse, repurposing, and/or recycling programs.

18.2.5.2 Water Systems

With respect to water efficiency, individuals do have somewhat more information than they do with food. Toilets that use less water to flush are labeled as such, and the amount of water used in showering and bathing is plain to see (reducing one's footprint). One can decide to shower in a way that conserves water either due to environmental values or—in high-poverty communities—due to the cost of heating and consuming water. Clothes washers and dishwashers have settings for lower water use and lower temperature water use. Installing composting toilets and recirculating *greywater* further reduces water use, but only higher-income households may have the economic means to do this. Having a backyard garden or other porous surface reduces the runoff to the municipal sewer system, and collecting rainwater for watering the lawn and garden reduces the demand for municipal water.

18.2.5.3 Energy Systems

Like water, with energy efficiency, individuals have somewhat more information than they do with food. Appliances and light bulbs come with energy ratings, and the length of time an appliance runs is evident. Thermostats control the energy used for heating and air conditioning. Cars are labeled with energy efficiency, and individuals control their energy consumption by choosing to: drive highly efficient vehicles, use carpools or mass transportation, bike, and/or walk. Energy consumption can be reduced by enjoying the recreational amenities of *local* lakes and rivers and coasts, rather than traveling to distant sites. Residents can retrofit their residences to reduce energy requirements for heating and cooling, installing **cool roofs** and vegetation in the yard to reduce the need for air conditioning (also sequestering carbon, to mitigate greenhouse gas emissions). Rooftop solar electric and solar thermal both reduce the need for air conditioning and generate energy with no carbon emissions. Backyard wind energy is now feasible for individuals. All of these are ways of maintaining the flow of critical energy while reducing one's footprint. As noted, they vary in their affordability and their impact on infrastructure. One factor justifying the household expense of installing the means for alternative energy production is the amount of money that the local electric utility pays for electricity fed into the grid; this is largely determined by the private sector with some influence from the government and civil society (policy).

18.2.5.4 FEW Systems

To illustrate the urban FEW nexus, the model individual walks or rides a community bicycle to the river that runs through the city and catches a fish that s/he takes home and cooks for dinner using photovoltaic energy. The exemplary urban farm provides a porous land surface and water storage cistern to receive wastewater from the food processing facility, and the wet and solid waste from the food processing plant provides a source of nutrients for the farm. The substitution of these water and nutrient sources for more energy-intensive municipal water supply and commercial fertilizer by the farm, and the reduction in energy demand for wastewater treatment by the processing facility, both provide benefits for energy and water security; as does the location of food production close to the urban population for food security. At the same time, the open area provided by urban farms may offer opportunities for siting wind energy generators and solar energy collectors.

18.3 Urban FEW Supply Chain Management

Various projections of demand for food, energy, and water to sustain a projected nine billion people globally by 2050—most of whom will reside in cities—include increases of more than 50%, 60%, and 55% in these vital consumable resources, respectively (the criticality criterion). Moreover, the sources of these essential resources are inextricably and reciprocally linked as discussed at length in Chap. 2. The increasing complexity of the FEW nexus requires a more sustainable approach that allows for the co-management of these resources.

In this section, we will discuss how supply chains can be one approach for understanding and managing the FEW nexus in cities. We will first discuss the concepts of *supply chain management (SCM)* and **sustainable supply chain management (SSCM)** in relation to food, energy, and water supply. Then, we will compare and contrast different types of FEW supply chains—centralized versus decentralized and integrated versus separated systems—illustrating how different approaches can affect the sustainability of the FEW nexus. The socio-economic drivers of supply chains are discussed at length in Chap. 5 and were discussed in Sect. 18.2 above.

18.3.1 Supply Chain Management from Resource to End-User at the FEW Nexus

Supply chains at the FEW nexus are complex and dynamic networks that exist in the context of:

- typically increasing population demands;
- dynamic interactions between food, energy, and water sectors; and
- changing and variable producer and consumer, civil society, and governmental demands with respect to FEW security and environmental, economic, and social impacts.

Appropriate management of FEW supply chains in cities is particularly critical in large populations with high densities that are substantially separated geographically, economically, or politically from the resources needed to support them, necessitating heavy infrastructure. In the case of cities with shrinking populations, SCM may be particularly challenging due to unanticipated changes in demands and capacities. Although rural populations in the developed world currently may rely largely on the same supply chains that support cities, the potential exists for different models to be employed in urban and rural areas that could increase overall supply chain sustainability. In contrast, in the less developed parts of the world, rural populations rely on FEW supply chains that are somewhat different from those in urban areas; in rural households, there is much more self-provisioning of food, energy, and water than in urban areas.

Supply chain management (SCM) encompasses the entire value chain and addresses materials and supply management from the extraction of raw materials to the end of products' useful lives, as well as disposal, recycling, repurposing, and reuse. It is an integrative approach to planning and controlling material flows from suppliers to end-users, and enhancing system performance by the coordination of manufacturing, logistics, and materials management within the whole supply chain network.

SCM has been widely applied in the food industry, from agricultural and industrial food production to distribution, retail, consumption, and waste disposal in order to ensure food safety and security. The application of SCM in energy and water supply is understandably much more limited than for food products—despite the similar importance and need for affordability of these resources, as discussed in

Sect. 1.3—for a variety of the following reasons. Market options may be limited, such as in the case of a single regional water supplier drawing from a single source like a reservoir. Firms may have few market options, such as in the case of a firm supplying power to the electric grid using technology that requires a single primary energy source. When bringing an alternative water or energy source on-line, the timeframe is much longer, and the cost for implementation is much greater than for implementing an alternative in the food chain.

A number of SCM practices evaluate energy and water use as metrics to assess environmental impacts, but SCM practices relating to water supply or the water industry itself are very few; two emerging examples are the purification of wastewater so that it can be recycled as drinking water, and the transmission of wastewater so that it can be used for irrigation. More common is the investigation of waste biomass-to-energy SCM, but the cost and operational complexity of logistics make the successful, energetic utilization of waste biomass challenging. Nevertheless, the SCM evaluation process can help justify the implementation of alternatives and the establishment of incentives, as discussed in Sect. 5.2.

18.3.2 Sustainable Supply Chain Management (SSCM) of the FEW Nexus

Understanding supply chains has become critical for understanding food, energy, and water sustainability because of the complexity of the interactions between these resources that are essential for human existence and the various physical and socio-economic factors discussed throughout this book. *SSCM* is a set of managerial practices that include the following:

1. Impact assessment within a framework of triple bottom line outcomes (i.e., social, environmental, and economic); and
2. Consideration of all stages across the entire value chain for each product and the entire product life cycle.

Integrating *life cycle assessment (LCA)* and other metrics into SCM (see Chap. 13) provides quantitative accounting for the environmental, economic and social effects of products, processes, or services (i.e., the total footprint) by assessing the following:

1. the energy and material inputs required;
2. the wastes released to the environment; and
3. the potential environmental, economic, and social impacts of the energy, materials, and wastes involved.

The use of LCA in supply chain network analysis has become an actively researched area and is being increasingly applied in academic, industrial, and governmental fields for evaluation of anticipated environmental, economic, and social impacts.

Possible computational approaches and challenges for taking this analysis from the product (micro-) scale to the city (meso-) scale are illustrated in Sect. 16.5.

An important distinction should be noted about energy usage as a sustainability criterion. While it is typical to consider energy as the element being optimized in efficiency models, it is often the energy byproduct (such as air and wastewater emissions including lead and mercury) that is the most critical to control for the safety and health of a community. Due to the non-linear relationship between energy consumption and emissions, there is no assurance that limitations on energy consumption afford specific reductions in energy emissions. The actual source of electricity (fossil fuel or nuclear or renewable resource) being consumed at any time and location in producing a product is not transparent to the consumer—even if the number of kilowatt-hours of energy associated with the production of a **consumer good** is known—such that even those products requiring lower energy to produce may have greater negative environmental impact if produced using polluting technologies. New data analytics technologies are bringing such transparencies to the market, as discussed further in Chap. 22, making SSCM increasingly practicable.

An important consideration in SSCM is the need for democratic decision-making, which requires transparent and accessible communication of technical information to elected officials, civil society groups, the public, and the development of governance approaches that include collaboration between experts and ordinary citizens.

18.3.3 Approaches to Sustainable FEW Supply in Cities

Given the variety in geographical location, age, population, and socio-economic development of cities, the demands for FEW supplies in different cities vary dramatically. To meet various demands, different FEW supply approaches are necessary. In this section, different types of FEW supplies will be discussed: centralized versus decentralized/local systems, and separated versus integrated systems.

18.3.3.1 Centralized Versus Decentralized FEW Supply Systems

Characterized by a profound process of industrialization and globalization, today, FEW supplies are primarily provided through **centralized systems**, especially in developed markets, such as industrial agriculture, centralized animal feeding operations (CAFOs), and large water treatment and energy generation facilities (heavy infrastructure). While on an individual facility level, centralization results in intensification and specialization of production and may result in more efficient operations, the reliance of centralized systems on uniform industrially produced external inputs and nonrenewable resources, and the related social and environmental costs of resource extraction and industrial production and distribution, raise questions about the sustainability of centralized systems (heavy footprint).

In contrast to centralized systems, decentralized systems have the potential to provide diverse, resilient, and sustainable FEW supplies. Examples of decentralized FEW supplies include urban gardens, urban agriculture, green stormwater infrastructure (e.g., green roofs, rain gardens, cisterns, rain barrels), and decentralized energy systems (e.g., individual rooftop solar supply, mini- or community grid, and distributed individual energy services). The **short supply chains (SSC)** characteristic of decentralized systems can be more environmentally friendly by reducing the number of intermediaries and the transportation distance between FEW commodities and consumers (lighter footprint). For example, supply chain analysis and/or life cycle assessment of decentralized systems may compare the energy efficiency of local food systems to long supply chains and demonstrate that urban agriculture has the better energy performance and can also reduce greenhouse gas emissions.

There is much debate about the overall relative sustainability and efficiency of centralized versus decentralized systems. Although decentralized systems reduce transportation distance and intermediaries, the consequent decrease in initial energy consumption may be offset by other energy uses (such as for food storage). Similarly, socio-economic benefits related to **decentralized FEW system** employment (more local jobs demanding skilled workers that receive higher wages) that may be lacking in centralized systems—due to resource extraction and production distant from the city—may result in drawbacks relating to commodity pricing and stability. Conversely, although decentralized FEW supply chain performance may be undermined by a weak structural optimization of supply logistics, centralized supply sometimes results in severely under-served populations (represented by the lack of affordable fresh food in urban centers, the lack of dependable water in low income areas of cities in developing countries, and the lack of energy following disasters caused by weather).

Certainly, the development of decentralized FEW supplies in cities faces numerous obstacles due to the scarcity, expense or fragmentation of land, and the challenge of integrating innovative practices with existing infrastructure (see the examples in Sect. 18.5 below). The sometimes controversial social, economic, and environmental impacts of urban FEW innovations include odor (from waste stream recapture and larger farm animals), noise (from wind generators and chickens), and perceived aesthetics and safety (from tall vegetation in gardens and stormwater management systems). Future research on both centralized and decentralized systems is warranted using more holistic assessments and considering design logistics within a framework of sustainability's triple bottom line, which considers social, economic, and environmental impacts, as discussed further in Chap. 22.

18.3.3.2 Integrated Versus Separated Systems

The complexity and historical structure of energy, water, and food systems typically result in separate management of their various elements, despite their numerous interwoven interactions. Examples of separate systems include groundwater pumping to supply a city, conventional tillage of an agricultural commodity by a farm for a global market, and fossil-fuel based power generation by a regional power provider. Efficiency across multiple sectors by multiple players rather than the productivity of isolated firms is, however, a necessity for the long-term sustainability of all three systems in the context of resource scarcity, increasing population pressure, and other stressors, as described throughout this book.

Examples of integrated urban FEW systems include:

- waste reclamation (i.e., recovering energy from food waste and wastewater, and recycling wastewater and *greywater* for irrigation in a **closed supply loop**);
- biomass-based energy (i.e., converting residues and waste to energy); and
- large-scale water infrastructure projects such as multi-use water reservoirs for energy generation, irrigation, and cooling water.

Separate systems may have advantages, including less intensive planning and coordination (lighter infrastructure), lower initial capital expenditures, public familiarity, associated economic and political structures (the policy criterion), and well-established existing support infrastructure.

Integrated systems may have advantages including: less resource extraction and associated environmental impacts and costs (such as when waste streams from one sector are used as resources for another sector), resource security (availability, safety, and affordability) in the case of scarce or diminishing resources, local job creation, and micro-scale practices such as green roofs and rain gardens have relatively low barriers to entry which may encourage broad adoption.

The challenges for implementing integrated systems may need to be addressed using a variety of approaches. For example, comprehensive data-based assessments may be necessary to evaluate potential benefits and risks and resolve stakeholder concerns about transitions from separated to integrated systems (such as for recycled water in garden irrigation, which could introduce pollutants such as heavy metals, pathogens, and chemicals of emerging concern into soils and food chains). Government, private enterprises, and civil society may need to engage more fully in theorizing and developing the political frameworks to achieve multiple sustainable development goals while minimizing the risks of adverse cross-sectoral impacts (policy heavy). Successful micro-scale pilot projects may need to be implemented widely before proof-of-concept and market forces can drive broad implementation.

Examples of integrative practices that started as pilot projects and are now more widely accepted include green roofs, which provide potential synergies in the FEW nexus by improving energy efficiency for buildings, reducing stormwater runoff and associated pollution, and capturing rainwater for potential reuse, while potentially supporting food production and/ or recreation (a lighter footprint). Similar to green roofs, rain gardens can be used to direct rainwater for urban food production and stormwater management, reducing the energy consumption required for both irrigation and water treatment. Waste-to-resource practices convert municipal and/or food waste to energy, fertilizer, and reclaimed water that can be used to supply other portions of the FEW nexus in cities.

When these separate actions take place within a broader vision for integrated management and governance across FEW sectors, systems and scales may organically come together to create substantive change with relatively low up-front investment. For instance, **U.S. EPA's Net Zero Strategy** has gradually built a culture of improving the environment, saving money, and helping communities become more sustainable and resilient via conserving water, reducing energy use, and eliminating solid waste at individual military and non-military communities.

18.3.4 Principles for Understanding Urban FEW Supply

There are multiple ways to describe the integration between two or three components of the FEW nexus, as diagrammed in Sect. 1.4. Figure 18.1 additionally describes the dynamics of food, energy, and water specific to urban areas, relating to their availability, affordability, and safety.

Among many parameters affecting the feasibility and sustainability of the urban FEW supplies, three fundamental principles are critical:

1. *Multi-objective*—The urban FEW nexus is inherently interdisciplinary and complex, encompassing social, environmental, and economic demands with multiple objectives. For these multi-objective-oriented systems, a systematic optimization strategy, such as **multi-criteria decision analysis**, is necessary to design urban FEW systems for meeting various demands.

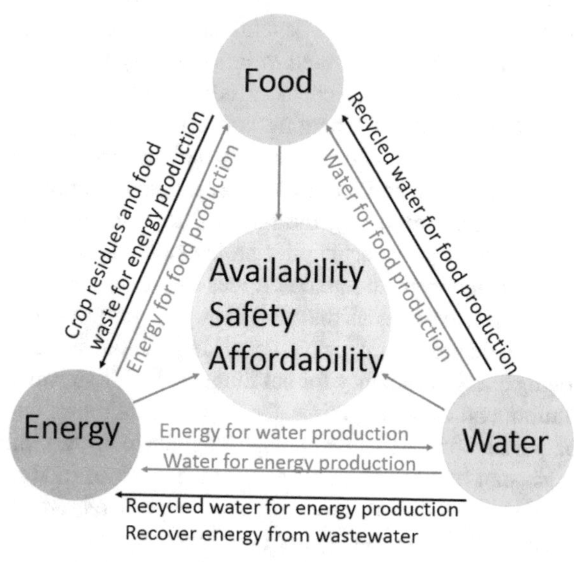

Fig. 18.1 The concept of the FEW nexus and potential integration of sustainable practices. Availability, safety, and affordability lie at the center of the FEW nexus. The grey arrows and corresponding contents indicate interactions between food, energy, and water, and the black arrows and corresponding contents indicate potential FEW integration

2. **Dynamic system accommodation**—Urban FEW systems are dynamic given geospatial/temporal variances, human population shifts, cultural diversity, social/economic development, and dependencies beyond the urban context. Therefore, comprehensive strategies that capture these dynamics and enable corresponding accommodation are imperative. For example, GIS-based analysis and predictive modeling in FEW system design and management that considers interactions between urban and rural systems would be necessary.
3. **Efficiency and sustainability planning**—In addition to the availability, safety, and affordability focus of the FEW nexus, two other important considerations are efficiency and sustainability. Urban FEW systems have to be operated efficiently to improve overall sustainability. For this purpose, life cycle assessment and other metrics are necessary planning tools.

18.4 Integrated FEW Crises Due to Poverty and Exclusion

To illustrate the challenge of providing effective governance of the complex systems at the FEW Nexus, the causes and effects of the public drinking water crisis that occurred in Flint, MI, starting in 2014 will likely continue to be analyzed for many years. When the State of Michigan took over the administration of the city via an appointed emergency financial administrator (policy heavy),[2] a change in the source of drinking water from Lake Huron to the corrosive Flint River was initiated as a cost-saving measure (the affordability criterion). The high cost of energy to pump drinking water over 60 miles from Lake Huron to the City's distribution system (heavy infrastructure) was a factor in these decisions. The change was implemented despite warnings from the supervisor in the City's water treatment facility to the Michigan Department of Environmental Quality and from leadership in the Governor's Office that the timeframe and safeguards for the switch were inadequate to protect public health (the criticality criterion).

Due to a variety of chemical, engineering, and operational issues, the water from the corrosive new water source leached lead from the water pipes into the drinking water in homes, schools, and businesses, causing high blood lead levels in the population, particularly among children. The change also increased bacterial counts in water, leading to a Legionnaire's disease outbreak. These crisis conditions contributed to impaired public health as they made food preparation, at-home food gardening, and personal hygiene extremely challenging for many households. Subsequently, high lead levels in publicly supplied water have been found in over 100 US cities, illustrating the potential for this issue to affect public health more broadly (heavy footprint).

[2] Implemented under the provisions of Michigan Public Act 4 of 2011 (since replaced by 436 of 2012); see also the previous Public Act 72 of 1990 for another emergency financial management policy.

The arguments for more effective urban governance of food, energy, and water systems are twofold:

1. *Cities offer unique perspectives for FEW*, both because urban system integration can demonstrate solutions for global crises and also because addressing the crises that cities themselves present for nexus security is essential for specific populations in the present time. Cities around the world are experiencing unprecedented problems at scales never before seen in terms of population growth and density, sprawl, polarization by income including extreme poverty, disinvestment, and exclusion or segregation by race, ethnicity, and national identity. Population migration, the flight of refugees from areas experiencing environmental distress or conflict, or from regions experiencing long-standing ethnic conflicts such as in the West Bank and the Gaza Strip, or locations experiencing economic distress and conflict such as Central America, are specific cases of this problem. We use *exclusion* to mean the denial of economic, political, and/or cultural resources and basic necessities as well as spatial exclusion and segregation. In the USA, such exclusion is experienced by many communities along the lines of race and ethnicity, including African American, Native American, and immigrant or ethnic minority communities.
2. *The experiences of communities—whether defined socially or spatially—that are subject to impoverishment and exclusion need to be the starting point for addressing integrated FEW crises.* Developing solutions that are equitable and sustainable is not just about stating the problem, but also about asking how examining the problems and their sources help us identify pathways that lead to positive outcomes for these communities and thus for society as a whole. **Positive models** that are available to us from some cities—such as Curitiba for equitable transportation as discussed in Sect. 18.5.3 and Belo Horizonte for equitable food systems—also need to be investigated for their suitability and appropriateness for all outcomes.

18.4.1 Examples of Urban Nexus Insecurity Impacts

From the perspective of households and neighborhoods as well as communities, nexus security issues need to be addressed in the form of housing, transportation, utilities, and other infrastructure that form the structure in which FEW resources come together and are accessed by households. Decisions and policies related to land and its use, political fragmentation, and the presence or absence of intergovernmental cooperation all affect how these structures and services are provided or accessed (the policy criterion). The dynamics of social and spatial exclusion within and between communities can result in a denial of FEW resources or degradation of their quality.

Residential segregation by race and ethnicity in US cities, for example, leads to pernicious disparities in everything from education to health and employment. Such segregation is not simply the result of personal prejudices or the unfortunate byproduct of economic decisions. Study after study has shown how government policies and regulations have created and reinforced racial exclusion over an extended time period (policy heavy). This exclusion results in FEW nexus insecurities.

For example, housing insecurity causes nexus insecurity with implications for health and reduces individual options to acquire a stable and sustainable future in economic, political, and cultural realms. Housing insecurity is a common phenomenon across global cities, whether it is due to lack of affordability, inadequate services and amenities for survival nearby, limited urban employment options, or the vulnerabilities posed by tenuous tenure in squatter settlements or in refugee, migrant, or detention camps. Housing insecurity can also challenge the delivery of water, electricity, and other sources of power needed for household survival and operation, either due to lack of household resources to maintain service or due to lack of system resources to provide broad access triggered by lack of rate-paying customers.

Lack of adequate, consistent, and clean water supplies can lead to food insecurity and disease due to a greater dependence on processed, packaged foods; inability to cook food or maintain sanitary conditions; and an inability to plan (the criticality criterion). This can have especially dire knock-on and lifetime effects for affected children. The poor end up paying more in both intangible and relative terms in their efforts to correct either situation. Nexus insecurity thus can challenge households' efforts to counter poverty, imposing constraints on members' ability to realize their potential through education, employment, and other forms of social achievement.

18.4.2 Factors Contributing to and Approaches for Addressing Urban Nexus Insecurity

Spatial exclusion of the form historically experienced by African American and Native American communities and other impoverished communities of color has created disparities along many dimensions, including Nexus insecurity. Nexus solutions to resolve the impacts of exclusion require strategies that target both places as well as people (policy heavy). An example of the former might be to enact policies for inclusionary housing such that all residential developments set aside a portion for households with incomes that are below the community's median value. The latter category might include rental vouchers for low-income households that will help them afford to rent apartments in better-off communities.

In Detroit, nexus insecurities are creating a water crisis: tens of thousands of households have experienced or are threatened with water shutoffs. Water shutoffs cripple personal care, dietary planning and preparation, and sanitation. Children miss school or face cognitive and social difficulties due to impacts related to personal hygiene and dirty clothes.

Similarly, in Flint, lead contamination of water has caused significant long-term physical and psychological hardships for residents, devalued tens of thousands of homes, and resulted in enormous public expenditures and mountains of plastic waste.

In both cases, and in similar others, racial exclusion was the basis upon which these more proximate dynamics have created Nexus insecurity.

Solutions to racial disparities in FEW security require government intervention in the market place (policy heavy). Civil society institutions are also developing programs based on public and philanthropic support to meet FEW needs using the unique assets of such communities and to build capacity of different kinds. Technological solutions that are affordable and appropriate need to be considered in such efforts. Small-scale examples of FEW integrated, bottom-up solutions in Detroit include stormwater/rainwater capture for urban gardens, compost, and energy production using food waste from nearby restaurants, breweries, and farmers' markets; and solar-powered water pumps for agriculture in Detroit.

Efforts to address social disparities in FEW security need to be based on an analysis of the historical patterns of exclusion, including the role that government policy and regulation played in such patterns. They need to be informed by efforts that such communities are putting in place to address needs and solve problems, and they need to involve members from excluded communities in leadership roles. Lessons from such efforts potentially have wider applications in communities here and elsewhere.

Additional factors contributing to FEW nexus insecurities in cities in the USA and around the world are as follows:

- Unprecedented growth (population and geographic)
- Income polarization
- Sprawl

These impacts are creating crises with FEW impacts at the household and community levels. In Midwestern US "rustbelt" cities, the FEW crisis cascade is exemplified by Flint; there are different manifestations elsewhere. Resilience planning will likely take various forms in other cities.

18.5 Case Studies: Nexus Innovation at the City Level

The real excitement, of course, happens when the private sector, public sector, civil society, and individual agency come together to influence the linkages between food, energy, and water. The examples that follow illustrate current transformative actions at the FEW nexus in cities. These examples transcend classic examples of transformative change[3] and attempt to illustrate centralized and decentralized actions across various scales using both integrated and separate practices. In places where the drivers of the FEW nexus align around the shared objective of greater sustainability, examples such as these can be found.

[3] One such classic example is the successful adoption of solar cookers in tropical cities that were formerly dependent on traditional charcoal stoves. Solar cookers proved to be well suited to the long, slow cooking needed for the dried fish, vegetables, and rough grains of local diets. Charcoal stoves were inefficient with unhealthy fumes, and contributed to deforestation and associated negative impacts on local lakes and rivers as well as carbon sequestration.

18.5.1 Portland, OR, USA

Portland can be characterized as a consistently growing city since the 1990s, with an estimated population of 639,863 in 2015, representing an increase of 15.7% since 2000. As a growing USA city with a strong municipal government, Portland's work at the FEW nexus is driven by innovative process-oriented municipal planning and policy aimed at protecting the environment and limiting the negative impacts of climate change. The city's evolution into a leader in urban sustainability has occurred in the context of successful top-down environmental planning at the municipal and regional levels that integrates large-scale citizen engagement strategies to garner public support from the bottom-up and bring innovative policies into effect.

Examples of work at the FEW nexus in Portland include a myriad of municipal plans and strategies published between 1993 and the present. Portland is an Innovator City within the C40 network of major world cities committed to combatting climate change and was recognized at the 2016 C40 Cities Awards for excellence in the category of Climate Action Plans & Inventories.

In the area of its urban food system, Portland has a Sustainable Food Program and an Urban Food Zoning Code (policy heavy) that aims to achieve triple-bottom-line goals around food access and health at the individual and community levels. The City of Portland's website hosts a community-supported agriculture (CSA) map, farmers' market locations and schedules, information about the food zoning code, and a searchable database of sustainable food resources in Portland and nationally. In collaboration with Multnomah County, the city established the Portland/Multnomah Food Policy Council in 2002, a citizen-led advisory council that is advisor to the municipal and county-level governments on matters related to food policy in the region.

In the sector of water, Portland has made water conservation a priority through conservation measures in city parks and green spaces and through planning and programs for watershed management at the municipal level (reduced footprint). Over eighty city parks in Portland utilize climate-based irrigation systems that conserve water and significantly reduce the total volume of water used by the City. Portland City Council endorsed the Planning and Sustainability Office's River Concept document in 2006, which is being used to guide the development of a River Plan. The City's Office of Environmental Services issues watershed report cards and runs a Community Watershed Stewardship Program that offers funding to residents and neighborhood groups for projects that improve neighborhood and watershed ecology through planting native plants and community gardens, and that support youth education and workforce development around environmental issues.

The gamut of Portland's sustainable energy initiatives extends across sectors and scales from solar and wind energy incentives for residents and businesses to green building standards, including the promotion of green roofs, district energy, and many municipal energy use reduction strategies (lighter infrastructure). Portland's Green Building Policy exempts sustainable building additions such as solar panels, small wind turbines, and green roofs from design review under existing zoning codes, and developers and businesses that install green roofs in the central city can achieve a **floor area ratio (FAR) bonus** for doing so.

The city's Water Bureau offers residents free water efficiency devices. Customized Energy Saver Kits covering both energy and water use can be obtained from the Energy Trust of Oregon. Between 2009 and 2012, the city's Office of Planning and Sustainability partnered with local neighborhood organizations on the "Solarize Portland!" campaign that installed solar panels and weatherization on nearly one thousand urban residences.

Along with successfully creating policies and implementing programs in the areas of food systems, water, and energy separately, the City of Portland has several initiatives that work at the integration of the FEW nexus. In particular, its Green Building Policy promotes energy efficiency through green roofs that reduce energy usage on a facility level relating to heating and cooling and at the meso-level through water recapture that reduces water treatment energy needs. Food is integrated into the two-way energy–water nexus of green roofs on a facility level by planting food-producing landscapes such as community and market gardens on the tops of buildings, which may also be used as functional recreational space.

Along with its garbage collection and recycling service, the City also runs a composting service that collects residential and commercial food scraps and sends them to Oregon composting facilities. The finished compost is then sold to landscaping companies and made available to residents. Further strengthening of the FEW nexus could be achieved by creating connections between the municipal composting program and EcoRoof policies to expand the use of food-producing EcoRoofs.

Another promising FEW nexus initiative is the fruit of a private-public partnership between Lucid Energy, the City of Portland, and private investors. The full-scale project aims to produce sustainable energy from the City's gravity-fed municipal water pipes—a constant and previously untapped source of micro-hydropower that avoids the damaging environmental impacts of large-scale river dams, including their negative effects on agricultural floodplain cultivation. According to the producer, four 42-in. turbines produce about 1100-Mwh of electricity each year, approximately enough to power 150 homes. Utilization of new technologies such as this is dependent on compatible topography and the availability of initial installation capital and other factors, but those installed in Portland are projected to result in net energy savings over the project lifetime.

18.5.2 Detroit, MI, USA

Due to its industrial past and history of extreme depopulation and disinvestment, Detroit can be characterized as a Rust Belt or Legacy city. Its population decline extends from its peak of over 1.8 million in the 1950s to an estimated 672,795 in 2015. Detroit's housing vacancy rate, combined with the age of its water infrastructure and other systems that were designed for more than double its current population and industrial capacity, have created a substantial economic burden for ongoing system maintenance and a constraint for technological innovation. Until more recently, mainly non-municipal actors at the grassroots level have driven innovative work at the FEW nexus in Detroit. Many of the main players involved in developing Detroit's sustainable systems at the FEW nexus are Civil Society (Social Impact and private) entities which are working in areas of sustainability and food systems, as well as individuals within the City's urban agricultural network such as entrepreneurial market growers and community gardeners, and leaders within City departments and regional agencies such as the Great Lakes Water Authority (part of the Detroit Water and Sewerage Department before it was separated from DWSD as part of the city's 2014 bankruptcy settlement).

Examples of innovations at the FEW nexus in Detroit are numerous but generally small-scale. The Detroit Black Community Food Security Network (DBCFSN) is innovating at the FEW nexus at its D-Town Farm located in the municipally owned Rouge Park on Detroit's Northwest side. Along with producing food and promoting food justice and autonomy within the Black community (the affordability criterion), DBCFSN has developed a system for water catchment for use on its extensive urban farm, creates compost from its agricultural waste, and has installed solar panels to power a water pump for rainwater reuse on the farm (decreased footprint). The DBCFSN is also engaged in developing a consumer food co-operative, The Detroit People's Food Co-op, on a campus that will offer a community kitchen and affordable housing. Such developments enhance FEW equity, both by improving access to food and housing but also cultivating African-American leadership on community-based solutions.

The Earthworks Urban Farm at the Capuchin Soup Kitchen on the East Side of Detroit has integrated FEW nexus innovations into its operations in a way similar to D-Town Farm. The two and a half acre Earthworks Urban Farm grows 14,000 pounds of USDA Certified Organic produce per year for use at the Capuchin Soup Kitchen, distribution to the surrounding community, and for sale at its farm stand. The farm offers a training program in agriculture for adults and an after-school program for children, among other initiatives. Earthworks employs three water collection tanks, two below ground and one above, to gather rainwater from gutters on top of the farm's hoop house. This water is then pumped from the tanks for use in the irrigation of crops by a solar-powered water pump.

Though small-scale, D-Town Farms' and Earthworks' systems use FEW innovation to realize several important environmental, municipal, and organizational benefits that can serve as examples for other individuals, businesses, and organizations, and that should be considered in the creation of future city policies regarding sustainability. For example, the farms' catchment systems offer water for food production at a higher quality and lower cost than municipal water services can provide; the collection of water from farm structures limits the amount of stormwater runoff entering an already overstressed municipal combined sewer infrastructure, saving additional energy in wastewater treatment. Locally, food produced from the farms supports the local community and strengthens the urban food system (the criterion of criticality), with regional energy savings by reducing food miles traveled. Finally, because these solutions are developed from within the community itself, they increase FEW capacity and resilience among stakeholders.

In 2013, Detroit passed its Urban Agriculture Ordinance, which gave legitimacy to existing urban agricultural operations occurring at the City's grassroots, amended the City's zoning code to define urban agricultural uses, and generally permitted urban agricultural activities less than one acre in size (the policy criterion). The City realized the need to address agricultural uses in its zoning code beginning in 2009 when several large-scale commercial urban farms were proposed within City boundaries by wealthy business interests, one of which was Hantz Farms. The ordinance was developed with involvement by grassroots participants in Detroit's urban agriculture sector and municipal staff, led by City Planning Commission staff member Kathryn Underwood and the Urban Agriculture Work Group. The Urban Agriculture Work Group was composed of urban agriculturalists, academics, county Cooperative Extension officials, and representatives from city- and state-level agencies.

While Detroit's Urban Agriculture Ordinance did not explicitly promote innovation at the FEW nexus, it gave urban agriculture in Detroit much-needed legitimacy and strengthened prospects for developing a healthy and thriving local food system for Detroit residents. The Urban Agriculture Ordinance is silent on the challenges of municipal water access for agriculture in the City but does include standards for water catchment systems on urban gardens and farms, supporting grassroots farmers who wish to innovate.

The potential for further development of supportive policy at the FEW nexus continues through the establishment of the Detroit Office of Sustainability in 2017 and the development of a Sustainability Action Agenda for the City through a public engagement process in 2018.

In recent years, the City of Detroit and its associated water and energy utilities have made strides toward greater sustainability at the FEW nexus through several municipal and municipally sanctioned planning and infrastructure initiatives. The City of Detroit partnered with the private energy company Detroit Renewable Power and its subsidiary Detroit Thermal to burn municipally collected waste to produce waste-to-energy electricity for the City, as well as district heating in the Midtown and downtown areas, with the steam produced in this process (reduced footprint). However, the benefits of waste-to-energy production and district heating must be weighed against the air quality concerns of burning municipal waste within the dense City center (questioning the reduced footprint) and the overall cost-benefits of burning versus recycling. Due to a track record of consistent emissions violations, the waste-to-energy facility remained controversial throughout its lifetime, and was permanently shut down in 2019.

At the water-energy nexus, energy optimization studies have long been part of water utility operation in Detroit due to its flat topography, requiring pumping to maintain surface drainage and distribute drinking water, as well as flap-gate operation to minimize structural intrusion by river water during wet weather. Recently, the Detroit Water and Sewerage Department's green infrastructure initiatives have resulted in better watershed management in the City, including collaboration with the quasi-governmental Detroit Land Bank Authority to promote strategic green infrastructure interventions on Detroit's vacant lands, and construction of park-based stormwater management basins for water quality improvement related to combined sewer overflows (CSOs) after major storm events. Detroit's Drainage Charge Program initiated in 2016 is designed to ensure all City parcels equitably share the cost of drainage infrastructure based on the amount of their impervious surface that contributes run-off to the system; rate credits are offered for run-off capture devices such as cisterns, detention/retention basins, bio-retention, and other water storage (the policy criterion). While the connection of these water storage systems to the FEW Nexus is not directly promoted, the legality, feasibility, and practicality of implementing water storage systems, and the contributions of water storage systems to the FEW Nexus, have already been established through work within the grassroots urban agriculture community.

The strategic vision for future FEW nexus initiatives in Detroit may continue to be influenced by the 2012 Detroit Future City Framework. Initiated by major foundations and quasi-governmental authorities including the Detroit Economic Growth Corporation and the Kresge, Kellogg, and Knight Foundations, it was completed as part of a citywide public-engagement effort and offers visions of what a sustainable future might look like, encompassing the areas of economic growth, land use, city systems, neighborhoods, land and building assets, and civic capacity. The framework is not a master plan or policy of the City of Detroit, but rather may be viewed as a strategic framework and vision from the nongovernmental and civil society sectors to provide context for what is possible for a more sustainable future in the City of Detroit. The framework includes a discussion of large-scale green and blue infrastructure integration within the urban landscape that would include innovations at the FEW Nexus.

18.5.3 Curitiba, Brazil

Curitiba is well known as one of the most environmentally innovative and sustainable cities in the world. Through the work of a pragmatic mayor, proactive municipal government, and the expertise and foresight of professional city planners, Curitiba transitioned from an ecologically unsustainable and expanding third world city to the eco-city that we know it as today.

Curitiba epitomizes innovation at the FEW Nexus through several large-scale programs and planning projects headed by former Mayor Jaime Lerner, who served as mayor of Curitiba for three terms. The City's Green Exchange Program is perhaps the best example of an integrated FEW innovation. The Green Exchange Program allows Curitiba residents to exchange trash and recyclables collected from the City streets for tokens that can be used to obtain fresh produce from local farmers as well as transit tickets for Curitiba's highly efficient bus rapid transit system. The Green Exchange Program was initiated by Lerner in 1989 in response to unsanitary conditions in the City's slums due to the accumulation of trash because of the inability of municipal collection vehicles to navigate the steep unpaved roads of the **favelas**. Lerner determined how much it would cost for the trucks to collect the trash from the *favelas*, and instead rerouted this money toward purchasing produce from local peri-urban farmers growing food just outside of the city limits that could be exchanged for the trash that the *favela* residents would bring down from their neighborhoods themselves. In this way, the Green Exchange Program simultaneously supported Curitiba's food system by supporting local farmers and food access for low-income residents, while also addressing sanitary concerns such as the accumulation of garbage and the resultant risk of water contamination and spread of disease (potentially water-borne) within Curitiba's most impoverished communities.

Another area in which Lerner's pragmatism resulted in innovations at the FEW Nexus is stormwater management through Curitiba's city parks and green spaces. When Lerner took office in 1971, Curitiba maintained one public park in the City center, which amounted to about two square feet of green space per City resident. In response to flooding concerns, the City had access to federal funding to channelize the City's five rivers. However, Lerner and his administration took a different approach. Instead of enclosing the City's rivers in cement channels, the City developed 28 public parks and gardens harnessing the power of natural floodplains, many using stormwater retention ponds and other green infrastructure projects as their centerpieces. By developing public park space for large-scale stormwater management, Curitiba provided its residents with functional recreational assets while also reducing stormwater treatment energy demands and flooding risk in the context of climate change.

The final major aspect of Curitiba's FEW innovations is its use of sustainable land use and transit planning to reduce transportation energy use and ensure that all residents are able to achieve sufficient mobility to access basic needs including food and health care (the affordability criterion). As air pollution and traffic congestion continued to be significant concerns during the mid-twentieth century, the City made the decision to create a system of transportation axes (the policy criterion) to funnel traffic in and out of the City efficiently while promoting the use of public transit through a highly efficient and well-used bus rapid transit system (reducing the footprint). The plan called for one axis to serve motorists coming into the City, the middle axis to be dedicated to rapid transit bus service traveling both ways, and the third axis serving motorists leaving the city. The City paid transit providers by the mile, which created incentives to fill as many seats as possible. Equity was supported through the distribution of transit tokens in return for trash collected by community members. The City's Master Plan called for the highest density development concentrated along these corridors, with density tapering off gradually with increased distance from them. The creation of the structural axes encouraged Curitiba's residents to use public transit and to settle near the City center, promoting density. Over time, these actions played an important role in limiting suburban sprawl, which resulted in energy savings through reductions in fuel usage and also avoided unnecessary expansion of the City's energy and water infrastructures.

18.5.4 *Tianjin Eco-City*

Tianjin Eco-City is a joint project between the governments of the People's Republic of China and Singapore located just outside of the Chinese city of Tianjin on the coast of the Bohai Sea. Begun in 2007, the goal of the two nations is to create a new sustainable eco-city from the ground-up, in response to rapid urbanization and environmental degradation in the region. FEW innovation in the Tianjin Eco-City project is driven by binational governmental cooperation supported by a wide range of private-sector partnerships.

In the master planning process for the Tianjin Eco-City, FEW nexus opportunities were identified to develop a new city that would serve as a sustainable, healthy environment for its residents, and as an industrial hub of the green economy for the two countries involved. One aspect of the Eco-City that relates directly to the FEW nexus is its water-energy resource loop that treats wastewater via anaerobic biodigestion to extract methane gas to be used for energy production.

Similar to the planning that Curitiba undertook in the late twentieth century, the Tianjin Eco-City Master Plan calls for high-density development nearest to major transportation axes, to promote population density near the City center and high levels of public transit use among residents (the policy criterion). The Plan also calls for the City being structured into discrete but connected cumulative units of varying scales, including eco-cells, eco-communities, and eco-districts from smallest to largest. Five major eco-districts will surround a green-blue eco-core to give the City its overarching structure. The green-blue eco-core is envisioned as both a space for recreation, transportation, and stormwater management.

18.6 Conclusion

The above case studies—both domestic and international, including Portland, a growing American city, and Detroit, a post-industrial legacy city, Curitiba, a 'retrofit' city, and Tianjin Eco-City, a new city imagined from the ground up—provide only a glimpse into the possibilities of integrated urban FEW innovations. These cases provide an illustration of integrated FEW supply solutions, both centralized and decentralized, implemented at different scales from the micro (facilities: urban farms in Detroit) to the meso (city systems: Curitiba's stormwater parks and transit system and Portland's energy-producing pipes) and macro (city/metropolitan region: Tianjin Eco-City) levels, driven by different actors including the private sector, public sector, and civil society.

Key Points

- Urban systems that are largely physically separated from the agricultural resources needed to support them exhibit dynamics and drivers that are inherently different from the dynamics and drivers in rural systems. Food, energy, and water supply chain management in cities are particularly critical in large population centers with high densities that are substantially separated geographically, economically, or politically from the FEW resources needed to support them.

- The socio-political-economic drivers of the FEW nexus in cities function within the *socio-ecological system* (see Sect. 9.1.1), at multiple interacting levels of social organization (international and national, state and regional, local, facility, household, and individual), driven by three types of institutions—*private sector* institutions, the state and the *public sector* institutions, and social impact organizations or *civil society* (also known as nongovernmental organizations).
- Decisions by food wholesalers, grocery retailers, policy-makers, and others have helped to create relatively large urban areas characteristic of many US cities where access to affordable and nutritious fresh foods is limited—often characterized by relatively low household incomes—called *food deserts*.
- Public-sector water agencies have, in some cases, successfully ensured water and wastewater service availability even to those households that lack sufficient income to pay for the services; this is sometimes framed as a discussion of water as a human right.
- In urban areas, housing security—defined broadly to include location, quality of amenities, affordability, and security of tenure—is crucial for FEW security at the household level.
- From a biophysical perspective, one of the strongest potential drivers of the FEW nexus in urban areas is global climate change. Impacts of climate change already being seen are more frequent and severe storms and flooding and droughts, impacting energy, drinking water, and food supply.
- The FEW nexus in cities is strongly shaped by the agency of individuals and households in pursuing the interests they have chosen based on their particular mix of ideology, knowledge, and resources.
- Multi-criteria decision analysis, dynamic system accommodation, and efficiency and sustainability planning are necessary for sustainable supply chain management of the FEW nexus in a city.
- While centralized supply chains result in intensification and specialization of production and may result in more efficient operations, the reliance of centralized systems on uniform industrially produced external inputs and nonrenewable resources, and the related social and environmental costs of resource extraction and industrial production and distribution, raise questions about the sustainability of centralized systems. In contrast to centralized systems, decentralized systems have the potential to provide a diverse, resilient, and sustainable FEW supply chain.
- While separately managed supply chains have advantages including less intensive planning and coordination, lower initial capital expenditures, public familiarity, associated economic and political structures, and well-established existing support infrastructure, integrated supply chain management has advantages including less resource extraction and associated environmental impacts and costs, resource security in the case of scarce or diminishing resources, local job creation, and relatively low barriers to entry which may encourage broad adoption.

- The experiences of communities—whether defined socially or spatially—that are subject to impoverishment and exclusion need to be the starting point for addressing integrated FEW crises. Developing solutions that are equitable and sustainable is not just about stating the problem but also about asking how examining the problems and their sources helps the USA to identify pathways that lead to positive outcomes for these communities and thus for society as a whole.
- Portland Oregon's work at the FEW nexus is driven by innovative process-oriented municipal planning and policy aimed at protecting the environment and limiting the negative impacts of climate change by engaging public, private, civil society, and individual agency to implement changes at a variety of scales.
- Many of the main players involved in developing Detroit, Michigan's sustainable systems at the FEW nexus are Civil Society (Social Impact and private) entities which are working in areas of sustainability and food systems, as well as individuals within the City's urban agricultural network, and leaders within City departments and regional agencies.
- Curitiba, Brazil innovates at the FEW nexus through several large-scale programs and planning projects, including a public exchange of recyclables and trash for tokens to obtain fresh produce and bus transportation that largely benefits low-income individuals, and the promotion of urban density to reduce energy usage for transportation and water supply.
- FEW innovation in the Tianjin Eco-City project is driven by binational governmental cooperation supported by a wide range of private-sector partnerships. In the master planning process for the Tianjin Eco-City, FEW nexus opportunities were identified to develop a new city that would serve as a sustainable, healthy environment for its residents, and as an industrial hub of the green economy for the two countries involved.

Discussion Points and Exercises

1. Discuss why it is imperative that we understand the food–energy–water nexus in cities, specifically. How does the FEW nexus in cities compare to that in rural, suburban, and peri-urban areas?
2. Discuss the role that the private sector and economy play as a driver of the food–energy–water nexus in cities. Draw an example of the private sector's role in the nexus from one of the case studies discussed in the chapter (see Sect. 18.5) or provide an example from your own experience.
3. Discuss the role that the public sector and governance play as a driver of the food–energy–water nexus in cities. Draw an example of the public sector's role in the nexus from one of the case studies discussed in the chapter (see Sect. 18.5) or provide an example from your own experience.
4. Discuss the role that civil society plays as a driver of the food–energy–water nexus in cities. Draw an example of civil society's role in the nexus from one of the case studies discussed in the chapter (see Sect. 18.5) or provide an example from your own experience.
5. Discuss the role that the knowledge, attitudes, and practices of individuals plays as a driver of the food–energy–water nexus in cities. Draw an example of

the role of individual knowledge, attitudes, and practices from one of the case studies discussed in the chapter (see Sect. 18.5) or provide an example from your own experience.

6. In the USA, which driver of the food–energy–water nexus (private sector, public sector, civil society, or individual knowledge, attitudes, and practices) do you think plays the largest role in cities, and why? If you had to focus on just two of these drivers, which two would you choose in order to have the greatest positive impact on FEW sustainability? Explain your reasoning.
7. If you were tasked with creating an Urban FEW Nexus Management Plan for your city, discuss how you would address each of the drivers discussed in the chapter, and describe a strategy of sustainable FEW management that could apply to each.
8. Refer to Exercises 2.2–2.9. Consider the urban FEW nexus as a complex system. Describe:

   (a) the boundaries of the system;
   (b) the main components;
   (c) the structural arrangement of the components;
   (d) the most significant functional interactions between components;
   (e) external factors and their interactions with the system;
   (f) the most significant (distributed) controls on the system including human actions;
   (g) issues that might alter the stability of the system; and
   (h) an emergent property of the system.

9. Diagram your answer to Exercise 18.8.
10. Which framing of the FEW nexus in the literature (Sect. 1.4) most closely resembles your diagram of the urban FEW nexus from your answer to Exercise 18.9? Which of the framings do you believe is most applicable to cities, and why?
11. How might international law (Chap. 6) better ensure that food, energy, and water resources are managed more sustainably in cities globally? Which international governing body do you think is most fit to consider the FEW nexus, and why? You may also choose more than one.
12. How might domestic US law (Chap. 8) better ensure that food, energy, and water resources are managed more sustainably in cities in the USA? Which level of government (local, state, or federal) do you think has the greatest influence on FEW considerations in American cities, and why? Which local, state, or federal agency or agencies do you think should be tasked with managing resources at the urban FEW nexus? Explain your answer.
13. Refer to Sect. 9.1. Identify a provisioning, a regulating, a supporting, and a cultural ecosystem service relevant to cities. In what ways can cities be designed to optimize and capitalize upon urban ecosystem services at the food–energy–water nexus? Provide an example.
14. Go play the SymbioCity game at http://www.btslearning.com/app/eBS/symbiocity/index.asp. How does this sustainable city simulation game address the

food–energy–water nexus? Does it address all three sectors? Does the game consider food, energy, and water systems as separate or integrated systems in cities? How so? If you were to add another component of urban sustainability to the game, what would it be, and why?

References

Abelman, J. (2017). *Working Group: From urban agriculture to new urban commons – Productive landscapes and infrastructural ecologies*. http://apru-scl.uoregon.edu/2017/03/08/from-urban-agriculture-to-new-urban-commons-productive-landscapes-and-infrastructural-ecologies/.

Alstone, P., Gershenson, D., & Kammen, D. M. (2015). Decentralized energy systems for clean electricity access. *Nature Climate Change, 5*(4), 305–314.

Anderson, O. (2017). Kentucky. In A. Hachadourian, C. Hillstrom (Eds.), *3 Creative solutions to protect water, wildlife, and a warming city nation of change*. http://www.yesmagazine.org/issues/50-solutions/3-creative-solutions-to-protect-water-wildlife-and-a-warming-city-20170205.

Aubry, C., & Kebir, L. (2013). Shortening food supply chains: A means for maintaining agriculture close to urban areas? The case of the French metropolitan area of Paris. *Food Policy, 41*, 85–93.

Baatz, E. (1995). The Chain Gang: Successful management of the supply chain can save a company both money and time, but CIO-100 winners are taking it a step further. *CIO, 8*, 46–46.

Bakker, K. (2003). Archipelagos and networks: Urbanization and water privatization in the South. *The Geographical Journal, 169*(4), 328–341.

Barnard, P. L., Hoover, D., Hubbard, D. M., et al. (2017). Extreme oceanographic forcing and coastal response due to the 2015–2016 El Niño. *Nature Communications, 8*, 14365.

Barthel, S., & Isendahl, C. (2013). Urban gardens, agriculture, and water management: Sources of resilience for long-term food security in cities. *Ecological Economics, 86*, 224–234.

Becchis, F., Asquer, A., & Russolillo, D. (Eds.). (2016). *The political economy of local regulation: Theoretical frameworks and international case studies*. London: Palgrave Macmillan.

Beveridge, R., Moss, T., & Naumann, M. (2017). Sociospatial understanding of water politics: Tracing the multidimensionality of water reuse. *Water Alternatives, 10*(1), 22–40.

Bizikova, L., et al. (2013). *The water-energy-food security nexus: Towards a practical planning and decision-support framework for landscape investment and risk management*. Winnipeg: International Institute for Sustainable Development.

Bouzarovski, S. (2015). *Retrofitting the city: Residential flexibility, resilience and the built environment*. London: IB Tauris.

Bouzit, M., Das, S., & Cary, L. (2017). Valuing treated wastewater and reuse: Preliminary implications from a meta-analysis. *Water Economics and Policy, 4*, 1–26.

Breitenbach, S. (2017). States see value in backing 'food hubs' for farmers, consumers. *Stateline*. http://www.pewtrusts.org/en/research-and-analysis/blogs/stateline/2017/01/19/states-see-value-in-backing-food-hubs-farmers-consumers.

Brodt, S., et al. (2013). Comparing environmental impacts of regional and national-scale food supply chains: A case study of processed tomatoes. *Food Policy, 42*, 106–114.

Brown, P. (2017). Wind turbines head for homes again. *environmentalresearchweb*, 2 Feb 2017. http://climatenewsnetwork.net/wind-turbines-head-homes/.

Budds, J., & McGranahan, G. (2003). *Privatization and the provision of urban water and sanitation in Africa, Asia, and Latin America*. London: International Institute for Environment and Development (IIED).

Coley, D., Howard, M., & Winter, M. (2009). Local food, food miles, and carbon emissions: A comparison of farm shop and mass distribution approaches. *Food Policy, 34*(2), 150–155.

Conklin, A. R. (2017). It ain't easy being green (infrastructure). *Aquatic Sciences Chronicle, 1*, 6–9.

Cousins, F. (2017). The US is suffering from a very real water crisis that few are acknowledging. *DeSmogBlog*, 24 Jan 2017. https://www.desmogblog.com/2017/01/24/america-suffering-very-real-water-crisis-few-are-acknowledging.

Coyle, B. D., & Ellison, B. (2017). Will consumers find vertically farmed produce "out of reach"?. Choices Quarter 1 Available online: http://www.choicesmagazine.org/choices-magazine/theme-articles/transformations-in-the-food-system-nutritional-and-economic-impacts/will-consumers-find-vertically-farmed-produce-out-of-reach.

Dahl, K. A., Fitzpatrick, M. F., & Spanger-Siegfried, E. (2017). Sea level rise drives increased tidal flooding frequency at tide gauges along the U.S. East and Gulf Coasts: Projections for 2030 and 2045. *PLoS One*. https://doi.org/10.1371/journal.pone.0170949.

Daniels, P. (2017). Designing a renewable food system. *Stanford Social Innovation Review*, 10 Jan 2017.

Dickson, J. D. (2017). Farming nonprofit aims to grow North End co-op. *Detroit News*, 26 Feb 2017.

Elliott, J. M., & Ali, S. R. (2007). *The state and local government political dictionary*. Rockville: Wildside Press.

EPA, U.S. (2017). Cited March 2017. Available from: https://www.epa.gov/water-research/promoting-sustainability-through-net-zero-strategies.

Farley, G. A. (1997). Discovering supply chain management: A roundtable discussion. *APICS—The Performance Advantage, 7*(1), 38–39.

Feldberg, S. (2017). Farm of the future: What grows in Las Vegas stays in Las Vegas. *National Public Radio The Salt*, 23 Jan 2017.

Fleming, L. N. (2018). State OKs permit letting Nestle draw more water in west Michigan. *The Detroit News*. Online at https://www.detroitnews.com/story/news/politics/2018/04/02/nestle-michigan-permit-water-withdrawal/33489833/.

Frazier, I. (2017). The vertical farm: Growing crops in the city without soil or natural light. *The New Yorker*, 9 Jan 2017.

German Advisory Council on Global Change. (2016). Humanity on the move: Unlocking the transformative power of cities. http://www.wbgu.de/fileadmin/templates/dateien/veroeffentlichungen/hauptgutachten/hg2016/hg2016_en.pdf.

Global Water Forum. (2012). Water outlook to 2050: The OECD call for early and strategic action. Available from http://www.globalwaterforum.org/2012/05/21/water-outlook-to-2050-the-oecd-calls-for-early-and-strategic-action/.

Goodman, P., & Goodman, P. (1947). *Communitas*. Chicago: University of Chicago Press.

Gould, K. A., & Lewis, T. L. (2016). *Green gentrification: Urban sustainability and the struggle for environmental justice*. New York: Routledge.

Grassi, M., Rizzo, L., & Farina, A. (2013). Endocrine disruptors compounds, pharmaceuticals and personal care products in urban wastewater: Implications for agricultural reuse and their removal by adsorption process. *Environmental Science and Pollution Research, 20*(6), 3616–3628.

Haarstad, H. (2016). Who is driving the 'smart city' agenda? Assessing smartness as a governance strategy for cities in Europe. In A. Jones et al. (Eds.), *Services and the green economy* (pp. 199–218). London: Palgrave Macmillan.

Hagelaar, G. J., & Van der Vorst, J. G. (2001). Environmental supply chain management: Using life cycle assessment to structure supply chains. *The International Food and Agribusiness Management Review, 4*(4), 399–412.

Hoff, H. (2011). *Understanding the nexus. Background paper for the Bonn2011 Conference: The water, energy and food security nexus*. Stockholm: Stockholm Environment Institute.

ICIC. (n.d.). Recommendations for the City of Boston. www.icic.org.

Ilieva, R. T. (2016). *Urban food planning: Seeds of transition in the global north*. New York: Routledge.

Jarosz, L. (2008). The city in the country: Growing alternative food networks in Metropolitan areas. *Journal of Rural Studies, 24*(3), 231–244.

Jeffries, N. (2017). Organic matters. *Circulate*. http://circulatenews.org/2017/03/organic-matters/.

Johannessen, A., & Wamsler, C. (2017). What does resilience mean for urban water services? *Ecology and Society, 22*(1), 1. https://doi.org/10.5751/ES-08870-220101.

Lafer, G. (2017). *The one percent solution: How corporations are remaking America, one state at a time*. Ithaca: Cornell University Press.

Lazarova, V., & Bahri, A. (2004). *Water reuse for irrigation: Agriculture, landscapes, and turf grass*. Boca Raton, FL: CRC Press.

Leck, H., et al. (2015). Tracing the water-energy-food nexus: Description, theory and practice. *Geography Compass, 9*(8), 445–460.

Lee, H. L., & Billington, C. (1992). Managing supply chain inventory: Pitfalls and opportunities. *Sloan Management Review, 33*(3), 65.

Lerma Montero, I. (2017). Chapter 11: Landfill culture: Some implications to degrowth. In *Transitioning to a post-carbon society*. London: Palgrave Macmillan.

Loftus, A. J., & McDonald, D. A. (2001). Of liquid dreams: A political ecology of water privatization in Buenos Aires. *Environment and Urbanization, 13*(2), 179–199.

Ma, S., Goldstein, M., Pitman, A. J., et al. (2017). Pricing the urban cooling benefits of solar panel deployment in Sydney, Australia. *Scientific Reports, 7*, 43938. http://www.nature.com/articles/srep43938.pdf.

Mack, E. A., & Wrase, S. (2017). A burgeoning crisis? A nationwide assessment of the geography of water affordability in the United States. *PLoS One, 12*(1). https://doi.org/10.1371/journal.pone.0169488.

Mansuy, J. (2016). Food waste: The actions of public and private actors globally. https://ssrn.com/abstract=2903253.

Massey, B. (2017). D.C.'s urban farms wrestle with gentrification and displacement. *Civil Eats*.

Masten, S., Davies, S., & McElmurry, S. P. (2016). Flint water crisis: What happened and why? *Journal of the American Water Works Association, 108*(12), 22–34.

Matos, S., & Hall, J. (2007). Integrating sustainable development in the supply chain: The case of life cycle assessment in oil and gas and agricultural biotechnology. *Journal of Operations Management, 25*(6), 1083–1102.

Middleton, C., & Allen, S. (2014). The (re) discovery of "the Nexus": Political economies and dynamic sustainabilities of water, energy and food security in Southeast Asia. In *Asia Pacific Sociological Association (APSA) conference "Transforming societies: Contestations and convergences in Asia and the Pacific"*, Chiang Mai, Thailand.

Milman, O. (2017). Atlantic City and Miami Beach: Two takes on tackling the rising waters. *The Guardian*, 20 Mar 2017.

Mix, T. L., Raridon, A., & Croff, J. M. (2016). "There is just a stigma here": Historical legacies, food justice, and solutions-based approaches toward urban community resilience. In B. Caniglia, M. Vallee, & B. Frank (Eds.), *Resilience, environmental justice and the city* (p. 79). New York: Routledge.

Molinero, S. (2017). Green business models: The key enabler for smart cities in Europe. *IISD SDG Knowledge Hub*, 31 Jan 2017. http://sdg.iisd.org/commentary/guest-articles/green-business-models-the-key-enabler-for-smart-cities-in-europe/.

Morgan, K., & Murdoch, J. (2000). Organic vs. conventional agriculture: Knowledge, power, and innovation in the food chain. *Geoforum, 31*(2), 159–173.

Murphy, C. (1999, May). *1999 cultivating Havana: Urban agriculture and food security in the years of crisis*. Oakland, CA: IFDP.

Murray, A. (2017). Can green infrastructure really solve Pittsburgh's stormwater problems?. http://www.alleghenyfront.org/can-green-infrastructure-really-solve-pittsburghs-stormwater-problems/.

Nastiti, A., Meijerink, S. V., Oelmann, M., et al. (2017). Cultivating innovation and equity in co-production of commercialized spring water in peri-urban Bandung, Indonesia. *Water Alternatives, 10*(1), 160–180.

Nogrady, B. (2017). Urban heat islands: Cooling things down with trees, green roads, and fewer cars. *The Guardian*.

Norman, K., Sargent, R., & Fanshaw, B. (2016). Shining cities 2016: How smart local policies are expanding solar power in America. *Environment America and Frontier Group*.

Padawangi, R. (2017). Building knowledge, negotiating expertise: Participatory water supply advocacy and service in globalizing Jakarta. *East Asian Science, Technology and Society, 11*(1), 71–90.

Pimentel, D., et al. (2008). Reducing energy inputs in the US food system. *Human Ecology, 36*(4), 459–471.

Ralson, K., & Newman, C. (2015, October). *A look at what's driving lower purchases of school lunches*. Washington, DC: United States Department of Agriculture, Economic Research Service. Amber Waves.

Ray, R., Fisher, D. R., & Fisher-Maltese, C. (2016). School gardens in the city: Does environmental equity help close the achievement gap? *DuBois Review: Social Science Research On Race, 13*(2), 379–395.

Reese, A. M. (2016). The roots of food inequalities. Gravy, Winter 2016.

Renting, H., & Wiskerke, J. S. C. (2010). New emerging roles for public institutions and civil society in the promotion of sustainable local agro-food systems. In *9th European IFSA symposium*, Vienna, Austria.

Riemenschneider, C., et al. (2016). Pharmaceuticals, their metabolites, and other polar pollutants in field-grown vegetables irrigated with treated municipal wastewater. *Journal of Agricultural and Food Chemistry, 64*(29), 5784–5792.

Rizet, C., et al. (2008). Chaînes logistiques et consommation d'énergie: cas des meubles et des fruits et légumes.

Roberts, W. (2017). Three ways to urban agricultures: Digging into three very different offerings from three new books. *Medium*, 30 Mar 2017.

Robinson, J. L. (2017). Chapter 9: Rethinking the politics of water: Risk, resilience and the rights of future generations. In B. S. Caniglia, M. Vallee, & B. F. Frank (Eds.), *Resilience, environmental justice and the city* (pp. 159–176). New York: Routledge.

Romero-Lankao, P., McPhearson, T., & Davidson, D. (2017). The food-energy-water nexus and urban complexity. *Nature Climate Change, 7*, 233–235.

Rueb, E. S. (2017). How New York city gets its electricity. *New York Times*. New York 101.

Schepelmann, P. (2016). Regional governance for climate protection: Transition management in North-Rhine Westphalia, Germany.

Schlich, E., et al. (2006). La consommation d'énergie finale de différents produits alimentaires: un essai de comparaison. *Le Courrier de l'environnement de l'INRA, 53*(53), 111–120.

Skocpol, T. (1980). Political response to capitalist crisis: Neo-Marxist theories of the state and the case of the new deal. *Politics and Society, 10*(2), 155–201.

Sofia, G., Roder, G., Dalla Fontana, G., et al. (2017). Flood dynamics in urbanised landscapes: 100 years of climate and humans' interaction. *Scientific Reports, 7*, 40527.

Solar Power in America. (2016). *Frontier Group and Environment America Research and Policy Center*. http://environmentamerica.org/sites/environment/files/reports/EA_shiningcities2016_scrn.pdf.

Stern, P., Kalof, L., Dietz, T., et al. (1995). Values, beliefs, and pro-environmental action: Attitude formation toward emergent attitude objects. *Journal of Applied Social Psychology, 25*(18), 1611–1636.

Stutz, B. (2017). A vulnerable community braces for the impacts of sea level rise. *Yale Environment 360*. http://e360.yale.edu/features/a-vulnerable-community-braces-for-the-impacts-of-sea-level-rise.

Swyngedouw, E. (2005). Dispossessing H_2O: The contested terrain of water privatization. *Capitalism Nature Socialism, 16*(1), 81–98.

Tachet R, Santi P et al. (2016). Revisiting street intersections using slot-based systems. PLoS One, 11(3): e0149607.

Tan, K. C. (2001). A framework of supply chain management literature. *European Journal of Purchasing & Supply Management, 7*(1), 39–48.
Tran, Q. K., Schwabe, K. A., & Jassby, D. (2016). Wastewater reuse for agriculture: Development of a Regional Water Reuse decision-support Model (RWRM) for cost-effective irrigation sources. *Environmental Science & Technology, 50*(17), 9390–9399.
Travers, J. (2017). Growing agrihoods: The next frontier in urban revitalization. *Civil Eats*, 30 Jan 2017.
Trienekens, J. H., et al. (2012). Transparency in complex dynamic food supply chains. *Advanced Engineering Informatics, 26*(1), 55–65.
UEMI. (2017). *Urban Electric Mobility Initiative*. http://www.uemi.net/.
United Nations. (2015). Sustainable development goals: 17 goals to transform our world. Available from http://www.un.org/sustainabledevelopment/sustainable-development-goals/.
United Nations Demographic Yearbook. (2016). *Table 6 Technical Notes*. Available from: https://unstats.un.org/unsd/demographic-social/products/dyb/documents/dyb2016/Notes06.pdf.
Walton, S. V., Handfield, R. B., & Melnyk, S. A. (1998). The green supply chain: Integrating suppliers into environmental management processes. *Journal of Supply Chain Management, 34*(1), 2–11.
Wellinghoff, J., & Weissman, S. (2015). The right to self-generate as a grid-connected customer. *Energy Law Journal*. http://scholarship.law.berkeley.edu/cgi/viewcontent.cgi?article=1010&context=cleepubs.
White, R., & Stirling, S. A. (2014). Resilience in sustainable lifestyles research project. https://www.researchgate.net/profile/Andy_Stirling/publication/282358849_Resilience_in_Sustainable_Lifestyles_Research_Project_Report/links/560e5c2308ae6b29b498658c.pdf.
World Energy Council. (2013). World energy scenarios: Composing energy futures to 2050. Available from http://www.worldenergy.org/wp-content/uploads/2013/09/World-Energy-Scenarios_Composing-energy-futures-to-2050_Executive-summary.pdf.
Zahran, S., McElmurry, S., et al. (2018). Assessment of the legionnaires' disease outbreak in flint Michigan. *Proceedings of the National Academy of Sciences of the United States of America*. www.pnas.org/cgi/doi/10.1073/pnas.1718679115.
Zeuli, K., & Niijhuis, A. (2017). The resilience of America's urban food systems: Evidence from five cities. http://icic.org/wp-content/uploads/2017/01/Rockefeller_ResilientFoodSystems_FINAL_post.pdf?x96880.
Zeuli, K., Niijhuis, A., & Murphy, P. (2015). Resilient food systems, resilient cities: Recommendations for the City of Boston.
Zhuikov, M. (2017). Attacking urban water contamination through science and policy. *Aquatic Sciences Chronicle, 1*, 1–3.

Further Reading

Brears, R. C. (2019). *The green economy and the water–energy–food nexus*. London: Palgrave Macmillan.
Chirisa, I., & Bandauko, E. (2015). African cities and the water-food-climate-energy nexus: An agenda for sustainability and resilience at a local level. *Urban Forum, 26*(4), 391–404. Springer Netherlands..
Duvernoy, I. (2018). Alternative voices in building a local food policy: Forms of cooperation between civil society organizations and public authorities in and around Toulouse. *Land Use Policy, 75*, 612–619.
Fan, J.-L., Kong, L.-S., Wang, H., & Zhang, X. (2019). A water-energy nexus review from the perspective of urban metabolism. *Ecological Modelling, 392*, 128–136.
Gondhalekar, D., & Ramsauer, T. (2017). Nexus city: Operationalizing the urban water–energy–food nexus for climate change adaptation in Munich, Germany. *Urban Climate, 19*, 28–40.

Hawken, P. (Ed.). (2017). *Drawdown: The most comprehensive plan ever proposed to reverse global warming*. UK: Penguin.

Heard, B. R., Miller, S. A., Liang, S., & Xu, M. (2017). Emerging challenges and opportunities for the food–energy–water nexus in urban systems. *Current Opinion in Chemical Engineering, 17*, 48–53.

Irazábal, C. (2017). *City making and urban governance in the Americas: Curitiba and Portland*. New York: Routledge.

Kurian, M. (2017). The water–energy–food nexus: Trade-offs, thresholds and transdisciplinary approaches to sustainable development. *Environmental Science & Policy, 68*, 97–106.

Lord, C., & Norquist, K. (2010). Cities as emergent systems: Race as a rule in organized complexity. *Environmental Law, 40*, 551–597.

Mallach, A. (2018). *The divided city: Poverty and prosperity in urban America*. Washington, DC: Island Press.

Newell, J., & Ramaswami, A. (Eds.). (2017). Focus on urban food–energy–water systems: Interdisciplinary, multi-scalar and cross-sectoral perspectives. *Environmental Research Letters, 12*.

Owen, A., Scott, K., & Barrett, J. (2018). Identifying critical supply chains and final products: An input-output approach to exploring the energy–water-food nexus. *Applied Energy, 210*, 632–642.

Pothukuchi, K., Arrowsmith, M., & Lyon, N. (2018). Hydraulic fracturing: A review of implications for food systems planning. *Journal of Planning Literature, 33*(2), 155–170.

Romero-Lankao, P., McPhearson, T., & Davidson, D. J. (2017). The food–energy–water nexus and urban complexity. *Nature Climate Change, 7*(4), 233.

Schlör, H., Venghaus, S., & Hake, J.-F. (2018). The FEW-Nexus city index–measuring urban resilience. *Applied Energy, 210*, 382–392.

Spahn, A. (2018). "The first generation to end poverty and the last to save the planet?"—Western individualism, human rights and the value of nature in the ethics of global sustainable development. *Sustainability, 10*(6), 1853.

Vogt, C., Zimmermann, M., & Brekke, K. (2014). *Operationalizing the urban NEXUS: Towards resource-efficient and integrated cities and metropolitan regions*. Germany: GIZ and ICLEI.

Chapter 19
Watersheds at the Nexus

Robert B. Richardson

19.1 Introduction

The interdependence of food, energy, and water systems is often revealed in watersheds or catchment areas where land use, population growth, and increasing demand for food, energy, and water exert pressure on watershed ecosystem services that can lead to the environmental degradation and the depletion of natural capital. The impacts of global climate change add to the complexity of the food–energy–water nexus, which highlights the importance of an integrated approach to watershed governance, management, and planning.

Watersheds are a critical scale at which to examine the nexus of food, energy, and water systems, and depending upon the scale and type of watershed, the intersection of these systems pose an array of challenges for governance of natural resources.

From a systems perspective, watersheds include not only the movement of water into a drainage basin but also the aquatic ecosystems and surrounding landscapes. Changes to the hydrological cycle—including both water withdrawals and releases—affect the structure and function of aquatic ecosystems by altering streamflow characteristics and natural rates of sedimentation. Changes in land use, such as the clearing of forested land for agriculture, will lead to increased erosion, sedimentation, and nutrient runoff that degrade both terrestrial and aquatic ecosystem services through soil loss and water pollution.

R. B. Richardson (✉)
Department of Community Sustainability, College of Agriculture & Natural Resources, Michigan State University, East Lansing, MI, USA
e-mail: rbr@msu.edu

P. Saundry, B. L. Ruddell (eds.), *The Food-Energy-Water Nexus*, AESS Interdisciplinary Environmental Studies and Sciences Series,
https://doi.org/10.1007/978-3-030-29914-9_19

There are inherent challenges related to the downstream impacts of upstream land uses, including the production of food and energy. In many cases, water is an input to the production of food (e.g., irrigation for agriculture, water for livestock) and energy (e.g., hydroelectric dams), and these activities may negative have impacts on water quality and aquatic ecosystem services, such as habitat for fish and other organisms.

The challenges associated with resource access rights, designated land uses, and protection of ecosystem services become more difficult when watersheds share multiple jurisdictional borders and different systems of governance and protection. Each authority may have unique policies that are driven by economic development priorities as well as local perceptions and cultural norms that are often disconnected from the structure and function of ecosystems. Addressing these challenges requires interdisciplinary scientific approaches in research as well as a synthesis and integration of cross-jurisdictional and interdisciplinary perspectives for effective cross-border watershed governance. The capacity for such integration varies widely across economic, social, cultural, and political contexts.

This chapter examines the food–energy–water nexus through three case studies of watersheds that are subject to different pressures from interactions at the nexus. Each of the watersheds also shares multiple national or regional borders and requires systems of cross-border governance and resource management. The three case studies include:

(a) the Great Lakes of North America, which is the largest freshwater system in the world, but faces increasing pressure from nutrient runoff from agriculture, mercury contamination from coal-fired power plants, and significant withdrawals of water for thermoelectric energy production;
(b) the Amazon River Basin in South America, which is the largest river system on Earth, but faces increasing pressure from the effects of deforestation, the development of hydroelectric dams, and the resulting collapse of fisheries; and
(c) the Lake Victoria Basin in East Africa, one of the most species-rich lakes in the world in terms of diversity, but faces increasing pressure from population growth, the effects of deforestation, rapidly increasing eutrophication, and declining productivity of fisheries.

In each case, the pressures on the watersheds and their ecosystem services are related to the linkages in the food–energy–water nexus, but because of the differences in geological, ecological, economic, and sociopolitical contexts, the challenges for cross-border governance vary widely. Each case study describes the biophysical context; the watershed challenges at the nexus of food, energy, and water systems; and a description of transboundary governance of each watershed. Capacity for effective governance is assessed for each watershed across a range of criteria including institutional capacity, coordination, distributional issues, social mobility, and the political-economic context.

19.2 The Great Lakes Region of North America

19.2.1 Background

The Great Lakes region of North America is part of the St. Lawrence River Watershed, which is among the 20 largest drainage basins in the world. The watershed drains to the Atlantic Ocean and spans two countries, including eight states in the USA and two provinces in Canada, as well as 60 major cities. The North American Great Lakes Basin is situated at an altitude that ranges between 74 m (Lake Ontario) and 183 m (Lake Superior) (ILEC 2005). A map of the North American Great Lakes Basin is presented below in Fig. 19.1.

The Laurentian Great Lakes of North America are the largest surface freshwater system on the Earth, containing approximately 84% of the continent's surface fresh water, and about 21% of the world's supply of surface fresh water. The Great Lakes drain through the St. Lawrence River, which flows about 1200 km^2 before emptying into the Gulf of St. Lawrence, the largest estuary in the world. While the drainage area of the Great Lakes covers more than 520,000 km^2, the drainage area of the entire Laurentian system (including the St. Lawrence River) is approximately one million km^2. In 2013, the Great Lakes supplied over 42 billion gallons of water per day for residential, agricultural and industrial use to more than 33 million people

Fig. 19.1 North American Great Lakes Basin. (Source: Great Lakes Information Network)

Table 19.1 Physical features of the North American Great Lakes

Characteristic	Lake Superior	Lake Michigan	Lake Huron	Lake Erie	Lake Ontario	Total
Average depth (m)	147	85	59	19	86	
Maximum depth (m)	406	282	229	64	244	406
Volume (km^3)	12,100	4920	3540	484	1640	22,684
Water area (km^2)	82,100	57,800	59,600	25,700	18,960	244,160
Land drainage area (km^2)	127,700	118,000	134,100	78,000	64,030	521,830
Shoreline length (km)	4385	2633	6157	1402	1146	17,017
Retention time (years)	191	99	22	3	6	

Sources: ILEC (2005) and EPA (2016)

(Great Lakes Commission 2014). Beyond this reliance on the Great Lakes for water supply, residents of the basin and visitors from far afield also realize recreational, ecological and cultural benefits from the basin's resources. Physical features of the North American Great Lakes Basin are presented in Table 19.1.

Evaporation on the Great Lakes varies seasonally, where evaporation rates are highest in late fall and early winter when weather conditions are colder. Evaporation is not directly driven by warm air temperatures, but rather by warm water temperatures.

The North American Great Lakes support fisheries that generate approximately $4.3 billion per year, and the Lakes provide water to more than 40 million people, including five major urban centers (i.e., Chicago, Detroit, Toledo, Cleveland, and Buffalo), and approximately 10% of the US population, and 30% of the Canadian population. The Basin-wide average population density rate is 43 persons per km^2 (ILEC 2005), but population density is greatest in and around the approximately 60 major cities in the Basin.

With a lengthy history of proposals for large-scale diversions of water from the Great Lakes, US states and Canadian provinces have worked for many years to respond to this risk. The Great Lakes Charter adopted in 1985 was the first formal agreement in the basin for how diversions would be restricted, and this ultimately led to the 2005 Great Lakes-St. Lawrence River Basin Sustainable Water Resources Agreement and the 2008 Great Lakes-St. Lawrence River Basin Compact (see Sect. 8.1.1). Through these agreements, the Great Lakes states and Canadian provinces have committed to water conservation and water use efficiency goals and formal limits on large diversions. However, these regional agreements are vulnerable to national-level pressures in the face of water scarcity elsewhere.

19.2.2 *Watershed Issues at the Food–Energy–Water Nexus*

The Great Lakes Basin faces threats from a range of anthropogenic stressors, including land use and landscape change, as well as invasive species, point source pollution, non-point source pollution, atmospheric deposition, and climate change.

The greatest threat to water quality is related to surface water contamination by municipal discharge, industrial pollution, and nutrient runoff from agricultural landscapes.

Although the Great Lakes are part of a single system, the challenges at the nexus of food, energy, and water systems are different for each lake. Lake Superior is characterized by low nutrient concentrations, high oxygen content, and a sparse growth of algae and other organisms. It is the largest, deepest, and coldest of the Great Lakes, and most of its basin is forested and sparsely populated, with little agricultural activity, and thus it has had relatively lower anthropogenic impacts.

The other four lakes exist in basins with deeper, more fertile soils, and they have undergone significant land use changes, from the clearing of forested land to intensive agriculture, industrialization, and urbanization. Lake Michigan is the second largest of the Great Lakes, and it ranges from less developed areas in the north to the most urbanized regions of the Great Lakes in the southern basin, including Chicago and Milwaukee.

The more temperate southern region of the Great Lakes Basin is well suited for agriculture because of a more favorable climate and more fertile soils. However, nutrient runoff from farms in Indiana, Michigan, and Ohio has led to severe contamination in Lakes Michigan, Huron, and Erie, including areas that have experienced algal blooms from nitrogen and phosphorous loading, widespread algae die-offs, and compromised public water utility systems (see Sect. 4.5.1 for a case study on incorporating behavioral heterogeneity in FEW systems in the Lake Erie watershed). Public health advisories about consumption of toxin-contaminated fish are common throughout the region.

Water quality issues are further aggravated by altered hydrology. In a water-abundant region, the ecological systems reflect both water abundance and seasonal fluctuations in-stream flows. Because the ecosystems have evolved in a water-rich environment, their health is intricately connected to water availability and specific hydrologic regimes. As a result, drops in lake water levels and stream flows are more harmful than they might be in areas that have evolved under more water-stressed conditions. Drawdowns in groundwater for agricultural irrigation and municipal consumption are one stressor, but climate uncertainty is adding further complexity because the region could end up with a warmer and wetter climate that leads to increased river flows, flooding, channel adjustments, bank erosion, and increased sedimentation.

Contamination and hydrologic changes compromise the ability of ecosystems to deal with other stressors. As a result, the lakes are particularly susceptible to invasive species introductions. Since 1840, more than 180 invasive species have been identified in the North American Great Lakes, each of which has disrupted the balance of aquatic ecosystems (EC and EPA 2014). Because the Great Lakes are historically nutrient poor, small increases in nutrient loadings lead to effects that are of a greater magnitude. On the other hand, because the Great Lakes Basin is water-abundant, small changes in flows lead to small responses that are difficult to detect or appreciate until a threshold has been crossed.

Related to the food–water nexus, excessive inputs of phosphorus from agricultural runoff have resulted in eutrophication and algal growth in some areas of the Great Lakes. Efforts that began in the 1970s to reduce phosphorus loadings were largely successful, but in some locations, phosphorus loads have been increasing again, and an increasing share of the phosphorus is a dissolved form that is biologically available to fuel nearshore algal blooms (EC and EPA 2009).

Contaminated sediments are a significant and persistent threat to the Great Lakes basin. Although the discharge of toxic chemicals to the Great Lakes has declined in recent decades, persistent high concentrations of contaminated sediments in the bottoms of rivers and harbors pose potential risks to aquatic organisms, wildlife, and humans (EPA 2016). Degraded harbor and tributary areas in the Great Lakes basin have been identified and labeled as Areas of Concern (AOCs). The Great Lakes Water Quality Agreement between the USA and Canada defines AOCs as areas where significant impairment of beneficial uses has occurred as a result of human activities at the local level (GLWQP 2012).

The problem of these buried contaminated sediments is a legacy of decades of urban and agricultural non-point source runoff, industrial and municipal discharges, and combined sewer overflows, all with some linkages to the food–energy–water nexus. To address these impairments, each AOC developed a Remedial Action Plan that identifies contaminated bottom sediments as a significant problem that must be addressed to restore beneficial uses (EPA 2016).

Since the 1970s, concentrations of historically regulated contaminants such as polychlorinated biphenyls (PCBs), dichlorodiphenyltrichloroethane (DDT), and mercury have generally declined in most monitored fish species. However, mercury levels in fish have been slowly increasing since 1990, after years of steady decline (EC and EPA 2009). Atmospheric deposition from coal-powered power plant and municipal waste incinerator emissions is the largest source of mercury pollution to the Great Lakes. Emissions from coal-fired power plants attributed to as much as 57% of the mercury pollution (Evers et al. 2011).

Mercury in the environment does not break down over time, and due to its bioaccumulative nature, even small concentrations of mercury in water can have large impacts on food webs. Mercury concentrations can be magnified by one to ten million times in fish and fish-eating birds like loons, reducing their growth and reproductive success (Evers et al. 2011).

At the energy–water nexus, increasing biofuels production, decommissioning of coal power plants, and shifts to nuclear energy are changing the stresses on the water in the Great Lakes. This is complicated by particular invasive species that have different effects on the lakes, including Zebra mussels clogging water pipes and lampreys altering fish populations.

Additionally, legacy toxic contamination associated with the history of industrial activity in the region remains a problem, and the US states and Canadian provinces continue to work toward resolving these issues under the Binational Toxics Strategy, adopted in 1997.

19.2.3 Watershed Governance

Given the multiple units of government with interests in the North American Great Lakes Basin, numerous treaties, laws, conventions, compacts, and other agreements have implemented over more than a century of watershed management.

Early efforts for regional collaboration between the USA of America and Canada resulted in the International Boundary Waters Treaty of 1909. This treaty which was ratified by the USA and Great Britain (which had dominion over Canada at the time), and established the International Joint Commission (IJC) consisting of Americans and Canadians to oversee any issue related to shared waters on the border between the USA and Canada.

The IJC works to prevent and resolve disputes between the USA and Canada, particularly those related to transboundary waters and pursues the common good of both countries as an independent and objective advisor to the two governments. The IJC has jurisdiction over issues related to fishing rights, water diversion, shipping, and other transportation rights, the building of dams and bridges, other shared water uses, and concerns over water pollution.

The IJC is a science-based, binational organization that works through consensus and relies on the impartial judgment of three appointed commissioners. Experts from both countries serve on technical boards for the Commission and carry out studies and fieldwork in response to requests from the countries. In 1912, the IJC conducted the largest spatial water quality assessment ever undertaken to investigate sources of bacterial threats to drinking water. While recommendations to prevent pollution and infrastructure alterations were implemented, after 100 years point and non-point sources of contamination remain a concern.

Water quality in the Great Lakes is governed by the Great Lakes Water Quality Agreement (GLWQA), which is a commitment between the USA and Canada to restore and protect the waters of the Great Lakes (EPA 2017). The Agreement provides a framework for identifying binational priorities and implementing actions that improve water quality. The USA and Canada first signed the Agreement in 1972, and it has been amended in 1983, 1987, and 2012 to enhance water quality programs that ensure the "chemical, physical, and biological integrity of the Waters of the Great Lakes" (GLWQP 2012, p. 5).

The 2012 agreement will facilitate the USA and Canadian action on threats to Great Lakes water quality and includes strengthened measures to anticipate and prevent ecological harm. New provisions address aquatic invasive species, habitat degradation and the effects of climate change, and support continued work on existing threats to people's health and the environment in the Great Lakes Basin such as harmful algae, toxic chemicals, and discharges from vessels.

Efforts continue at all levels of government to address water quality problems driven by multiple factors including the transformation of the landscape and critical cumulative impacts due to climate change.

Finally, the North American Great Lakes are also governed by the Great Lakes-St. Lawrence River Basin Water Resources Compact of 2008, which protects the lakes from consumptive uses and diversions. The Great Lakes-St. Lawrence River Basin Water Resources Compact became State and federal law in 2008 and established the Great Lakes-St. Lawrence River Basin Water Resources Council. Each of the eight Great Lakes State legislatures in the USA ratified the Compact and Congress provided its consent for this historic accord. The Compact details how the States will work together to manage and protect the Great Lakes-St. Lawrence River Basin. It also provides a framework for each State to enact programs and laws protecting the Basin.

In general, governance of the Great Lakes Basin is characterized by high levels of institutional capacity and coordination in an industrialized political economy. While the Boundary Waters Treaty of 1909 and the Great Lakes-St. Lawrence River Basin Water Resources Compact of 2008 create a partially formed architecture for transboundary management of the waters of the Great Lakes Basin, voluntary, cooperative, and informal relationships may be required to adequately protect the ecosystems and quality of life in the Great Lakes region.

Extensive networks of informal governance organizations compete and collaborate to affect the hydro-ecological systems of the North American Great Lakes. Nevertheless, legal protections and regulations will be increasingly essential as the effects of global climate change increase the challenges for all stakeholders.

19.3 The Amazon River Basin of South America

19.3.1 Background

The Amazon Basin is the part of South America drained by the Amazon River and its tributaries. South America's Amazon River is the largest river system on Earth in terms of discharge of water, and much of the basin is facing increasing pressure from infrastructure development and the resulting collapse of fisheries. With an average drainage rate of approximately 200,000 m^3/s, the Amazon is larger than the next six largest rivers combined, and it is equivalent to approximately 18% of the world's river input to the oceans (Richey et al. 1989).

The Amazon River was previously thought to be the second-longest river in the world, after the Nile in sub-Saharan Africa. The longest upstream extension of the Amazon River has long been thought to be the source of the Río Apurímac drainage in the Andes Mountains of southwestern Peru. However, recent research has traced the origin of the Amazon River to the headwaters of the Mantaro River, which joins Río Apurímac and other tributaries to form the Ucayali River, which in turn meets with the Marañón River in Peru as the main tributary of the Amazon. These findings imply that when the source location is defined as the most distant source of water in the Amazon basin, the new source extends the length of the Amazon River by about 284 km (176 miles), making it 105 km (65 miles) longer than the Nile. This brings the total length of the Amazon River to 6800 km (4225 miles) long. The total length of the Nile River is 6695 km (4160 miles).

Fig. 19.2 Map of the Amazon River Basin. (Source: Wikimedia Commons)

The Amazon drainage basin covers an area of about 7.5 million km^2 (approximately 2.9 million mi^2) or roughly 40% of the South American continent. The Amazon Basin includes drainage areas in eight countries, including Bolivia, Brazil, Colombia, Ecuador, Guyana, Peru, Suriname, and Venezuela. A map of the Amazon River Basin is presented in Fig. 19.2.

The Amazon Basin is considered one of the world's most important ecological systems, primarily because it includes the largest remaining area of tropical rainforest and one of Earth's greatest collections of biological diversity (Foley et al. 2007). Most of the Basin is covered in dense tropical forest; at 5,500,000 km^2 (approximately 2,100,000 mi^2) in size, Amazonia is the largest rainforest in the world, and the forests of the Amazon Basin play a critical role in the global carbon cycle. Throughout the Basin, the rivers and forests are also important sources of livelihoods for hundreds of indigenous groups and forest-dependent communities.

The Brazilian Amazon is significant because of both environmental and sociodemographic characteristics. It accounts for well over two-thirds of the Amazonian Basin in spatial terms (approximately four million km^2), and its population has grown rapidly through in-migration during the past 50 years. The Brazilian Amazon accounts for more than a quarter of the Earth's rainforests, and within the Amazon Basin, tens of millions of people depend on ecosystem services provided by the forest. Fish in the Amazon River and its tributaries are an important source of protein in the region. Annual floods replenish nutrients in floodplain areas that are used for agriculture. The rivers of the Basin are important vectors for transportation, logging and timber production are major industries in many cities and towns, and collection of non-timber forest products provide important sources of income to support rural livelihoods. The rainforest also helps reduce the risk of fire, helps reduce air pollution, and sequesters carbon dioxide that would otherwise contribute to global greenhouse gas emissions.

19.3.2 Watershed Issues at the Food–Energy–Water Nexus

The ecosystems of the Amazon River Basin provide a wide range of ecosystem goods and services, including habitat, regulation of hydrological flows, erosion control, and food provision, especially migratory fish. Per capita, fish consumption is high in the Amazon Basin relative to other parts of the world. Average per capita fish consumption was estimated to be 94 kg/year among riverine populations in the Amazon Basin, and 40 kg/year among urban populations, rates that are 5.8 and 2.5 times the world average, respectively (Isaac and Almeida 2011). Other important ecosystem services that support and sustain life include food provisioning, water supply, nutrient cycling, soil formation, water regulation, and climate regulation, among others (Foley et al. 2007).

There are numerous challenges at the nexus of food, energy, and water systems in the Amazon River Basin. The river itself supports fisheries that are important sources of food, but it also provides the source of energy for hydroelectric dams, and of water for the irrigation of food and feed crops. The production of food and energy have contributed to economic development in countries throughout the Basin, but those activities have also led to widespread environmental degradation, including deforestation, sedimentation, nutrient runoff, and increased greenhouse gas emissions. Environmental degradation has been further exacerbated by overfishing.

The Amazon River Basin is being rapidly degraded and faces a range of threats, including the impacts of hydroelectric dams, mining, overfishing, and deforestation (see Sects. 13.2.3 and 20.2.1). Much of the Basin is facing increasing pressure from infrastructure development and the resulting collapse of fisheries. Developing countries in the region are rapidly pursuing hydropower infrastructure construction, among other large infrastructure projects, to improve their position in the global economy. Infrastructure development is also an important component of the economic development strategies of these countries, including increasing capacity to produce food, generate energy, provide opportunities for employment and income, and improve the overall quality of life of their citizens.

There has been an unprecedented expansion of hydropower development throughout the Amazon River Basin, and current plans suggest that the pace of development is likely to continue (Anderson et al. 2018). Brazil has 256 large dams in operation or planned, and the country generates nearly 65% of its electricity from hydropower. The country is home to some of the largest dams in the world, including the 14,000-MW Itaipu Dam on the Paraná River, as well as the controversial Belo Monte Dam project currently under construction. There is evidence that expansion of hydropower development in the Andean Amazon has been underestimated. A recent study documented 142 dams in operation or under construction and 160 dams in various stages of planning in Colombia, Ecuador, Peru, and Bolivia (Anderson et al. 2018). There has been little coordination between these countries in their hydroelectric planning, and there is growing concern that they are engaged in a hydropower race where there will be winners and losers. The pace is so fast that institutions and research organizations can barely meet the requirements of the environmental impact statements required, and where data sharing has been guarded by the construction companies due to litigation and protests by civil society and ethnic minorities affected by these dams. The problem is only exacerbated in those cases taking place near the border areas.

The hydrologic consequences of large-scale dams and reservoirs are extensive, as humans have appropriated over 50% of the world's available freshwater runoff, expected to exceed 70% by 2025. Sharp declines in available freshwater drive changes in seasonal river discharge, downstream freshwater habitat, loss of floodplains and even coastal erosion and salinity changes. The consequences for ecosystem structure and composition (e.g., habitat fragmentation, loss of aquatic faunal diversity), and function (e.g., nutrient flows, primary production) are severe. Reservoirs can also be significant sources of greenhouse gas emissions during bacterial decomposition of flooded peatlands and upland forests (Giles 2006; Fearnside and Pueyo 2012). Tropical hydroelectric dams, such as those throughout the Amazon River Basin, emit significant amounts of greenhouse gases, especially methane. Emissions from tropical hydroelectric dams are often underestimated or ignored in greenhouse gas accounting, and they can often exceed emissions of fossil fuels for decades (Fearnside and Pueyo 2012).

The human costs of large dams are also significant, especially for those displaced by reservoirs are only the most visible victims of large dams. The social, cultural, economic, and political disruptions of the involuntary resettlement of displaced populations have costs that are routinely underestimated or ignored (Tortejada et al. 2012; Égré and Senécal 2003). For example, in the Amazon, during the construction of the Tucuruí Dam in 1989, the area of the reservoir ended up being much larger than planned, flooded more communities than planned and led to the resettlement of many more people than originally estimated (Fearnside 1999). Millions more have lost land and homes to the construction of canals, roads, irrigation projects, and industrial development that accompany dams. Others have lost access to clean water, sources of food, and other natural resources in the dammed area. Dams have also been associated with increases in communicable diseases.

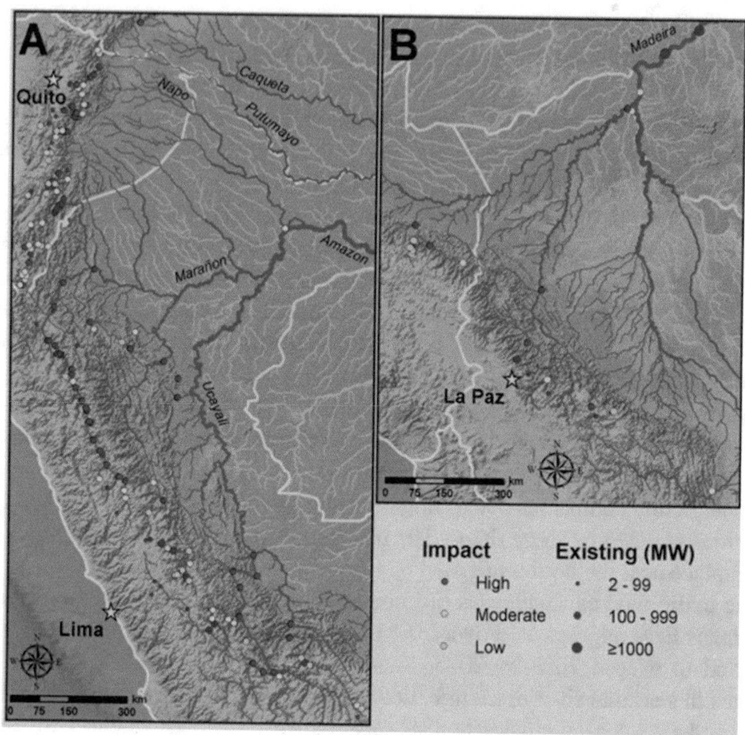

Fig. 19.3 Results of ecological impact analysis of planned dams in the Andean Amazon. (**a**) Results for tributaries originating in the Colombian, Ecuadorian, and northern Peruvian Andes; (**b**) results for tributaries originating in the Bolivian and southern Peruvian Andes. (Source: Finer and Jenkins, 2012)

A recent study involved an ecological impact analysis of planned hydroelectric dams across all six major Andean tributaries of the Amazon River (Caqueta, Madeira, Napo, Marañon, Putumayo, and Ucayali), which span five countries, namely, Bolivia, Brazil, Colombia, Ecuador, and Peru. The framework for the ecological impact analysis focused on river connectivity and forest loss caused by dam-related infrastructure. Based on the classification framework used in the study, 71 (47%) of the planned dams were identified as high impact, 51 (34%) as moderate impact, and 29 (19%) as low impact. A map depicting the results of the ecological impact analysis is presented in Fig. 19.3.

In addition to the negative effects of hydroelectric dams, deforestation is exacerbating the adverse impacts on the rivers of the Amazon Basin, and it is driven largely by clearing land for agriculture. Soybean production and cattle ranching are important agricultural activities in the region, and rising demand for both soybeans and beef in global markets have expanded agricultural production. Deforestation in the floodplains along the mainstream of the lower Amazon River has led to increased soil erosion, which has altered water quality and clarity and caused sedimentation. Deforestation can also increase the variability of water levels between the dry and rainy seasons, affecting plant and wildlife communities.

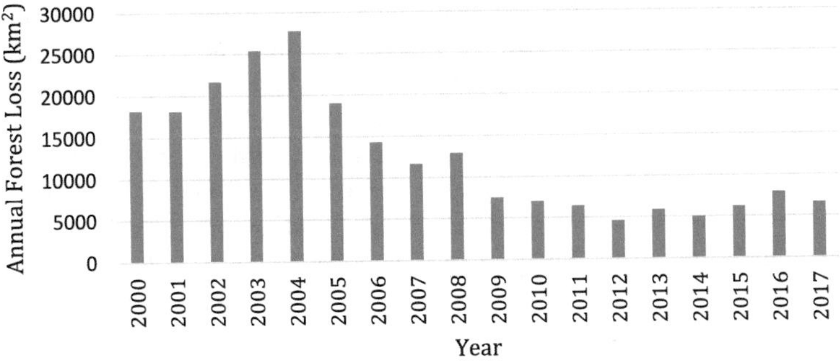

Fig. 19.4 Annual forest loss in the Brazilian Amazon, 2000–2017. (Source: National Institute for Space Research)

Nearly 20% of the forests of the Brazilian Amazon has been cleared since 1970 (Foley et al. 2007). Annual forest loss in the Brazilian Amazon was at its peak in 1995 at nearly 30,000 km^2, and it has fluctuated over the years, rising again to more than 25,000 km^2 in 2003 and 2004, and falling to a historic low of approximately 4570 km^2 in 2012. However, after years of decline, deforestation has risen sharply since 2015, in part due to illegal logging. According to data from the National Institute for Space Research in Brazil, 7989 km^2 of land was cleared between August 2015 and July 2016, 29% higher than the previous year. Annual forest loss in the Brazilian Amazon since 2000 is presented in Fig. 19.4.

Evidence from recent research indicates that land use in the Amazon extends far beyond deforestation of large areas; selective logging and other canopy damage is also pervasive in the region (Foley et al. 2007). Deforestation can cause collateral damage to surrounding forests through enhanced drying of the forest floor, increased incidence of forest fires, and lowered productivity of biomass. The loss of healthy forests can degrade critical ecosystem services, such as carbon sequestration in woody biomass and soils, regulation of water balance and hydrological flows, regulation of regional climate patterns, and amelioration of infectious diseases.

19.3.3 Watershed Governance

The primary formal agreement for governance in the Amazon River Basin is the Treaty for Amazonian Cooperation (also Amazon Cooperation Treaty), signed on July 1978 by Bolivia, Brazil, Colombia, Ecuador, Guyana, Peru, Suriname, and Venezuela, is a legal instrument that recognizes the transboundary nature of the river basin (ACTO 2017). The ACT reaffirms the sovereignty of the countries of the Amazon Basin and encourages, institutionalizes, and guides regional cooperation between them.

It also supports increased scientific and technological research, information exchange, natural resources conservation, freedom of navigation, preservation of cultural heritage, development of transport and communication infrastructure, and enhanced tourism and trade along the borders. The main purpose of the ACT is to promote the development of the Amazon Basin while reconciling the dual goals of economic growth and environmental preservation.

Despite the long-standing ACT, there has been little shared information or cooperation resulting from this treaty to coordinate between the eight countries within the Basin. Concern is growing about a hydropower race in which there will be winners and losers. The pace of project development is so fast that institutions and research organizations can barely meet the requirements of the environmental impact statements, and data sharing has been guarded by the construction companies due to litigation and protests by civil society and ethnic minorities affected by these hydroelectric dam projects. The challenges for cooperation and problems from conflict are only exacerbated in projects developed in cross-border areas.

Thus, institutional capacity for transboundary governance is considered to be low, in part because of low levels of coordination, in a political context that is characterized by centralized governments. The complexity of watershed dynamics is increased by high levels of income inequality and sociocultural heterogeneity.

Land use governance in the Amazon River Basin has undergone significant changes at multiple scales in recent decades. There have been some successful cross-border, national, and municipal governance initiatives. First, the Latin America Water Funds Partnership provides “Water Funds” that help strengthen integrated watershed management and governance of water resources through the financing of long-term conservation initiatives. Funds also provide technical and financial assistance for water management and support local stakeholders in implementing and facilitating information exchange between communities and relevant water management actors (LAWFP 2017).

These funds are leveraging the concept of payment for ecosystem services, and several more such programs are in development throughout the Amazon Basin. The Amazon Waters Initiative, created by the Wildlife Conservation Society to identify important areas for fisheries across the Amazon, is producing useful data needed for effective cross-boundary governance. The Amazon Waters Initiative seeks to promote a vision of the Amazon Basin, in which the region is valued not just for its rich tropical forests and its importance for carbon storage, but for its role as the world’s greatest and most diverse freshwater system.

At the national level, law enforcement capacity has increased, and downstream industries linked to commodity chains responsible for deforestation have begun to monitor some of their suppliers’ impacts on forests. There are also examples of “Green Municipalities” supported by public–private partnerships (such as the Nature Conservancy and The Ministry of the Environment of Brazil) that are helping to reduce illegal deforestation and track and plan for deforestation driven by development and conversion to farmland. Community-based collaborative fisheries management combined with regulation and capacity building for local governments is leading to the recovery of some fishing stocks.

19.4 Lake Victoria Basin in East Africa

19.4.1 Background

Lake Victoria is located in East Africa and is one of several freshwater lakes situated along the Great Rift Valley. The shore of Lake Victoria borders Uganda, Kenya, and Tanzania, and the Lake Victoria Basin also includes tributaries that flow from Rwanda and Burundi. It is situated at an altitude of approximately 1134 m (ILEC 2005). A map of the Lake Victoria Basin is presented in Fig. 19.5.

Lake Victoria is the second-largest freshwater lake in the world by surface area, which is estimated at 68,800 km^2 (Bootsma and Hecky 1993). Lake Victoria Basin receives inflow from 17 tributaries, the largest of which is the Kagera River, which flows out of Burundi along the Rwanda–Burundi, Rwanda–Tanzania, and Tanzania–Uganda borders before eventually emptying into Lake Victoria in Uganda. However, these tributaries contribute less than 20% of the water that enters the Lake, with the remainder provided by rainfall. The Nile River flows out of Lake Victoria near Jinja, Uganda, on the northern shore. Table 19.2 presents several morphometric and hydrological data for Lake Victoria and its catchment area.

Half of the shoreline area borders Uganda, and about one-third of the shoreline is adjacent to Tanzania. Descriptive statistics for the distribution of lake surface area, shoreline area, and tributary area for the five countries in the Lake Victoria Basin are presented in Table 19.3.

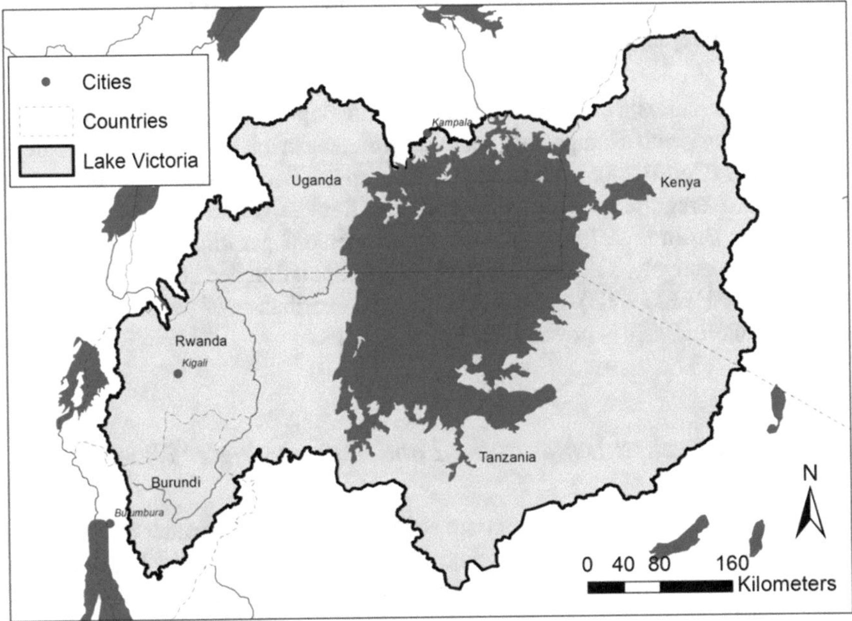

Fig. 19.5 Map of Lake Victoria Basin. (Source: UNEP 2014)

Table 19.2 Morphometric and hydrological data for Lake Victoria

Characteristic	Measurement
Catchment area	195,000 km^2
Lake area	68,800 km^2
Basin area	193,000 km^2
Maximum depth	92 m
Mean depth	40 m
Volume	2760 km^3
Water volume	2750 km^3
Shore length	3440 km
Surface elevation	1134 m
Average rainfall	1791 mm/year
Average evaporation rate	1551 mm/year

Sources: Bootsma and Hecky (1993) and ILEC (2005)

Table 19.3 Lake Victoria surface area, shoreline, and basin area per country

Country	Lake surface area		Shoreline		Tributary	
	km^2	%	km^2	%	km^2	%
Kenya	4113	6	550	17	38,913	21.5
Tanzania	33,756	49	1150	33	79,570	44.0
Uganda	31,001	45	1750	50	28,857	15.9
Rwanda	0	0	0	0	20,550	11.4
Burundi	0	0	0	0	13,060	7.2
Total	68,870	100	3450	100	180,950	100.0

Source: Shepherd et al. (2000)

Lake Victoria Basin supports a rapidly growing population currently estimated at over 35 million, with average population growth rates in the Basin of approximately 3% per year (UNEP 2006). Population density in the Basin is one of the highest in the world; average population density on the Kenyan, Tanzanian, and Ugandan sides of the Basin is 297 persons per km^2, 97 persons per km^2, and 635 persons per km^2, respectively. The Basin-wide average population density rate is 155 persons per km^2 (ILEC 2005). A map depicting the distribution of population density in Lake Victoria Basin is presented below in Fig. 19.6.

19.4.2 Watershed Issues at the Food–Energy–Water Nexus

The Lake Victoria Basin is an important source of fish and freshwater for domestic consumption and agricultural production. The catchment areas of the Lake serve as sinks for agricultural, domestic, and industrial runoff of numerous effluents and residues, which pose a threat to the habitat of fish species. The scientific value of

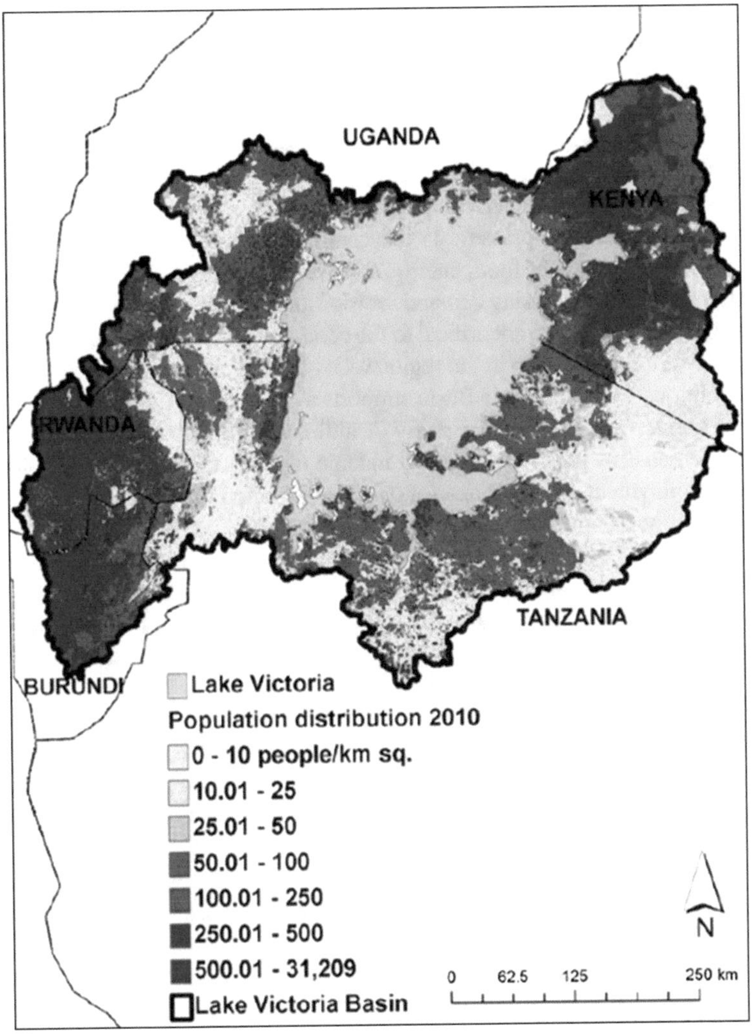

Fig. 19.6 Distribution of population density of Lake Victoria Basin (Bremner et al. 2013)

these lakes is underscored by the richness in fish and invertebrate species, many of which are endemic.

However, many of the watersheds around Lake Victoria and other lakes are highly populated, and conflicts among user groups have increased the demands on the lakes' resources. Consequently, collapses of fish stocks have been observed in most of the lakes due to overfishing. Introductions of exotic fish species such as the Nile perch (*Lates niloticus*), though followed by increases in fishing pressure, have been accompanied by a severe decline and in some cases extinction of native fish species.

Some of the lakes have been invaded by the water hyacinth (*Eichhornia crassipes*). Agricultural activities, deforestation, and the removal of other vegetation in the catchment areas have increased siltation and contributed to the loss of habitats and biodiversity. Increased nutrient inputs from agriculture, sewage and industrial discharges and combustion processes contribute to problems of eutrophication. There are also increased threats of toxic pollution from industrial waste discharge, mining, pesticides, and oil residues and spills.

Arguably, population pressure is the central driver of watershed management challenges at the nexus of food, energy, and water systems in Lake Victoria Basin, primarily due to the increasing demand for food and energy by a growing population.

Fishing and agriculture are critical to the economies of the countries in the Basin. Over 80% of the population in the region relies on land, agriculture, and fishing for their livelihoods. Lake Victoria Basin supports a population of over 35 million people who derive their livelihood directly or indirectly from the resources within the basin. The fisheries produce an annual income of more than $US 600 million and provide employment opportunities for over three million people (Njiru et al. 2008). Approximately 22 million people in the region rely on fish for consumption (Bremner et al. 2013). Yet the region is marked by high rates of rural poverty and food insecurity, which are related to stagnant agricultural productivity, soil degradation, desertification, livestock diseases, declining fisheries, and biodiversity losses. Environmental stresses adversely impact on the ecosystems of the Basin, as well as the region's economy and livelihoods.

There are three primary challenges related to the food–energy–water nexus in the Lake Victoria Basin, and each is associated with unsustainable natural resource uses.

The first is pollution and eutrophication, which is driven by excessive nutrient runoff that also contributes to water quality degradation, oxygen depletion, and fish mortality.

The second is deforestation, which is driven by both the clearing of land for agriculture and the dependence on wood fuels for energy. Deforestation contributes to erosion and sedimentation that degrades water quality in the Lake.

The third is overfishing and declining fish catch, which is driven in part by the other two challenges, along with the dependence on fisheries for food security, livelihoods, and household income.

Pollution and eutrophication in Lake Victoria have led to problems associated with water quality, oxygen depletion, algal blooms, and fish mortality. The primary nutrients of concern are nitrogen and phosphorous, and they originate from three major sources: discharges of sewage and other untreated wastewater, stormwater runoff, and nonpoint-source pollution, mainly from agricultural activities (ILEC 2005).

Deforestation has been documented as a critical environmental concern across all countries in the Basin. Extraction of wood fuels for cooking and heating is an important factor in deforestation throughout the region. All countries an unsustainable energy deficit, as fuelwood is being consumed faster than it is being replaced (LVBC 2007). Most rural households rely on firewood that is collected, and more than 80% of urban households in sub-Saharan Africa use charcoal as their main source of cooking energy.

Charcoal is also a major source of livelihoods for rural households in areas with access to urban markets. More than 96% of Rwandans depend on wood for domestic energy, and 81% of the country's energy consumption is from wood (LVBC 2007).

Throughout the region, wood is increasingly scarce and requires more time and effort to collect. In Kenya and Tanzania, people have to go longer distances to obtain firewood, requiring the use of lower-quality biomass fuels. Urbanization has dramatically increased the consumption of charcoal in cities, and as the population increases, fuelwood utilization is expected to increase, further constraining fuel supply.

Deforestation in the region can also be attributed to clearing land for agriculture. Most of the population of the countries in the Basin is engaged in smallholder agriculture for their livelihoods, and food insecurity and malnutrition remain persistent challenges.

Since subsistence crop production dominates the agricultural economy, the main driver of food insecurity and malnutrition is the limited ability to grow enough food to feed the growing population. Ninety percent of the population of Rwanda is engaged in subsistence farming, and they are not able to grow sufficient amounts of food (LVBC 2007).

In Uganda, the food system is characterized by declining per capita production, soil degradation from unsustainable agricultural practices, and highly variable rainfall. Low agricultural productivity in the region is caused in part by a loss of soil fertility, lack of water for irrigation, and a lack of access to agricultural inputs such as fertilizer and seeds.

Overfishing and declining fish catch is driven in part by the other two challenges, which demonstrates a kind of causal feedback loop, where deforestation, pollution, and eutrophication are contributing to a decline in fish populations, and the dependence on fisheries for food security, livelihoods, and household income by a growing human population puts additional pressure on a declining resource. Increasing fishing pressure is a consequence of population growth, increasing demand for fish for consumption as food, and a growing commercial fishing sector. An audit by the government of Tanzania found that between 2009 and 2011, yields of Nile perch on Lake Victoria fell about 5% (URT 2013).

Lake Victoria is Africa's most important source of inland fishery production, mainly due to the introduction of Nile perch, which transformed Lake Victoria from a primarily artisanal fishery that supported subsistence consumption to a multimillion dollar commercial fishery in just over a decade. Annual fish catch in Lake Victoria increased from approximately 30,000 metric tons in the late 1970s and peaked at 560,000 metric tons in the early 1990s (Njiru et al. 2008). Annual catch has since declined to approximately 500,000 metric tons, and the declining catch has been attributed in part to overfishing.

Both Nile perch and Nile tilapia were introduced to Lake Victoria in the 1950s in order to boost the fishing industry, but it also contributed to the loss of many endemic fish species, particularly cichlids (ILEC 2005; LVFO 2017). Approximately 75% of the Nile perch catch is exported, primarily to Europe and the USA, which contributes to economic growth and foreign exchange. Nile tilapia mainly serves the domestic and regional markets, which supports food security, employment, and household income.

There is evidence that global climate change may also contribute to the decline in fish yields in the region. A study of algae records from the bottom of nearby Lake Tanganyika found that over the last century, warming waters have driven down fish yield in the Lake by 30% (O'Reilly et al. 2003). As surface water is warmed by the sun and air, levels of dissolved oxygen in deep waters in the Lake gradually decline, forcing fish to swim closer to the surface. Nutrients from the depths are prevented from reaching plankton near the surface, leading to a diminishing supply of food for fish. This finding has implications for the importance of considering the dynamic nature of climate change when studying the food–energy–water nexus in watersheds, particularly in semi-tropical regions.

19.4.3 Watershed Governance

The riparian countries in the Basin are taking measures to address these three issues of eutrophication, deforestation, and overfishing, including the creation of the Lake Victoria Basin Commission (LVBC) under the East African Community of six countries, facilitates free trade and cooperation between member states. The LVBC has the objective of restoring the health of the Lake to better support biodiversity and **human welfare**. LVBC was formed as an apex institution responsible for all initiatives in the Basin, including a number of conservation and restoration initiatives and programs designed to eradicate poverty, promote economic growth, and reverse environmental degradation.

A Strategic Action Plan for the Lake Victoria Basin was adopted by the East African Community in April 2007 (LVBC 2007). It is an important point of departure for improved natural resources management and development and includes elements such as investments in reforestation, implementation of policies for fisheries and, reduction of point source pollution. However, many components in the action plan focus on addressing the symptoms of problems related to the food–energy–water nexus in the Basin, rather than on the underlying causes of unsustainable management of natural resources and degraded ecosystems, which is ultimately a lack of alternative livelihoods, a lack of alternative sources of energy, and weak monitoring and enforcement of existing laws and policies.

The challenge of transboundary governance for the Lake Victoria Basin is made difficult by the fact that the riparian areas and watershed lands of most of these lakes lie within more than one political jurisdiction. This challenge is characterized by low levels of coordination in a fragmented political system, in the context of a diverse, rural economy. Water resource and environmental management organizations often do not understand the complementary nature of their responsibilities and do not work together effectively toward shared watershed goals.

Across political boundaries, different political environments, levels of economic development, social norms, and administrative settings lead to different approaches to lake basin management, with detrimental effects on the environment and development status of Basin communities.

19.5 Conclusions

Food, energy, and water systems provide critical resources that sustain and fulfill life on Earth. All humans must have access to these elements of life in a modern society to be empowered with the tools to engage with each other, to self-organize to meet the challenges of climate change, and to educate their children for the future. The sustainable management of these resources and the ecosystems that depend on them is also needed to ensure future prosperity and quality of life.

However, these systems almost always occupy a transboundary space at local, national, regional, and transnational scales. Across these boundaries, there is tremendous variation in the political, economic, and social contexts within which food, energy, and water are governed, including differences in the capacity for self-organization, relative dependence on government, and political traditions. Further aggravating these issues, population and climate change are altering the pressures on components of the systems, leading to conflict over access and use of finite resources.

Clearly, there is a growing interest in—and concern for—the food–energy–water nexus, but the exact nature of this nexus and the new issues it raises are poorly understood. It is clear that agricultural and energy systems depend on water as inputs, and the production of food and energy affects the quality of water. But the boundaries of the food–energy–water nexus are not yet well defined, and the appropriate scale for examining the nexus is not well understood. From the case studies in this chapter, it is clear that land and land use are inextricably linked to the food–energy–water nexus, but it is not explicitly considered as a dimension of the nexus. These case studies highlight the tensions over land use for food and biofuel production, and water quality issues associated with runoff from land use.

Furthermore, there are two competing forces that are shaping national priorities about watershed governance. First, there is a growing concern for the well-being of local communities and issues related to sustainability and equity that contribute to one agenda. Second, there is an influencing role of multinational corporations in shaping national priorities for public policy, and this view would argue that that the problems we face in food, water, and energy systems are too complex and would best be addressed by empowering large-scale agricultural producers, constructing large dams, and generating ever-larger investments in infrastructure in order to produce and use more of all three dimensions of the nexus. The challenges identified in these case studies suggest that perhaps the entire conception of the food–energy–water nexus should be reconsidered.

It is clear that there are tremendous differences in the political, economic, and social contexts within which the food–energy–water nexus manifests in regions where water resources are abundant. In water-abundant areas, it may be more difficult to convince people that they should organize and develop collaborative governance of their water resources and watersheds. Such challenges call for building consensus on what should be changed at both the production and consumption end of food, energy, and water systems in order to ensure their sustainability. Furthermore, these challenges require a shift in thinking about the governance of these systems, away from the incessant pursuit of increased production, and toward more efficient production and use of food, energy, and water resources.

In an effort to synthesize across the various governance issues identified in these three case studies, capacity for effective governance was assessed for each watershed across a range of criteria including institutional capacity, scale, inclusiveness, coordination, coordination, distributional issues, sociocultural heterogeneity, political system, social mobility, and the political-economic context. An overview of this assessment is presented in Table 19.4.

Watersheds are a critical scale at which to examine the nexus of food, energy, and water systems, and the intersection of these systems pose an array of challenges for governance of natural resources, particularly where transboundary waters require high levels of institutional capacity and coordination for effective governance. The interdependence of these systems is often revealed in watersheds or catchment areas where land use, population growth, and increasing demand for food, energy, and water exert pressure on watershed ecosystem services that can lead to environmental degradation and the depletion of natural capital.

The challenges associated with resource access rights, designated land uses, and protection of ecosystem services become more difficult when watersheds share multiple jurisdictional borders and different systems of governance and protection. The impacts of global climate change add to the complexity of the food–energy–water nexus, which highlights the importance of an integrated approach to watershed governance, management, and planning. Emerging challenges call for a new framework for transboundary governance of the food–energy–water nexus in watersheds that are characterized by abundance.

Table 19.4 Overview of assessment of governance issues in each case study

Issue	Great Lakes	Amazon River Basin	Lake Victoria
Institutional capacity	High	Low	Medium
Scale	National-regional	Local-global	Local-regional
Inclusiveness	High	Low	Medium
Coordination	High	Low	Low
Distributional issues	Low	Highly unequal	Moderate
Heterogeneity	Medium	High	High
Political system	Federalism	Centralized	Fragmented
Social mobility	High	Low	Medium
Political economy	Industrialized	Commodity export	Rural-agrarian

In defining a new governance framework for the future of the food–energy–water nexus, it will be important to consider how to restructure the nexus toward four objectives: (1) increased efficiency in the production and use of water and energy; (2) a shift toward energy independence and diversification of sources of energy; (3) increased efficiency in the production, distribution, and consumption of food; and (4) reduced nutrient losses and other agricultural runoff, which currently cause algal blooms and loss of water quality downstream from agricultural fields. Even in the case of abundant water resources, these considerations pose considerable challenges for transboundary governance of food–energy–water resources. Building institutional capacity for transboundary governance that is inclusive, equitable, and well-coordinated is likely to be more effective in the context of abundant water resources such as these three large watersheds.

Key Points

- Watersheds are a critical scale at which to examine the nexus of food, energy, and water systems, and depending upon the scale and type of watershed, the intersection of these systems pose an array of challenges for governance of natural resources.
- The interdependence of these systems is often revealed in watersheds or catchment areas where land use, population growth, and increasing demand for food, energy, and water exert pressure on watershed ecosystem services that can lead to environmental degradation and the depletion of natural capital.
- The challenges associated with resource access rights, designated land uses, and protection of ecosystem services become more difficult when watersheds share multiple jurisdictional borders and different systems of governance and protection.
- Addressing these challenges requires interdisciplinary scientific approaches in research as well as synthesis and integration of cross-jurisdictional and interdisciplinary perspectives for effective cross-border watershed governance.
- An examination of the food–energy–water nexus through three case studies of watersheds reveals that the capacity for cross-border watershed governance varies widely across economic, social, cultural, and political contexts.
- Emerging challenges call for a new framework for transboundary governance of the food–energy–water nexus in watersheds that are characterized by abundance.
- Building institutional capacity for transboundary governance that is inclusive, equitable, and well-coordinated is likely to be more effective in the context of abundant water resources such as these three large watersheds.

Discussion Points and Exercises

1. Describe how watersheds are a critical scale at which to examine the nexus of food, energy, and water systems. How does the intersection of these systems pose an array of challenges for governance of natural resources, particularly where transboundary waters require high levels of institutional capacity and coordination for effective governance?

2. How are institutional capacity, coordination, distributional issues, social mobility, and the political-economic context related to effective governance of watersheds?
3. How is the interdependence of food, energy, and water systems revealed in watersheds or catchment areas?
4. How do land use, population growth, and increasing demand for food, energy, and water exert pressure on watershed ecosystem services?
5. How do land uses, population growth, and increasing demand for food, energy, and water exert pressure on ecosystem services and contribute to environmental degradation and the depletion of natural capital in watersheds?
6. Describe how these pressures on watersheds and their ecosystem services are related to linkages in the food–energy–water nexus.
7. How do these pressures pose challenges for cross-border governance of watersheds at the nexus of food, energy, and water systems?
8. How do the impacts of global climate change add to the complexity of the food–energy–water nexus in the context of watersheds? Describe the implications of an integrated approach to watershed governance, management, and planning.
9. Describe the challenges related to the downstream impacts of upstream land uses in the context of watersheds. How are these challenges related to linkages in the food–energy–water nexus?
10. In the case of the Great Lakes of North America, describe how the pressures on the watersheds and their ecosystem services are related to the linkages in the food–energy–water nexus.
11. In the case of the Great Lakes of North America, what are the challenges for cross-border governance of watersheds? Describe how each challenge is associated with unsustainable natural resource uses.
12. In the case of the Amazon River Basin in South America, describe how the pressures on the watersheds and their ecosystem services are related to the linkages in the food–energy–water nexus.
13. In the case of the Amazon River Basin in South America, what are the challenges for cross-border governance of watersheds? Describe how each challenge is associated with unsustainable natural resource uses.
14. In the case of the Lake Victoria Basin in East Africa, describe how the pressures on the watershed and its ecosystem services are related to the linkages in the food–energy–water nexus.
15. In the case of the Lake Victoria Basin in East Africa, what are the challenges for cross-border governance of watersheds? Describe how each challenge is associated with unsustainable natural resource uses.
16. How do differences in the political, economic, and social contexts of watersheds affect decision-making related to the food–energy–water nexus in regions where water resources are abundant?
17. How do the challenges associated with resource access rights, designated land uses, and protection of ecosystem services become more difficult when watersheds share multiple jurisdictional borders and different systems of governance and protection?

18. The chapter concluded that two competing forces that are shaping national priorities about watershed governance across the three case studies. What are these competing forces? What are their roles in contributing to public policy?

References

ACTO (Amazon Cooperative Treaty Organization). (2017). *Amazon Cooperative Treaty*. Available at: http://otca.info/portal/tratado-coop-amazonica.php#.

Anderson, E. P., Jenkins, C. N., Heilpern, S., Maldonado-Ocampo, J. A., Carvajal-Vallejos, F. M., Encalada, A. C., Rivadeneira, J. F., Hidalgo, M., Cañas, C. M., Ortega, H., Salcedo, N., Maldonado, M., & Tedesco, P. A. (2018). Fragmentation of Andes-to-Amazon connectivity by hydropower dams. *Science Advances, 4*(1), 1642.

Bootsma, H. A., & Hecky, R. E. (1993). Conservation of the African Great Lakes: A limnological perspective. *Conservation Biology, 7*(3), 644–656.

Bremner, J., Lopez-Carr, D., Zvoleff, A., & Pricope, N. (2013). Using new methods and data to assess and address population, fertility, and environment links in the Lake Victoria Basin. In *Proceedings of the 2013 International Union for the Scientific Study of Population (IUSSP)*, Busan, South Korea, August 16–21, 2013.

EC and EPA (Environment Canada and the U.S. Environmental Protection Agency). (2009). *State of the Great Lakes 2009*. Cat. No. En161-3/1-2009E-PDF. EPA 905-R-09-031. Available at: http://binational.net/.

EC and EPA (Environment Canada and the U.S. Environmental Protection Agency). (2014). *State of the Great Lakes 2011*. Cat. No. En161-3/1-2011E-PDF. EPA 950-R-13-002. Available at: http://binational.net/.

Égré, D., & Senécal, P. (2003). Social impact assessments of large dams throughout the world: Lessons learned over two decades. *Impact Assessment and Project Appraisal, 21*(3), 215–224.

EPA (Environmental Protection Agency). (2016). *Physical features of the Great Lakes*. Available at: https://www.epa.gov/greatlakes/.

EPA (Environmental Protection Agency). (2017). *Great Lakes Water Quality Agreement (GLWQA)*. Available at: https://www.epa.gov/glwqa/.

Evers, D. C., Wiener, J. G., Driscoll, C. T., Gay, D. A., Basu, N., Monson, B. A., Lambert, K. F., Morrison, H. A., Morgan, J. T., Williams, K. A., & Soehl, A. G. (2011). *Great Lakes mercury connections: The extent and effects of mercury pollution in the Great Lakes region*. Gorham, Maine: Biodiversity Research Institute. Report BRI 2011-18. 44p.

Fearnside, P. M. (1999). Social impacts of Brazil's Tucuruí Dam. *Environmental Management, 24*(4), 483–495.

Fearnside, P. M., & Pueyo, S. (2012). Greenhouse-gas emissions from tropical dams. *Nature Climate Change, 2*(6), 382–384.

Finer, M., & Jenkins, C. N. (2012). Proliferation of hydroelectric dams in the Andean Amazon and implications for Andes-Amazon connectivity. *PLoS One, 7*(4): e35126.

Foley, J. A., Asner, G. P., Costa, M. H., Coe, M. T., DeFries, R., Gibbs, H. K., Howard, E. A., Olson, S., Patz, J., Ramankutty, N., & Snyder, P. (2007). Amazonia revealed: forest degradation and loss of ecosystem goods and services in the Amazon Basin. *Frontiers in Ecology and the Environment, 5*(1), 25–32.

Giles, J. (2006). Methane quashes green credentials of hydropower. *Nature, 44*, 524–525.

GLWQP (Great Lakes Water Quality Protocol of 2012). (2012). *Protocol amending the agreement between Canada and the United States of America on Great Lakes Water Quality*. Washington, DC. Available at: https://binational.net/glwqa-aqegl/.

Great Lakes Commission. (2014). *Annual report of the Great Lakes regional water use database*. Ann Arbor, MI: Great Lakes Commission. Available at: http://projects.glc.org/waterusedata/pdf/wateruserpt2013.pdf.

ILEC (International Lake Environment Committee). (2005). Managing lakes and their basins for sustainable use. A report for lake basin managers and stakeholders. Kusatsu, Japan: International Lake Environment Committee Foundation.

Isaac, V. J., & Almeida, M. C. (2011). *El consumo de pescado en la Amazonia brasileña (Fish consumption in the Brazilian Amazon)*. COPESCAALC Documento Ocasional. No. 13. Rome: Food and Agriculture Organization of the United Nations. Available at: http://www.fao.org/docrep/014/i2408s/i2408s.pdf.

LAWFP (Latin American Water Funds Partnership). (2017). *What is a water fund?* Available at: http://waterfunds.org/esp/what-is-a-water-fund/.

LVBC (Lake Victoria Basin Commission). (2007). *Strategic action plan for the Lake Victoria Basin*. Kisumu, Kenya: Lake Victoria Basin Commission.

LVFO (Lake Victoria Fisheries Organization). (2017). *Lake Victoria Fisheries: An introduction*. Jinja, Uganda: Lake Victoria Fisheries Organization. Available at: http://www.lvfo.org/.

Njiru, M., Kazungu, J., Ngugi, C. C., Gichuki, J., & Muhoozi, L. (2008). An overview of the current status of Lake Victoria fishery: Opportunities, challenges and management strategies. *Lakes & Reservoirs: Research & Management, 13*, 1–12.

O'Reilly, C. M., Alin, S. R., Plisnier, P.-D., Cohen, A. S., & McKee, B. A. (2003). Climate change decreases aquatic ecosystem productivity of Lake Tanganyika, Africa. *Nature, 424*, 766–768.

Richey, J. E., Nobre, C., Deser, C., & C. (1989). Amazon River discharge and climate variability: 1903 to 1985. *Science, 246*(4926), 101–103.

Shepherd, K., Walsh, M., Mugo, F., Ong, C., Hansen, T. S., Swallow, B., Awiti, A., Hai, M., Nyantika, D., Ombao, D., Grunder, M., Mbote, F., & Mungai, D. (2000). *Improved land management in the Lake Victoria basin: Linking land and lake, research and extension, catchment and lake basin*. Nairobi, Kenya: International Centre for Research in Agroforestry (ICRAF) and Kenya Ministry of Agriculture and Rural Development.

Tortejada, C., Altinbilek, D., & Biswas, A. K. (Eds.). (2012). *Impacts of large dams: A global assessment*. Dordrecht: Springer.

UNEP (United Nations Environment Programme). (2006). *Lake Victoria Basin environment outlook: Environment and development*. Nairobi, Kenya: UNEP.

UNEP (United Nations Environment Programme). (2014). *Development of tools to incorporate climatic variability and change in particular floods and droughts into basin planning processes*. Available at: http://fdmt.iwlearn.org/en.

URT (United Republic of Tanzania). (2013). *Performance audit of the management of fisheries in Lake Victoria*. Dar es Salaam, Tanzania: Ministry of Livestock and Fisheries Development.

Further Reading

Amazon Waters. (2018). *The initiative (Amazon Waters Initiative)*. Lima, Peru: Wildlife Conservation Society. Available at: http://amazonwaters.org/the-initiative.

Bazilian, M., Rogner, H., Howells, M., Hermann, S., Arent, D., Gielen, D., Steduto, P., Mueller, A., Komor, P., Tol, R. S. J., & Yumkella, K. K. (2011). Considering the energy, water and food nexus: Towards an integrated modelling approach. *Energy Policy, 39*(12), 7896–7906.

Benson, D., Gain, A. K., & Rouillard, J. J. (2015). Water governance in a comparative perspective: From IWRM to a 'nexus' approach? *Water Alternatives, 8*(1), 756–773.

Foley, J. A., Asner, G. P., Costa, M. H., Coe, M. T., DeFries, R., Gibbs, H. K., Howard, E. A., Olson, S., Patz, J., Ramankutty, N., & Snyder, P. (2007). Amazonia revealed: Forest degrada-

tion and loss of ecosystem goods and services in the Amazon Basin. *Frontiers in Ecology and the Environment, 5*(1), 25–32.
Lake Victoria Basin Commission. (2007). *Strategic action plan for the Lake Victoria Basin*. Kisumu, Kenya: Lake Victoria Basin Commission.
Lake Victoria Basin Commission and GRID-Arendal. (2017). *Lake Victoria Basin: Atlas of our changing environment*. Arendal, Norway: GRID-Arendal. Available at: http://www.grida.no/publications/328.
Lake Victoria Fisheries Organization. (2017). *Lake Victoria Fisheries: An introduction*. Jinja, Uganda: Lake Victoria Fisheries Organization. Available at: http://www.lvfo.org/.
Moran, E. F., Brondízio, E. S., & VanWey, L. K. (2005). Population and environment in Amazônia: Landscape and household dynamics. In B. Entwisle & P. C. Stern (Eds.), *Population, land use, and environment: Research directions*. Washington, DC: National Academies Press.
Ogutu-Ohwayo, R., Hecky, R. E., Cohen, A. S., & Kaufman, L. (1997). Human impacts on the African Great Lakes. *Environmental Biology of Fishes, 50*(2), 117–131.
Tollefson, J. (2016). Deforestation spikes in Brazilian Amazon. *Nature, 540*, 182.

Chapter 20
Managing Human Conflicts

Lara B. Fowler and Robert T. Caccese

20.1 Introduction

Finding constructive ways to address and manage conflicts that arise in the FEW system is critical. Impacts to food, energy, and water can cause **conflict**, particularly where human populations outstrip local supplies or distribution systems needed to bring in supplies. Shortages aren't the only source of impact: a flood can also have cascading impacts through the food, energy, and water systems (see, e.g., Kiger 2013).

Conflicts can arise in a number of ways. As noted in the introduction, human dimensions of FEW systems are affected by perceptions, cultural and economic motivations, laws and policies operating at different scales, and disputes between various parties with different values, power, and aspirations. Separate or different policies for food, energy, and water can lead to conflicts between communities with a primary interest in one of the food, energy or water components. While international policies usually involve diplomacy and trade arrangements, nations can and have resorted to a military conflict to address food, energy, and water crises or have targeted such systems for attack.

Along with these human dynamics, climate change and extreme events are **threat multipliers**: underlying social instabilities are exacerbated by floods,

L. B. Fowler
Penn State Law, Penn State Institutes of Energy and the Environment, The Pennsylvania State University, University Park, PA, USA
e-mail: lbf10@psu.edu

R. T. Caccese (✉)
Pennsylvania Fish and Boat Commission, Harrisburg, PA, USA

P. Saundry, B. L. Ruddell (eds.), *The Food-Energy-Water Nexus*, AESS Interdisciplinary Environmental Studies and Sciences Series,
https://doi.org/10.1007/978-3-030-29914-9_20

droughts, heat waves or other variation in weather patterns. For example, the Syrian Civil War has been linked in part to a multiyear drought and related crop failures that escalated underlying but already existing tensions into a civil war.

While impacts to food, energy, and water systems can be significant, challenges to these integrated systems can also be the catalyst to bring people together. While the rhetoric of conflict may be strong, reality can also be different, particularly in the case of water. In 1965, Elinor Ostrom coined the term "**common pool resources**" as those that have a "high level of interdependency with other individual interests" within the common pool that people would come together to protect (Ostrom 1965) (see also Ostrom and Hess 2007). More recently, Dr. Aaron Wolf and colleagues have systematically analyzed the rhetoric of "water wars" and found that water is far more often used as a catalyst for cooperation than actual war (Wolf 2007).

Such findings may be critical in navigating conflicts over the FEW dynamics now and into the future. This chapter provides examples where conflict exists for some or all parts of the FEW system, then outlines tools used to manage such conflicts and examples of the tool application.

A note to addressing human conflicts related to food, energy, and water. While people may wish to find a permanent "resolution" to such conflicts, a final resolution is seldom possible; challenges in the FEW systems are only continuing to grow. However, designing a robust system for *managing* these challenges can—and has—helped decrease the potential for future conflicts and provide people with ways of adapting to the changes they face. Collaborative solutions can lead to more collaboration, a dynamic that may prove entirely necessary given the rapidly changing impacts to the FEW system.

20.2 Existing Conflicts Over Food, Energy, and Water Systems

There are numerous examples of conflicts over food, energy, and water ranging from local to international. As discussed elsewhere in this book, a shortage of water may also result in a shortage of power production (either diminished hydro-electric production or an inability to cool a power plant); at the same time, food production is likely to be limited in that same region (see, e.g., Puckett 2016). The more widespread an impact to the system caused by something like as a regional drought, the broader the impacts to the FEW system and the more likelihood of conflict.

The metrics discussed in Chap. 13, data in Chap. 14, and modeling tools in Chap. 15 can all help elucidate potential challenges and provide opportunities for running scenarios to see where conflicts might result. Such tools can also help run scenarios to find solutions in addressing such challenges. The section below highlights examples of domestic (within country) conflicts and regional conflicts that cross state or country boundaries.

20.2.1 Examples of Domestic or Regional Conflicts

Some conflicts remain within a single country or region within a country. For example, the recent 5-year drought in California (2012–2017) had a huge impact throughout the FEW system. Persistent drought and reduced winter snowpack resulted in record low flows, saltwater intrusion, depleted groundwater aquifers, and subsiding land levels. Drought reduced hydroelectric production by 57,000 GWh from 2011–2015. Natural gas combustion replaced the lost hydroelectric potential, costing ratepayers about $2 billion dollars (Gleick 2015). Food production also suffered; producers had limited or no ability to irrigate crops. In many areas, they turned to pumping groundwater, thus increasing electricity costs and exacerbating the decline of groundwater levels despite a new statewide groundwater law enacted in 2014. The impacts of withdrawals remain as groundwater levels had not yet recovered into 2019.

This California drought highlighted significant tension between different communities such as between the agricultural community and those working to protect endangered fish species. Because of litigation and agency regulations, water contracted for delivery in Central and Southern California was left in streams to provide flow for endangered salmon and Delta smelt. Producers bought water from other sources, drilled wells, or let fields to go fallow. In 2016, California traded drought for floods (see e.g., Stockton 2017).

Even with more precipitation, California has swung between both flood and dry conditions. In 2017 and 2018, periods of extremely dry conditions and sparks from electricity lines led to catastrophic wildfires that killed numerous people and led to widespread property damage. The 2017 wildfire season inflicted the most damage to California communities, topped only by the 2018 wildfire season. The continued extremes have placed a significant strain on California's statewide system for food, energy, and water and will continue to do so into the future. In early 2019, California's largest utility, PG&E, declared bankruptcy over its link to recent fires and as fire season began, started shutting down the electricity system to prevent fire, thus raising considerable uncertainty about energy supplies (Figs. 20.1 and 20.2).

Impacts to food, energy, and water may also play out in a regional context that cross state or country boundaries (Brown et al. 2015). For example, the Ogallala aquifer stretches from Nebraska south to Texas, providing a crucial water supply for corn, wheat, and cattle production in a 20 billion U.S. dollars/year industry (Parker 2016). This area is heavily dependent on irrigation made possible after World War II because of both inexpensive energy and better pumping technology. However, excessive pumping has depleted the water table to record low levels faster than the system is being recharged (Fig. 20.3).

Producers have been left with hard choices: use less water to slow depletion and bear the economic loss in the near term or continue pumping at normal levels to maintain production but face a longer-term total loss of water (Bloomberg 2016). More energy is required to drill deeper into the aquifer, water is harder to obtain, and the impact on food security is heightened (see, e.g., Steward et al. 2013).

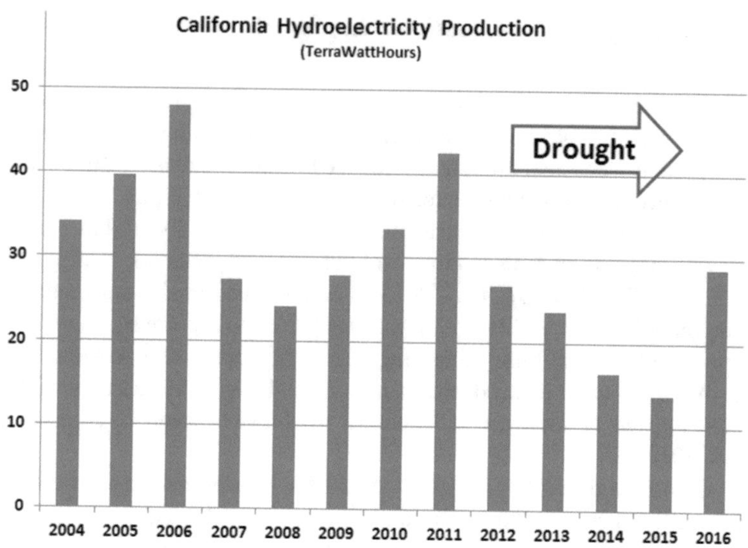

Fig. 20.1 Impact of California drought on hydroelectricity production (map derived from data in Gleick (2015); for another example, see also Sect. 13.3 on the 2011 Texas drought)

Problems of higher temperature and drought are especially exacerbated in states like Texas and Kansas where crop irrigation is entirely dependent on groundwater. Because this region supplies nearly one-fifth of the US agricultural harvest of wheat, corn, and cotton (Little 2009), the decline in this aquifer impacts not only the US but beyond.

Conflicts over water scarcity are not just a western US issue. For example, there have been legal fights over the Apalachicola-Chattahoochee-Flint (ACF) River Basin in Georgia, Alabama, and Florida (Fig. 20.4) since 1990. In the most recent legal battle brought to the U.S. Supreme Court, Florida attributed its lack of water and crashing oyster populations in Apalachicola Bay to increasing water use and withdrawal in Georgia. Georgia argued the need for fresh water for the City of Atlanta and irrigation for its farmers. Despite millions spent on legal fees, this case has not yet been resolved (see, e.g., Reilly 2017).

Although Alabama is not a party to the U.S. Supreme Court case, it has been part of other litigation dating to the 1990s. Alabama uses water from the ACF Basin for agriculture, fisheries, power generation, and recreation. In recent years, Alabama has suffered impacts from severe to extreme droughts that greatly impacted agricultural production, energy supplies, and drinking water. Figure 20.5 below shows the percent of Alabama in mild (yellow) to extreme (dark red) drought below since the year 2000.

These types of issues are not limited to the US. In parts of southern Africa, strong economic and military nations control natural resources for potential political gain and agenda; access to water is a key piece of this. For example, the Limpopo River flows through South Africa, Botswana, Zimbabwe, and Mozambique.

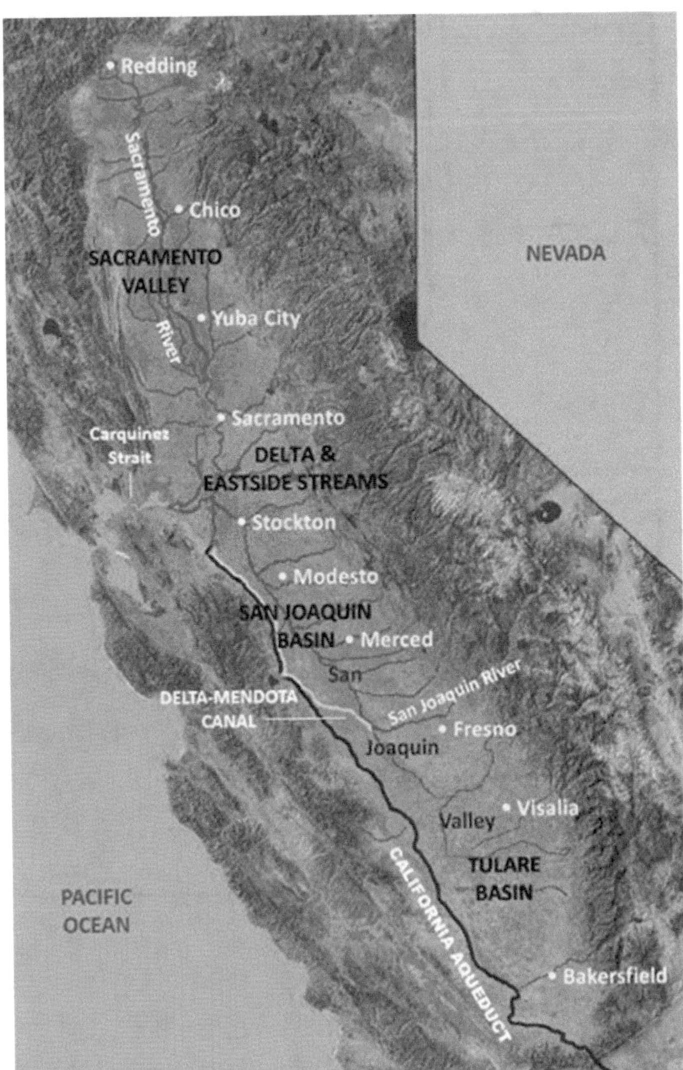

Fig. 20.2 Image of the Central Valley of California: water delivery to Southern California travels through the Sacramento-Bay Delta region. (Credit: Central Valley photo courtesy of the U.S. Geological Survey Department of the Interior/USGS)

Of the 44 dams on the river, 26 are located in South Africa. Many of the trans-boundary agreements signed by these nations favor South Africa, holding the border to be the high-water mark of the river. Another regional river, the Orange River, faces tensions around the creation of an Orange-Senqu Commission for river management; such tension has prevented countries like Namibia from using the river during times of drought while water flows through South African dams to produce power (Kings 2016). When smaller nations in this region ask for renegotiated

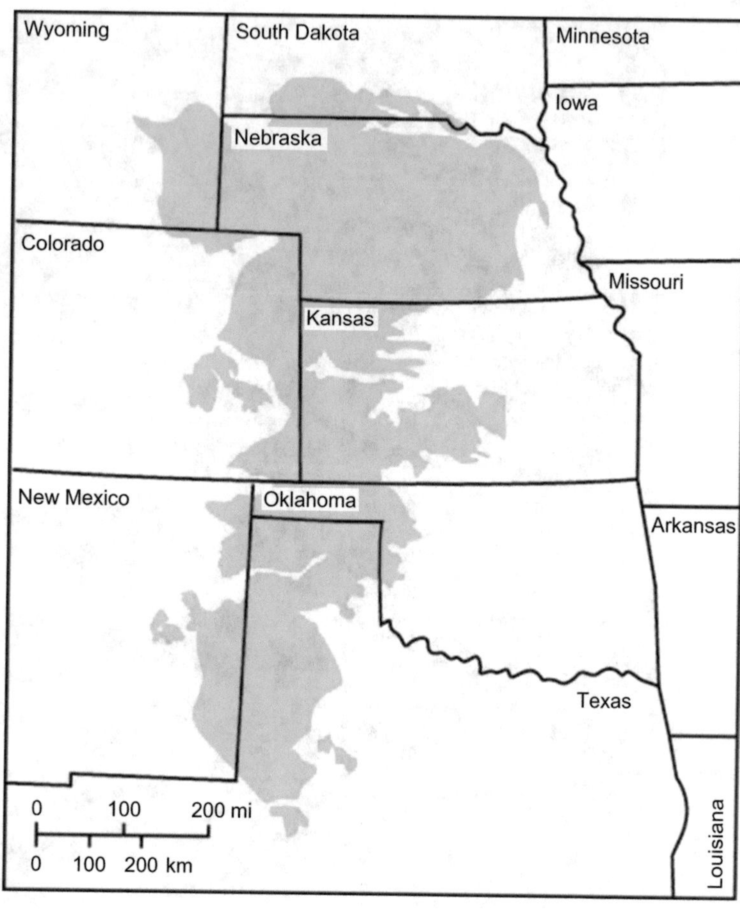

Fig. 20.3 The Ogallala (High Plains) Aquifer underlies 451,000 km² in eight states. (Credit: Kansas Geological Survey)

water agreements, stronger sovereigns have mobilized troops in particular regions to secure water supply facilities. Section 19.4 includes a case study on other parts of Africa as well (Fig. 20.6).

Elsewhere, pollution from industrial runoff and dumping in Zimbabwe into the Muene River upstream has impacted water supplies in Mozambique; this problem has worsened exponentially under drought conditions. Both countries share the river as a border. Low water flow in times of drought increases the concentration of contaminants and chemicals suspended in the water column as they are carried downstream into Chicamba Dam, a critical water supply source in Mozambique (Kings 2016). Moreover, outbreaks in cholera and fish die-offs have become commonplace.

Fig. 20.4 Map of the ACF River Basin. (Credit: U.S. Geological Survey Department of the Interior/USGS)

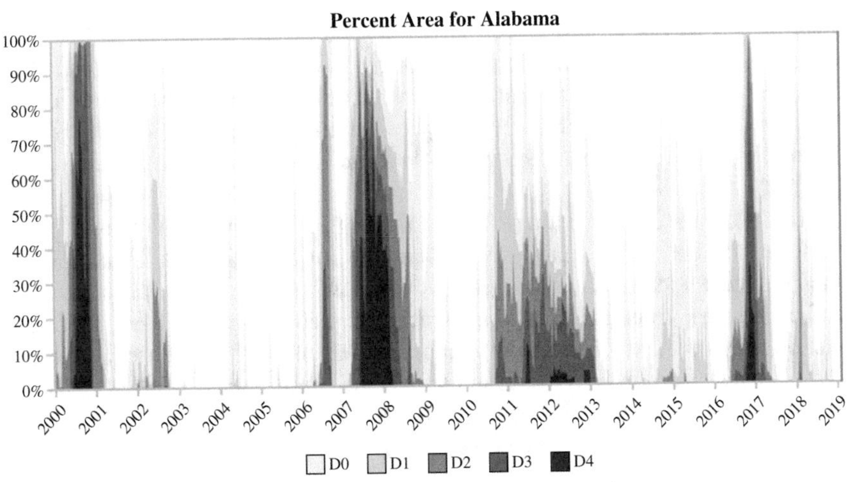

Fig. 20.5 Percentage of Alabama in drought (2000–2018). (Credit: National Drought Monitor (2019))

Similarly, food security problems have arisen between Botswana and Namibia as a direct result of sovereignty issues related to Sedudu Island in the Chobe River. The island's contested legal status to ownership is unresolved, with the Chobe River serving as the border between both countries. A drought forced Namibian fishermen

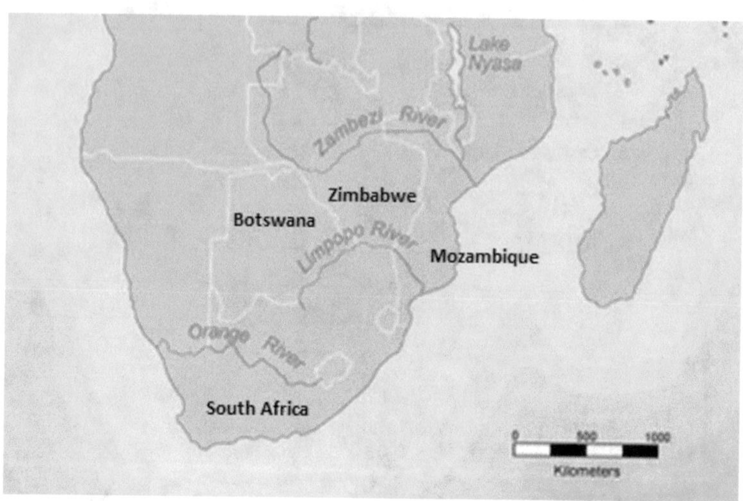

Fig. 20.6 Map showing the broad route of various rivers in southern Africa (Credit: https://garza-mercadocapital.com/wp-content/uploads/2019/04/b637e471a7c59d9829b189a50e6d07d8.jpg)

to fish toward the Botswana side of the river or from the island itself. As a result, several Namibians were shot by the Botswana army or arrested for fishing illegally (Kings 2016).

Elsewhere, widespread drought affected much of northeastern Brazil from 2012–2017. This impacted agriculture, biofuel crops, and reservoirs used for water supply and energy production. In 2014, Brazil's largest city with a population of 20 million people, São Paulo, had less than 20 days water supply and a reservoir 97% empty. While rain solved the immediate issue, deforestation in the Amazon may exacerbate drought and continue to impact the city and country's water supply (Ritter 2018). See also Sect. 19.3 on the Amazon River Basin for a more in-depth discussion.

For rapidly developing countries in Asia, energy development can present problems for food and water sectors. Water demands for dams and fossil fuel sources often coincide with an increasing population. However, as water availability (i.e., surface streams, groundwater aquifers) becomes strained across sectors, food production may be at risk because of competing demand from the energy industry. These are big issues in both China and India. On a smaller scale, this has also been the case in Sri Lanka, a small island nation that depends on hydropower for its electricity. The recent drought forced electricity to be generated from coal and oil sources, with hydropower contributing about 10% compared to the normal 40%. Drought also dropped paddy rice production from 800,000 to 300,000 **hectares** due to lacking irrigation water (Perera 2017). Reservoir levels were only a third full and restrictions were in place to only allow withdrawal for drinking purposes.

These types of issues are playing out all over the world, with underlying dynamics in any given region made tenser by extremes in weather conditions: drought, floods, heat, intense storms, and so on.

20.2.2 Underlying Dynamics for the FEW Nexus

As discussed elsewhere in this book, there are a number of underlying dynamics that make managing conflicts in the FEW nexus challenging. One dynamic is the challenge of managing natural boundaries such as waterways versus political boundaries such as those between states or countries. As noted in the introduction, decision-making on food, energy, and water starts with the 7.6 billion individuals (as of early 2018) on the planet. Decisions are made at every level of social organization, both public (i.e., household, village, town, city, county, state/province, region, nation, international, and global) and private (i.e., small businesses like farmers, large companies, non-profit and non-governmental organizations, and many more). See Chaps. 4 (human behavior); 12 (scale); and 17 (applying science to practice). Much like a stone thrown into a pond, any one decision can have a ripple effect that flows through the FEW system.

Likewise, FEW systems nearly always include a large number of independent actors of different kind (individual, public and private) of different sizes and power with different objectives and different values and cultures. Problems between different actors may be said to be "tractable" when their goals overlap and compromise is desired or pushed. Where there is no overlap in the objectives of the different actors, no willingness to compromise, or big differences in framing the challenges, problems may be viewed as intractable.

Another dynamic results from imbalances in power. This occurs where one actor, or set of actors, has significantly greater power than others; this can often lead to outcomes that prioritize their objectives over those of others. Such power may result from status or position; wealth and level of development; political power or number of followers, or other sources of power, both formal and informal. In the illustration below, one source of "power" is insurance: if someone plants a crop earlier than dictated by insurance, they bear the risk of that decision by themselves. Yet adapting to extremes in conditions may mean also adapting financial incentives (or disincentives) to address on-the-ground conditions (see, e.g., Illgner 2016; Munich RE 2017).

Another dynamic is that values and who should participate in the allocation of resources changes over time. For example, the USA and Canada are parties to the Columbia River Treaty. The agreement, signed in 1964 for 60 years, is primarily designed for flood control and power generation between the countries (Columbia Basin Trust n.d.). Currently, negotiations are underway about the treaty's future, which expires in 2024 unless extended. When the treaty was signed in 1964, environmental considerations were not part of the discussions. However, in present-day negotiations where values have changed, Native American tribes from each country have lobbied for reliable water flows for fish populations (i.e., salmon, steelhead) in any signed extension to preserve their cultural identity and traditions (Columbia River Inter-Tribal Fish Commission 2017). Guaranteeing this requires new management techniques and cooperation between agricultural producers, energy developers, and political officials. Power generation benefits received by Canada from US production are set to be re-negotiated, along with the possibility of raising reservoir levels on the Canadian side to protect US cities from floods.

Taken together, managing conflicts that may arise with the FEW system is a challenge, one made more significant due to population growth and the impact of climate change.

Illustration of Complex Decisions Made at the Individual Level
Global dynamics often play out at very local levels and in individual decisions. Early in 2012, Ethan Cox, an Illinois farmer, had a choice to make. After 2 years of extreme rain and floods in 2010 and 2011, many Midwestern farmers were confident that another season of rainfall would produce bountiful crops. However, Mr. Cox realized based on past experience, scant snowfall during the winter of 2012 might mean a drought during the growing season. With unseasonably warm temperatures in March, Mr. Cox planted his crops earlier than his neighbors to take advantage of whatever soil moisture remained and gave his crops a better chance to survive. In addition, using no-till methods on his land reduced erosion and allowed cover crops to shade his seedlings. By planting earlier than timelines set by agricultural professionals, however, Mr. Cox made himself ineligible to receive insurance payments if he lost his crops from drought or intense storms. Ultimately, Mr. Cox had a high crop yield compared to his neighbors because of his adaptation techniques to use less water and energy to grow corn and soybeans (McGraw 2015). This kind of story illustrates the difficult decisions being made daily: decisions that affect every part of the FEW system.

20.2.3 Climate Change as a "Threat Multiplier"

While the climate has always changed, a warming climate across the globe is leading to more extremes in weather, sea level rise, and in some places, weather systems that stall in place like Hurricane Harvey in 2016 which dropped more than 50 in. of rain in a concentrated area (see Chap. 11, climate change).

Such extremes are acting like a "**threat multiplier**" that heightens underlying tensions and dynamics and may create new threats upon the nexus. In 2011, António Guterres, former High Commissioner for Refugees, noted in an address to the UN Security Council that "climate change is the defining challenge of our times: a challenge which interacts with and reinforces the other global megatrends such as population growth, urbanization, and growing food, water and energy insecurity. It is a challenge which is adding to the scale and complexity of human displacement, and a challenge that has important implications for the maintenance of international peace and security" (Guterres 2011).

The *2011 Arab Spring* is a prime example of climate threats to multiple sectors of society. Across the region, extreme and prolonged drought forced people to the cities to find work, leading to violence, poverty, and instability (Johnstone and Mazo 2011). In Syria, the ongoing civil unrest and refugee crisis is partially attributed to a severe drought that added to political instability. From 2006–2011, more than 60% of Syria experienced devastating drought, crop failure, and loss of livestock. With the increased population in urban areas, water supplies became strained and made living conditions worse. Coupled with governance issues, the war forced millions to leave Syria for other Middle Eastern nations or Europe (Werrell and Femia 2013). Within Syria, critical infrastructure for water, food, and energy have been destroyed and are not easily rebuilt without significant investment, making repatriation of displaced Syrians that much more difficult.

Mass migrations or forced migrations have occurred due to both economic hardship and poor conditions for producing food, and as a result, larger societal changes have occurred in major regions of the world. Socio-economic and political stressors in response to new climate conflicts and violence have threatened the stability of government structures in responding appropriately. For example, West African countries such as Mali, Niger, and Gambia have experienced vast migrations of young men searching for limited jobs in neighboring Libya and Algeria. As temperatures rise and rain becomes increasingly unpredictable, life as a farmer has all about disappeared. Instead, men chance migration through the Sahara Desert and jihadist-ridden territory to look for new means of income to support families, oftentimes dying along the way or returning home with nothing (Sengupta 2016).

In turn, such migrations impact receiving countries. In places like Jordan's Za'atari refugee camp, thousands of refugees rely on water trucked in daily. Lack of adequate sanitation forces residents to discard waste water onto the bare desert inside the camp, jeopardizing groundwater supplies. Providing water to more people has further stressed an already scarce resource (Farishta 2014).

Stories in the news are being backed up by systematic research. In a study released in 2019, researchers from the International Institute for Applied Systems Analysis (IAASA) linked climate, conflict, and migration, noting that "Climate change will not cause conflict and subsequent asylum-seeking flows everywhere. But in a context of poor governance and a medium level of democracy, severe climate conditions can create conflict over scarce resources" (Abel et al. 2019).

The idea of **climate refugees** is relatively new. While there is no legal definition, the issue of migration caused by climate impacts is being considered by the United Nations High Commission for Human Rights because of displacement caused by disasters and climate change (see https://www.unhcr.org/climate-change-and-disasters.html). The number of people displaced by such disasters is growing: an estimated 24 million people have been displaced each year since 2008, with an estimated 143 million people displaced by climate impacts in sub-Saharan Africa, South Asia, and Latin America by 2050 (McDonnell 2018). How to address such refugees in international law remains an open question. Climate as a threat multiplier and cause of mass migration is not the only challenge made more severe climate change.

Areas like the Arctic which have previously been seen as peaceful have already started to see conflicts emerge. As the Arctic sea ice melts, parts of the Arctic Ocean have started to remain ice-free. New fishing grounds and vessel passage routes have emerged, but questions remain as to how this new "frontier" will affect fish populations, sovereignty issues and the Law of the Sea, and human migration across this landscape. Likewise, conflicting policy among nations regarding energy exploration is already starting to develop, with Russia planting its flag on the bottom of the Arctic Ocean and starting to conduct both oil and gas exploration and military operations.

20.3 Toolkit to Address and Manage Conflicts

How to address and manage these types of challenges and conflicts is a critical question. Traditional sector-specific approaches to conflict resolution make it difficult to develop solutions across the nexus. In many cases, regulations or court adjudications define who wins, focus on a specific problem, and often end with parties in contentious positions.

In contrast, creativity in addressing and managing conflicts that address all elements of the food–energy–water nexus is possible using a range of tools. Methods of collaborative governance and **alternative dispute resolution** (ADR, also called "appropriate dispute resolution") can provide ways forward (see, e.g., Moffitt and Bordone 2005). ADR processes incorporate the human dimension of environmental problems directly by designing problem-specific plans to help stakeholders arrive at informed decisions in a way that can account for legal, scientific, and political ambiguity and risk (Fowler and Shi 2016).

Much as challenges to the FEW system come from many sources, ways to address them also come from many sources. Managing issues within the nexus does not revolve around one single method or solution. Instead, tools to manage conflict range from an individual to international, informal to formal, legislative to judicial. This section explores such tools and provides examples of how they have been used.

Because the relationship between the parts of the food, energy, and water nexus is still developing, traditional practices of conflict resolution, negotiation, collaborative governance, the use of "serious games," facilitation, mediation, and arbitration can offer innovative ways to solve issues between sectors. Even judicial resolution of conflicts—at local, national, and international levels—is important to consider. Finally, international treaties can be important tools in helping manage potential conflicts (see Chap. 6).

20.3.1 Traditional Dispute Resolution Practices

Traditional dispute resolution mechanisms are as old as human society. Village elders, religious leaders, and others have historically been called on to help resolve disputes between people (see, e.g., Barrett and Barrett 2004; Dempsey and Coburn 2010). Modern day usage of traditional dispute resolution mechanisms can be quite helpful in addressing local to global disputes. Such mechanisms have been written about in numerous scholarly articles with country or region-specific details. On the global level, the traditional South African dispute resolution mechanism of "**indaba**" was successfully used to bring more than 200 countries together during the Paris Climate negotiations in 2015.

The indaba process involves breaking a large group of people into smaller groups, allowing participants to rotate during discussions to ensure all voices are heard evenly. In most cases, a facilitator is active in overseeing the dialogue. Each participant is given an opportunity to answer a question presented to the group without interruption. The caveat is that participants must answer stating personal positions, adhering to certain thresholds that shouldn't be crossed. In addition, each participant must provide a solution to the problem that can be accepted by the group. After all group members speak, the large group may reconvene and consider the offered solution as part of a potential compromise (Rocket 2016). Decision-making power is limited to a few leaders who consider the groups' recommendations together.

In the case of the Paris discussions, a number of breakthroughs occurred in short order, sometimes as short as thirty (30) minutes. Notably, the inclusion of all countries—each given an opportunity to share their perspective—helped break down barriers between smaller nations and world powers.

Ultimately, the indaba discussions reduced about nine hundred (900) points of down to roughly three hundred (300). Agreements on some issues could not be solved, as the subject of climate mitigation among sovereigns is a thorny issue by itself (Felix 2015). However, global diplomats welcomed the "participatory, yet fair" method which offered a helpful tool for complex issues encompassing a wide range of sectors. Although the targets set during these discussions are far from met and some parties like the USA have threatened to withdraw, these agreements set the stage for remarkable progress on global issues. More on international agreements is explored in Sect. 20.3.8 below.

20.3.2 Negotiation

Another key tool for addressing potential conflicts is **negotiation**, which is simply "the back-and-forth communication designed to reach an agreement when you and the other side have some interests that are shared and others that are opposed" (Fisher and Ury 1981). In their seminal work on negotiation, *Getting to Yes*, Fisher and Ury discuss the difference between positions and interests: "Your **position** is something you have decided upon. Your **interests** are what caused you to so decide." In the food, energy, water nexus, people will often start a conversation by stating their position: I want to build a dam. I don't want you to irrigate. You have to provide energy for my project. One way to open up a conversation and start to understand better what is happening is through better understanding the interests at play: why does someone want to build a dam? Why should someone stop irrigating? What

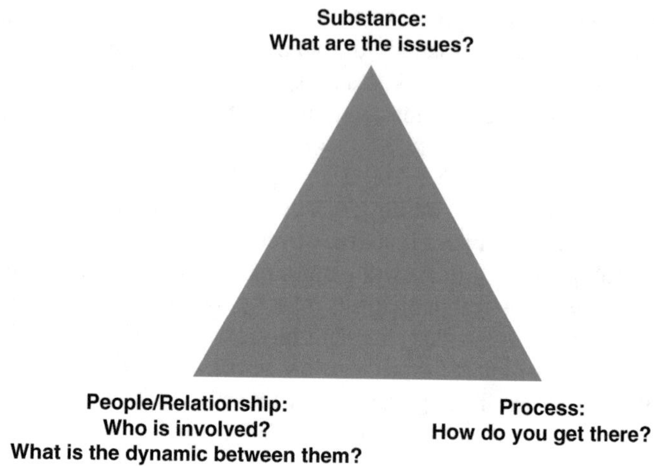

Fig. 20.7 Dynamics in conflict (Source: Carpenter and Kennedy 2001)

do people value? Digging deeper and seeking a better understanding of interests involved may help prevent conflict from spiraling up in the first place (Fig. 20.7).

Another way to think about building more constructive negotiations is to think about what is happening in a negotiation. Often people want to talk about the *substance* of an issue: how to ensure reliable energy supply. However, their ability to do so is often hampered by distrust of either the *people* engaged or the *process* used (Carpenter and Kennedy 2001). Failure to recognize conflicts in these areas can result in a "spiral of conflict" where what might have been a smaller dispute spirals into a much bigger one (Carpenter and Kennedy 2001). Spending time to think about both people and process can be critical in moving a negotiation on FEW topics forward in a more fruitful way.

One example negotiation that has led to real on-the-ground changes in the management system comes from Kansas. Kansas is well known for its rich history as an agricultural state. Both the Ogallala (also known as the High Plains Aquifer) and Great Bend Prairie aquifers underlie the surface with a vast network of underground streams and pools; water is managed under the prior appropriation doctrine as a real property right (Fig. 20.8). Center-pivot irrigation systems introduced in the 1950s resulted in an exponential increase of water withdrawal for irrigation. Through the 1960s and 1970s, state officials acknowledged serious depletion of aquifers from excessive pumping and over-allocation, resulting in an adverse effect on wetlands. Because groundwater is hydrologically connected with surface water, pumping groundwater for irrigation caused surface levels in the wildlife-rich areas to recede.

To address the declining water levels, the State of Kansas created groundwater management districts and Intensive Groundwater Use Control Areas to conserve water, along with opportunities for local areas to develop their own solutions. The creation of local enhanced management areas (LEMAs) allows water rights

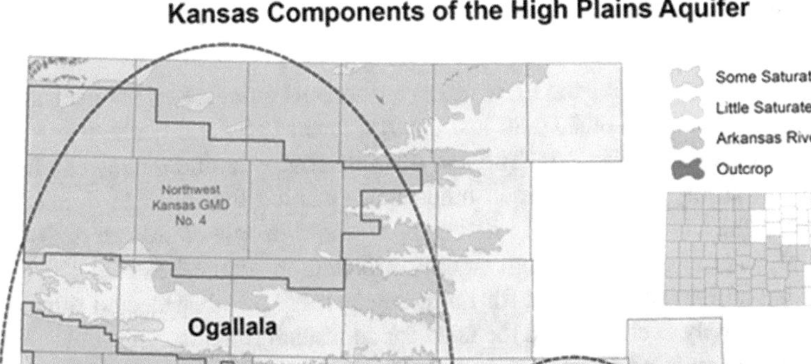
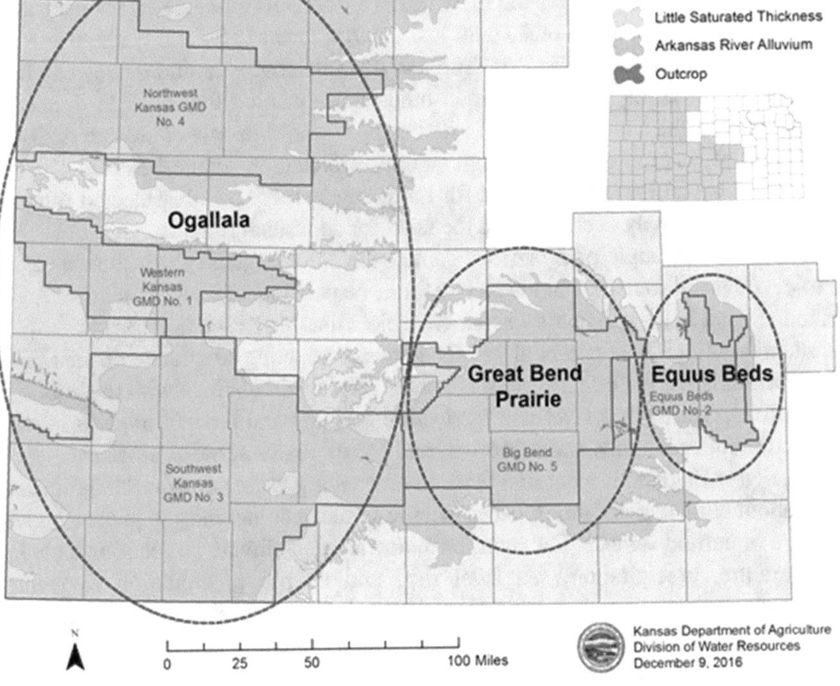

Fig. 20.8 Kansas portions of the Ogallala (High Plains) Aquifer. (Copyright 2017 KDA and/or its suppliers. All rights reserved)

holders to use voluntary mechanisms to restore groundwater to appropriate levels and recharge rates. What makes LEMAs different than most laws is the amount of collaboration between parties to achieve results. Concerned water users submit a proposal to a groundwater management district (GMD) board, which then works with the State Water Resources Department to develop an official management plan. Once agreed upon, the plan goes to the Legislature and Governor to finally become law. A need for such a plan must exist with reasonable geographic boundaries and public interest needs (Owen 2016).

Negotiated local agreements have led to good success: as of November 2016, pumping reduction is ahead of schedule for a number of areas (Golden 2016). Much like in Elinor Ostrom's work on governing the commons, people in these areas came together to create LEMAs to shepherd their own resources (Owen 2016). The success is based on communities working together to reduce water use through efficiency; self-enforcement has proven more successful than enforcement by a state agency. Such efforts at conservation ensure continued production of food and help reduce energy consumption.

20.3.3 Serious Games

Another tool used to examine the FEW nexus are **serious games**. Serious games are those that have a purpose other than just entertainment (MIT n.d.); see also Sect. 17.3.4 on immersive experiences. The use of in person or online games can help people try scenarios, work with each other, and examine the multiple dynamics involved with the FEW system. In one example where people could run different scenarios of water allocation through an online allocation "game," elected officials who participated realized the need for more engagement of other elected officials in local water supply decisions and the need for additional funding.

In another example of a "serious game," the New England Climate Adaptation Project engaged four coastal towns in Maine, New Hampshire, Massachusetts, and Rhode Island in a participatory research project that tested science-based role-play simulations to discuss potential mechanisms for adapting to climate change. This project examined the effectiveness of such role-play simulations with the public to adapt to issues like sea level rise or riverine flooding and prevent impacts to their infrastructure, economy, water sources, and coastline (Susskind et al. 2015).

The New England project first focused on creating usable down-scaled information about climate risks and impacts specific for each community to present a clear picture of current science. For each town, downscaled climate projections included temperature, precipitation, sea level rise, and intense precipitation frequency. Researchers then translated projections into a formal Summary Risk Assessment that contained local risks and potential adaptation options (Susskind et al. 2015).

At the same time, researchers conducted multiple interviews to gather information about stakeholders, local awareness of climate impacts, and ideas for solutions. Once assessments were completed, more than 500 people participated in role-play simulations and workshops. The simulations used the results of local climate risks and political dynamics specific to each community. Roles such as town planner, business owner, and public works director were assigned to participants, who then used the climate projections and public feedback to form responses and management plans (Susskind et al. 2015).

Post-event interviews and data indicated that many communities in coastal areas are not ready or equipped to manage risks from climate impacts. The events also showed several other key outcomes: increased community concern and support for collective action in adapting to rising seas and flooded rivers; enhanced community capacity to collaborate and empathize with each other; better social learning; and catalyzed community action (Susskind et al. 2015; Rumore 2017). Participants indicated the value of custom climate projections for a community in obtaining grant funding, engaging other communities, and developing local policy based on projected environmental conditions.

The use of such "serious games" allows people to respond to scenarios, try out different outcomes, and build a sense of what could happen. Working with somewhat hypothetical scenarios and perhaps different roles than they might normally have also allows participants to think about perspectives other than their own.

There are other ways to do this as well. In Japan, for example, some processes will designate a few people to represent the future to ensure future generational needs are also addressed.

20.3.4 Collaborative Governance, Facilitation, and Stakeholder Engagement

On the legislative or quasi-legislative side where laws and regulations are being written, communities have started to various models of **collaborative governance** to bring people together. There are a variety of mechanisms, including "deliberative democracy, e-democracy, public conversations, participatory budgeting, citizen juries, study circles, collaborative policymaking, and other forms of deliberation and dialogue among groups of stakeholders or citizens" (Blomgren-Bingham et al. 2005; Blomgren-Bingham 2009). All of these mechanisms seek to bring ordinary citizens into the legislative or rulemaking process, often by empaneling or inviting people in who might not normally be involved with crafting regulations. For example, a collaborative policy making process may bring multiple **stakeholders** or people who may be affected by the outcome of a process together to help inform this process. This can be done through more formal mechanisms such as formal facilitated or mediated process (see below) or can be done more informally. Participatory budgeting brings everyday people into the process of building budgets. Voters approved using this for New York City's budgeting process in the 2018 election, for example. There are additional resources for collaborative governance listed at the end of this chapter.

One key way to bring people together is through a facilitated process. **Facilitation** is the use of a neutral third party to help a group of people talk with each other and hopefully achieve a mutual understanding of interests and values. The use of third-party facilitators can be used in the settings described above to help identify potential stakeholders, convene a discussion, set ground rules to help participants engage with each other, and address potential disagreements or difficulties as they arise. A facilitated discussion does not necessarily involve conflict but occurs when a person or group decides that it would be useful to have assistance in having more productive conversations.

How these tools fit together is illustrated in three case studies below: one case study on flood impacts in Washington State's Chehalis River Basin; one on post-hurricane recovery in Louisiana, and one on managing drought in Australia's Murray-Darling River Basin.

20.3.4.1 Example #1: The Chehalis River Basin of Washington State

The Chehalis River Basin offers examples both of collaborative governance and the use of facilitation. The Chehalis River Basin is the second largest watershed in Washington State. In recent decades, the Basin has had several significant floods, including a record-breaking flood in 2007. The impacts of such floods have damaged infrastructure, food production, and the ability to provide food, water, sanitation, and energy. Floods are one concern, but the basin also suffers from drought and degraded habitat for aquatic species, notably salmon. This, in turn, affects Native tribes reliant on fish for food and culture.

After the 2007 flood, then-Governor Christine Gregoire allocated state funding to create a local "Chehalis Basin Flood Authority" to find local solutions to the impacts caused by the catastrophic flooding. This authority included elected officials, representatives from state and federal government, NGOs, Tribes, citizens, and the interested public. Parties understood that doing nothing was not an option. Because river boundaries cross political boundaries, however, collaborative work at a local level was critical to break the Basin's nearly century-long cycle of flood, study, blame, and repeat.

Challenges with such a diverse group of parties included funding issues, distrust between various stakeholders, and strong emotional ties to the basin. Facilitated flood policy workshops allowed people to voice their concerns and interests so others could understand their perspectives. Meeting locations were rotated within the Basin and tours provided people a chance to talk with each. Over time, finding common ground on various management options eased tension between community members and elected officials. A small but diverse work group eventually developed a plan to reduce flood damage and restore aquatic species simultaneously (Governor's Chehalis Basin Work Group 2014).

The more informal work group prepared recommendations for the state Legislature, resulting in the Governor officially convening the Chehalis Basin Work Group to work on large-scale capital projects, land use management, and effective flood warning systems. The Work Group included six Basin representatives, elected officials, citizens, and tribal leaders (Chehalis Basin Strategy 2016). State agencies collaborated with the Work Group members to seek stakeholder input and design a comprehensive environmental strategy for the basin. Including all relevant stakeholders was vital because potential solutions and funding options were not possible without constructive dialogue (Washington Department of Ecology 2015).

Building relationships at workshops and initiating tough conversations allowed participants to address issues one at a time through the help of various facilitators. The combined efforts of the locally driven effort resulted in significant funding; the 2013–2015 biennium capital budget included $28.2 million dollars to implement the Work Group's recommendations (William D. Ruckelshaus Center 2015). Today, work continues to implement proposed solutions for flood risk reduction and ecosystem restoration (Fig. 20.9).

20.3.4.2 Example #2: Louisiana Speaks

Work post-Hurricane Katrina offers another way stakeholder engagement can be implemented. After Hurricane Katrina and the less discussed Hurricane Rita in 2005, Louisiana developed a hands-on approach to repair the battered coast and help over a million displaced people. The "Louisiana Speaks" initiative aimed to

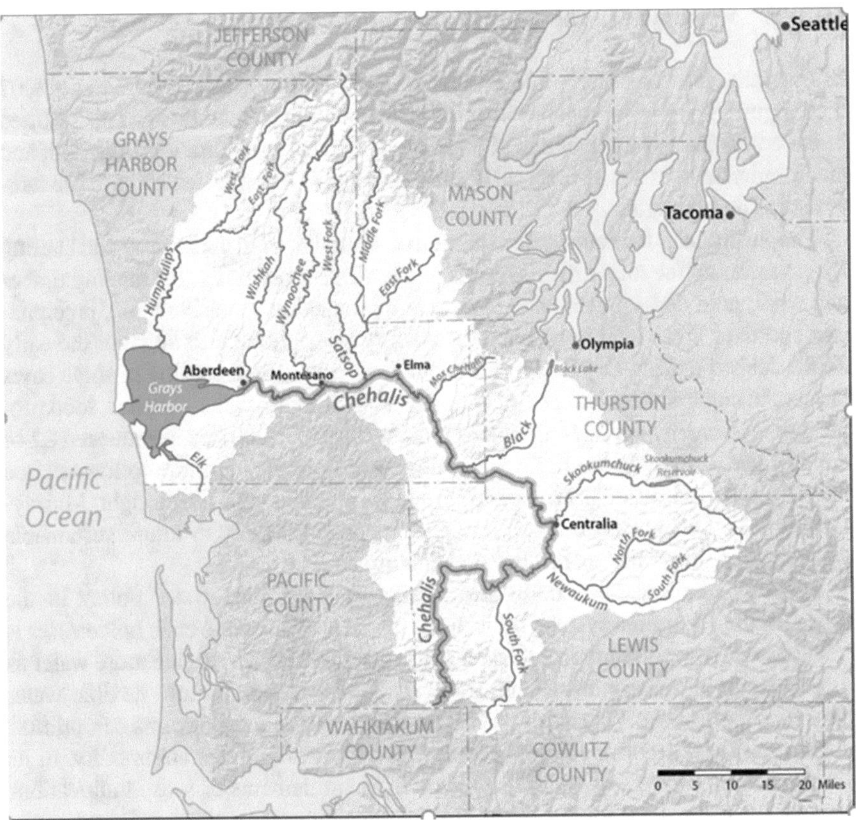

Fig. 20.9 Chehalis River Basin. (Credit: "Map of the Chehalis River watershed in Washington, USA." by Shannon1 through Wikipedia, licensed under CC BY-SA 4.0)

create a sustainable, long-term vision for Louisiana by combining local, state, and federal entities with various stakeholders and citizens to rethink Louisiana's future. The ultimate product was a fifty (50) year growth plan focused on planning at the building, neighborhood, parish, and regional scale (Center for Planning Excellence 2016) (see also previous chapters that address "scale").

This final plan relied heavily upon the public participation at workshops, notably citizens' opinions, concerns, and ideas for new management. The open public process resulted in over 1,000 workshops over 25,000 square miles with more than 27,000 people participating in various ways.

Responses were dominated by the need to restore coastal wetlands and regional infrastructure while reinventing historic communities. The Louisiana Recovery Authority used this engagement to create a comprehensive regional plan in 2007 focused on three main vision areas: sustainable recovery, smarter growth, and regional focus (Louisiana Recovery Authority 2007). Ultimately, the plan became the leading recovery, redevelopment, and planning effort in US history (Center for Planning Excellence 2016).

20.3.4.3 Example #3: The Murray-Darling River Basin of Australia

In Australia, the "Big Dry"- a long running drought- eventually resulted in a 2007 reconfiguration of water regulation in the Murray-Darling River Basin. This resulted in the comprehensive management of water resources, agricultural production, and environmental uses. As of early 2019, this structure is being tested by extreme temperatures and drought.

The Basin spans the southeastern portion of Australia. Both the Murray and Darling Rivers make up the major river systems and have a history of use for hauling timber and wool; gold development; and agricultural production. An arid region, precipitation and river flows tend to be unpredictable. A major drought throughout the early 2000s dubbed the Big Dry decimated agricultural production, causing in some cases farmers to commit suicide. Lack of water affected drinking water supplies, food production, and energy supplies as well as local wetlands. A 2007 audit indicated 22 of the 23 river systems in the basin were in poor ecological shape due to low surface flows, saltwater intrusion from depleted groundwater tables, and the drought (Murray-Darling Basin Commission 2008). Officials theorized the key to future sustainable water reserves depended on healthy ecosystems and river systems.

Reacting to these dire conditions, regulators overhauled water policy in the Basin. In the Australian system, the volume of water allocated to each holder/user is based on water availability in a watershed. In wet years, users receive more water as a percentage of their right, less in dry years. Coupled with a more flexible water allocation policy, the Water Act of 2007 added water market reforms to optimize water use. These reforms focused on the true cost of water as a commodity. In its simplest form, farmers or water users who hold entitlements or season allocations wait for the Basin Authority to release annual allocations for each user. Based on the allocation, the water user decides whether it will be more profitable to grow crops or sell their water right. As the price of water fluctuates, users have relied on sales of water allocations as a major revenue source (Richter et al. 2016). In addition, the Water Act of 2007 requires managing surface water and groundwater together while providing environmental flows as much priority as the actual consumption of water by users.

The Australian government also earmarked $12.6 billion AUD over 10 years starting in 2007 to improve environmental conditions across the basin through a program called "Water for the Future." This program depended on two methods: (1) direct buy-backs of water entitlements from holders, and (2) efforts to improve on-farm efficiency through a focus on rural water use and infrastructure projects. Water saved from increased efficiency is directed to river systems or used for flooding wetland areas to initiate natural flow and water recharge (Richter et al. 2016). Until recently, changes in regulatory policy have seen positive environmental responses in the Basin and economic stability for farmers active in growing crops. However, more recently in 2019, extremely high temperatures and low flows have led to widespread fish kills and greater impacts to the Murray-Darling Basin. Adapting to new extremes conditions may also require revisiting water regulations.

).3.5 *Mediation*

lediation also relies on the use of a neutral third party—*a mediator*—to help peo-e have a constructive conversation. However, formal mediation tends to occur fter a conflict has arisen and parties are in a dispute. In the USA, all 50 states have mediation privilege that generally protects anything created for the purpose of the nediation from being introduced into formal court proceedings. The job of a media-or is to work with the parties to design a process, assess the issues in conflict, larify the interests and issues, generate and then evaluate options, develop potential greements, and work with parties to determine potential implementation steps (see, .g., Susskind and Ozawa 1983; Susskind et al. 1999). At the end of the process, it s ultimately up to the parties involved to decide how and whether to proceed or reach and implement an agreement, not the mediator.

Both mediation and facilitation rely on a cooperative approach by the neutral third party to stimulate dialogue and hopefully build productive relationships among stakeholders. Vital to this process is the identification of issues among parties and how to address them (Fowler and Shi 2016). One benefit unique to both methods is the control stakeholders possess over the process and outcome while the third-party neutral facilitates hopefully more productive engagement and outcomes.

For the food–energy–water nexus, mediation is an essential alternative to the traditional judicial system where backlogs often occur, and knowledge of environ-mental matters can be limited. Moreover, mediation allows parties to convene at times of their choosing and flexibility in which issues to investigate. Mediation bridges the gap between the complexity of environmental conflicts and the jurisdic-tional reach of traditional courts. Furthermore, the method provides decision-making processes in a participatory format, allowing the mediator to help parties evaluate the strength and weaknesses of the options. Finally, mediation presents an opportunity to create "social capital" among participants by working to improve relationships (Fowler and Shi 2016).

An example of where mediation has made a difference is in the ability to store then use additional groundwater in the greater Los Angeles area. Two groundwater basins—the Central and West Coast Groundwater Basins—run from the City of Los Angeles to the City of Long Beach (Fig. 20.10). Subject of much litigation in the 1950s and 1960s due to declining groundwater levels, the parties took themselves to court and ultimately reached settlements on who could extract how much water and from where (Witherall and Simes 2014). Elinor Ostrom wrote about this history as an example of people managing "common pool resources" for themselves (Ostrom 1965). In the early 2000s, parties to these settlements wanted to re-examine who could store and then use additional groundwater; they hired mediators to help with this process. After countless discussions and negotiations, the parties agreed on new governance mechanisms to address groundwater storage and use in these basins; these were eventually approved by the court in amendments to the original judg-ments (Sprague 2013).

Fig. 20.10 West Coast and Central Basin Groundwater Basins. (Map credit to the Water Replenishment District of Southern California)

20.3.6 Arbitration

In contrast to mediation, **arbitration** involves the use of a neutral third party or panel to make a decision about a dispute. Sometimes called private judging, arbitration is an adjudicative form of resolution where an arbitrator renders a final decision on a dispute between the parties. This decision is legally binding and is often used in contract disputes (Fowler and Shi 2016). The use of arbitration is also heavily used in international energy disputes, which may have significant consequences for national or international energy policy. At the end of 2016, an estimated 42% of all arbitration cases handled by one firm arose from the energy sector, with the "Energy Charter Treaty" (see treaties below) the "most frequently invoked international investment agreement" (Kluwer Arbitration Blog 2018). (For more on the Energy Charter Treaty, see discussion in Sect. 7.4.1.)

Billions of dollars are being decided in international arbitration decisions, often in ways that are opaque to policymakers. Decisions made in arbitration usually cannot be appealed or reviewed by any other body. Like other decisions made by actors throughout the FEW system, a decision made in arbitration about one case can ripple through the system. For example, after changes in various tax and regulatory incentives, the Czech Republic and Spain faced claims from investors in the solar sector and Poland from investors in the renewable energy sector (Franzetti 2017). After Germany decided to stop using nuclear power production after the 2011

Fukushima Daichi nuclear disaster in Japan, it too faced claims in arbitration for this decision. The outcome of any of these arbitration cases can be quite costly and affect national policy, but not necessarily in a way that is readily seen or understood.

20.3.7 Adjudication in the Courts

Courts at various levels are also used to determine the outcomes in the context of the FEW nexus. In the USA at the state level, state agencies focused on water or energy often have *administrative law judges (ALJ)* who may hear a preliminary dispute. There are also specialized courts that may hear environmental cases such as an environmental hearings board or pollutions control hearing board. Two states—Montana and Colorado—have specialized water courts. Washing State has a specialized Energy Facilities Siting Commission to review potential new energy projects. Appeal from decisions by a specialized tribunal like these go to a trial court (often called a superior court); judges at a trial court hear both civil and criminal cases and generally do not specialize in environmental issues. Appeals from this level are reviewed by an appellate court and finally to a state's supreme court for what is usually final determination of a case. Processing an environmental case through this set of courts can take many years, which is one reason that parties will sometimes try to resolve a dispute outside the court system.

There is also a parallel system at the federal level (see Chap. 8). A conflict brought under a federal law like the Clean Water Act can be heard by an administrative law judge within the federal agency charged with implementing the law. For the Clean Water Act, this would be the U.S. Environmental Protection Agency (EPA). EPA has ALJs who "conduct hearings and render decisions in proceedings between the EPA and persons, businesses, government entities, and other organizations that are or are alleged to be regulated under environmental laws" (U.S. EPA n.d.). Appeals of a decision at this level go through one of the 94 U.S. District Courts in the relevant state or area, then to one of the 13 U.S. Circuit Courts of Appeals, and potentially to the U.S. Supreme Court. The U.S. Supreme Court is asked to review around 10,000 cases per year but accepts fewer than 100 each year, so the odds are that the decision issued by the U.S. Court of Appeals will be a final judgment or outcome of a case (Fig. 20.11).

An example set of cases that have fundamentally altered water, energy, and food production relate to the management of both the Columbia River (Pacific Northwest) and California's Bay Delta region. In both regions, declines in endangered fish species have been challenged in court, where federal judges decided that insufficient water for declining fish populations requires reallocation of water otherwise used to generate energy along the Columbia or irrigation water in California. Management of large river systems in both regions depends on the outcome of these on-going and challenging cases.

For disputes between states, which can arise in the case of water allocation between states, the U.S. Supreme Court hears a case directly and acts as a trial court.

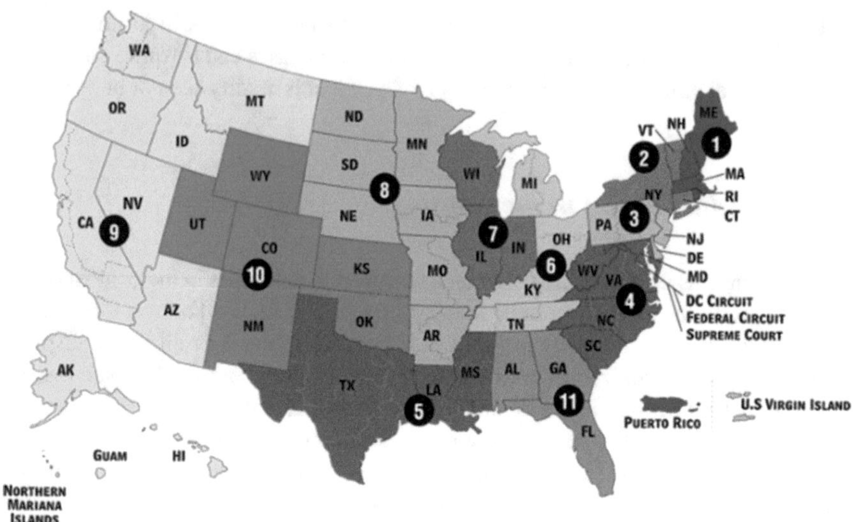

Fig. 20.11 US Courts of Appeals by region (e.g., California and other western states are in the Ninth Circuit Court of Appeals). See https://www.uscourts.gov/sites/default/files/u.s._federal_courts_circuit_map_1.pdf)

In practice, the U.S. Supreme Court assigns a "*special master*" to collect evidence from the parties and issue a decision; the U.S. Supreme Court then reviews the finding. As of 2019, there are six U.S. Supreme Court cases pending over water issues between states, including two in the eastern USA and four in the western USA. The *Florida* vs. *Georgia* case is discussed above.

Beyond the USA, each country has its own court system to decide cases. Whether or how these touch FEW Nexus questions is an area of potential research. Beyond the judicial system, international agreements are another way that potential conflicts in the FEW system can be addressed.

20.3.8 International Agreements: Goals, Treaties, and Conventions

International agreements can be a way to address potential, evolving or existing conflicts between countries. There are a number of international agreements that affect or address various aspects of food, energy and particularly water (see generally Chap. 6 on international governance). International agreements include goals, treaties, and conventions. Each of these is explained below with examples to help illustrate their role.

International *goals* such as the *Sustainable Development Goals* (SDGs) can be adopted as a resolution by the United Nations (UN). The UN adopted the SDGs after a 3-year consultation process involving 83 national surveys and engagement of

Fig. 20.12 UN Sustainable Development Goals (Source: https://www.un.org/sustainabledevelopment/sustainable-development-goals/)

over seven million people through a 2015 resolution. This resolution formally agreed to the 2030 Agenda for Sustainable Development and its 17 Sustainable Development Goals, adopting these as part of the "Road to Paris" for the 2015 Climate Agreement discussed above.

Although not legally binding, the Sustainable Development Goals are supposed to set out a "blueprint" for achieving critical international goals related to a wide range of topics. In addition to other goals that touch on aspects of the FEW nexus (Fig. 20.12), the SDGs specifically include the FEW nexus: Zero Hunger (SDG 2); Clean Water and Sanitation (SDG 6); and Clean and Affordable Energy (SDG 7). See Sect. 3.7 for more on the Sustainable Development Goals.

A *treaty* is a formally ratified agreement between countries. Ratification means a country has formally agreed to the treaty; in the USA, for example, the U.S. Senate must ratify all treaties (UN n.d.). Treaties are and will continue to be an important international tool because watersheds (either surface or groundwater) generally do not follow political boundaries. A treaty can be between two countries (a *bilateral treaty*) or more than two countries (a *multilateral treaty*). There are numerous international treaties over water in particular, with an estimated 400 water-related treaties enacted between 1820 and 2007, not including those related to navigation; of these, approximately 86% are bilateral and just 14% multilateral (Lubner 2015). Although a number of people say that fights over water will result in the next war, researchers like Aaron Wolf have examined the rhetoric of "water wars" versus the reality and found that the need to share scarce water resources is far more likely to lead to cooperation (Wolf 2007).

A *convention* includes representatives from many countries who work together to develop a general agreement about procedures or actions they will take on specific topics. The Paris Agreement to address climate change is the most recent action taken to enact the UN Framework Convention on Climate Change (UNFCC), adopted in 1992 and ratified by 197 parties. The 2015 Paris Agreement sets out agreements by various countries to meet Nationally Determined Contributions (NDCs) to reduce greenhouse gas emissions. While the original 1992 UNFCC addresses climate change, other conventions have been adopted that deal more specifically with water, such as the Convention on the Protection and Use of Transboundary Watercourses and International Lakes (adopted in 1997) and the Convention on the Non-Navigation of International Water Sources (adopted in 2014).

A full explanation of international agreements in their various forms is beyond the scope of this book. However, they can be useful as a way to help decrease potential conflict.

20.4 Conflict Resolution Toolkit

From the above examples, a number of ways exist for developing solutions to challenges throughout the FEW system. For individuals entering the field, the "toolkit" discussed above provides several ways to address nexus challenges; this is not an exclusive list. Tools discussed here include the following:

- Creative negotiation techniques, including those that draw on traditional dispute resolution methods (i.e., Paris Accord/indaba);
- Negotiated development of regulations that merge human and environmental needs (i.e., Kansas LEMAs, "Louisiana Speaks" Regional Plan, Murray-Darling water law amendments);
- The use of serious games and role plays to address threats (i.e., the NECAP Project);
- A variety of collaborative governance tools, including facilitated workshops with the public (i.e., Chehalis Basin, "Louisiana Speaks," NECAP Project);
- Mediation to bring parties together through the use of a neutral third party (i.e., the Central and West Coast Groundwater Basins in the Los Angeles area);
- Arbitration to decide issues like energy contracts (i.e., changes in European laws);
- Court judgments on water issues (i.e., Columbia River issues); and
- International goals, agreements, and treaties (i.e., the Paris Climate Agreement, Sustainable Development Goals).

There are other tools beyond the scope of this chapter that may be helpful to consider as well such as the financial incentives used to leverage public conservation of resources in the Murray-Darling Basin of Australia. More reading is provided below.

20.5 Conclusion/Summary

Conflict in all or parts of the food–energy–water nexus continues to be an evolving realm, with a continuous influx of new parties, challenges, and creative solutions. Although struggles within the USA are common, the reach of this nexus relationship is global in scope.

Climate impacts and weather variability act to increase the intensity of current threats and introduce new issues to each sector. Furthermore, weak governance institutions, transboundary issues, population increases, and cultural differences provide contextual challenges in developing viable solutions.

The scales of conflicts range from individual to community, regional, and international levels, each with its own unique set of challenges and factors. Opportunities to address conflict at varying scales offer creative ways to blend management directives with sustainable solutions.

Vital to the success of developing solutions is reliable collaboration among involved parties. Dispute resolution techniques, such as facilitation and mediation, present decision-makers and interested stakeholders with a platform to hear interests, concerns, and creative options for solving challenges. Complicated challenges in the nexus will require continuous stakeholder engagement and a connection of science to law, policy, metrics, and modeling systems. These are areas where training may be helpful, as may be hiring someone with process-related skills.

Key Points

- Conflicts among food, energy, and water sectors range from individuals to global in scale.
- Minor changes to one sector can have profound impacts on the operations of the other two due to their interrelatedness.
- Changes in climate can exasperate current conflicts and act as a threat multiplier for the occurrence and intensity of future disputes.
- Weak institutional and governance factors, increasing populations, and transboundary issues act as additional human challenges to addressing conflicts within the nexus.
- Dispute resolution techniques, such as facilitation, mediation, and arbitration, offer avenues for solutions as opposed to litigation or escalating conflicts.
- A key attribute to reducing conflict is constructively engaging stakeholders to participate in improving nexus relations; this takes process-based skills which can be learned and are invaluable to anyone working within the nexus.
- Connecting decision-making with appropriate scientific factors or regulatory policy is crucial for viable solutions to nexus conflicts (Jacobs 2002).
- Solution types can be specific to a particular scale or reach across a plethora of continuums including, but not limited to, negotiations, public involvement in regulatory policy development, community preparedness workshops, and global agreements to implement change.
- Conflict in the food, energy, and water nexus continues to be a subject area worth understanding better both within the USA and internationally. Each sector has a

clear effect on the others, but defining the scale of connection between all three has been challenging, unpredictable, and oftentimes, uncharted. More importantly, developing viable solutions to cope with conflict in the nexus is critical to future management, policy development, and human interaction with the environment.

Discussion Points and Exercises

1. We learned about the challenges in California with respect to water allocation, drought, and flooding. Based on the solutions discussed above, how would you improve water management in California to balance human and environmental needs? What specific conflict resolution tools might you use?
2. With respect to climate change, how can governments ensure the FEW nexus stays in balance? What tools can be used to adapt to climate changes while ensuring security for food, energy, and water sources? (See, e.g., U.S. DOE 2015).
3. Many areas of the world remain vulnerable to imbalances in the nexus, especially in areas of the Middle East and Africa where food and water security is of concern. What factors should be prioritized in resolving transboundary conflicts between nations in these areas? Why did you choose these factors?
4. A number of solutions have been developed at multiple scales for reducing human conflict within the nexus. Which solution do you think can be used at all scales successfully? Why?
5. In many parts of the world, weak governance structures lead to conflict over nexus factors. How can alternative dispute resolution processes help overcome this lack of framework?
6. You are an ambassador for a country that is part of a multi-nation global environmental agreement to conserve freshwater supplies. One of the signatory nations has proposed to amend the agreement to build more dams along a river that creates the border with another nation also party to the agreement. The border nation opposes additional construction of dams for fear of receiving less water from the shared river source. As a result, it has threatened to leave the environmental agreement altogether on national security grounds. How might this issue be managed where the interests of both countries are taken into account?
7. After learning about human conflict and the nexus, how can modeling tools be used to help resolve a conflict over competing demands for food, energy, and water supplies?
8. Based on the roles of economics, infrastructure, and ecosystems play on the nexus, how can human conflict exasperate or mitigate these factors?
9. Human conflict often emerges from competing demands for limited resources. Based on earlier chapters addressing data challenges, how would you resolve conflicts where data gaps make it difficult to develop policy or for leaders to make important decisions?
10. Modeling tools can be effective predictors for future climate scenarios, as well as water and food supplies. However, models can produce different results,

potentially with big ramifications. How would you remedy varying model results in conflicts between shared transboundary resources?

11. As an employee for a state environmental agency specializing in water management, your state is in the midst of determining whether a quantified drought plan is necessary for future water supply decisions. To make an informed decision, you have been asked to research drought statistics for your state from the USA Drought Monitor. How will you incorporate data and statistics about drought in your state to allow decision-makers to make the best decision as to water use during shortages while avoiding conflict among competing users?
12. As a consultant in the northwestern USA, your company is attempting to construct a large reservoir on a major river that would generate hydroelectric power for the region. However, Chinook salmon use this same river for spawning and migration patterns annually. How can you use datasets and flow requirements for salmon life cycles from the U.S. Fish and Wildlife Service to resolve a conflict among environmental advocates and proponents for the dam?

References

Abel, G. J., et al. (2019). Climate, conflict and forced migration. *Global Environmental Change, 54*, 239–249. https://doi.org/10.1016/j.gloenvcha.2018.12.003.

Barrett, J. T., & Barrett, J. P. (2004). *A history of alternative dispute resolution: The story of a political, cultural, and social movement*. San Francisco: Jossey-Bass.

Blomgren-Bingham, L. (2009). Collaborative governance: Emerging practices and the incomplete legal framework for public and stakeholder voice. *Journal of Dispute Resolution, 2009*(2). Available at: https://scholarship.law.missouri.edu/jdr/vol2009/iss2/2.

Blomgren-Bingham, L., Nabatchi, T., & O'Leary, R. (2005). The new governance: Practices and processes for stakeholder and citizen participation in the work of government. *Public Administrative Review, 65*(5). Available online at: http://citeseerx.ist.psu.edu/viewdoc/download?doi=10.1.1.455.8574&rep=rep1&type=pdf.

Bloomberg, J. W. (2016). *Dilemma for U.S. farmers: Which crops will lose the least money?* Retrieved November 18, 2016, from http://www.stltoday.com/business/local/dilemma-for-u-s-farmers-which-crops-will-lose-the/article_7e3869fb-e297-5526-a669-556d43574323.html.

Brown, M. E., et al. (2015) *Climate change, global food security, and the U.S. Food System*. U.S. Global Change Research Program Executive Summary. Retrieved December 13, 2016, from http://www.usda.gov/oce/climate_change/FoodSecurity2015Assessment/FullAssessment.pdf.

Carpenter, S. L., & Kennedy, W. J. D. (2001). *Managing public disputes: A practical guide for professionals in business, government, and citizens' groups*. San Francisco: Jossey-Bass.

Center for Planning Excellence. (2016). *"Louisiana Speaks" regional plan*. Retrieved December 6, 2017, from http://www.cpex.org/louisiana-speaks/.

Chehalis Basin Strategy. (2016). *Background*. Retrieved December 2, 2016, from http://chehalis-basinstrategy.com/background/.

Columbia Basin Trust. (n.d.) *An overview: Columbia River treaty*. Columbia River Trust. Retrieved November 27, 2016, from https://thebasin.ourtrust.org/wp-content/uploads/delightful-downloads/CRT_Overview.pdf.

Columbia River Inter-Tribal Fish Commission. (2017). *Columbia River Treaty*. Retrieved January 2, 2017, from https://www.critfc.org/tribal-treaty-fishing-rights/policy-support/columbia-river-treaty/.

Dempsey, J., & Coburn, N. (2010). *Traditional dispute resolution and stability in Afghanistan.* U.S. Institutes of Peace Peacebrief. Available online at: https://www.usip.org/sites/default/files/PB%2010%20Traditional%20Dispute%20Resolution%20and%20Stability%20in%20Afghanistan.pdf.

Farishta, A. (2014). The impact of Syrian refugees on Jordan's water resources and water management planning. Master's thesis, Columbia University.

Felix, B. (2015). *Climate talks turn to South African indaba process to unlock deal.* Available via Reuters. Retrieved December 12, 2016, from http://www.reuters.com/article/us-climatechange-summit-indaba-idUSKBN0TT29C20151210.

Fisher, R., & Ury, W. L. (1981). *Getting to yes: Negotiating agreement without giving in.* New York: Penguin Books.

Fowler, L. B., & Shi, X. (2016). Human conflicts and the food, energy, and water nexus: Building collaboration using facilitation and mediation to manage environmental disputes. *Journal of Environmental Studies and Sciences, 6*, 104–122. https://doi.org/10.1007/s13412-016-0373-x.

Franzetti, E. (2017). Recent investor-state arbitral awards in the energy sector, energy across borders: Geopolitics, conflict and international arbitration. Handout provided at Penn State's 2017 energy days.

Fritz, A. (2017). *The Oroville Dam spillway failed miserably, so California is blowing it up.* Retrieved July 13, 2017, from https://www.washingtonpost.com/news/capital-weather-gang/wp/2017/06/06/the-oroville-dam-emergency-spillway-failed-miserably-so-california-is-blowing-it-up/?utm_term=.6cc40c6b4f23.

Gillis, J. (2016). *Flooding of coast, caused by global warming, has already begun.* Retrieved December 4, 2016, from https://www.nytimes.com/2016/09/04/science/flooding-of-coast-caused-by-global-warming-has-already-begun.html.

Gleick, P. (2015). *Impacts of California's ongoing drought: Hydroelectricity generation.* Pacific Institute. Retrieved July 13, 2017, from http://pacinst.org/wp-content/uploads/2015/03/California-Drought-and-Energy-Final1.pdf.

Golden, B. (2016). *Monitoring the impacts of Sheridan County 6 local enhanced management area.* Retrieved December 17, 2016, from https://agriculture.ks.gov/docs/default-source/lemas/2016-12-15_lema-report-formal-review.pdf?sfvrsn=4.

Governor's Chehalis Basin Work Group. (2014). *Chehalis Basin strategy 2014 recommendation report.* Retrieved December 4, 2016, from http://chehalisbasinstrategy.com/wp-content/uploads/2015/09/ChehalisBasinWorkGroupRecommendationsReport_Final_0004.pdf.

Guterres, A. (2011). Statement by Mr. António Guterres, United Nations High Commissioner for Refugees, United Nations Security Council Briefing "Maintenance of International Peace and Security: "New Challenges to International Peace and Security and Conflict Prevention". Retrieved January 31, 2019, from https://www.unhcr.org/admin/hcspeeches/4ee21edc9/statement-mr-antonio-guterres-united-nations-high-commissioner-refugees.html.

Illgner, R. (2016). Using financial incentives for environmental compliance. Oral presentation at 2016 annual American Water Resources Association conference, Florida Hotel & Conference Center, Orlando, 13–17 Nov 2016.

Jacobs, K. (2002). *Connecting science, policy, and decision-making: A handbook for researchers and science agencies.* NOAA Office of Global Programs. Retrieved January 4, 2017, from https://leopoldleadership.stanford.edu/sites/default/files/Jacobs_2001-02_Connecting.Science.Decisionmaking.pdf.

Johnstone, J., & Mazo, J. (2011). Global warming and the Arab Spring. *Survival, 53*(2), 11–17.

Kiger, P. J. (2013). *Colorado flooding imperils oil and gas sites, causes spill.* Retrieved December 7, 2016, from http://news.nationalgeographic.com/news/energy/2013/09/130919-colorado-flooding-causes-oil-spill/.

Kings, S. (2016). *Climate change is testing southern Africa water agreements.* Retrieved December 3, 2016, from http://www.climatechangenews.com/2016/12/02/climate-change-is-testing-southern-africa-water-agreements/.

Kluwer Arbitration Blog. (2018). *The three hottest energy arbitrations of 2017.* Available online at: http://arbitrationblog.kluwerarbitration.com/2018/01/04/three-hottest-energy-arbitrations-2017/.

Little, J. B. (2009). *The Ogallala Aquifer: Saving a vital U.S. water source*. Retrieved January 30, 2019, from https://www.scientificamerican.com/article/the-ogallala-aquifer/#googDisableSync.

Louisiana Recovery Authority. (2007). Louisiana speaks regional plan, vision and strategies for recovery and growth in south Louisiana. Louisiana Recovery Authority. Retrieved December 6, 2016, from https://static1.squarespace.com/static/536d55f1e4b07afeea8cef61/t/54500a8de4b0d2e64620a808/1414531725899/LA_Speaks_FINAL.pdf.

Lubner, V. E. (2015). The impact of international water treaties on transboundary water conflicts: A study focused on large transboundary lakes. Theses and dissertations. Paper 817. University of Wisconsin-Milwaukee.

Massachusetts Institutes of Technology. (n.d.). *Definition of serious games*. http://gamelab.mit.edu/tag/serious-games/.

McDonnell, T. (2018). *The refugees the world barely pays attention to*. Retrieved January 30, 2019, from https://www.npr.org/sections/goatsandsoda/2018/06/20/621782275/the-refugees-that-the-world-barely-pays-attention-to.

McGraw, S. (2015). *Betting the farm on a drought: Stories from the front lines of climate change*. Austin: University of Texas Press.

Moffitt, M. L., & Bordone, R. C. (Eds.). (2005). *The handbook of dispute resolution*. Cambridge, MA: Program on Negotiation at Harvard Law School.

Munich, R. E. (2017). *Natural catastrophic losses at their highest for four years*. Retrieved January 5, 2017, from https://www.munichre.com/en/media-relations/publications/press-releases/2017/2017-01-04-press-release/index.html.

Murray-Darling Basin Commission. (2008). *Sustainable rivers audit: A report on the ecological health of rivers in the Murray-Darling Basin, 2004–2007*. Murray-Darling Basin Commission. Retrieved January 4, 2017, from https://www.mdba.gov.au/sites/default/files/archived/mdbc-sra-reports/1-SRA_Report_1_tech_full.pdf.

National Drought Monitor. (2019). *History of Alabama droughts, 2000 to 2018*. Available online at: https://www.drought.gov/drought/states/alabama.

Ostrom, E. (1965). *Public entrepreneurship: A case study in ground water management*. Available online at: https://dlc.dlib.indiana.edu/dlc/handle/10535/3581.

Ostrom, E., & Hess, C. (2007). Private and common property rights, encyclopedia of law economics. Available at: https://surface.syr.edu/cgi/viewcontent.cgi?article=1024&context=sul.

Owen, C. C. (2016). Local enhanced management areas or "What's not the matter with Kansas?" Oral presentation at 2016 annual American Water Resources Association conference, Florida Hotel & Conference Center, Orlando, 13–17 Nov 2016.

Parker, L. (2016). *What happens to the U.S. Midwest when the water's gone?* Retrieved December 13, 2016, from http://www.nationalgeographic.com/magazine/2016/08/vanishing-midwest-ogallala-aquifer-drought/.

Perera, A. (2017). *With drought looming, Sri Lanka tries something new: Preparing*. Reuters. Retrieved January 18, 2017, from http://www.reuters.com/article/us-sri-lanka-drought-resilience-idUSKBN1520X6.

Puckett, K. (2016). *Report says climate change will cut farm, ranch earnings*. Retrieved December 14, 2016, from http://www.greatfallstribune.com/story/news/local/2016/02/25/report-says-climate-change-will-cut-farm-ranch-earnings/80962412/.

Reilly, A. (2017). *Special master hands Ga. a victory in Southeast water war*. Available via E&E NEWS. Retrieved July 5, 2017, from https://www.eenews.net/stories/1060050120/.

Richter, B., et al. (2016). *Water share, using water markets and impact investment to drive sustainability*. The Nature Conservancy. Retrieved January 3, 2017, from https://thought-leadership-production.s3.amazonaws.com/2016/08/16/13/41/58/5e9b26b2-5c77-40f6-81fd-03e0c3de78a9/WaterShareReport.pdf.

Ritter, K. (2018). *São Paulo heading to another dry spell*. Available via Circle of Blue. Retrieved January 31, 2019, from http//www.circleofblue.org/2018/water-climate/drought/sao-paulo-heading-to-another-dry-spell/).

Rocket, M. (2016). *Indaba negotiation strategy to reach big decisions*. Available via AVATAR. Retrieved December 13, 2016, from http://www.avatar.co.nz/blog/indaba-negotiation-strategy-to-reach-big-decisions.html.

Rumore, D. (2017). *Using serious games to address climate risk*. University of Utah Environmental Dispute Resolution Blog. Available online at: https://www.law.utah.edu/using-serious-games-to-help-communities-make-progress-on-serious-problems/.

Sengupta, S. (2016). *Heat, hunger and war force Africans onto a 'Road on Fire.'* New York Times. Retrieved December 15, 2016, from https://www.nytimes.com/interactive/2016/12/15/world/africa/agadez-climate-change.html?smid=tw-share.

Sprague, M. (2013). *Judge approves water storage plan for southeast area*. Whittier Press Telegram. Retrieved February 2, 2019, from https://www.presstelegram.com/2013/12/18/judge-approves-water-storage-plan-for-southeast-area/.

Steward, D. R., et al. (2013). Tapping unsustainable groundwater stores for agricultural production in the High Plains Aquifer of Kansas, projections to 2110. *Proceedings of the National Academy of Sciences of the United States of America, 110*(37), E3477–E3486. https://doi.org/10.1073/pnas.1220351110.

Stockton, N. (2017). *California floods its fields to keep its cities from flooding*. Available at WIRED. Retrieved January 10, 2017, from https://www.wired.com/2017/01/california-flooding-keep-cities-flooding/.

Susskind, L., & Ozawa, C. (1983). Mediated negotiation in the public sector, 27 *Am. Behavioral Scientist* 255, 256.

Susskind, L. E., Levy, P., & Thomas-Larmer, J. (1999). *Negotiating environmental agreements: How to avoid escalating confrontation, needless costs and unnecessary litigation*. Cambridge, MA: MIT-Harvard Public Disputes Program.

Susskind, L. E., et al. (2015). *Managing climate risks in coast communities: Strategies for engagement, readiness and adaptation*. New York: Anthem Press.

U.S. Department of Energy Office of Energy Policy and Systems Analysis. (2015). *Climate change and the U.S. energy sector: Regional vulnerabilities and resilience solutions*. U.S. Department of Energy. Retrieved December 11, 2016, from https://energy.gov/sites/prod/files/2015/10/f27/Regional_Climate_Vulnerabilities_and_Resilience_Solutions_0.pdf.

U.S. Environmental Protection Agency (EPA). (n.d.). Filings, Procedures, Orders and Decisions of EPA's Administrative Law Judges, from https://www.epa.gov/alj.

United Nations. (n.d.). *Glossary*. Available online at: https://treaties.un.org/pages/Overview.aspx?path=overview/glossary/page1_en.xml.

Washington Department of Ecology. (2015). *Restoring habitat and reducing flood damage: Working together in the Chehalis River Basin*. Retrieved July 14, 2017, from http://www.ecy.wa.gov/news/2015/133.html.

Werrell, C. E., & Femia, F. (2013). The Arab Spring and climate change. In *A climate and security correlations series*. Center for Climate and Security. Retrieved December 12, 2016, from https://climateandsecurity.files.wordpress.com/2012/04/climatechangearabspring-ccs-cap-stimson.pdf.

William D. Ruckelshaus Center. (2015). *Chehalis Basin strategy*. Retrieved December 2, 2016, from http://ruckelshauscenter.wsu.edu/projects/current-projects/chehalisfloodingfish/.

Witherall, A., & Simes, J. (2014). *Los Angeles Basin groundwater adjudication summary*. U.S. Bureau of Reclamation. Available online at: https://www.usbr.gov/lc/socal/basinstudies/LA%20Adjudication%20Dec%202014.pdf.

Wolf, A. T. (2007). Shared waters: Conflict and cooperation. *Annual Review of Environment and Resources*. Available online at: https://www.annualreviews.org/doi/10.1146/annurev.energy.32.041006.101434.

Further Reading

Bingham, L. B. (2009). Collaborative governance: Emerging practices and the incomplete legal framework for public and stakeholder voice. *Journal of Dispute Resolution, 2009*, 269–324.

Fisher, R., & Ury, W. (1991). *Getting to yes-negotiating agreement without giving in* (2nd ed.). UK: Penguin Books.

Fowler, L. B., & Shi, X. (2016). Human conflicts and the food, energy, and water nexus: Building collaboration using facilitation and mediation to manage environmental disputes. *Journal of Environmental Studies and Science, 6*, 104–122.

Lewicki, R. J., Gray, B., & Elliot, M. (Eds.). (2003). *Making sense of intractable-environmental conflicts: Concepts and cases*. Washington, DC: Island Press.

Moffitt, M. L., & Bordone, R. C. (Eds.). (2013). *The handbook of dispute resolution*. Cambridge, MA: Program of Negotiation at Harvard Law School.

Rogers, N., Bordone, R. C., Sander, F. E. A., & McEwen, C. A. (2018). *Designing systems and processes for managing disputes* (2nd ed.). New York: Wolters Kluwer.

Susskind, L. E. (1994). *Environmental diplomacy: Negotiating more effective global agreements*. New York: Oxford Press.

Susskind, L. E., et al. (2015). *Managing climate risks in coast communities: Strategies for engagement, readiness and adaptation*. New York: Anthem Press.

Swain, A., & Ojendal, J. (Eds.). (2018). *Routledge handbook of environmental conflict and peacebuilding*. New York: Routledge.

Waldo, J. C., Fowler, L. B., & West, C. W. (2012). Mediation and voluntary resolution of environmental, natural resources, and land use conflicts. In A. W. Mackie et al. (Eds.), *Washington real property deskbook series. Vol. 6: Land use development* (4th ed., pp. 18–30). Washington: Washington State Bar Association.

Chapter 21
Opportunities at the Nexus

Beth Kinne, Darrin Magee, Bruce A. McCarl, Rabi Mohtar, Robert B. Richardson, Benjamin L. Ruddell, Peter Saundry, and Lara Treemore-Spears

21.1 Introduction

Management of the FEW Nexus is essential for the successful development and support of humanity. Fortunately, opportunities at the Nexus are many. As problems in one area often cause further problems in other areas, so improvements in one area often lead to benefits in other areas. Every chapter in this textbook emphasizes opportunities, problems, and tools for addressing the opportunities; we will summarize some of these key high-level FEW Nexus opportunities here.

B. Kinne · D. Magee
Environmental Studies, Hobart and William Smith Colleges, Geneva, NY, USA
e-mail: kinne@hws.edu; magee@hws.edu

B. A. McCarl
Department of Agricultural Economics, College of Agriculture and Life Sciences, Texas A&M University, College Station, TX, USA
e-mail: mccarl@tamu.edu

R. Mohtar
Faculty of Agricultural and Food Sciences, American University of Beirut, Beirut, Lebanon

Department of Biological and Agricultural Engineering, Texas A&M University, College Station, TX, USA
e-mail: mohtar@aub.edu.lb; mohtar@tamu.edu

R. B. Richardson
Department of Community Sustainability, College of Agriculture & Natural Resources, Michigan State University, East Lansing, MI, USA
e-mail: rbr@msu.edu

B. L. Ruddell
School of Informatics, Computing, and Cyber Systems, Northern Arizona University, Flagstaff, AZ, USA
e-mail: Benjamin.Ruddell@nau.edu

P. Saundry, B. L. Ruddell (eds.), *The Food-Energy-Water Nexus*, AESS Interdisciplinary Environmental Studies and Sciences Series,
https://doi.org/10.1007/978-3-030-29914-9_21

We began this book by noting that FEW systems provide critical consumable commodities, require massive infrastructure, are currently footprint-heavy, must be extremely accessible and affordable, and are the focus of extensive governance at all levels of human society. Food, energy, and water security are critical to healthy, prosperous, and stable human societies. Recognition of this is reflected by the prominence of food, energy, and water in the 2015 Sustainable Development Goals (Sect. 3.7) and many of "grand challenges" in science and engineering (see Sect. 1.3.3). Achieving FEW security requires the integrated management of FEW systems sustainably.

FEW systems are profoundly influenced by demographics and societal development, human behavior, economics and trade, air pollution, ecosystems, climate, and climate change. One or more of these factors are major parts of and Nexus research project and practice.

For all of these reasons, people are the center of all framings (macroscopes) of the nexus, six of which were presented in Sect. 1.4. Thus, opportunities at the nexus that change human behavior are as important as opportunities to improve science and technology.

In Chap. 17, we explored the application of nexus science to real-world problems, or practice, as carried out by both scientists and non-scientists referred to as practitioners. We noted the centrality of decision-making and how stakeholders can come together as Communities of Practice to utilize science-based tools that enhance their ability to make decisions by maximizing areas of agreement and minimizing areas of conflict. We explored a number of valuable tools that can be used to make science useful to decision-making processes, including:

- Data integration
- Integrated assessment modeling
- Methods of visual analytics
- Online platforms
- Immersive decision environments
- Tools for addressing decision-making under uncertainty
- Tradeoff analysis
- Communities of practice

The many case studies throughout this book, and in particular Chaps. 18–20, highlight the varied contexts where food, energy, and water systems interact and the many opportunities to apply science to practice. The case studies in this book also illustrate the changes to FEW systems which can be made. Many possible changes considered as positive or progressive, emphasize sustainability. A positive change

P. Saundry (✉)
Energy Policy and Climate, Advanced Academic Programs, Krieger School of Arts and Sciences, Johns Hopkins University, Washington, DC, USA
e-mail: psaundr1@jhu.edu

L. Treemore-Spears
Health Urban Waters, College of Engineering, Wayne State University, Detroit, MI, USA
e-mail: treemorespears@wayne.edu

sustains and regenerates the biophysical environment. Progressive changes provide opportunities for rewarding labor, investment, and consumption in the economic system. Positive opportunities impel and support social systems that are inclusive, open, equitable, and just in terms of race, ethnicity, gender, and age.

In this chapter, we shift from "what is" to "what might be." We begin by exploring some criteria for identifying real-world challenges that provide greater impetus and opportunity for applying integrated FEW science to real-world practice. There is the greatest opportunity for FEW Nexus applications to improve outcomes where there is a specific combination of scarcity, competition, externalities, and shared benefits. This often involves the invention of mechanisms for cooperation, reallocation of resources, sharing of private data, transactions across system boundaries, and the limited but proper role of government and law and, more broadly, the community of science and practice in the FEW nexus. We conclude with a number of case studies of Nexus work and practice that epitomize the opportunities that exist.

21.2 Situations Favorable to the Application of Nexus Science to Practice

In Sect. 1.5.2, we noted that certain situations provide greater impetus and opportunity for applying integrated FEW science to real-world practice. We now revisit the three examples of this in the context of the entire book.

21.2.1 Acute Scarcity of Two or More FEW Commodities

The aphorism "no crisis should go to waste" encapsulates opportunity for significant change that becomes possible following a natural or man-made catastrophe. Rebuilding after a storm (e.g., Puerto Rico), or creating new structures of governance after a war (e.g., the United Nations) are familiar examples. Situations of food, energy, and water insecurity are also opportunities for change. Scarcity primarily refers to the physical availability and the physical, legal, and economic access attributes of food, energy, and water security (see Sect. 3.2). These case studies provide opportunities to learn from systems approaches to resource management to build resilience.

Droughts are occurring with some frequency. Examples of droughts that have been noted in this book include California (2011–2017) (see Sects. 11.4.4 and 20.2.1), Texas (2011) (see Sect. 13.3), southern Africa (see Sect. 20.2.1), northeastern Brazil (2012–2017) (see Sect. 20.2.1), Sri Lanka (see Sect. 20.2.1), and the Murray-Darling River Basin of Australia (see Sect. 20.3.4). In each of these locations, water is critical for energy (typically for power plant cooling) and food production (especially irrigation) as well as direct consumption (which can experience increased contamination in low flows). Droughts frequently bring siloed interests in food, energy, and water into conflict. Thus, many of the case studies in Chap. 20 (Managing Human Conflicts) are triggered by droughts. As shown in Chap. 11, climate change will result in more droughts in the future.

Chapter 20 also notes that FEW conflicts are often outbreaks based on long-term simmering tensions between different interests and that droughts are an opportunity to recognize and address long-term FEW tensions and develop integrated solutions.

Because food and energy are heavily traded (Chap. 7), local disruptions in food and energy production can be mitigated by imports. However, imports can be physically denied (e.g., the oil crisis of 1973–1974), and be too expensive for many in need. These situations often highlight issues of systemic, long-term poverty, and other forms of social exclusion. In Chap. 3 (Development), we noted the importance of this issue in developing countries. In Sect. 18.4, we described the significance of poverty and exclusion in cities. In such situations, responses can recognize interactions with other components and tensions between them. For example, solutions to local disruptions in food and energy flows can include greater demands on water, arable land, and biomass, each of which creates tensions between FEW interests.

Acute shortages can also be a result of infrastructure failure. Short-term failures of electric power grids are not uncommon. Rivers are examples of natural infrastructure that move food, energy, and water. One important aspect of droughts is their ability to impact FEW transportation on rivers. During the Great Plains Drought of 2012, low water levels on the Mississippi River limited barge traffic on the river moving coal and crops.

Finally, it should be recognized that flooding, in addition to drought, can also bring about acute shortages of food, energy, and water, through direct impacts and through infrastructure impacts.

Whatever the cause of acute scarcity, it is frequently associated with a level of social unrest that motivates action by governments and international organizations. Acute scarcities were part of the motivation of the Sustainable Development Goals (Sect. 3.7). Responding to acute shortages and achieving the SDGs for food, energy, and water simultaneously is a strong motivation for integrating food, energy, and water management. In particular, repeated shortages highlight the limitation of single-sector solutions and make clear the need for integrated management.

21.2.2 Significant Externalities Arising from FEW Decisions and Stakeholder Actions

Throughout this book were have noted many examples of externalities. In Sect. 5.3.1, we noted examples such as the following:

1. applications of nitrogen fertilizers on food crops impacting local rivers and aquifers but not being reflected in the costs of the fertilizer appliers or their resultant crop product price;

2. pollutants entering aquifers due to infiltration of produced fracking water; and
3. greenhouse gas emissions coming largely from affluent, high energy-use societies, while the adverse externality (climate change effects) fall disproportionately on poor, low energy-use societies (Chap. 11).

Climate change is the highly visible global externality that is driving major shifts in the governance and practice of energy systems (primarily), but also in food and water systems. For example, coal-fired electricity generation is water-intensive compared to natural gas and renewables. Shifts away from coal-fired power generation in the USA are reducing water use. In a second example, emissions of greenhouse gases associated with food systems are leading to efforts to change agricultural practices, bolster soil carbon stocks, change food choices (especially meat-intensive diets), and reduce food waste.

There are many examples of the degradation of ecosystems and ecosystem services as externalities. Soil erosion degrading soil quality, and agricultural productivity is one example. The case study of erosion control services and conservation agriculture (Sect. 9.4) illustrates a solution with cross-cutting benefits. A shift to conservation agriculture creates benefits for food (increased productivity), as well as for water (e.g., less runoff means better flood regulation and more irrigation storage), and for energy (e.g., water available for hydropower production or more traditional biomass available).

Air pollution, water pollution, contamination of arable land, and biodiversity loss, are broad classes of externalities related to FEW systems. However, externalities only motivate action when there is a public reaction against that externality. The reaction usually begins in communities that are adversely affected by the externality, but it must include all parties to the problem in order to succeed in creating positive change.

Societies have considerable experience in developing governance strategies for externalities, ranging from rules on behavior and technology (command-and-control regulation) to market-based regulation such as pricing the pollutant or the commodity that is being used in a manner that leads to the externality (see Chap. 5). While there is little experience in addressing the FEW nexus with these strategies, sector-specific experiences provide enough confidence to many to view regulating externalities as a tool for addressing some challenges at the FEW nexus.

21.2.3 Potential Benefits to Many Communities from Coordinated Actions

Throughout this book, we have explored many instances of cooperation. Cooperation generally occurs where parties see potential benefits for themselves, even if that benefit is just the avoidance or minimization of harm. For example, Chap. 6 (International Governance) described the extensive international system established following World War II to facilitate international cooperation largely to avert a repeat of the harm caused by two world wars and the Great Depression. That system is largely responsible for the cooperation embodied in the Sustainable Development Goals (Sect. 3.7) and climate change (Chap. 11).

Chapter 5 (Economics) provides several tools to identify who gains and who loses under alternative scenarios, an important way to understand the incentives needed to attain cooperation. As noted in that Chapter, market incentives can be created to promote cooperation. Examples of incentives include assignment of private rights to property in place of property held in common, taxes on pollutants and inefficiencies, or the provision of subsidies for systemically beneficial practices.

The extensive system of cooperative international trading in FEW commodities described in Chap. 7 is largely a result of joining economic tools of cooperation with international governance structures supporting cooperation. Cooperative trading of FEW commodities supports FEW security but also achieves mutual economic benefits through the application of comparative advantage.

Chapter 10 (Infrastructure) noted that FEW infrastructures are sources of cooperation or conflict regionally and between nation-states. Chapters 18–20 of the book argue for the centrality of cities, watersheds, and conflicts as the most important contexts of application of FEW Nexus thinking. Cities are hubs of wealth, consumption, and knowledge, and have the power to dramatically alter human behavior and system function through their many economic and social connections. Watersheds are hydro-political units that integrate water governance, land use, food production, transportation, and water management, and are therefore a significant opportunity for FEW Nexus applications. Human conflicts at all scales often touch on (or are caused by) the FEW Nexus, and the presence of conflict presents great opportunities and risks for the application of Nexus thinking.

21.3 Opportunities

The core opportunity of work at the food–energy–water nexus is the opportunity to develop and improve tools in science, engineering, communication, stakeholder collaboration, decision-making, policy, governance, and conflict management that support the achievement of nexus goals. Many case studies and illustrations of this opportunity set are included in this book to inspire solutions.

New science and practice must be understood in the context of both the biophysical environment and the socio-economic-political environment. It is important to recognize that gaps in our understanding of both exist for four primary reasons. First, it is frequently, and inaccurately, assumed that sufficient use-inspired research has already been completed to support sound science-based decision-making leading to practical interventions. Second, the interactions between human activities and biophysical systems have had both positive and negative consequences for different interests. Third, the complexity of coupled human-natural systems makes separation of causes and effects difficult—every effect is also a cause, and every cause is also an effect. Finally, the scale and complexity of biophysical systems make it difficult to forecast accurately all the impacts of human attempts to influence a given biophysical system. Similarly, the complexity of multifaceted socio-economic-political systems makes it difficult to forecast all the impacts of particular changes in FEW governances on human behavior accurately.

Recognizing each of these gaps is an opportunity for future work that fills them and enables ever more effective solutions to Nexus challenges.

We will now review several interwoven themes where opportunities exist to overcome those challenges.

21.3.1 Communities of Science and Practice

Integrated research, capacity building, outreach, education, and informed private, public, and civil sectors are essential for the development of Nexus solutions. Solutions can be applied and tested locally and then used beyond regional and national borders. Such solutions require interdisciplinary cooperation, inclusivity, and transparency among stakeholders. Scientifically enabled policy, monitoring, assessment, and cooperation must complement the Sustainable Development Goals (SDGs) to make it possible to achieve them. We noted in Sect. 17.5, the value of developing effective communities of science and practice that bring diverse scientists, engineers, and non-science stakeholders together to address nexus challenges.

Interdisciplinary science has traditionally been challenging because of the narrow disciplinary training and incentives provided by academic and governmental institutions. Public funding of research has likewise been dominated by narrowly disciplinary silos for funding. While the value of interdisciplinary research has been long recognized, programs that funded such research have often been short-lived compared to the decade or more that it takes to build effective interdisciplinary communities of science. However, academic and government program opportunities funding for Nexus research have emerged. For example, in the USA, the National Science Foundation (NSF) launched a program on Innovations at the Nexus of Food, Energy and Water Systems (INFEWS) in 2016 (NSF (n.d.)) in the words of NSF:

> *The INFEWS program seeks to support research that conceptualizes FEW systems broadly and inclusively, incorporating social and behavioral processes (such as decision making and governance), physical processes (such as built infrastructure and new technologies for more efficient resource utilization), natural processes (such as biogeochemical and hydrologic cycles), biological processes (such as agroecosystem structure and productivity), and cyber-components (such as sensing, networking, computation and visualization for decision-making and assessment). Investigations of these complex systems may produce discoveries that cannot emerge from research on food or energy or water systems alone. It is the synergy among these components in the context of sustainability that will open innovative science and engineering pathways to produce new knowledge, novel technologies, and innovative predictive capabilities.*
>
> *The overarching goal of the INFEWS program is to catalyze well-integrated, convergent research to transform understanding of the FEW Nexus as integrated social, engineering, physical, and natural systems in order to improve system function and management, address system stress, increase resilience, and ensure sustainability. The NSF INFEWS activity is designed specifically to attain the following goals:*

1. *Significantly advance our understanding of the food–energy–water system of systems through quantitative, predictive and computational modeling, including support for relevant cyberinfrastructure;*
2. *Develop real-time, cyber-enabled interfaces that improve understanding of the behavior of FEW systems and increase decision support capability;*
3. *Enable research that will lead to innovative and integrated social, engineering, physical, and natural systems solutions to critical FEW systems problems;*
4. *Grow the scientific workforce capable of studying and managing the FEW system of systems, through education and other professional development opportunities.*

NSF has also engaged other parts of the US government, like the U.S. Department of Agriculture, in INFEWS funding opportunities. While the longevity of INFEWS and FEWS nexus research grants that will sustain a Nexus Community of Science is uncertain, other programs support work at the nexus from a variety of sector-based and cross-cutting perspectives such as sustainability. Thus, the opportunity for public and private funders around the world to sustain the emergent Nexus Community of Science is an important one.

In Sect. 17.5, we described the example of the Sustainable Water–energy–food Nexus Working Group of Water Future as a global nexus community of science focused on water research in support of international scientific collaboration to drive solutions to the world's water problems. Many opportunities to form similar communities exist. FEW Nexus Communities of Practice are immature at this time but emerging and evolving. We argue that communities of practice provided with appropriate tools can most effectively develop and apply solutions to nexus challenges. Providing communities of practice with useful decision-making tools and platforms frame most of the specific opportunities that follow.

Establishing a global FEW Nexus Community of Science Practice (CoSiP) would provide the foundation of a general stakeholder community that will provide a global platform to effectively address the substantial existing knowledge gaps in science, education, and governance. Such a community would also enable integrated research efforts and improve capacity building, outreach, and education efforts. Proposals for a FEW Nexus Community of Science and Practice seek to transcend regional and national borders to promote inclusive, transparent, interdisciplinary cooperation and intergovernmental in approaches between all stakeholders. Its philosophy would be supportive and complementary to the United Nations Sustainable Development Goals and be used to encourage scientifically enabled policy, monitoring, assessments, and cooperation. Indeed, locally relevant work would provide the foundation for identifying solutions to common, global problems. To achieve these ends, a FEW Nexus Community of Science Practice might offer a global platform for the nexus debate and will bridge between science, policy-making, and the general stakeholder community by:

1. Establishing a shared data platform (with national and international components) that serves all three sectors (water, energy, and food);
2. Identifying data needs and shortcomings through the evaluation of existing libraries and their ability, or lack thereof, to support multi-scale, transdisciplinary research;

3. Defining the interconnectivity of few nexus systems through a set of comparative local- and regional-scale pilot projects that test data and implement solutions in multiple locations (regional, national, and international);
4. Developing a common accounting framework that supports holistic, regional and national resource management approaches;
5. Promoting the development of innovative, high tech solutions to effectively relieve the stresses and address the challenges posed by the stressors; and
6. Establishing a set of "governance indicators" for monitoring the role and effectiveness of governance in management practices in both developed and developing countries.

These activities flow into the opportunities for future nexus work described below. Since FEW research is fundamentally applied (and also Use-Inspired), communities of science and practice can, therefore, be viewed as a foundational structure to support and sustain nexus projects. The opportunities for local, regional, and global communities of science and practice exist.

21.3.2 Defining Questions

In Chap. 12, we explored many aspects of the key to effective use-inspired FEW Nexus science is defining questions that integrate research (especially basic research) with valuable applications (see Pasteur's Quadrant in Sect. 12.2.1). Communities of science and practice provide significant help in defining nexus questions that most effectively align scientific research with the needs of stakeholders and decision-makers. Thus, in Chap. 4, we described the importance of considering the role of human behavior and adaptation in FEW systems.

However, it is crucial to remember the conversation between those that conduct research and those that apply it is a two-way conversation. While science that misses the mark of decision-maker needs is not useful (and probably not sufficiently use-inspired), a decision-maker's demand that science answer value and purpose questions lying beyond the scope of empirical science is also not useful. The FEW nexus exists in the context of both the biophysical environment and the socio-economic-political environment (see Sect. 12.2.2).

Decision science, the interdisciplinary study of human decision-making at the individual, collective, and institutional levels, can be extremely useful in connecting science to applications (see Sects. 4.3, 12.2.3, and 17.2). Decision science incorporates theories and techniques from psychology, behavioral economics, and statistics, among others, to investigate how people make decisions. This understanding can help define biophysical reach questions that will address decision-maker needs. Because the FEW Nexus is fundamentally an interdisciplinary and systems concept (and not a basic disciplinary science), researchers at the Nexus should aspire to both outcomes.

Examples of biophysical research areas relevant to decision-maker needs include:

1. Improving our understanding of soil processes, phenomena and interactions relating to soil organic matter, microbes, nutrients, and moisture to changing climatic conditions, can lead to methods that increase food yields in ways that require lower energy, fertilizer, and water inputs and with fewer negative environmental, greater farm labor safety, higher quality food, and with greater economic returns to farm operators.
2. Scaling up the integration of biophysical processes into infrastructure that supports food production (e.g., pollinator conservation and restoration, multifunctional landscapes, and urban agriculture) or productive use of wasted resources (e.g., nutrient recapture from waste streams, and heating services from cooling water in power plants [cogeneration].

Examples of socio-economic-political research areas include the following:

1. The design of stakeholder engagement processes that lead to the co-production of knowledge and science and ultimately more informed policy and management solutions.
2. Improving our understanding of the behavioral reactions of individuals to changes in FEW systems.
3. Improving our understanding of the multidimensional benefits and costs of actions in alternative uses so that scarce resources can be reallocated to "higher-value" users.
4. Non-market valuation of important nexus concerns such as **public welfare**, ecosystem values, environmental damage, and cultural values.

In Sect. 12.3, we explored the challenge of scale selection in defining questions. Nexus students should align scales of space and time in biophysical processes with those in FEW governance, resource management, and decision-making so that they can be synergistic rather than discordant. This is particularly challenging because the system processes (institutional and physical) involved in the FEW Nexus operate at varied—but specific—scales, so trade-offs between micro-, meso-, and macro-scale framings are required.

It is possible to ask many research and practice questions at the FEW nexus, but as observed by physicist Lisa Randall (2011):

> *An almost indispensable skill for any creative person is the ability to pose the right questions. Creative people identify promising, exciting, and, most important, accessible routes to progress—and eventually formulate the questions correctly.*

21.3.3 Metrics

In Chap. 13, we described the role of metrics as a bridge between science and decision-making, and ultimately behavior. Metrics are selected as means of measuring things that society values and which can be backed by science. Metrics facilitate effective stakeholder communication, engagement, and decision-making. This was illustrated in Chap. 3, where we reviewed an array of metrics used to measure progress toward the food, energy, and water objectives of the Sustainable Development Goals.

Because of the importance of metrics in thinking about issues by the public and decision-makers, society is itself molded by the utilization of metrics. As a result, different stakeholders with different objectives often favor different metrics measured at different spatial and temporal scales. The evolution of metrics from those used in the Millennium Development Goals to those used for the Sustainable Development Goals reflect the values and objectives of a larger and more diverse group of stakeholders as well as a more ambitious set of objectives.

Because the choice of metric constrains data collection and modeling, the choice of metric has major scientific implications, too. Getting the metrics right is very important. The central challenge of choosing metrics is to accurately reflect desired social outcomes for both the near-term and the long-term while maximizing the ability of science to provide them. Near-term desired social outcomes involve decision-making based on what we currently understand and value. Long-term desired social outcomes require recognizing that what we understand and value will change, and future decision-makers will be locked-in to a greater or lesser extent by the prior decisions. The student of FEW systems should guide the choice of metrics in a way that both educates the public and decision-makers about near-term decision-making and encourages them to think about future options for decision. The process of choosing metrics in FEW systems provides an important opportunity for science and non-science stakeholders to engage in discussions that profoundly shape science, communication, education, and decision-making. Thus, metrics are more than a tool to measure; they are an opportunity to frame future decisions and actions.

21.3.4 Data

In Chap. 14, we noted that adequate data tends to be the limiting factor on the quality of our estimation, modeling, understanding, decision-making, and prediction. While there is a lot of data about FEW systems, it is often challenging to locate, access, or use given critical gaps. Common data scales often do not match the scales of the decision-makers' questions. FEW systems data cover a myriad of highly specialized public and private applications, and these are voluminous, complex, and diverse with respect to data structure and standard, as well as the repositories that handle each application. Data quality, management, and rules are essential concerns for FEW systems. Fusing data for different parts of the FEW system remains a serious challenge.

However, significant advancements in management are underway. Improvements in creating and deploying low-cost sensors for onsite and remote sensing, combined with wired and wireless connectivity, and in fast computing power, give us ever better abilities to design, collect, curate, share, integrate, and utilize high-quality data.

These advancements create significant opportunities to more effectively obtain and employ data in support of better understanding and managing of FEW systems and forecasting how systems will respond to internal and external changes.

Notable examples of opportunities to advance FEW systems data include "smart" agricultural, water, and manufacturing operations that generate data using sensors, detailed supply chain data, systems databases that describe all aspects of FEW systems processes in a coherent environment, public–private and private-private partnerships on data sharing between organizations, and differential privacy tools to allow appropriate and safe access to data by various parties with various levels of access and trust.

21.3.5 Models

In Chap. 15, we reviewed the state of the art in modeling for FEW systems and the challenges in developing integrated modeling tools. In particular, we emphasize the challenge imposed upon modeling by the independent, siloed, decision-making of different actors who often prefer single-system models tailored to the details of their "silo" with only minimal consideration of the other connected systems.

However, as noted above, certain situations provide significant impetus for developing integrated models to support integrated FEW management solutions—acutely scarcities, consequential externalities, and compelling potential benefits. In such situations, successful models are based on the most important and shared needs of stakeholders, tailored to the spatial and temporal requirements of science and decision-making, and addressing system vulnerabilities and resilience to human and natural stressors. *Integrated Assessment Models* are whole-system models that aim to evaluate the systemic effects of policies and trends. Communities of science and practice can play an important role in shaping effective models. In addition to system and optimization models, much of the nexus modeling is about quantifying and analyzing trade-offs. These tools are scale and stakeholderdependent (Daher and Mohtar 2015; Miralles-Wilhelm 2016).

While models of FEW systems that can project outcomes under different possible scenarios of the future are needed for decision support, there are significant challenges and opportunities for advancing in this area of knowledge. In this book, we noted and have described a number of powerful models, including the following:

- Pacific Northwest National Laboratory (PNNL)—Global Change Assessment Model (GCAM)
- USA Environmental Protection Agency (EPA)—Automated Geospatial Watershed Assessment Tool (AGWA)
- Food & Agriculture Organization of the UN (FAO)—Land & Water Division (NRL): Diagnostic, Financial, and Institutional Tool for Investment in Water for Agriculture
- Stockholm Environment Institute (SEI) WEAP (Water Evaluation and Planning System)

- Water–energy–food Nexus Tool 2.0
- Stockholm Environment Institute (SEI)—LEAP (Long Range Energy Alternatives Planning System)
- United Nations Statistics Division—The System of Environmental-Economic Accounting (SEEA)
- The WBCSD (World Business Council on Sustainable Development) Global Water Tool
- UK DECC (Department of Energy & Climate Change) United Kingdom: 2050 Pathways Calculator
- MuSIASEM—Multi-Scale Integrated Analysis of Societal & Ecosystem Metabolism—The Flow-Fund Model
- The International Atomic Energy Agency (IAEA)—Climate, Land-use, Energy, and Water (CLEW)
- Stockholm Environment Institute (SEI)—REAP (Resource and Energy Analysis Programme)
- Agriculture and Agri-food Canada—BIMAT (Biomass Inventory mapping and Analysis Tool)

Such models illustrate opportunities for modeling, including simulation and trade-offs of FEW systems, with a focus on one or more of these primary resources at different scales.

21.3.6 Computing

In Chap. 16, we describe how the increasing volume, velocity, and variety of data required to analyze FEW systems creates challenges for traditional computational tools. Advances in experimental computer and software engineering and design applied to experimental algorithms provide an array of tools that can be selectively deployed on diverse models and data sets. The combination of systems modeling, big data, and high-performance computing power is particularly powerful because of its potential to unlock a new class of rapid interactive and exploratory immersive decision-making processes that are informed by a complete set of systems connections. This is an intuitive way for decision-makers to immerse themselves in systems and explore the connections—and it is a fundamentally transformative capability made possible by advanced computing power.

However, without careful integration of the different types of science involved in FEW system analysis, these tools can often be used as "black boxes" without looking into what is going on "under the hood." Thus, future efforts should focus on bringing the developers and domain scientists together to develop prescriptive solutions instead of over-the-counter ones that will improve algorithm efficiencies as well as the understanding of the effects of various scenarios for specific use cases.

21.3.7 Communication

There is rarely a linear flow of science to decision-making. Decision-makers utilize science to a lesser or greater degree but are influenced by their own experiences and values, as well as the objectives of any community or constituency that they represent, before making choices and judgments. Thus, the relationship between science and decisions is a complex one. Where the products of science are more aligned with the processes and needs of decision-makers, they are more influential. Where science is more effectively communicated to and understood by decision-makers, it is more influential. How risk and uncertainty are understood and perceived can strongly influence the impact of science. Thus, two-way communication is a critical part of the application of science to practice.

Communication is a serious challenge to successful work in communities of science (i.e., scientists from different disciplines working together) and in communities of practice (both scientist/non-scientist communication, and communication between non-science stakeholders representing different communities). The core challenge of communication is to achieve two process goals: First, establish a common language and understanding of nexus issues that support communication and collaborative problem solving and solution development by diverse stakeholders, and second, implement processes that facilitate communication between different stakeholders. Fortunately, there are several examples of how to address this challenge, such as a research question, a metric, a modeling outcome, or a decision-making support tool.

Throughout this book, we have seen many examples of communication processes;

1. University-based research initiatives frequently hold stakeholder workshops. For example, The Texas A&M University System Water–energy–food Nexus Initiative, which focuses on Decision Support for Water Stressed FEW Nexus Decisions, held a 2018 "Stakeholder Information Sharing and Engagement Workshop" involving "over 70 stakeholders drawn from the water, energy, and food sectors in San Antonio and surrounding region." Facilitated small-group sessions were held to obtain stakeholder input on research questions to be asked, and on limitations and opportunities for stakeholder engagement on WEF nexus-related work (Rosen et al. 2018).
2. The Food and Agricultural Organization of the United Nations (FAO) operates multi-stakeholder processes at the level of countries to "decide what issues to focus on and what actions to take" and is "fundamentally about participatory decision-making and information sharing at the country level." One rationale is, "If local people take ownership of all stages and levels of decision-making, development activities are more likely to build on local strengths, meet local needs and priorities, and foster self-determination and sustainability." (FAO website).
3. The International Joint Commission (see Sect. 19.2.3) which oversees issue related to shared waters on the border between the USA and Canada has a Great Lakes Science Advisory Board which engages a diverse set of scientists to provide advice on research and scientific matters, including science priorities

and research coordination. The IJC also conducts a binational poll to understand stakeholder concerns and aspirations for water resources.

4. The Intergovernmental Panel on Climate Change (IPCC) (Chap. 11) is tasked with engaging the scientific community to synthesize and communicate the current state of scientific understanding in three areas (physical science, impacts, adaptation and vulnerability; and mitigation). Each of the reports includes a Summary for Policymakers (SPM), which is drafted first by scientists and then reviewed by governments who provide feedback. A second draft by scientists is later discussed, sentence-by-sentence, in a meeting that includes delegates from government and observer organizations and scientists.

Communication between different stakeholders often benefits from the use of a third-party neutral facilitator to help people have more productive conversations and meetings, as illustrated in Chap. 20 (Managing Human Conflicts). As noted in Sect. 17.1, the two-way, iterative engagement between producers and users of scientific information builds trust, facilitates social learning, and increases the credibility, saliency, legitimacy of research.

It is often surprising to non-scientists that scientists with different disciplinary backgrounds have great difficulty communicating with each other. Scientific disciplines develop languages that have exact meanings to the practitioners of each discipline. Disciplinary scientists can become intellectually siloed within disciplinary academic departments and profession advancement decisions based on publications in disciplinary journals where adherence to shared definitions of terms is essential. Thus, interdisciplinary science requires agreement on a common language and reference shared by scientists in different disciplines. Better interdisciplinary communication between scientists is therefore essential for FEW research.

Further, this common language must be shared and understood by practitioners and decision-makers to facilitate communication with them. Often scientists can adopt the general terms of practitioners and decision-makers. In Sect. 11.3, we described an approach developed and used by the Intergovernmental Panel on Climate Change to communicate scientific uncertainty and risk, using terms that are generally accessible (see Table 11.1). Given that understanding and addressing uncertainty and precision of information and decisions is essential for making high-quality decisions (Sects. 1.5 and 17.4), this approach has valuable lessons to the communication of FEW system science.

Visualizations of FEW nexus data, projections or predictions, and other scientific results (see Sect. 17.3.3) can serve as a decision support system to decision-makers and stakeholders with less knowledge about the underlying interconnected components. When using visualizations, it is important to identify the best means of visualization and modeling systems to represent stakeholder interests and provide stakeholders with the greatest understanding and decision-support. Online platforms (see Sects. 17.3.4 and 17.6.2) and immersive decision environments (see Sect. 17.3.5) are examples of powerful communication tool we provide significant opportunities for more effective communication. Visual communication is a powerful common language that nearly all humans share, across disciplines and other boundaries.

21.3.8 Collaborative Solutions

Throughout this book, we have seen challenges at the nexus in terms of balancing demands for food, energy, and water against a wide array of consequences—environmental, human, economic, cultural, and other impacts. FEW management is usually a matter of weighing trade-offs. Successful solutions meet societal demands for food, energy, and water while minimizing adverse effects. In Chap. 20 (Managing Human Conflicts) we described how conflicts among food, energy, and water sectors range from individuals to global in scale with minor changes to one sector having profound impacts in other sectors. However, defining the scale of connection between all three has been challenging, unpredictable, and oftentimes, uncharted. Conflicts can be exacerbated by climate change, ecosystem degradation, weak institutions and governance, population growth, and transboundary issues.

In Chap. 20, we described how constructively engaging stakeholders to participate in improving nexus relations and reducing conflict and develop solutions that all, or most stakeholders can agree with. How stakeholders are engaged can be specific to a particular scale or reach across a plethora of continuums including, but not limited to, negotiations, public involvement in regulatory policy development, community preparedness workshops, and global agreements to implement change.

Developing viable solutions to cope with conflict in the nexus is critical to future management, policy development, and human interaction with the environment. In our considerations of communities of practice, defining questions, and communications, we have already addressed many opportunities that help create a collaborative framework for Nexus projects that lead to solutions likely to be acceptable to a larger set of stakeholders.

There are, however, challenges in the development of cooperative solutions:

1. Identifying potential win-win situations;
2. Convincing parties that benefits are both real and worth of their engaged cooperation;
3. Not allowing the interests of important stakeholders without political and economic power to be marginalized by powerful stakeholders; and
4. Governance systems (e.g., institutions and treaties) that are siloed in a manner that artificially limits, rather than supports, cooperation across the separate sectors.

Fortunately, there are many opportunities to advance collaborative solutions and many tools to utilize.

Collaborative governance (see Sect. 20.3.4) is a class of processes that advance collaborative policy and regulation as a solution. Collaborative governance engages stakeholders in making, implementing, and enforcing public policy. Techniques of collaborative governance include "deliberative democracy, e-democracy, public conversations, participatory budgeting, citizen juries, study circles, collaborative policymaking, and other forms of deliberation and dialogue among groups of stakeholders or citizens" (Blomgren-Bingham et al. 2005).

Collaborative geodesign (see Sect. 17.6.1) is an example of a collaborative tool that engages stakeholders in landscape design using a tool that interactively lets them test and receive feedback for different design decisions. The benefit of this approach is that the stakeholders can immediately see the impact of their design decisions on biophysical and social indicators. Collaborative geodesign is a step forward to initiate discussions among stakeholders and domain scientists. Thus, the actual parties (i.e., stakeholders) that are affected by the decisions have the opportunity to communicate their concerns with the scientific community as well as the policymakers to make better and more realistic design decisions.

21.4 Case Studies in Opportunity

21.4.1 Watershed Integration Case Study

Globally, watersheds are diverse in terms of their scale, resource uses, and governance structures and they are also subject to different pressures from interactions at the FEW nexus. Watersheds often share multiple municipal, regional, or national borders, and this characteristic suggests the need for systems of cross-border governance and resource management. The challenges for cross-border governance vary widely because of the differences in geological, ecological, economic, and sociopolitical contexts. While the challenges are real, so are the opportunities to use watersheds to solve Nexus problems. Certain geographic and socioeconomic conditions provide greater opportunity, momentum, and political will for applying the integrated scientific study of the FEW nexus to real-world practice at the watershed scale.

In the context of watersheds, building institutional capacity for transboundary governance that is inclusive, equitable, and well-coordinated is likely to be more effective in the context of abundant water resources (not scarce resources)—such as the success of the Great Lakes Compact in the Great Lakes region of North America (Sects. 8.1.1 and 19.2), where significant success has been achieved. However, there are also numerous governance issues that may impede coordination in addressing FEW nexus challenges under conditions of resource abundance—specifically, the issues identified above in Table 19.4, including:

- Institutional capacity for effective decision-making;
- Scale of the watershed;
- Inclusiveness in decision-making;
- Coordination in integrated action;
- Distributional issues related to benefits or negative externalities;
- Heterogeneity among stakeholders and their objectives;
- Political system, and associated trust in its efficacy;
- Social mobility across socioeconomic strata; and
- Political economy and alignment of relations with law, custom, and government.

There exist many opportunities to advance FEW nexus science and application in the context of watershed management, especially in the following areas:

- Building communities of science and practice, especially where watersheds require cross-border coordination and governance;
- Defining scientific questions that address the needs of decision-makers and the capabilities of the scientific community;
- Using participatory processes of choosing metrics and indicators to engage stakeholders in both the scientific and non-scientific communities in discussions the advance knowledge, communication, education, and decision-making;
- Using new abilities to collect, integrate and utilize vast amounts of data for robust science at the nexus of FEW systems in watersheds;
- Developing better models that capture the complex interactions of FEW systems and accurately project future outcomes under defines changes to the system;
- Connecting upstream and downstream communities;
- Utilizing significant advances in computing and data analytics to develop models and machine-learning technologies in a manner that generates useful results to researchers, stakeholders, and decision-makers;
- Delivering more effective communication about FEW systems and the trade-offs inherent in decision-making regarding policies and other actions to change systems for more desirable outcomes;
- Developing and deploying collaborative solutions with diverse groups of stakeholders from both the scientific and non-scientific communities;
- Achieving global food, energy, and water security, in all their aspects, for all people, in a sustainable manner that does not undermine the functional integrity of ecosystems.

Issues related to the FEW are complex, particularly in watersheds that are characterized by resource abundance, and that share cross-border governance. FEW nexus approaches offer many opportunities for the twenty-first-century researcher, student, and practitioner to explore trade-offs at the nexus of FEW systems governance, particularly under conditions of resource abundance.

Looking ahead, now that you are equipped with a systems perspective and toolkit, how can approaches to understanding integrated food–energy–water systems help address problems at the nexus of FEW systems? How can such approaches contribute to solutions under conditions of resource abundance? How can such approaches be useful in a watershed context that is characterized by cross-border governance? How can a new framework for transboundary governance of the nexus of FEW systems in watersheds address emerging challenges, even in conditions of resource abundance?

21.4.2 Environmental Governance Case Study

In 1994, US President Bill Clinton signed Executive Order 12898, requiring federal agencies and grantees to consider environmental justice in their decision-making. While EO 12898 has its limitations and could be revoked by another president at any time, few dispute its importance in raising awareness about environmental justice. In a similar fashion, any president could issue an executive order requiring consideration of food–energy–water nexus impacts, perhaps as part of NEPA.

NEPA has its limitations, chiefly because it mandates procedural steps but not substantive outcomes. In other words, the environmental impacts of various alternative project proposals must be considered in the process, but the sponsoring federal agency is not required to select the proposal option that is least detrimental to the environment. However, NEPA precedent provides some latitude for the executive branch to take a more substantive interpretation of the law. Forcing FEW considerations into NEPA by executive order, then, could be highly instrumental in bringing nexus analysis to the forefront of US policy.

A second means of incorporating FEW consideration into US policy would be to reform the Clean Water Act and Safe Drinking Water Act to require nonpoint source pollution prevention. Doing so would strengthen existing voluntary initiatives to install riparian buffers or sediment traps, or to incorporate manure into farmland using ecologically appropriate methods.

Still another option would be to form a National Council on the FEW Nexus, like the Council on Environmental Quality. The Council could guide key agencies regulating energy, food, and water in coordinating policies, identifying unintended consequences, and reducing inefficiencies and conflicts among FEW policy and laws. Similar structures could also enhance the coordination of the individual FEW sectors.

At the state and local levels, laws and land use regulations designed to protect specific economic sectors, increase energy security, or support local agriculture can directly conflict with environmental goals. Right to Farm laws designed to protect farmers from nuisance suits can hamper water conservation and pollution reduction. At the same time, better financial support for county-based Soil and Water Conservation Districts and similar institutions involved in outreach about best practices in agriculture could decrease the externalities of agriculture on water resources. In addition, policies supporting technical approaches that make farms more energy-efficient and reduce the carbon footprint of food production could help. Expanding these policies to include explicit consideration of energy and water flows and costs would improve the models and increase the resilience of food regulations.

Changes in political administrations can impact environmental regulations. Regulations may be perceived as overly burdensome to industry and detrimental to the economy. FEW regulations are no exception; in early 2018, the EPA sought public comment on whether it should clarify or revise its interpretation that discharges to surface waters via groundwater should be subject to regulation by the CWA. This debate strikes at the larger question about how to best incorporate changing scientific knowledge into law and policy. We invariably assign rights based on our current understanding of the world. When science proves that understanding to be inaccurate, significant legal, political, and practical challenges result.

One benefit of federalism is that when states fail to act, the federal government may step in, and vice versa. When federal administrators decrease protections or support for water, renewable energy, or climate change mitigation that would protect FEW resources, state governments may step in to counter those moves. California's persistent engagement in climate discussions, even as the Trump Administration withdrew at a national level, is one important example. Other actors such as basin commissions, watershed associations, energy cooperatives, and third-party certification programs can fill the gap with forward-looking initiatives that target specific problems. For these efforts to bear fruit, policymakers must be willing to listen to the scientific community, and scientists must be able to communicate the results of their work in language accessible to policymakers and the general public.

21.4.3 *Data Fusion Case Study*

The FEWSION™ project (https://fewsion.us) is a data fusion effort funded in 2016 by the National Science Foundation's interdisciplinary INFEWS program (Innovations at the Nexus of Food Energy and Water Systems). FEWSION brings together a large number of academic and government data sources to describe the commodity flows in the US FEW system in a single seamless dataset. This data enables place-based researchers and those studying a single component of the system to place their work within the broader perspective of the entire FEW system.

This data fusion requires expertise in a large number of distinct datasets and data formats, along with expertise in the data science tools for upscaling, downscaling, cross-walking, and harmonizing voluminous and heterogeneous datasets into a single data structure. This process involves ingesting a large number of Level-0 (raw source) datasets, their transformation into a single large Level-1 (coherent integrated) dataset, and then the application of quality control tools to produce reliable Level-2 (inspected and quality controlled) datasets. This process is accomplished using a scientific workflow implemented using Python language, allowing the reproducibility of the dataset. The resulting dataset size is measured in Petabytes, and its calculation required high-performance computing (HPC). This data resource features documentation, a data model, metadata, a codebase, and both publicly available extracts of the data and also privately controlled source datasets, along with visualization and data download services.

The FEWSION Database™ 1.0 includes some of the following data types:

- 43 Commodity flow categories, based on the "SCTG+FEWSION" code scheme
  - Food and Beverages (for people)
  - Agricultural Products
  - Fuels (Natural Gas, Diesel, Gasoline, Coal)
  - Electricity
  - Water Use
  - Surface water flows and transfers
  - … and all other major commodity types
- Flows between 3,143 US Counties and 8 Foreign Regions
- Seven transportation modes (Pipeline, Power Grid, Rail/Train, Road/Truck, Water/Ship, Air/Plane, Mixed)
- 2012 annual data

The FEWSION Database™ 1.0 utilizes some of the following data inputs:

- U.S. Census Population Data
- U.S. Census Economic Census
- Bureau of Labor Statistics
- U.S. Geological Survey (Water use census, surface flows)
- U.S. Department of Agriculture National Agricultural Statistics
- U.S. Department of Agriculture Economic Research Service
- U.S. Census Commodity Flow Survey
- Oak Ridge National Laboratory/U.S. Department of Transportation Freight Analysis Framework
- U.S. Energy Information Administration
- U.S. Environmental Protection Agency
- U.S. Department of Homeland Security
- U.S. Department of Agriculture CropScape
- DHS HIFLD Open Data
- National Renewable Energy Laboratory ReEDS Energy & Power Flow Data
- National Renewable Energy Laboratory ReEDS Water Withdrawal and Consumption Data
- U.S. Foreign Trade Data
- Global Water Productivity Data
- Water Footprint Network
- Academic surface water flow models
- Academic surface water transfer statistics
- Academic electrical power flow models

FEWSION provides an online publicly accessible visualization and data search and retrieval system called FEW-View™. FEW-View™ 1.0 allows a user to select commodity types and units, choose locations, and visually map the supply chains. Users can benchmark and compare their community's FEW usage or footprints with other US communities. Users can view analytics that describes their supply chain network-like resilience or circularity metrics. Users can print out reports for their communities' supply chains, and can directly download the data that they see on their screen. The map interface looks like this (Fig. 21.1).

Visual analysis and exploration is one of the most effective strategies for orienting both technical analysts and stakeholders within a systems context. People have a limited capacity to grasp systems of connections, but people are relatively adept at visual comprehension and exploratory analysis. However, before a user can employ this kind of analysis tool, the user must be trained. Even relatively simple interfaces require significant training and experience. In order to streamline the user's onboarding to the tool, FEW-View™ utilizes a combination of science art and narrative storytelling, followed by preconfigured scenario maps, to ease the user into the interface. An example scenario follows below (Fig. 21.2).

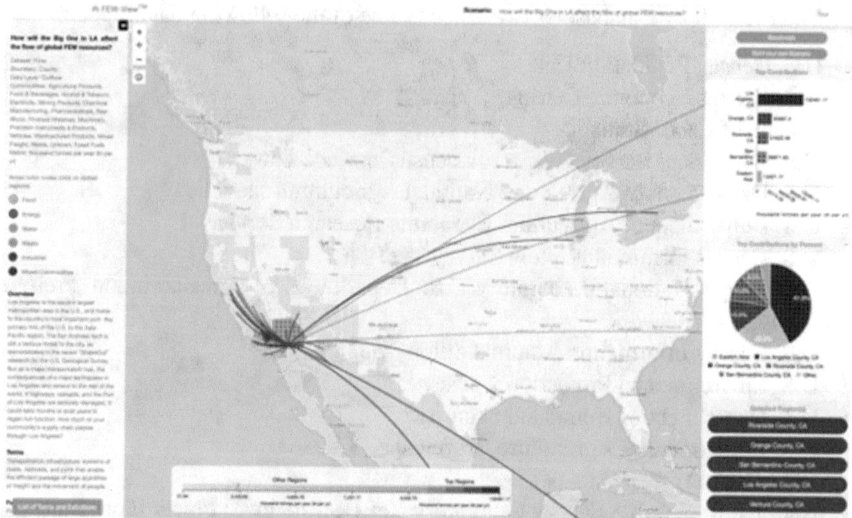

Fig. 21.1 Screenshot from the FEW-View™ 1.0 mapping interface. Used with permission from the authors

Key Points

- Certain situations provide greater impetus and opportunity for applying integrated FEW science to real-world practice, such as:
  - Acute scarcity of one or more commodity.
  - Significant externalities from FEW decisions and actions.
  - Potential benefits to many communities from coordinated actions.
- There exist many opportunities to advance FEW nexus science and application, especially in the following areas:
  - Building communities of science and practice;
  - Defining scientific questions that meet the needs of decision-makers and the capabilities of the scientific community;
  - Using the process of choosing metrics to engage science and non-science stakeholders in discussions that profoundly shape science, communication, education, and decision-making;
  - Using new abilities to collect, integrate and utilize vast amounts of data for stronger science;
  - Developing better models that capture the complex interactions of few systems and accurately project future outcomes under defined changes to the system;
  - Utilize significant advances in computing to use data and run models in a manner that generates useful results to stakeholders and decision-makers;
  - Carry out more effective communication about few systems and the choices in policies and other actions to change systems for more desirable outcomes;

Fig. 21.2 Screenshot from the FEW-View™ 1.0 onboarding scenario interface, for Hawaii's food and fuel supply chains

  - Develop and deploy collaborative solutions with diverse science and non-science stakeholders; and
  - Achieve food, energy, and water security, in all their aspects, for all people, in a sustainable manner.
  - Share data, tools, and knowledge and disseminate using the ever-increasing e-space.

- A requirement to include food–energy–water nexus considerations in analysis carried out under the U.S. National Environmental Policy Act (NEPA) could increase consideration of externalities of decisions on nexus resources.
- Data fusion, modeling, and visualization systems (like FEWSION) provide a useful and accessible interface between scientists, stakeholders, decision-makers, and the public—if they are carefully implemented with the user community in mind.

Discussion Points and Exercises

1. Now that you are armed with a FEW systems perspective and toolkit, how will you use these Nexus opportunities to make the world a better place?

References

Blomgren Bingham, L., Nabatchi,T., O'Leary, R. (2005). The New Governance: Practices and Processes for Stakeholder and Citizen Participation in the Work of Government https://doi.org/10.1111/j.1540-6210.2005.00482.x.

Daher, B. T., & Mohtar, R. H. (2015). Water-energy-food (WEF) Nexus Tool 2.0: guiding integrative resource planning and decision-making, Water International, https://doi.org/02508060.2015.1074148.

Randall, L. (2011). *Knocking on heaven's door: How physics and scientific thinking illuminate the universe and the modern world.* USA: HarperCollins Publishers.

Rosen, R. A., Daher, B., & Mohtar, R. H. (2018). *Water-energy-food nexus stakeholder information sharing and engagement workshop*. Texas A&M University. Available at https://www.water-energy-food.org/resources/resources-detail/workshop-report-water-energy-food-nexus-stakeholder-information-sharing-and-engagement-workshop/.

Further Reading

An, L. (2012). Modeling human decisions in coupled human and natural systems: Review of agent-based models. *Ecological Modelling, 229*(2012), 25–36.

Bazilian, M., Rogner, H., Howells, M., Hermann, S., Arent, D., Gielen, D., Steduto, P., Mueller, A., Komor, P., & Tol, R. S. J. (2011). Considering the energy, water and food nexus: Towards an integrated modelling approach. *Energy Policy, 39*, 7896–7906.

Garrick, D., & Hall, J. W. (2014). Water security and society: Risks, metrics, and pathways. *Annual Review of Environment and Resources, 39*, 611–639.

Hoolohan, C., Larkin, A., McLachlan, C., Falconer, R., Soutar, I., Suckling, J., et al. (2018). Engaging stakeholders in research to address water–energy–food (WEF) nexus challenges. *Sustainability Science, 13*(5), 1415–1426.

Ingwersen, W., et al. (2014). Integrated metrics for improving the life cycle approach to assessing product system sustainability. *Sustainability, 6*(3), 1386–1413.

Kehrer, J., & Hauser, H. (2013). Visualization and visual analysis of multifaceted scientific data: A survey. *IEEE Transactions on Visualization and Computer Graphics, 19*(3), 495–513.

Kraucunas, I., et al. (2015). Investigating the nexus of climate, energy, water, and land at decision-relevant scales: The Platform for Regional Integrated Modeling and Analysis (PRIMA). *Climatic Change, 129*(3–4), 573–588.

Kurian, M., Ardakanian, R., Veiga, L. G., & Meyer, K. (2016). *Resources, services and risks: How can data observatories bridge the science-policy divide in environmental governance?* New York: Springer.

Mercure, J. F., Pollitt, H., Bassi, A. M., Viñuales, J. E., & Edwards, N. R. (2016). Modelling complex systems of heterogeneous agents to better design sustainability transitions policy. *Global Environmental Change, 37*(2016), 102–115.

Miralles-Wilhelm, F. (2016). Development and application of integrative modeling tools in support of food–energy–water nexus planning—A research agenda. *Journal of Environmental Studies and Sciences, 6*, 3. https://doi.org/10.1007/s13412-016-0361-1.

Mohtar, R. (2018). Opportunities in the water–energy–food nexus approach: Innovatively driving economic development, social wellbeing, and environmental sustainability. *Science of the Total Environment Special Issue*. Available online at https://www.sciencedirect.com/journal/science-of-the-total-environment/special-issue/10K1703WP3V.

Mohtar, R. H., & Daher, B. (2016). Water–energy–food nexus framework for facilitating multi-stakeholder dialogue. *Water International*. https://doi.org/10.1080/02508060.2016.1149759.

Mohtar, R. H., & Lawford, R. (2016). Present and future of the water–energy–food nexus and the role of the community of practice. *Journal of Environmental Studies and Sciences, 6*, 192. https://doi.org/10.1007/s13412-016-0378-5.

Mohtar, R. H. (2017). Making the leap into the nexus: Changing nexus. Presentation at the Watershed conference, Vatican City, 22 Mar 2017.

None. (2016). Identifying interdisciplinary data science approaches and challenges to enhance understanding of interactions of food systems with energy and water systems. http://www.spatial.cs.umn.edu/few/few_report_draft.pdf.

NSF (website) Innovations at the Nexus of Food, Energy and Water Systems (INFEWS), National Science Foundation. Available at https://www.nsf.gov/funding/pgm_summ.jsp?pims_id=505241.

Roy, B. (1993). Decision science or decision-aid science? *European Journal of Operational Research, 66*(2), 184–203.

Serrat-Capdevila, A., Valdes, J. B., & Gupta, H. V. (2011). Decision support systems in water resources planning and management: Stakeholder participation and the sustainable path to science-based decision making. In C. Jao (Ed.), *Efficient decision support systems—Practice and challenges from current to future* (pp. 423–440). Rijeka, Croatia: InTech.

Smith, D. W., Welch, M., Bennett, K. E., Padgham, J., & Mohtar, R. (2017). Building a WEF Nexus Community of Practice (NCoP). *Current Sustainable/Renewable Energy Reports, 4*, 168–172. Springer.. https://doi.org/10.1007/s40518-017-0080-6.

Stapp, W. B. (2000). Watershed education for sustainable development. *Journal of Science Education and Technology, 9*, 183. https://doi.org/10.1023/A:1009430215477.

Strachan, N., Fais, B., & Daly, H. (2016). Reinventing the energy modelling–policy interface. *Nature Energy, 1*, 16012.

Tevar, A. D., Aelion, H. M., Stang, M. A., & Mendlovic, J. (2016). The need for universal metrics in the energy–water-food nexus. *Journal of Environmental Studies and Sciences, 6*(1), 225–230.

Appendix A: Institutions

Amazon Cooperation Treaty Organization (ACTO): International organization created under the 1978 Treaty for Amazonian Cooperation (also Amazon Cooperation Treaty) as a secretariat to facilitate its aims. Also La Organización del Tratado de Cooperación Amazónica (OTCA) Website: http://www.otca-oficial.info/home

American Center for Life Cycle Assessment: Nonprofit organization "providing education, awareness, advocacy, and communications to build capacity and knowledge of environmental LCA. ACLCA membership consists of industry, academia, government, consulting, and NGOs." Website: www.aclca.org

American Farm Bureau Federation (or Farm Bureau): National organization representing farm and ranch families in the USA. State-level Farm Bureaus operate under the national organization. Website: https://www.fb.org/

American Petroleum Institute (API): Trade association representing 600 oil and natural gas companies. Website: https://www.api.org/

American Society of Civil Engineers (ASCE): International professional association of people from 177 countries working in civil engineering. Website: https://www.asce.org/

American Water Works Association (AWWA): Professional association of people and institutions associated with water treatment issues. Primarily US Website: https://www.awwa.org/

California Energy Commission: State government agency responsible for energy policy and planning within the state of California. Website: https://www.energy.ca.gov/

CGIAR (formerly the Consultative Group for International Agricultural Research): International Research partnership organization with 15 research centers (see International Food Policy Research Institute) that address issues of food security, health, and nutrition, rural poverty, environment, and sustainability. Website: https://www.cgiar.org/

P. Saundry, B. L. Ruddell (eds.), *The Food-Energy-Water Nexus*, AESS Interdisciplinary Environmental Studies and Sciences Series,
https://doi.org/10.1007/978-3-030-29914-9

Coalition for Publishing Data in the Earth and Space Sciences (CPDESS): International organization that "connects Earth and space science publishers and data facilities to help translate the aspirations of open, available, and useful data from policy into practice." Website: https://copdess.org/

Consortium of Universities for the Advancement of Hydrologic Science, Inc. (CUAHSI): "Research organization representing more than 130 U.S. universities and international water science-related organizations. CUAHSI receives support from the National Science Foundation (NSF) to develop infrastructure and services for the advancement of water science in the United States." Website: https://www.cuahsi.org/

Data.gov: Website of US federal government that provides public access to data from a wide array of federal and nonfederal sources. Website: https://www.data.gov/

DataOne (Observational Network for Earth): Project supported by the U.S. National Science Foundation that manages a "distributed framework and sustainable cyberinfrastructure that meets the needs of science and society for open, persistent, robust, and secure access to well-described and easily discovered Earth observational data." Website: https://www.dataone.org/

Delaware River Basin Commission: Regional organization that facilitates the coordination of policy and activities related to the Delaware River Basin (USA) by state and federal agencies and other stakeholders. Website: http://www.state.nj.us/drbc/

Earth Observing System (EOS): Program of NASA which includes "a coordinated series of polar-orbiting and low inclination satellites for long-term global observations of the land surface, biosphere, solid Earth, atmosphere, and oceans." Website: https://eospso.nasa.gov/

East African Community (EAC): Intergovernmental organization representing six countries of the African Great Lakes region (Burundi, Kenya, Rwanda, South Sudan, Tanzania, and Uganda) that facilitates free trade and cooperation between member states. Website: https://www.eac.int/

Environment Canada (fully: Environment and Climate Change Canada or ECCC): Agency of the national government of Canada responsible for environmental policies and programs. Website: https://www.canada.ca/en/environment-climate-change.html

EPA Conflict Prevention and Resolution Center: Part of the U.S. Environmental Protection Agency that provides "services and expertise in alternative dispute resolution (ADR), environmental conflict resolution, consensus-building, and collaborative problem-solving." Website: https://www.epa.gov/adr

European Union (EU): A political and economic union of 28 European countries (as of mid-2019; 27 should the United Kingdom leave). Website: https://europa.eu/

Federal Energy Regulatory Commission (FERC): Independent agency of the US federal government that regulates the interstate transmission of natural gas, oil, and electricity and related issues. It is particularly important with respect to electricity issues under the Federal Power Act and natural gas under the Natural Gas Act. Website: https://www.ferc.gov/

Federal Geographic Data Committee (FGDC): A structure within the US federal government that provides "executive, managerial, and advisory direction and oversight for geospatial decisions and initiatives across the Federal government." Website: https://www.fgdc.gov/

Food and Agriculture Organization of the United Nations (FAO): Specialized agency of the United Nations System which "works in partnership with governments and other development actors at global, regional and national levels to develop supportive policy and institutional environments" to lead "international efforts to defeat hunger." The FAO includes work on fisheries and agriculture. Website: https://www.fao.org

Foreseer: A scenario generation tool developed by the Use Less Group based in the Department of Engineering, University of Cambridge to study the nexus of water, energy and land resources. Website: https://www.foreseer.group.cam.ac.uk

Future Earth: International research organization with 20 Global Research Projects focused on sustainability. See Water Future (Sustainable Water Future Programme). Website: http://www.futureearth.org

Global Environment Facility (GEF): Intergovernmental organization with 183 member states addressing global environmental issues. Website: https://www.thegef.org/

Global Footprint Network: Think tank that develops and promotes tools based upon measuring and using an Ecological Footprint as a metric of sustainability. Website: www.footprintnetwork.org

Global Open FAIR Initiative: Organization that aims to implement the FAIR data principles, making data Findable, Accessible, Interoperable and Reusable. It offers an open and inclusive ecosystem for consortia of individuals, institutions, and organizations working together. Website: https://www.go-fair.org/

Global Water Partnership: A "global action network with over 3000 Partner organizations in 183 countries" that "provides knowledge and builds capacity to improve water management at all levels: global, regional, national and local," particularly through integrated water resources management. Website: www.gwp.org

Great Lakes Science Advisory Board: Scientific advisory body to the International Joint Commission and Great Lakes Water Quality Board. Website: https://www.ijc.org/en/sab

Government Accountability Office (GAO): Independent, non-partisan agency under the US Congress. That conducts analysis on "how taxpayer dollars are spent and provides Congress and federal agencies with objective, reliable information to help the government save money and work more efficiently." Website: https://www.gao.gov/

Integrated Solutions for Water, Energy, and Land (ISWEL): Partnership of International Institute for Applied Systems Analysis, the Global Environment Facility (GEF), and the United Nations Industrial Development Organization (UNIDO) exploring "cost-effective solutions to jointly meet water, land and energy demands under different development and climate pathways. The project takes a global approach, but it also zooms into two large transboundary basin facing multiple developments and environmental challenges: The Zambezi and the Indus." Website: http://www.iiasa.ac.at/web/home/research/iswel/ISWEL.html

International Atomic Energy Agency (IAEA): Intergovernmental organization that works "for scientific and technical co-operation in the nuclear field. It works for the safe, secure and peaceful uses of nuclear science and technology." While not under the direct control of the United Nations, the IAEA report to both the UN General Assembly and Security Council. Website: https://www.iaea.org/

International Energy Agency (IEA): Intergovernmental organization with 30 member states to whom it provides "data, analysis, and solutions on all fuels and all technologies; helping governments, industry, and citizens make good energy choices." Website: https://www.iea.org/

International Food Policy Research Institute (IFPRI): International organization that "provides research-based policy solutions to sustainably reduce poverty and end hunger and malnutrition in developing countries." One of 15 CGIAR research centers. Website: http://www.ifpri.org/

International Fund for Agricultural Development (IFAD): Specialized agency of the United Nations system which provides low-interest loans and grants to developing countries to finance innovative agricultural and rural development programs and projects. Website: https://www.ifad.org

International Institute for Applied Systems Analysis (IIASA): International think tank providing "insights and guidance to policymakers worldwide by finding solutions to global and universal problems through applied systems analysis in order to improve human and social wellbeing and to protect the environment." Website: http://www.iiasa.ac.at/

International Joint Commission (IJC): Bi-national intergovernmental organization of Canada and the USA prevents established under the 1909 Boundary Waters Treaty and also supporting the 1978 Great Lakes Water Quality Agreement which facilitates cooperation on boundary waters and resolves disputes. Website: http://www.ijc.org/

International Law Association (ILA): International organization whose objectives are "the study, clarification and development of international law, both public and private, and the furtherance of international understanding and respect for international law." Website: http://www.ila-hq.org/

International Law Commission (ILC): A group of 34 international law experts that assist the Unite Nations in the development and codification of international law.

International Maritime Organization (IMO): Specialized agency of the United Nations System related to shipping. Website: https://www.imo.org

International Monetary Fund (IMF): Specialized agency of the United Nations System to promote financial stability by having countries cooperate in how they managed their currency exchange rates and financial flows; and to support nations which face problems in their payments of international debts, thereby avoiding broader international financial crises. Website: https://www.imf.org

Lake Victoria Basin Commission (LVBC): Specialized institution of the East African Community (EAC) mandated to "coordinate sustainable development and management of the Lake Victoria Basin." Website: https://www.lvbcom.org/

Latin America Water Funds Partnership (LAWFP): International organization created by the Inter-American Development Bank, FEMSA Foundation, the Global Environment Facility and The Nature Conservancy to "contribute to the water security in Latin America and the Caribbean, through the creation and strengthening of Water Funds" that support integrated watershed management and governance of water resources. Website: http://waterfunds.org/esp/

Magic-Nexus Project: Moving Towards Adaptive Governance in Complexity: Informing Nexus Security (MAGIC)-Nexus is a European Union project to advance "Integrated approaches to food security, low-carbon energy, sustainable water management, and climate change mitigation." Website: www.magic-nexus.eu

Massachusetts Department of Environmental Protection (MassDEP): State government agency responsible for environmental regulation and oversight within the state of Massachusetts. Website: https://www.mass.gov/orgs/massachusetts-department-of-environmental-protection

Ministry of the Environment (Ministério do Meio Ambiente) of Brazil: Agency of the national government of Brazil responsible for environmental policies and programs. Website: http://www.mma.gov.br/

NASA (National Aeronautics and Space Administration): Agency of the US federal government responsible for the civilian space program, aeronautics, and aerospace research. Website: https://www.nasa.gov/

Nexus: The Water, Energy & Food Security Resource Platform: An online collection of nexus resources maintained by the German Federal Ministry of Economic Cooperation and Development and the European Union. Website: www.water–energy–food.org

OPEC (Organization of Petroleum Exporting Countries): Intergovernmental organizations of 14 countries facilitate coordination of their petroleum policies. Website: https://www.opec.org

Open GIS Consortium (OGC): International organization advancing open standards for geographic information system (GIS) data: Website: http://www.opengeospatial.org/

Organization for Economic Cooperation and Development (OECD): Intergovernmental organization of 35 members (as of mid-2018) drawn from wealthier and more developed countries which seeks to "promote policies that will improve the economic and social well-being of people around the world" by providing a forum for analysis, dialogue, and collaboration. Website: https://www.oecd.org

Perspectivity: International organization of individuals and professional consultants, dedicated to help foster dialogue and build regenerative organizations and societies. It maintains a number of online "challenges" which guide users through explorations of issues, include the FEW nexus, where they make decisions and see the consequences. Website: www.theperspectivitychallenge.org

Privacy International: International organization promoting the right to privacy in various forms, especially data, information technology, and cyber issues. Website: https://privacyinternational.org/

SIM4NEXUS: Project funded by the European Union to advance understanding of issues related to the nexus of water, land, food, energy, and climate and "predict society-wide impacts of resource use and relevant policies on sectors such as agriculture, water, biodiversity and ecosystem services through a model-based analysis." Website: www.sim4nexus.eu

Society for Environmental Toxicology and Chemistry (SETAC): International professional association of individuals and institutions "dedicated to the study, analysis, and solution of environmental problems, the management, and regulation of natural resources, research and development, and environmental education." Website: www.setac.org

Susquehanna River Basin Commission: Regional organization that facilitates the coordination of policy and activities related to the Susquehanna River Basin (USA) by state and federal agencies and other stakeholders. Website: https://www.srbc.net/

Tennessee Valley Authority (TVA): A corporation owned by the US federal government that operates electric power facilities on the Tennessee River system, sells electricity and provides flood control, navigation, and land management services to the region. Website: https://www.tva.gov/

The Nature Conservancy (TNC): International environmental organization with a mission to "conserve the land and water upon which all life depends." Website: https://www.nature.org

US Federal Government: The national government of the USA as defined by the US Constitution with three independent co-equal branches. The executive branch headed by the President; the legislative branch (Congress) composed of the US Senate and US House of Representatives; and, the judicial branch headed by the Supreme Court of the USA. Powers not vested in the federal government are reserved for the states or the people directly.

U.S. Agency for International Development (USAID): Agency of the US federal government which leads international development and humanitarian efforts. Website: https://www.usaid.gov

U.S. Army Corps of Engineers (USACE): Agency within the Department of Defense of the US federal government involved in major projects, mostly civilian and within the USA, such as dams, canals, and flood protection. USACE operates approximately half of federal hydropower capacity. Website: https://www.usace.army.mil/

U.S. Bureau of Reclamation (USBR): Agency within the Department of Interior of the US federal government involved mostly in oversees water resource management projects including hydropower and irrigation in the western USA. USBR operates over one-third of federal hydropower capacity. Website: https://www.usbr.gov/

U.S. Census Bureau: Agency of the US federal government responsible for carrying out a national census every 10 years and for producing data about the American people and economy. Website: https://www.census.gov/

U.S. Department of Agriculture (USDA): Department of the US federal government composed of 29 agencies focused on food, agriculture, natural resources, rural development, nutrition, and related issues. Website: https://www.usda.gov/

U.S. Department of Energy (DOE): Department of the US federal government composed of agencies, offices and national laboratories focused on energy, environment, and nuclear issues. Website: https://www.energy.gov/

U.S. Department of Health and Human Services (HHS): Department of the US federal government composed of agencies, offices, and research institutes focused on medicine, public health, and social services. Website: https://www.hhs.gov/

U.S. Department of Homeland Security (DHS): Department of the US federal government composed of agencies focuses on threats to US territory. Website: https://www.dhs.gov/

U.S. Department of Interior (DOI): Department of the US federal government composed of agencies and offices focused on the management and conservation of natural and cultural resources. Website: https://www.doi.gov/

U.S. Department of Labor (DOL): Department of the US federal government composed of agencies and programs focused on federal labor laws. Website: https://www.dol.gov/

U.S. Department of Transportation (DOT): Department of the US federal government composed of agencies and programs focused on transportation planning, projects, and safety. Website: https://www.transportation.gov/

U.S. Energy Information Administration (EIA): Agency within the Department of Energy of the US federal government that researches, collects, and disseminates information on energy topics. Website: https://www.eia.gov

U.S. Environmental Protection Agency (EPA): Agency of the US federal government that develops and enforces environmental regulations in a wide range of areas and conducts research. Website: https://www.epa.gov

U.S. Food and Drug Administration (FDA): Agency within the Department of Health and Human Services of the US federal government responsible for "protecting the public health by ensuring the safety, efficacy, and security of human and veterinary drugs, biological products, and medical devices; and by ensuring the safety of our nation's food supply, cosmetics, and products that emit radiation." Website: https://www.fda.gov

U.S. Geological Survey (USGS): Agency with the Department of Interior of the US federal government monitors and provides data and scientific understanding on the availability and condition of water, energy minerals, and other natural resources. Website: https://www.usgs.gov

U.S. Institute for Conflict Resolution: Institute within the Udall Foundation, an independent agency of the US federal government, with a mission to "provide impartial collaboration, consensus-building, and conflict resolution services." Website: https://www.udall.gov/ourprograms/institute/institute.aspx

Union of Concerned Scientists (UCS): A US-based science advocacy organization with programs focused on applying science to issues of energy, environment, food, nuclear war, and equity. Website: https://www.ucsusa.org

UN-Energy: Interagency mechanism that coordinates the activities of many UN agencies and offices on energy. Website: https://www.un-energy.org/

UN-Water: Interagency mechanism that coordinates the activities of many UN agencies and offices on water. Website: http://www.unwater.org/

United Nations System: A set of institutions established following World War II to maintain international peace and facilitating international cooperation and governance. The UN has six "principal organs"—the General Assembly, the Economic and Social

Council, the International Court of Justice, the Security Council, the Trusteeship Council, and the UN Secretariat. There are also a large number of agencies and organizations established by and reporting to different parts of the UN. Website: https://www.un.org/

United Nations Children's Fund (UNICEF): An organization of the United Nations system that addresses issues relevant to the health and well-being of children. Website: https://www.unicef.org/

United Nations Development Programme (UNDP): An organization of the United Nations system that addresses issues of development and "works to eradicate poverty and reduce inequalities through the sustainable development of nations." Website: https://www.undp.org

United Nations Environment Programme (UNEP): An organization of the United Nations system that facilitates the setting of a "global environmental agenda, promotes the coherent implementation of the environmental dimension of sustainable development within the United Nations system, and serves as an authoritative advocate for the global environment." Website: https://www.unenvironment.org/

United Nations Economic and Social Council (ECOSOC): Part of the United Nations system which facilitates international dialogue and action on economic, social and environmental aspects of sustainable development. Website: https://www.un.org/ecosocECOSOC includes five regional commissions, including the Economic and Social Commission for Western Asia (ESCWA). Website: https://www.unescwa.org/

United Nations Educational, Scientific and Cultural Organization (UNESCO): Specialized agency of the United Nations System focused on educational, scientific, and cultural activities. Website: https://en.unesco.org/

Office of the High Commissioner for Human Rights (OHCHR): Office within the United Nations system that "provides assistance in the form of technical expertise and capacity-development in order to support the implementation of international human rights standards on the ground" and acts as the lead UN entity of Human Rights. Website: https://www.ohchr.org/

United Nations Human Rights Council (HRC): Intergovernmental organization within the United Nations system responsible for "strengthening the promotion and protection of human rights around the globe and for addressing situations of human rights violations and make recommendations on them." Website: https://www.ohchr.org/EN/HRBodies/HRC/Pages/Home.aspx

United Nations Industrial Development Organization (UNIDO): Specialized agency of the United Nations System that "promotes industrial development for poverty reduction, inclusive globalization, and environmental sustainability." Website: https://www.unido.org/

United Nations System of Environmental Economic Accounting (SEEA): A initiative within the United Nations systems that brings together economic and environmental statistical information "into a common framework to measure the condition of the environment, the contribution of the environment to the economy and the impact of the economy on the environment." Website: www.seea.un.org

United Nations University (UNU): Academic and research arm of the United Nations system with campuses in many countries. Website: https://unu.edu/UNU includes an Institute for Integrated Management of Material Fluxes and of Resources (UNU-FLORES) which advances a nexus approach to the sustainable management of environmental resources and maintains a Nexus Tools Platform. Website: https://flores.unu.edu and https://ntptools.ara.uberspace.de

Upper Mississippi River Basin Association: Regional organization that facilitates the coordination of policy and activities related to the Upper Mississippi River Basin (USA) by state and federal agencies and other stakeholders. Website: http://www.umrba.org

World Bank Group (WBG): Specialized agency of the United Nations systems components primarily aimed at supporting development. Included within the WBG are:

- International Bank for Reconstruction and Development (loans to middle-income countries)
- International Finance Corporation (work with the private sector)
- International Development Association (interest-free loans or grants to low-income countries)
- International Centre for Settlement of Investment Disputes
- Multilateral Investment Guarantee Agency (insurance for investments)

- Website: https://www.worldbank.org/

World Health Organization (WHO): Specialized agency of the United Nations systems that focuses on issues of public health. Website: https://www.who.int/

World Meteorological Organization (WMO): Specialized agency of the United Nations systems that focuses on issues of meteorology, climatology, and operational hydrology. Website: https://public.wmo.int

Water Footprint Network: International organization that facilitates "collaboration between companies, organizations, and individuals to solve the world's water crises by advancing fair and smart water use." http://waterfootprint.org

Water Future (Sustainable Water Future Programme): Global research platform operated through Future Earth "with expertise and innovation in water research, policy, security, and sustainability." Website: http://www.futureearth.org/projects/water-future-sustainable-water-future-programme

Western States Water Council (WSWC): Regional organization that facilitates information sharing and cooperation between 18 western states of the USA and with the federal government on water issues. Website: http://www.westernstateswater.org/WSWC maintains the Water Data Exchange (WaDE) as a platform for member states to share "water supply, water use, and water administration datasets with each other, with federal partners, and with the public." Website: http://www.westernstateswater.org/wade

Wildlife Conservation Society: International environmental organization seeking to "conserve the world's largest wild places in 16 priority regions, home to more than 50% of the world's biodiversity." Website: https://www.wcs.org/

World Council on Dams: International organization that existed between 1997 and 2001 that developed guidelines for dam building that would provide greater protection to dam-affected people and the environment, and more equitable distribution of benefits.

World Energy Council: International organization that is a network of individuals and institutions involved in energy that facilitates dialogue and cooperation to advance affordable, stable, and environmentally sensitive energy system. Website: https://www.worldenergy.org/

World Food Programme: A program of the United Nations system that delivers food assistance in emergencies and works with communities to improve nutrition and build resilience. Website: https://www1.wfp.org/

World Resources Institute (WRI): International think tank that develops solutions to sustainable development on issues of climate, energy, food, forests, water, and cities and transport. Website: https://www.wri.org/

World Trade Organization: An organization to facilitate international negotiations and conflict resolution related to trade. A successor organization to the General Agreement on Tariffs and Trade (GATT). Website: https://www.wto.org/

Appendix B: Treaties, Declarations, and Laws

Notes:

UNTS = United Nations Treaty Series, a publication produced by the Secretariat of the United Nations containing all treaties and international *agreements* registered or filed and recorded by the Secretariat since 1945, according to Article 102 of the Charter of the United Nations.

ILM = International Legal Materials, a publication produced by the American Society of International Law that reproduces primary international legal documents.

Agreement on Agriculture (AoA): Treaty of the World Trade Organization (WTO)/General Agreement on Tariffs and Trade (GATT) negotiated during the Uruguay Round (1986–1995). The treaty addresses domestic support for agriculture, market access, and export subsidies. Entry into force 1 January 1995. 1867 UNTS 410.

Agreement on the Application of Sanitary and Phytosanitary Measures (SPS): Treaty of the World Trade Organization (WTO)/General Agreement on Tariffs and Trade (GATT) negotiated during the Uruguay Round (1986–1995). The treaty addresses food safety and animal and plant health (phytosanitation). Entry into force 1 January 1995. 1867 UNTS 493.

Agreement on Technical Barriers to Trade (TBT): Treaty of the World Trade Organization (WTO)/General Agreement on Tariffs and Trade (GATT) negotiated during the Uruguay Round (1986–1995). The treaty addresses technical regulations, standards, and conformity assessment procedures. Entry into force 1 January 1995. 1868 UNTS 120.

Agreement on Trade-Related Aspects of Intellectual Property Rights (TRIPs): Treaty of the World Trade Organization (WTO)/General Agreement on Tariffs and Trade (GATT) negotiated during the Uruguay Round (1986–1995). The treaty addresses domestic support for agriculture, market access, and export subsidies. Entry into force 1 January 1995. 1869 UNTS 299; 33 ILM 1197 (1994).

P. Saundry, B. L. Ruddell (eds.), *The Food-Energy-Water Nexus*, AESS Interdisciplinary Environmental Studies and Sciences Series,
https://doi.org/10.1007/978-3-030-29914-9

Amazon Cooperation Treaty: See Treaty for Amazonian Cooperation.

Boundary Waters Treaty of 1909: Bilateral treaty between the USA and Canada to collaborate and resolve disputes related to boundary waters, including the Great Lakes, utilizing the International Joint Commission. Available at: https://www.ijc.org/en/boundary-waters-treaty-1909/

Charter of the United Nations: Foundational treaty of the United Nations. Entry into force 24 October 1945.

Convention on Nuclear Safety: Treaty of International Atomic Energy Agency establishing safety rules at nuclear power plants negotiated after the 1986 Chernobyl accident. Entry into force 24 October 1996. 1963 UNTS 293; 33 ILM 1514 (1994).

Convention on the Elimination of all Forms of Discrimination against Women: International treaty adopted by the United Nations General Assembly which defines what constitutes discrimination and establishes an agenda for countries to address such discrimination. Entry into force 9 March 1981. 1249 UNTS 13; 19 ILM 33 (1980). Available at: https://www.un.org/womenwatch/daw/cedaw/cedaw.htm

Constitution of the Food and Agriculture Organization: International treaty defining the purposes and operational rules of the Food and Agricultural Organization of the United Nations. Entry into force 16 October 1945. 40 AJIL Supp. 76. Available at: http://www.fao.org/3/x5584e/x5584e0i.htm

Convention on the Law of the Non-Navigational Uses of International Watercourses (also U.N. Watercourses Convention—UNWC): International treaty adopted by the United Nations General Assembly related to the uses and conservation of all waters that cross international boundaries. (1997) Entry into force 17 August 2014. U.N. Doc A/51/869; 36 ILM 700 (1997).

Convention on the Protection and Use of Transboundary Watercourses and International Lakes (also Water Convention—ECEWC): International treaty of the United Nations Economic Commission for Europe (ECE) related to transboundary waters. Entry into force 6 October 1996, 1936 UNTS 269; 31 ILM 1312 (1992). Available at http://www.unece.org/env/water/text/text.htm

Convention on the Rights of the Child: International treaty adopted by the United National General Assembly related to the rights of children (persons under the age of 18, unless the laws of a particular country set the legal age for adulthood younger). Entry into force 2 September 1990. 1577 UNTS 3; 28 ILM 1448 (1989). Available at https://www.ohchr.org/en/professionalinterest/pages/crc.aspx

Convention on Third Party Liability in the Field of Nuclear Energy: Treaty of the Organization for Economic Cooperation and Development (OECD) related to liability and compensation for damage caused by nuclear power production. Amended on several occasions. Entry into force 1 April 1968. 956 UNTS 251. Available at https://www.oecd-nea.org/law/nlparis_conv.html

European Energy Charter (also 1991 Energy Charter and 2015 International Energy Charter): International agreement establishing principles for international energy cooperation which led to the Energy Charter Treaty. Entry into force 16 April 1998. 2080 UNTS 95; 34 ILM 360 (1995).

Energy Charter Treaty (ECT): See European Energy Charter.

International Treaty on Plant Genetic Resources for Food and Agriculture (ITPGR): Treaty of the Food and Agriculture Organization of the United Nations (FAO) related to the conservation and sustainable use of plant genetic resources and the fair and equitable sharing of the benefits arising from their use. Entry into force 29 June 2004. 2400 UNTS 303.

General Agreement on Tariffs and Trade (GATT): A framework for facilitation international negotiations to regulate trade that operated from 1947 to 1995 and treaties associated with it. The World Trade Organization in 1995 succeeded GATT. Initial entry into force 1 January 1948. 55 UNTS 194. 1994 GATT: 1867 UNTS 187, 33 ILM 1153 (1994).

General Assembly resolution 1803 (XVII) of 14 December 1962 (Declaration on Permanent Sovereignty over Natural Resources): Resolution adopted by the United Nations General Assembly based upon a recommendation by the Commission on Human Rights on the right of peoples and nations to permanent sovereignty over their natural wealth and resources. G.A. Res. 1803 (XVII), U.N. Doc. A/RES/1803(XVII) (14 December 1962). Available at https://www.ohchr.org/Documents/ProfessionalInterest/resources.pdf

General Assembly resolution 62/292 of 28 July 2010 (The Human Right to Water and Sanitation): Resolution adopted by the United Nations General Assembly which asserted "the right to safe and clean drinking water and sanitation as a human right that is essential for the full enjoyment of life and all human rights." GA/10967, United Nations. Available at https://www.un.org/en/ga/search/view_doc.asp?symbol=A/RES/64/292

Great Lakes-St. Lawrence River Basin Sustainable Water Resources Agreement: A 2005 agreement entered into by eight US states and the Canadian provinces of Ontario and Quebec to manage the waters of the basin as a connected, hydrological whole. Available at http://www.glslregionalbody.org/Docs/Agreements/Great_Lakes-St_Lawrence_River_Basin_Sustainable_Water_Resources_Agreement.pdf

Great Lakes-St. Lawrence River Basin Water Resources Compact: A legally binding agreement affecting eight US states bordering the Great Lakes (Illinois, Indiana, Michigan, Minnesota, New York, Ohio, Pennsylvania, and Wisconsin) approved by the state legislatures and enacted as federal law. The Compact details how the States will work together to manage and protect the Great Lakes-St. Lawrence River Basin. It also provides a framework for each State to enact programs and laws protecting the Basin. Implemented through the Great Lakes-St. Lawrence River Water Resources Regional Body (Regional Body) and the Great Lakes-St. Lawrence River Basin Water Resources Council (Council). http://www.glslcompactcouncil.org/

Great Lakes Water Quality Protocol of 2012: Protocol amending the agreement between Canada and the USA of America on Great Lakes Water Quality. Washington, DC. Available at: https://binational.net/2012/09/05/2012-glwqa-aqegl/

International Covenant on Civil and Political Rights: International treaty adopted by the United Nations General Assembly related to civil and political rights. Entry into force 23 March 1976. GA Res. 2200A (XXI), 21 UN GAOR

Supp. (No. 16) at 52, U.N. Doc. A/6316 (1966); 999 UNTS 171; 6 ILM 368 (1967). Available at https://www.ohchr.org/EN/ProfessionalInterest/Pages/CCPR.aspx

International Covenant on Economic, Social and Cultural Rights (ICESCR): International treaty adopted by the United Nations General Assembly related to rights to food, work, housing, and education. Entry into force 3 January 1976. GA Res. 2200A (XXI), 21 UN GAOR Supp. (No. 16) at 49, U.N. Doc. A/6316 (1966); 993 UNTS 3; 6 ILM 368 (1967). Available at https://www.ohchr.org/EN/ProfessionalInterest/Pages/CESCR.aspxCESCR General Comment No. 12 (1999) The Right to Adequate Food (Art. 11) Adopted at the Twentieth Session of the Committee on Economic, Social and Cultural Rights, on 12 May 1999 (Contained in Document E/C.12/1999/5). Available at https://www.refworld.org/pdfid/4538838c11.pdfCESCR General Comment No. 15 (2002) The Right to Water (arts. 11 and 12) Adopted at the Twenty-Ninth Session of the Committee on Economic, Social and Cultural Rights, on 11–12 November 2002 (Contained in Document E/C.12/2002/11). Available at https://www.refworld.org/pdfid/4538838d11.pdf

European Energy Charter: See European Energy Charter.

Kyoto Protocol to the United Nations Framework Convention on Climate Change: International agreement committing certain nations to reductions in emissions of greenhouse gases. Entry into force 16 February 2005. UN Doc FCCC/CP/1997/7/Add.1, Dec. 10, 1997; 37 ILM 22 (1998). Available at https://unfccc.int/resource/docs/convkp/kpeng.pdf

MARPOL (International Convention for the Prevention of Pollution from Ships): Treaty of the International Maritime Organization related to pollution from ships. Entry into force 2 October 1983. 12 ILM 1319 (1973); TIAS No. 10,561; 34 UST 3407;1340 UNTS 184.

Paris Agreement: International agreement under the United Nations Framework Convention on Climate Change (UNFCCC) wherein countries submitted Intended Nationally Determine Commitments (INDCs) related to emissions of greenhouse gases. Adopted at the 21st Conference of Parties (COP21) to the UNFCCC (30 November to 13 December 2015). Entry into force 4 November 2016. Available at https://treaties.un.org/pages/ViewDetails.aspx?src=TREATY&mtdsg_no=XXVII-7-d&chapter=27&clang=_en

Protocol for Sustainable Development of Lake Victoria Basin: This 2003 Protocol established the Lake Victoria Basin Commission (LVBC), under the authority of the 1999 Treaty for the Establishment of the East African Community, which was signed by the partner states of the Republic of Kenya, Republic of Uganda and the United Republic of Tanzania. The Treaty obliges the Partner States to cooperate in relation to Lake Victoria Basin in a "coordinated and sustainable manner." The Protocol authorized the LVBC to develop a management for the conservation and the sustainable utilization of the resources of the Basin. Entry into force 29 November 2003. Available at: https://www.internationalwaterlaw.org/documents/regionaldocs/Lake_Victoria_Basin_2003.pdf

Ramsar Convention on Wetlands of International Importance especially as Waterfowl Habitat (also Wetlands Convention): Treaty of the United Nations

Educational, Scientific and Cultural Organization (UNESCO) related to the protection of many values of wetlands. Entry into force 21 December 1975. 996 UNTS 245; 11 ILM 963 (1972). Available at https://www.ramsar.org/

Treaty for Amazonian Cooperation (also Amazon Cooperation Treaty): Treaty between eight South American countries (Bolivia, Brazil, Colombia, Ecuador, Guyana, Peru, Suriname, Venezuela) to cooperate the in development and conservation of the Amazon basin. Facilitated by Amazon Cooperative Treaty Organization. Entry into force 12 August 1980. 1202 UNTS 51. Available at http://otca.info/portal/tratado-coop-amazonica.php#

Treaty for the Establishment of the East African Community: This Treaty was signed in November 1999 and entered into effect in July 2000, following its ratification by the original three Partner States, the Republic of Kenya, Republic of Uganda and the United Republic of Tanzania. The accord established the East African Community, whereby all participating nations agreed to establish more cooperative commercial and political relations for more than 130 million citizens of the Partner States. The Republic of Rwanda and the Republic of Burundi became full members of the East African Community (EAC) in July 2007. http://eacj.eac.int/?page_id=33

United Nations Framework Convention on Climate Change (UNFCCC): International treaty adopted at the 1992 United Nations Conference on Environment and Development (UNCED), which established an intergovernmental structure to address climate change. Entry into force 21 March 1994, 1771 UNTS 107; S. Treaty Doc No. 102–38; U.N. Doc. A/AC.237/18 (Part II)/Add.1; 31 ILM 849 (1992). Available at https://unfccc.int/files/essential_background/background_publications_htmlpdf/application/pdf/conveng.pdf

US Federal Law.

Notes:

USC = United States Code, a compilation of federal law in the USA.

CFR = U.S. Code of Federal Regulations (such as CFR Title 40: Protection of Environment) that includes the mandatory agency rules and requirements implementing US laws.

The Federal Register (FR) includes proposed US agency rules, notices, executive orders, and other official notices that may have changed the actual implementation of the CFR since the last time it was officially published.

Guidance and Policy specific to individual laws and regulations are typically available from the regulatory agency website. These items may change quickly in response to court decisions and other agency changes.

State food, energy, and water-related laws and regulations can be more restrictive, but not more permissive than, the federal government.

Clean Air Act: Federal law directed at air pollution. Existing clean air law has been amended many times, but especially in 1970, 1977, and 1990. The Act is primarily implemented through the U.S. Environmental Protection Agency and State Implementation Plans (SIPs). 42 USC § 7401 et seq.

Clean Water Act: Federal law directed at water pollution. Existing clean water law has been amended many times, but especially in 1970, 1972, and 1977. Changes in 1990 incorporated provisions of the Great Lakes Water Quality Agreement of 1978. The Act is primarily implemented through the U.S. Environmental Protection Agency and State Implementation Plans (SIPs). 33 USC § 1151 et seq.

Colorado River Compact: A 1922 agreement among seven US states (Arizona, California, Colorado, Nevada, New Mexico, Utah, and Wyoming) regarding the allocation of the water rights from the Colorado River basin. Combined with several federal laws, legal decisions, and agreements, the compact is part of the "law of the river."

Energy Policy Act of 2005: Legislation that established many policies to address energy issues including a Renewable Fuel Standard, tax incentives for many types of energy, and energy efficiency (Pub.L. 109–58) affects 42 USC. § 15801 et seq., § 13201 et seq, and 16 USC § 2601 et seq.

Farm Bill: Legislation passed approximately every 5 years on a wide range of agricultural and food issues.

Food Safety Modernization Act (FSMA) of 2011: Legislation that impacted the regulation of food by the U.S. Food and Drug Administration. It shifted the focus of regulatory efforts to the prevention of foodborne illness by addressing multiple steps in the production and supply chain (Pub.L. 111–353) affecting 21 USC § 301 et seq.

Louisiana-Mississippi Tangipahoa River Waterway Compact: 1988 agreement between the states of Louisiana and Mississippi regarding the Tangipahoa River.

Mississippi River Interstate Pollution Phase-out Compact: Proposed interstate compact regarding the Mississippi River.

National Environmental Policy Act (NEPA): 1970 legislation that established the White House Council on Environmental Quality (CEQ) and the requirement all federal agencies consider every significant aspect of the environmental impact of an action before proceeding with it. This occurs through Environmental Impact Statements. (Pub.L. 91–190) 42 USC § 4321 et seq.

National Pollutant Elimination Discharge System (NPDES): Permitting system under the Clean Water Act for regulating point sources of pollution discharged into "Waters of the USA." 33 USC § 1342.

Oil Pollution Act of 1990: Legislation passed following the 1989 Exxon Valdez oil spill in Prince William Sound, Alaska. It addressed many issues of liability and responsibility for oil spills and required oil tankers to have double hulls. (Pub.L. 101–380) 33 USC § 2701.

Resource Conservation and Recovery Act (RCRA): Federal law directed at the disposal of solid and hazardous wastes. Existing law was significantly revised in 1974 (Pub.L. 93–523). The law has been amended at least seven times since 1974. The Act is primarily implemented through the U.S. Environmental Protection Agency and State Implementation Plans (SIPs). 42 USC § 6901 et seq.

Safe Drinking Water Act (SDWA): Federal law directed at public drinking water supplies. The law was developed in 1965 but was overhauled in 1976

(Pub.L. 94–580). The Act is primarily implemented through the U.S. Environmental Protection Agency and State Implementation Plans (SIPs). 42 USC § 300 et seq.

Universal Declaration of Human Rights (UDHR): Resolution adopted by the United Nations General Assembly on 10 December 1948. The Declaration consists of 30 articles that address a range of human rights. Although legally non-binding, its common usage leads many to consider it customary international law. It began the process leading to the International Bill of Human Rights. U.N. Doc A/810 at 71 (1948).

Vienna Convention on Civil Liability for Nuclear Damage: Treaty of International Atomic Energy Agency establishing standards of financial protection against damages caused by certain peaceful uses of nuclear energy. Entry into force 12 November 1977. 1062 UNTS 265; 2 ILM 727 (1963). Amended in 1997 with 2241 UNTS 270.

Vienna Convention on the Law of Treaties: Treaty developed by the International Law Commission that establishes rules for international treaties. Entry into force 27 January 1980. 1155 UNTS 331; 8 ILM 679.

WTO Agreement: **Marrakesh Agreement**: Treaty of the World Trade Organization (WTO)/General Agreement on Tariffs and Trade (GATT) negotiated during the Uruguay Round (1986–1995) that establishing the World Trade Organization. Entry into force 1 January 1995. 1867 UNTS 154.

Appendix C: Data Sources

Chapter 3: Development

CEOS (2018), Satellite Earth Observations in Support of the Sustainable Development Goals. CEOS. http://eohandbook.com/sdg/

CIA (2018). The World Factbook. U.S. Central Intelligence Agency. https://www.cia.gov/library/publications/the-world-factbook/

Data.gov (2018), Global Development. Data.gov. https://www.data.gov/development/

DI (2018), The Development Data Hub. http://data.devinit.org/

EIA (2018) International Energy Outlook. Energy Information Administration Available online at https://www.eia.gov/outlooks/ieo/

GapMinder: https://www.gapminder.org/

Google (2018), Public Data Search: World Development Indicators. https://www.google.com/publicdata/explore?ds=d5bncppjof8f9_

GPSDD (2018), Global Partnership for Sustainable Development Data. http://www.data4sdgs.org/

Holden, J. (2016). The top 10 sources of data for international development research. The Guardian, 16 March 2016. https://www.theguardian.com/global-development-professionals-network/2016/mar/16/the-top-10-sources-of-data-for-international-development-research

OECD (2018), International Development Statistics (IDS) online databases. OECD. http://www.oecd.org/dac/stats/idsonline.htm

OECD (2018), OECD International Development Statistics. OECD. https://www.oecd-ilibrary.org/development/data/oecd-international-development-statistics_dev-data-en

Out World in Data: https://ourworldindata.org/

UN (2018), Global Human Development Report 2018 Statistical Update. http://hdr.undp.org/en/2018-update

P. Saundry, B. L. Ruddell (eds.), *The Food-Energy-Water Nexus*, AESS Interdisciplinary Environmental Studies and Sciences Series,
https://doi.org/10.1007/978-3-030-29914-9

UN Department of Economic and Social Affairs, Population Division (2017), World Population Prospects 2017. Data and related materials available online at https://esa.un.org/unpd/wpp/

UN Food and Agriculture Organization, Prevalence of undernourishment. Available online at https://data.worldbank.org/indicator/SN.ITK.DEFC.ZS

World Bank (2018) World Development Indicators. Data and related materials available online at https://datacatalog.worldbank.org/dataset/world-development-indicators or https://data.worldbank.org/products/wdi

World Bank (2017) Global Tracking Framework 2017. Available online at http://www.worldbank.org/en/topic/energy/publication/global-tracking-framework-2017

World Bank (2018), World Bank Open Data. https://data.worldbank.org/

World Health Organization, Global Database on Child Growth and Malnutrition. Data and related materials available online at http://www.who.int/nutgrowthdb/en/

World Health Organization/UNICEF Joint Monitoring Programme (JMP) for Water Supply, Sanitation and Hygiene website https://washdata.org/

World Food Programme, Food security analysis website (http://www1.wfp.org/food-security-analysis).

World Resource Institute, ResourceWatch website https://resourcewatch.org/

Chapter 4: Human Behavior and Adaptation

EPA's Alternative Fuels Data Center (https://afdc.energy.gov/data/categories/afvs-and-hevs).

Chapter 8: US Governance

American Bar Association. Waters of the U.S. Rule. Available online at https://www.americanbar.org/groups/environment_energy_resources/resources/wotus/wotus-rule.html

Draper & Draper, LLC. U.S. Transboundary Water Database. Available online at http://www.draperllc.com/water-disputes-map. (Includes an interactive map and database of interstate water compacts, conflicts, and court cases.)

Energy Information Administration power plant data (form 860: https://www.eia.gov/electricity/data/eia860/, form 923: https://www.eia.gov/electricity/data/eia923/).

Environmental Protection Administration. National Pollutant Discharge Elimination System (NPDES) State Program Information. Available online at https://www.epa.gov/npdes/npdes-state-program-information

Environmental Protection Administration. Sole Source Aquifers. Available online at https://epa.maps.arcgis.com/apps/webappviewer/index.html?id=9ebb047ba3ec41ada1877155fe31356b

USA Geological Water Use Survey (quinquennial water survey: https://water.usgs.gov/watuse/).

U.S. Fish and Wildlife Service, Digest of Federal Resource Laws of Interest to the U.S. Fish and Wildlife Service. Available online at https://www.fws.gov/laws/lawsdigest/compact.html.

Chapter 7: Trade

Enerdata, Global Energy Statistical Yearbook 2018. https://yearbook.enerdata.net/total-energy/world-import-export-statistics.html, Accessed February 20th, 2019.

EnergyData.info, The World Bank, https://energydata.info/, Accessed February 20th, 2019.

FAOSTAT, Food and Agriculture Organization of the United Nations. http://www.fao.org/faostat/en/, Accessed February 20th, 2019.

IEA, Statistics. https://www.iea.org/statistics/index.html?country=WORLD&year=2016&category=Energy%20supply&indicator=TPESbySource&mode=chart&dataTable=BALANCES, Accessed February 20th, 2019.

International Trade Centre, Trade Map. http://www.intracen.org/itc/market-info-tools/trade-statistics/, Accessed February 20th, 2019.

OECD, Energy Data, https://data.oecd.org/energy.htm, Accessed February 20th, 2019.

OECD Summary of IEA World Energy Statistics and Balances. https://www.oecd-ilibrary.org/energy/data/iea-world-energy-statistics-and-balances_enestats-data-en, Accessed February 20th, 2019.

Ortiz-Ospina, D. Beltekian, and M. Roser, Trade and Globalization, Our World in Data. https://ourworldindata.org/trade-and-globalization, Accessed February 20th, 2019.

World Bank Data Bank: https://data.worldbank.org/

Chapter 8: US Governance

American Bar Association. Waters of the U.S. Rule. Available online at https://www.americanbar.org/groups/environment_energy_resources/resources/wotus/wotus-rule.html

Draper & Draper, LLC. U.S. Transboundary Water Database. Available online at http://www.draperllc.com/water-disputes-map. (Includes an interactive map and database of interstate water compacts, conflicts, and court cases.)

Energy Information Administration power plant data (form 860: https://www.eia.gov/electricity/data/eia860/) (form 923: https://www.eia.gov/electricity/data/eia923/).

Environmental Protection Administration. National Pollutant Discharge Elimination System (NPDES) State Program Information. Available online at https://www.epa.gov/npdes/npdes-state-program-information

Environmental Protection Administration. Sole Source Aquifers. Available online at https://epa.maps.arcgis.com/apps/webappviewer/index.html?id=9ebb047ba3ec41ada1877155fe31356b

USA Geological Water Use Survey (quinquennial water survey: https://water.usgs.gov/watuse/).

U.S. Fish and Wildlife Service, Digest of Federal Resource Laws of Interest to the U.S. Fish and Wildlife Service. Available online at https://www.fws.gov/laws/lawsdigest/compact.html.

Chapter 9: Ecosystems and Ecosystem Services

The Economics of Ecosystems and Biodiversity—Valuation Database—http://doc.teebweb.org/wp-content/uploads/2017/03/teeb_database_teebweb.xlsx

World Resource Institute, ResourceWatch website https://resourcewatch.org/data/explore

Chapter 11: Climate Change

European Commission (2018) EDGAR: Emissions Database for Global Atmospheric Research. Available online at http://edgar.jrc.ec.europa.eu/

UN Food and Food and Agriculture Organization (2018). FAO yearbook. Fishery and Aquaculture Statistics 2016. Rome 104pp.

UN Food and Agriculture Organization, Prevalence of undernourishment. Available online at https://data.worldbank.org/indicator/SN.ITK.DEFC.ZS

World Bank (2017) Global Tracking Framework 2017. Available online at http://www.worldbank.org/en/topic/energy/publication/global-tracking-framework-2017

World Resource Institute, ResourceWatch website https://resourcewatch.org/data/explore

Chapter 15: Modeling

Shared Socioeconomic Pathways: global database for scenario projections for food production, power generation and water resources: https://tntcat.iiasa.ac.at/SspDb/dsd?Action=htmlpage&page=welcome

GEOGLAM Crop Monitor: https://cropmonitor.org

Chapter 16: Computing

Computational Tools:

- Stackoverflow Q&A
- Anaconda Python Package Manager
- R Website
- Tutorials on a variety of Tools/Approaches: Medium
- GitHub Idea Portal: GIST
- QGIS Geographic Information System Software
- ARCGIS Geographic Information System Software
- Spatial Database Management System

Visualization Tools (R or Python)

- ggplot,
- matplotlib
- mapnik
- folium
- plotly

Datasets:

- Google Dataset Search
- NASA Open Data Portal
- ESA Open Data Portal
- GEOGLAM Crop Monitor

Appendix D: Educational Resources

General

Bloomberg U.S. Land Use Visualization, https://www.bloomberg.com/graphics/2018-us-land-use/

USEPA Watershed Academy, https://www.epa.gov/watershedacademy

USEPA Water Topics Index, https://www.epa.gov/environmental-topics/water-topics

HydroViz Energy-Water Nexus, http://nexus.hydroviz.org/Lessons/Index/National/Nexus

FEWSION Visualization, https://fewsion.us/visualization/

FEWSION Education, https://fewsion.us/education/

USGS Water Science School, https://water.usgs.gov/edu/

Johns Hopkins Foodspan Curriculum, Unit 1 Intro: http://www.foodspanlearning.org/lesson-plans/unit-1-meet-the-food-system/index.html#section1

Johns Hopkins Foodspan Curriculum Unit 2 Supply Chain: http://www.foodspanlearning.org/lesson-plans/unit-2-farmers-factories-and-food-chains/

Johns Hopkins Foodspan Curriculum, Unit 3 Consumers: http://www.foodspanlearning.org/lesson-plans/unit-3-consumers-and-communities/

Nourish Curriculum, http://www.nourishlife.org/teach/food-system-tools/

USEIA Energy Kids, https://www.eia.gov/kids/

USDOE Energy Literacy Curriculum, https://www.energy.gov/eere/education/energy-literacy-essential-principles-and-fundamental-concepts-energy-education

NEA Clean Energy Education, http://www.nea.org/tools/clean-energy-education.html

USEPA Energy and the Environment, https://www.epa.gov/energy

Global Footprint Network Ecological Footprint Calculator, https://www.footprintcalculator.org/

USEPA Carbon Footprint Calculator, https://www3.epa.gov/carbon-footprint-calculator/

P. Saundry, B. L. Ruddell (eds.), *The Food-Energy-Water Nexus*, AESS Interdisciplinary Environmental Studies and Sciences Series,
https://doi.org/10.1007/978-3-030-29914-9

GRACE Communications Foundation Water Footprint Calculator, https://www.watercalculator.org/

It's Fresh Food Waste Calculator, http://www.itsfresh.com/food-waste-calculator/

USEIA Radio Programs, https://www.eia.gov/pressroom/radio/

USEIA International Energy Outlook, https://www.eia.gov/outlooks/ieo/

OpenEI Energy Information, https://openei.org/wiki/Information

NREL BITES (carbon emissions reduction tool), https://bites.nrel.gov/

CLEAN Climate and Energy Education, https://cleanet.org/index.html

Chapter 3: Development

Borr, M. (2018), Examination of Environmental Change, Vulnerability, and Human Migrations. Center for Global Studies at Penn State. http://cgs.la.psu.edu/teaching-resources/community-resources/miiies-international-intercultural-module-examination-of-environmental-change-vulnerability-and-human-migrations

GCS (2018), Teaching Resources. The Center for Global Studies at Penn State. http://cgs.la.psu.edu/teaching-resources

Klein, J.D. (2017), BUILDING A BETTER WORLD: SIX STRATEGIES FOR ENGAGING THE SUSTAINABLE DEVELOPMENT GOALS IN THE CLASSROOM (with links to SDG teaching resources). November 14th, 2017. http://www.p21.org/news-events/p21blog/2282-building-a-better-world-six-strategies-for-engaging-the-sustainable-development-goals-in-the-classroom

NEA (2018), Global Education Resources. National Education Association. http://www.nea.org/home/37409.htm

Selcher, W.A. (2018), The WWW Virtual Library: International Affairs Resources. https://internationalaffairsresources.com/

TeachSDGS (2018), Assets: Instructional Materials, Free Courses, Tools, Resources & Publications. Teach SDGs. http://www.teachsdgs.org/assets.html

WLL (2018), World's Largest Lesson. http://worldslargestlesson.globalgoals.org/

World Bank (2018), GINI Index. https://data.worldbank.org/indicator/si.pov.gini?view=map

Chapter 14: Data

Data Carpentry, https://datacarpentry.org/

GapMinder. https://www.gapminder.org/, Accessed December 4th, 2018.

GitHub Guides, *GitHub*. https://guides.github.com/, Accessed December 4th, 2018.

Best Practices for Data Management, *DataOne*, https://www.dataone.org/best-practices, Accessed December 4th, 2018.

Data Best Practices, *Stanford Libraries*. https://library.stanford.edu/research/data-management-services/data-best-practices, Accessed December 4th, 2018.

USDA NASS Data Visualization, https://www.nass.usda.gov/Data_Visualization/index.php

USDA NASS Cropscape Visualization, https://nassgeodata.gmu.edu/CropScape/

USGS WaterWatch, https://waterwatch.usgs.gov/

USEIA International Energy Outlook, https://www.eia.gov/outlooks/ieo/

NASA Worldview, https://worldview.earthdata.nasa.gov/

PRISM Current Month U.S. Climate, http://prism.oregonstate.edu/mtd/

iSciences Water, Climate, and Human Security, https://www.isciences.com/

Bloomberg U.S. Land Use Visualization, https://www.bloomberg.com/graphics/2018-us-land-use/

HydroViz Energy-Water, http://nexus.hydroviz.org/Lessons/Index/National/Nexus

WRI Aqueduct, https://www.wri.org/our-work/project/aqueduct

ArcGIS Open Data Hub, http://hub.arcgis.com/pages/open-data

PolicyMap, https://www.policymap.com/

FEWSION Project FEW-View, https://fewsion.us/visualization/

Google Earth Engine Public Datasets, https://earthengine.google.com/datasets/

Pardee RAND Food-Energy-Water Index, https://www.prgs.edu/pardee-initiative/food-energy-water.html

US EPA NPDES Tools, https://www.epa.gov/npdes/npdes-tools

Water Energy Food (WEF) Nexus Tool 2.0, www.wefnexustool.org

WRI ResourceWatch, https://resourcewatch.org

Chapter 15: Modeling

An Introduction to Integrated Assessment of Climate-Energy–water-Land-Economy Systems; all course materials: lectures, videos, exercises.

Chapter 16: Computing

Online Training Portals:

- EDX
- Coursera
- Udemy

Chapter 17: Applying Science to Practice

FEWSION for Community Resilience (F4R), https://fewsion.us/f4r/, Accessed March 13th, 2019.

HydroViz Energy-Water Nexus, http://nexus.hydroviz.org/Lessons/Index/National/Nexus, Accessed March 13th, 2018.

INFEWS-ER (2019), Innovations at the Nexus of Food, Energy, and Water Systems Educational Resources, https://infews-er.net/, Accessed March 13th, 2019.

InTeGrate (2018), Unit 2: Community-Based Participatory Solutions, https://serc.carleton.edu/integrate/teaching_materials/food_energy_water/student_materials/unit_2.html, Accessed March 13th, 2019.

Ruddell, B.L., Miller, J., Rushforth, R.R., Salla, R., and Sọktoeva, E. (2018), FEW-View 1.0, the FEWSION Visualization System, https://dev2.dtn.asu.edu/dev/fewsion/public/, Accessed May 21st, 2018.

Ruddell and Konar, et al. (2017), Master Class: Food, Energy, and Water Systems in a Global Economy https://www.cuahsi.org/community/news-events/master-class-food-energy-and-water-systems-in-a-global-economy, Accessed March 13th, 2019.

UNFAO (2019), Water-Energy-Food Nexus Rapid Appraisal, http://www.fao.org/energy/water-food-energy-nexus/water-energy-food-nexus-ra/en/, Accessed March 13th, 2019.

Chapter 18: Cities at the Nexus

EPA 2018. Net Zero Strategy. https://www.epa.gov/water-research/promoting-sustainability-through-net-zero-strategies

ICLEI 2019. Local Governments for Sustainability. http://icleiusa.org/

National Academies of Sciences, Engineering, and Medicine 2016. *Pathways to Urban Sustainability: Challenges and Opportunities for the United States*. Washington, DC: The National Academies Press. https://doi.org/10.17226/23551.

SymbioCity 2019. "Sustainability by Sweden." http://www.symbiocity.se/

SymbioCity Sustainable City Simulation Game. http://www.btslearning.com/app/eBS/symbiocity/index.asp

Wolfson, Richard. *Energy, environment, and climate, Third Ed*. WW Norton & Company, Inc., 2017.

Chapter 19: Watersheds at the Nexus

An Educator's Guide to the Meaningful Watershed Educational Experience. Chesapeake Bay Program, Annapolis, MD. Available at: https://www.cbf.org/document-library/education/teachers-guide-to-meaningful-watershed-education-experience.pdf

Rock Your Watershed! A Game of Chance and Choice, Iowa State University Extension and Outreach. Available at: http://water-rocks.herokuapp.com/game/index/

Stapp, W. B., Wals, A. E., Moss, M. R., & Goodwin, J. E. (eds.) (1996). *International Case Studies on Watershed Education*. Dubuque, IA: Kendall/Hunt Publishing Company.

The Watershed Game, Distance Learning and Educational Web Adventures. Available at: http://games.bellmuseum.umn.edu/watershed/

The Watershed: What is a Watershed?, Central Sierra Environmental Resource Center. Available at: https://www.cserc.org/sierra-fun/games/watershed-game/

Tools and Resources to Protect Watersheds, United States Environmental Protection Agency. Available at: https://www.epa.gov/hwp/tools-and-resources-protect-watersheds/

Watershed Education, Pennsylvania Department of Conservation and Natural Resources. Available at: http://www.docs.dcnr.pa.gov/stateparks/watersheded/overview/index.htm

Chapter 20: Managing Human Conflicts

American Bar Association Section on Dispute Resolution: https://www.americanbar.org/groups/dispute_resolution/

Association for Conflict Resolution: https://acrnet.org/

Mediate.com: https://mediate.com/

Oregon State University's Program in Water Management and Transformation Data Set on Transboundary Waters: https://transboundarywaters.science.oregonstate.edu/content/data-and-datasets

Program on Negotiation at Harvard Law School: https://www.pon.harvard.edu/

Transboundary Water Governance: http://www.watergovernance.org/

University Network for Collaborative Governance: https://www.kitchentable.org/uncg

University of Utah Environmental Dispute Resolution Program: https://www.law.utah.edu/blogs/edr-blog/

Glossary

Absolute Metric A metric independent of context concerning budget, constraint, peer, or history, e.g., a 1000-MW baseload power plant may consume 200 gallons of water per megawatt hour (intensive), equivalent to 4.8 million gallons of water per day (extensive).

Acre An imperial unit of area, equivalent to 4047 m^2 or around 0.4 Ha.

Acre-Foot A commonly used unit of water withdrawals or consumption, especially in irrigation in the Western U.S. It is the volume of water to cover an area of one acre to a depth of one foot, equal to 1233 m^3.

Afforestation Planting trees, saplings, and seeds on land devoid of trees to establish a forest. Compare to Reforestation.

Agent-based Models (ABMs) Computational models that model how the decisions of individual agents affect the system outcome(s) of interest. ABMs typically allow for types of individuals with different decision rules, and for interaction between these different types.

Agents Natural or human entities who control stocks, and who produce, consume, originate, terminate, and take inputs and outputs concerning flows. Agents exhibit sentient decision-making behavior, both rational and irrational.

Aggregation A common method of de-identifying PII or PCII or other sensitive data by reporting the space-time location, stock, flow, etc. for a group instead of an individual.

Agricultural Drought Drought conditions as measured by soil moisture. Compare to hydrological drought and meteorological drought.

Agricultural Food Commodity A staple crop (e.g., wheat, corn, rice, barley, maize, sugar, fruits, and vegetables) or animal products (e.g., fish, meat, and dairy products) that is the basis for foodstuffs.

Alluvium Sediment of sand, silt, gravel, and clay that is deposited by streams and rivers and often forms very fertile soil in a river valley or delta.

Alternative Dispute Resolution (ADR) Sometimes called appropriate dispute resolution, ADR is simply "A process adopted to end a problem before taking legal

P. Saundry, B. L. Ruddell (eds.), *The Food-Energy-Water Nexus*, AESS Interdisciplinary Environmental Studies and Sciences Series,
https://doi.org/10.1007/978-3-030-29914-9

action." (Black's Law Dictionary online). It can be a range of tools that include negotiation, mediation, and other tools used to address conflicts.

Anthropocene The time period where the human activity has been the dominant influence on climate and the environment, including ecosystems, water systems, the cryosphere and atmosphere.

Anticipatory Adaptation A strategy to increase the resiliency (vulnerability to disruption) of a system by adapting the system in anticipation of possible disruptions. The strategy requires monitoring and foresight.

Apache Hadoop A set of open-source software utilities that facilitate using a network of many computers to solve problems involving massive amounts of data and computation.

Applied Research Research creates near-term utility and value.

Aquifer Layers of porous and permeable rock or sand below the surface of the earth where water is stored.

Arbitration "The investigation and determination of a matter or matters of difference between contending parties, by one or more unofficial persons, chosen by the parties, and called "arbitrators," or "referees." (Black's Law Dictionary online). Sometimes called "private judging," a third party neutral or panel of neutrals hear the conflict between parties, take evidence, and render a decision in a case.

Artificial Neural Network A class of algorithms that were designed by mimicking the neural connections inside animal brains. They are designed to learn from data that will later be used for predicting the outcomes in similar situations.

Assimilation A modeling technique that blends theoretical model estimates with observations to produce an optimally accurate dataset.

Auditability of Data A requirement that the quality and provenance of data can be verified by tracking it upstream through the data life cycle to its source.

AWS Amazon Web Services is a set of tools incorporated in Amazon's cloud platform to provide fast, reliable and efficient computation, storage and analytics services.

Baseload Power Plants Power plants that operate continuously to meet the minimum amount of daily electricity demand. Baseload power plants are often nuclear or coal, though sometimes natural gas combined-cycle plants are used for baseload.

Basic Research Research that creates new fundamental knowledge and concepts in the short term and leads to long-term value.

Basin An area of land where precipitated water collects into a common body of water, e.g., river or lake.

Behavioral Heterogeneity The recognition that people may be motivated differently and that their behavior may differ in systematic ways, rather than assuming that people behave in the same way when faced with particular stimuli.

Behavioral Spillover When engaging in a target behavior is linked to another seemingly unrelated behavior. Behavioral spillover can be positive (an increase in the target behavior is causing an increase in another behavior) or negative (an increase in one behavior causing a decrease in another behavior).

Benefit–Cost Analysis A method with defined procedures of calculating the benefits of a particular action as well as the associated costs, which end up subtracting the costs from benefits to see if an action is desirable or undesirable from a strictly economic viewpoint.

Benefits Transfer The use of economic information from one setting in another setting. An example is taking an estimated value of a fish in Oregon and using it in evaluating a California project.

Benefits (Human Welfare) Positive outcomes in well-being from the fulfillment of needs and wants.

BioCro A model for perennial biomass feedstocks, including Miscanthus, switchgrass, and willow, which captures vegetation response to atmospheric and climate change.

Biodiversity The diversity of biological life in a particular ecosystem, region, component of the earth (e.g., terrestrial or marine), or the earth as a whole. See Ecosystem Diversity.

Biofuels Fuels created through biochemical processes such as fermentation and distillation, most commonly from plant matter but sometimes from other wastes. Examples include biodiesel and bioethanol.

Biogeophysical The biological and physical components, processes, and interactions that occur in a defined location, area, or generally. See Biophysical and Geophysical.

Biophysical The components and interactions between biotic (living) and abiotic (non-living) things, such as species and ecosystems and their interactions with non-living physical processes.

Biosphere The parts of the earth system composed of living organisms. This includes parts of the land, oceans, and atmosphere and is sometimes referred to as the sum of all ecosystems on the earth.

Biota Biological life, usually refers to just plants and animal life, but also includes fungi, archaea, and bacteria.

Black Swan Event An extremely consequential and unique one-of-a-kind event; this kind of event cannot be predicted with sufficient accuracy.

Boomerang Effect Observed when the promotion of a particular behavioral or attitudinal change has an effect opposite of what was intended. See Rebound Effect.

Bottom-Up Approaches, methods, and models that involve building up from small units. Bottom-up models are constructed from their subparts. Bottom-up methodologies involve direct measurement at the process scale, and subsequent aggregation and gap filling from many process observations to estimate "mass balance" at larger scales. See Top-Down.

Boundaries (System) The limits in space and time a system which defines the parts, external factors and which interactions must be considered.

Buffering A design strategy that provides reliability, robustness, and/or resilience by storing enough of a commodity to sustain operations during disruptions; Inventory is buffering.

Bushel An imperial unit of volume used primarily for food production. It is equivalent to 35.2391 L.

Cairns Group A group of 19 agricultural exporting countries that work to liberalize international trade in agricultural commodities, especially through the World Trade Organization.

Calories A unit of measure of energy content (normally for food) equivalent to the energy required to raise the temperature of a gram of water by 1 °C. There can be some confusion as, when capitalized; the unit Calories refers to the energy required to raise the temperature of one kilogram of water by 1 °C, i.e., one Calorie = one kilocalorie = 1000 calories.

Capacity The peak rate at which a system (e.g., a power plant) can operate. Compare to Utilization.

Capital Cost The cost of building a system.

Capital Goods Goods that are used to produce other goods that are sold to consumers (see consumer goods).

Capitalist Economic System Assumes that government acts to benefit the interests of one or more specific capitals (extractive, agricultural, industrial, commercial, financial) and/or to benefit the long-term interests of capitalism as a mode of social organization.

Carbon Capture and Storage/Sequestration (CCS) The capture of carbon, typically from the waste stream of a large point source of carbon dioxide (e.g., power plant), and its long-term storage to mitigate climate change and ocean acidification.

Carbon Dioxide (CO_2) A molecule of one carbon atom and two oxygen atoms that is an important greenhouse gas generated when combusting hydrocarbon fuels, such as coal.

Carrying Capacity The estimated maximum population of a species that an environment can sustain indefinitely.

Cascading Failure A situation where a failure in one part of a system causes failures in other parts of the system. For example, a failure in "upstream" infrastructure leading to failures in the "downstream" infrastructure that is dependent on the upstream component.

Catchment Basin (also Watershed or Drainage Basin) A geographical delineation characterized by the fact that all runoff within the basin will eventually drain to a given water body. Catchment basins are nested. For example, the catchment of a small tributary is within the catchment of a larger river, and both of those are within the catchment basin of the lake or large river to which they drain. Also called a *watershed* or *drainage basin.*

Celsius (C) A metric unit temperature scale defined by 0 °C as the freezing point of water and 100 °C as the boiling point of water at sea level and standard temperature and pressure. It is commensurate with the Kelvin scale but shifted such that 0 °C = 273.15 K.

Centralized FEW System A system in which food, energy, and water resources are efficiently produced in a centralized or concentrated location or hub in high volumes or quantities, and then transported to dispersed consumers. Examples industrial agriculture and CAFOs (concentrated animal feeding operations), electricity produced in a large power plant and then delivered to consumers across the grid, and municipal water and sewerage systems that collect water from a centralized input source such a river, deliver it to consumers via vast pipe systems, and then return it to a single centralized WWTP.

Circularity The degree to which an economy, or flows of goods, services, or information, cycle and feed back to the original source; for example, if half of a community's food is produced locally, the circularity of that food economy is 0.5.

City An urbanized area of human settlement, defined by the United Nations as a locality with more than 50,000 inhabitants. Definitions of cities vary substantially worldwide, with some countries defining a city as a locality with at least 200 residents.

Civil Society The societal space in which collective action around shared values, interests, and purposes occurs. Civil society includes non-profit entities such as charities, community groups, professional and faith-based organizations, social movements, advocacy groups, trade societies, and other non-governmental and non-commercial human associations.

Clean Water Act (U.S.) The U.S. federal statute that governs discharges of pollutants and removal of dredged material from USA waters. Its precursor was the Federal Water Pollution Control Act, passed in 1948, but in 1972 the act was significantly restructured and renamed the Clean Water Act. It can be found at 33 U.S.C. 1251 et seq.

Clean Water Rule of 2015 (U.S.) The U.S. Environmental Protection Agency's interpretation of the jurisdiction of the Clean Water Act to include all waters that have a "significant nexus" with navigable waters, including those that are hydrologically connected but have no visual surface connection.

Climate Change (Anthropogenic) Long-term climate change driven by changes in the concentration of greenhouse gases within the Earth's atmosphere made by human actions. The Earth's climate has changed dramatically over its history. Human activities are the main driver of current climate change.

Climate Refugee Migration caused by climate impacts because of displacement caused by disasters and climate change (see https://www.unhcr.org/climate-change-and-disasters.html).

Climate Variability Variations in climate due to natural drivers as opposed to anthropogenic drivers.

Climate-Weather Research and Forecasting (CWRF) Model A regional climate model developed at the University of Maryland, which simulates the interactions between land, atmosphere, ocean; convection and microphysics; and clouds, aerosols, and radiation, yielding improved forecast skills on climate timescales.

Closed Supply Loop A closed-loop supply chain that maximizes efficiency by converting waste streams from one sector or process into resources for another sector or process.

Collaborative Geodesign The involvement of stakeholders in landscape design using a tool which interactively lets them test and receive feedback for different design decisions. The benefit of this approach is that the stakeholders can immediately see the impact of their design decisions on biophysical and social indicators. Collaborative geodesign is a step forward to initiate discussions among stakeholders and domain scientists. Thus, the actual parties (i.e., stakeholders) that are affected by the decisions have the opportunity to communicate their concerns with the scientific community as well as the policymakers to make better and more realistic design decisions.

Collaborative Governance A "broad range of processes through which citizens and stakeholders collaborate to make, implement, and enforce public policy" (Blomgren-Bingham 2009).

Combined Sewer Overflows (CSOs) In cities with a combined sewer system, occurs when precipitation overloads the wastewater processing facility during large storm events, resulting in raw sewage being discharged into the receiving body of water.

Combined Sewer System A sewer system, often found in older cities, that collects, transports, and processes both stormwater runoff and sanitary wastewater inputs together.

Commodities Basic physical consumable goods that are bought and sold in large quantities.

Common Pool Resource Resources have "two attributes of importance for economic activities: (1) it is costly to exclude individuals from using the good either through physical barriers or legal instruments and (2) the benefits consumed by one individual subtract from the 9 benefits available to others." (Ostrom and Hess 2007).

Community of Practice A group of people who share a domain of interest and learn to improve action regarding that interest through regular interaction.

Comparative Advantage The advantage that an individual, company, or nation has in its ability to complete a particular activity, such as making a product, compared to others. The advantage may exist for many reasons, including the availability of natural resources, capital goods, human resources, or technological capacity; or because of economies of scale, location, customer preferences, subsidies, or helpful policies.

Compensation Principle A criterion that says a particular action is desirable if the net benefits accruing to those gaining from the action exceed the net cost to those losing because of the action.

Computable General Equilibrium (CGE) Models Economic models that are used to estimate how an economy might react to changes in policies, technologies or other external factors, such as climate change.

Computational Complexity Mathematical models of computation to define how much resources an algorithm needs.

Conflict Generally, a significant disagreement between parties. The Dictionary of Conflict Resolution includes more than 20 pages of definitions; the difference between a "dispute" and a "conflict" may be one of magnitude where "conflicts are often seen as broader (involving more people), deeper (extending beyond surface issues into questions of value, identity, fear, or need), and more systematic (reaching beyond a single interaction or claim)" (Moffitt and Bordone 2005).

Conservation Agriculture A set of agricultural practices, especially soil management practices, which sustain the natural soil fertility, ecosystem functions, and species diversity of farmland.

Consumer Goods Goods that are bought and used by consumers as opposed to capital goods that are used to produce other goods.

Consumer Surplus The difference between the maximum price a consumer is willing to pay to acquire a given quantity of goods less the price that is paid summed over all quantities of the good consumed. It reflects consumer satisfaction of paying lower prices for goods.

Consumption (Water) The difference between water withdrawn from a watershed and the quantity of water that is returned to the same watershed in liquid form. Generally, water consumption is due to evaporation and evapotranspiration or its embodiment in some product (e.g., food). See Withdrawal (Water).

Cool Roofs A roof of a building that is designed to reflect more sunlight and absorb less heat than a standard roof, typically utilizing either highly reflective paint, tiles or shingles, sheet coverings, or protective coatings.

Cornucopian A view of the challenges of providing resources for a population that emphasizes the ability of humans to develop solutions, especially technological solutions, that provide more resources (in contrast to a neo-Malthusian view).

Correlative Rights A legal doctrine whereby the owners of adjoining areas of land, or the governments of adjacent regions, recognize each other's rights to shared resources like a water-body or aquifer that extends across the border between the two owners or governments.

Cost–Benefit Analysis A technique designed to determine the feasibility of a project or plan by quantifying its costs and benefits.

Couplings The points of contact between different systems or subsystems.

Cryosphere The frozen water parts of the earth system including all places composed of ice and snow such as ice caps; ice sheets; glaciers; permafrost; frozen seas, lakes and rivers; icebergs; and snowfields.

Cubic Foot (ft^3, cu-ft, or CF) Cubic foot is an imperial unit of volume, equivalent to 28.3168 L. Natural gas is often measured in million cubic feet (MCF) or trillion cubic feet (TCF).

Cubic Meter (m^3) A metric unit of volume, defined as a cube with sides of one meter. A cubic meter is equivalent to 1000 L or 220 gallons.

Cultural Ecosystem Services Ecosystem services that have physical, emotional, artistic, recreational, religious, or other values important to the social behavior of human communities.

Customary International Law Obligations between nations rooted in established or "customary" practices.

Data Useful facts and assumptions, and in the digital era, these facts and assumptions are usually (but not always) quantitative and numeric in nature.

Data Ethics Principles that stipulate that data collection and use must yield benefits to the object of the data (not the subjects conduction analysis) that outweigh the risks to the object of the data.

Data Life Cycle The design, collection, quality control, metadata description, curation, discovery, integration, and analysis of data.

Data Protection The law, and/or means of enforcement of the law, protecting private or sensitive data and exercising the right to data privacy.

Data Quality The standard of data relative to its intended use. Data quality involves the validity, completeness, precision (accuracy), resolution, and provenance of the data.

Data Utility The value and employability of data to solve a problem or answer a question and utility is often balanced against (in tension with) privacy.

Data Validity The correctness and quality of the data in hand.

Database A collection of information organized to provide efficient storage, update, and retrieval. To perform these tasks, specialized software, namely, Database Management Systems are used.

Decentralized FEW System A system in which food, energy, and water resources are produced in many smaller-scale and dispersed locations, closer to the consumer. Examples include urban and peri-urban agriculture and backyard food production, localized electrical mini-grids utilizing solar, wind, or geothermal energy, and well-water systems. Decentralized systems are characterized by their greater potential for resilience than centralized systems.

Decision Science The interdisciplinary study of human decision-making at the individual, collective, and institutional levels. Decision science incorporates theories and techniques from psychology, behavioral economics, and statistics, among others, to investigate how people make decisions.

Decision Support System for Agrotechnology Transfer (DSSAT) A community-developed software application program that comprises dynamic crop growth simulation models for over 40 crops.

Decoupling A design strategy that reduces connections and interdependencies between systems and subsystems; this reduces the vulnerability of connected infrastructures to unwanted changes and to cascading failures.

Degrowth The overall down-scaling of production and consumption to reduce the environmental impact of human activities.

Demography The statistical study of a population, its composition, attributes and changes over time.

Destination (Economics) The location in space and time, and the system agent, where a good or service is delivered; usually this agent consumes the good or service, uses it to produce something else, or stores the goods.

Differential Privacy A database technology that allows users to query only those fields and joins that are permitted based on their credentials.

Disaggregation (Modeling) A technique that attempts to reverse aggregation and achieve finer resolution using assumptions which trade reduced validity for increased precision.

Discount Rate The rate that converts effects occurring in future times so they are on the same basis and can be compared. Use of the discount rate converts future dollars into ones equivalent with present dollars.

Discrete Data Data collected by regular/irregular intervals through time/space. Note that natural phenomena occur in a continuous way. However, analytics on discrete data is easier and more manageable than continuous data.

Distributed Infrastructure Infrastructure that is organized with a decentralized structure comprised of many small systems to minimize the cost and risk associated with centralized systems. Compare to Decoupling.

Distributional (Distributive) Effects How policies may differentially affect different portions of the population, such as urban and rural residents, or high- and low-income populations.

Disturbance (Ecological) An event which causes a significant temporary or long-term change in ecosystem functioning. Natural disturbances include intense fires, floods, storms, diseases, and droughts. Anthropogenic disturbances include the introduction of non-native species, global climate change, large-scale land use changes, and major species predation.

Drought Worse than Drought of Record (DWDR) Drought condition under LCRA's 2015 Water Management Plan based on inflows, drought duration, and combined storage, where an ongoing drought has a real likelihood of becoming a new Drought of Record. A DWDR declaration by LCRA's Board of Directors would trigger action to cut off interruptible stored water and implement mandatory pro rata curtailment of water for firm demands.

Drought A period of abnormally low precipitation which results in a shortage of water.

Dynamic System Accommodation Comprehensive strategies that capture and adapt to the complex and dynamic nature of FEW systems and allow for adaptation to change.

Dynamics Changes through time, often expressed as rates of change.

Ecological Process Any change or reaction which occurs within ecosystems, either physical, chemical or biological. Ecosystem processes include decomposition, production, nutrient cycling, and fluxes of nutrients and energy.

Economies of Scale The ability to produce goods at a lower cost per unit by producing large amounts of goods either within a single process or in amalgamation with related goods.

Ecosystem Accounting The process of constructing formal accounts for ecosystems.

Ecosystem Diversity The diversity of ecosystems in a particular region, a component of the earth, or the earth as a whole. Ecosystem diversity is a type of biodiversity. See Biodiversity.

Ecosystem Function A subset of the interactions between ecosystem structure and processes that underpin the capacity of an ecosystem to provide goods and services.

Ecosystem Health A state or condition of an ecosystem that expresses attributes of biodiversity within "normal" ranges, relative to its ecological stage of development. Ecosystem health depends inter alia on ecosystem resilience and resistance.

Ecosystem Integrity Implies completeness or wholeness and infers capability in an ecosystem to maintain all its components as well as functional relationships when disturbed.

Ecosystem Management An approach to maintaining or restoring the composition, structure, function, and delivery of services of natural and modified ecosystems for the goal of achieving sustainability. It is based on an adaptive, collaboratively developed vision of desired future conditions that integrates ecological, socio-economic, and institutional perspectives, applied within a geographic framework, and defined primarily by natural ecological boundaries.

Ecosystem Services The direct and indirect contributions of ecosystems to human well-being. The concept "ecosystem goods and services" is synonymous with ecosystem services.

Ecosystem A dynamic complex of plant, animal, and microorganism communities and their non-living environment interacting as a functional unit. For practical purposes, it is important to define the spatial dimensions of concern.

Efficiency and Sustainability Planning Urban and environmental planning processes and strategies that go beyond the consideration of the availability, safety, and affordability of FEW resources to also consider efficiency and sustainability factors.

El Niño–Southern Oscillation (ENSO) The warm phase of an irregular fluctuation in sea surface temperature compared to the atmosphere in the central and east-central Equatorial Pacific Ocean. The cool phase is called La Niña. Events typically occur every 2–7 years and usually last for about 12 months, but sometimes longer. ENSO is associated with significant weather variations in the Americas.

Embedded Energy The energy used to create a product or provide a service, the consequences of which (e.g., greenhouse gas emissions) are implicitly "embedded" within the product. See virtual water.

Energy Balance The difference between energy input and energy output of a system. This is often applied to the earth system and to specific human energy systems, such as biofuels where it refers to the difference between all of the energy inputs to delivering a unit of biofuel and the energy obtained from that unit.

Energy The ability to do work which usually delivers services such as the operation of machines, mobility, heating and cooling, lighting, and commercial and industrial activity.

Energy Return on Investment (EROI) A metric to determine the relative effectiveness of energy delivery technologies by comparing the amount of energy delivered to the amount of energy that society must invest in the technology.

Energy Security The ability of people to have access to adequate supplied of reliable and relatively inexpensive energy for consumption. See Food Security and Water Security.

Environmental Flows Flows in streams that are maintained for the purpose of protecting the biological health of the stream ecosystem.

Erosion The physical removal of particles, such as by a stream, river or glacier. Eroded particles may then be transported and deposited by the same or a different process than the one that caused the erosion.

Establishment (Scale) A type of spatial micro-scale that revolves around any location encompassing one specific activity, e.g., a power plant generates electricity. Establishment scale is often called the "address" scale or "customer" scale of an individual building, residence, or facility.

Exclusion The denial of economic, political and/or cultural resources as well as spatial exclusion and segregation; may be experienced due to race, ethnicity, identity, economic and other factors.

Expert Determination Expert Determination is a contextually specific method of de-identifying PII data to remove the risk of identification while maximizing the utility of the data.

Extensive Metrics Metrics dependent on scale or level of consumption, e.g., a larger power plant will consume more water for cooling than a smaller power plant of the same design.

External Factors Factors outside the boundary of a system that influences the system.

Externalities The impacts of a system on factors beyond those being studied.

Externality A typically negative effect arising from an activity such as the production or consumption of goods and services by one group which damages another group without appropriate compensation being paid. For example, water pollution by an upstream industry damaging downstream water users.

Extreme Affordability A good or service that can be afforded by everyone, even the radically poor in financial means.

Facilitation The use of a third-party neutral to help people have more productive conversations. A facilitator can often be used to help run more effective meetings, for example.

Factor Input Valuation Methods Methods to estimate the value of ecosystem services based upon their ability to substitute for an input that is valued in an existing market, for when the ecosystem service contributes to a measurable marketed output.

Fahrenheit (F) An imperial temperature scale where 32 °F is the freezing point of water and 212 °F is the boiling point of water at sea level and standard atmospheric pressure.

FAIR Principles (Data Management) Principles that emphasize findability, accessibility, interoperability, and reuse.

Fallacy of Composition The observed fact that the consequences of action by a whole industry are different from the results if just an individual implements that action. For example, the adoption of a new technology may be profitable for an individual company within an industry, whereas the adoption of the technology by the whole industry drives down prices and lowers industry profits.

FASOMGHG FASOMGHG is a dynamic, multi-period, intertemporal, price-endogenous, mathematical programming model depicting land transfers and other resource allocations between and within the agricultural and forest sectors in the USA.

Favela A word in Brazilian Portuguese referring to low-income and historically informal urban slum areas in Brazil that often consist of dense multilevel housing constructed on steep hillsides and built from salvaged materials.

Federation (Data) A strategy for linking multiple repositories to enable sharing and search of diverse databases while allowing the originators of the data to maintain control and ownership of data if they desire.

FEW System Models The mathematical relationships between food, energy, and water systems that capture their spatial and/or temporal dynamics, as well as feedbacks between them.

FEW-Everything System (FEWe) A FEW system that is connected to every other aspect of the Coupled Natural Human System and has fuzzy boundaries.

Firm (Economics) A single economic entity that operates at multiple establishments.

Firm Yield (Water) The amount of water that a reservoir could have produced annually if it had been in place during the worst drought of record.

Floor Area Ratio (FAR) Bonus A zoning tool used to control building bulk by setting a ratio of building mass to the area of the building's lot. The award of a FAR bonus is often used as an economic incentive for developers to provide additional public benefits or amenities to the community.

Flows Accounted quantities of material, good, currency, and so on that move from one space-time location to another; flows may be human (e.g., currency payment, oil by pipeline) or natural (e.g., water in a river).

Food Desert An area that lacks access to the range of affordable and healthy foods that make up a full and healthy diet, including fresh fruits and vegetables and whole grains. Food deserts often occur in urban, rural, minority, and low-income areas devoid of large full range grocery stores and food markets.

Food A nutritious substance that is consumed to sustain life and support growth in the young.

Food Miles The distance that food is transported from where it is produced to where it is consumed.

Food Policy Council A council composed of diverse stakeholders engaged in the local food system, usually established through governmental or grassroots action, that is tasked with examining the food system and providing policy recommendations to local and regional governments on how to improve that system.

Food Security The physical, social and economic access to sufficient safe and nutritious food that meets their dietary needs and food preferences for an active and healthy life. See Energy Security and Water Security.

Foot (ft) An imperial unit of distance, equivalent to 0.3048 m.

Footprints (Environmental) The direct and indirect biophysical impacts of consumption of natural resources on the earth, usually natural resources that have a planetary boundary on their availability.

Forecast A quantitative prediction, usually with a margin of uncertainty.

Function The behavior of a system, which is often purposeful. Function is constrained by structure and serves to support and feed structure.

Gallon (gal or G) An imperial unit of volume normally reserved for liquids, equivalent to 4.54609 L (UK gallon) or 3.785411784 L (US gallon). Establishment water capacity or use (e.g., for water treatment plants or by power plants) is often stated as million gallons per day (MGD). Fuel economy is often stated in miles per gallon (mpg).

Gender Development Index (GDI) An index that calculates separate Human Development Indices for males and females.

Gender Inequality Index (GII) An index that utilizes metrics differentiated by gender on health, empowerment, and labor.

General Agreement on Tariffs and Trade (GATT) A framework for facilitation international negotiations to regulate trade that operated from 1947 to 1995. It was succeeded by the World Trade Organization.

General Circulation Model (GCM) A mathematical model of the earth's climate represented in a three-dimensional grid of cells defined by vertical layers and horizontal elements over the globe. The number and thickness of layers above and below the surface of the earth are defined within each GCM according to its scientific objective.

Geodesign A collaborative design process that involves multiple stakeholders to find an optimal solution for a problem that has a spatial component. For example, a land use geodesign project may include farmers, city planners as well as water scientists to reach consensus on how to find an optimal solution for all stakeholders.

GEOGLAM (Group on Earth Observations Global Agricultural Monitoring Initiative) A global scale project launched by the Group of Twenty (G20) Agriculture Ministers in 2011.

Geophysical The physical components and interactions such as those involving the earth's atmosphere, water, and geology.

Global Change Assessment Model (GCAM) An integrated assessment tool for exploring consequences and responses to global change. Climate change is a global issue that impacts all regions of the world and all sectors of the global economy.

Global Scale Modeling of an activity on a global scale involves data aggregation at spatial scales ranging from regions, states, or nations up to the planetary boundary.

Global Warming Potential (GWP) The amount of heat a greenhouse gas traps in the earth's atmosphere compared to that of carbon dioxide. Because each greenhouse gas has a unique lifespan in the atmosphere, GWP is specific to a particular duration.

Globalization The process by which organizations, commerce, culture, and other entities have greater international connections and impact.

GOSSYM An agricultural simulation model for cotton crop growth and yield, developed by the USA Department of Agriculture.

Governance The processes by which communities of people develop and achieve specific ends. As such, governance refer to the policies, laws, institutions, and actions of governments at all levels. Governance also refers to rules and policies that are developed by through groups of governmental bodies (e.g., an international treaty) and outside of formal governmental institutions (e.g., through private corporations or civil society organizations).

GPU (Graphics Processing Unit) In modern computers, GPU devices are designed to do mathematical calculations in parallel to render images for a display device. Since GPU units have a highly parallel structure and efficient matrix manipulation and convolution ability, it makes them well suited for neural network algorithms that require high number of parallel matrix calculations.

Great Lakes Compact A 2005 agreement between the states of Illinois, Indiana, Michigan, Minnesota, New York, Ohio, Wisconsin and the Commonwealth of Pennsylvania to enact concurrent legislation to protect, restore, conserve and manage the waters of the basin for the benefit of their citizens through the enactment of coordinated policies and programs. Pursuant to Article I, section 10 of the U.S. Constitution, required the consent of Congress. (Full name: Great Lakes-St Lawrence River Basin Water Resources Compact.)

Green Infrastructure Hard infrastructure that emphasizes the use of natural ecological systems instead of concrete and steel.

Greenhouse Gases (GHG) Gases which trap outgoing thermal (infrared) radiation from the Earth and serve to increase the temperature within the Earth's atmosphere.

Grey Infrastructure Hard infrastructure that emphasizes the use of efficient materials and technologies that are the product of human technology.

Greywater System A wastewater recirculation system for reusing water that drains from bathroom sinks, showers, tubs and washing machines for flushing toilets, irrigation, and other non-potable uses; a system not intended for reusing wastewater from toilets which is known as blackwater.

Gross Domestic Product (GDP) The market value of goods and services produced and sold to final consumption within a national economy over a specified period (typically 1 year).

Gross National Income (GNI) A measure of wealth based on the sum of a nation's Gross Domestic Product and the net income it receives from outside its geographic borders.

Gross National Product (GNP) The market value of goods and services produced (i.e., the "output") by the citizens and corporations of a country regardless of where that activity takes place.

Groundwater Water that is stored in or flows through subsurface aquifers.

Habitat The physical and biological characteristics of an area that support the survival and reproduction of a particular species, including the food required by that organism, the shelter provided by vegetation and other structures that allow it's young to survive, appropriate water sources, and other features.

Hard Infrastructure Physical systems that form infrastructure, such as pipes or roads. Compare to Soft Infrastructure. See also Green Infrastructure, Grey Infrastructure, and Natural Infrastructure.

Hard Law International treaties, conventions, and agreements which have legal binding obligations with specific or "hard" consequences if breached.

Harmful Algal Blooms (HABs) High concentrations of various strains of cyanobacteria that, under specific conditions, produce toxins that are harmful to people and animals.

Hectare (Ha) Hectare is a metric unit of area equivalent to 10,000 m^2, or a square of 100 m on each side.

Helsinki Rules on the Uses of the Waters of International Rivers are non-binding international guidelines of the use of transboundary rivers' groundwater which were codified in 1966 by the International Law Association. The Helsinki Rules were superseded in 2004 by the Berlin Rules.

Hotspots Critical thematic topics or threatened locations for which the nexus approach has the potential of strong societal impact. Note that the term "hotspot" defined here is different than the statistical "hotspot" term which is used for statistically significant clusters geolocated activities/events (e.g., crime/disease hotspots).

Human Development Index An index of development that aggregates data on life expectancy at birth, expected years of schooling, mean years of schooling, and Gross National Income per capita.

Hydrocarbons Molecular chemicals composed of chains of carbon atoms with associated hydrogen atoms. They are often used as fuels, such as methane (CH_4) and ethanol (C_2H_6O).

Hydrological Drought Drought conditions as measured by water found in streams, lakes, reservoirs, and groundwater. Compare to agricultural drought and meteorological drought.

Hypothetical Bias A bias that exists when people report different behavior in surveys or experiments when these decisions are hypothetical, then they would when faced with the same decision in their daily lives.

Indaba A traditional South African dispute resolution mechanism that relies on group identification and discussion of ideas from all participants.

Indicator Information based on measured data used to represent a particular attribute, characteristic, or property of a system.

Induced Innovation When the price of one input increases sharply relative to the price of other inputs this stimulates developing innovative technologies that reduce the use of the high price input.

Inelastic Demand Curve A relationship between demand and supply where significant changes in price does not lead to a corresponding significant change in demand. The more inelastic the demand, the steeper the curve and the less quantity will react to a price change.

Inequality-adjusted Human Development Index (IHDI) An index which modifies the three main metrics within the Human Development Index in accordance with the degree of inequality in that index.

Information The answer to a question.

Information Theory The quantification of questions and answers.

Informed Consent The written explicit consent granted by a private party to allow public release of private data or its use for a specifically defined purpose.

Infrastructure The physical and organizational apparatus that handles high-volume goods and services that require heavily capitalized, large-scale, durable, reliable, shared, interdependent, and specialized systems that are highly efficient and achieve low marginal costs. Compare to Hard, Soft, Green, Natural, Grey, and Distributed Infrastructure.

Inputs and Outputs The flows of goods and services in and out of a process; these may include raw materials, finished products and services, and wastes; input-output coefficients describe these.

Institution (Organizations) A significant and established organization in a society or culture, especially of a public character.

Institutions (Rules) The rules that guide how people within societies live, work, and interact with each other. Formal institutions are written or codified rules. Examples of formal institutions would be the constitution, the judiciary laws, the organized market, and property rights. Informal institutions are rules governed by social and behavioral norms of the society, family, or community, also referred to as organizations.

In-stream Flows Legally required minimum flows of water in streams or rivers. In most states, water rights for in-stream flows can be held only by the state.

Insurance A strategy to mitigate the risk involved in making capital-intensive investments (e.g., for infrastructure) by pooling the risk of many similar systems using carefully calculated payments into a pool of funds that cover loss.

Integrated Assessment Models Whole-system models that aim to evaluate the systemic effects of policies and trends.

Integrated Development Environment (IDE) Software that facilitates software development for programmers by providing bundles of code editing, building, testing and debugging tools. In addition, they often have a variety of features such as syntax highlighting, code formatting, code completion, error diagnostics and reporting. Modern IDEs also help synchronize team work, versioning and deployment process in an efficient way.

Integrated FEW System A system in which the interconnectedness and interactions of the three great consumable sectors, food, energy, and water are considered together.

Integrated Water Resource Management The coordinated development and management of water, land, and related resources to maximize economic and social welfare.

Intensive Metrics Metrics that are independent of scale or level of consumption, e.g., a power plant may consume X gallons of water per kilowatt hour of electricity generated.

Interdependency (Infrastructure) Infrastructure property where one type of infrastructure is substantially dependent on another type of infrastructure to provide its inputs. Compare to Cascading failure.

Interests The reason or values about *why* someone might decide on a position. See also definition on position. Getting to Yes by Fisher and Ury (1981) discusses interest-based negotiation as a way to achieve potential positive outcomes beyond assuming a negotiation is a win-lose zero-sum game.

Intermediate Producers Producers who take a commodity and consume it as input into the production of a higher-value commodity, usually via manufacturing operations.

Internal Interactions The interactions that describe how parts of a system influence each other directly or indirectly.

International Treaty "An international agreement concluded between States in written form and governed by international law"—1980 Vienna Convention on the Law of Treaties.

Interval-Scale A micro-temporal data that records the rapid changes in a process's function, often at scales of seconds or minutes.

Intrinsic Value The value of someone or something in and for itself, irrespective of its utility for someone else.

Inventory Data Data that tracks and accounts for how much of a product is available at an establishment, on order, or en route, and is a central component of private sector supply chain management.

ISO The International Organization for Standardization (or Organisation Internationale de Normalization), which maintains many essential metadata standards.

Isoquant Curve A curve that shows all the combinations of inputs that yield the same level of output. Thus, for example, the following two points would appear on the same isoquant: suppose 2 acres of land and 10 h of labor produces a given amount of output, while one gets an equal output when using 1.5 acres of land and 15 h of labor. This reveals possible substitutions of inputs as prices change.

Joule (J) A metric unit of energy, defined as the energy required to move a mass of 100 g a distance of 1 m in Earth's gravitational field, approximately 10 m/s^2, roughly the energy to lift an apple from the floor onto a desk.

Knightian Unknowns Design factors, error factors, or risk factors that cannot in principle be quantified with sufficient precision and accuracy to make a decision; unknowable unknowns; for instance, if you have a probability distribution or estimate of precision and accuracy, this is a constrained unknown and is *not* a Knightian uncertainty. Compare to Knowable Unknowns.

Knock-on Effect Something such as an event, process, or action that causes another event to happen, but indirectly.

Knowable Unknowns Factors that are known to exist and can be quantified in principle but which have not yet been measured with sufficient precision and accuracy to be estimated with sufficient confidence to make a decision.

Knowledge System A social system for thinking, remembering, and communicating knowledge.

Kriging In geostatistics, kriging is an interpolation technique that is used to predict values at locations without observations using the weighted average of the values from locations with observations. These weights are calculated by a variogram model which takes the distance into account. Thus, closer locations will have a higher weight than distant locations.

LEAP (Model) The Long-range Energy Alternatives Planning modeling system is a software tool for energy policy analysis and climate change mitigation assessment developed at the Stockholm Environment Institute.

Least Developed Countries (LDCs) Countries defined by the UN as "low-income countries confronting severe structural impediments to sustainable development" which are "highly vulnerable to economic and environmental shocks and have low levels of human assets." Forty-seven countries were classified as LDCs in 2018.

Life Cycle Assessment (LCA) A framework to evaluate the environmental (and increasingly other) impacts of a good or service over its full life cycle from raw material extraction, through operation and end-of-life, e.g., disposal.

Lock-in A property of infrastructure wherein human and natural ecosystems build up around the readily available and low-cost services provided by an infrastructure and the system subsequently becomes very expensive to adapt.

Loess A deposit of silty or loamy particles concentrated in a geographic area primarily by wind. The materials for loess formation may originate from glacial sources such as large braided river valleys, non-glacial sources like volcanic ash eruptions, or any large dry flat area over which strong winds can develop sufficiently to transport fine particles, such as deserts and playa lakes.

Lower Colorado River Authority (LCRA) An organization that manages the allocation of water within the Colorado River Basin in Texas.

Machine Learning Machine learning is the name of a set of methods of data science that aims to analyze data to learn, identify patterns and make informed decisions by using mathematical and statistical concepts.

Macro-scale A spatial or temporal scale that summarizes the totals or averages of a metric, without attempting to preserve the patterns of variation between granular individuals.

Macroscope A tool or framework to sense and perceive a key aspect of a complex system.

Maladaptation Action taken ostensibly to avoid or reduce vulnerability to climate change by one group that increases the vulnerability of others.

Marginal Cost The cost of one additional unit of system output, without including fixed costs like capital and maintenance.

MARKAL (Model) A energy systems model that represents the evolution over a period of usually 40–50 years of a specific energy system at the national, regional, state or province, or community level. It was developed in a cooperative multinational project over a period of almost two decades by the Energy Technology Systems Analysis Program (ETSAP) of the International Energy Agency.

Measure (or Measurement) Information which refers to the actual measurement of a state, quantity or process derived from observations or monitoring.

Mediation The use of a third-party neutral (a mediator) to help people work through conflict but where the parties retain the ability to decide the outcome. Mediation more often is used after a dispute arises and sometimes after litigation has already started. All of the US states have mediation privilege statutes that protect communications made for purposes of a mediation process.

Mercantilism An economic system that utilizes international trade to generate wealth and usually including policies such as tariffs, subsidies, and quota to maximize exports and minimize imports.

Meso-scale A spatial or temporal scale that aggregates together many micro-scale objects to preserve privacy and reduce excessive detail, but without destroying the major pattern and information content of the data. Meso-scale is where top-down and bottom-up methods usually meet and is the finest scale at which aggregated census-style data is often available without running afoul of private or sensitive data protection procedures.

MESSAGE (Modeling) A modeling framework for medium- to long-term energy system planning, energy policy analysis, and scenario development developed by the International Institute of Applied Systems Analysis (IIASA). It provides core inputs for major international assessments and scenarios studies, such as the Intergovernmental Panel on Climate Change (IPCC), the World Energy Council (WEC), the German Advisory Council on Global Change (WBGU), the European Commission, and the Global Energy Assessment (GEA).

Metadata "Data about data" which describe a dataset's authorship, contents, format, resolution, timing, provenance, sources, methods, a license of use, globally unique identification, etc.

Meteorological Drought Drought conditions as measured by the level of precipitation. Compare to agricultural drought and hydrological drought.

Micro-scale A spatial or temporal scale that resolves the individual granular components and processes within a system.

Mode of Transportation The type of infrastructure used to move a good or service, for example, a truck running on a road.

Models A mathematical or intellectual construct that describes a system's structure and function. Models are often computerized and quantitative—but may be conceptual. Models capture hypotheses. Models can describe and project.

Modern Renewable Energy Renewable energy sources excluding traditional biomass fuels, and including hydropower, wind, solar, and modern biomass/biofuel—along with (arguably) third generation and newer nuclear sources.

Multi-criteria Decision Analysis A decision-making framework tool that can be applied to complex problems that require stakeholders to understand, evaluate, and choose among multiple different alternatives and their trade-offs.

Multidimensional Poverty Index (MPI) An index that measures "overlapping deprivations suffered by individuals at the same time" in terms of health (nutrition and child mortality), education (years of schooling and children enrolled) and standards of living (cooking fuel, toilet, water, electricity, floor, and assets).

Nation State A self-governing territory occupied by people under the same government and laws.

National Environmental Policy Act (NEPA) A 1970 federal statute that requires an environmental impact assessment, and potentially an environmental impact statement, for any major federal action with significant environmental impacts. (NEPA, 42 U.S.C. 4321 et seq.)

National Pollution Discharge Elimination System (NPDES) A system by which any discharge of pollutants into a navigable body of water must be authorized by a permit.

Natural Capital An economic metaphor for the limited stocks of physical and biological resources found on earth.

Natural Gas Liquids Petroleum liquids that are produced along with natural gas when natural gas is the primary objective.

Natural Infrastructure Hard infrastructure that operates largely (not entirely) without the need for direct human decisions, designs, or investments.

Natural Resources Materials produced by nature that have value to humanity, including arable land, fish, coal, oil, natural gas, wind, solar radiation, and water.

Need-to-Know (Data) A data access minimization principle applied to classified and categorized national security data, such that only those persons with a valid need to know are allowed access to sensitive data.

Negotiation "The back-and-forth communication designed to reach an agreement when you and the other side have some interests that are shared and others that are opposed" (Fisher and Ury 1981).

Neoliberal Economics An economic theory that emphasizes the benefits of economic competition with limited regulation by governments. However, neoliberalism has typically recognized a positive role for some government regulations in contrast with classical liberal economics which advocated for the smallest possible regulation by government.

Neo-Malthusian A view of the challenges of providing resources for a population that emphasizes efforts to limit population growth (in contrast to a cornucopian view).

Net Energy Analysis (NEA) A methodological framework for accounting the energy used in the extraction, processing, transportation, conversion, and use of energy commodities.

Network Level The level of multiple interacting facilities within the same basin, region, or country.

Non-consumptive (Water) Water use that does not permanently remove water from the watershed. Examples of non-consumptive use include water used in hydroelectric power plants; the return flows from irrigation; and, in-stream or environmental flows. See Consumption and withdrawal.

Non-point Source Pollution Pollution that enters a water body through diffuse channels, such as agricultural runoff. Non-point sources are comparatively difficult to regulate.

Normative Behavioral Models Models are models of what behavior people *should* exhibit in a particular scenario, regardless of what behavior people actually exhibit. (The opposite of positive behavioral models.)

North American Free Trade Agreement (NAFTA) A trade agreement entered into in 1994 among the USA, Mexico, and Canada to remove tariffs and streamline trade among the three countries.

No-Till Cropping An agricultural practice of growing crops or pasture (grass/plants for grazing animals) without tilling (plowing, rototilling, harrowing, rolling, etc.) the soil in order to increase soil fertility and resilience, water infiltration and retention of nutrients and biota.

Ontology (Data) A set of concepts that name and describe the categories of objects in a dataset and the relationships between these types.

Origin (Economics) The location in space and time, and the system agent, that produced and ships a good or service.

Origins and Destinations (Economics) The space-time locations where a good or service was transported from, and to, respectively; a flow emanates from an origin and arrives at a destination.

Pareto Optimal (Economics) A theoretical state where the distribution of resources is such that it is impossible to reallocate resources benefit one individual without being detrimental to one or more other individuals.

PCII (Data) A government-enforced national security categorization originating after the September 11th, 2001 attacks in the USA, such that some food energy or water infrastructure data concerns security-critical protected critical infrastructure information.

Pedolith A soil structure which has been impacted by biogeochemical processes have added, removed, or changed its components.

Pee-Cycling The conversion of human liquid waste (urine) to sterilized nutrient-rich fertilizer that can be re-used for agricultural and ecological purposes.

Peri-urban Areas immediately surrounding a city that have mixed urban and rural characteristics.

Petroleum A blend of hydrocarbon liquids that can be separated into a variety of valuable products through distillation.

Petrostate A nation whose economy is highly dependent on the production and sale of petroleum.

PII (Data) A categorization as personally identifiable information that is covered by data privacy and data protection law.

Planetary Boundary(ies) A concept used to define a 'safe operating space for humanity' by defining science-based limits on environmental impacts at the planetary scale, beyond which society might be negatively affected.

Pluralist Model of Government A model of government in which no single set of interests in the society has sufficient power to influence governmental action to a significant extent. Organizations and processes of government determine the public good and act accordingly.

PODIUM (Modeling) A modeling framework to develop scenarios of water and food supply at river basin, sub-national and national level, developed by the International Water Management Institute.

Point-Source Pollution Pollution that enters a water body through a discreet point, such as a pipeline. Point sources are regulated by the NPDES system.

Policy The rules, regulations, investments, and actions of an organization that are designed to achieve important outcomes.

Pollutant Discharge The discharge of a chemical, sediment, or other material that reduces the quality of the receiving water body.

Position Something that a person or group has decided on. See also interests.

Positive Behavioral Models Models of behavior people actually *do* exhibit in a particular scenario, regardless of the behavior that a specific model predicts. (See Normative Behavioral Models.)

Pound (lb) An imperial unit of mass, equivalent to 0.454 kg.

Practice "The actual application or use of an idea, belief, or method, as opposed to theories relating to it" (Oxford Dictionaries). In this book, practice refers to the application of Nexus research to real-world problems.

Practitioners Scientists and nonscientists involved in the practice of application of Nexus science to real-world problems.

Prediction A forecast of what is expected to happen. Compare to Projection.

Preemption (US Government) The US doctrine that federal law supersedes conflicting state law, based in the Supremacy Clause in Article VI of the US Constitution. At the state level, preemption allows state law to supersede conflicting local (town and city) laws. Preemption may be express or implied.

Press Events Events that occur gradually and alter the system incrementally, such as rising temperatures and sea levels, or gradual population decline. Press events do not require people to react in a short amount of time. (In contrast, see Pulse Events.)

Primacy (US Government) The authority of a US state to implement federal laws after gaining federal approval of the state implementation plan.

Primary Data Data collected by researchers from the population of interest (e.g., through surveys or experiments), for the purpose(s) of testing a specific research question.

Primary Sector The natural resource extraction sector of the economy.

Private Data Privileged information that has not been made publicly available, and to which the public has no legal right of access unless the private party explicitly grants informed consent.

Private International Law International laws related to individuals and organizations, like corporations, when they move across international borders and operate in different countries.

Private Sector The portion of the economy that is not under direct government control, comprised of individuals and companies, and characterized by the intention of producing a profit.

Process Scale A type of spatial and temporal micro-scale that resolves the space and time scale of operation of a process and involves the creation, transformation, or transportation of a specific product, good, or service, and implicates the inputs and outputs of the process.

Producer Surplus The difference between the total cost a producer incurs to supply goods for and the revenue made when the goods are sold.

Product Preferences The preference of particular populations for particular products not based on the essential function of the products (e.g., a dietary preference independent of nutritional value).

Production Input A terms used in economics to assign a value to inputs required to achieve a certain level of production or output.

Profitability Trade-off Managing current assets and current liabilities in such a way so that profitability will be optimum.

Projection An estimate of how a system is likely to behave under certain conditions and assumptions. Compare to Prediction.

Provisioning Ecosystem Services The products obtained from ecosystems, including, for example, genetic resources, food and fiber, and freshwater.

Public Goods A good that an individual will generally not produce since they do not gain the full benefits of production and rather is something that needs a public role in the production. Technically, economists define this as a good that is both non-excludable and non-rivalrous in that individuals cannot be effectively excluded from use and where use by one individual does not reduce availability to others.

Public International Law International laws governing the relationships between states as opposed to their private citizens and private organizations (e.g., non-governmental organizations and corporations).

Public Sector The portion of the economy under direct governmental control, including public services, public enterprises, and public policy.

Pulse Events Events that occur suddenly and are likely to affect people and communities over a short period of time and are more difficult to predict (e.g., sudden weather events or economic shocks). (In contrast, see press events.)

Race to the Bottom A theory that unregulated economic competition results in a focus of cost-reduction at the expense of all other factors such as environmental protection, and worker pay and safety.

Rationality assumption Agents are often assumed to behave rationally, such that their decisions are always those that maximize a particular objective (usually profit or utility). This assumption is often violated in human decision-making.

Rebound Effect An unintended response to changes in the social system, such as the introduction of new technology or a policy, that results in a reduction in benefits from that technology of policy.

Redundancy A design strategy that provides reliability, robustness, or resilience using multiple and backup systems. See Reliability, Robustness, and Resilience.

Reforestation Planting of native trees in an area with decreased density/numbers of trees. Compare to Afforestation.

Region An area of land that has some common features, which could be natural, such as climate or landscape, or social, such as language or religion.

Regulating Ecosystem Services The benefits obtained from the regulation of ecosystem processes, including, for example, the regulation of climate, water, and some human diseases.

Relative Metrics Metrics that take account of context concerning budget, constraint, peer, or history, (e.g., a power plant may consume water at 90% of its operational capacity).

Reliability The percentage of time a system is free from failure.

Renewable Fuel Standards (US) A US federal program originating in the Energy Policy Act of 2005 and expanded under the Energy Independence and Security Act of 2007 that requires transportation fuel sold in the USA to contain minimum volumes of renewable fuels.

Replacement Cost The costs incurred by replacing ecosystem services with artificial technologies.

Replacement Value The value obtained by replacing ecosystem services with artificial technologies.

Repository A generalized library service that curates data and that ideally provides FAIR data management services.

Resilience The capability of a system to adapt to or recover from a disruption or harmful event without losing essential functions; however, resilience allows for adaptation of structure. For ecosystems, resiliency describes the ability of an ecosystem to recover from disturbance without human intervention.

Resolution Resolution is the level of spatial, temporal, or categorical detail at which a dataset is aggregated.

Resource Any physical or virtual entity of limited availability that provides a benefit.

Responses Human actions, including policies, strategies, and interventions, to address specific issues, needs, opportunities, or problems. In the context of ecosystem management, responses may be of legal, technical, institutional, economic, and behavioral nature and may operate at various spatial and time scales.

Revealed Preference Method An economic practice of analyzing the strength of a preference that individuals have for certain outcomes, like reduced environmental harm, by studying what they are willing to buy and pay. See also stated preference method.

Right-to-Farm Laws (US) Laws enacted by local municipalities in agricultural regions that prohibit neighbors or the general public from bringing nuisance suits against farmers for smells, noises, or other inconveniences associated with standard animal husbandry or food production processes.

Risk Risk is an estimate of impact or damage, and is a function of three factors: (1) the likelihood (or frequency) of an event, (2) the vulnerability of a system to the event (or, how protected a system is from the event), and (3) severity (or potential impact, or exposure) of an event if it occurs and overwhelms the system's protection; for events with high likelihood, high vulnerability, and high severity, risk is high for a system.

Risk Multiplier A factor, such as climate change, which increases preexisting risks.

Robustness A design strategy that selects systems which perform well under a wide range of possible futures, reducing the need to precisely anticipate future conditions.

Route The precise mode of transport, pathway, and waypoints followed by a flow that transports a good or service from origin to destination.

Rule of Fifteen A common legal threshold in US State law specifying that utility customer data and other PII data must be de-identified through aggregation into groups of not less than fifteen individuals, any one of which comprises not more than fifteen percent of the group's total.

Rural An area of low-density human population and smaller population aggregates outside of urban and peri-urban areas that usually are characterized by significant primary production (farming, fishing, forestry, mining) and tourism.

Safe Harbor (Data) A legal protection of specifically named PII data against release unless an expert determination is made that this release does not create a significant risk of identification and/or harm to the person(s) involved.

Scale The measurable dimensions of phenomena or observations. Expressed in physical units, such as meters, years, population size, or quantities moved or exchanged. In observation, scale determines the relative fineness and coarseness of different detail and the selectivity among patterns these data may form (MA 2005a).

Scenario A set of assumptions about a system such as its key inputs, conditions, and functioning.

Secondary Data Data that has been previously collected by an entity other than current researchers (e.g., census data).

Secondary Sector The manufacturing sector of the economy.

Security Through Obscurity Security through Obscurity is a tactic, often unintentional or implicit, of protecting data by minimizing its findability, accessibility, and interoperability.

Security The ability of people to have affordable, reliable, and high-quality access to their basic FEW needs so that they can live healthy and productive lives unconstrained by existential resource limitations; security implicates attributes of availability; access; utilization; and, stability reliability.

Sedimentation The deposition of particles, typically after having been suspended in and/or transported by water, such as in a river delta or floodplain. The size and sorting of the deposited particles typically depend upon water velocity, with larger rocks transported by faster water and generally settling out more quickly as water velocity decreases, with silts and clays staying in suspension longer and settling out in quieter water further downstream such as lakes and large rivers.

Sensitive Data Data for which there are significant sensitivities or risks involved in disclosure.

Sensor Systems Networks of sensors used to make large volumes of observations and measurement automatically.

Separated FEW Systems Considering the three food, energy, and water systems or sectors individually.

Serious Games Games, including role plays, that include a purpose other than just entertainment; see, for example, MIT's Serious Game Lab: http://gamelab.mit.edu/tag/serious-games/.

Short Supply Chains (SSCs) Supply chains that are more environmentally friendly due to reduced transportation distance and fewer intermediaries.

Social Costs and Benefits Costs and benefits as seen from the perspective of society as a whole. These differ from private costs and benefits in being more inclusive (all costs and benefits borne by some members of society are taken into account) and in being valued at social opportunity cost rather than market prices, where these differ—sometimes termed "economic" costs and benefits.

Social Impact or Civil Society Organization A human organization that is organized around a mission of collective action, shared values, interests, or purposes. These include nonprofit entities such as charities, identity and faith-based organizations, community groups, professional and trade organizations, social movements, advocacy groups, and other entities that are neither governmental nor for-profit in purpose.

Socio-ecological System A complex interdependent system of biogeophysical and social factors interacting in a nonlinear multidimensional ordered manner that may exhibit unpredictable behavior.

Socio-Ecological-Technical Systems (SETS) Socio-Ecological-Technical Systems combine hard and soft infrastructures.

Soft Infrastructure Human institutions and social systems that form infrastructure, for example an irrigation district or elected board of officials. Compare to Hard Infrastructure.

Soft Law Agreements between parties, including nation-states, without legally binding components.

Soil A layer of relatively loose minerals that have been weathered from the earth's crust and have undergone a variety of other physical, chemical, and biological processes making them suitable to support plant growth. Typically composed of a mixture of minerals, organic matter, gases, liquids and microorganisms, different soils may exhibit specific textural, structural, and chemical properties. Soil is typically distinguished from its underlying parent material or subsoil by the presence of organic matter and other materials that can support life.

Sovereignty The right and power of entity to self-government, independently from other governments.

Spatial Datasets Datasets with location information. Although often confused with "geospatial," spatial datasets do not need to have a geographic location. Thus, any coordinate system that defines a location in regards to other objects in space may be considered spatial.

Spatial Scale It defines the extent of the area, length or distance of a spatial data object.

Stakeholder A person, group or organization that has a stake in the outcome of a particular activity.

Stakeholders People and organizations in the real-world system and who have something to gain and/or lose from an initiative or conversation.

Stated Preference Method An economic practice of measuring individuals' value for environment quality directly, by asking them to state their preference for the environment. See also revealed preference method. See also Revealed Preference Method.

Stocks Stocks are accounted quantities of material, good, currency, and so on that exist at a specific space-time location; stocks may be human (e.g., bank account, gasoline tank) or natural (e.g., aquifer, coal seam).

Storage An accumulation of a commodity or input, usually for the purpose of absorbing changes in supply or demand rates. Compare to Buffer.

Stranded Capital The circumstance where capital invested in assets (e.g., a power plant) or resources (e.g., coal) loses its value because the asset or resource is no longer required and cannot be adapted to other valuable purposes. Compare to lock-in.

Strategic Petroleum Reserve A reserve of petroleum managed by the U.S. Department of Energy stored at four sites in Texas and Louisiana intended to maintain supplies to the USA during severe energy supply interruptions.

Structure (System Science) The relationships between different parts of a system and between its parts and external systems. Structure establishes the potential for function and the pathways of functional interaction. Structure is typically fixed or slowly varying, especially when based upon infrastructure, and creates and constrains the pathways and options by which a system can achieve its purposes.

Substitutability The extent to which human-made capital can be substituted for natural capital (or vice versa).

Supply Chain Management (SCM) An integrative approach to planning and controlling material flows from suppliers to end-users that aims to enhance system performance through the coordination of manufacturing, logistics, and materials management within the entire supply chain network. SCM encompasses the entire value chain and addresses materials and supply management from the extraction of raw materials to the end of a product's useful life as well as disposal, recycling, and reuse.

Supply Chains The sequences of steps in the production and delivery of goods and services.

Supporting Ecosystem Services Ecosystem services that are necessary for the maintenance of all other ecosystem services. Some examples include biomass production, production of atmospheric oxygen, soil formation and retention, nutrient cycling, water cycling, and provisioning of habitat (MA 2005a).

Sustainability A characteristic or state whereby the needs of the present and local population can be met without compromising the ability of future generations or populations in other locations to meet their needs (MA 2005a).

Surface Water Freshwater that flows on the surface of the earth, such as in streams and lakes.

Survey A census method for collecting data, especially usage, production, or transportation data.

Sustainable Supply Chain Management (SSCM) A set of managerial practices that include impact assessment within a framework of triple bottom line outcomes (i.e., social, environmental, and economic) and a consideration of all stages across the entire value chain for each product and the entire product life cycle.

Sustainable Use (Ecosystems) The use of ecosystems in a way that benefits present generations while maintaining the potential to meet the needs and aspirations of future generations.

SWAT (Model) The Soil and Water Assessment Tool is a small watershed to river basin-scale model used to simulate the quality and quantity of surface and groundwater and predict the environmental impact of land use, land management practices, and climate change. SWAT is widely used in assessing soil erosion prevention and control, non-point source pollution control and regional management in watersheds.

System A system is a set of things that connected in a way that creates some unified whole.

Technology Adoption The uptake of new technologies, such as electric vehicles or solar energy, as a strategy for adapting to changes in natural FEW systems.

Temporal/Timescale Mismatches When the costs and benefits of a particular action occur at different temporal scales, often leading to trade-offs between short-term costs and long-term gains.

Tertiary Sector The service sector of the economy.

Texas Commission on Environmental Quality (TCEQ) The state agency that issues surface water rights.

Thermal Discharge A discharge of warmer water into a receiving water body that changes the temperature of the receiving water body.

Thermoelectric Power Plants Thermoelectric power plants use fuels (either chemical or nuclear) to generate heat to drive a turbine to produce electricity.

Threat Multiplier Used to refer to the likely impact that climate change will have on underlying social instabilities that can be exacerbated by floods, droughts, heat waves, or other variations in weather patterns.

Ton Ton is an imperial measure of mass. There is some ambiguity, as the term may refer to short tons, used in the US equivalent to 907.1847 kg, or to long tons, formerly used in the UK and British Commonwealth, equivalent to 1016.047 kg.

Tonne A metric unit of mass equivalent to 1000 kg.

Top-Down Approaches, methods, and models that begin with a whole system and deconstruct it into essential components. Top-down methods often involve estimation or approximate measurement from afar (often, from space!) and using methods that intrinsically enforce a "mass balance" for the aggregate measure. See Bottom-Up.

Trade Secrets Private data that a business elects to hold private to avoid providing an unfair advantage to competitors.

Trade The buying and selling of goods and services.

Traditional Biomass Wood, charcoal, leaves, agricultural residue, animal/human waste and urban waste (in contrast to modern renewables such as hydropower, wind, solar, geothermal, and modern biomass.) Traditional biomass is considered renewable.

Traditional Dispute Resolution Mechanisms Community-based structures such as tribal councils or village and/or religious leaders who help resolve disputes. In Afghanistan, for example, "community councils (often called *shuras or jirgas*) generally consist of community elders and other respected elders sitting together to reach an equitable resolution of disputes and to reconcile the disputants, their families and the community as a whole" (Dempsey and Coburn 2010).

Transaction Costs Expenses incurred when buying or selling a good or service. It can be thought of as the costs of middlemen/intermediaries or the direct financial costs of brokerage services.

U.S. EPA Net Zero Strategy An initiative of the U.S. Environmental Protection Agency that aims to assist US communities, municipalities, water utilities, developers, and the military in achieving Net Zero and Net Positive Energy, Net Zero Waste, and Net Zero Water sustainability goals. Net Zero means achieving a balance between water demand and availability, consuming only as much energy as is produced and producing additional energy from renewable sources, and eliminating solid wastes sent to landfills.

Undernourishment Receiving dietary energy from usual food consumption that is below “minimum energy requirement norms” for an individual’s age, gender, body weight, and level of activity as defined by the WHO/UNICEF.

Underweight Defined by the WHO as “less than two standard deviations below the median weight for age groups in the international reference population.”

Urban Refers to higher-density non-agricultural population centers, typically associated with cities and their metropolitan areas.

U.S. Federal Reserve System The Central Banking System of the USA that manages the currency, money supply, and interest rates of the USA under the leadership of an independent Federal Reserve Board.

Use-inspired Research Research that combines the benefits of applied and basic research and is Pascal’s Quadrant of a graph of quest for fundamental knowledge against considerations of use by society.

Utility An organization (sometimes publicly owned, sometimes privately owned by highly regulated) that provides an essential public service involving major infrastructure such as electricity, potable water, and sanitation.

Utility Customer Data Data on the amounts of energy, water, etc. consumed by an individual, household, or organization, and/or the amount of money billed for the utility service.

Utility Maximization A theoretical framework used to model human behavior in economics. Some assumptions of utility maximization that people have a well-defined utility function, for example, may not be accurate.

Utilization The rate at which the system is functioning, operating, or producing. Compare to Capacity.

Values-laden Technology Technology that embodies and facilitates specific human preferences and values (nearly all technology does this to some degree).

Virtual Water The water that was consumed in the production of a good or service, the consequences of which (e.g., groundwater depletion) are implicitly “embedded” within the product or service. See embedded energy.

Visual Analytics Visual Analytics is an outgrowth of the fields of information visualization and scientific visualization that focuses on analytical reasoning facilitated by interactive visual interfaces.

Vulnerability The degree to which an entity is susceptible to damage resulting from a harmful event.

Vulnerability to Extreme Events A property of the largest infrastructures; these infrastructures persist for a long time and will therefore eventually encounter a very large event that exceed s the designed capabilities of the system.

Water A substance that is the fluid basis of animal and plant life.

Water Management Plan (WMP) A framework for managing the water resources of a region.

Water Security The ability to have access adequate and reliable quantities of clean water for consumption, proper sanitation, food and goods production, and sustainable health care.

Water Trust A typically not-for-profit organization dedicated to advancing water programs such as efficient use, maintaining natural flows, and public water rights. In developing nations, Water trusts often support access, sanitation, and other key aspects of water security.

Watershed (also Catchment Basin or Drainage Basin) A geographical delineation characterized by the fact that all runoff within the basin will eventually drain to a given water body. Catchment basins are nested. For example, the catchment of a small tributary is within the catchment of a larger river, and both of those are within the catchment basin of the lake or large river to which they drain.

Watershed Associations Organizations, which may or may not be incorporated, dedicated to preserving or improving management of a watershed or catchment basin. They may engage in public outreach and education, lobbying for legal protections for the watershed, and engaging citizens in data collection and restoration efforts. Most members of watershed associations live or work in the watershed.

Watt (W) A metric unit of power defined as the flow of 1 J/s. Power plant capacities range from a few MW for a very small plant to 22,500 MW (22.5 GW) for Three Gorges Dam, the largest power plant in the world.

Watt-hour (Wh) A metric unit of energy defined a power flow of 1 W for one hour, equivalent to 3600 J. Electric power production or demand is often defined in kilowatt hours (kWh) for residential, megawatt-hours (MWh) for power plant generation, or terawatt hours (TWh) for global scales.

WEAP (Software) The water evaluation and planning system is a software tool that takes an integrated approach to water resources planning, developed by the Stockholm Environment Institute.

Weathering The physical and chemical environmental processes associated with breaking materials down into their component parts, through the activity of the earth's atmosphere, water cycle and biological organisms; examples include water freezing in rock cracks to break it apart and limestone dissolving from exposure to rainfall because atmospheric carbon dioxide dissolves in water to produce carbonic acid.

Withdrawal (Water) Water that is removed from a watershed for human purposes without consideration of whether that water is returned to the same watershed in liquid form. See Consumption (Water).

XML (Computing) The extensible markup language is a generic standard for structuring metadata (and less commonly, data).

Index

P. Saundry, B. L. Ruddell (eds.), *The Food-Energy-Water Nexus*, AESS Interdisciplinary Environmental Studies and Sciences Series,
https://doi.org/10.1007/978-3-030-29914-9

D

E

Printed in the United States
by Baker & Taylor Publisher Services